Fungal Pathogenesis in Plants and Crops: Molecular Biology and Host Defense

Fungal Pathogenesis in Plants and Crops: Molecular Biology and Host Defense

Editor:

Carlos Tello Lacal, Ph.D

Delve Publishing

www.delvepublishing.com

Fungal Pathogenesis in Plants and Crops: Molecular Biology and Host Defense

Editor: Carlos Tello Lacal, Ph.D

708 3rd Avenue, 6th Floor

New York, NY 10017

United States of America

www.delvepublishing.com

ISBN: 978-1-68095-898-0

Library of Congress Control Number: 2016951501

About the Editor

Carlos Tello Lacal, Ph.D

Carlos Tello (Valencia, Spain, 1981) studied Agricultural Engineering with specialization in Biotechnology at the Universidad Politécnica de Valencia, where he graduated with a BSc thesis on the characterization of an Arabidopsis mutant with increased tolerance to seed ageing and salt stress. He moved afterwards to Sevilla to complete his MSc and PhD at the Instituto de Recursos Naturales y Agrobiología, where he studied the mechanisms employed by plants in their adaptation to abiotic stress and more specifically the role of the Arabidopsis Na^+/H^+ antiporter SOS1 in the regulation of Na^+ and K^+ homeostasis. Since 2014, he is based in Zürich and develops several freelance science-related jobs. He has recently retrained himself in the field of cell culture in bioreactors by attending an advanced training course at the Zürcher Hochschule für Angewandte Wissenschaften

Contents

List of Contributors

Eva Miedes
Centro de Biotecnología y Genómica de Plantas, Universidad Politécnica Madrid, Madrid, Spain
Departamento Biotecnología, Escuela Técnica Superior Ingenieros Agrónomos, Universidad Politécnica Madrid, Madrid, Spain

Ruben Vanholme
Department of Plant Systems Biology, VIB (Flanders Institute for Biotechnology), Gent, Belgium
Department of Plant Biotechnology and Bioinformatics, Ghent University, Gent, Belgium

Wout Boerjan
Department of Plant Systems Biology, VIB (Flanders Institute for Biotechnology), Gent, Belgium
Department of Plant Biotechnology and Bioinformatics, Ghent University, Gent, Belgium

Antonio Molina
Centro de Biotecnología y Genómica de Plantas, Universidad Politécnica Madrid, Madrid, Spain
Departamento Biotecnología, Escuela Técnica Superior Ingenieros Agrónomos, Universidad Politécnica Madrid, Madrid, Spain

Daniel V. Savatin
Department of Biology and Biotechnology "Charles Darwin", Sapienza-University of Rome, Rome, Italy

Giovanna Gramegna
Department of Biology and Biotechnology "Charles Darwin", Sapienza-University of Rome, Rome, Italy

Vanessa Modesti
Department of Biology and Biotechnology "Charles Darwin", Sapienza-University of Rome, Rome, Italy

Felice Cervone
Department of Biology and Biotechnology "Charles Darwin", Sapienza-University of Rome, Rome, Italy

Shaowu Meng
Fungal Genomics Laboratory, Center for Integrated Fungal Research, North Carolina State University, Raleigh, NC 27695, USA,
Current address: Hayes Laboratory, Lineberger Comprehensive Cancer Center, School of Medicine, CB# 7295, University of North Carolina at Chapel Hill, Chapel Hill, NC 27599-7295, USA and

Trudy Torto-Alalibo
Virginia Bioinformatics Institute, Virginia Polytechnic and State University, Blacksburg, VA 24061, USA,

Marcus C Chibucos
Virginia Bioinformatics Institute, Virginia Polytechnic and State University, Blacksburg, VA 24061, USA,
Current address: Institute for Genome Sciences, University of Maryland School of Medicine, Baltimore, MD 21201, USA

Brett M Tyler
Virginia Bioinformatics Institute, Virginia Polytechnic and State University, Blacksburg, VA 24061, USA,

Ralph A Dean
Fungal Genomics Laboratory, Center for Integrated Fungal Research, North Carolina State University, Raleigh, NC 27695, USA,

Carrie Selin
Department of Plant Science, University of Manitoba, Winnipeg, MB, Canada

Teresa R. de Kievit
Department of Microbiology, University of Manitoba, Winnipeg, MB, Canada

Mark F. Belmonte
Department of Biological Sciences, University of Manitoba, Winnipeg, MB, Canada

W. G. Dilantha Fernando
Department of Plant Science, University of Manitoba, Winnipeg, MB, Canada

Mark S. Dixon,
The Sainsbury Laboratory, John Innes Centre, Norwich Research Park, Colney Lane, Norwich NR4 7UH, United Kingdom

Catherine Golstein,
The Sainsbury Laboratory, John Innes Centre, Norwich Research Park, Colney Lane, Norwich NR4 7UH, United Kingdom

Colwyn M. Thomas,
The Sainsbury Laboratory, John Innes Centre, Norwich Research Park, Colney Lane, Norwich NR4 7UH, United Kingdom

Erik A. van der Biezen,
The Sainsbury Laboratory, John Innes Centre, Norwich Research Park, Colney Lane, Norwich NR4 7UH, United Kingdom

Jonathan D. G. Jones
The Sainsbury Laboratory, John Innes Centre, Norwich Research Park, Colney Lane, Norwich NR4 7UH, United Kingdom

Upinder S. Gill,
Plant Biology Division, The Samuel Roberts Noble Foundation, Ardmore, OK 73401.

Seonghee Lee,
Plant Biology Division, The Samuel Roberts Noble Foundation, Ardmore, OK 73401.

Kirankumar S. Mysore
Plant Biology Division, The Samuel Roberts Noble Foundation, Ardmore, OK 73401.

Simeon O Kotchoni
Department of Plant Molecular Biology, Institute of Botany, Kirschallee 1, University of Bonn, D-53115 Germany

Emma W Gachomo
Institute for Plant Diseases, Nussallee 9, University of Bonn, D-53115 Germany

Tünde Pusztahelyi
Central Laboratory, Faculty of Agricultural and Food Sciences and Environmental Management, University of Debrecen, Debrecen, Hungary

Imre J. Holb
Faculty of Agricultural and Food Sciences and Environmental Management, Institute of Horticulture, University of Debrecen, Debrecen, Hungary
Department of Plant Pathology, Centre for Agricultural Research, Plant Protection Institute, Hungarian Academy of Sciences, Debrecen, Hungary

István Pócsi
Department of Biotechnology and Microbiology, Faculty of Science and Technology, University of Debrecen, Debrecen, Hungary

Mansoor Karimi Jashni
Laboratory of Phytopathology, Wageningen University and Research Centre, Wageningen, Netherlands
Department of Plant Pathology, Tarbiat Modares University, Tehran, Iran

Rahim Mehrabi
Laboratory of Phytopathology, Wageningen University and Research Centre, Wageningen, Netherlands
Cereal Research Department, Seed and Plant Improvement Institute, Karaj, Iran

Jérôme Collemare
Laboratory of Phytopathology, Wageningen University and Research Centre, Wageningen, Netherlands
UMR1345, IRHS-INRA, Beaucouzé, France

Carl H. Mesarich
Laboratory of Phytopathology, Wageningen University and Research Centre, Wageningen, Netherlands
Bioprotection Technologies, The New Zealand Institute for Plant and Food Research Limited, Mount Albert Research Centre, Auckland, New Zealand

Pierre J. G. M. de Wit
Laboratory of Phytopathology, Wageningen University and Research Centre, Wageningen, Netherlands

Philippe Jeandet
Laboratory of Stress, Defenses and Plant Reproduction, Research Unit "Vines and Wines of Champagne", UPRES EA 4707, Department of Biology and Biochemistry, Faculty of Sciences, University of Reims, 51687 Reims cedex 02, France

Claire Hébrard
Laboratory of Stress, Defenses and Plant Reproduction, Research Unit "Vines and Wines of Champagne", UPRES EA 4707, Department of Biology and Biochemistry, Faculty of Sciences, University of Reims, 51687 Reims cedex 02, France

Marie-Alice Deville
Champagne Deville, 13 rue Carnot, Verzy 51380, France

Sylvain Cordelier
Laboratory of Stress, Defenses and Plant Reproduction, Research Unit "Vines and Wines of Champagne", UPRES EA 4707, Department of Biology and Biochemistry, Faculty of Sciences, University of Reims, 51687 Reims cedex 02, France

Stéphan Dorey
Laboratory of Stress, Defenses and Plant Reproduction, Research Unit "Vines and Wines of Champagne", UPRES EA 4707, Department of Biology and Biochemistry, Faculty of Sciences, University of Reims, 51687 Reims cedex 02, France

Aziz Aziz
Laboratory of Stress, Defenses and Plant Reproduction, Research Unit "Vines and Wines of Champagne", UPRES EA 4707, Department of Biology and Biochemistry, Faculty of Sciences, University of Reims, 51687 Reims cedex 02, France

Jérôme Crouzet

Laboratory of Stress, Defenses and Plant Reproduction, Research Unit "Vines and Wines of Champagne", UPRES EA 4707, Department of Biology and Biochemistry, Faculty of Sciences, University of Reims, 51687 Reims cedex 02, France

Xiaotang Di
Molecular Plant Pathology, Faculty of Science, Swammerdam Institute for Life Sciences, University of Amsterdam, Amsterdam, Netherlands

Frank L. W. Takken
Molecular Plant Pathology, Faculty of Science, Swammerdam Institute for Life Sciences, University of Amsterdam, Amsterdam, Netherlands

Nico Tintor
Molecular Plant Pathology, Faculty of Science, Swammerdam Institute for Life Sciences, University of Amsterdam, Amsterdam, Netherlands

Hege H. Divon
Department of Plant and Environmental Sciences, Norwegian University of Life Sciences, A˚ s, Norway;
Norwegian Institute for Agricultural and Environmental Research - Bioforsk, Plant Health and Plant Protection Division, A˚ s, Norway; and

Robert Fluhr
Department of Plant Science, Weizmann Institute of Science, Rehovot, Israё

Martin B. Dickman
Norman Borlaug Center, Texas A&M University, College Station, Texas, United States of America,
Department of Microbial Pathogenesis and Immunology, College of Medicine, Texas A&M Health Science Center, College Station, Texas, United States of America,

Paul de Figueiredo
Norman Borlaug Center, Texas A&M University, College Station, Texas, United States of America,
Department of Microbial Pathogenesis and Immunology, College of Medicine, Texas A&M Health Science Center, College Station, Texas, United States of America,
Department of Veterinary Pathobiology, College of Veterinary Medicine, Texas A&M University, College Station, Texas, United States of America,
Department of Plant Pathology and Microbiology, Texas A&M University, College Station, Texas, United States of America

Yi Fang
Nano Electrochemistry Laboratory, College of Engineering, University of Georgia, Athens, GA 30602, USA

Ramaraja P. Ramasamy
Nano Electrochemistry Laboratory, College of Engineering, University of Georgia, Athens, GA 30602, USA

Abdelbasset El Hadrami
Department of Plant Science, University of Manitoba, 222, Agriculture Building, Winnipeg, Manitoba, R3T 2N2, Canada
OMEX Agriculture Inc., P.O. Box 301, 290 Agri Park Road, Oak Bluff, Manitoba, R0G 1N0, Canada

Ahmed F. El-Bebany
Department of Plant Science, University of Manitoba, 222, Agriculture Building, Winnipeg, Manitoba, R3T 2N2, Canada
Department of Plant Pathology, Faculty of Agriculture, Alexandria University, El-Shatby, Alexandria, 21545, Egypt

Zhen Yao
Department of Plant Science, University of Manitoba, 222, Agriculture Building, Winnipeg, Manitoba, R3T 2N2, Canada

Lorne R. Adam
Department of Plant Science, University of Manitoba, 222, Agriculture Building, Winnipeg, Manitoba, R3T 2N2, Canada

Ismailx El Hadrami
Laboratoire de Biotechnologies, Protection et Valorisation des Ressources Végétales (Biotec-VRV), Faculté des Sciences Semlalia, B.P. 2390, Marrakech, 40 000, Morocco

Fouad Daayf
Department of Plant Science, University of Manitoba, 222, Agriculture Building, Winnipeg, Manitoba, R3T 2N2, Canada

Mathias Choquer
UMR 1290 BIOGER-CPP, INRA, Route de Saint-Cyr, Versailles, France; and

Elisabeth Fournier
UMR 1290 BIOGER-CPP, INRA, Route de Saint-Cyr, Versailles, France; and

Caroline Kunz
UMR 217 Interaction Plantes-Pathoge` nes, INRA/AgroParisTech/Universite´ Pierre et Marie Curie, Paris, France

Caroline Levis
UMR 1290 BIOGER-CPP, INRA, Route de Saint-Cyr, Versailles, France; and

Jean-Marc Pradier
UMR 1290 BIOGER-CPP, INRA, Route de Saint-Cyr, Versailles, France; and

Adeline Simon
UMR 1290 BIOGER-CPP, INRA, Route de Saint-Cyr, Versailles, France; and

Muriel Viaud
UMR 1290 BIOGER-CPP, INRA, Route de Saint-Cyr, Versailles, France; and

Jian-Hua Wang
Molecular Biotechnology Laboratory of Triticeae Crops, Huazhong Agricultural University, Wuhan 430070, Hubei, China; Email:

Mbacke Ndoye
Molecular Biotechnology Laboratory of Triticeae Crops, Huazhong Agricultural University, Wuhan 430070, Hubei, China; Email:

Jing-Bo Zhang
Molecular Biotechnology Laboratory of Triticeae Crops, Huazhong Agricultural University, Wuhan 430070, Hubei, China; Email:
College of Plant Science and Technology, Huazhong Agricultural University, Wuhan 430070, Hubei, China

He-Ping Li
Molecular Biotechnology Laboratory of Triticeae Crops, Huazhong Agricultural University, Wuhan 430070, Hubei, China; Email:

Yu-Cai Liao
Molecular Biotechnology Laboratory of Triticeae Crops, Huazhong Agricultural University, Wuhan 430070, Hubei, China; Email:
College of Plant Science and Technology, Huazhong Agricultural University, Wuhan 430070, Hubei, China

Yi Chung Chen
Plant-Microbe Interactions Laboratory; School of Agriculture and Food Sciences; The University of Queensland; Brisbane, Queensland, Australia

Brendan N Kidd
Plant-Microbe Interactions Laboratory; School of Agriculture and Food Sciences; The University of Queensland; Brisbane, Queensland, Australia

Lilia C Carvalhais
Plant-Microbe Interactions Laboratory; School of Agriculture and Food Sciences; The University of Queensland; Brisbane, Queensland, Australia

Peer M Schenk
Plant-Microbe Interactions Laboratory; School of Agriculture and Food Sciences; The University of Queensland; Brisbane, Queensland, Australia

Shijie Zhang
NWAFU-Purdue Joint Research Center, College of Plant Protection, Northwest A&F University, Shaanxi, People's Republic of China,

Jin-Rong Xu
NWAFU-Purdue Joint Research Center, College of Plant Protection, Northwest A&F University, Shaanxi, People's Republic of China,
Department of Botany and Plant Pathology, Purdue University, West Lafayette, Indiana, United States of America

Andrés Pinzón
Mycology and Phytopathology Laboratory, Department of Biological Sciences, Universidad de los Andes, Bogotá, Colombia,
Bioinformatics center, Colombian EMBnet node, Biotechnology Institute, National University of Colombia, Bogotá, Colombia,

Emiliano Barreto
Bioinformatics center, Colombian EMBnet node, Biotechnology Institute, National University of Colombia, Bogotá, Colombia,

Adriana Bernal
Mycology and Phytopathology Laboratory, Department of Biological Sciences, Universidad de los Andes, Bogotá, Colombia,

Luke Achenie
Department of Chemical Engineering, Virginia Polytechnic Institute and State University, Blacksburg Virginia, USA,

Andres F González Barrios
Grupo de Diseño de Productos y Procesos, Department of Chemical Engineering, Los Andes University, Bogotá, Colombia and

Raúl Isea
Fundación IDEA, Centro de Biociencias, Hoyo de la puerta, Baruta 1080, Venezuela

Silvia Restrepo
Mycology and Phytopathology Laboratory, Department of Biological Sciences, Universidad de los Andes, Bogotá, Colombia,

Melania Figueroa
Department of Plant Pathology, University of Minnesota, St. Paul, MN, USA
Stakman-Borlaug Center for Sustainable Plant Health, University of Minnesota, St. Paul, MN, USA

Claudia V. Castell-Miller
Department of Plant Pathology, University of Minnesota, St. Paul, MN, USA
Stakman-Borlaug Center for Sustainable Plant Health, University of Minnesota, St. Paul, MN, USA

Feng Li
Department of Plant Pathology, University of Minnesota, St. Paul, MN, USA
Stakman-Borlaug Center for Sustainable Plant Health, University of Minnesota, St. Paul, MN, USA

Scot H. Hulbert
Department of Plant Pathology, Washington State University, Pullman, WA, USA
Department of Crop and Soil Sciences, Washington State University, Pullman, WA, USA

James M. Bradeen
Department of Plant Pathology, University of Minnesota, St. Paul, MN, USA
Stakman-Borlaug Center for Sustainable Plant Health, University of Minnesota, St. Paul, MN, USA

Evelyn Vollmeister
Institute for Microbiology, Heinrich Heine University D ¨usseldorf, D ¨usseldorf, Germany;

Kerstin Schipper
Institute for Microbiology, Heinrich Heine University D ¨usseldorf, D ¨usseldorf, Germany;

Sebastian Baumann
Institute for Microbiology, Heinrich Heine University D ¨usseldorf, D ¨usseldorf, Germany;
Max Planck Institute for Terrestrial Microbiology, Marburg, Germany

Carl Haag
Institute for Microbiology, Heinrich Heine University D ¨usseldorf, D ¨usseldorf, Germany;

Thomas Pohlmann
Institute for Microbiology, Heinrich Heine University D ¨usseldorf, D ¨usseldorf, Germany;
Max Planck Institute for Terrestrial Microbiology, Marburg, Germany

Janpeter Stock
Institute for Microbiology, Heinrich Heine University D ¨usseldorf, D ¨usseldorf, Germany;

Michael Feldbrugge
Institute for Microbiology, Heinrich Heine University D ¨usseldorf, D ¨usseldorf, Germany;
Max Planck Institute for Terrestrial Microbiology, Marburg, Germany

Graeme J. Kettles
Department of Plant Biology and Crop Science, Rothamsted Research, Harpenden, UK

Kostya Kanyuka
Department of Plant Biology and Crop Science, Rothamsted Research, Harpenden, UK

Preface

Fungal pathogens constitute one of the major threats to global crop production, with an estimated annual yield loss of 15% being attributed to them. The main strategies employed by farmers to prevent, reduce or fight fungal infection range from the use of resistant crop cultivars to the treatment with fungicides or the establishment of crop rotation practices. The life cycles and infectious strategies of the main crop fungal pathogens, as well as the mechanisms triggered by plants in response to these pathogenic agents, emerge as essential areas of study requiring furhter investigation in order to optimize fungal fight practices. In this book, a wide range of peer-reviewed research papers fully covering the above-mentioned topics have been compiled.

The first chapter describes the cell wall, the initial barrier fungal pathogens need to overcome in order to colonize their host, particularly how modifications in its composition influence plant resistance to infection. Chapter 2 provides an overview of the different signaling pathways triggered in response to plant cell wall wounding, since wounded tissue constitutes an important way of entry into the plant employed by pathogens. The main features of the infection process by fungi and oomycetes in terms of gene ontology are reviewed in Chapter 3. Chapter 4 discusses the roles of the effectors released by biotrophic and hemibiotrophic fungal pathogens in the infectious process and the mechanisms enabling their release by fungi and uptake by plant hosts. The functioning of the complex plant-pathogen recognition process is illustrated in Chapter 5 by studying the case of tomato Cf-2-mediated resistance against leaf moulds. The process continues to be described in Chapter 6, where the differences between host and non-host resistance mechanisms are presented. Plants react against the presence of fungal pathogens by producing a plethora of compounds with antifungal activity, such as reactive oxygen species, secondary metabolites, proteases and phytoalexins, whose roles in fungal fight are reviewed in Chapters 7 to 10. Chapter 11 investigates the effects of the main phytohormones in shaping the plant-pathogen interaction by depicting the example of the soil fungus Fusarium oxysporum. Going back to the infectious process, Chapter 12 examines the use of plant nutrient supply by the growing phytopathogenic fungi. In Chapter 13, the control by plants of the programmed cell death process in response to fungal pathogens is analyzed. The different laboratory-based techniques employed in plant disease detection, ranging from polymerase chain reaction (PCR) to fluorescence in situ hybridization (FISH) or enzyme-linked immunosorbent assay (ELISA) among others, are reviewed in Chapter 14. Chapter 15 discusses the latest trends for studying plant-pathogen interactions in the proteomics era. The last eight chapters feature practical cases involving the most important fungal crop pathogens. Chapter 16 describes the infectious process of Botrytis cinerea, the genetical variability observed in both strains and virulence factors and the development of high-throughput genomics and proteomics analyses aimed at deciphering its host interaction mechanisms. In Chapter 17, the application of molecular studies to decipher the population structure and genetic diversity of the Fusarium graminearum species complex is reviewed. Chapter 18 briefly describes the molecular responses triggered in plant roots in response to Fusarium oxysporum and the use of biological control agents and genetic manipulation techniques against this pathogen. The production of effectors and its delivery into the host plant is once again described in Chapter 19, this time exemplified by the case of Magnaporthe oryzae. Chapter 20 illustrates the study of the Phytophtora infestans - Solanum tuberosum pathosystem by employing a computational model-based approach. In Chapter 21, the use of Brachypodium distachyon as a model system to underpin resistance mechanisms to rust fungi belonging to the Puccinia complex is discussed. Chapter 22 explores the whole life cycle and infectiuous process corresponding to the maize pathogen Ustilago maydis. Finally, the molecular interactions between Zymoseptoria tritici and its host crop wheat (Triticum aestivum L.) are dissected in Chapter 23.

The present book intends to be beneficial for college students, researchers, teachers, and farmers in their approach to the field of plant fungal pathogenesis.

Editor
Carlos Tello Lacal, Ph.D

INTRODUCTION

Plant pathology (also phytopathology) is the scientific study of diseases in plants caused by pathogens (infectious organisms) and environmental conditions (physiological factors). Organisms that cause infectious disease include fungi, oomycetes, bacteria, viruses, viroids, virus-like organisms, phytoplasmas, protozoa, nematodes and parasitic plants. Not included are ectoparasites like insects, mites, vertebrate, or other pests that affect plant health by consumption of plant tissues. Plant pathology also involves the study of pathogen identification, disease etiology, disease cycles, economic impact, plant disease epidemiology, plant disease resistance, how plant diseases affect humans and animals, pathosystem genetics, and management of plant diseases.

Control of plant diseases is crucial to the reliable production of food, and it provides significant reductions in agricultural use of land, water, fuel and other inputs. Plants in both natural and cultivated populations carry inherent disease resistance, but there are numerous examples of devastating plant disease impacts (see Irish potato famine, chestnut blight), as well as recurrent severe plant diseases (see rice blast, soybean cyst nematode, citrus canker). However, disease control is reasonably successful for most crops. Disease control is achieved by use of plants that have been bred for good resistance to many diseases, and by plant cultivation approaches such as crop rotation, use of pathogen-free seed, appropriate planting date and plant density, control of field moisture, and pesticide use. Across large regions and many crop species, it is estimated that diseases typically reduce plant yields by 10% every year in more developed settings, but yield loss to diseases often exceeds 20% in less developed settings.

Fungal pathogens of plants and animals

Feature	Plants	Animals
Number	Many	Few
Importance	25-30% crop losses	Immunocompromised hosts
Life style & environmental niche	Often saprophytes; complex *in planta* – biotroph; necrotroph	Often soil saprophytes; not obligate
Experimental systems	Many; often shallow focus; crops	Few; different host cell & immunosuppression regimes
Host specificity	Often; coevolution	Not often
Reproduction	Sexual & asexual	Asexual (ascomycetes)
Dispersal/ transmission	Between hosts	Not usually between hosts
Pathogen requirements	Complex set of physical & molecular barriers to overcome	Inocula must get in, must survive 37°C, avoid or subvert immune responses

Fungal pathogens of plants

Puccinia graminis fsp. tritici
Stem rust of wheat
(R. Park)

Sclerotinia sclerotiorum
Stem rot of canola (M Rodriguez)

Magnaporthe oryzae
Rice blast

Leptosphaeria maculans
Blackleg of canola

- Recognise and attach to host surface
- Germinate (spores)
- Enter host
- Colonise & derive nutrition (survive)
- Avoid or subvert host defence responses
- Reproduce
- Exit and disperse and find new host

Steps are developmentally regulated; mutants arrested at particular stages. Some processes (recognition, pH regulation, oxidative burst, signalling) occur several times during disease. Plant symbionts (eg. mycorrhizae) have similar requirements

Overview

Interplay of signaling systems in plant-pathogen interactions is the key component in the development of diseases in plants. Signaling between fungal pathogens and plants begins when they meet on the plant surface. The evidence is strong that plant-fungal interactions begin, and perhaps their outcomes (susceptibility or resistance) are determined, during the very first few minutes of association and perhaps even earlier. If the signals at the plant surface are perceived as favorable by the fungi, the fungal spores germinate and either directly penetrate the cuticle or penetrate the cuticle by producing appressorium and infection peg. Both the physical and chemical plant signals may be involved in the penetration process. The plant signals may be used by the fungus to trigger the expression of the genes necessary to penetrate into the host.

Plants have evolved sophisticated detection and response systems that decipher pathogen signals and induce synthesis of various antifungal compounds and fortify their cell wall components to defend themselves. For successful pathogenesis. The pathogens have to overcome these defense mechanisms by producing appropriate signals. On perception of pathogen signals. Several signaling pathways are activated in plants mounting resistance response to invading pathogens. The interplay of the plant and pathogen signals allows the plant to fine-tune defense responses and the pathogen to succumb to the toxic environment in the plants or to overcome these defense mechanisms. Resulting in disease susceptibility or resistance. Susceptibility and resistance are opposite sides of the same coin and indicate the winning candidate in the battle between plant and pathogen. Resistance and susceptibility are often determined at the single-cell level. These complex signaling systems in pathogens and plants during pathogenesis in susceptible and resistant conditions arc described in this book. In this chapter. The molecular events occurring during the fungal development in host plants on perception of plant signals are described.

Signaling and Transduction Systems in "First Touch" And Adhesion of Fungal Spores

First Touch or Initial Contact Triggers the Infection Process

A physical contact of the pathogen. What is called "first touch." it has been shown to initiate various signaling and transduction systems in the fungal infection process. First touch or initial surface contact may induce the fungal spores to germinate and differentiate to form infection structures. Such as appressoria. Any solid-surface contact is sufficient for inducing spore germination in Colletoirichunt spp. Conidia of Colletttrichum glorimporinitles are induced to germinate and differentiate to form appressoria by different plant signals. How-ever. contact with a hard surface is necessary for the plant signals to induce appres-sorium formation C. gluravrotinkles conidia formed appressoria on both a hydrophilic cover glass and a hydrophobic polystyrene petri dish when exposed to plant-surface wax or ethylene signals. On the other hand, on soft hydrophilic or hydrophobic substrates, such as 2% agar or petrolatum. Respectively. Only germination occurred. All hard surfaces tested induced appressorium development, including both hydrophobic and hydrophilic surfaces of Gelliond, whereas soft surfaces such as water agar 12%1 failed to induce appressorium formation in C. sloes: valet It appears that a chain of molecular events that ultimately lead to differentiation of the germ tubes into appressoria is initiated upon contact with a hard surface. The fungal spores, which land on the host surface, first recognize certain substratum characteristics within seconds of making contact. On the first touch, the spore releases copious extracellular matrix (ECM) in advance of germination. The ECM production in the barley powdery mildew pathogen Blumeria gruntinis luwdei was associated with changes in spore surface features and the ECM was shown to be released biphasically: an initial release was evident after 2 min. and a second release after IS min (Nicholson et al., 1988). In fact, the release of ECM from the conidia of B. gran:kis f. sp. honk! Could be observed within 20 s of first touch demonstrated hydrolytic enzyme activity in the region corresponding to conidial ECM. Within 3 min of spore deposition. The ECM continued to accumulate at the conidium-substratum interface for up to —60 min after contact.

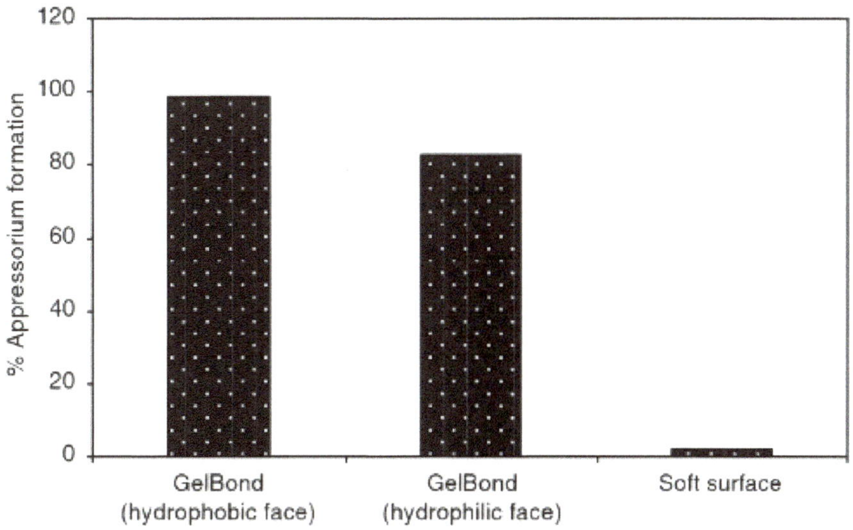

Figure.1 Rctalhc cf88-am of hanl hydrophobic or hydrophilic) and %team on appremorium formation in C✓leintrklimm glorovvfloides.

Pathogenic Fungi & Sites of Entry

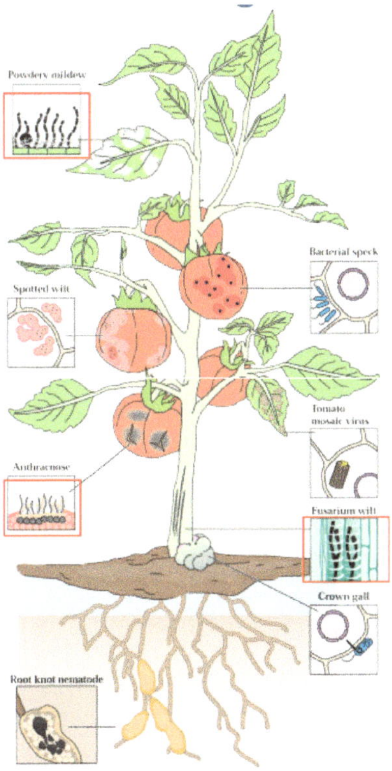

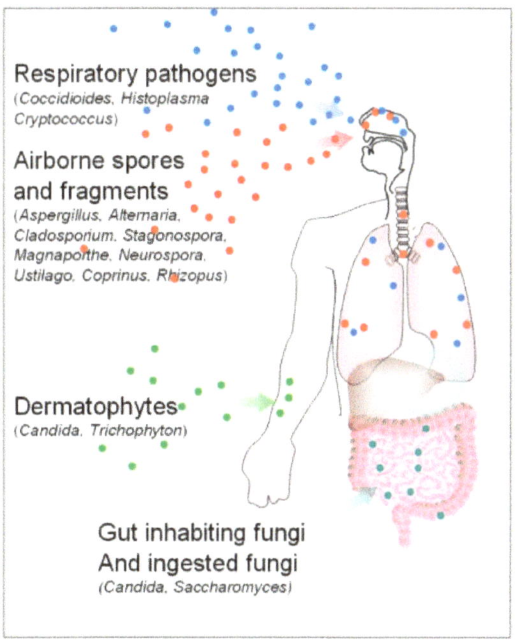

Bowyer et al. BMC Genomics. 2006; 7: 251.

Fungal Plant Pathogens and Symptomology

Fungi are spore-forming, non-chlorophytic, eukaryotic (cells having true nuclei) organisms and most of the true fungi are filamentous and branched. Most of the over 100,000 species of fungi are saprophytes. However, over 20,000 species of fungi are parasites and cause disease in crops and plants (USEPA 2005). Fungal parasites are by far the most prevalent plant pathogenic organism. All plants are attacked by one species or another of phytopathogenic fungi. Individual species of fungi can parasitize one or many different kinds of plants. The body, or thallus, of most of the higher fungi is called a mycelium (pl. mycelia). The mycelium is made up of thread-like structures called hyphae (sing. hypha). Hyphae grow only from their tips, and under favorable conditions can grow indefinitely. Hyphae, most often, are partitioned by cross walls called septa (sing. septum). Septate hyphae are divided into individual cells by these internal walls. Septa usually have a central pore which allows small organelles, and in some cases nuclei, to pass from one cell to another. The hyphae of the classes Ascomycetes, Basidiomycetes and Deuteromycetes are septate and the hyphae of class Oomycetes and Zygomycetes are nonseptate or coenocytic. Coencytic hyphae have multiple nuclei in a common cytoplasm.

Hyphae produce specialized simple or branched projections called haustoria (sing. Haustorium). Haustoria penetrate the host tissue and act as nutrient absorbing organs. Haustoria contain nuclei and a concentrated number of mitochondria. The haustoria do not rupture the host cell protoplasmic membrane but invaginates itself into the cell thus greatly increasing the absorptive surface of the fungus.

Reproduction of fungi is primarily by means of spores. Spores are reproductive bodies that consist of one or a few cells. In function they are analogous to the seeds of green plants. Spores are produced by sexual and asexual reproduction. Sexual reproduction involves the union of two compatible nuclei as produced by meiosis. For many phytopathogenic fungi the sexual cycle occurs only once during each growing season. Sexually produced spores include oospores, zygospores, ascospores and basidiospores characteristic of the phyla Oomycota, Zygomycota, Ascomycota and Basidiomycota respectively. Asexual reproduction usually occurs by means of mitosis which produce mitotic spores, mycelial fragmentation, fission and budding. With asexual reproduction the repeating cycles of infection can continue throughout the growing season. Asexual spores may be classified as oidia (formed by fragmentation of hyphae into individual cells), conidia (borne on tips or sides of specialized branches of hyphae) and sporangiospores (a nonmotile spore born in a sporangim or case). There are many types and different characteristics of fruiting bodies, spores and mycelium. Fungi are both classified and identified by these features.

The classification of pathogenic fungi is important for identifying and diagnosing plant disease. Each kingdom, phylum, class, order, genus and specie has its own identifying characteristics. Many of the diagnostic factors in identifying different fungi are very subtle and without proper lab equipment and skill are difficult to identify down to the species level. If we can identify the type of fruiting body and/or spores that are associated with a particular phylum, we can narrow down our search considerably. Most often, we rely on visual signs and symptoms for diagnosis of fungal disease.

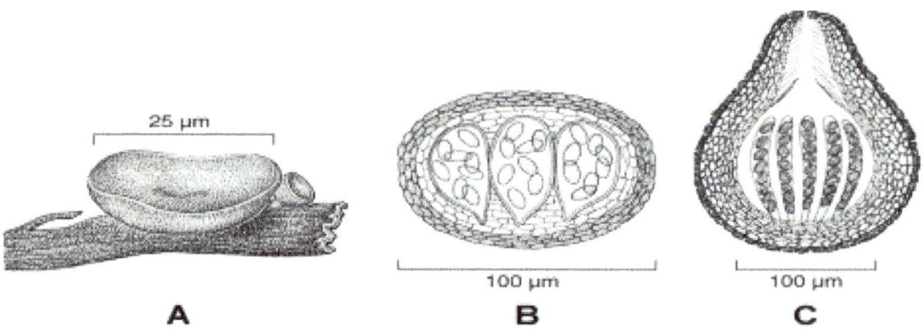

The primary phyla dealt with in plant pathology and disease diagnosis are listed below. This outline is very brief and only includes a few of the more important pathogens and problems. Also included are some of the more important characteristics of each phylum. For further clarification on fungal taxonomy, you can refer to any number of resources at the Extension office. Also, the internet is an excellent source of information on this subject.

Diagnostic Signs and Symptoms of Fungal Infections

Fungi can cause general or localized signs and/or symptoms. In the majority of cases, fungal infections cause general necrosis of host tissue and often cause stunting, distortions and abnormal changes in plant tissue and organs. The most distinctive and easily identifiable characteristics of fungal infections are the physical presence of signs of the pathogen. Signs include hyphae, mycelia, fruiting bodies and spores of the fungal pathogen are significant clues to proper identification and diagnosis of a disease.

The fruiting bodies of fungi range from microscopic to macroscopic. They come in many shapes and configurations and have their individual characteristics. The fruiting bodies, along with spores, and mycelium, in most cases can lead to an accurate identification of the disease. The following symptoms are common in fungal infections whether alone or in combination with other fungal pathogens.

Leaf Spots are very common in both biotic and abiotic plant disorders. Fungal leaf spots often take the form of localized lesions consisting of necrotic and collapsed tissue. Leaf spots can vary in size and are generally round and concentric, but can be ovoid or elongated on both leaves and stems of the host. The typical fungal leaf spot will have a "bulls-eye-like" appearance consisting of roughly concentric rings that may display zones of different colors such as yellow, red or purple, and will often have a tan center. As the spots develop, they are not restricted by the leaf veins as can be the case in bacterial leaf spots. Fungal leaf spots will usually have a dry texture but are not dry and papery.

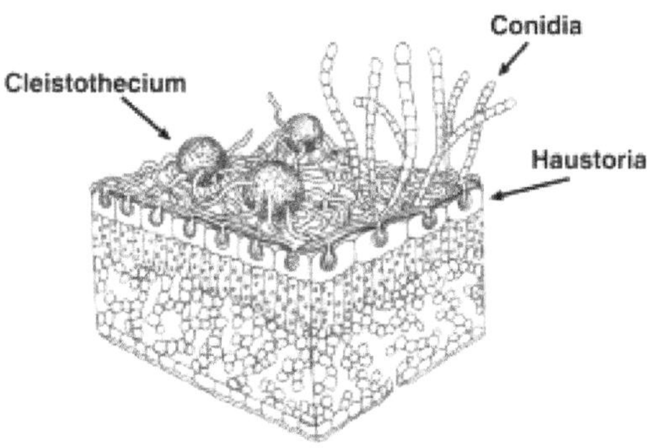

The Top 10 fungal pathogens in molecular plant pathology

Recently, the journal Molecular Plant Pathology considered which viruses would appear in a 'Top 10' of plant viruses on the basis of their perceived importance, scientifically or economically, in terms of the views of the contributors to the journal. This was carried out as many papers, reviews and grant applications claim that a particular plant virus is of huge importance, and this is probably rightly so. As a result of the interest generated in this exercise for plant viruses, a similar survey was instigated for plant-pathogenic fungi, and with a similar mechanism to the virus review. All authors, reviewers, editorial board members and senior editors of the journal Molecular Plant Pathology, with an interest in fungi, were contacted and asked to nominate three plant-pathogenic fungi they would expect to see in a list of the most scientifically/economically important fungal pathogens. The review, by its very nature, is similar in format and layout to the Top 10 virus review. The survey generated 495 votes from the international community, and allowed the generation of a Top 10 fungal plant pathogen list for the journal Molecular Plant Pathology (see Table 1).

Table 1. Top 10 fungal plant pathogens.

Rank	Fungal pathogen	Author of fungal description
1	*Magnaporthe oryzae*	Ralph Dean
2	*Botrytis cinerea*	Jan A. L. van Kan
3	*Puccinia* spp.	Zacharias A. Pretorius
4	*Fusarium graminearum*	Kim Hammond-Kosack
5	*Fusarium oxysporum*	Antonio Di Pietro
6	*Blumeria graminis*	Pietro Spanu
7	*Mycosphaerella graminicola*	Jason J. Rudd
8	*Colletotrichum* spp.	Marty Dickman
9	*Ustilago maydis*	Regine Kahmann
10	*Melampsora lini*	Jeff Ellis

The table represents the ranked list of fungi as voted for by plant mycologists associated with the journal *Molecular Plant Pathology*.

The fungus making the strongest appearance in the voting was Magnaporthe oryzae, in first place with almost double the number of votes of Botrytis cinerea, in second place. All voters highlighted the economic importance of M. oryzae; with over one-half of the world's population relying on rice as the main source of calories, this pathogen can have devastating effects; however, many also highlighted how this pathogen has developed into a model system for the study of plant–pathogen interactions.

Botrytis cinerea, in second place, clearly has an impact in many areas because of the broad host range, causing severe damage, both pre- and post-harvest. It may be one of the few entries in the Top 10 that can claim a potential beneficial use, through its role in some aspects of wine production, with some benefits also claimed for the ninth entry, Ustilago maydis, in infected cobs.

Puccinia spp., with many voters grouping the three wheat-infesting rust diseases together, are grouped together in third position. These fungal pathogens were already a serious disease threat, but have increased in impact with the emergence of race Ug99, now posing a serious challenge to wheat production.

In fourth and fifth places are two Fusarium species, but with contrasting host ranges, with F. graminearum causing significant damage predominantly to cereals and a few noncereal species, and F. oxysporum having a wide host range, with severe losses in crops as diverse as tomato, cotton and banana. However, pathogens of cereals continue to have a significant presence in the Top 10 with Blumeria graminis in sixth position and Mycosphaerella graminicola in seventh.

Colletotrichum spp., in eighth position, are members of an important genus of plant pathogens, for which the taxonomy may be described as being in a state of flux. The proposed number of species within the genus ranges from 29 to over 700 depending on taxonomic interpretation. Colletotrichum spp. have long served as a model system for hemibiotrophic pathogens, with a short biotrophic stage, followed by a switch to tissue ramification and necrotrophic development.

It is interesting that Ustilago maydis and Melampsora lini make up ninth and tenth positions, respectively, with many voters indicating strong scientific rather than economic reasons for their inclusion. Both are now developed as model systems, which are readily tractable, and provide vital insights into the molecular basis of plant immunity and infection processes.

Although the aim of this review article was to identify the Top 10 most important plant-pathogenic fungi according to the contributors to Molecular Plant Pathology, we are very much aware that importance and priorities can vary locally across continents and disciplines. We are also aware that not all fungi can make it into any Top 10, because of obvious numerical limits, although such fungi can still be regarded as hugely important. We therefore felt it appropriate to cite honourable mentions for fungi just missing out on the Top 10 list, including Phakopsora pachyrhizi and Rhizoctonia solani, both clearly important. In future years, when another survey of the Top 10 is carried out, it will be interesting to see whether fungi such as Phakopsora pachyrhizi, the causal agent of Asian soybean rust, makes an appearance, as it is clearly a disease that has only recently appeared in certain parts of the world and will potentially increase in importance.

This review contains single-page descriptions of the Top 10 plant-pathogenic fungi to introduce the reader to each of them, with illustrative figures and key references for further reading. The intention of this review is to trigger discussion and debate amongst the plant mycology community, as well as to lay down a bench-mark, as it will be interesting to see how perceptions change in future years and which fungi enter and leave the list.

CHAPTER 1

THE ROLE OF THE SECONDARY CELL WALL IN PLANT RESISTANCE TO PATHOGENS

Eva Miedes, Ruben Vanholme, Wout Boerjan and Antonio Molina

Keywords: Cell wall, plant pathogen, plant immunity, xylan, cellulose, lignin

INTRODUCTION

In their natural environments, plants are under continuous threat of biotic stresses caused by pathogenic bacteria, fungi, viruses, and oomycetes, that compromise plant survival and reproduction (Panstruga et al., 2009). Given that green plants are the ultimate source of energy for most other organisms, it is not surprising that plants have evolved a plethora of resistance mechanisms which are either constitutively present or induced after pathogen attack (Glazebrook, 2005; Panstruga et al., 2009). An important defense element common to all plants is the cell wall.

All the plant cells that are in developmental expansion have a constantly remodeled primary cell wall that mainly consists of carbohydrate-based polymers (classified as cellulose, hemicelluloses and pectins) and hydroxyproline-rich *O*-glycoproteins, such as extensins and arabinogalactan proteins (AGPs; Carpita and McCann, 2000). In addition, those cells that have completed their cellular expansion and need to reinforce their structure for functional reasons (e.g., to form vessel or fiber cells) generate a secondary cell wall that is mainly composed of cellulose, hemicelluloses (mostly xylans) and lignin (Cosgrove, 2005; Sarkar et al., 2009). Besides having multiple essential functions during plant development, plant cell walls also play important roles in preventing pathogen invasion. First, cell walls act as a passive barrier; local or extensive breakdown of the wall matrix is typically required for the progression of pathogen infection (Cantu et al., 2008; Hematy et al., 2009). Second, the cell wall is a reservoir of antimicrobial compounds, which are released during cell wall degradation (García-Olmedo et al., 2001; Schulze-Lefert, 2004; Vorwerk et al., 2004). Moreover, plants have a dedicated cell wall integrity (CWI) maintenance mechanism similar to that existing in fungi, which initiates responses to regulate CWI during plant development and in response to external stimuli (Wolf et al., 2012; Engelsdorf and Hamann, 2014). Impairment of CWI by pathogen attack or wounding results in the release of plant signaling molecules, the so-called Damage-Associated Molecular Patterns (DAMPs; Vorwerk et al., 2004; Cantu et al., 2008). DAMPs can modulate plant innate immune responses upon recognition by plant Pattern Recognition Receptors (PRRs), through molecular mechanisms that are similar to those regulating the activation of immune responses by Pathogen-Associated Molecular Patterns (PAMPs) derived from microbial pathogens (reviewed by Dodds and Rathjen, 2010; Macho and Zipfel, 2014; Malinovsky et al., 2014). The recognition of DAMPs and PAMPs by PRRs activates protein kinase cascades, which regulate downstream immune responses

Citation: Miedes E, Vanholme R, Boerjan W and Molina A (2014) the role of the secondary cell wall in plant resistance to pathogens. Front. Plant Sci. 5:358. doi: 10.3389/fpls.2014.00358

that can lead, among others, to cell-wall reinforcement (Ringli, 2010; Ferrari et al., 2013; Engelsdorf and Hamann, 2014; Malinovsky et al., 2014). Notably, in addition to its role in protecting plants against infection, the plant cell wall can also act as a source of nutrients from the pathogen point of view, thereby promoting pathogen growth and development (Cantu et al., 2008; Hematy et al., 2009).

Cell walls are also considered as a valuable feedstock for the production of second generation biofuels and bio-based chemicals. These so-called lignocellulosic feedstocks can either derive from agricultural and industrial practices, such as maize stover, straw and sugarcane bagasse, or from dedicated crops, such as fast growing grasses and trees, grown for the purpose of generating large volumes of lignocellulosic biomass. In both cases, engineering cell wall composition is a promising strategy to ease lignocellulosic biomass conversion toward fuels and chemicals in industrial processes (Simmons et al., 2010; Ong et al., 2014). Importantly, the changes in cell-wall composition needed for industrial biomass processing should not conflict with the principal biological roles of the cell wall as a supportive and protective structure. Indeed, some cell wall modifications result in negative repercussions on biomass yield (Bonawitz and Chapple, 2013). A better understanding of the processes underlying the yield-penalty in plants with modified cell walls has led to successful engineering strategies to recover the biomass yield, while maintaining the anticipated cell wall modifications (Petersen et al., 2012; Yang et al., 2013; Bonawitz et al., 2014). Likewise, cell wall modifications should not have negative repercussions on crop susceptibility toward pathogens. Clearly, a better understanding of the processes underlying the interactions between pathogens and the cell wall will support the development of plants with optimized lignocellulosic characteristics, without negatively affecting disease resistance.

A relatively large number of studies have described the influence of plant cell-wall modifications on pathogen infection (Cantu et al., 2008; Bellincampi et al., 2014; Malinovsky et al., 2014). Contra-intuitively, "weakening" the cell wall by knocking out essential genes involved in cell-wall biosynthesis sometimes leads to enhanced resistance toward specific pathogens. In this review, we summarize the consequences of secondary cell-wall modifications on pathogenic infection and link them with our current knowledge on the role of the cell wall in plant resistance to pathogens. Because lignin is both stress-induced and developmentally deposited in the secondary thickened cell wall and because it is a major target for lignocellulosic biomass engineering, we have put special emphasis on the effect of altering lignin amount and composition on pathogen infection and spread.

THE EFFECT OF ALTERING CELL WALL POLYSACCHARIDES ON RESISTANCE TO PATHOGENS

The contribution of the secondary cell wall to plant immunity has been mainly demonstrated through the characterization of plant mutants impaired in secondary wall composition (Cantu et al., 2008; Underwood, 2012). For instance, the resistance of a set of *Arabidopsis thaliana* (*Arabidopsis* herein) mutants defective in cellulose synthase (CESA) subunits required for secondary cell wall formation (i.e., CESA4, CESA7, and CESA8) toward a series of pathogens has been tested. These mutants make less cellulose, which results in collapsed xylem vessels, and therefore they are called *ir*regular *x*ylem mutants (*irx5*, *irx3*, and *irx1*, respectively). These mutants showed enhanced resistance to different pathogens, including the necrotrophic fungi*Plectosphaerella cucumerina* and *Botrytis cinerea*, the vascular bacterium *Ralstonia solanacearum*, and the biotrophic bacterium *Pseudomonas syringae* (Table 1; Hernández-Blanco et al., 2007). In line with these results, an *Arabidopsis* mutant defective in the MYB46 transcription factor that directly regulates the expression of genes required for secondary cell wall formation, including lignin and cellulose biosynthesis (among which *CESA4, CESA7*, and *CESA8*), also showed enhanced resistance to necrotrophic fungi (Ramírez et al., 2011). The disease resistance phenotype of *irx1*, *irx3*, *irx5*, and *myb46* mutants was in part explained by the constitutive activation of plant immune responses rather than by alterations of the passive wall barrier. In these mutants, the abscisic acid signaling pathway was constitutively active and antimicrobial peptides and tryptophan-derived metabolites accumulated to a higher extent than in wild-type plants (Hernández-Blanco et al., 2007; Sánchez-Vallet et al., 2010). Plant resistance to pathogens is also altered in *Arabidopsis* mutants affected in CESAs subunits required for cellulose biosynthesis of the primary cell wall, such as the CESA3 defective *isoxaben resistant*(*ixr1*)/*constitutive expression of VSP* (*cev1*) mutants (Ellis et al., 2002). The *ixr1/cev1* mutant alleles are more resistant than wild-type plants to *B. cinerea*, *P. syringae*, and *Erysiphe cichoracearum* (Ellis et al., 2002), whereas their resistance to *R. solanacearum* and *P. cucumerina*does not differ from that of wild-type plants, which contrasts with the resistance phenotype of the secondary cell wall cellulose mutants, *irx1, irx3* and *irx5* (Hernández-Blanco et al., 2007). In*ixr1/cev1* plants, the ethylene and jasmonic acid, but not the abscisic acid signaling pathway, are constitutively activated. These results with *Arabidopsis cesA* mutants illustrate that specific immune responses can be activated by alteration of the CWI of either the primary or the secondary wall (Ellis et al., 2002; Hernández-Blanco et al., 2007).

Table 1: Resistance phenotype of plants with alterations in secondary cell wall structure/composition.

Gene name (mutant/transgenic)	Plant species	Pathogen tested	Phenotype[1]	Reference
[illegible]	Arabidopsis thaliana	[illegible]	R	[illegible]
WAT1 (wat1)	Arabidopsis thaliana	R. solanacearum, P. cucumerina, Xanthomonas campestris, Verticillium dahliae, V. alboatrum	R	Denancé et al., 2013
DET3 (det3)	Arabidopsis thaliana	P. cucumerina	R	[illegible]
[illegible]	Arabidopsis thaliana	P. cucumerina	R	Delgado-Cerezo et al., 2012
[illegible]	Arabidopsis thaliana	[illegible]	R	[illegible]
IRX6 (irx6)	Arabidopsis thaliana	P. cucumerina	R	Delgado-Cerezo et al., 2012
[illegible]	Arabidopsis thaliana	[illegible]	R	[illegible]
[illegible]	Arabidopsis thaliana	[illegible]	S	[illegible]
[illegible]	Arabidopsis thaliana	[illegible]	S	[illegible]
		[illegible]	R[2]	[illegible]
[illegible]	Arabidopsis thaliana	[illegible]	S	[illegible]
[illegible]	Arabidopsis thaliana	[illegible]	S	[illegible]
[illegible]	Arabidopsis thaliana	[illegible]	R	[illegible]
[illegible]	Nicotiana tabacum	[illegible]	R	[illegible]
[illegible]	[illegible]	[illegible]	R	[illegible]
[illegible]	Nicotiana tabacum	[illegible]	R	[illegible]
[illegible]	Medicago sativa	[illegible]	R	[illegible]
[illegible]	Triticum monococcum	[illegible]	S	[illegible]
[illegible]	[illegible]	[illegible]	S	[illegible]
[illegible]	Sorghum bicolor	[illegible]	R	[illegible]
[illegible]	Sorghum bicolor	[illegible]	R	[illegible]
[illegible]	[illegible]	[illegible]	R	[illegible]

[illegible]

A severe reduction in secondary wall thickness of fibers, but not that of xylem vessels, as it occurs in the *Arabidopsis WALLS ARE THIN 1* (*wat1*) mutant, also increased resistance to vascular plant pathogens, such as the bacteria *R. solanacearum* and *Xanthomonas campestris* pv. *campestris*, the fungi *Verticillium dahliae* and *Verticillium alboatrum*, and the necrotrophic fungus *P. cucumerina*(Denancé et al., 2013). *WAT1* encodes a tonoplast localized indole acetic acid (auxin) transporter (Pesquet et al., 2005; Ranocha et al., 2010, 2013). Auxin content was found to be lower in roots, but not in leaves of the *wat1* mutant than in those of wild-type plants. In contrast, salicylic acid content was higher in the roots of the *wat1* mutant than in those of wild-type plants. Introduction in *wat1*plants of *NahG*, the bacterial gene coding for a salicylic acid-degrading hydroxylase, restored full susceptibility to the bacteria (Denancé et al., 2013). These data and those obtained by comparative transcriptomic analyses of *wat1* and wild-type plants suggest that *wat1*-mediated resistance is again not caused by altering the strength of the wall as a passive barrier, but that it is dependent on the activation of immune responses, mainly localized in the vascular system, which are partially dependent on the salicylic acid pathway. This defense response has been described as "vascular immunity" (Denancé et al., 2013).

Alteration of glucoronoxylans and xyloglucans or modifications in the content of wall xylose, which is the major sugar component of these polysaccharides, also impacts resistance to pathogens in*Arabidopsis*. For example, plants with enhanced levels of wall-bound xylose, as it occurs in the *de-etiolated3* (*det3*) and *irx6* mutants (Brown et al., 2005; Rogers et al., 2005) or with alterations in the structure of xyloglucan, as in the *xyl1-2* mutant (Sampedro et al., 2010), show an enhanced resistance to the necrotrophic fungus *P. cucumerina* (Delgado-Cerezo et al., 2012; Table 1). In contrast, impairment of the *ERECTA* (*ER*) gene encoding a PRR resulted in a reduced content of xylose besides other cell wall alterations in *Arabidopsis* (Sánchez-Rodríguez et al., 2009). The *er*mutant was found to be more susceptible than wild-type plants to several pathogens, such as the necrotrophic fungus *P. cucumerina*, the vascular bacterium *R. solanacearum* and the vascular oomycete *Pythium irregulare* (Table 1; Godiard et al., 2003; Llorente et al., 2005; Adie et al., 2007). The enhanced susceptibility to *P. cucumerina* and the cell wall features of the *er* mutant, including its reduced xylose content, were restored to wild-type levels by mutations in*SUPPRESSOR OF ERECTA 1* and 2 (*SER1* and *SER2)*, further suggesting a link between cell wall xylose content and resistance to pathogens (Sánchez-Rodríguez et al., 2009). Although several defense genes are constitutively up-regulated in the *ser1* and *ser2* mutants, the precise molecular basis of their resistance has not yet been fully elucidated and the *SER* genes have not been characterized yet (Sánchez-Rodríguez et al., 2009). *Arabidopsis* mutants in the Gβ and Gγ1/γ2 subunits of the heterotrimeric G protein (i.e., *agb1* single and *agg1 agg2* double mutants, respectively) also have a reduced content of xylose in their cell walls and are hypersusceptible to the necrotrophic fungi *P. cucumerina* and *Alternaria brassicicola*, the biotrophic bacterium *P. syringae* and the vascular fungus *Fusarium oxysporum* (Table 1; Llorente et al., 2005; Trusov et al., 2010; Klopffleisch et al., 2011; Delgado-Cerezo et al., 2012; Liu et al., 2013; Lorek et al., 2013;Torres et al., 2013). Interestingly, the reduced resistance of *agb1* single and *agg1 agg2* double mutants was found to be independent of defense pathways required for resistance to these pathogens, such as those regulated by abscisic acid, salicylic acid, jasmonic acid and ethylene, and those that regulate the biosynthesis of tryptophan-derived metabolites (Delgado-Cerezo et al., 2012; Lorek et al., 2013; Torres et al., 2013). It has been suggested that the reduced resistance in the *agb1* and *agg1 agg2* mutants is rather the direct consequence of a weakened cell-wall and a defective production of reactive oxygen species (ROS) upon pathogen infection (Delgado-Cerezo et al., 2012; Jiang et al., 2012; Liu et al., 2013; Lorek et al., 2013). Together, these data suggest that shifts in the xylose content of the cell wall, e.g., by altering the glucoronoxylan and xyloglucan content, are responsible, at least in part, for the altered susceptibility of some *Arabidopsis*secondary cell wall mutants to pathogens.

Cell wall polysaccharides such as xylan, (gluco) mannan and xyloglucan can be acetylated. Four*Reduced Wall Acetylation* genes (*RWA1–RWA4*) are involved in the acetylation of xylan during secondary wall biosynthesis. The expression of these genes is regulated by SND1, a transcriptional master switch of secondary wall biosynthesis (Lee et al., 2011). Remarkably, the *Arabidopsis rwa2*mutant, that has ~20% lower levels of polysaccharide *O*-acetylation but no obvious alteration in growth and development, is more resistant than wild-type plants to the necrotrophic fungus *B. cinerea* (Manabe et al., 2011). The relevance of the degree of xylan acetylation in plant resistance to pathogens is further supported by the enhanced resistance to the necrotrophic fungi *B. cinera* and*Bipolaris sorkiniana* of transgenic *Arabidopsis* and *Brachypodium distachyon*, respectively, that have a reduced xylan acetylation due to overexpression of a xylan acetylesterase from *Aspergillus nidulans* (*AnAXE*; Table 1; Pogorelko et al., 2013). These data indicate that the degree of acetylation of specific secondary cell wall polymers might be a determinant of susceptibility to particular pathogens. In addition to RWA proteins, members of the trichome birefringence (TBR) and TBR-like (TBL) protein families are also involved in the *O*-acetylation of wall polysaccharides (Gille et al., 2011). The *Arabidopsis powdery mildew resistant5* (*pmr5*) mutant, impaired in a TBL member, has a decrease in cell-wall esterification as demonstrated by Fourier transform infrared (FTIR) analysis, but it has yet to be demonstrated whether *pmr5* cell walls have an altered polysaccharide *O*-acetylation (Vogel et al., 2004; Gille and Pauly, 2012). The *pmr5* mutant is more resistant than wild-type plants to powdery mildew fungi (i.e., *E. cichoracearum* and *E. orontii),*whereas its resistance to the bacterium *P. syringae* or the oomycete *Peronospora parasitica* (re-named *Hyaloperonospora arabidopsidis)* was similar to that of wild-type plants (Vogel et al., 2004). Taken together, these data indicate that a decrease in cell wall acetylation in *Arabidopsis*resulted in an enhanced resistance to several fungi, but the molecular mechanisms explaining this resistance phenotype have yet to be elucidated.

Primary cell wall remodeling can also impact pathogen resistance, as exemplified by the enhanced resistance to some pathogens of mutants defective in the CESA subunits required for primary cell wall cellulose biosynthesis (i.e., *ixr1/cev1*; Ellis et al., 2002; Hernández-Blanco et al., 2007). Similarly, modification of the biosynthesis and/or structure (e.g., degree of methylesterification or acetylation) of wall pectins can affect pathogen resistance (Vogel et al., 2002, 2004; Lionetti et al., 2007; Raiola et al., 2011; Volpi et al., 2011; Bethke et al., 2014). The complex contribution of pectin amount/structure to the regulation of plant innate immunity has been nicely reviewed in several recent publications that also describe the different virulence mechanism used by pathogens to modify or degrade pectins in order to favor plant colonization (Ferrari et al., 2012; Lionetti et al., 2012; Bellincampi et al., 2014).

PHENYLPROPANOID AND LIGNIN BIOSYNTHESIS

Lignin is an aromatic polymer that is mainly deposited in secondary thickened cell walls where it provides strength and imperviousness. In monocot and dicot plants, lignin is mainly made from the monolignols coniferyl and sinapyl alcohol that give rise to the guaiacyl (G) and syringyl (S) units in the lignin polymer, respectively. *p*-Coumaryl alcohol, that gives rise to the *p*-hydroxyphenyl (H) units in the lignin polymer, is a minor monolignol that is slightly more abundant in monocot than in dicot cell walls. Lignin of gymnosperms is typically composed of G units and low levels of H units, but lacks S units. In several plant species, the traditional monomers are incorporated into the lignin in acylated forms. For instance, kenaf lignin is rich in sinapyl acetate-derived units, lignin of grasses has a high content of sinapyl *p*-coumarate-derived lignin units and poplar lignin incorporates sinapyl *p*-hydroxybenzoate (Morreel et al., 2004; Del Río et al., 2007; Lu and Ralph, 2008; Hatfield et al., 2009). In addition, plants do accept a range of other phenolics as lignin monomers. For example, lignin in wheat straw has relatively high levels of the flavonoid tricin (Del Río et al., 2012), whereas phenylpropanoid aldehydes and acids are found in the lignin of a range of wild-type and genetically engineered plants (Kim et al., 2000; Dauwe et al., 2007; Vanholme et al., 2012a; Van Acker et al., 2013, 2014).

The lignin biosynthetic pathway is relatively well described in many species including *Arabidopsis*, tobacco, alfalfa and poplar, and is generally divided in two branches: (i) the general phenylpropanoid pathway from Phe to feruloyl-CoA and (ii) the monolignol-specific pathway from feruloyl-CoA to the monolignols (Figure 1). At least eleven enzymes are involved in the biosynthesis of the monolignols from Phe: phenylalanine ammonia-lyase (PAL), cinnamate 4-hydroxylase (C4H), 4-coumarate:CoA ligase (4CL), hydroxycinnamoyl-CoA shikimate/quinate hydroxycinnamoyl transferase (HCT), *p*-coumarate 3-hydroxylase (C3H), caffeoyl shikimate esterase (CSE), caffeoyl-CoA *O*-methyltransferase (CCoAOMT), cinnamoyl-CoA reductase (CCR), ferulate 5-hydroxylase (F5H), caffeic acid *O*-methyltransferase (COMT), and cinnamyl alcohol dehydrogenase (CAD; Figure 1; Boerjan et al., 2003; Bonawitz and Chapple, 2010; Vanholme et al., 2010, 2013). After their biosynthesis, the monolignols are transported to the cell wall where they are oxidized by laccases and/or peroxidases to monolignol radicals. Subsequently, these monolignol radicals couple in a combinatorial fashion with the formation of various types of chemical bonds of which the ether (8-O-4), resinol (8–8), and coumaran (8–5) bonds are the most prominent ones (Boerjan et al., 2003; Ralph et al., 2004; Vanholme et al., 2012a).

Figure 1. The phenylpropanoid and monolignol biosynthetic pathway in*Arabidopsis* **(adapted from Vanholme et al., 2010, 2013).**

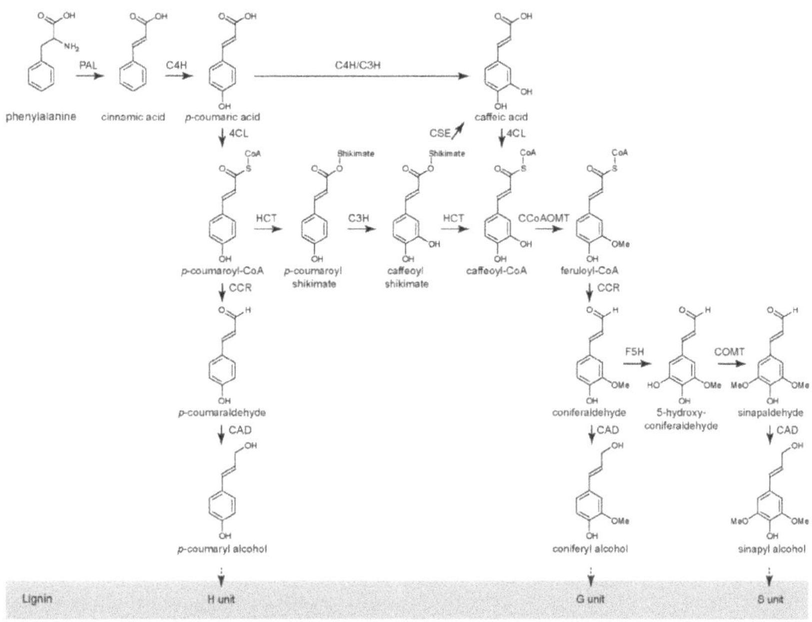

The biosynthesis and deposition of lignin in secondary cell walls is developmentally programmed and it is generally accepted that lignin provides a physical barrier against initial pathogen colonization (Buendgen et al., 1990; Bonello et al., 2003). In addition, lignin or lignin-like phenolic polymers are synthesized and rapidly deposited in cell walls in response to biotic and abiotic stresses and to cell wall structure perturbations (Caño-Delgado et al., 2003; Tronchet et al., 2010; Sattler and Funnell-Harris, 2013). The deposition of lignin in infected cells may prevent the spread of toxins and enzymes of the pathogen into the host and at the same time also the transfer of water and nutrients from the host cells to the pathogen (Smith et al., 2007). Both biotic and abiotic stresses have been shown to induce the expression of genes of the phenylpropanoid pathway in different tissues and cell cultures of several plants species, resulting in the enhanced accumulation of the corresponding enzymes, increased enzymatic activities and wall lignification (Kliebenstein et al., 2002; Bhuiyan et al., 2007; Zhao et al., 2009). For example, in Chinese cabbage (*Brassica rapa L.* ssp. *pekinensis)* infected with the necrotrophic bacterium *Erwinia carotovora* subsp. *carotovora*, transcriptomic analyses revealed that 12 genes putatively encoding enzymes involved in lignin biosynthesis were up-regulated (Zhang et al., 2007), and in *Camelina sativa* the expression of *CsCCR2* was induced after inoculation with the necrotrophic fungus *Sclerotinia sclerotiorum* (Eynck et al., 2012). In cell suspension cultures of *Linum usitatissimum* treated with different fungal PAMPs, the expression of genes encoding PAL, CCR, and CAD was elevated, PAL activity was enhanced and monolignol-derived compounds accumulated (Hano et al., 2006). Similarly, in suspension cultured cells of a blast-resistant rice genotype (*Oryza sativa* L. cv. Gigante Vercelli) treated with cell wall hydrolysates from the fungal pathogen *Magnaporthe oryzae*, up-regulation of *PAL* genes was observed (Giberti et al., 2012). Also in wheat lignification acts as a defense response against pathogens: for example, S-rich lignin was found to accumulate during the hypersensitive reaction of wheat to *Puccinia graminis* infection (Menden et al., 2007), and S lignin was made in wheat sheath epidermal cells infected with *F. proliferatum* (Bishop et al., 2002). In contrast, no changes in the lignin content occurred in wheat leaves infected with Wheat Streak Mosaic Virus (Kofalvi and Nassuth, 1995).

In addition to its function in the biosynthesis of lignin, the phenylpropanoid pathway is required for the synthesis of numerous other phenolic compounds, such as stilbenes, coumarins, (neo-)lignans, phenylpropanoid conjugates, and flavonoids (Lo and Nicholson, 1998; Yu et al., 2000, 2005; Dixon et al., 2002; Naoumkina et al., 2010). Many of these compounds are considered to be phytoalexins, i.e., antimicrobial compounds implicated in plant defense (Daayf et al., 2012;König et al., 2014). Therefore, impairing steps of the phenylpropanoid pathway can result in either accumulation or reduced abundance of these compounds, often resulting in pleiotropic effects on plant resistance (Weiergang et al., 1996; Ruuhola and Julkunen-Tiitto, 2003; Dicko et al., 2005;Pan et al., 2006; Lozovaya et al., 2007). Phenolic compounds are also important in plant-pathogen recognition. For example, the expression of *Agrobacterium tumefaciens Virulence* (*Vir*) genes, needed for infection, is induced by phenolic compounds and infection of the host plant cannot take place in the absence of these compounds (Lee et al., 1992, 1995; Maury et al., 2010).

EFFECT OF PHENOLIC CONTENT AND LIGNIN MODIFICATIONS ON PATHOGEN RESISTANCE

Evidence for a role for lignin and soluble phenolics in plant defense has been obtained from the analysis of the pathogen resistance of transgenic plants and mutants with contrasting lignin amount or composition. For example, in tomato, the total content of soluble phenolics and lignin were significantly higher in varieties that were resistant to the vascular bacterium *R. solanacearum*than in susceptible ones, and this enhanced resistance was associated to a greater accumulation of lignin in roots upon bacterial infection, a process that was triggered by salicylic acid (Mandal et al., 2011, 2013). In tobacco, plants down-regulated for *PAL* had reduced levels of chlorogenic acid and exhibited more rapid and extensive lesion development than wild-type plants upon infection with the fungal pathogen *Cercospora nicotianae* (Maher et al., 1994). The increased disease susceptibility in this *PAL*-suppressed line was suggested not to result from the inhibition of the pathogen-induced response, but rather from the decrease in the developmental accumulation of chlorogenic acid (Maher et al., 1994). However, the lignin content in this *PAL*-suppressed line was not determined and therefore it cannot be excluded that the enhanced susceptibility was caused by reduced lignin content or a weaker cell wall (Maher et al., 1994). In accordance with the latter studies, transgenic tobacco plants constitutively overexpressing *PAL* genes showed a higher tolerance toward *C. nicotianae* and *Phytophthora parasitica* pv. *nicotianae* (Way et al., 2002,2011; Shadle et al., 2003). Notably, *COMT* and *CCoAOMT* antisense tobacco lines were more resistant to *Agrobacterium tumefaciens* infection and showed a reduced tumor area and mass relative to wild-type plants (Maury et al., 2010). The phenolic compounds secreted by these antisense plants upon wounding did not induce the expression of the bacterial *Vir* genes equally well as those secreted from wounded wild-type plants (Maury et al., 2010). In other words, the*Agrobacterium* did not recognize its host because of the difference in soluble phenolics.

In cotton (*Gossypium hirsutum*), quantitative analysis of resistance to the wilt fungus *V. dahliae*revealed an association between increased lignification in the stems upon infection and resistance against wilt (Xu et al., 2011). In line with these data, overexpression of the cotton *DIRIGENT1* gene, which enhances lignification, blocks the spread of *V. dahliae* (Shi et al., 2012). In alfalfa (*Medicago sativa*), down-regulation of the *HCT* gene leads to plants with reduced lignin levels, constitutive defense responses and enhanced tolerance to the fungal pathogen *Colletotrichum trifolii*. This activation of defense responses was hypothesized to be triggered by bioactive cell wall fragments released from the secondary cell wall (Gallego-Giraldo et al., 2011b). In melon (*Cucumis melo*), lignin accumulation upon infection was found to increase faster and to a higher level in lines resistant to the powdery mildew fungus *Podosphaera fusca* than in susceptible lines, and this differential accumulation correlated with enhanced PAL levels (Romero et al., 2008). Lignin composition seems to play an important role in pathogen resistance in flax, as RNAi-mediated suppression of a *CAD* gene increased flax susceptibility to the vascular fungus *F. oxysporum*(Wróbel-Kwiatkowska et al., 2007).

Also in grasses the effect of modifying lignin biosynthesis on plant susceptibility has been investigated. For instance, in wheat *(Triticum monococcum)*, silencing the monolignol biosynthesis genes *TmPAL*, *TmCOMT*, *TmCCoAOMT*, and *TmCAD* led to super-susceptibility of leaf tissues to the fungus*B. graminis* f. sp. *tritici*, the causal agent of powdery mildew disease (Bhuiyan et al., 2009). The increased accumulation of mono- and diferulates in the cell walls of oat and wheat upon infection with *P. coronate* sp. *avenae* and *Agrobacterium* sp., respectively, has been associated with resistance toward these pathogens (Ikegawa et al., 1996; Parrott et al., 2002). In transgenic rice overexpressing the *NPR1 HOMOLOG 1* (*NH1*), a suppressor mutant screening was performed and a mutation in the *SUPPRESSSOR OF NH1-MEDIATED LESION FORMATION AND RESISTANCE* (*SNL6*) gene, which encodes a CCR-like protein, was selected. *snl6* mutants had a lower lignin content and a reduced resistance to the bacterium *X. oryzae* pv. *oryzae* (Bart et al., 2010). Mutations in *BROWN MIDRIB 6* (*BMR6*) and *BMR12* in sorghum (*Sorghum bicolor* L.) allowed the development of forage and grain lines with a reduced lignin content and modified lignin composition (Oliver et al., 2005). The *bmr6* and *bmr12* mutants, that are defective in CAD and COMT proteins, respectively, restricted the growth of different *Fusarium* spp. (*F. thapsinum*,*F. proliferatum*, and *F. verticillioides*), but not that of *Gibberella fujikuroi* (Bout and Vermerris, 2003; Sattler et al., 2009; Funnell-Harris et al., 2010). It is unknown whether the alteration of lignin composition or the accumulation of phenolic compounds is causative to the enhanced resistance of these sorghum mutants to *Fusarium* sp.

In trees, the contribution of lignin amount or composition on susceptibility to pathogens has been also investigated. For example, in eucalyptus, the deposition of lignin in necrophylatic periderm in the early stages of infection by *Mycosphaerella* explains the greater resistance of *Eucalyptus nitens*as compared with *E. globulus* (Smith et al., 2007). Comparative metabolite profiling of xylem tissue of *Ulmus minor* and *Ulmus minor* × *Ulmus pumila* after inoculation with *Ophiostoma novo-ulmi*showed that the hybrid has a faster defense response, which is characterized by an increase in the amount of lignin (Martin et al., 2007). Similarly, the infection of *Pinus nigra* by *Sphaeropsis sapinea* induces an increase in the deposition of lignin that was associated to resistance (Bonello and Blodgett, 2003). Interestingly, in hybrid poplar (*Populus tremula* × *Populus alba*) no increased disease incidence was observed in field-grown antisense *COMT* and *CAD* lines relative to that observed in wild-type trees, nicely showing that altered lignin biosynthesis does not necessarily negatively impact resistance to pathogens (Pilate et al., 2002; Halpin et al., 2007).

In summary, a general positive correlation between lignin amount and pathogen resistance has been observed, in particular when the plant-pathogen interaction concerns vascular pathogens, such as *Fusarium* sp., *Xanthomonas* sp. or *Verticilium* sp., that generally spread through the secondary-thickened xylem. In the majority of the examples analyzed, the impact of lignin modification on the regulation of other defense responses has not been studied, and it is not yet possible to conclude whether the role of lignin in resistance is merely passive or active by regulating specific immune responses.

CONTRIBUTION OF LIGNIN TO PATHOGEN RESISTANCE IN *ARABIDOPSIS*

Several lines of evidence support a role for lignin in immunity of *Arabidopsis* to pathogens. The expression of some lignin-biosynthesis genes was induced, and the amount of lignin increased, by treating *Arabidopsis* with hormones (i.e., salicylic acid, abscisic acid or jasmonic acid) that regulate plant defense (Mohr and Cahill, 2007; Chen et al., 2009; Gallego-Giraldo et al., 2011a). Similarly, infection of *Arabidopsis* with particular pathogens, such as the bacteria *P. syringae* pv. *tomato* and *X. campestris,* resulted in increased expression of lignin-biosynthesis genes and in higher lignin levels (Mohr and Cahill, 2007; Quentin et al., 2009). Experiments also hinted to specific stress-related roles for the different gene family members involved in certain enzymatic conversions. For example, the *Arabidopsis CCR2* gene has been suggested to participate in the hypersensitive response to *X. campestris* as its expression was up-regulated after inoculation with this bacterium, in contrast to *CCR1* which was preferentially expressed during development (Lauvergeat et al., 2001).

The analysis of *Arabidopsis* mutants defective in lignin biosynthesis and of transgenic plants overexpressing lignin biosynthesis genes has contributed to unravel the role of lignin in plant immunity. For example, two *pal1/2/3/4* quadruple mutants with 20% and 25% residual lignin levels and 25% residual salicylic acid levels, showed a stunted growth and were hypersusceptible to *P. syringae*. In addition, the total salicylic acid levels in the quadruple mutants after infection were about 50% of those in wild type, suggesting that pathogen-induced, salicylic acid-mediated resistance might be partially impaired in this mutant (Huang et al., 2010). The *Arabidopsis comt* mutant was found to be slightly more susceptible than wild-type plants to *P. syringae* pv. *tomato DC3000*, but also to *B. cinerea*, *A. brassicicola* and *X. campestris* pv. *campestris*, and *B. graminis* f. sp. *hordei*, that is a barley pathogen that does not colonize *Arabidopsis* plants (Quentin et al., 2009). Unexpectedly, asexual sporulation of the oomycete *H. arabidopsidis*, causing the downy mildew disease, was impaired in the *comt* mutant (Quentin et al., 2009). This phenotype was not correlated with an increased salicylic and jasmonic acid-dependent defense, but with a higher frequency of oomycete sexual reproduction within *comt* mutant tissues (Quentin et al., 2009). It was further proven that *comt* mutants accumulated soluble 5-hydroxyferuloyl-malate and that this compound promoted sexual oomycete reproduction *in vitro* (Quentin et al., 2009). *Arabidopsis f5h1* mutants showed an increased susceptibility to the fungal pathogen *S. sclerotiorum* and to the vascular fungus *Verticillium longisporum* (Huang et al., 2009; König et al., 2014). The *f5h1* mutants have similar amounts of lignin as compared to wild type, but lack S units in the lignin and are sinapate ester deficient (Meyer et al., 1998; Vanholme et al., 2012b; König et al., 2014). Because sinapate esters inhibit fungal growth *in vitro*, their absence in *f5h1* might explain the mutant's increased susceptibility toward fungal pathogens (König et al., 2014). Interestingly, the *UGT72E2* over-expressing *Arabidopsis* line, in which lignin was not altered but the soluble phenylpropanoid coniferin accumulated, was less susceptible to *V. longisporum* (König et al., 2014). All together these results again strongly support that not only the lignin polymer but also the soluble phenolic pool plays a significant role in the defense of plants against pathogens. As described above for crops and other plant species, a general positive correlation between lignin amount and resistance toward pathogens has been observed in *Arabidopsis* too. However, a deeper characterization of the impact of the lignin alterations on *Arabidopsis* immune responses and its relation with the content and profile of phenolics is needed to elucidate the molecular mechanisms explaining the differential responses of the mutants to pathogen infection (Table 1).

In *Arabidopsis*, the plant's response to the perturbation of lignin has been studied in a collection of mutants, each mutated in a single gene of this pathway, by combining transcriptomics and metabolomics (Vanholme et al., 2012b; Figure 1). These analyses revealed that *c4h*, *4cl1*, *ccoaomt1*, and *ccr1* mutants, that produced less lignin, upregulated the shikimate, methyl-donor, and phenylpropanoid pathways (i.e., the pathways supplying the monolignols), whereas, *f5h1* and *comt* mutants, that provoked lignin compositional shifts, downregulated the very same pathways (Vanholme et al., 2012b). Moreover, some of these mutant alleles revealed subtle differences in the metabolic and gene expression profiles that might contribute to differential resistance responses to pathogens (Vanholme et al., 2012b). This collection of mutants represents a unique tool to further characterize the specific contribution of lignin biosynthesis to resistance against different types of pathogens in *Arabidopsis*.

EFFECTS OF MODIFYING LIGNIN AND SECONDARY CELL WALL STRUCTURE ON PLANT PATHOGEN RESISTANCE: HYPOTHESES AND OPEN QUESTIONS

Several hypotheses can be formulated to explain the resistance or the susceptibility observed in mutants and transgenic plants affected in the amount and/or composition of lignin and, more broadly, the secondary cell wall (Figure 2): (i) The perturbation of lignin or secondary cell wall structure modifies the physical barrier that pathogens must overcome to invade the plant. This perturbation can lead to enhanced resistance or susceptibility as these pathogens might lack the enzymes required for a proper degradation of this novel physical barrier, or wall degradation by pathogens is now facilitated, respectively; (ii) The reduced amount of lignin or the modification of lignin or secondary cell wall composition impacts the strength of the secondary cell wall resulting in collapsed xylem. A drop in vascular conduction might negatively contribute to plant colonization by vascular pathogens; (iii) The reduction of lignin amount and the modification of the secondary cell wall can loosen the wall, facilitating the constitutive or pathogen-induced release of cell wall DAMPs, which might trigger immune responses resulting in enhanced resistance to pathogens; (iv) The perturbation of the lignin pathway could lead to the accumulation of soluble phenolic compounds that are either toxic to some pathogens (e.g., resulting in a reduced virulence), or serve as a new carbon or nutrient source for pathogens that then will grow better (e.g., resulting in enhanced virulence); (v) Similarly, the perturbation of the lignin pathway could also lead to the accumulation or decrease of soluble phenolics that are plant-pathogen recognition compounds, which would result in an enhanced susceptibility or resistance, respectively. Although these hypotheses might explain some of the published phenotypes, other molecular explanations cannot be excluded and a deeper molecular and biochemical characterization is required for a better understanding of the contribution of the secondary cell wall to pathogen resistance.

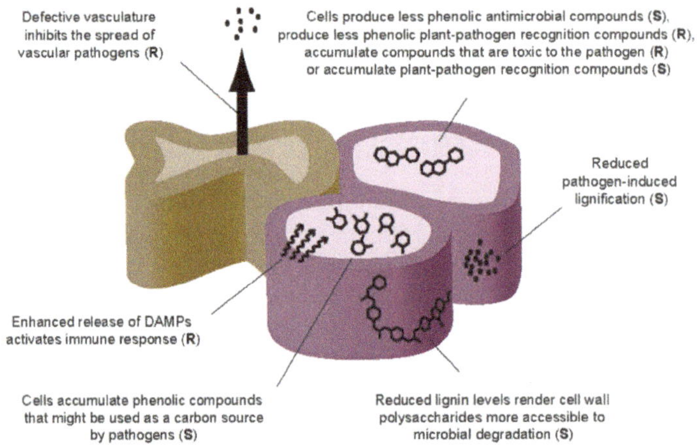

Figure 2. Model illustrating lignin and secondary cell wall modifications resulting in enhanced resistance (R) or enhanced susceptibility (S) to pathogens.

PERSPECTIVES

Lignin negatively impacts the conversion of lignocellulosic biomass into fermentable sugars, making it one of the most important limiting factors in the processing of plant biomass to pulp and biofuels (Chen and Dixon, 2007; Dien et al., 2009; Van Acker et al., 2013). Hence, modifications of the plant secondary cell wall can contribute to the improvement of biomass processing in paper mills and bio-refineries. However, one critical question for lignocellulosic feedstock development is whether engineering the secondary cell wall, including its lignin content and composition, will affect plant defense against pathogens. In this review, we have summarized the role of the cell wall in plant resistance toward pathogens. We conclude that plants with altered secondary cell walls may either have an enhanced or a reduced resistance toward pathogens, or no effect at all, depending on the alterations made and the pathogens tested. Because our current knowledge on the role of the cell wall (primary or secondary) in defense against pathogens is still fragmentary, it is difficult to predict how specific alterations of the cell wall will influence a plant's resistance toward pathogens. A deeper investigation of the role of the plant cell wall in pathogen resistance and the biochemical networks underlying this resistance is required.

CONFLICT OF INTEREST STATEMENT

The authors declare that the research was conducted in the absence of any commercial or financial relationships that could be construed as a potential conflict of interest.

ACKNOWLEDGMENTS

We gratefully acknowledge funding through the Spanish Ministry of Economy and Competitiveness (MINECO, grants BIO2012-32910 and PIM2010PKB-00749), the European Commission's Directorate-General for Research within the 7th Framework Program (FP7/2007–2013) under the grant agreement No. 311804 (SUNLIBB) and No. 270089 (MULTIBIOPRO), Stanford University's Global Climate and Energy Project (GCEP, "Lignin management: optimizing yield and composition in lignin-modified plants"), and the Multidisciplinary Research Partnership "Biotechnology for a Sustainable Economy" (01MRB510W) of Ghent University. Ruben Vanholme is indebted to the Research Foundation-Flanders for a postdoctoral fellowship. Eva Miedes was a JdC postdoctoral fellow from the Spanish MINECO.

REFERENCES

1. Adie, B. A., Perez-Perez, J., Perez-Perez, M. M., Godoy, M., Sanchez-Serrano, J. J., Schmelz, E. A.,et al. (2007). ABA is an essential signal for plant resistance to pathogens affecting JA biosynthesis and the activation of defenses in *Arabidopsis*. *Plant Cell* 19, 1665–1681. doi: 10.1105/tpc.106.048041
2. Bart, R. S., Chern, M., Vega-Sánchez, M. E., Canlas, P., and Ronald, P. C. (2010). Rice Snl6, a cinnamoyl-CoA reductase-Like gene family member, is required for NH1-mediated immunity to *Xanthomonas oryzae* pv. *oryzae*. *PLoS Genet.* 6:e1001123. doi: 10.1371/journal.pgen.1001123
3. Bellincampi, D., Cervone, F., and Lionetti, V. (2014). Plant cell wall dynamics and wall-related susceptibility in plant–pathogen interactions. *Front. Plant Sci.* 5:228. doi: 10.3389/fpls.2014.00228
4. Bethke, G., Grundman, R. E., Sreekanta, S., Truman, W., Katagiri, F., and Glazebrook, J.

(2014). *Arabidopsis* PECTIN METHYLESTERASEs contribute to immunity against *Pseudomonas syringae*. *Plant Physiol.* 164, 1093–1107. doi: 10.1104/pp.113.227637

5. Bhuiyan, N., Liu, W., Liu, G., Selvaraj, G., Wei, Y., and King, J. (2007). Transcriptional regulation of genes involved in the pathways of biosynthesis and supply of methyl units in response to powdery mildew attack and abiotic stresses in wheat. *Plant Mol. Biol.* 64, 305–318. doi: 10.1007/s11103-007-9155-x
6. Bhuiyan, N. H., Selvaraj, G., Wei, Y., and King, J. (2009). Gene expression profiling and silencing reveal that monolignol biosynthesis plays a critical role in penetration defence in wheat against powdery mildew invasion. *J. Exp. Bot.* 60, 509–521. doi: 10.1093/jxb/ern290
7. Bishop, D. L., Chyatterton, N. J., Harrison, P. A., and Hatfield, R. D. (2002). Changes in carbohydrate partitioning and cell wall remodelling with stress-induced pathogenesis in wheat sheaths. *Physiol. Mol. Plant Pathol.* 61, 53–63. doi: 10.1016/S0885-5765(02)90416-9
8. Boerjan, W., Ralph, J., and Baucher, M. (2003). Lignin biosynthesis. *Annu. Rev. Plant Biol.* 54, 519–546. doi: 10.1146/annurev.arplant.54.031902.134938
9. Bonawitz, N. D., and Chapple, C. (2010). The genetics of lignin biosynthesis: connecting genotype to phenotype. *Annu. Rev. Genet.* 44, 337–363. doi: 10.1146/annurev-genet-102209-163508
10. Bonawitz, N. D., and Chapple, C. (2013). Can genetic engineering of lignin deposition be accomplished without an unacceptable yield penalty? *Curr. Opin. Plant Biol.* 24, 336–343. doi: 10.1016/j.copbio.2012.11.004
11. Bonawitz, N. D., Kim, J. I., Tobimatsu, Y., Ciesielski, P. N., Anderson, N. A., Ximenes, E.,et al. (2014). Disruption of mediator rescues the stunted growth of a lignin-deficient *Arabidopsis* mutant. *Nature* 15, 509, 376–380. doi: 10.1038/nature13084
12. Bonello, P., and Blodgett, J. T. (2003). *Pinus nigra–Sphaeropsis sapinea* as a model pathosystem to investigate local and systemic effects of fungal infection of pines. *Physiol. Mol. Plant Physiol.* 63, 249–261. doi: 10.1016/j.pmpp.2004.02.002
13. Bonello, P., Storer, A. J., Gordon, T. R., Wood, D. L., and Heller, W. (2003). Systemic effects of *Heterobasidion annosum* on ferulic acid glucoside and lignin of presymptomatic ponderosa pine phloem, and potential effects on bark-beetle-associated fungi. *J. Chem. Ecol.* 29, 1167–1182. doi: 10.1023/A:1023833707382
14. Bout, S., and Vermerris, W. (2003). A candidate-gene approach to clone the sorghum Brown midrib gene encoding caffeic acid O-methyltransferase. *Mol. Genet. Genomics* 269, 205–214.
15. Brown, D. M., Zeef, L. A. H., Ellis, J., Goodacreb, R., and Turner, S. R. (2005). Identification of novel genes in Arabidopsis involved in secondary cell wall formation using expression profiling and reverse genetics. *Plant Cell* 17, 2281–2295. doi: 10.1105/tpc.105.031542
16. Buendgen, M. R., Coors, J. G., Grombacher, A. W., and Russell, W. A. (1990). European corn borer resistance and cell wall composition of three maize populations. *Crop Sci.* 30, 505–510. doi: 10.2135/cropsci1990.0011183X003000030005x
17. Caño-Delgado, A., Penfield, S., Smith, C., Catley, M., and Bevan, M. (2003). Reduced cellulose synthesis invokes lignification and defense responses in *Arabidopsis thaliana*. *Plant J.* 34, 351–362. doi: 10.1046/j.1365-313X.2003.01729.x
18. Cantu, D., Vicente, A. R., Labavitch, J. M., Bennett, A. B., and Powell, A. L. (2008). Strangers in the matrix: plant cell walls and pathogen susceptibility. *Trends Plant Sci.* 13, 610–617. doi: 10.1016/j.tplants.2008.09.002
19. Carpita, N., and McCann, M. (2000). "The cell wall," in *Biochemistry and Molecular Biology of Plants*, eds B. B. Buchanan, G. Wilhelm, and R. L. Jones (Rockville, IL: American Society of Plant Physiologists), 52–108.
20. Chen, F., and Dixon, R. A. (2007). Lignin modification improves fermentable sugar yields for biofuel production. *Nat. Biotechnol.* 25, 759–761. doi: 10.1038/nbt1316
21. Chen, Z., Zheng, Z., Huang, J., Lai, Z., and Fan, B. (2009). Biosynthesis of salicylic acid in plants. *Plant Signal. Behav.* 4, 493–496. doi: 10.4161/psb.4.6.8392
22. Cosgrove, D. J. (2005). Growth of the plant cell wall. *Nat. Rev. Mol. Cell Biol.* 6, 850–861. doi: 10.1038/nrm1746
23. Daayf, F., El Hadrami, A., El-Bebany, A. F., Henriquez, M. A., Yao, Z., Derksen, H.,et al. (2012). "Phenolic compounds in plant defense and pathogen counter-defense mechanisms," in *Recent Advances in Polyphenol Research*, eds V. Cheynier, P. Sarni-Manchado, and S. Quideau (Oxford, UK: Wiley-Blackwell). doi: 10.1002/9781118299753.ch8
24. Dauwe, R., Morreel, K., Goeminne, G., Gielen, B., Rohde, A., Van Beeumen, J.,et al. (2007). Molecular phenotyping of lignin-modified tobacco reveals associated changes in cell–wall metabolism, primary metabolism, stress metabolism and photorespiration. *Plant J.* 52, 263–285. doi: 10.1111/j.1365-313X.2007.03233.x
25. Del Río, J. C., Marques, G., Rencoret, J., Martínez,Á. T., and Gutiérrez, A. (2007). Occurrence of naturally acetylated lignin units. *J. Agric. Food Chem.* 55, 5461–5468. doi: 10.1021/jf0705264
26. Del Río, J. C., Rencoret, J., Prinsen, P., Martínez,Á. T., Ralph, J., and Gutiérrez, A. (2012). Structural characterization of wheat straw lignin as revealed by analytical pyrolysis, 2D-NMR, and reductive cleavage methods. *J. Agric. Food Chem.* 60, 5922–5935. doi: 10.1021/jf301002n
27. Delgado-Cerezo, M., Sánchez-Rodríguez, C., Escudero, V., Miedes, E., Fernández, P. V., Jordá, L.,et al. (2012). *Arabidopsis*heterotrimeric G-protein regulates cell wall defense and resistance to necrotrophic fungi. *Mol. Plant* 5, 98–114. doi: 10.1093/mp/ssr082
28. Denancé, N., Ranocha, P., Oria, N., Barlet, X., Rivière, M. P., Yadeta, K. A.,et al. (2013). *Arabidopsis* wat1 (walls are thin1) mediated resistance to the bacterial vascular pathogen, *Ralstonia solanacearum*, is

accompanied by cross-regulation of salicylic acid and tryptophan metabolism. *Plant J.* 73, 225–239. doi: 10.1111/tpj.12027

29. Dicko, M. H., Gruppen, H., Barro, C., Traore, A. S., Van Berkel, W. J. H., and Voragen, A. G. J. (2005). Impact of phenolic compounds and related enzymes in sorghum varieties for resistance and susceptibility to biotic and abiotic stresses. *J. Chem. Ecol.* 31, 2671–2688. doi: 10.1007/s10886-005-7619-5

30. Dien, B. S., Sarath, G., Pedersen, J. F., Sattler, S. E., Chen, H., Funnell-Harris, D. L.,et al. (2009). Improved sugar conversion and ethanol yield for forage sorghum (*Sorghum bicolor* L. Moench) lines with reduced lignin contents. *Bioenergy Res.* 2, 153–164. doi: 10.1007/s12155-009-9041-2

31. Dixon, R. A., Achnine, L., Kota, P., Liu, C. J., Reddy, M. S., and Wang, L. (2002). The phenylpropanoid pathway and plant defence – a genomics perspective. *Mol. Plant Pathol.* 3, 371–390. doi: 10.1046/j.1364-3703.2002.00131.x

32. Dodds, P. N., and Rathjen, J. P. (2010). Plant immunity: towards an integrated view of plant–pathogen interactions. *Nat. Rev. Genet.* 11, 539–548. doi: 10.1038/nrg2812

33. Ellis, C., Karafyllidis, I., Wasternack, C., and Turner, J. G. (2002). The *Arabidopsis* mutant cev1 links cell wall signaling to jasmonate and ethylene responses. *Plant Cell* 14, 1557–1566. doi: 10.1105/tpc.002022

34. Engelsdorf, T., and Hamann, T. (2014). An update on receptor-like kinase involvement in plant cell wall integrity maintenance.*Ann. Bot.* doi: 10.1093/aob/mcu043 [Epub ahead of print].

35. Eynck, C., Séguin-Swartz, G., Clarke, W. E., and Parkin, I. A. (2012). Monolignol biosynthesis is associated with resistance to*Sclerotinia sclerotiorum* in *Camelina sativa*. *Mol. Plant Pathol.* 13, 887–899. doi: 10.1111/j.1364-3703.2012.00798.x

36. Ferrari, S., Savatin, D. V., Sicilia, F., Gramegna, G., Cervone F., and DeLorenzo, G. (2013). Oligogalacturonides: plant damage associated molecular patterns and regulators of growth and development. *Front. Plant Sci.* 4:49. doi: 10.3389/fpls.2013.00049

37. Ferrari, S., Sella, L., Janni, M., Favaron, F., and D'Ovidio, R. (2012). Transgenic expression of polygalacturonase-inhibiting proteins in *Arabidopsis* and wheat increases resistance to the flower pathogen *Fusarium graminearum*. *Plant Biol.* 14, 31–38. doi: 10.1111/j.1438-8677.2011.00449.x

38. Funnell-Harris, D. L., Pedersen, J. F., and Sattler, S. E. (2010). Alteration in lignin biosynthesis restricts growth of *Fusarium*spp. in brown midrib sorghum. *Phytopathology* 100, 671–681. doi: 10.1094/PHYTO-100-7-0671

39. Gallego-Giraldo, L., Escamilla-Trevino, L., Jackson, L. A., and Dixon, R. A. (2011a). Salicylic acid mediates the reduced growth of lignin down-regulated plants. *Proc. Natl. Acad. Sci. U.S.A.* 108, 20814–20819. doi: 10.1073/pnas.1117873108

40. Gallego-Giraldo, L., Jikumaru, Y., Kamiya, Y., Tang, Y., and Dixon, R.A. (2011b). Selective lignin down-regulation leads to constitutive defense response expression in alfalfa (*Medicago sativa* L.). *New Phytol.* 190, 627–639. doi: 10.1111/j.1469-8137.2010.03621.x

41. García-Olmedo, F., Rodríguez-Palenzuela, P., Molina, A., Alamillo, J. M., López-Solanilla, E., Berrocal-Lobo, M.,et al. (2001). Antibiotic activities of peptides, hydrogen peroxide and peroxynitrite in plant defence. *FEBS Lett.* 498, 219–222. doi: 10.1016/S0014-5793(01)02456-5

42. Giberti, S., Bertea, C. M., Narayana, R., Maffei, M. E., and Forlani, G. (2012). Two phenylalanine ammonia lyase isoforms are involved in the elicitor-induced response of rice to the fungal pathogen *Magnaporthe oryzae*. *J. Plant Physiol.* 169, 249–254. doi: 10.1016/j.jplph.2011.10.008

43. Gille, S., de Souza, A., Xiong, G., Benz, M., Cheng, K., Schultink, A.,et al. (2011). O-acetylation of *Arabidopsis* hemicellulose xyloglucan requires AXY4 or AXY4L, proteins with a TBL and DUF231 domain. *Plant Cell* 23, 4041–4053. doi: 10.1105/tpc.111.091728

44. Gille, S., and Pauly, M. (2012). O-acetylation of plant cell wall polysaccharides. *Front. Plant Sci.* 3:12. doi: 10.3389/fpls.2012.00012

45. Glazebrook, J. (2005). Contrasting mechanisms of defense against biotrophic and necrotrophic pathogens. *Annu. Rev. Phytopathol.* 43, 205–227. doi: 10.1146/annurev.phyto.43.040204.135923

46. Godiard, L., Sauviac, L., Torii, K. U., Grenon, O., Mangin, B., and Grimsley, N. H.,et al. (2003). ERECTA, an LRR receptor-like kinase protein controlling development pleiotropically affects resistance to bacterial wilt. *Plant J.* 36, 353–365. doi: 10.1046/j.1365-313X.2003.01877.x

47. Halpin, C., Thain, S. C., Tilston, E. L., Guiney, E., Lapierre, C., and Hopkins, D. W. (2007). Ecological impacts of trees with modified lignin. *Tree Genet. Genomes* 3, 101–110. doi: 10.1007/s11295-006-0060-2

48. Hano, C., Martin, I., Fliniaux, O., Legrand, B., Gutierrez, L., Arroo, R. R.,et al. (2006). Pinoresinol–lariciresinol reductase gene expression and secoisolariciresinol diglucoside accumulation in developing flax (*Linum usitatissimum*) seeds. *Planta* 224, 1291–1301. doi: 10.1007/s00425-006-0308-y

49. Hatfield, R. D., Marita, J. M., Frost, K., Grabber, J., Ralph, J., Lu, F.,et al. (2009). Grass lignin acylation: p-coumaroyl transferase activity and cell wall characteristics of C3 and C4 grasses. *Planta* 229, 1253–1267. doi: 10.1007/s00425-009-0900-z

50. Hematy, K., Cherk, C., and Somerville, S. (2009). Host–pathogen warfare at the plant cell wall. *Curr. Opin. Plant Biol.* 12, 406–413. doi: 10.1016/j.pbi.2009.06.007

51. Hernández-Blanco, C., Feng, D. X., Hu, J., Sánchez-Vallet, A., Deslandes, L., Llorente, F.,et al. (2007). Impairment of cellulose synthases required for *Arabidopsis* secondary cell wall formation enhances disease resistance. *Plant Cell* 19, 890–903. doi: 10.1105/tpc.106.048058

52. Huang, J., Bhinu, V. S., Li, X., Bashi, Z. D., Zhou, R., and Hannoufa, A. (2009). Pleiotropic changes in *Arabidopsis* f5h and sct mutants revealed by large-scale gene expression and metabolite analysis. *Planta* 230, 1057–1069. doi: 10.1007/s00425-009-1007-2

53. Huang, J., Gu, M., Lai, Z., Fan, B., Shi, K., Zhou Y. H.,et al. (2010). Functional analysis of the Arabidopsis PAL gene family in plant growth, development, and response to environmental stress. *Plant Physiol.* 153, 1526–1538. doi: 10.1104/pp.110.157370

54. Ikegawa, T., Mayama, S., Nakayashiki, H., and Kato, H. (1996). Accumulation of diferulic acid during the hypersensitive response of oat leaves to *Puccinia coronate* f. sp. *avenae* and its role in the resistance of oat tissues to cell wall degrading enzymes. *Physiol. Mol. Plant Pathol.* 48, 245–256. doi: 10.1006/pmpp.1996.0021

55. Jiang, K., Frick-Cheng, A., Trusov, Y., Delgado-Cerezo, M., Rosenthal, D. M., Lorek, J.,et al. (2012). Dissecting *Arabidopsis* G-β signal transduction on the protein surface. *Plant Physiol.* 159, 975–983. doi: 10.1104/pp.112.196337

56. Kim, H., Ralph, J., Yahiaoui, N., Pean, M., and Boudet, A.-M. (2000). Cross-coupling of hydroxycinnamyl aldehydes into lignins. *Org. Lett.* 2, 2197–2200. doi: 10.1021/ol005906o

57. Kliebenstein, D. J., Lim, J. E., Landry, L. G., and Last, R. L. (2002). *Arabidopsis* UVR8 regulates ultraviolet-B signal transduction and tolerance and contains sequence similarity to human regulator of chromatin condensation 1. *Plant Physiol.*130, 234–243. doi: 10.1104/pp.005041

58. Klopffleisch, K., Phan, N., Augustin, K., Bayne, R. S., Booker, K. S., Botella, J. R.et al. (2011). *Arabidopsis* G-protein interactome reveals connections to cell wall carbohydrates and morphogenesis. *Mol. Syst. Biol.* 7, 532. doi: 10.1038/msb.2011.66

59. Kofalvi, S. A., and Nassuth, A. (1995). Influence of wheat streak mosaic virus infection on phenylpropanoid metabolism and the accumulation of phenolics and lignin in wheat. *Physiol. Mol. Plant Pathol.* 47, 365–377. doi: 10.1006/pmpp.1995.1065

60. König, S., Feussner, K., Kaever, A., Landesfeind, M., Thurow, C., Karlovsky, P.,et al. (2014). Soluble phenylpropanoids are involved in the defense response of *Arabidopsis* against *Verticillium longisporum*. *New Phytol.* 202, 823–837. doi: 10.1111/nph.12709

61. Lauvergeat, V., Lacomme, C., Lacombe, E., Lasserre, E., Roby, D., and Grima-Pettenati, J. (2001). Two cinnamoyl-CoA reductase (CCR) genes from *Arabidopsis thaliana* are differentially expressed during development and in response to infection with pathogenic bacteria. *Phytochemistry* 57, 1187–1195. doi: 10.1016/S0031-9422(01)00053-X

62. Lee, C., Teng, Q., Zhong, R., and Ye, Z. H. (2011). The four *Arabidopsis* REDUCED WALL ACETYLATION genes are expressed in secondary wall-containing cells and required for the acetylation of xylan. *Plant Cell Physiol.* 52, 1289–1301. doi: 10.1093/pcp/pcr075

63. Lee, K., Dudley, M. W., Hess, K. M., Lynn, D. G., Joerger, R. D., and Binns, A. N. (1992). Mechanism of activation of*Agrobacterium* virulence genes: identification of phenol-binding proteins. *Proc. Natl. Acad. Sci. U.S.A.* 89, 8666–8670. doi: 10.1073/pnas.89.18.8666

64. Lee, Y. W., Jin, S., Sims, W. S., and Nester, E. W. (1995). Genetic evidence for direct sensing of phenolic compounds by the VirA protein of *Agrobacterium tumefaciens*. *Proc. Natl. Acad. Sci. U.S.A.* 92, 12245–12249. doi: 10.1073/pnas.92.26.12245

65. Lionetti, V., Cervone, F., and Bellincampi, D. (2012). Methyl esterification of pectin plays a role during plant–pathogen interactions and affects plant resistance to diseases. *J. Plant Physiol.* 169, 1623–1630. doi: 10.1016/j.jplph.2012.05.006

66. Lionetti, V., Raiola, A., Camardella, L., Giovane, A., Obel, N., Pauly, M.,et al. (2007). Overexpression of pectin methylesterase inhibitors in *Arabidopsis* restricts fungal infection by *Botrytis cinerea*. *Plant Physiol.* 143, 1871–1880. doi: 10.1104/pp.106.090803

67. Liu, J., Ding, P., Sun, T., Nitta, Y., Dong, O., Huang, X.,et al. (2013). Heterotrimeric G proteins serve as a converging point in plant defense signaling activated by multiple receptor-like kinases. *Plant Physiol.* 161, 2146–2158. doi: 10.1104/pp.112.212431

68. Llorente, F., Alonso-Blanco, C., Sanchez-Rodriguez, C., Jorda, L., and Molina, A. (2005). ERECTA receptor-like kinase and heterotrimeric G protein from *Arabidopsis* are required for resistance to the necrotrophic fungus *Plectosphaerella cucumerina*.*Plant J.* 43, 165–180. doi: 10.1111/j.1365-313X.2005.02440.x

69. Lo, S. C. C., and Nicholson, R. L. (1998). Reduction of light-induced anthocyanin accumulation in inoculated sorghum mesocotyls: implications for a compensatory role in the defense response. *Plant Physiol.* 116, 979–989. doi: 10.1104/pp.116.3.979

70. Lorek,J.,Griebel,T.,Jones,A.M.,Kuhn,H.,andPanstruga, R. (2013). The role of *Arabidopsis* heterotrimeric G-protein subunits in MLO2 function and MAMP-triggered immunity. *Mol. Plant Microbe. Interact.* 26, 991–1003. doi: 10.1094/MPMI-03-13-0077-R

71. Lozovaya, V. V., Lygin, A. V., Zernova, O. V., Ulanov, A. V., Li, S. X., Hartman, G. L.,et al. (2007). Modification of phenolic metabolism in soybean hairy roots through down regulation of chalcone synthase or isoflavone synthase. *Planta* 225, 665–679. doi: 10.1007/s00425-006-0368-z

72. Lu, F., and Ralph, J. (2008). Novel tetrahydrofuran structures derived from β-β-coupling reactions involving sinapyl acetate in kenaf lignins. *Org. Biomol. Chem.* 6, 3681–3694. doi: 10.1039/b809464k

73. Macho, A. P., and Zipfel, C. (2014). Plant PRRs and the activation of innate immune signaling. *Mol. Cell.* 54, 263–272. doi: 10.1016/j.molcel.2014.03.028

74. Maher, E. A., Bate, N. J., Ni, W., Elkind, Y., Dixon, R. A., and Lamb, C. J. (1994). Increased disease susceptibility of transgenic tobacco plants with suppressed levels of preformed phenylpropanoid products. *Proc. Natl. Acad. Sci. U.S.A.* 91, 7802–7806. doi: 10.1073/pnas.91.16.7802

75. Malinovsky, F. G., Fangel, J. U., and Willats, W. G. T. (2014). The role of the cell wall in plant immunity. *Front. Plant Sci.* 5:178. doi: 10.3389/fpls.2014.00178

76. Manabe, Y., Nafisi, M., Verhertbruggen, Y., Orfila, C., Gille, S., Rautengarten, C.,et al. (2011). Loss-of-function mutation of REDUCED WALL ACETYLATION in *Arabidopsis* leads to reduced cell wall acetylation and increased resistance to *Botrytis cinerea*. *Plant Physiol.* 155, 1068–1078. doi: 10.1104/pp.110.168989

77. Mandal, S., Das, R. K., and Mishra, S. (2011). Differential occurrence of oxidative burst and antioxidative mechanism in compatible and incompatible interactions of tomato and *Ralstonia solanacearum*. *Plant Physiol. Biochem.* 49, 117–123. doi: 10.1016/j.plaphy.2010.10.006

78. Mandal, S., Kar, I., Mukherjee, A. K., and Acharya, P. (2013). Elicitor-induced defense responses in *Solanum lycopersicum*against *Ralstonia solanacearum*. *Sci. World J.* 25:561056. doi: 10.1155/2013/561056

79. Martin, J. A., Solla, A., Woodward, S., and Gil, L. (2007). Detection of differential changes in lignin composition of elm xylem tissues inoculated with *Ophiostoma novoulmi* using Fourier transform-infrared spectroscopy. *For. Pathol.* 37, 187–191. doi: 10.1111/j.1439-0329.2007.00495.x

80. Maury, S., Delaunay, A., Mesnard, F., Cronier, D., Chabbert, B., Geoffroy, P.,et al. (2010). O-methyltransferase(s)-suppressed plants produce lower amounts of phenolic vir inducers and are less susceptible to *Agrobacterium tumefaciens* infection. *Planta*232, 975–986. doi: 10.1007/s00425-010-1230-x

81. Menden, B., Kohlhoff, M., and Moerschbacher, B. M. (2007). Wheat cells accumulate a syringyl-rich lignin during the hypersensitive resistance response. *Phytochemistry* 68, 513–520. doi: 10.1016/j.phytochem.2006.11.011

82. Meyer, K., Shirley, A. M., Cusumano, J. C., Bell-Lelong, D. A., and Chapple, C. (1998). Lignin monomer composition is determined by the expression of a cytochrome P450-dependent monooxygenase in *Arabidopsis*. *Proc. Natl. Acad. Sci. U.S.A.*95, 6619–6623. doi: 10.1073/pnas.95.12.6619

83. Mohr, P. G., and Cahill, D. M. (2007). Suppression by ABA of salicylic acid and lignin accumulation and the expression of multiple genes, in *Arabidopsis* infected with *Pseudomonas syringae* pv. tomato. *Funct. Integr. Genomics* 7, 181–191. doi: 10.1007/s10142-006-0041-4

84. Morreel, K., Ralph, J., Kim, H., Lu, F., Goeminne, G., Ralph, S.,et al. (2004). Profiling of oligolignols reveals monolignol coupling conditions in lignifying poplar xylem. *Plant Physiol.* 136, 3537–3549. doi: 10.1104/pp.104.049304

85. Naoumkina, M. A., Zhao, Q., Gallego-Giraldo, L., Dai, X., Zhao, P. X., and Dixon, R. A. (2010). Genome-wide analysis of phenylpropanoid defense pathways. *Mol. Plant Pathol.* 11, 829–846. doi: 10.1111/j.1364-3703.2010.00648.x

86. Oliver, A. L., Pedersen, J. F., Grant, R. J., Klopfenstein, T. J., and Jose, H. (2005). Comparative effects of the sorghum bmr-6 and bmr-12 genes: II. Grain yield, stover yield, and stover quality in grain sorghum. *Crop Sci.* 45, 2240–2245. doi: 10.2135/cropsci2004.0660

87. Ong, R. G., Chundawat, S. P. S., Hodge, D. B., Keskar, S., and Dale, B. E. (2014). "Linking plant biology and pretreatment: understanding the structure and organization of the plant cell wall and interactions with cellulosic biofuel production," in *Plants and BioEnergy, Advances in Plant Biology*, eds M. C. McCann, M. S. Buckeridge, and N. C. Carpita (New York, NY: Springer), 231–253.

88. Pan, Q., Zhan, J., Liu, H., Zhang, J., Chen, J., Wen, P.,et al. (2006). Salicylic acid synthesized by benzoic acid 2-hydroxylase participates in the development of thermotolerance in pea plants. *Plant Sci.* 171, 226–233. doi: 10.1016/j.plantsci.2006.03.012

89. Panstruga, R., Parker, J. E., and Schulze-Lefert, P. (2009). SnapShot: plant immune response pathways. *Cell* 136, 978–U976. doi: 10.1016/j.cell.2009.02.020

90. Parrott, D. L., Anderson, A. J., and Carman, J. G. (2002). *Agrobacterium* induces plant cell death in wheat (*Triticum aestivum*L.). *Physiol. Mol. Plant Pathol.* 60, 59–69. doi: 10.1006/pmpp.2002.0378

91. Pesquet, E., Ranocha, P., Legay, S., Digonnet, C., Barbier, O., Pichon, M.,et al. (2005). Novel markers of xylogenesis in zinnia are differentially regulated by auxin and cytokinin. *Plant Physiol.* 139, 1821–1839. doi: 10.1104/pp.105.064337

92. Petersen, P. D., Lau, J., Ebert, B., Yang, F., Verhertbruggen, Y., Kim, J. S.,et al. (2012). Engineering of plants with improved properties as biofuels feedstocks by vessel-specific complementation of xylan biosynthesis mutants. *Biotechnol. Biofuels* 26, 84. doi: 10.1186/1754-6834-5-84

93. Pilate, G., Guiney, E., Holt, K., Petit-Conil, M., Lapierre, C., Leplé, J.-C.,et al. (2002). Field and pulping performances of transgenic trees with altered lignification. *Nat. Biotechnol.* 20, 607–612. doi: 10.1038/nbt0602-607

94. Pogorelko, G., Lionetti, V., Fursova, O., Sundaram, R. M., Qi, M., and Whitham, S. A.,et al. (2013). *Arabidopsis* and*Brachypodium distachyon* transgenic plants expressing *Aspergillus nidulans* acetylesterases have decreased degree of polysaccharide acetylation and increased resistance to pathogens. *Plant Physiol.* 162, 9–23. doi: 10.1104/pp.113.214460

95. Quentin, M., Allasia, V., Pegard, A., Allais, F., Ducrot, P. H., Favery, B.,et al. (2009). Imbalanced lignin biosynthesis promotes the sexual reproduction of homothallic oomycete pathogens. *PLoS Pathog.* 5:e1000264. doi: 10.1371/journal.ppat.1000264

96. Raiola, A., Lionetti, V., Elmaghraby, I., Immerzeel, P., Mellerowicz, E. J., Salvi, G.,et al. (2011). Pectin methylesterase is induced in *Arabidopsis* upon infection and is necessary for a successful colonization by necrotrophic pathogens. *Mol. Plant Microbe Interact.* 24, 432–440. doi: 10.1094/MPMI-07-10-0157

97. Ralph, J., Lundquist, K., Brunow, G., Lu, F., Kim, H., Schatz, P. F.,et al. (2004). Lignins: natural polymers from oxidative coupling of 4-hydroxyphenylpropanoids. *Phytochem. Rev.* 3, 29–60. doi: 10.1023/B:PHYT.0000047809.65444.a4

98. Ramírez, V., Agorio, A., Coego, A., García-Andrade, J., Hernández, M. J., Balaguer, B.,et al. (2011). MYB46 modulates disease susceptibility to *Botrytis cinerea* in *Arabidopsis*. *Plant Physiol.* 155, 1920–1935. doi: 10.1104/pp.110.171843

99. Ranocha, P., Denancé, N., Vanholme, R., Freydier, A., Martinez, Y., Hoffmann, L.,et al. (2010). Walls are thin 1 (WAT1), an*Arabidopsis* homolog of *Medicago truncatula* NODULIN21, is a tonoplast-localized protein required for secondary wall formation in fibers. *Plant J.* 63, 469–483. doi: 10.1111/j.1365-313X.2010.04256.x

100. Ranocha, P., Dima, O., Nagy, R., Felten, J., Corratgé-Faillie, C., Novak, O.,et al. (2013). *Arabidopsis* WAT1 is a vacuolar auxin transport facilitator required for auxin homeostasis. *Nat. Commun.* 4, 2625. doi: 10.1038/ncomms3625

101. Ringli, C. (2010). Monitoring the outside: cell wall-sensing mechanisms. *Plant Physiol.* 153, 1445–1452. doi: 10.1104/pp.110.154518

102. Rogers, L. A., Dubos, C., Surman, C., Willment, J., Cullis, I. F., Mansfield, S. D.,et al. (2005). Comparison of lignin deposition in three ectopic lignification mutants. *New Phytol.* 168, 123–140. doi: 10.1111/j.1469-8137.2005.01496.x

103. Romero, D., Rivera, M. E., Cazorla, F. M., Codina, J. C., Fernández-Ortuño, D., Torés, J. A.,et al. (2008). Comparative histochemical analyses of oxidative burst and cell wall reinforcement in compatible and incompatible melon-powdery mildew (*Podosphaera fusca*) interactions. *J. Plant Physiol.* 165, 1895–1905. doi: 10.1016/j.jplph.2008.04.020

104. Ruuhola, T., and Julkunen-Tiitto, R. (2003). Trade-off between synthesis of salicylates and growth of micropropagated *Salix pentandra*. *J. Chem. Ecol.* 29, 1565–1588. doi: 10.1023/A:1024266612585

105. Sampedro, J., Pardo, B., Gianzo, C., Guitián, E., Revilla, G., and Zarra, I. (2010). Lack of α-xylosidase activity in *Arabidopsis*alters xyloglucan composition and results in growth defects. *Plant Physiol.* 154, 1105–1115. doi: 10.1104/pp.110.163212

106. Sánchez-Rodríguez, C., Estevez, J. M., Llorente, F., Hernandez-Blanco, C., Jorda, L., Pagan, I.,et al. (2009). The ERECTA Receptor-Like Kinase regulates cell wall mediated resistance to pathogens in *Arabidopsis thaliana*. *Mol. Plant Microbe. Interact.* 22, 953–963. doi: 10.1094/MPMI-22-8-0953

107. Sánchez-Vallet, A., Ramos, B., Bednarek, P., López, G., Piślewska-Bednarek, M., Schulze-Lefert, P.,et al. (2010). Tryptophan-derived secondary metabolites in *Arabidopsis thaliana* confer non-host resistance to necrotrophic *Plectosphaerella cucumerina*fungi. *Plant J.* 63, 115–127. doi: 10.1111/j.1365-313X.2010.04224.x

108. Sarkar, P., Bosneaga, E., and Auer, M. (2009). Plant cell walls throughout evolution: towards a molecular understanding of their design principles. *J. Exp. Bot.* 60, 3615–3635. doi: 10.1093/jxb/erp245

109. Sattler, S. E., and Funnell-Harris, D. L. (2013). Modifying lignin to improve bioenergy feedstocks: strengthening the barrier against pathogens? *Front Plant Sci.* 4:70. doi: 10.3389/fpls.2013.00070

110. Sattler, S. E., Saathoff, A. J., Haas, E. J., Palmer, N. A., Funnell-Harris, D. L., Sarath, G.,et al. (2009). A nonsense mutation in a cinnamyl alcohol dehydrogenase gene is responsible for the sorghum brown midrib 6 phenotype. *Plant Physiol.* 150, 584–595. doi: 10.1104/pp.109.136408

111. Schulze-Lefert, P. (2004). Knocking on the heaven's wall: pathogenesis of and resistance to biotrophic fungi at the cell wall.*Curr. Opin. Plant Biol.* 7, 377–383. doi: 10.1016/j.pbi.2004.05.004

112. Shadle, G. L., Wesley, S. W., Korth, K. L., Chen, F., Lamb, C., and Dixon, R. A. (2003). Phenylpropanoid compounds and disease resistance in transgenic tobacco with altered expression of l-phenylalanine ammonia-lyase. *Phytochemistry* 64, 153–161. doi: 10.1016/S0031-9422(03)00151-1

113. Shi, H., Liu, Z., Zhu, L., Zhang, C., Chen, Y., Zhou, Y.,et al. (2012). Overexpression of cotton (*Gossypium hirsutum*) dirigent1 gene enhances lignification that blocks the spread of *Verticillium dahliae*. *Acta Biochim. Biophys. Sin.* 44, 555–564. doi: 10.1093/abbs/gms035

114. Simmons, B. A., Loqué, D., and Ralph, J. (2010). Advances in modifying lignin for enhanced biofuel production. *Curr. Opin. Plant Biol.* 13, 313–320. doi: 10.1016/j.pbi.2010.03.001

115. Smith, A. H., Gill, W. M., Pinkard, E. A., and Mohammed, C. L. (2007). Anatomical and histochemical defence responses induced in juvenile leaves of *Eucalyptus globulus* and *Eucalyptus nitens* by *Mycosphaerella* infection. *For. Pathol.* 37, 361–373. doi: 10.1111/j.1439-0329.2007.00502.x

116. Torres, M. A., Morales, J., Sánchez-Rodríguez, C., Molina, A., and Dangl, J. L. (2013). Functional interplay between *Arabidopsis*NADPH oxidases and heterotrimeric G protein. *Mol. Plant Microbe Interact.* 26, 686–694. doi: 10.1094/MPMI-10-12-0236-R

117. Tronchet, M., Balagué, C., Kroj, T., Jouanin, L., and Roby, D. (2010). Cinnamyl alcohol dehydrogenases C and D, key enzymes in lignin biosynthesis, play an essential role in disease resistance in *Arabidopsis*. *Mol. Plant Pathol.* 11, 83–92. doi: 10.1111/j.1364-3703.2009.00578.x

118. Trusov, Y., Jordá, L., Molina, A., and Botella, J. R. (2010). "G proteins and plant innate immunity," in *Integrated G Proteins Signaling in Plants*, eds S. Yalovsky, F. Baluška, and A. Jones (Berlin, Heidelberg: Springer),221–250.

119. Underwood, W. (2012). The plant cell wall: a dynamic barrier against pathogen invasion. *Front. Plant Sci.* 3:85. doi: 10.3389/fpls.2012.00085

120. Van Acker, R., Leplé, J.-C., Aerts, D., Storme, V., Goeminne, G., Ivens, B.,et al. (2014). Improved saccharification and ethanol yield from field-grown

transgenic poplar deficient in cinnamoyl-CoA reductase. *Proc. Natl. Acad. Sci. U.S.A.* 111, 845–850. doi: 10.1073/pnas.1321673111

121. Van Acker, R., Vanholme, R., Storme, V., Mortimer, J. C., Dupree, P., and Boerjan, W. (2013). Lignin biosynthesis perturbations affect secondary cell wall composition and saccharification yield in *Arabidopsis thaliana*. *Biotechnol. Biofuels* 6:46. doi: 10.1186/1754-6834-6-46

122. Vanholme, R., Cesarino, I., Rataj, K., Xiao, Y., Sundin, L., Goeminne, G.,et al. (2013). Caffeoyl shikimate esterase (CSE) is an enzyme in the lignin biosynthetic pathway in *Arabidopsis*. *Science* 341, 1103–1106. doi: 10.1126/science.1241602

123. Vanholme, R., Demedts, B., Morreel, K., Ralph, J., and Boerjan, W. (2010). Lignin biosynthesis and structure. *Plant Physiol.*153, 895–905. doi: 10.1104/pp.110.155119

124. Vanholme, R., Morreel, K., Darrah, C., Oyarce, P., Grabber, J. H., Ralph, J.,et al. (2012a). Metabolic engineering of novel lignin in biomass crops. *New Phytol.* 196, 978–1000. doi: 10.1111/j.1469-8137.2012.04337.x

125. Vanholme, R., Storme, V., Vanholme, B., Sundin, L., Christensen, J. H., Goeminne, G.,et al. (2012b). A systems biology view of responses to lignin biosynthesis perturbations in *Arabidopsis*. *Plant Cell* 24, 3506–3529. doi: 10.1105/tpc.112.102574

126. Vogel, J. P., Raab, T. K., Schiff, C., and Somerville, S. C. (2002). PMR6, a pectate lyase-like gene required for powdery mildew susceptibility in *Arabidopsis*. *Plant Cell* 14, 2095–2106. doi: 10.1105/tpc.003509

127. Vogel, J. P., Raab, T. K., Somerville, C. R., and Somerville, S. C. (2004). Mutations in PMR5 result in powdery mildew resistance and altered cell wall composition. *Plant J.* 40, 968–978. doi: 10.1111/j.1365-313X.2004.02264.x

128. Volpi, C., Janni, M., Lionetti, V., Bellincampi, D., Favaron, F., and D'Ovidio, R. (2011). The ectopic expression of a pectin methyl esterase inhibitor increases pectin methyl esterification and limits fungal diseases in wheat. *Mol. Plant Microbe. Interact.* 24, 1012–1019. doi: 10.1094/MPMI-01-11-0021

129. Vorwerk, S., Somerville, S., and Somerville, C. (2004). The role of plant cell wall polysaccharide composition in plant disease resistance. *Trends Plant Sci.* 9, 203–209. doi: 10.1016/j.tplants.2004.02.005

130. Way, H. M., Birch, R. G., and Manners, J. M. (2011). A comparison of individual and combined l-phenylalanine ammonia lyase and cationic peroxidase transgenes for engineering resistance in tobacco to necrotrophic pathogens. *Plant Biotechnol. Rep.* 5, 301–308. doi: 10.1007/s11816-011-0183-2

131. Way, H. M., Kazan, K., Mitter, N., Goulter, K. C., Birch, R. G., and Manners, J. M. (2002). Constitutive expression of a phenylalanine ammonia-lyase gene from *Stylosanthes humilis* in transgenic tobacco leads to enhanced disease resistance but impaired plant growth. *Physiol. Mol. Plant Pathol.* 60, 275–282. doi: 10.1016/S0885-5765(02)90407-8

132. Weiergang, I., Hipskind, J. D., and Nicholson, R. L. (1996). Synthesis of 3-deoxyanthocyanidin phytoalexins in sorghum occurs independent of light. *Physiol. Mol. PlantPathol.* 49, 377–388. doi: 10.1006/pmpp.1996.0060

133. Wolf, S., Hématy, K., and Höfte, H. (2012). Growth control and cell wall signaling in plants. *Annu. Rev. Plant Biol.* 63, 381–407. doi: 10.1146/annurev-arplant-042811-105449

134. Wróbel-Kwiatkowska, M., Starzycki, M., Zebrowski, J., Oszmiański, J., and Szopa, J. (2007). Lignin deficiency in transgenic flax resulted in plants with improved mechanical properties. *J. Biotechnol.* 128, 919–934. doi: 10.1016/j.jbiotec.2006.12.030

135. Xu, L., Zhu, L., Tu, L., Liu, L., Yuan, D., Jin, L.,et al. (2011). Lignin metabolism has a central role in the resistance of cotton to the wilt fungus *Verticillium dahliae* as revealed by RNA-Seq-dependent transcriptional analysis and histochemistry. *J. Exp. Bot.* 62, 5607–5621. doi: 10.1093/jxb/err245

136. Yang, F., Mitra, P., Zhang, L., Prak, L., Verhertbruggen, Y., Kim, J. S.,et al. (2013). Engineering secondary cell wall deposition in plants. *Plant Biotechnol. J.* 11, 325–335. doi: 10.1111/pbi.12016

137. Yu, C. K., Springob, K., Schmidt, J., Nicholson, R. L., Chu, I. K., Yip, W. K.,et al. (2005). A stilbene synthase gene (SbSTS1) is involved in host and nonhost defense responses in sorghum. *Plant Physiol.* 138, 393–401. doi: 10.1104/pp.105.059337

138. Yu, O., Jung, W., Shi, J., Croes, R. A., Fader, G. M., McGonigle, B.,et al. (2000). Production of the isoflavones genistein and daidzein in non-legume dicot and monocot tissues. *Plant Physiol.* 124, 781–794. doi: 10.1104/pp.124.2.781

139. Zhang, S. H., Yang, Q., and Ma, R. C (2007). *Erwinia carotovora* ssp. carotovora infection induced "defense lignin" accumulation and lignin biosynthetic gene expression in Chinese cabbage (*Brassica rapa* L. ssp. pekinensis). *J. Integr. Plant Biol.*49, 993–1002. doi: 10.1111/j.1672-9072.2007.00478.x

140. Zhao, J., Buchwaldt, L., Rimmer, S. R., Sharpe, A., Mcgregor, L., Bekkoui, D.,et al. (2009). Patterns of differential gene expression in *Brassica napus* cultivars infected with *Sclerotinia sclerotiorum*. *Mol. Plant Pathol.* 10, 635–649. doi: 10.1111/j.1364-3703.2009.00558.x

CHAPTER 2 WOUNDING IN THE PLANT TISSUE

Daniel V. Savatin, Giovanna Gramegna, Vanessa Modesti and Felice Cervone

Keywords: Wounding, oligogalacturonides, wound response, systemin, jasmonate

INTRODUCTION

The sessile condition exposes plants to any possible environmental stress. Injury, one of the most frequent stress conditions that plants must face, may cause both loss of nutrients and entry of microbes. Therefore, plants have evolved sophisticated mechanisms to promptly respond to wounding, rapidly heal the tissue and prevent microbe infections. Unlike animals, plants do not have mobile cells specialized for defense, but each plant cell has the capability to activate protective mechanisms upon injury sensing. The capacity of cells to activate defense responses upon "danger" sensing and recognition of no self microbe-associated molecular patterns (MAMPs) and/or endogenous damage-associated molecular patterns (DAMPs) is characteristic of the plant innate immunity (Akira et al., 2006). Defense responses activated by wounding are similar and overlapping with those activated by MAMPs and DAMPs, indicating that both injury and pathogens are limited by plants in a similar manner. Most of our knowledge on wounding derives from studies in which plants are mechanically damaged. Mechanical injury activates defenses that are similar to those induced by herbivores and insects (Reymond et al., 2000; Arimura et al., 2005; Rehrig et al., 2014), although the damage caused by herbivores has peculiar characteristics and mechanical wounding is necessary but not sufficient to trigger the full response activated by insects (Maffei et al., 2007). Differences between mechanical wounding and attacks by herbivores and insects will be discussed in other reviews of this issue (Krautz et al., 2014).

Plants contrast wounding with both constitutive structures, such as epicuticular films and crystals of wax, and secretory conduits for latices or resins, that restrict the access of opportunistic microorganisms to the tissue, as well as wound-induced responses that, unlike the constitutive defenses, are energetically costly (Leon et al., 2001; Bonaventure and Baldwin, 2010) and thereby regulated and triggered only when required. The primary events of the response to wounding occur at the injured site (local response) while the undamaged tissues respond later (systemic response), upon perception of mobile signals that communicate the existence of a critical condition (Farmer and Ryan, 1992). Extracellular signals such as cell wall-derived oligogalacturonides (OGs) and peptides, like systemin, have been characterized as typical signals of wounding (Roberts, 1992). Genes involved in biosynthesis of jasmonic acid (JA) and ethylene (ET) as well as genes for general stress responses (oxidative stress, dehydration stress, heat-shock proteins, etc.) are rapidly induced (Reymond et al., 2000; Delessert et al., 2004). Later,

Citation: Savatin DV, Gramegna G, Modesti V and Cervone F (2014) Wounding in the plant tissue. Front. Plant Sci. 5:470. doi: 10.3389/fpls.2014.00470

events of protein turnover and transport processes involving aquaporins, lipid transfer proteins, ABC transporters, sugar, and peptide transporters occur. Finally, the modulation of primary metabolism (carbohydrate and lipid metabolism, nitrogen assimilation) and the expression of genes involved in the biosynthesis of secondary metabolites with repellent or anti-digestive activity [i.e., glucosinolates, cyanogenic glucosides, alkaloids, phenolics, and proteinase inhibitors (PI)] may occur. Transcriptional profiling analyses performed on 8.200 *Arabidopsis* genes revealed that approximately 8% of these genes are differential expressed after wounding and about 20% of the wounding-regulated genes encode proteins involved in signal transduction, such as members in the AP2, WRKY, and MYB families (Cheong et al., 2002). A large fraction of the wound-responsive genes are also responsive to pathogens, suggesting that signaling pathways activated by these stimuli are shared (see below). This clearly indicates that an important reprograming of gene expression occurs in plants to defend the damaged tissue, which represents an easy passage for pathogen invasion. Receptors and signal transduction elements usually involved in pathogen response as well as several putative disease resistance genes (*R* genes) are up-regulated by wounding (Cheong et al., 2002). This review is focused on what is currently known about the putative signals that are released upon wounding, on the mechanism of their perception and transduction and on plant defense responses activated upon injury sensing.

SENSING THE WOUNDING THROUGH THE CELL WALL INTEGRITY

The cuticle, composed of cutin and cuticular waxes, covers the outermost epidermal cell wall in the aerial portions of plants (Riederer and Schreiber, 2001; Nawrath, 2006). The integrity of cuticle and cell wall (CWI) is affected by wounding and may be sensed by the plant cells. Studies on*Arabidopsis* plants expressing cutinase (Sieber et al., 2000) and on mutants impaired in cuticle biosynthesis have indicated that a more permeable cuticle allows the passage of diffusates with growth-inhibiting activity against phytopathogenic fungi (Bessire et al., 2007; Chassot et al., 2007). A breach in the cuticle caused by wounding also favors the diffusion of elicitors that, therefore, have an easier access to the cell surface, while an intact cuticle may prevent their passage from the surface. An early recognition of elicitors may lead to a prompt and efficient activation of the immune responses (L'Haridon et al., 2011; Benikhlef et al., 2013).

The cell wall is a dynamic structure that play a critical role in growth and development as well as in preventing wounding and pathogen attack (Bellincampi et al., 2014). The perception of an altered CWI is a key event during wounding (Nuhse, 2012; Wolf et al., 2012). Subtle modifications caused by physical perturbations such as light touch, soft mechanical stress, wind or contact with insects may be sensed at the level of the plasma membrane through stretch-activated mechano-sensitive channels that increase the intracellular Ca^{2+} concentration and trigger further signal transduction events (Nakagawa et al., 2007; Haswell et al., 2008; Benikhlef et al., 2013). Expression of the*Arabidopsis* calcium channels *MATING INDUCED DEATH 1* (MID1)-COMPLEMENTING ACTIVITY 1 (MCA1) and MCA2 in yeast complements the lethal effect of loss-of-function mutations in the *MID1* and *MID2* genes encoding stretch-activated calcium channels (Kanzaki et al., 1999;Ketela et al., 1999) and promote calcium influx upon mechanical stimulation (Nakagawa et al., 2007; Yamanaka et al., 2010). Putative cation channels belonging to the GLUTAMATE RECEPTOR-LIKE (GLR) family, which mediate calcium influxes in response to MAMPs (Kwaaitaal et al., 2011), are also required for the expression of several JA-inducible genes upon wounding in*Arabidopsis* (Mousavi et al., 2013).

Several *Arabidopsis* receptor like kinases (RLKs), among which those belonging to the*Catharanthus roseus* receptor-like kinase 1 family that includes THESEUS 1 (THE1), HERCULES 1, and FERONIA, have been proposed as possible sensors of CWI (Hematy et al., 2007; Guo et al., 2009). Loss of THE1 function attenuates the growth defects and ectopic lignification phenotype caused by a mutation in the *CELLULOSE SYNTHASE 6* (Hematy et al., 2007). Moreover, THE1 is involved in the accumulation of reactive oxygen species (ROS) and lignin deposition induced by isoxaben, an inhibitor of cellulose synthesis (Denness et al., 2011). This evidence clearly shows that defects in the cell wall caused by a disturbance of cellulose biosynthesis are sensed through THE1.

The monitoring of the status of pectin contributes to the sensing of CWI alterations (De Lorenzo et al., 2011). Plants carrying mutations that significantly alter pectin integrity, such as *quasimodo 2* or*tumorous shoot development 2*, exhibit constitutive induction of defense responses (Krupkova et al., 2007; Mouille et al., 2007). However, minor modifications in the methylation status, which occur in transgenic plants overexpressing the inhibitors of pectin methylesterases or in KO mutants of *PECTIN METHYLESTERASE 3*, do not influence the expression of defense genes (Lionetti et al., 2007, 2010, 2012; Raiola et al., 2011). Possible indicator of an altered pectin integrity is the presence of OGs, a well-known class of DAMPs that, similarly to MAMPs, act as danger signals for the activation of the immune responses (Boller and Felix, 2009; De Lorenzo et al., 2011). OGs are released from the plant cell walls upon partial degradation of homogalacturonan, the main component of pectin, by wound-induced hydrolytic enzymes or, during infections, by microbial hydrolytic enzymes. The size of OGs is critical for their elicitor activity, being OGs with a degree of polymerization (DP) between 10 and 15 most active while shorter oligomers are inactive. OGs induce in several plant species a wide range of defense responses, including production of ROS, nitric oxide, phytoalexins, glucanase, chitinase, and callose (Bellincampi et al., 2000; Galletti et al., 2008; Rasul et al., 2012; Ferrari et al., 2013). In tomato, OGs, probably generated by the action of a wound-inducible plant-derived polygalacturonase (PG; Bergey et al., 1999), induce the accumulation of PI (Ryan and Jagendorf, 1995). OGs may act only locally, because their oligoanionic nature confers them a limited mobility in the tissues (Baydoun and Fry, 1985). In*Arabidopsis*, both wounding and OG treatment induce a strong local resistance against the necrotrophic fungus *Botrytis cinerea* that is independent of salicylic acid (SA)- and JA-mediated signaling (Chassot et al., 2007; Ferrari et al., 2007). OGs also antagonize auxin responses (Branca et al., 1988; Bellincampi et al., 1996; Ferrari et al., 2008; Savatin et al., 2011), but the auxin-

OG antagonism is uncoupled from their activity as defense elicitors. Indeed, the *Arabidopsis* mitogen-activated protein (MAP) kinase kinase kinases ANPs have been identified as elements in the OG-mediated induction of defenses, but do not play a major role in the inhibition of the auxin-induced gene expression (Savatin et al., 2014).

OG sensing in *Arabidopsis* may involve wall-associated kinases (WAKs; Brutus et al., 2010; Kohorn and Kohorn, 2012). WAKs are RLKs consisting of an extracellular domain, containing epidermal growth factor repeats, a transmembrane domain and a cytoplasmic Ser/Thr kinase domain. The extracellular domain of WAK1 and WAK2 also contains an N-terminal portion that binds pectin *in vitro* (Decreux and Messiaen, 2005; Kohorn et al., 2009). OGs with a DP > 9 bind reversibly WAK1 and the binding increases when OGs are present as dimers in a calcium-mediated "egg box" conformation (Decreux and Messiaen, 2005; Cabrera et al., 2008). Five *WAK* genes are clustered on chromosome 1 and additional 21 *WAK-like* genes (*WAKL*) are present in *Arabidopsis* (Verica et al., 2003). *WAK1*, *WAK2*, *WAKL5*, and *WAKL7* are induced by wounding (Wagner and Kohorn, 2001; Verica et al., 2003). In rice, *OsWAK1* is also induced by mechanical wounding as well as by SA and methyl-JA (MeJA) but not by abscisic acid (ABA; Li et al., 2008). A role of WAK2 in the wound response was suggested by the analysis of gene expression in plants overexpressing WAK2 fused to a TAP epitope (Kohorn et al., 2012).

A lectin receptor kinase-I.9 (DORN1), which plays a role in the perception of extracellular ATP, is also involved in the wound responses. Indeed, ectopic expression of DORN1 enhances expression of genes co-regulated by wounding and ATP (Choi et al., 2014). A maize wound-induced gene encoding a leucine-rich RLK (WPK1) is involved in JA- and phytochrome-mediated signaling (He et al., 2005). In tobacco, a leucine-rich repeat RLK (WRK) is involved in the JA-dependent wound signaling and acts upstream of the SA- and wound-induced protein kinases SIPK and WIPK, respectively (Seo et al., 1995; Zhang and Klessig, 1998a,b; Takabatake et al., 2006). WRK expression increases 15 min after wounding (Ito et al., 2002). WRK orthologs are present in dicots (*Arabidopsis* and tomato) but not in monocots (rice and wheat; Takabatake et al., 2006).

WOUND-ASSOCIATED DAMPS

Peptides that function as DAMPs have been isolated in wounded tissues. Systemin, a 18-aminoacid peptide, was identified in tomato after wounding or insect attack as a cleavage product released into the apoplast from prosystemin, i.e., a larger cytoplasmic precursor protein that accumulates in the cytosol of phloem parenchyma cells (Jacinto et al., 1997; Narvaez-Vasquez and Ryan, 2004;Schilmiller and Howe, 2005). Sensing of systemin activates the biosynthesis of JA, which, in turn, activates defenses responses in neighboring cells (Orozco-Cardenas et al., 1993). The systemin receptor was identified as the tomato homolog of the brassinosteroid receptor BRI1, SR160 (Scheer and Ryan, 1999), but more recent findings argued against this evidence (Hind et al., 2010). Hydroxyproline-rich systemins (HypSys) that trigger plant immunity during herbivore or pathogen attack (Heiling et al., 2010; Bhattacharya et al., 2013) have been identified in Solanaceae (Pearce et al., 2001, 2007, 2009; Pearce and Ryan, 2003; Bhattacharya et al., 2013) and in sweet potato (Chen et al., 2008). HypSys peptides, as systemin, are processed from precursor proteins which are induced by wounding (Narvaez-Vasquez et al., 2005).

A peptide, Pep1, was identified in *Arabidopsis* for its capability of inducing alkalinization in suspension-cultured cells. Pep1 is a 23-amino acid peptide released from the C-terminus of a 92 amino acid precursor protein, PROPEP1, which is induced by wounding, MeJA and ET. *PROPEP1*belongs to a gene family of eight members. The family members *PROPEP2* and *PROPEP3*, and, to a lesser extent, *PROPEP1* are strongly induced by microbial pathogens such as *B. cinerea*,*Phytophthora infestans*, and *Pseudomonas syringae* as well as by various MAMPs and DAMPs elicitors, including NPP1, HrpZ, flg22, and OGs (Craigon et al., 2004; Toufighi et al., 2005; Denoux et al., 2008). PROPEPs are localized in the cytosol and the tonoplast and may function in the amplification/modulation of elicitor-triggered responses rather than being signals responsible for the initiation of the defense responses (Huffaker et al., 2006; Bartels et al., 2013). Homologues of AtPeps have been identified in maize. ZmPep1 regulates disease responses whereas ZmPep3 triggers the biosynthesis of JA and ET and induces the production of anti-herbivore volatiles (Huffaker et al., 2011, 2013). AtPeps are perceived by two RLKs (PEPR1 and PEPR2), which share structural and functional similarity to the MAMP receptors FLS2 and EFR (Yamaguchi et al., 2006;Krol et al., 2010). PEPR1 and PEPR2 are induced by wounding and MeJA but not by SA and 1-aminocyclopropane-1-carboxylic acid (ACC) synthase (Yamaguchi et al., 2010). They are also differentially induced by DAMPs (AtPeps and OGs) and MAMP (flg22 and elf18; Zipfel et al., 2004,2006; Denoux et al., 2008; Yamaguchi et al., 2010).

Cutin monomers, that are formed as a breakdown of the cuticle, have been proposed as signal molecules for the induction of disease resistance in cereals, i.e., barley and rice (Schweizer et al., 1994). Fungal pathogens such as *Erysiphe graminis* and *Magnaporthe grisea* are able to produce and secrete cutinases that facilitate the formation of cutin monomers in the infection site. Pretreated barley leaves with cutin monomers display acquired protection against *E. graminis*(Schweizer et al., 1996b) and evidences that free cutin monomers can be recognized by plant cells as endogenous stress-related signals were obtained in cultured potato cells (Schweizer et al., 1996a).

SIGNAL TRANSDUCTION UPON WOUNDING

Many events triggered by wounding have been uncovered and are discussed here.

Electric Signals

The involvement of electrical signals in the local and systemic alert in plants was postulated in 1992, when it was found that mechanical wounding in tomato cotyledons causes the transmission of a potential action to the first unwounded leaf concomitantly with the induction of PI proteins at the site of injury (Wildon et al., 1992). More recently, it has been shown that mechanical wounding at the tips of *Arabidopsis* leaves generates, within a few seconds, wound-activated surface potential (WASP) changes that are consequent to a plasma membrane depolarization. The WASP signal first moves from tips toward the center of the rosette leaves and then to a restricted and selected number of distal leaves. For example, wounding at the tip of leaf no. 8 causes WASP changes of the same amplitude and duration in leaves no. 5, 11, 13, and 16 but not in other leaves. Both JA and JA-responsive gene expression increases with a total of 313 genes up-regulated both locally and systemically. GLR proteins, putative cation channels, are required for WASP propagation leading to defense gene expression (Mousavi et al., 2013). Similar WASP effects on JA levels and defense gene expression have been reported in tomato plants upon wounding (Herde et al., 1996). Electric signals that propagate over distances of 100 cm from the wounded site are generated in avocado trees (Oyarce and Gurovich, 2011).

Ion Fluxes

One of the earliest responses (0.5–2 min) activated by the elicitors of plant defenses is the membrane depolarization following the influx of H^+ and Ca^{2+} and a concomitant efflux of K^+ and nitrate across the membrane (Nurnberger et al., 2004; Mithofer et al., 2005). Injury-induced ion fluxes occur in both dicots such as *Vicia faba* and monocots such as *Hordeum vulgare*(Zimmermann et al., 2009). Calcium spikes are critical for downstream signaling, since the physiological concentration of cytosolic calcium very rapidly increases after "danger" sensing (Lecourieux et al., 2006; Kudla et al., 2010; Reddy et al., 2011). In plants, as in animals, calcium is a well-known second messenger. Plants discriminate among the various stimuli by generating "calcium signatures" that are characteristic in terms of sub-cellular localization, amplitude, duration and frequency (Sanders et al., 2002). Intracellular peaks of calcium are detected in both epidermis and vascular cells proximal to the injury within 6 s (Beneloujaephajri et al., 2013). Signatures are decoded by three major types of sensor proteins: calmodulins (CAMs) and CAM-like proteins, calcineurin B-like proteins (CBL) and calcium-dependent protein kinases (CDPKs), a class of calcium sensors bearing both protein kinase and CAM-like domains in a single polypeptide (Luan et al., 2002; Harper and Harmon, 2005; Luan, 2009). Different studies highlight the role of calcium sensors in plant immunity. For example, the *Arabidopsis* CAM binding protein (CBP) 60 g contributes to flg22-induced accumulation of SA and is involved in resistance against *P. syringae*(Wang et al., 2009); the rice CBL-interacting protein kinases (CIPKs) 14 and 15 are involved in various MAMP-induced immune responses (Kurusu et al., 2010); the potato and tobacco CDPKs participate in the activation of the oxidative burst (Ludwig et al., 2005; Kobayashi et al., 2007,2012). In *Arabidopsis*, CALCIUM-DEPENDENT PROTEIN KINASE 3 (CPK3) and CPK13 are required for defense gene induction upon feeding by the generalist herbivore *Spodoptera littoralis*(Kanchiswamy et al., 2010). CPK3 is also activated by flg22 in *Arabidopsis* protoplasts suggesting that it is involved in MAMP signaling as well (Boudsocq et al., 2010). In tomato, LeCDPK2 contributes to wound-triggered ET production by phosphorylating and activating the ET biosynthesis enzyme ACC SYNTHASE 2 (Kamiyoshihara et al., 2010). In *Nicotiana attenuata*, CDPK4 and CDPK5 are negative regulators of JA synthesis; plants with silenced expression of these two CDPKs are more resistant to larvae of *Manduca sexta* and exhibit enhanced responses to mechanical wounding (Yang et al., 2012). MeJA, touching and mechanical wounding enhance a calcium-activated CDPK activity that induces systemic wound responses also in maize (Szczegielniak et al., 2012).

Reactive Oxygen Species

The production of ROS is a highly conserved process among aerobic organisms and is involved in defense and development processes of plants. ROS are emerging as signal molecules in plant immunity activation in response to both pathogens and wounding (Mittler et al., 2011; Suzuki and Mittler, 2012). In tomato, hydrogen peroxide is detected within 1 h after wounding and increases at 4–6 h both locally and in the upper unwounded leaves. OGs generated by a plant PG probably act as mediators of this process. A tomato mutant unable to properly respond to wounding neither induce PG nor generates hydrogen peroxide and is more susceptible to larvae of *Manduca sexta*(Orozco-Cardenas and Ryan, 1999). Given its toxicity, hydrogen peroxide must be tightly regulated to work as a signal molecule, and this is achieved through a complex mechanism involving calcium, protein phosphorylation, and production of ROS-scavenging enzymes that determine its steady-state levels in the cell. Wound-induced apoplastic hydrogen peroxide is produced by transmembrane NADPH oxidases (RBOHs) and by peroxidases, which also have a role in detoxification of other ROS (Minibayeva et al., 2014). The C-terminal region of plant RBOHs contains cytosolic FAD- and NADPH-binding domains and six conserved transmembrane domains while the cytosolic N-terminal region contains two EF-hand motifs which bind calcium (Kobayashi et al., 2007; Oda et al., 2010; Proels et al., 2010; Kimura et al., 2012; Drerup et al., 2013). Indeed, the wound-induced oxidative burst is dependent on calcium spikes and occurs also in the absence of the stimulus through artificially increasing the calcium levels in the cells (Monshausen et al., 2007; Takeda et al., 2008; Kimura et al., 2012). On the other hand, wound-related production of hydrogen peroxide is abolished by pretreatments with the calcium channel blocker verapamil or calcium chelators EGTA and oxalate (Beneloujaephajri et al., 2013). A calcium-dependent protein kinase CPK5 phosphorylates RBOHD and, probably, represents the link between calcium accumulation and ROS production. CPK5 phosphorylates *in vitro* and *in vivo* the N-terminal serine residues S39, S148, S163, and S347 of AtRBOHD (Dubiella et al., 2013) while, in a contradicting report, ROS production triggered by pathogen infection is reduced in *cpk1 cpk2* double mutant plants (Gao et al., 2013). RBOHD forms complexes with EFR and FLS2 as well as with the plasma membrane-associated kinase BOTRYTIS-INDUCED KINASE 1 (BIK1), which is also required for the protection conferred by wounding against pathogens (Laluk et al., 2011). BIK1 directly interacts with and phosphorylates different residues of RBOHD in response to elicitors (Kadota et al., 2014). In addition, RBOHF activity is regulated both through direct binding of Ca^{2+} to EF-hands and through calcium-dependent phosphorylation by CBL1/9-CIPK26 complexes (Drerup et al., 2013).

Mitogen-Activated Protein Kinases (MAPKs)

Mitogen-activated protein kinase cascades amplify several abiotic and biotic stimuli leading to appropriate physiological responses (Rodriguez et al., 2010). They consist of a core module of three kinases that perform sequential phosphorylation reactions: a MAP kinase kinase kinase (MAP3K) activates a MAP kinase kinase (MAP2K), which activates a MAPK. Involvement of MAPKs in wounding has been widely described in various plant species (Nakagami et al., 2005). In*Arabidopsis*, wounding activates MEKK1, MPK3, MPK19 and, consequently, MEKK1 phosphorylates MKK1 (Hadiarto et al., 2006). Wounding also activates MPK4 and MPK6 and plants overexpressing the PP2C-type phosphatase AP2C1, which dephosphorylates and inactivates MPK4 and MPK6, do not respond to wounding (Ichimura et al., 2000). On the contrary, *ap2c1*mutants display enhanced responses to wounding and are more resistant to phytophagous mites (*Tetranychus urticae*; Schweighofer et al., 2007). On the other hand, MPK8, which is activated through direct binding of CAMs in a Ca^{2+}-dependent manner and through a MKK3-mediated phosphorylation, negatively regulate the expression of RBOHD and ROS homeostasis triggered by wounding (Nemoto et al., 2011). In tobacco, the SA-induced protein kinase kinase SIPKK and MPK4, which are orthologs of *Arabidopsis* MKK1/MKK2 and MPK4, respectively, are required for wound-induced expression of JA-responsive genes, being MPK4 activated by SIPKK (Gomi et al., 2005). Moreover, WIPK and SIPK, which are orthologs of the *Arabidopsis* MPK3 and MPK6, respectively, are also involved in wounding signaling (Seo et al., 2007).

Hormones

Wounding induces *de novo* synthesis of JA, ABA, and ET, which are known to activate a network of interconnected pathways that coordinate host defense responses (Peña-Cortés et al., 1995; Bergey et al., 1996; Bouquin et al., 1997). JA accumulates in wounded plants and activates expression of various defense genes such as those encoding PI, thionin, and enzymes involved in secondary metabolism (Creelman and Mullet, 1997). Jasmonates, including the active form jasmonoyl-isoleucine (JA-Ile), derive from plastidial fatty acids through at least 10 intermediates and the involvement of three cellular compartments (Staswick and Tiryaki, 2004; Browse, 2009; Fonseca et al., 2009; Schaller and Stintzi, 2009). In *Arabidopsis*, wounding at leaf no. 8 promptly (90 s) induces an increase of JA amount in leaf no. 13, which shares a connected vasculature with leaf no. 8 (Dengler and Kang, 2001). Plant 13-lipoxygenases (13-LOXs) catalyze the first event in JA synthesis, i.e., the dioxygenation of fatty acids (Andreou and Feussner, 2009). In *Arabidopsis*, LOX2 is required for the JA synthesis proximal to the wound (Glauser et al., 2009) while LOX6 is required for JA and JA-Ile accumulation in the wounded as well as in the distal unwounded leaves. The conversion of JA to JA-Ile takes 50 s in the wounded leaf and about 100 s in the distal connected leaf no. 13 (Chauvin et al., 2013). JA and JA-Ile accumulation in response to wounding have been demonstrated to be dependent on WASPs (see above, Mousavi et al., 2013). Another important step in JA synthesis is the accumulation of the JA precursor 12-oxo-phytodienoic acid (OPDA) catalyzed by allene oxidase (AOS), which is induced by tissue injury (Leon et al., 2001). In potato, two putative AOS genes, StAOS1 and StAOS2, are differentially induced upon wounding and are required for OPDA and JA accumulation both in wounded and in unchallenged tissues (Taurino et al., 2014). A possible link between JA signaling and CWI alterations is suggested by the analysis of the *Arabidopsis cev1* and *cob* mutants, which have defects in cellulose synthesis and deposition and produce higher amount of JAs (Ellis et al., 2002; Ko et al., 2006). On the other hand, plants with a reduced expression of StAOS1 and StAOS2, and, consequently, a lower amount of OPDA, display reduced PME activity, increased methyl esterification level of pectins and an increased susceptibility to an hypovirulent strain of *Dickeya dadantii* (Taurino et al., 2014). In tomato, JA is preferentially generated in vascular bundles and accumulates in the midrib of leaves (Stenzel et al., 2003). AOS and lipoxygenases are located in the companion-cell–sieve-element complex of the vascular bundle (Hause et al., 2003). Since systemin accumulates in phloem parenchyma cells (Narvaez-Vasquez and Ryan, 2004) and activates the octadecanoid pathway for JA biosynthesis, it may be hypothesized that perception of systemin on the surface of companion cells initiates the synthesis of JA that is rapidly transported along the phloem (Schilmiller and Howe, 2005).

Abscisic acid is a stress hormone that mediates plant responses to drought and salinity (Finkelstein, 2013) as well as the expression of wound-induced *PROTEINASE INHIBITOR II*(*PIN2*) gene in tomato and potato (Peña-Cortés et al., 1995). Moreover, ABA positively regulates programmed death in *Arabidopsis* leaf cells surrounding the wounding site to confine injury and/or pathogen infections (Bostock and Stermer, 1989). Spreading of programmed cell death from wounded sites is repressed by the transcription factor MYB108 or BOTRYTIS SENSITIVE1 (BOS1;Mengiste et al., 2003), and plants lacking this element exhibit mis-regulated cell death after wounding (Cui et al., 2013).

Ethylene production upon wounding has been documented (O'Donnell et al., 1996; Bouquin et al., 1997; Liu et al., 1997). Among the early wound-induced genes there are several ACC synthase genes and many of ET response transcription factors, i.e., EREBPs (Cheong et al., 2002). In tomato, ET and wound signaling, mediated by systemin and JA, have been reported to independently act on resistance against *B. cinerea* (Diaz et al., 2002). ET and JA, besides mediating inducible defenses in response to wounding, have been also proposed to function in the trade-off between growth and defense and the associated changes in resource allocations (Onkokesung et al., 2010).

Late Responses

Early intra- and inter-cellular events activated around the wounded site are required for late responses such as deposition of callose, suberin, lignin, and synthesis of various phenolics that may function both as a physical barrier and as antimicrobial substances. An *Arabidopsis* callose synthase, PMR4, is required for wound-induced callose formation (Jacobs et al., 2003). Callose is a (1→3)-β-D-glucan synthetized in all types of plant tissues in response to wounding (Chen and Kim, 2009). Its accumulation is dependent on the oxidative burst (Daudi et al., 2012; O'Brien et al., 2012) and occurs at the level of the cell wall either at wounded penetration sites or during attempted infections of fungi (Bellincampi et al., 2014). Callose may also prevent the spread of viruses through plasmodesmata (Benitez-Alfonso et al., 2011). Activity of both copper amine oxidases (CuAO) and flavin-containing amine oxidases (PAO), hydrogen peroxide-producing enzymes responsible for the oxidative de-amination of polyamines, appears to be important in wound healing in tobacco plants (Tisi et al., 2008). CuAO mediates also the enhanced accumulation of cell wall phenolics, observed on wound surface in tobacco plants over-expressing a fungal endopolygalacturonase, which show constitutively activated defenses. This observation suggests an important role of polyamine catabolism-derived hydrogen peroxide in the response activated by a compromised pectin integrity (Cona et al., 2014). Reconstruction of damaged tissues often takes place upon wounding and involves vascular and/or other cells that may divide and differentiate to reunite the existing tissues. Moreover, active biosynthesis and accumulation of pectic substances has been described in the cell wall of the reunion region in the cortex in cucumber and tomato hypocotyls (Asahina et al., 2002). The transcription factors RAP2.6L and ANAC071 are induced by ET and JA, differentially expressed around the injury site and are essential for tissue reunion of*Arabidopsis* wounded flowering stems (Asahina et al., 2011).

CONCLUSION

Pathogens often utilize wounded tissues for their entry into the plant. Wounding is rapidly perceived through an efficient surveillance mechanism of tissue integrity followed by cell-to-cell communication and long-distance signaling. Every cell is able to rapidly produce and propagate different alert messages, such as WASPs and ROS waves (Figure 1), which rapidly prime the rest of the plant to set up defenses against the potential danger. Propagation occurs over long distances, between different parts of the same plant and even between different individuals through volatile molecules production (Komarova et al., 2014). In the recent years several elements involved in sensing and signaling of wounding have been identified showing that the defense-related responses activated by wounding are comparable and almost overlapping with those activated after elicitor sensing. Thus, injury triggers a similar level of alert as a pathogen does, indicating that a breach in the physical barriers of the plant needs to be efficiently defended.

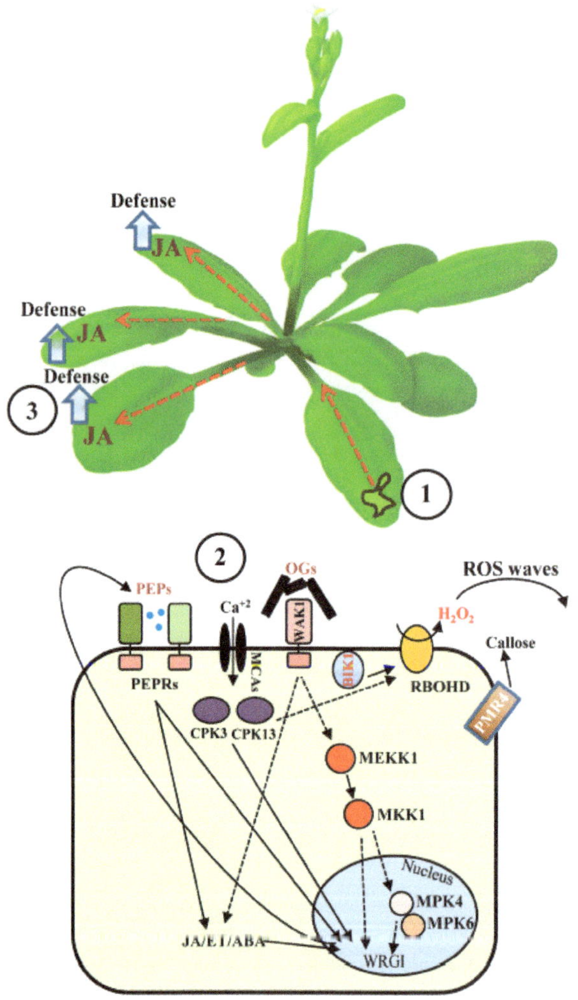

Figure 1. Local and systemic responses induced by wounding in *Arabidopsis*.Wounding of *Arabidopsis* leaves (1) is sensed through mechano-sensitive elements or by recognition of damage-associate molecular patterns, such as OGs or PEPs, which are perceived at the plasma membrane level. (2) Elements involved in wound signaling include calcium channels, MAPK cascades, CDPKs, and other kinases. Cell-to-cell communication is achieved by H_2O_2 waves produced by the trans-membrane NADPH oxidase RBOHD. Alert messages are generated and systemically propagated to undamage tissues through WASPs (red dashed lines) and other signals, such as JA. (3) Dashed lines indicate still partially uncharacterized roles of MAPKs or hypothetical cascades. WRGI: wound-regulated gene induction.

CONFLICT OF INTEREST STATEMENT

The authors declare that the research was conducted in the absence of any commercial or financial relationships that could be construed as a potential conflict of interest.

ACKNOWLEDGMENTS

This work was supported by the ERC (ERC_AdG_233083 grant awarded to Felice Cervone) and by the Istituto Pasteur – Fondazione Cenci-Bolognetti. We thank Prof. Giulia De Lorenzo for helpful discussions.

REFERENCES

1. Akira, S., Uematsu, S., and Takeuchi, O. (2006). Pathogen recognition and innate immunity. *Cell* 124, 783–801. doi: 10.1016/j.cell.2006.02.015
2. Andreou, A., and Feussner, I. (2009). Lipoxygenases – structure and reaction mechanism. *Phytochemistry* 70, 1504–1510. doi: 10.1016/j.phytochem.2009.05.008
3. Arimura, G., Kost, C., and Boland, W. (2005). Herbivore-induced, indirect plant defenses. *Biochim. Biophys. Acta* 1734, 91–111. doi: 10.1016/j.bbalip.2005.03.001
4. Asahina, M., Azuma, K., Pitaksaringkarn, W., Yamazaki, T., Mitsuda, N., Ohme-Takagi, M.,et al. (2011). Spatially selective hormonal control of RAP2.6L and ANAC071 transcription factors involved in tissue reunion in *Arabidopsis*. *Proc. Natl. Acad. Sci. U.S.A.* 108, 16128–16132. doi: 10.1073/pnas.1110443108
5. Asahina, M., Iwai, H., Kikuchi, A., Yamaguchi, S., Kamiya, Y., Kamada, H.,et al. (2002). Gibberellin produced in the cotyledon is required for cell division during tissue reunion in the cortex of cut cucumber and tomato hypocotyls. *Plant Physiol.* 129, 201–210. doi: 10.1104/pp.010886
6. Bartels, S., Lori, M., Mbengue, M., van Verk, M., Klauser, D., Hander, T.,et al. (2013). The family of Peps and their precursors in*Arabidopsis*: differential expression and localization but similar induction of pattern-triggered immune responses. *J. Exp. Bot.*64, 5309–5321. doi: 10.1093/jxb/ert330
7. Baydoun, E. A. H., and Fry, S. C. (1985). The immobility of pectic substances in injured tomato leaves and its bearing on the identity of the wound hormone. *Planta* 165, 269–276. doi: 10.1007/BF00395050
8. Bellincampi, D., Cardarelli, M., Zaghi, D., Serino, G., Salvi, G., Gatz, C.,et al. (1996). Oligogalacturonides prevent rhizogenesis in rol B transformed tobacco explants by inhibiting auxin-induced expression of the rol B gene. *Plant Cell* 8, 477–487. doi: 10.1105/tpc.8.3.477
9. Bellincampi, D., Cervone, F., and Lionetti, V. (2014). Plant cell wall dynamics and wall-related susceptibility in plant–pathogen interactions. *Front. Plant Sci.* 5:228. doi: 10.3389/fpls.2014.00228
10. Bellincampi, D., Dipierro, N., Salvi, G., Cervone, F., and De Lorenzo, G. (2000). Extracellular H_2O_2 induced by oligogalacturonides is not involved in the inhibition of the auxin-regulated rolB gene expression in tobacco leaf explants. *Plant Physiol.* 122, 1379–1385. doi: 10.1104/pp.122.4.1379
11. Beneloujaephajri, E., Costa, A., L'Haridon, F., Metraux, J. P., and Binda, M. (2013). Production of reactive oxygen species and wound-induced resistance in *Arabidopsis thaliana* against *Botrytis cinerea* are preceded and depend on a burst of calcium. *BMC Plant Biol.* 13:160. doi: 10.1186/1471-2229-13-160
12. Benikhlef, L., L'Haridon, F., Abou-Mansour, E., Serrano, M., Binda, M., Costa, A.,et al. (2013). Perception of soft mechanical stress in *Arabidopsis* leaves activates disease resistance. *BMC Plant Biol.* 13:133. doi: 10.1186/1471-2229-13-133
13. Benitez-Alfonso, Y., Jackson, D., and Maule, A. (2011). Redox regulation of intercellular transport. *Protoplasma* 248, 131–140. doi: 10.1007/s00709-010-0243-4
14. Bergey, D. R., Howe, G. A., and Ryan, C. A. (1996). Polypeptide signaling for plant defensive genes exhibits analogies to defense signaling in animals. *Proc. Natl. Acad. Sci. U.S.A.* 93, 12053–12058. doi: 10.1073/pnas.93.22.12053
15. Bergey, D. R., Orozco-Cardenas, M., De Moura, D. S., and Ryan, C. A. (1999). A wound- and systemin-inducible polygalacturonase in tomato leaves. *Proc. Natl. Acad. Sci. U.S.A.* 96, 1756–1760. doi: 10.1073/pnas.96.4.1756
16. Bessire, M., Chassot, C., Jacquat, A. C., Humphry, M., Borel, S., Petetot, J. M.,et al. (2007). A permeable cuticle in *Arabidopsis*leads to a strong resistance to *Botrytis cinerea*. *EMBO J.* 26, 2158–2168. doi: 10.1038/sj.emboj.7601658
17. Bhattacharya, R., Koramutla, M. K., Negi, M., Pearce, G., and Ryan, C. A. (2013). Hydroxyproline-rich glycopeptide signals in potato elicit signaling associated with defense against insects and pathogens. *Plant Sci.* 207, 88–97. doi: 10.1016/j.plantsci.2013.03.002
18. Boller, T., and Felix, G. (2009). A renaissance of elicitors: perception of microbe-associated molecular patterns and danger signals by pattern-recognition receptors. *Annu. Rev. Plant Biol.* 60, 379–406. doi: 10.1146/annurev.arplant.57.032905.105346
19. Bonaventure, G., and Baldwin, I. T. (2010). Transduction of wound and herbivory signals in plastids. *Commun. Integr. Biol.* 3, 313–317. doi: 10.4161/cib.3.4.11834
20. Bostock, R. M., and Stermer, B. A. (1989). Perspectives on wound healing in resistance to pathogens. *Annu. Rev. Phytopathol.*27, 343–371. doi: 10.1146/annurev.py.27.090189.002015
21. Boudsocq, M., Willmann, M. R., McCormack, M., Lee, H., Shan, L., He, P.,et al. (2010). Differential innate immune signaling via Ca^{2+} sensor protein kinases. *Nature* 464, 418–422. doi: 10.1038/nature08794
22. Bouquin, T., Lasserre, E., Pradier, J., Pech, J. C., and Balague, C. (1997). Wound and ethylene induction of the ACC oxidase melon gene CM-ACO1 occurs via two direct and independent transduction pathways. *Plant Mol. Biol.* 35, 1029–1035. doi: 10.1023/A:1005902226054

23. Branca, C., De Lorenzo, G., and Cervone, F. (1988). Competitive inhibition of the auxin-induced elongation by α-D-oligogalacturonides in pea stem segments. *Physiol. Plant.* 72, 499–504. doi: 10.1111/j.1399-3054.1988.tb09157.x

24. Browse, J. (2009). The power of mutants for investigating jasmonate biosynthesis and signaling. *Phytochemistry* 70, 1539–1546. doi: 10.1016/j.phytochem.2009.08.004

25. Brutus, A., Sicilia, F., Macone, A., Cervone, F., and De Lorenzo, G. (2010). A domain swap approach reveals a role of the plant wall-associated kinase 1 (WAK1) as a receptor of oligogalacturonides. *Proc. Natl. Acad. Sci. U.S.A.* 107, 9452–9457. doi: 10.1073/pnas.1000675107

26. Cabrera, J. C., Boland, A., Messiaen, J., Cambier, P., and Van Cutsem, P. (2008). Egg box conformation of oligogalacturonides: the time-dependent stabilization of the elicitor-active conformation increases its biological activity. *Glycobiology* 18, 473–482. doi: 10.1093/glycob/cwn027

27. Chassot, C., Nawrath, C., and Metraux, J. P. (2007). Cuticular defects lead to full immunity to a major plant pathogen. *Plant J.*49, 972–980. doi: 10.1111/j.1365-313X.2006.03017.x

28. Chauvin, A., Caldelari, D., Wolfender, J. L., and Farmer, E. E. (2013). Four 13-lipoxygenases contribute to rapid jasmonate synthesis in wounded *Arabidopsis thaliana* leaves: a role for lipoxygenase 6 in responses to long-distance wound signals. *New Phytol.* 197, 566–575. doi: 10.1111/nph.12029

29. Chen, X. Y., and Kim, J. Y. (2009). Callose synthesis in higher plants. *Plant Signal. Behav.* 4, 489–492. doi: 10.4161/psb.4.6.8359

30. Chen, Y. C., Siems, W. F., Pearce, G., and Ryan, C. A. (2008). Six peptide wound signals derived from a single precursor protein in *Ipomoea batatas* leaves activate the expression of the defense gene sporamin. *J. Biol. Chem.* 283, 11469–11476. doi: 10.1074/jbc.M709002200

31. Cheong, Y. H., Chang, H. S., Gupta, R., Wang, X., Zhu, T., and Luan, S. (2002). Transcriptional profiling reveals novel interactions between wounding, pathogen, abiotic stress, and hormonal responses in *Arabidopsis*. *Plant Physiol.* 129, 661–677. doi: 10.1104/pp.002857

32. Choi, J., Tanaka, K., Cao, Y. R., Qi, Y., Qiu, J., Liang, Y.,et al. (2014). Identification of a plant receptor for extracellular ATP.*Science* 343, 290–294. doi: 10.1126/science.343.6168.290

33. Cona, A., Tisi, A., Ghuge, S. A., Franchi, S., De Lorenzo, G., and Angelini, R. (2014). Wound healing response and xylem differentiation in tobacco plants over-expressing a fungal endopolygalacturonase is mediated by copper amine oxidase activity.*Plant Physiol. Biochem.* 82, 54–65. doi: 10.1016/j.plaphy.2014.05.004

34. Craigon, D. J., James, N., Okyere, J., Higgins, J., Jotham, J., and May, S. (2004). NASCArrays: a repository for microarray data generated by NASC's transcriptomics service. *Nucleic Acids Res.* 32, D575–D577. doi: 10.1093/nar/gkh133

35. Creelman, R. A., and Mullet, J. E. (1997). Biosynthesis and action of jasmonates in plants. *Annu. Rev. Plant Physiol. Plant Mol. Biol.* 48, 355–381. doi: 10.1146/annurev.arplant.48.1.355

36. Cui, F., Brosche, M., Sipari, N., Tang, S., and Overmyer, K. (2013). Regulation of ABA dependent wound induced spreading cell death by MYB108. *New Phytol.* 200, 634–640. doi: 10.1111/nph.12456

37. Daudi, A., Cheng, Z. Y., O'Brien, J. A., Mammarella, N., Khan, S., Ausubel, F. M.,et al. (2012). The apoplastic oxidative burst peroxidase in *Arabidopsis* is a major component of pattern-triggered immunity. *Plant Cell* 24, 275–287. doi: 10.1105/tpc.111.093039

38. De Lorenzo, G., Brutus, A., Savatin, D. V., Sicilia, F., and Cervone, F. (2011). Engineering plant resistance by constructing chimeric receptors that recognize damage-associated molecular patterns (DAMPs). *FEBS Lett.* 585, 1521–1528. doi: 10.1016/j.febslet.2011.04.043

39. Decreux, A., and Messiaen, J. (2005). Wall-associated kinase WAK1 interacts with cell wall pectins in a calcium-induced conformation. *Plant Cell Physiol.* 46, 268–278. doi: 10.1093/pcp/pci026

40. Delessert, C., Wilson, I. W., Van Der, S. D., Dennis, E. S., and Dolferus, R. (2004). Spatial and temporal analysis of the local response to wounding in *Arabidopsis* leaves. *Plant Mol. Biol.* 55, 165–181. doi: 10.1007/s11103-004-0112-7

41. Dengler, N., and Kang, J. (2001). Vascular patterning and leaf shape. *Curr. Opin. Plant Biol.* 4, 50–56. doi: 10.1016/S1369-5266(00)00135-7

42. Denness, L., McKenna, J. F., Segonzac, C., Wormit, A., Madhou, P., Bennett, M.,et al. (2011). Cell wall damage-induced lignin biosynthesis is regulated by a reactive oxygen species- and jasmonic acid-dependent process in *Arabidopsis*. *Plant Physiol.* 156, 1364–1374. doi: 10.1104/pp.111.175737

43. Denoux, C., Galletti, R., Mammarella, N., Gopalan, S., Werck, D., De Lorenzo, G.,et al. (2008). Activation of defense response pathways by OGs and Flg22 elicitors in *Arabidopsis* seedlings. *Mol. Plant* 1, 423–445. doi: 10.1093/mp/ssn019

44. Diaz, J., ten Have, A., and van Kan, J. A. (2002). The role of ethylene and wound signaling in resistance of tomato to *Botrytis cinerea*. *Plant Physiol.* 129, 1341–1351. doi: 10.1104/pp.001453

45. Drerup, M. M., Schlucking, K., Hashimoto, K., Manishankar, P., Steinhorst, L., Kuchitsu, K.,et al. (2013). The Calcineurin B-like calcium sensors CBL1 and CBL9 together with their interacting protein kinase CIPK26 regulate the *Arabidopsis* NADPH oxidase RBOHF. *Mol. Plant* 6, 559–569. doi: 10.1093/mp/sst009

46. Dubiella, U., Seybold, H., Durian, G., Komander, E., Lassig, R., Witte, C. P.,et al. (2013). Calcium-dependent protein kinase/NADPH oxidase activation circuit is required for rapid defense signal propagation. *Proc. Natl. Acad. Sci. U.S.A.* 110, 8744–8749. doi: 10.1073/pnas.1221294110

47. Ellis, C., Karafyllidis, I., Wasternack, C., and Turner, J. G. (2002). The *Arabidopsis* mutant cev1 links cell wall signaling to jasmonate and ethylene responses. *Plant Cell* 14, 1557–1566. doi: 10.1105/tpc.002022

48. Farmer, E. E., and Ryan, C. A. (1992). Octadecanoid precursors of jasmonic acid activate the synthesis of wound-inducible proteinase inhibitors. *Plant Cell* 4, 129–134. doi: 10.1105/tpc.4.2.129

49. Ferrari, S., Galletti, R., Denoux, C., De Lorenzo, G., Ausubel, F. M., and Dewdney, J. (2007). Resistance to *Botrytis cinerea*induced in *Arabidopsis* by elicitors is independent of salicylic acid, ethylene, or jasmonate signaling but requires PHYTOALEXIN DEFICIENT3. *Plant Physiol.* 144, 367–379. doi: 10.1104/pp.107.095596

50. Ferrari, S., Galletti, R., Pontiggia, D., Manfredini, C., Lionetti, V., Bellincampi, D.,et al. (2008). Transgenic expression of a fungal endo-polygalacturonase increases plant resistance to pathogens and reduces auxin sensitivity. *Plant Physiol.* 146, 669–681. doi: 10.1104/pp.107.109686

51. Ferrari, S., Savatin, D. V., Sicilia, F., Gramegna, G., Cervone, F., and De Lorenzo, G. (2013). Oligogalacturonides: plant damage-associated molecular patterns and regulators of growth and development. *Front. Plant Sci.* 4:49. doi: 10.3389/fpls.2013.00049

52. Finkelstein, R. (2013). Abscisic acid synthesis and response. *Arabidopsis Book* 11:e0166. doi: 10.1199/tab.0166

53. Fonseca, S., Chico, J. M., and Solano, R. (2009). The jasmonate pathway: the ligand, the receptor and the core signalling module. *Curr. Opin. Plant Biol.* 12, 539–547. doi: 10.1016/j.pbi.2009.07.013

54. Galletti, R., Denoux, C., Gambetta, S., Dewdney, J., Ausubel, F. M., De Lorenzo, G.,et al. (2008). The AtrbohD-mediated oxidative burst elicited by oligogalacturonides in *Arabidopsis* is dispensable for the activation of defense responses effective against *Botrytis cinerea*. *Plant Physiol.* 148, 1695–1706. doi: 10.1104/pp.108.127845

55. Gao, X. Q., Chen, X., Lin, W. W., Chen, S. X., Lu, D. P., Niu, Y. J.,et al. (2013). Bifurcation of *Arabidopsis* NLR immune signaling via Ca^{2+}-dependent protein kinases. *PLoS Pathog.* 9:e1003127. doi: 10.1371/journal.ppat.1003127

56. Glauser, G., Dubugnon, L., Mousavi, S. A., Rudaz, S., Wolfender, J. L., and Farmer, E. E. (2009). Velocity estimates for signal propagation leading to systemic jasmonic acid accumulation in wounded *Arabidopsis*. *J. Biol. Chem.* 284, 34506–34513. doi: 10.1074/jbc.M109.061432

57. Gomi, K., Ogawa, D., Katou, S., Kamada, H., Nakajima, N., Saji, H.,et al. (2005). A mitogen-activated protein kinase NtMPK4 activated by SIPKK is required for jasmonic acid signaling and involved in ozone tolerance via stomatal movement in tobacco.*Plant Cell Physiol.* 46, 1902–1914. doi: 10.1093/pcp/pci211

58. Guo, H. Q., Li, L., Ye, H. X., Yu, X. F., Algreen, A., and Yin, Y. H. (2009). Three related receptor-like kinases are required for optimal cell elongation in *Arabidopsis thaliana*. *Proc. Natl. Acad. Sci. U.S.A.* 106, 7648–7653. doi: 10.1073/pnas.0812346106

59. Hadiarto, T., Nanmori, T., Matsuoka, D., Iwasaki, T., Sato, K., Fukami, Y.,et al. (2006). Activation of *Arabidopsis* MAPK kinase kinase (AtMEKK1) and induction of AtMEKK1-AtMEK1 pathway by wounding. *Planta* 223, 708–713. doi: 10.1007/s00425-005-0126-7

60. Harper, J. F., and Harmon, A. (2005). Plants, symbiosis and parasites: A calcium signalling connection. *Nat. Rev. Mol. Cell Biol.*6, 555–566. doi: 10.1038/nrm1679

61. Haswell, E. S., Peyronnet, R., Barbier-Brygoo, H., Meyerowitz, E. M., and Frachisse, J. M. (2008). Two MscS homologs provide mechanosensitive channel activities in the *Arabidopsis* root. *Curr. Biol.* 18, 730–734. doi: 10.1016/j.cub.2008.04.039

62. Hause, B., Hause, G., Kutter, C., Miersch, O., and Wasternack, C. (2003). Enzymes of jasmonate biosynthesis occur in tomato sieve elements. *Plant Cell Physiol.* 44, 643–648. doi: 10.1093/pcp/pcg072

63. He, G. Z., Tarui, Y., and Iino, M. (2005). A novel receptor kinase involved in jasmonate-mediated wound and phytochrome signaling in maize coleoptiles. *Plant Cell Physiol.* 46, 870–883. doi: 10.1093/pcp/pci092

64. Heiling, S., Schuman, M. C., Schoettner, M., Mukerjee, P., Berger, B., Schneider, B.,et al. (2010). Jasmonate and ppHsystemin regulate key malonylation steps in the biosynthesis of 17-hydroxygeranyllinalool diterpene glycosides, an abundant and effective direct defense against herbivores in *Nicotiana attenuata*. *Plant Cell* 22, 273–292. doi: 10.1105/tpc.109.071449

65. Hematy, K., Sado, P. E., Van Tuinen, A., Rochange, S., Desnos, T., Balzergue, S.,et al. (2007). A receptor-like kinase mediates the response of *Arabidopsis* cells to the inhibition of cellulose synthesis. *Curr. Biol.* 17, 922–931. doi: 10.1016/j.cub.2007.05.018

66. Herde, O., Atzorn, R., Fisahn, J., Wasternack, C., Willmitzer, L., and Peña-Cortés, H. (1996). Localized wounding by heat initiates the accumulation of proteinase inhibitor II in abscisic acid-deficient plants by triggering jasmonic acid biosynthesis.*Plant Physiol.* 112, 853–860.

67. Hind, S. R., Malinowski, R., Yalamanchili, R., and Stratmann, J. W. (2010). Tissue-type specific systemin perception and the elusive systemin receptor. *Plant Signal. Behav.* 5, 42–44. doi: 10.4161/psb.5.1.10119

68. Huffaker, A., Dafoe, N. J., and Schmelz, E. A. (2011). ZmPep1, an ortholog of *Arabidopsis* elicitor peptide 1, regulates maize innate immunity and enhances disease resistance. *Plant Physiol.* 155, 1325–1338. doi: 10.1104/pp.110.166710

69. Huffaker, A., Pearce, G., and Ryan, C. A. (2006). An endogenous peptide signal in *Arabidopsis* activates components of the innate immune response. *Proc. Natl. Acad. Sci. U.S.A.* 103, 10098–10103. doi: 10.1073/pnas.0603727103

70. Huffaker, A., Pearce, G., Veyrat, N., Erb, M., Turlings, T. C. J., Sartor, R.,et al. (2013). Plant elicitor peptides are conserved signals regulating direct and indirect antiherbivore defense. *Proc. Natl. Acad. Sci. U.S.A.* 110, 5707–5712. doi: 10.1073/pnas.1214668110

71. Ichimura, K., Mizoguchi, T., Yoshida, R., Yuasa, T., and Shinozaki, K. (2000). Various abiotic stresses rapidly activate*Arabidopsis* MAP kinases ATMPK4 and ATMPK6. *Plant J.* 24, 655–665. doi: 10.1046/j.1365-313x.2000.00913.x

72. Ito, N., Takabatake, R., Seo, S., Hiraga, S., Mitsuhara, I., and Ohashi, Y. (2002). Induced expression of a temperature-sensitive leucine-rich repeat receptor-like protein kinase gene by hypersensitive cell death and wounding in tobacco plant carrying the N resistance gene. *Plant Cell Physiol.* 43, 266–274. doi: 10.1093/pcp/pcf031

73. Jacinto, T., McGurl, B., Franceschi, V., Delano-Freier, J., and Ryan, C. A. (1997). Tomato prosystemin promoter confers wound-inducible, vascular bundle-specific expression of the β-glucuronidase gene in transgenic tomato plants. *Planta* 203, 406–412. doi: 10.1007/s004250050207

74. Jacobs, A. K., Lipka, V., Burton, R. A., Panstruga, R., Strizhov, N., Schulze-Lefert, P.,et al. (2003). An *Arabidopsis* callose synthase, GSL5, is required for wound and papillary callose formation. *Plant Cell* 15, 2503–2513. doi: 10.1105/tpc.016097

75. Kadota, Y., Sklenar, J., Derbyshire, P., Stransfeld, L., Asai, S., Ntoukakis, V.,et al. (2014). Direct regulation of the NADPH oxidase RBOHD by the PRR-associated kinase BIK1 during plant immunity. *Mol. Cell* 54, 43–55. doi: 10.1016/j.molcel.2014.02.021

76. Kamiyoshihara, Y., Iwata, M., Fukaya, T., Tatsuki, M., and Mori, H. (2010). Turnover of LeACS2, a wound-inducible 1-aminocyclopropane-1-carboxylic acid synthase in tomato, is regulated by phosphorylation/dephosphorylation. *Plant J.* 64, 140–150. doi: 10.1111/j.1365-313X.2010.04316.x

77. Kanchiswamy, C. N., Takahashi, H., Quadro, S., Maffei, M. E., Bossi, S., Bertea, C.,et al. (2010). Regulation of *Arabidopsis*defense responses against *Spodoptera littoralis* by CPK-mediated calcium signaling. *BMC Plant Biol.* 10:97. doi: 10.1186/1471-2229-10-97

78. Kanzaki, M., Nagasawa, M., Kojima, I., Sato, C., Naruse, K., Sokabe, M.,et al. (1999). Molecular identification of a eukaryotic, stretch-activated nonselective cation channel. *Science* 285, 882–886. doi: 10.1126/science.285.5429.882

79. Ketela, T., Green, R., and Bussey, H. (1999). *Saccharomyces cerevisiae* Mid2p is a potential cell wall stress sensor and upstream activator of the PKC1-MPK1 cell integrity pathway. *J. Bacteriol.* 181, 3330–3340.

80. Kimura, S., Kaya, H., Kawarazaki, T., Hiraoka, G., Senzaki, E., Michikawa, M.,et al. (2012). Protein phosphorylation is a prerequisite for the Ca^{2+}-dependent activation of *Arabidopsis* NADPH oxidases and may function as a trigger for the positive feedback regulation of Ca^{2+} and reactive oxygen species. *Biochim. Biophys. Acta* 1823, 398–405. doi: 10.1016/j.bbamcr.2011.09.011

81. Ko, J. H., Kim, J. H., Jayanty, S. S., Howe, G. A., and Han, K. H. (2006). Loss of function of COBRA, a determinant of oriented cell expansion, invokes cellular defense responses in *Arabidopsis thaliana*. *J. Exp. Bot.* 57, 2923–2936. doi: 10.1093/jxb/erl052

82. Kobayashi, M., Ohura, I., Kawakita, K., Yokota, N., Fujiwara, M., Shimamoto, K.,et al. (2007). Calcium-dependent protein kinases regulate the production of reactive oxygen species by potato NADPH oxidase. *Plant Cell* 19, 1065–1080. doi: 10.1105/tpc.106.048884

83. Kobayashi, M., Yoshioka, M., Asai, S., Nomura, H., Kuchimura, K., Mori, H.,et al. (2012). StCDPK5 confers resistance to late blight pathogen but increases susceptibility to early blight pathogen in potato via reactive oxygen species burst. *New Phytol.*196, 223–237. doi: 10.1111/j.1469-8137.2012.04226.x

84. Kohorn, B. D., Johansen, S., Shishido, A., Todorova, T., Martinez, R., Defeo, E.,et al. (2009). Pectin activation of MAP kinase and gene expression is WAK2 dependent. *Plant J.* 60, 974–982. doi: 10.1111/j.1365-313X.2009.04016.x

85. Kohorn, B. D., and Kohorn, S. L. (2012). The cell wall-associated kinases, WAKs, as pectin receptors. *Front. Plant Sci.* 3:88. doi: 10.3389/fpls.2012.00088

86. Kohorn, B. D., Kohorn, S. L., Todorova, T., Baptiste, G., Stansky, K., and McCullough, M. (2012). A dominant allele of*Arabidopsis* pectin-binding wall-associated kinase induces a stress response suppressed by MPK6 but not MPK3 mutations.*Mol. Plant* 5, 841–851. doi: 10.1093/mp/ssr096

87. Komarova, T. V., Sheshukova, E. V., and Dorokhov, Y. L. (2014). Cell wall methanol as a signal in plant immunity. *Front. Plant Sci.* 5:101. doi: 10.3389/fpls.2014.00101

88. Krautz, R., Arefin, B., and Theopold, U. (2014). Damage signals in the insect immune response. *Front. Plant Sci.* 5:342. doi: 10.3389/fpls.2014.00342

89. Krol, E., Mentzel, T., Chinchilla, D., Boller, T., Felix, G., Kemmerling, B.,et al. (2010). Perception of the *Arabidopsis* danger signal peptide 1 involves the pattern recognition receptor AtPEPR1 and its close homologue AtPEPR2. *J. Biol. Chem.* 285, 13471–13479. doi: 10.1074/jbc.M109.097394

90. Krupkova, E., Immerzeel, P., Pauly, M., and Schmulling, T. (2007). The TUMOROUS SHOOT DEVELOPMENT2 gene of*Arabidopsis* encoding a putative methyltransferase is required for cell adhesion and co-ordinated plant development. *Plant J.*50, 735–750. doi: 10.1111/j.1365-313X.2007.03123.x

91. Kudla, J., Batistic, O., and Hashimoto, K. (2010). Calcium signals: the lead currency of plant information processing. *Plant Cell*22, 541–563. doi: 10.1105/tpc.109.072686

92. Kurusu, T., Hamada, J., Nokajima, H., Kitagawa, Y., Kiyoduka, M., Takahashi, A.,et al. (2010). Regulation of microbe-associated molecular pattern-induced hypersensitive cell death, phytoalexin production, and defense gene expression by calcineurin B-like protein-interacting protein kinases, OsCIPK14/15, in rice cultured cells. *Plant Physiol.* 153, 678–692. doi: 10.1104/pp.109.151852

93. Kwaaitaal, M., Huisman, R., Maintz, J., Reinstadler, A., and Panstruga, R. (2011). Ionotropic glutamate receptor (iGluR)-like channels mediate MAMP-induced calcium influx in *Arabidopsis thaliana*. *Biochem. J.* 440, 355–365. doi: 10.1042/BJ20111112

94. L'Haridon, F., Besson-Bard, A., Binda, M., Serrano, M., Abou-Mansour, E., Balet, F.,et al. (2011). A permeable cuticle is associated with the release of reactive oxygen species and induction of innate immunity. *PLoS Pathog.* 7:e1002148. doi: 10.1371/journal.ppat.1002148

95. Laluk, K., Luo, H. L., Chai, M. F., Dhawan, R., Lai, Z. B., and Mengiste, T. (2011). Biochemical and genetic requirements for function of the immune response regulator BOTRYTIS-INDUCED KINASE1 in plant growth, ethylene signaling, and PAMP-triggered immunity in *Arabidopsis*. *Plant Cell* 23, 2831–2849. doi: 10.1105/tpc.111.087122

96. Lecourieux, D., Ranjeva, R., and Pugin, A. (2006). Calcium in plant defense-signaling pathways. *New Phytol.* 171, 249–269. doi: 10.1111/j.1469-8137.2006.01777.x

97. Leon, J., Rojo, E., and Sanchez-Serrano, J. J. (2001). Wound signaling in plants. *J. Exp. Bot.* 52, 1–9. doi: 10.1093/jexbot/52.354.1

98. Li, H., Zhou, S. Y., Zhao, W. S., Su, S. C., and Peng, Y. L. (2008). A novel wall-associated receptor-like protein kinase gene, OsWAK1, plays important roles in rice blast disease resistance. *Plant Mol. Biol.* 69, 337–346. doi: 10.1007/s11103-008-9430-5

99. Lionetti, V., Cervone, F., and Bellincampi, D. (2012). Methyl esterification of pectin plays a role during plant-pathogen interactions and affects plant resistance to diseases. *J. Plant Physiol.* 169, 1623–1630. doi: 10.1016/j.jplph.2012.05.006

100. Lionetti, V., Francocci, F., Ferrari, S., Volpi, C., Bellincampi, D., Galletti, R.,et al. (2010). Engineering the cell wall by reducing de-methyl-esterified homogalacturonan improves saccharification of plant tissues for bioconversion. *Proc. Natl. Acad. Sci. U.S.A.* 107, 616–621. doi: 10.1073/pnas.0907549107

101. Lionetti, V., Raiola, A., Camardella, L., Giovane, A., Obel, N., Pauly, M.,et al. (2007). Overexpression of pectin methylesterase inhibitors in *Arabidopsis* restricts fungal infection by botrytis cinerea. *Plant Physiol.* 143, 1871–1880. doi: 10.1104/pp.106.090803

102. Liu, J. H., Lee-Tamon, S. H., and Reid, D. M. (1997). Differential and wound-inducible expression of 1-aminocylopropane-1-carboxylate oxidase genes in sunflower seedlings. *Plant Mol. Biol.* 34, 923–933. doi: 10.1023/A:1005801420172

103. Luan, S. (2009). The CBL-CIPK network in plant calcium signaling. *Trends Plant Sci.* 14, 37–42. doi: 10.1016/j.tplants.2008.10.005

104. Luan, S., Kudla, J., Rodriguez-Concepcion, M., Yalovsky, S., and Gruissem, W. (2002). Calmodulins and calcineurin B-like proteins: calcium sensors for specific signal response coupling in plants. *Plant Cell* 14, S389–S400. doi: 10.1105/tpc.001115

105. Ludwig, A. A., Saitoh, H., Felix, G., Freymark, G., Miersch, O., Wasternack, C.,et al. (2005). Ethylene-mediated cross-talk between calcium-dependent protein kinase and MAPK signaling controls stress responses in plants. *Proc. Natl. Acad. Sci. U.S.A.*102, 10736–10741. doi: 10.1073/pnas.0502954102

106. Maffei, M. E., Mithofer, A., and Boland, W. (2007). Before gene expression: early events in plant-insect interaction. *Trends Plant Sci.* 12, 310–316. doi: 10.1016/j.tplants.2007.06.001

107. Mengiste, T., Chen, X., Salmeron, J., and Dietrich, R. (2003). The BOTRYTIS SUSCEPTIBLE1 gene encodes an R2R3MYB transcription factor protein that is required for biotic and abiotic stress responses in *Arabidopsis*. *Plant Cell* 15, 2551–2565. doi: 10.1105/tpc.014167

108. Minibayeva, F., Beckett, R. P., and Kranner, I. (2014). Roles of apoplastic peroxidases in plant response to wounding.*Phytochemistry* doi: 10.1016/j.phytochem.2014.06.008 [Epub ahead of print].

109. Mithofer, A., Ebel, J., and Felle, H. H. (2005). Cation fluxes cause plasma membrane depolarization involved in beta-glucan elicitor-signaling in soybean roots. *Mol. Plant Microbe Interact.* 18, 983–990. doi: 10.1094/MPMI-18-0983

110. Mittler, R., Vanderauwera, S., Suzuki, N., Miller, G., Tognetti, V. B., Vandepoele, K.,et al. (2011). ROS signaling: the new wave?*Trends Plant Sci.* 16, 300–309. doi: 10.1016/j.tplants.2011.03.007

111. Monshausen, G. B., Bibikova, T. N., Messerli, M. A., Shi, C., and Gilroy, S. (2007). Oscillations in extracellular pH and reactive oxygen species modulate tip growth of *Arabidopsis* root hairs. *Proc. Natl. Acad. Sci. U.S.A.* 104, 20996–21001. doi: 10.1073/pnas.0708586104

112. Mouille, G., Ralet, M. C., Cavelier, C., Eland, C., Effroy, D., Hematy, K.,et al. (2007). Homogalacturonan synthesis in*Arabidopsis thaliana* requires a Golgi-localized protein with a putative methyltransferase domain. *Plant J.* 50, 605–614. doi: 10.1111/j.1365-313X.2007.03086.x

113. Mousavi, S. A. R., Chauvin, A., Pascaud, F., Kellenberger, S., and Farmer, E. E. (2013). GLUTAMATE RECEPTOR-LIKE genes mediate leaf-to-leaf wound signalling. *Nature* 500, 422–426. doi: 10.1038/nature12478

114. Nakagami, H., Pitzschke, A., and Hirt, H. (2005). Emerging MAP kinase pathways in plant stress signalling. *Trends Plant Sci.* 10, 339–346. doi: 10.1016/j.tplants.2005.05.009

115. Nakagawa, Y., Katagiri, T., Shinozaki, K., Qi, Z., Tatsumi, H., Furuichi, T.,et al. (2007). *Arabidopsis* plasma membrane protein crucial for Ca^{2+} influx and touch sensing in roots. *Proc. Natl. Acad. Sci. U.S.A.* 104, 3639–3644. doi: 10.1073/pnas.0607703104

116. Narvaez-Vasquez, J., Pearce, G., and Ryan, C. A. (2005). The plant cell wall matrix harbors a precursor of defense signaling peptides. *Proc. Natl. Acad. Sci. U.S.A.* 102, 12974–12977. doi: 10.1073/pnas.0505248102

117. Narvaez-Vasquez, J., and Ryan, C. A. (2004). The cellular localization of prosystemin: a functional role for phloem parenchyma in systemic wound signaling. *Planta* 218, 360–369. doi: 10.1007/s00425-003-1115-3

118. Nawrath, C. (2006). Unraveling the complex network of cuticular structure and function. *Curr. Opin. Plant Biol.* 9, 281–287. doi: 10.1016/j.pbi.2006.03.001

119. Nemoto, K., Seto, T., Takahashi, H., Nozawa, A., Seki, M., Shinozaki, K.,et al. (2011). Autophosphorylation profiling of*Arabidopsis* protein kinases using the cell-free system. *Phytochemistry* 72, 1136–1144. doi: 10.1016/j.phytochem.2011.02.029

120. Nuhse, T. S. (2012). Cell wall integrity signaling and innate immunity in plants. *Front. Plant Sci.* 3:280. doi: 10.3389/fpls.2012.00280

121. Nurnberger, T., Brunner, F., Kemmerling, B., and Piater, L. (2004). Innate immunity in plants and animals: striking similarities and obvious differences. *Immunol. Rev.* 198, 249–266. doi: 10.1111/j.0105-2896.2004.0119.x

122. O'Brien, J. A., Daudi, A., Butt, V. S., and Bolwell, G. P. (2012). Reactive oxygen species and their role in plant defence and cell wall metabolism. *Planta* 236, 765–779. doi: 10.1007/s00425-012-1696-9

123. O'Donnell, P. J., Calvert, C., Atzorn, R., Wasternack, C., Leyser, H. M. O., and Bowles, D. J. (1996). Ethylene as a signal mediating the wound response of tomato plants. *Science* 274, 1914–1917. doi: 10.1126/science.274.5294.1914

124. Oda, T., Hashimoto, H., Kuwabara, N., Akashi, S., Hayashi, K., Kojima, C.,et al. (2010). Structure of the N-terminal regulatory domain of a plant NADPH oxidase and its functional implications. *J. Biol. Chem.* 285, 1435–1445. doi: 10.1074/jbc.M109.058909

125. Onkokesung, N., Baldwin, I. T., and Galis, I. (2010). The role of jasmonic acid and ethylene crosstalk in direct defense of*Nicotiana attenuata* plants against chewing herbivores. *Plant Signal. Behav.* 5, 1305–1307. doi: 10.4161/psb.5.10.13124

126. Orozco-Cardenas, M., McGurl, B., and Ryan, C. A. (1993). Expression of an antisense prosystemin gene in tomato plants reduces resistance toward *Manduca sexta* larvae. *Proc. Natl. Acad. Sci. U.S.A.* 90, 8273–8276. doi: 10.1073/pnas.90.17.8273

127. Orozco-Cardenas, M., and Ryan, C. A. (1999). Hydrogen peroxide is generated systemically in plant leaves by wounding and systemin via the octadecanoid pathway. *Proc. Natl. Acad. Sci. U.S.A.* 96, 6553–6655. doi: 10.1073/pnas.96.11.6553

128. Oyarce, P., and Gurovich, L. (2011). Evidence for the transmission of information through electric potentials in injured avocado trees. *J. Plant Physiol.* 168, 103–108. doi: 10.1016/j.jplph.2010.06.003

129. Pearce, G., Bhattacharya, R., Chen, Y. C., Barona, G., Yamaguchi, Y., and Ryan, C. A. (2009). Isolation and characterization of hydroxyproline-rich glycopeptide signals in black nightshade leaves. *Plant Physiol.* 150, 1422–1433. doi: 10.1104/pp.109.138669

130. Pearce, G., Moura, D. S., Stratmann, J., and Ryan, C. A. (2001). Production of multiple plant hormones from a single polyprotein precursor. *Nature* 411, 817–820. doi: 10.1038/35081107

131. Pearce, G., and Ryan, C. A. (2003). Systemic signaling in tomato plants for defense against herbivores. Isolation and characterization of three novel defense-signaling glycopeptide hormones coded in a single precursor gene. *J. Biol. Chem.* 278, 30044–30050. doi: 10.1074/jbc.M304159200

132. Pearce, G., Siems, W. F., Bhattacharya, R., Chen, Y. C., and Ryan, C. A. (2007). Three hydroxyproline-rich glycopeptides derived from a single petunia polyprotein precursor activate defensin I, a pathogen defense response gene. *J. Biol. Chem.* 282, 17777–17784. doi: 10.1074/jbc.M701543200

133. Peña-Cortés, H., Fisahn, J., and Willmitzer, L. (1995). Signals involved in wound-induced proteinase inhibitor II gene expression in tomato and potato plants. *Proc. Natl. Acad. Sci. U.S.A.* 92, 4106–4113. doi: 10.1073/pnas.92.10.4106

134. Proels, R. K., Oberhollenzer, K., Pathuri, I. P., Hensel, G., Kumlehn, J., and Huckelhoven, R. (2010). RBOHF2 of barley is required for normal development of penetration resistance to the parasitic fungus *Blumeria graminis* f. sp. *hordei. Mol. Plant Microbe Interact.* 23, 1143–1150. doi: 10.1094/MPMI-23-9-1143

135. Raiola, A., Lionetti, V., Elmaghraby, I., Immerzeel, P., Mellerowicz, E. J., Salvi, G.,et al. (2011). Pectin methylesterase is induced in *Arabidopsis* upon infection and is necessary for a successful colonization by necrotrophic pathogens. *Mol. Plant Microbe Interact.* 24, 432–440. doi: 10.1094/MPMI-07-10-0157

136. Rasul, S., Dubreuil-Maurizi, C., Lamotte, O., Koen, E., Poinssot, B., Alcaraz, G.,et al. (2012). Nitric oxide production mediates oligogalacturonide-triggered immunity and resistance to *Botrytis cinerea* in *Arabidopsis thaliana*. *Plant Cell Environ.* 35, 1483–1499. doi: 10.1111/j.1365-3040.2012.02505.x

137. Reddy, A. S., Ali, G. S., Celesnik, H., and Day, I. S. (2011). Coping with stresses: roles of calcium- and calcium/calmodulin-regulated gene expression. *Plant Cell* 23, 2010–2032. doi: 10.1105/tpc.111.084988

138. Rehrig, E. M., Appel, H., Jones, A. D., and Schultz, C. J. (2014). Roles for jasmonate- and ethylene-induced transcription factors in the ability of *Arabidopsis* to respond differentially to damage caused by two insect herbivores. *Front. Plant Sci.* 5:407. doi: 10.3389/fpls.2014.00407

139. Reymond, P., Weber, H., Damond, M., and Farmer, E. E. (2000). Differential gene expression in response to mechanical wounding and insect feeding in *Arabidopsis*. *Plant Cell* 12, 707–720. doi: 10.1105/tpc.12.5.707

140. Riederer, M., and Schreiber, L. (2001). Protecting against water loss: analysis of the barrier properties of plant cuticles. *J. Exp. Bot.* 52, 2023–2032. doi: 10.1093/jexbot/52.363.2023

141. Roberts, K. (1992). Potential awareness of plants. *Nature* 360, 14–15. doi: 10.1038/360014a0

142. Rodriguez, M. C., Petersen, M., and Mundy, J. (2010). Mitogen-activated protein kinase signaling in plants. *Annu. Rev. Plant Biol.* 61, 621–649. doi: 10.1146/annurev-arplant-042809-112252

143. Ryan, C. A., and Jagendorf, A. (1995). Self defense by plants. *Proc. Natl. Acad. Sci. U.S.A.* 92:4075. doi: 10.1073/pnas.92.10.4075

144. Sanders, D., Pelloux, J., Brownlee, C., and Harper, J. F. (2002). Calcium at the crossroads of signaling. *Plant Cell* 14, S401–S417. doi: 10.1105/tpc.002899

145. Savatin, D. V., Ferrari, S., Sicilia, F., and De Lorenzo, G. (2011). Oligogalacturonide-auxin antagonism does not require posttranscriptional gene silencing or stabilization of auxin response repressors in *Arabidopsis*. *Plant Physiol.* 157, 1163–1174. doi: 10.1104/pp.111.184663

146. Savatin, D. V., Gigli, B. N., Marti, L., Fabbri, C., Cervone, F., and De Lorenzo, G. (2014). The *Arabidopsis* NPK1-related protein kinases ANPs are required for elicitor-induced oxidative burst and immunity. *Plant Physiol.* 165, 1188–1202. doi: 10.1104/pp.114.236901

147. Schaller, A., and Stintzi, A. (2009). Enzymes in jasmonate biosynthesis – structure, function, regulation. *Phytochemistry* 70, 1532–1538. doi: 10.1016/j.phytochem.2009.07.032

148. Scheer, J. M., and Ryan, C. A. (1999). A 160-kD systemin receptor on the surface of *Lycopersicon peruvianum* suspension-cultured cells. *Plant Cell* 11, 1525–1535. doi: 10.1105/tpc.11.8.1525

149. Schilmiller, A. L., and Howe, G. A. (2005). Systemic signaling in the wound response. *Curr. Opin. Plant Biol.* 8, 369–377. doi: 10.1016/j.pbi.2005.05.008

150. Schweighofer, A., Kazanaviciute, V., Scheikl, E., Teige, M., Doczi, R., Hirt, H.,et al. (2007). The PP2C-type phosphatase AP2C1, which negatively regulates MPK4 and MPK6, modulates innate immunity, jasmonic acid, and ethylene levels in *Arabidopsis.Plant Cell* 19, 2213–2224. doi: 10.1105/tpc.106.049585

151. Schweizer, P., Felix, G., Buchala, A., Müller, C., and Métraux, J. P. (1996a). Perception of free cutin monomers by plant cells.*Plant J.* 10, 331–341. doi: 10.1046/j.1365-313X.1996.10020331.x

152. Schweizer, P., Jeanguenat, A., Whitacre, D., Metraux, J. P., and Mosinge, E. (1996b). Induction of resistance in barley against*Erysiphe graminis* f.sp. *hordei* by free cutin monomers. *Physiol. Mol. Plant Pathol.* 49, 103–120. doi: 10.1006/pmpp.1996.0043

153. Schweizer, P., Jeanguenat, A., Mosinger, E., and Metraux, J. P. (1994). Plant protection by free cutin monomers in two cereal pathosystems. *Adv. Mol. Gen. Plant Microbe Interact.* 21, 371–374. doi: 10.1007/978-94-011-0177-6_55

154. Seo, S., Katou, S., Seto, H., Gomi, K., and Ohashi, Y. (2007). The mitogen-activated protein kinases WIPK and SIPK regulate the levels of jasmonic and salicylic acids in wounded tobacco plants. *Plant J.* 49, 899–909. doi: 10.1111/j.1365-313X.2006.03003.x

155. Seo, S., Okamoto, N., Seto, H., Ishizuka, K., Sano, H., and Ohashi, Y. (1995). Tobacco MAP kinase: a possible mediator in wound signal transduction pathways. *Science* 270, 1988–1992. doi: 10.1126/science.270.5244.1988

156. Sieber, P., Schorderet, M., Ryser, U., Buchala, A., Kolattukudy, P., Metraux, J. P.,et al. (2000). Transgenic *Arabidopsis* plants expressing a fungal cutinase show alterations in the structure and properties of the cuticle and postgenital organ fusions. *Plant Cell* 12, 721–737. doi: 10.1105/tpc.12.5.721

157. Staswick, P. E., and Tiryaki, I. (2004). The oxylipin signal jasmonic acid is activated by an enzyme that conjugates it to isoleucine in *Arabidopsis*. *Plant Cell* 16, 2117–2127. doi: 10.1105/tpc.104.023549

158. Stenzel, I., Hause, B., Maucher, H., Pitzschke, A., Miersch, O., Ziegler, J.,et al. (2003). Allene oxide cyclase dependence of the wound response and vascular bundle-specific generation of jasmonates in tomato – amplification in wound signalling. *Plant J.*33, 577–589. doi: 10.1046/j.1365-313X.2003.01647.x

159. Suzuki, N., and Mittler, R. (2012). Reactive oxygen species-dependent wound responses in animals and plants. *Free Radic. Biol. Med.* 53, 2269–2276. doi: 10.1016/j.freeradbiomed.2012.10.538

160. Szczegielniak, J., Borkiewicz, L., Szurmak, B., Lewandowska-Gnatowska, E., Statkiewicz, M., Klimecka, M.,et al. (2012). Maize calcium-dependent protein kinase (ZmCPK11): local and systemic response to wounding, regulation by touch and components of jasmonate signaling. *Physiol. Plant.* 146, 1–14. doi: 10.1111/j.1399-3054.2012.01587.x

161. Takabatake, R., Seo, S., Ito, N., Gotoh, Y., Mitsuhara, I., and Ohashi, Y. (2006). Involvement of wound-induced receptor-like protein kinase in wound signal transduction in tobacco plants. *Plant J.* 47, 249–257. doi: 10.1111/j.1365-313X.2006.02781.x

162. Takeda, S., Gapper, C., Kaya, H., Bell, E., Kuchitsu, K., and Dolan, L. (2008). Local positive feedback regulation determines cell shape in root hair cells. *Science* 319, 1241–1244. doi: 10.1126/science.1152505

163. Taurino, M., Abelenda, J. A., Rio-Alvarez, I., Navarro, C., Vicedo, B., Farmaki, T.,et al. (2014). Jasmonate-dependent modifications of the pectin matrix during potato development function as a defense mechanism targeted by *Dickeya dadantii*virulence factors. *Plant J.* 77, 418–429. doi: 10.1111/tpj.12393

164. Tisi, A., Angelini, R., and Cona, A. (2008). Wound healing in plants: cooperation of copper amine oxidase and flavin-containing polyamine oxidase. *Plant Signal. Behav.* 3, 204–206. doi: 10.4161/psb.3.3.5372

165. Toufighi, K., Brady, S. M., Austin, R., Ly, E., and Provart, N. J. (2005). The Botany array resource: e-northerns, expression angling, and promoter analyses. *Plant J.* 43, 153–163. doi: 10.1111/j.1365-313X.2005.02437.x

166. Verica, J. A., Chae, L., Tong, H. Y., Ingmire, P., and He, Z. H. (2003). Tissue-specific and developmentally regulated expression of a cluster of tandemly arrayed cell wall-associated kinase-like kinase genes

in *Arabidopsis*. *Plant Physiol.* 133, 1732–1746. doi: 10.1104/pp.103.028530

167. Wagner, T. A., and Kohorn, B. D. (2001). Wall-Associated kinases are expressed throughout plant development and are required for cell expansion. *Plant Cell* 13, 303–318. doi: 10.1105/tpc.13.2.303

168. Wang, L., Tsuda, K., Sato, M., Cohen, J. D., Katagiri, F., Glazebrook, J.,et al. (2009). *Arabidopsis* CaM binding protein CBP60g contributes to MAMP-induced SA accumulation and is involved in disease resistance against. *PLoS Pathog.* 5:e1000301. doi: 10.1371/journal.ppat.1000301

169. Wildon, D. C., Thain, J. F., Minchin, P. E. H., Gubb, I. R., Reilly, A. J., Skipper, Y. D.,et al. (1992). Electrical signaling and systemic proteinase inhibitor induction in the wounded plant. *Nature* 360, 62–65. doi: 10.1038/360062a0

170. Wolf, S., Hematy, K., and Hofte, H. (2012). Growth control and cell wall signaling in plants. *Annu. Rev. Plant Biol.* 63, 381–407. doi: 10.1146/annurev-arplant-042811-105449

171. Yamaguchi, Y., Huffaker, A., Bryan, A. C., Tax, F. E., and Ryan, C. A. (2010). PEPR2 is a second receptor for the Pep1 and Pep2 peptides and contributes to defense responses in *Arabidopsis*. *Plant Cell* 22, 508–522. doi: 10.1105/tpc.109.068874

172. Yamaguchi, Y., Pearce, G., and Ryan, C. A. (2006). The cell surface leucine-rich repeat receptor for AtPEP1, an endogenous peptide elicitor in *Arabidopsis*, is functional in transgenic tobacco cells. *Proc. Natl. Acad. Sci. U.S.A.* 103, 10104–10109. doi: 10.1073/pnas.0603729103

173. Yamanaka, T., Nakagawa, Y., Mori, K., Nakano, M., Imamura, T., Kataoka, H.,et al. (2010). MCA1 and MCA2 that mediate Ca^{2+}uptake have distinct and overlapping roles in *Arabidopsis*. *Plant Physiol.* 152, 1284–1296. doi: 10.1104/pp.109.147371

174. Yang, D. H., Hettenhausen, C., Baldwin, I. T., and Wu, J. (2012). Silencing *Nicotiana attenuata* calcium-dependent protein kinases, CDPK4 and CDPK5, strongly up-regulates wound- and herbivory-induced jasmonic acid accumulations. *Plant Physiol.*159, 1591–1607. doi: 10.1104/pp.112.199018

175. Zhang, S. Q., and Klessig, D. F. (1998a). The tobacco wounding-activated mitogen-activated protein kinase is encoded by SIPK.*Proc. Natl. Acad. Sci. U.S.A.* 95, 7225–7230. doi: 10.1073/pnas.95.12.7225

176. Zhang, S. Q., and Klessig, D. F. (1998b). Resistance gene N-mediated de novo synthesis and activation of a tobacco mitogen-activated protein kinase by tobacco mosaic virus infection. *Proc. Natl. Acad. Sci. U.S.A.* 95, 7433–7438. doi: 10.1073/pnas.95.13.7433

177. Zimmermann, M. R., Maischak, H., Mithofer, A., Boland, W., and Felle, H. H. (2009). System potentials, a novel electrical long-distance apoplastic signal in plants, induced by wounding. *Plant Physiol.* 149, 1593–1600. doi: 10.1104/pp.108.133884

178. Zipfel, C., Kunze, G., Chinchilla, D., Caniard, A., Jones, J. D. G., Boller, T.,et al. (2006). Perception of the bacterial PAMPEF-Tu by the receptor EFR restricts*Agrobacterium*-mediated transformation. *Cell* 125, 749–760. doi: 10.1016/j.cell.2006.03.037

179. Zipfel, C., Robatzek, S., Navarro, L., Oakeley, E. J., Jones, J. D., Felix, G.,et al. (2004). Bacterial disease resistance in*Arabidopsis* through flagellin perception. *Nature* 428, 764–767. doi: 10.1038/nature02485

CHAPTER 3 PROCESSES IN PATHOGENESIS BY FUNGAL AND OOMYCETE PLANT PATHOGENS

Shaowu Meng, Trudy Torto-Alalibo, Marcus C Chibucos, Brett M Tyler, and Ralph A Dean

COMMON PATHOGENESIS PROGRAMS OF FUNGI AND OOMYCETES

Oomycetes, although phylogenetically very distant, share many common morphological and physiological features with the true fungi [1, 2, 3]. For example, they have similar filamentous, branching, indeterminate bodies, and they acquire nutrition by secreting digestive enzymes and then absorbing the resultant breakdown products. More importantly, fungi and oomycetes share a unique capability compared with other microbial pathogens, namely that they are able to breach cuticles of host plants and establish infection rapidly [4]. Consequently, both are causal agents of many destructive plant diseases and are responsible for significant economic losses every year.

In this review, we summarize common mechanisms of pathogenesis displayed by oomycetes and fungi. Pathogenesis by a fungus or oomycete is a complex process. Briefly, it includes the following steps: dispersal and arrival of an infectious particle (usually a spore of some kind) in the vicinity of the host, adhesion to the host, recognition of the host (which may occur prior to adhesion), penetration into the host, invasive growth within the host, lesion development in the host, and finally production of additional infectious particles [5, 6] (see Figures 1, 2). In order to describe the entire process, we formulate a description of pathogenesis using standardized terms from the Gene Ontology (GO), including 256 new terms developed by members of the PAMGO (Plant-Associated Microbe Gene Ontology) consortium http://pamgo.vbi.vt.edu, an official interest group of the GO Consortium, as well as 38 extant GO terms that are placed in shaded boxes in Figures 3, 4, 5,6.

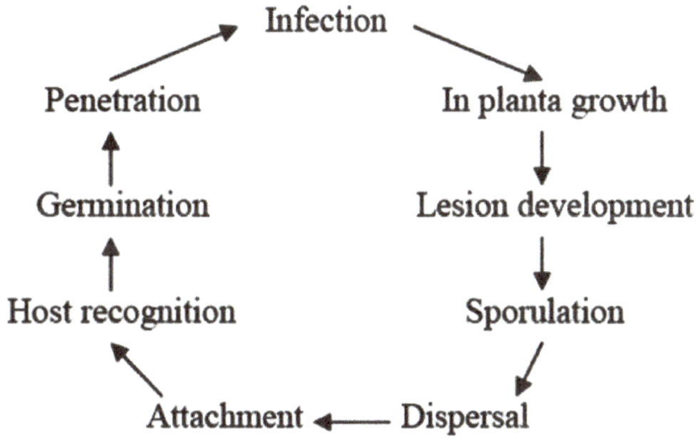

Figure 1: A generalized diagram displaying infection and disease cycle caused by fungi and oomycetes.

Citation: Meng, Trudy Torto-Alalibo, Marcus C Chibucos, Brett M Tyler, and Ralph A Dean, Processes In Pathogenesis by Fungal and Oomycete Plant Pathogens, DOI: 10.1186/1471-2180-9-S1-S7

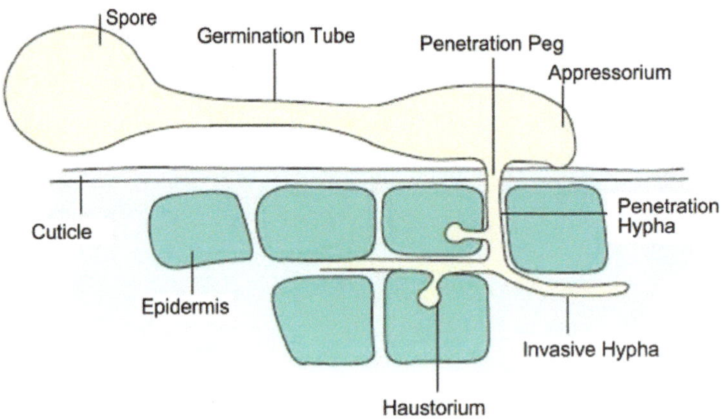

Figure 2: The infection process in fungal and oomycete pathogens. Modified by permission from Schumann, G. L., 1991, Plant diseases: Their biology and social impact, American Phytopathological Society, St. Paul, MN.

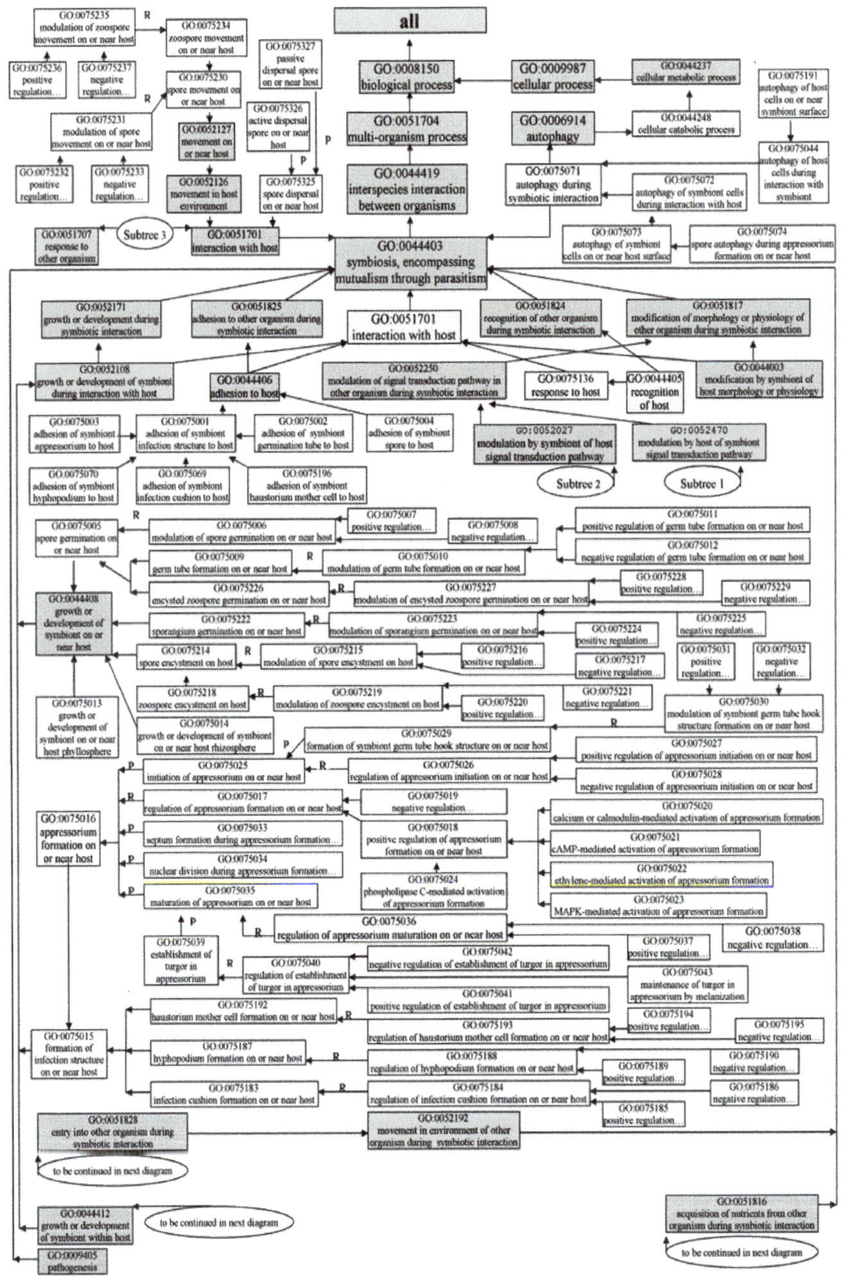

Figure 3: Gene Ontology terms for processes related to infection and disease (Part 1). Subtree 1 and 2 are depictured in Figure 5, and Subtree 3 is depictured in Figure 6. Shaded boxes indicate pre-existing GO terms, and unshaded boxes represent GO terms developed under the PAMGO project. "R" indicates "regulates relationship", "P" indicates "part of relationship", and null indicates "is a relationship" (see the Gene Ontology website athttp://www.geneontology.org for further information).

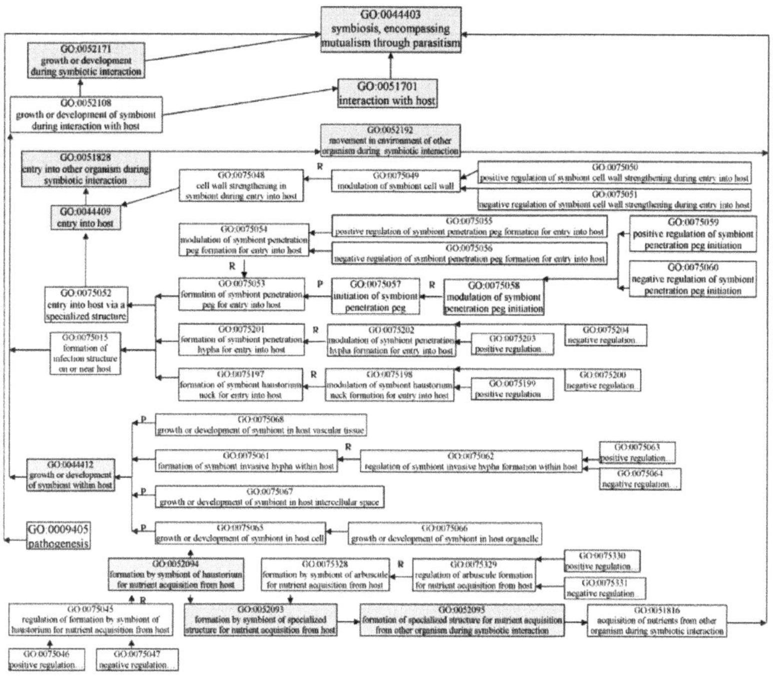

Figure 4: Gene Ontology terms for processes related to infection and disease (Part 2). Shaded boxes indicate pre-existing GO terms, and unshaded boxes represent GO terms developed under the PAMGO project. "R" indicates "regulates relationship", "P" indicates "part_of relationship", and null indicates "is_a relationship" (see the Gene Ontology website at http://www.geneontology.org for further information).

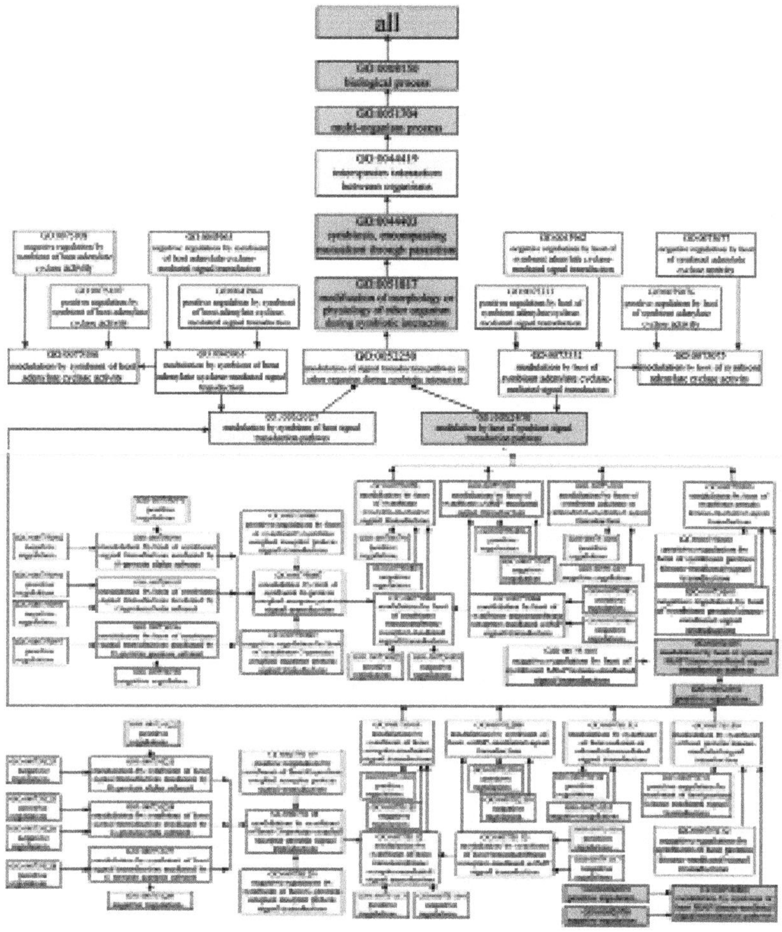

Figure 5: Gene Ontology terms for signal transduction processes related to infection and disease (Part 1). Subtree 1 consists of GO terms intending to annotate host gene products that stimulate signal transduction in symbiont. Subtree 2 represents the opposite perspective of Subtree 1. Shaded boxes indicate pre-existing GO terms, and unshaded boxes represent GO terms developed under the PAMGO project.

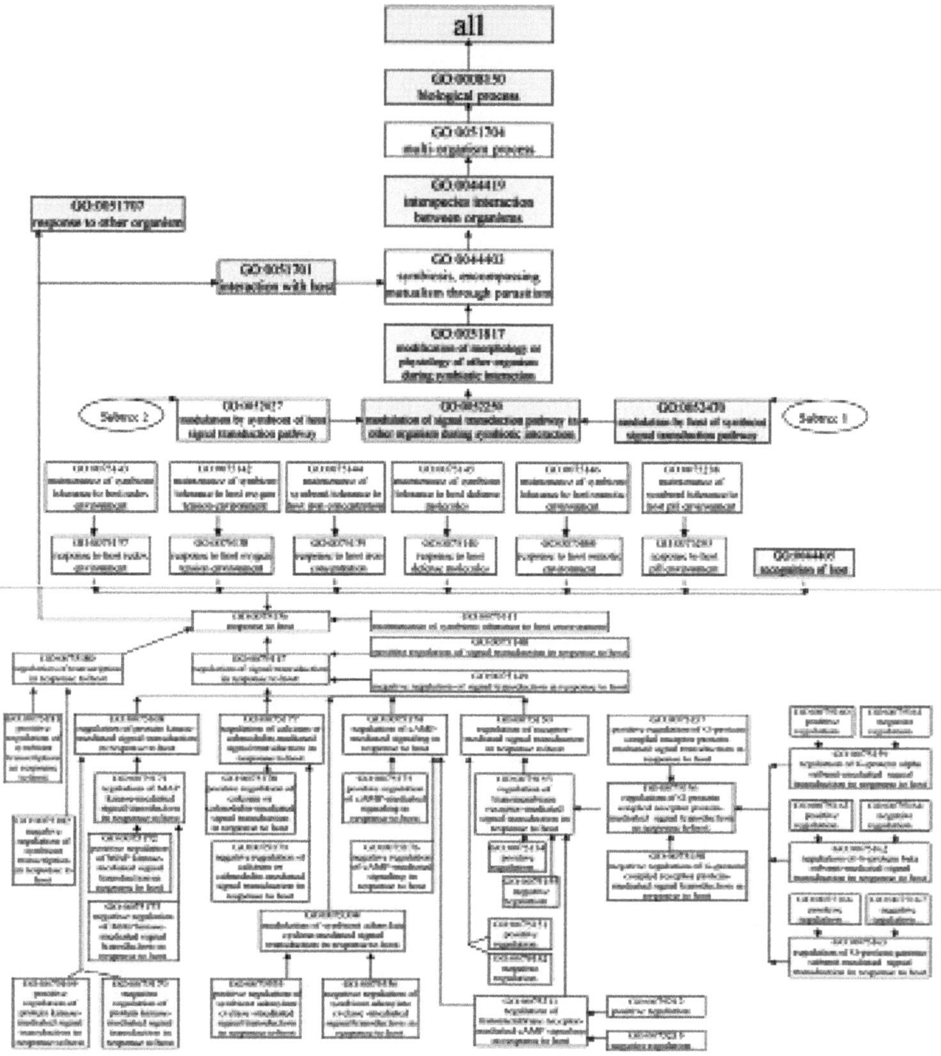

Figure 6: Gene Ontology terms for signal transduction processes related to infection and disease (Part 2). Subtree 3 consists of GO terms intending to annotate symbiont gene products that stimulate signal transduction in symbiont in response to host. Shaded boxes indicate pre-existing GO terms, and unshaded boxes represent GO terms developed under the PAMGO project.

All of the 256 new terms are placed within GO under the node "GO ID 0044403 symbiosis, encompassing mutualism through parasitism" (note the original, broad definition of "symbiosis" used in GO, which specifies use of the words "symbiont" and "host" as the smaller and larger, respectively, of the symbiotic interactants). Among these new terms, every term that starts with "modulation" or "regulation" has two child terms, one is "positive regulation of...", and the other is "negative regulation of..." Note that these child terms are general GO terms; "position regulation," for example, includes induction, upregulation, stimulation, etc. Four diagrams (see Figures 3, 4, 5, 6) encompassing the 256 new and 38 extant GO terms explicitly depict our description of pathogenesis, with an emphasis on appressorium formation and signal transduction. More details about each step are presented in the following sections.

SPORE DISPERSAL

Dispersal of spores is the most common process to initiate new infections [5], though direct infection by hyphae may occur. An example of the latter is the spread of ectomycorrhizal basidiomycetes in forest soils. Dispersal mechanisms can be grouped into two types: one is passive dispersal by wind, water or animal [7], and the other is active dispersal such as shooting ascospores through the boundary layer of air surrounding the fruiting body by forcible discharge [8]. Similarly, spores can be grouped into two types according to their motility. In fungi non-motile spores include sexual spores such as ascospores, rust urediniospores, sclerotia and conidiospores, while non-motile oomycete spores include oospores, sporangiospores and conidia. Motile spores with flagella, called zoospores, are ubiquitous among oomycetes and are also found in chytrid fungi [9]. Additionally, spores vary in requirements for dormancy. Some spores, such as zoospores, must encyst and differentiate to acquire qualities of dormancy before they become true spores [10]. Different spores vary in the length of the dormancy period; for example, some ascospores and oospores show extended dormancy, while others such as zoospores and ascomycete conidia are usually short-lived.

Three new GO terms including the term "GO ID 0075325 spore dispersal on or near host" were developed under the node "GO ID 0051701 ...interaction with host" to describe the mechanisms of spore dispersal. New terms describing active or passive dispersal mechanisms were placed as children of "spore dispersal on or near host" (see Figure 3).

Eight new GO terms describing spore motility were listed under the node "GO ID 0052127 movement on or near host". The term "GO ID 0075230 spore movement on or near host" is central to these eight terms with "GO ID 0075234 zoospore movement on or near host" as the principal child term (see Figure 3).

Similarly, eight new GO terms were added under "GO ID 0044408 growth or development of symbiont on or near host" to describe spore encystment. The term "GO ID 0075214 spore encystment on host" is central to these eight terms, with the term "GO ID 0075218 zoospore encystment on host" as the main child term (see Figure 3).

These 19 new terms are appropriate for annotating gene products known to be involved in spore dispersal. For example, inhibition of the vacuolar H+-ATPase by potassium nitrate causes a reduction in vacuole expulsion in zoospores of the oomycete *Phytophthora nicotianae* and leads to premature encystment [11]. Thus, H+-ATPase negatively regulates zoospore encystment and can be annotated with the new term «GO ID 0075221 negative regulation of zoospore encystment on host».

ADHESION TO THE HOST

Adhesion of spores to the host involves physical and chemical processes [3]. Typically, when spores reach the surface of a host tissue, they attach via adhesion molecules [5]. A germination tube then emerges from the spore or the encysted zoospore (see Figure 2). From the germination tube, a growth hypha or an infection structure such as an appressorium [12, 13, 14, 15, 16] develops, which also becomes firmly attached to the host surface via adhesion molecules. A variety of other infection structures such as hyphopodia [17, 18, 19], haustorium mother cells [20, 21, 22, 23], or infection cushions [24] are generated by fungal pathogens after germinating on the host surface. These all serve a common function of facilitating the pathogen's entry into the host tissue. It should be noted that the sporangia of many oomycetes may germinate directly to form an infection hypha, or else in the presence of abundant water they may differentiate, through specialized cleavage vesicles, into 10–30 zoospores that can individually disperse to initiate sites of infection [25].

Seven new GO terms under the parent, "GO ID 0044406 adhesion to host", were developed to describe in detail the biological process of adhesion to a host. The term "GO ID 0075001 adhesion of symbiont infection structure to host" is central to this section. Among the seven terms, five terms that describe adhesion of a specific infection structure, including appressorium, hyphopodium, haustorium mother cell, infection cushion, or germination tube, are children of "adhesion of symbiont infection structure to host" (see Figure 3).

To describe spore germination on or near host tissue, 16 new terms under the parent, "GO ID 0044408 growth or development of symbiont on or near host", were developed. The 16 terms cover spore germination, sporangium germination, encysted zoospore germination, and germ tube formation. The term "GO ID 0075005 spore germination on or near host" is central to this section. Major relationships among the sixteen terms are shown in Figure 3.

The 23 new GO terms in this section are useful for annotating pathogen gene products involved in adhesion to host tissue. For example, Car (cyst-germination-specific acidic repeat) proteins of the oomycete *Phytophthora infestans* are transiently expressed during germination of cysts (i.e., encysted zoospores) and during formation of appressoria, and they are localized at the surface of germlings. These proteins have considerable sequence homology to mucins and have an internal octapeptide repeat with the mucin consensus sequence TTYAPTEE. Therefore, like mucins, Car proteins should serve as a mucous cover protecting the germling and assisting in adhesion to the leaf surface [26]. Thus, the Car proteins can be annotated with the new terms «GO ID 0075226 encysted zoospore germination on or near host» and «GO ID 0075001 adhesion of symbiont infection structure to host», using the GO evidence code ISS (Inferred from Sequence or Structural Similarity).

SIGNAL TRANSDUCTION DURING RECOGNITION OF THE HOST

Signal transduction is an integral component of the host recognition process. Examples include protein kinase-mediated signal transduction [27], receptor-mediated signal transduction [28], G-protein coupled receptor protein signal transduction, G-protein subunit-mediated signal transduction [29], cAMP-mediated signal transduction [30], calcium or calmodulin-mediated signal transduction [31], and adenylate cyclase-mediated signal transduction [12].

In order to adequately describe signal transduction during symbiont interaction with its host, three sets of new terms were developed. Signal transduction pathways involved in the recognition between host and symbiont are generally quite extensively characterized and consequently 127 new terms were developed.

The first set of new terms is intended for annotation of host gene products that stimulate symbiont signal transduction (see Subtree 1, which includes terms under the node "GO ID 0052470 modulation by host of symbiont signal transduction pathway" in Figure 5). This set has 37 new terms. Five of these terms describing different types of signal transduction pathways are children of "GO ID 0052470" (see Subtree 1 in Figure 5).

The second set of new terms is intended for annotation of symbiont gene products that stimulate host signal transduction (see Subtree 2, which includes terms under the node "GO ID 0052027 modulation by symbiont of host signal transduction pathway" in Figure 5). This set has 36 new GO terms and has the same structure as the first set (see Subtree 2 in Figure 5). The terms in the second set are essentially the converse of the terms in the first set. For example, the term "GO ID 0075130 modulation by symbiont of host protein kinase-mediated signal transduction" in the second term set has a complementary term "GO ID 0075099 modulation by host of symbiont protein kinase-mediated signal transduction" in the first term set.

The third set of new terms is intended for annotation of symbiont gene products that stimulate symbiont signal transduction in response to the host (see Subtree 3, which includes terms under the nodes "GO ID 0051701 interaction with host" and "GO ID 0051707 response to other organism" in Figure 6). This set has 56 new GO terms. The new term "GO ID

0075136 response to host" is central to the 56 new terms. This term has 10 child terms that describe 10 symbiont responses to host organisms, such as "GO ID 0075147 regulation of signal transduction in response to host", "GO ID 0075140 response to host defense molecules", and "GO ID 0075180 regulation of transcription in response to host" etc. (see Figure 6). Eight of the 10 terms have their own child and lower level offspring terms, and each of those "response" terms has a child term such as "maintenance of symbiont tolerance to host ..." (see details in Figure 6).

The term "GO ID 0075147 regulation of signal transduction in response to host" has five children to describe different types of signal transduction, similar to the five child terms of "GO ID 0052470 modulation by host of symbiont signal transduction pathway" in the first set. Each of the five terms has child terms for positive regulation and negative regulation.

The three sets of new GO terms can be used to explicitly describe genes of signal transduction pathways involved in host recognition. For instance, the PMK1 gene of the rice blast fungus *Magnaporthe oryzae*encodes a mitogen-activated protein kinase (MAPK), which is a key component in the MAPK signaling cascade and is involved in appressorium formation and infectious growth [32]. Thus, the PMK1 protein can be annotated with the term "GO ID 0075171 regulation of MAP kinase-mediated signal transduction in response to host". Note that this gene product would not be annotated with "GO ID 0052435 modulation by host of symbiont MAP kinase-mediated signal transduction pathway" since this latter GO term is reserved to annotate host gene products. Similarly, this protein should not be annotated with "GO ID 0052080 modulation by symbiont of host MAP kinase-mediated signal transduction pathway" since PMK1 belongs to the symbiont's and not the host's signaling transduction pathway.

In addition, the modulation terms have children that describe more specific kinds of signal transduction. For example, «GO ID 0075168 regulation of protein kinase-mediated signal transduction in response to host» has a child «GO ID 0075171 regulation of MAP kinase-mediated signal transduction in response to host» (see details in Figure 6).

PENETRATION INTO THE HOST

Pathogens have evolved several mechanisms that include structural and/or enzymatic components in order to enter into their plant hosts [5]. Many fungi, such as *Alternaria alternata*, *Colletotrichum graminicola, M. oryzae*,*Pyrenophora teres*, and many oomycetes, such as *P. infestans* and *Phytophthora cinnamomi*, develop appressoria to directly penetrate plant cuticles [13, 33, 34, 35, 36, 37, 38]. An appressorium is a highly specialized structure that differentiates from the end of a symbiont germ tube. It is a swollen, dome-shaped or cylindrical organ, from which a narrow penetration peg emerges to rupture the plant cuticle and cell wall [33]. The penetration peg extends and forms a penetration hypha to penetrate through the epidermal cells and emerge into the underlying tissue [34, 35]. In some instances, penetration is driven by astoundingly high turgor pressures within the appressoria [36, 38]. Generation of turgor is due in part to a thick inner cell wall layer of melanin in mature appressoria. In other instances, cell wall degrading enzymes may play a primary role in or may facilitate the penetration process [39, 40, 41]. Appressoria produced by some fungi, such as rust fungi, do not penetrate directly through the cuticle, but gain entry through stomata [42].

Sixty-four new GO terms were developed to describe the biological process of penetration into the host, and they form two groups. The first group includes 43 new GO terms related to infection structures established on the outside of the host tissue, such as appressoria, hyphopodia, infection cushions, and haustorium mother cells. The second group has 21 new terms related to specialized structures that directly pierce the surface of the host, for example penetration pegs, penetration hyphae, and haustorium necks.

All of the 43 terms in the first group are children or lower level offspring of "GO ID 0052108 growth or development of symbiont during interaction with host". The core of this group is "GO ID 0075015 formation of infection structure on or near host". Twenty-eight terms in this group are related to appressorium formation. In particular, five of the 28 terms describe in detail the process of appressorium formation, namely "GO ID 0075025 initiation of appressorium on or near host", "GO ID 0075034 nuclear division during appressorium formation on or near host", "GO ID 0075033 septum formation during appressorium formation on or near host", "GO ID 0075035 maturation of appressorium on or near host", and "GO ID 0075017 regulation of appressorium formation on or near host" (see details in Figure 3).

Besides the child term "GO ID 0075016 appressorium formation on or near host", the term "GO ID 0075015 formation of infection structure on or near host" has three more detailed child terms: "GO ID 0075192 haustorium mother cell formation on or near host", "GO ID 0075187 hyphopodium formation on or near host", and "GO ID 0075183 infection cushion formation on or near host" (see details in Figure 3).

All of the 21 terms in the second group are children or lower level offspring of "GO ID 0044409 entry into host". The core of this group is "GO ID 0075052 entry into host via a specialized structure", which has three child terms related to penetration peg, penetration hypha, or haustorium neck for entry into the host (see details in Figure 4).

The 64 new terms can be used to annotate the gene products of penetration-related genes. For example, genes involved in melanin biosynthesis in the rice blast fungus, such as ALB1, RSY1 and BUF1, are required for appressorium function since mutants lacking these genes make appressoria, but are unable to penetrate susceptible rice leaves [43]; these can be annotated with the term "GO ID 0075053 formation of symbiont penetration peg for entry into host".

INVASIVE GROWTH WITHIN THE HOST

After successful penetration, invasive hyphae are formed that ramify through the host tissue [44, 45]. In some cases, special structures, such as a haustorium or an arbuscle, are formed in host cells for the symbiont to absorb nutrition [22, 23].

To describe invasive growth, 15 new GO terms were developed that are children or lower level offspring of "GO ID 0044412 growth or development of symbiont within host". The term "GO ID 0075065 growth or development of symbiont in host cell" has two children, "GO ID 0052094 formation by symbiont of haustorium for nutrient acquisition from host" and

"GO ID 0075066 growth or development of symbiont in host organelle". Additionally, arbuscules produced by mycorrhizal fungi are a type of structure functionally similar to haustoria, and thus "GO ID 0075328 formation by symbiont of arbuscule for nutrient acquisition from host" is a sibling of "GO ID 0052094" (see details in Figure 4).

The 15 new GO terms in this section meet the need to annotate pathogen genes that are involved in invasive growth. For example, the *MST12* gene in the rice blast fungus *M. grisea* was found to regulate infectious growth but not appressorium formation [46]. In particular, no obvious defects in vegetative growth, conidiation, or conidia germination were observed in *MST12* deletion mutants. Also, *MST12* mutants produce typical dome-shaped and melanized appressoria. When inoculated through wound sites, *MST12* mutants fail to cause spreading lesions and appear to be defective in infectious growth. As a result, *MST12* mutants are nonpathogenic [46]. Thus, the *MST12* gene can be annotated with the term "GO ID 0075061 formation of symbiont invasive hypha within host".

LESION DEVELOPMENT IN THE HOST

The eventual result of infection in most cases is lesion development. A lesion can be defined as any abnormality involving any tissue or organ due to any disease or any injury (cited from MedicineNet.com). Not surprisingly, there are many types of lesions including those caused by damage such as cold injury or insects' bites etc. It is difficult to define lesions objectively, as this requires a subjective judgment on what constitutes abnormal or damage and from what perspective, ranging for example from perturbation of a few cells to death of an entire tissue or organ. Similarly, formation of a lesion is not a specific process belonging to either the pathogen or the host and can be highly dependent on the environment. Therefore, at this time only one term, "GO ID 0009405 pathogenesis", is appropriate for genes involved in lesion formation.

OTHER NEW GO TERMS

Six new terms were placed jointly under the nodes "GO ID 0006914 autophagy" and "GO ID 0044403 symbiosis, encompassing mutualism through parasitism". The term "GO ID 0075071 autophagy during symbiotic interaction" is the core of the six terms, and it has two complementary children, "GO ID 0075044 autophagy of host cells during interaction with symbiont" and "GO ID 0075072 autophagy of symbiont cells during interaction with host". The latter term has a child "GO ID 0075073 autophagy of symbiont cells on or near host surface", which itself has a lower level child "GO ID 0075074 spore autophagy during appressorium formation on or near host" (see details in Figure 3).

The six autophagy-related GO terms are applicable to describe the functions of several genes in fungal pathogens during symbiotic interaction. For example, formation of a functional appressorium in the rice blast fungus requires autophagic cell death of the conidium, which is controlled by the *MgATG8* gene. Deletion of*MgATG8* results in impaired autophagy, arrested conidial cell death, and a nonpathogenic fungus [14]. Thus,*MgATG8* can be annotated with the new term "GO ID 0075074 spore autophagy during appressorium formation on or near host».

CONCLUSION

Two hundred fifty-six new GO terms were developed to annotate genes or gene products involved in common pathogenic processes in fungi and oomycetes, including spore dispersal, host adhesion, recognition, penetration, and invasive growth. These new GO terms provide the opportunity to apply a standard set of terms to annotate gene products of fungi, oomycetes, and their associated hosts, as well as those of other plant-associated pathogens and their hosts. The ability to compare and contrast these annotations for widely different plant-associated microbes and their hosts, using a standardized vocabulary, will greatly facilitate the identification of unique and conserved features of pathogenesis across different kingdoms. In addition, such comparisons should provide insight into the evolution of pathogenic processes.

ACKNOWLEDGEMENTS

All authors read and approved the final manuscript. We thank Candace Collmer, Michelle Gwinn Giglio, and the editor at The Gene Ontology Consortium Jane Lomax for their comments and suggestions in developing these PAMGO terms. This work is a part of PAMGO project, which is supported by the USDA NRI-CSREES (grant number 2005-35600-16370), and the National Science Foundation (grant number EF-0523736).

This article has been published as part of *BMC Microbiology* Volume 9 Supplement 1, 2009: The PAMGO Consortium: Unifying Themes In Microbe-Host Associations Identified Through The Gene Ontology. The full contents of the supplement are available online at http://www.biomedcentral.com/1471-2180/9?issue=S1.

COMPETING INTERESTS

The authors declare that they have no competing interests.

REFERENCES

1. Money NP: Why oomycetes have not stopped being fungi. Mycology Research. 1998, 102 (6): 767-768. 10.1017/S095375629700556X.
2. Latijnhouwers M, Wit PJGMD, Govers F: Oomycetes and fungi: similar weaponry to attack plants. Trends in Microbiology. 2003, 11 (10): 462-469. 10.1016/j.tim.2003.08.002.
3. Epstein L, Nicholson RL: Adhesion and adhesives of fungi and oomycetes. Biological Adhesives. Edited by: Smith AM, Callow JA. 2006, Springer-Verlag Berlin Heidelberg
4. Soanes DM, Richards TA, Talbot NJ: Insights from sequencing fungal and oomycete genomes: what can we learn about plant disease and the evolution of pathogenicity?. The Plant Cell. 2007, 19: 3318-3326. 10.1105/tpc.107.056663.
5. Agrios GN: Plant pathology. 2005, Elsevier Academic Press, London, UK, Fifth
6. Colditz F, Krajinski F, Niehaus K: Plant proteomics upon fungal attack. Plant Proteomics. Edited by: Šamaj J, Thelen J. 2007, Springer

7. Ingold GT: Dispersal in Fungi. 1953, Clarendon Press, Oxford; Oxford University Press, New York

8. Trail F: Fungal cannons: explosive spore discharge in the Ascomycota. FEMS Microbiol Lett. 2007, 276: 12-18. 10.1111/j.1574-6968.2007.00900.x.

9. James TY, Letcher PM, Longcore JE, Mozley-Standridge SE, Porter D, Powell MJ, Griffith GW, Vilgalys R: A molecular phylogeny of the flagellated fungi (Chytridiomycota) and description of a new phylum (Blastocladiomycota). Mycologia. 2006, 98 (6): 860-871. 10.3852/mycologia.98.6.860.

10. Griffin DH: Fungal Physiology. 1996, Published by Wiley_Default

11. Mitchell HJ, Hardham AR: Characterisation of the water expulsion vacuole in *Phytophthora nicotianae* zoospores. Protoplasma. 1999, 206: 118-130. 10.1007/BF01279258.

12. Choi W, Dean RA: The adenylate cyclase gene MAC1 of *Magnaporthe grisea* controls appressorium formation and other aspects of growth and development. Plant Cell. 1997, 9 (11): 1973-1983. 10.1105/tpc.9.11.1973.

13. Hardham AR: The cell biology behind *Phytophthora* pathogenicity. Australasian Plant Pathology. 2001, 30: 91-98. 10.1071/AP01006.

14. Veneault-Fourrey C, Barooah MK, Egan MJ, Talbot NJ: Autophagic fungal cell death is necessary for infection by the rice blast fungus. Science. 2006, 312 (5773): 580-583. 10.1126/science.1124550.

15. Talbot NJ: Fungal genomics goes industrial. Nature Biotechnology. 2007, 25 (5): 542-543. 10.1038/nbt0507-542.

16. Grenville-Briggs LJ, Anderson VL, Fugelstad J, Avrova AO, Bouzenzana J, Williams A, Wawra S, Whisson SC, Birch PRJ, Bulone V, West PV: Cellulose synthesis in *Phytophthora infestans* is required for normal appressorium formation and successful infection of potato. The Plant Cell. 2008, 20: 720-738. 10.1105/tpc.107.052043.

17. Onyile AB, Edwards HH, Gessner RV: Adhesive material of the hyphopodia of *Buergenerula spartinae*. Mycologia. 1982, 74 (5): 777-784. 10.2307/3792864.

18. Du M, Schard CL, Nuckles EM, Vaillancourt LJ: Using mating-type gene sequences for improved phylogenetic resolution of *Collectotrichum* species complexes. Mycologia. 2005, 97 (3): 641-658. 10.3852/mycologia.97.3.641.

19. Sukno SA, García VM, Shaw BD, Thon MR: Root infection and systemic colonization of maize by *Colletotrichum graminicola*. Applied and environmental microbiology. 2008, 74 (3): 823-832. 10.1128/AEM.01165-07.

20. Heath MC, Heath IB: Ultrastructural changes associated with the haustorial mother cell ceptum during haustorium formation in *Uromyces phaseoli* var. *vignae*. Protoplasma. 1975, 84: 297-314. 10.1007/BF01279359.

21. Glidewell DC, Mims CW: Ultrastructure of the haustorial apparatus in the rust fungus *Kunkelia nitens*. Botanical Gazette. 1979, 140 (2): 148-152. 10.1086/337071.

22. Mendgen K, Dressler E: Culturing *Puccinia coronata* on a cell monolayer of the *Avena saliva* coleoptile. Phytopath. 1983, 108: 226-234. 10.1111/j.1439-0434.1983.tb00583.x.

23. Wiethölter N, Horn S, Reisige K, Beike U, Moerschbacher BM: In vitro differentiation of haustorial mother cells of the wheat stem rust fungus, *Puccinia graminis* f. sp. *tritici*, triggered by the synergistic action of chemical and physical signals. Fungal Genetics and Biology. 2003, 38: 320-326. 10.1016/S1087-1845(02)00539-X.

24. Demirci E, Döken MT: Host penetration and infection by the anastomosis groups of *Rhizoctonia solani* Kühn isolated from potatoes. Tr J of Agriculture and Forestry. 1998, 22: 609-613.

25. Birch PRJ, Cooke DEL: Mechanisms of infection: Oomycetes. Encyclopedia of Plant and Crop Science. 2004, 1 (1): 697-700.

26. Görnhardt B, Rouhara I, Schmelzer E: Cyst germination proteins of the potato pathogen *phytophthora infestans* share homology with human mucins. Mol Plant-Micro Interact. 2000, 13 (1): 32-42. 10.1094/MPMI.2000.13.1.32.

27. Zhao X, Kim Y, Park G, Xu J-R: A mitogen-activated protein kinase cascade regulating infection-related morphogenesis in*Magnaporthe grisea*. The Plant Cell. 2005, 17: 1317-1329. 10.1105/tpc.104.029116.

28. Ligterink W, Kroj T, Nieden UZ, Hirt H, Scheel D: Receptor-mediated activation of a MAP kinase in pathogen defense of plants. Science. 1997, 276 (27): 2054-2057. 10.1126/science.276.5321.2054.

29. Miwa T, Takagi Y, Shinozaki M, Yun C-W, Schell WA, Perfect JR, Kumagai H, Tamaki H: Gpr1, a putative G-protein-coupled receptor, regulates morphogenesis and hypha formation in the pathogenic fungus *Candida albicans*. Eukaryotic Cell. 2004, 3 (4): 919-931. 10.1128/EC.3.4.919-931.2004.

30. Lafon A, Han K-H, Seo J-A, Yu J-H, d'Enfert C: G-protein and cAMP-mediated signaling in aspergilli: A genomic perspective. Fungal Genetics and Biology. 2006, 43 (7): 490-502. 10.1016/j.fgb.2006.02.001.

31. Praveen RJ, Reena G, Subramanyam C: Calmodulin-dependent protein phosphorylation during conidial germination and growth of *Neurospora crassa*. Mycol Res. 1997, 101: 1484-1488. 10.1017/S0953756297004255.

32. Xu J-R, Hamer JE: MAP kinase and cAMP signaling regulate infection structure formation and pathogenic growth in the rice blast fungus *Magnaporthe grisea*. Genes & Dev. 1996, 10: 2696-2706. 10.1101/gad.10.21.2696.

33. DeZwaan TM, Carroll AM, Valent B, Sweigard JA: *Magnaporthe grisea* Pth11p is a novel plasma membrane protein that mediates appressorium differentiation in response to inductive substrate cues. The Plant Cell. 1999, 11: 2013-2030. 10.2307/3871094.

34. Clergeot P-H, Gourgues M, Cots J, Laurans F, Latorse M-P, Pépin R, Tharreau D, Notteghem J-L, Lebrun M-H: *PLS1*, a gene encoding a tetraspanin-like protein, is required for penetration of rice leaf by the fungal pathogen *Magnaporthe grisea*. PNAS. 2001, 98 (12): 6963-6968. 10.1073/pnas.111132998.

35. Park G, Bruno KS, Staiger CJ, Talbot NJ, Xu J-R: Independent genetic mechanisms mediate turgor generation and penetration peg formation during plant infection in the rice blast fungus. Molecular Microbiology. 2004, 53 (6): 1695-1707. 10.1111/j.1365-2958.2004.04220.x.

36. Wang ZY, Jenkinson JM, Holcombe LJ, Soanes DM, Veneault-Fourrey C, Bhambra GK, Talbot NJ: The molecular biology of appressorium turgor generation by the rice blast fungus *Magnaporthe grisea*. Biochem Soc Trans. 2005, 33 (Pt 2): 384-388.

37. Blanco FA, Judelson HS: A bZIP transcription factor from *Phytophthora* interacts with a protein kinase and is required for zoospore motility and plant infection. Molecular Microbiology. 2005, 56 (3): 638-648. 10.1111/j.1365-2958.2005.04575.x.

38. Liu XH, Lu JP, Zhang L, Dong B, Min H, Lin FC: Involvement of a *Magnaporthe grisea* serine/threonine kinase gene, MgATG1, in appressorium turgor and pathogenesis. Eukaryot Cell. 2007, 6 (6): 997-1005. 10.1128/EC.00011-07.

39. Tonukari NJ, Scott-Craig JS, Walton JD: The *Cochliobolus carbonum SNF1* gene is required for cell wall-degrading enzyme expression and virulence on maize. Plant Cell. 2000, 12: 237-248. 10.1105/tpc.12.2.237.

40. Li D, Ashby AM, Johnstone K: Molecular evidence that the extracellular cutinase Pbc1 is required for pathogenicity of*Pyrenopeziza brassicae* on oilseed rape. Mol Plant-Micro Interact. 2003, 16: 545-552. 10.1094/MPMI.2003.16.6.545.

41. Aro N, Pakula T, Penttila M: Transcriptional regulation of plant cell wall degradation by filamentous fungi. FEMS Microbiology Reviews. 2005, 29: 719-739. 10.1016/j.femsre.2004.11.006.

42. Prats E, Llamas MJ, Jorrin J, Rubiales D: Constitutive coumarin accumulation on sunflower leaf surface prevents rust germ tube growth and appressorium differentiation. Crop science. 2007, 47: 1119-1124. 10.2135/cropsci2006.07.0482.

43. Chumley FG, Valent B: Genetic analysis of melanin-deficient, nonpathogenic mutants of *Magnaporthe grisea*. Mol Plant-Micro Interact. 1990, 3: 135-143.

44. Walker SK, Chitcholtan K, Yu YP, Christenhusz GM, Garrill A: Invasive hyphal growth: An F-actin depleted zone is associated with invasive hyphae of the oomycetes *Achlya bisexualis* and *Phytophthora cinnamomi*. Fungal Genetics and Biology. 2006, 43 (5): 357-365. 10.1016/j.fgb.2006.01.004.

45. Bassilana M, Blyth J, Arkowitz RA: Cdc24, the GDP-GTP exchange factor for Cdc42, is required for invasive hyphal growth of *Candida albicans*. Eukaryotic cell. 2003, 2 (1): 9-18. 10.1128/EC.2.1.9-18.2003.

46. Park G, Xue C, Zheng L, Lam S, Xu J-R: *MST12* regulates infectious growth but not appressorium formation in the rice blast fungus *Magnaporthe grisea*. Mol Plant-Micro Interact. 2002, 15 (3): 183-192. 10.1094/MPMI.2002.15.3.183.

CHAPTER 4

ROLE OF EFFECTORS IN PLANT-FUNGAL INTERACTIONS

Carrie Selin, Teresa R. de Kievit, Mark F. Belmonte, and W. G. Dilantha Fernando

Keywords: Effectors, biotrophs, necrotrophs, hemibiotrophs, plant pathogen interactions, plant resistance

INTRODUCTION

Fungal plant pathogens are economically important due to the threat they pose to the production and yield of agricultural crops. It has been estimated that global agricultural production suffers an average annual loss of 15% due to plant diseases (Oerke, 2006; Lo Presti et al., 2015; Schwessinger et al., 2015). To reduce and/or prevent fungal plant diseases, farmers depend on resistant crop cultivars or fungicide treatments. Furthermore, current agricultural cropping strategies rely primarily on the rotation of one cropping genotype within large areas of land, promoting the selection of fungal isolates capable of overcoming crop resistance. Such farming practices impose the need for continuous development and introduction of new genetic resistance traits into crop plants through breeding (Lo Presti et al., 2015).

The lifestyles of plant pathogenic fungi are highly diverse and utilize distinct strategies to interact with the host plant. For example, necrotrophic fungi infect and kill host tissue and extract nutrients from dead host cells. Biotrophic fungi colonize living host tissue and obtain nutrients from living tissue; whereas hemibiotrophic fungi display two phases during the infection process; first is an initial biotrophic phase followed by a necrotrophic stage (Lo Presti et al., 2015). Despite the diversity among these lifestyles, all fungi that colonize plants are recognized by the plant innate immune system, which elicits a host defense response. The innate immune perception triggers both local and systemic reactions, allowing a plant to respond to pathogen attack in a quick and localized manner over an extended period of time (Schwessinger and Ronald, 2012). For this reason, the plant utilizes a two-tier innate immune response that involves a combination of localized plasma membrane and intracellular receptors (Jones and Dangl, 2006; Dodds and Rathjen, 2010; Asai and Shirasu, 2015).

The initial tier of the innate response is generally triggered by molecules essential to the pathogen and cannot be modified without significant loss of viability. These molecules include abundant bacterial proteins, elongation factors Tu (Ef-Tu), flagellin, and fungal cell wall components such as polysaccharides and chitin (Win et al., 2012; Newman et al., 2013). These components are usually exposed and are referred to as pathogen-associated molecular patterns (PAMPs) and microbe-associated molecular patterns (MAMPs). During infection, PAMPs present within the plant apoplast and are recognized by membrane-localized pattern recognition receptors (PRRs), which initiates the first reaction of defense called PAMP-triggered immunity (PTI) (Jones and Dangl, 2006; Koeck et al.,

Citation: Selin C, de Kievit TR, Belmonte MF and Fernando WGD (2016) Role of Effectors in Plant-Fungal Interactions, Front. Microbiol. 7:600. doi: 10.3389/fmicb.2016.00600

2011). In *Arabidopsis thaliana*, the flagellin receptor, FLS2, a leucine rich repeat (LRR) receptor-like kinase, recognizes the highly conserved N-terminus of bacterial flagellin (flg22) and activates the MAP kinase pathway to induce expression of defense response genes (Gomez-Gomez and Boller, 2002). In addition, changes in hormone biosynthesis and plant cell reinforcement by callose deposition have been shown to occur upon PRR activation (Macho and Zipfel, 2014). Interestingly, oligomers of fungal chitin are recognized by the LysM-RLK (receptor-like kinase) AtCERK1 *A. thaliana* receptor via three extracellular domains (Miya et al., 2007; Liu et al., 2012). The resulting chitin-induced homodimerization of CERK1 has been shown to be essential for the activation of downstream signaling (Liu et al., 2012). In a recent study, however, a lysin motif receptor kinase (LYK) termed AtLYK5 was shown to be the primary chitin receptor, not AtCERK1 (Cao et al., 2014). Interestingly, the AtLYK5 appears to directly interact with AtCERK1, forming a chitin inducible complex to induce plant defense (Cao et al., 2014). Similary in rice, the CEBiP, a LysM-receptor like-protein (RLP), was also shown to directly bind chitin elicitors and interact with OsCERK1, a homolog of AtCERK1, in a chitin-dependent manner (Kaku et al., 2006;Shinya et al., 2012). Studies have shown that reduced expression of either *CEBiP* or rice *CERK1* in RNA interference (RNAi) lines results in an impaired response to chitin elicitors (Kaku et al., 2006;Shimizu et al., 2010). This suggests that both of these molecules are required for chitin signaling in rice.

To successfully facilitate infection, or to establish compatible interactions that lead to proliferation, fungi must be able to counteract PTI. To suppress the immune response and manipulate host cell physiology, plant pathogens secrete effector proteins (Stergiopoulos and de Wit, 2009; de Jonge et al., 2011; Giraldo and Valent, 2013). Although these secreted proteins are key players in suppressing PTI, they are also recognized by the plant surveillance system, which in turn triggers the second tier of immune response termed effector triggered immunity (ETI). Effectors that elicit an ETI response can be recognized by plant resistance proteins (R proteins), which are intracellular nucleotide-binding leucine rich repeat (NLR) receptors (Cui et al., 2015). Recognition of effector proteins via NLR receptors occurs through direct (receptor-mediated binding) or indirect (accessory protein-mediated) interactions (Dodds and Rathjen, 2010; Cui et al., 2015). Activation of ETI results in disease resistance and is usually associated with a hypersensitive cell death response (HR) localized at the infection site (Jones and Dangl, 2006). The strong HR response and resulting phenotype is a product of what is termed host specific gene-for-gene interactions, where an effector, coined Avr (avirulence), is recognized by the cognate R-protein produced by the host plant (Dodds and Rathjen, 2010). To date, ~83 effector proteins have been cloned and characterized from crop-infecting fungi and oomycetes (Table 1); 43 of which are encoded by *Avr* genes. In addition, a majority of cognate plant R-proteins associated with a specific Avr have also been identified (Stergiopoulos and de Wit, 2009; Gururania et al., 2012; Ali et al., 2014). Elucidation of the role of Avr effectors in virulence and the underlying mechanisms involved remains a challenge. Nevertheless, recent research is beginning to reveal the function of increasing numbers of fungal effectors bringing forward new technologies that may help address some of these knowledge gaps and improve our understanding in plant-pathogen interactions.

Table 1: Effectors of well-characterized biotrophic and hemibiotrophic plant pathogenic fungi[a] **and oomycetes that have been cloned and studied to date (Excludes toxins).**

Effector Protein[b]	Length[c]	N-terminal signal motif or conserved domains	Localization *in planta*	Expression	R-protein association; role in virulence	Genomic location	References
***Blumeria graminis f. sp hordei* (biotroph; host: barley)**							
AvrA10	286	Unknown	Cytoplasm (predicted)	Unknown	Interaction with Mla10	TE-rich region	Ridout et al., 2006; Shen et al., 2007
AvrK1	177	Unknown	Cytoplasm (predicted)	Unknown	Interaction with Mlk1	TE-rich region	Ridout et al. 2006
Cladosporium fulvum (biotroph; host: tomato)							
Avr 2	78	Unknown	Apoplast	*In planta*	Interaction with Cf2. Binds and inhibits tomato cysteine proteases Rcr3 and Pip3	Not reported	Dixon et al., 1996
Avr4	135	Unknown	Apoplast	*In planta*	Interaction with Cf4. Protects against chitinases	Not reported	Joosten et al., 1994; van den Burg et al., 2006
Avr4E	121	Unknown	Apoplast	*In planta*	Interaction with Cf4-E	Not reported	Westerink et al., 2004
Avr9	63	Unknown	Apoplast	*In planta*	Interaction with Cf9	Not reported	van Kan et al., 1991; van den Hooven et al., 2001
Extracellular proteins (Ecp)							
Ecp1	65	Unknown	Apoplast	*In planta*	Required for full virulence	Unknown	Lauge et al., 1997
Ecp2	143	Unknown	Apoplast	*In planta*	Required for full virulence	Unknown	Lauge et al., 1997
		Unknown	Apoplast	*In planta*	Unknown	Unknown	Lauge et al., 2000
Ecp5	[illegible]	Unknown	Apoplast	*In planta*	Unknown	Unknown	Lauge et al., 2000
Ecp6	199	Unknown	Apoplast	*In planta*	Reduces plant defense response by binding to chitin	Unknown	Bolton et al., 2008
Ecp7	100	Unknown	Apoplast	*In planta*	Unknown	Unknown	Bolton et al., 2008
***Fusarium oxysporum f. sp. lycopersici* (hemibiotroph; host: tomato)**							
Avr1 (Six4)	242	Unknown	Xylem	Induced during infection	Interaction with I-1	B-chromosome (TE-rich region)	Houterman et al., 2008
Avr2 (Six3)	163	RxLR motif-like	Xylem (translocated to cytoplasm)	Induced during infection	Interaction with I-2	B-chromosome (TE-rich region)	Houterman et al., 2009
Avr3 (Six1)	284	Unknown	Xylem	Stimulated by living cells	Interaction with I-3. Aggressiveness determinant.	B-chromosome (TE-rich region)	Rep et al., 2004
Secreted in xylem proteins (SIX)							

Six5	119	Unknown	Xylem	Likely induced during infection	Required for virulence; pairs with Avr2 for full I-2 resistance	Chromosome 14; upstream *Avr2*; shares promoter region	Ma et al., 2015
Six6	199	Unknown	Xylem	Induced during infection	Suppresses I-2 mediated cell death	B-chromosome	Gawehns et al., 2014
Leptosphaeria maculans **(hemibiotroph; host: oilseed rape)**							
AvrLm1	205	Unknown	Cytoplasm (predicted)	Induced during infection	Interaction with Rlm1 and LepR3	TE-rich, AT-Isochore	Gout et al., 2007; Šašek et al., 2012; Larkan et al., 2013
AvrLm2	232	Unknown	Cytoplasm (predicted)	Likely induced during infection	Interaction with Rlm2 (predicted)	TE-rich, AT-isochore; with unique feature harboring a second gene (head-to-head) forming a small GC-isochore	Ghanbarnia et al., 2015
AvrLm3	160	Unknown	Cytoplasm (predicted)	Induced during early infection	Interaction with Rlm3 (predicted)	TE-rich, AT-Isochore	Plissonneau et al., 2016
AvrLm4-7	143	(R/N)(Y/F)(R/S)E(F/W) motif along with RAWG motif located in nearby loop	Possibly apoplastic then translocated into plant cells	Induced during infection	Interaction with Rlm4 and Rlm7; strongly required for virulence;	TE-rich, AT-Isochore	Parlange et al., 2009; Blondeau et al., 2015; Novákova et al., 2015
Avrlm6	144	RxLR-like motif	Cytoplasm (predicted)	Induced during infection	Interaction with Rlm6	TE-rich, AT-Isochore	Fudal et al., 2007, 2009; Kale et al., 2010; Van de Wouw et al., 2010
Avrlm11	95	Unknown	Cytoplasm (predicted)	Induced during infection	Interaction with Rlm11	Dispensable minichromosome; AT-rich isochore	Balesdent et al., 2013
AvrlmJ1 (Avrlm5)	141	Unknown	Cytosplasm (predicted)	Induced during infection (predicted)	Interaction with Rlm5	AT-rich isochore	Van de Wouw et al., 2014, unpublished
Magnaporthe oryzae **(hemibiotroph; host: rice)**							
Avr-Pita 1	224	Unknown	Cytoplasm	*In planta*: accumulate in BIC	Interaction with Pi-ta	Telomeric	Orbach et al., 2000; Khang et al., 2010
PWL1	147	Unknown	Cytoplasm	*In planta:* accumulate in BIC	Unknown	Telomeric	Khang et al., 1995, 2010
PWL2	145	Unknown	Cytoplasm	*In planta:* accumulate in BIC	Unknown	TE-associated	Khang et al., 1995; Sweigard et al., 1995
ACE1	4035	Unknown	Fungal Appresorium	Expressed in appressoria	Encodes a polyketide synthase; Insertion of MINE retrotransposon into ACE1 gene responsible for virulence in strain 2/0/3	*M. oryzae* genome	Bohnert et al., 2004; Fudal et al., 2005
Avr-CO39	89	Unknown	Cytoplasm	Expression *in planta*	Interaction with CO39; Interaction with RGA4/RGA5	TE-associated	Li et al., 2009; Cesari et al., 2013
AvrPiz-t	108	Unknown	Unknown	Unknown	Interaction with Piz-t. Reduces flg-22 and chitin induced ROS generation; Targets U3 ubiquitin ligase from host for degredation for suppressing PTI	TE-associated	Zhou et al., 2006; Li et al., 2009; Park et al., 2012
AvrPia	85	Unknown	Cytoplasm	Expression *in planta*	Interaction with Pi-a	TE-associated	Yoshida et al., 2009
AvrPii	70	C2H2 zinc finger-like motif	Cytoplasm	Expression *in planta*	Interaction with Pi-i	Subtelomeric and TE-associated	Yoshida et al., 2009
AvrPik/km/kp	113	Unknown	Cytoplasm	Expression *in planta*	Interaction with Pi-k	Genome	Yoshida et al., 2009
AvrPib	75	Unknown	Cytoplasm (predicted)	Unknown	Interaction with Pi-b	TE-associated	Zhang et al., 2015
Secreted LysM protein							
SLP1	162	LysM domain	Apoplast	Expression *in planta*	Required for full virulence; Binds chitin and suppresses chitin-induced plant immune response	Not reported	Mentlak et al., 2012
MC69	54	No signal motifs identified	Apoplastic	Expression during early infection stage	Required for virulence	Chromosome 6	Saitoh et al., 2012
Biotrophy-associated secreted (BAS) proteins							
BAS1	115	Unknown	Cytoplasm	*In planta:* accumulate in BIC	Unknown	Unknown	Mosquera et al., 2009
BAS2	102	Unknown	Cell wall crossing points	*In planta:* accumulate in BIC	Unknown	Unknown	Mosquera et al., 2009
BAS3	113	Unknown	Cell wall crossing points	*In planta:* accumulate in BIC	Unknown	Unknown	Mosquera et al., 2009
BAS4	102	Unknown	BIC	*In planta*	Unknown	Unknown	Mosquera et al., 2009

***Melamspora lini* (biotroph; host: flax)**							
AvrL567	150	RxLR-like motif	Cytoplasm	Expressed in haustoria	Interaction with L5, L6, and L7 Not required for full virulence	Unknown	Dodds et al., 2004; Lawrence et al., 2010
AvrM	314	RxLR-like motif	Cytoplasm	Expressed in haustoria	Interaction with M	Unknown	Catanzariti et al., 2006; Kale et al., 2010
AvrP123	117	Unknown	Cytoplasm	Expressed in haustoria	Interaction with P1, P2 and P3	Unknown	Catanzariti et al., 2006
AvrP4	95	Unknown	Cytoplasm	Expressed in haustoria	Interaction with P4	Unknown	Catanzariti et al., 2006
***Puccinia graminis f. sp tritci* (biotroph; hosts: wheat and barley)**							
RGDBP	818	RGD motif	Apoplast	Constitutive	Interaction with Rpg1	Unknown	Nirmala et al., 2011
VPS9	744	Forms complex with RGDBP	Apoplast	Constitutive	Interaction with RpgA	Unknown	Nirmala et al., 2011
PGTAUSPE-10-1	Unknown	Unknown	Unknown	Expressed in haustoria	Possible interaction with Sr22	Unknown	Upadhyaya et al., 2014
***Ustilago hordei* (biotroph; host: barley)**							
UhAvr1	170	Not reported	Not reported	Not reported	Interaction with Uh1	TE-rich	Grewal et al., 2008
***Ustilago maydis* (biotroph; host: maize)**							
Pep1 (protein essential for penetration1)	178	Not reported	Apoplast	Induced during early infection	Required for full virulence; Directly interacts with POX12 and suppresses plant defense by scavenging ROS	Not reported	Doehlemann et al., 2009; Hemetsberger et al., 2012; Hof et al., 2014
Pit2 (protein involved in tumors 2)	118	Not reported	Apoplast	Induced during infection	Required for full virulence; Directly interacts with cysteine protease	Not reported	Doehlemann et al., 2011
Cmu1		Not reported	Apoplast	Induced during infection	Required for full virulence; A chorismate mutase that potentially cooperates with maize chorismate mutase to reduce available SA	Not reported	Eberhard et al., 1996; Djamei et al., 2011
Tin2 (Tumor inducing 2)	207	Not reported	Apoplast	Induced during infection	Required for full virulence; Interacts with maize TKK1 which positively influences anthocyanin production	Not reported	Brefort et al., 2009; Tanaka et al., 2014
See1	157	Not reported	Expressed in nucleus and cytoplasm	Strongly expressed in seedling	Required for full virulence	Not reported	Redkar et al., 2015
***Phytophthora infestans* (oomocyte; hosts: potato and tomato)**							
Avr2	118	RxLR motif	Cytoplasm	Induced during infection	Interaction with R2	TE-rich	Lokossou et al., 2009; Gilroy et al., 2011b
Avr3a	147	RxLR motif	Cytoplasm	Induced during infection	Interaction with R3a; INF1-meidated plant death	TE-rich	Armstrong et al., 2005; Bos et al., 2010; Gilroy et al., 2011a
Avr4	287	RxLR motif	Likely cytoplasm	Induced during infection	Interaction with R4	TE-rich	van Poppel et al., 2008, 2009
AvrBlb1; Ipi01	153	RxLR motif	Cytoplasm	Induced during infection	Interaction with RpiBlb1; Disrupts cell wall-plasma membrane adhesion	TE-rich	Pieterse et al., 1994; Senchou et al., 2004; Vleeshouwers et al., 2008
AvrBlb2	101	RxLR motif	Cytoplasm	Induced during infection	Interaction with RpiBlb2; Required for full virulence; blocks plant protease secretion	TE-rich	Oh et al., 2009; Bozkurt et al., 2011

SNE1 (Suppressor of Necrosis1)	291	RxLR motif	Cytoplasm	Induced during infection	Suppresses NLP-triggered and effector-triggered cell death	TE-rich	Kelley et al., 2010
Crinkling and necrosis inducing effectors (CRN)							
PiCRN2	456	Conserved FLAK (for Phe, Leu, Ala, Lys) motif	Cytoplasm	Induced during infection	Triggers plant cell death	TE-rich	Schornack et al., 2010
PiCRN8	599	Conserved FLAK motif	Cytoplasm	Induced during infection	Triggers plant cell death; C-terminal kinase-like domain exhibits kinase activity-suggests role in modification of host-cell signaling pathway	TE-rich	Schornack et al., 2010; Liu et al., 2011
PiCRN15	615	Conserved FLAK motif	Cytoplasm	Induced during infection	Triggers plant cell death	TE-rich	Schornack et al., 2010
PiCRN16	618	Conserved FLAK motif	Cytoplasm	Induced during infection	Triggers plant cell death	TE-rich	Schornack et al., 2010
Extracellular protease inhibitors (EPI)							
EPIC1	126	Cystatin-like domain	Apoplast	Induced during infection	Binds and inhibits tomato cysteine proteases Rcr3 and PIP1	Not reported	Tian et al., 2007; Song et al., 2009
EPIC2A	125	Cystatin-like domain	Apoplast	Unknown	Unknown	Not reported	Tian et al., 2007
EPIC2B	125	Cystatin-like domain	Apoplast	Induced during infection	Binds and inhibits tomato cysteine proteases Rcr3 and PIP1	Not reported	Tian et al., 2007; Song et al., 2009
EPIC3	131	Cystatin-like domain	Apoplast	Constitutive	Unknown	Not reported	Tian et al., 2007
EPIC4	173	Cystatin-like domain	Apoplast	Constitutive	Unknown	Not reported	Tian et al., 2007
EPI1	149	Cystatin-like domain	Apoplast	Induced during infection	Binds and inhibits tomato serine protease P69B	Not reported	Tian et al., 2004
EPI10	228	Cystatin-like domain	Apoplast	Induced during infection	Binds and inhibits tomato serine protease P69B	Not reported	Tian et al., 2005
***Phytophthora sojae* (oomocyete; host: soybean)**							
Avr1b-1	138	RxLR motif	Cytoplasm	Induced during infection	Interaction with Rps1b and Rps1k; Suppresses plant cell death	TE-rich	Shan et al., 2004; Dou et al., 2008b
Avr1a	122	RxLR motif	Cytoplasm	*In planta*	Interaction with Rps1a	TE-rich	Qutob et al., 2009
Avh331; Avr1k	279	RxLR motif	Cytoplasm	Induced during infection	Interaction with Rpsk; Strongly suppresses PAMP-triggered and effector-triggered cell death	TE-rich	Dou et al., 2008b, Kale et al., 2010
Avr3a/5	111-119	RxLR motif	Cytoplasm	Induced during infection	Interaction with Rps3a and Rps5; Suprresses effector-triggered cell death	TE-rich	Qutob et al., 2009; Dong et al., 2011; Wang et al., 2011
Avr3b	314	RxLR motif	Cytoplasm	Strongly induced in germinating cysts and during infection	Interaction with Rps3b; Suppresses effector-triggered cell death and ROS production	TE-rich	Dong et al., 2011
Avr3c	220	RxLR motif	Cytoplasm	Expressed in early infection stage	Interaction with Rps3c	TE-rich	Dong et al., 2009
Avr4/6	123	RxLR motif	Cytoplasm	Induced during infection	Interaction with Rps4 and Rps6	TE-rich	Dou et al., 2010
Avr1d	125	RxLR motif	Cytoplasm	Induced during infection	Interaction with Rps1d; Suppressed BAX (proapoptotic protein)-induced cell death	TE-rich	Na et al., 2013; Yin et al., 2013
Avh172	218-227	RxLR motif	Cytoplasm	Expressed in early infection stage	Suppresses effector-triggered cell death; Required for virulence	TE-rich	Wang et al., 2011
Avh238	134-142	RxLR motif	Cytoplasm	Induced during infection	Suppresses PAMP-triggered cell death; required for full virulence	TE-rich	Wang et al., 2011
PsCRN63	450	Conserved FLAK motif	Cytoplasm	Constitutive; Slightly induced during infection	Triggers plant cell death	TE-rich	Liu et al., 2011
PsCRN115	449	Conserved FLAK motif	Cytoplasm	Constitutive; Slightly induced during infection	Suppresses plant cell death	TE-rich	Liu et al., 2011
PsNPP1; NLPps	237	Not reported	Apoplast	Induced during transition to necrotrophy	Triggers plant cell death	TE-rich	Qutob et al., 2006

[a]*Plant pathogenic fungi listed in table are known to infect crop plants (excluded Arabidopsis thaliana).*
[b]*Includes proteins that have shown specific host-interaction phenotypes or indication of host cell entry.*
[c]*The number of amino acid residues within the unprocessed protein.*

In this review we focus on effector molecules of biotrophic and hemibiotrophic fungi, taking a close look at the mechanisms involved in release and uptake of effector molecules by the fungi and plant cells, respectively. We place emphasis on how effectors were discovered, and difficulties associated with determining the effector repertoire. We then discuss the role of specific effectors found in biotrophic and hemibiotrophic fungi and look at how new technology for generating direct mutations may provide a new avenue for elucidating the function of fungal effector proteins.

EFFECTOR DELIVERY MECHANISMS OF FUNGAL PATHOGENS

Effectors can be defined as molecules that alter host cell structure and function, facilitating infection (virulence factors or toxins) and/or triggering defense responses (avirulence factors: Avr). These proteins can be grouped into two classes based on their target sites in the host plant (Kamoun, 2006, 2007). Apoplastic effectors are secreted into the plant apoplast, where they interact with extracellular targets and surface receptors, whereas cytoplasmic effectors are translocated inside the plant cell (Djamei et al., 2011; Dong et al., 2011; Yaeno et al., 2011; Park et al., 2012). Regardless of the effector type, efficient delivery of effectors to the plant is required for the infection process. Since pathogenic fungi have developed distinct lifestyles, they have also established diverse effector delivery systems upon infection. Biotrophic and hemibiotrophic fungal pathogens feed and live on living host cells, and secrete effectors that are targeted for the host apoplast or cytoplasm using specialized infection structures such as appressoria or haustoria (Kemen et al., 2005; Catanzariti et al., 2006; Rafiqi et al., 2010;Koeck et al., 2011; Petre and Kamoun, 2014). Obligate biotrophs such as rust fungi or powdery mildew fungi are only capable of propagating on living host tissue. After penetration of the epidermal cell wall, a lobed haustorium develops within the meshophyll (Manners and Gay, 1983; Mackie et al., 1991; Harrison, 1998). The haustorium is completely surrounded by a membrane termed the extrahaustorial membrane, which appears to be an invagination of the plant plasma membrane (Garnica et al., 2014). Interestingly, the biotroph *Ustilago maydis* does not establish bulbous feeding structures similar to the haustoria-forming rust fungi. During penetration, the host plasma membrane invaginates and completely encloses the intracellular hyphae, establishing an extensive area of interaction where the exchange of molecules between the fungus and host occurs (Djamei and Kahmann, 2012; Lo Presti et al., 2015). Hemibiotrophic pathogens utilize a biotrophic phase early in the infection process followed by a necrotrophic phase killing host cells to complete their lifecycle. For the hemibiotrophic pathogen, *Magnaporthe oryzae*, two discrete secretion systems for delivery of apoplastic and cytoplasmic effectors are utilized (Giraldo et al., 2013). For cytoplasmic effector delivery, these proteins appear to accumulate in the biotrophic interface complex (BIC) near the tip of the first bulbous cell formed after host cell penetration. Apoplastic effectors, on the other hand, are not associated with BIC, and once secreted they are dispersed in the extracellular space between the fungal cell wall and the extra-invasive-hyphal membrane (Giraldo et al., 2013; Zhang and Xu, 2014). There have been no studies to elucidate effector delivery strategies employed by the pathogen *Leptosphaeria maculans*, a hemibiotroph that causes disease in *Brassica* plant species. However, one might speculate that it uses a mechanism similar to that of *M. oryzae*, where some of the cytoplasmic Avr effectors accumulate at the BIC. One particular review suggested that *L. maculans* was an apoplastic pathogen and unlike BIC-forming hemibiotrophs, it triggers a slower defense response where maximum expression of the effectors does not occur until after an initial endophytic growth phase (Stotz et al., 2014).

TARGETING OF EFFECTORS TO HOST CELLS

No system analogous to the bacterial secretion system has been identified in fungi. However, one common theme that has emerged in terms of fungal effector secretion is based on host targeting via N-terminal translocation domains, which are found after the general secretory signal peptide. Insights into the effector movement process have come from studies conducted with oomycete pathogens, which employ a similar infection strategy that are used by fungi (Koeck et al., 2011). In the oomycetes, secreted effector proteins share a common N-terminal host targeting domain, which contains common motifs such as RxLR (Arg-x-Leu-Arg), LxLFLAK or CRN (Crinkler motif) and CHxC amino acid sequences (Jiang et al., 2008). The N-terminal RxLR motif along with a downstream DEER (Asp-Glu-Glu-Arg) amino acid sequence in the Avr3a effector protein of*Phytopthera infestans* has been shown to be required for translocation into potato cells (Whisson et al., 2007; Dou et al., 2008a; Bos et al., 2010; Kale et al., 2010).

Identification of N-terminal signal motifs involved in cell entry is not as clearly defined for most fungi as it appears to be for oomycetes. In general, fungal effectors do not appear to share significant sequence similarity; a feature attributed to its rapid variation and host adaptation (Sperschneider et al., 2015a), although some exceptions have been reported. For example, the*Cladosporium fulvum* Ecp6 effector contains a LysM domain, which was shown to align with 16 putative *C. fulvum* Ecp6-like proteins from: *Aspergillus niger, Magnaporthe grisea, Mycosphaerella fijiensis, M. graminicola, Botrytis cinerea, Sclerotinia sclerotiorum, A. nidulans, A. oryzae, A flavus, C. lindemuthianum*, and *L. maculans* (Bolton et al., 2008). Also, a chorismate mutase effector, Cmu1, secreted by the maize pathogen *U. maydis*, illustrated a role in virulence, suggesting that some effectors may, based on protein sequence, have predicted functional roles (Djamei et al., 2011). It should be noted that evidence exists demonstrating conservation within the N-terminal sequence motifs of some fungi. In the barley powdery mildew fungus, *Blumeria gramanis* f. sp. *hordei*, for example, the effector proteins share an N-terminal [YFW]xC motif within 30 amino acids of the signal peptide (Godfrey et al., 2010). This particular motif has also been reported in other effectors found within rust fungi, but amino acid position within the proteins is less conserved (Duplessis et al., 2011). In the genus *Fusarium*, a set of effector proteins share a conserved [SG]PC[KR]P motif located immediately downstream of the N-terminal signal peptide. However, these motifs have yet to be functionally characterized and remain undetermined whether they are indeed fungal effector motifs (Manning et al., 2008). In *Melamspora lini*, the AvrL567 and AvrM effectors enter flax cells autonomously which is governed by N-terminal uptake domains, but the two proteins do not share conserved motifs or structures (Rafiqi et al., 2010).

Other features that may unite groups of effectors include structural similarities, for example the arrangement of secondary structural elements relative to each other, i.e., three-dimensional folds. Studies with the genus *Phytophthora* proteins AVR3a11 and PexRD2 suggested a three-helix bundle fold as the basic structural unit, which is formed by the repeating WY motifs. This particular structural element has been identified in more than 520 related RxLR effector proteins (Boutemy et al., 2011). Interestingly, a duplicated four helical motif with similarity to the WY domain of oomycete effectors was illustrated in the *M. lini* AvrM effector. In a recent study by Goritschnig et al. (2016), the WY-domain of an effector produced by *Hyloperonospora Arabidopsis* (ATR1) was shown to be directly recognized by the *Arabidopsis* NLR RPP1 through association with the C-terminal LRR (Goritschnig et al., 2016), suggesting that this feature may be important for NLR recognition. Other features such as β-sandwich structures have also been discovered in AvrL567 from *M. lini* (Wang et al., 2007) and in the *M. oryzae* effector Piz-t (Zhang et al., 2013), suggesting that structural conservation exists among some fungal effectors. Despite this commonality, a lack of significant effector protein conservation is expected to hamper the use of these features for predicting effector function.

IDENTIFYING AND PREDICTING EFFECTOR CANDIDATES

Prior to the advent of genomic sequencing, identification of many effectors was carried out by genetic map-based cloning. Examples include AvrPi-ta (Orbach et al., 2000), ACE1 (Bohnert et al., 2004), and Avr-CO39 (Farman and Leong, 1998) from *M. oryzae*; AvrLm1 (Gout et al., 2006), Avrlm4-7 (Parlange et al., 2009), AvrLm6 (Fudal et al., 2007) from *L. maculans*. The availability of transcriptomic and fungal genomic sequences has accelerated the number of effectors identified in the last decade. By screening for cDNA clones that associate with avirulence loci, Avr3a (Armstrong et al., 2005) and Avr4 (van Poppel et al., 2009) from *P. infestans* and AvrL567 from *M. lini* (Dodds et al., 2004) were identified and characterized. In some cases expressed sequence tags (EST) were screened bioinformatically to identify potential secreted proteins with specific N-terminal secretion signals (Vleeshouwers et al., 2008; Oh et al., 2009; Zhu et al., 2012). Interestingly, analysis of a cDNA library from barley tissue containing the ascomycete *B. graminis* f. sp. *hordei* illustrated that the majority of expressed fungal genes encoded unrelated small proteins containing N-terminal signal peptides and an N-terminal motif of [YFW]xC (Godfrey et al., 2010). Identification of effectors using these types of approaches defined structural properties of effector proteins allowing for bioinformatic predictions of putative effectors. Prediction of hundreds of potential RxLR effector genes in *P. sojae, P. ramorum, P. infestans* (Tyler et al., 2006; Jiang et al., 2008; Haas et al., 2009), *L. maculans* (Rouxel et al., 2011) and powdery mildew fungi were made possible through using a bioinformatic strategy.

Despite the identification of conserved sequence features among effectors, fungal effector prediction approaches remain problematic as they are largely based on relatively broad criteria, which principally rely on the presence of a secretion signal and the fact that the majority of effector proteins are small in size and cysteine-rich (Sperschneider et al., 2015a). These features have been used to mine predicted secretomes, and although it is helpful in reducing the number of candidates, the problem remains that not all secreted small cysteine rich proteins will function as an effector and conversely not all fungal effectors will be small and cysteine-rich. For instance, some cytoplasmic effectors translocated into the host cell are low in cysteine residues and are quite large in size. In *L. maculans*, AvrLm1 effector has only one cysteine residue (Gout et al., 2006), suggesting that the criteria of small and cysteine rich, while being valuable for apoplastic effectors, may have a tendency to exclude potential effector candidates (Sperschneider et al., 2015b).

New approaches for predicting both apoplastic and cytoplasmic effector candidates have been developed that do not rely solely on the presence of a secretion signal, small size and high cysteine residue content. Saunders et al. (2012) designed a thorough *in silico* analysis to identify the potential effector repertoire from the genome of two pathogenic rust fungi. The pipeline is founded on the observation that known effector proteins from filamentous pathogens have at least one of the following properties: (i) the presence of a secretion signal, (ii) being encoded by genes that are induced *in planta*, (iii) exhibiting similarity to haustorial proteins, (iv) small and cysteine rich, (v) the presence of a known effector motif or a nuclear localization signal, (vi) being encoded by genes with long intergenic regions, (vii) the presence of internal repeats, (viii) a lack of PFAM domains, excluding those associated with pathogenicity. The authors then used clustering algorithms to group protein families of rust pathogens and rank them according to the likelihood of being effectors. Using this approach they were able to identify approximately eight families of effector contenders. This pipeline was also utilized in a study conducted on *M. lini*, where 200 potential effector candidates were identified (Nemri et al., 2014), suggesting that this methodology may be useful in future effector predictions. A similar strategy was developed to predict effector candidates in *S. sclerotiorum*. This method involved the equivalent selection pipeline utilized by Saunders et al. (2012), as described above, in addition to selection of proteins belonging to duplicated gene families, and those that illustrated signatures of positive selection (Guyon et al., 2014). With this strategy, the authors were able to putatively identify 78 effector candidates from a predicted proteome of over 14,500 proteins (Guyon et al., 2014), suggesting that the additional criterion for selection within this pipeline could potentially narrow down the number of effector candidates, although more studies need to be carried out to ensure that potential effector proteins are not being missed. Taken together, these studies clearly illustrate that with the development of computational algorithms and enhancement of specific analytical criteria for protein selection pipelines, predicting fungal effectors is becoming more efficient. By looking beyond sequence similarity based methods and integrating relevant physiological effector functions such as gene expression *in planta*, genomic features, or taxonomic information, computational analytical algorithms will prove to be an extremely powerful tool for predicting effector candidates within a large proteome.

Proteomics is another important tool utilized for large-scale studies on proteins that are involved in plant-pathogen interactions. Characterization of a subset of proteins during plant-pathogen interactions has been able to provide a more direct view of cellular processes compared to DNA or RNA analysis and has resulted in the identification of both fungal and oomocyte effectors (Ricci et al., 1989; Kamoun et al., 1993; De Wit et al., 2002; Rose et al., 2002; Rep et al., 2004; Xu et al., 2007; Houterman et al., 2008, 2009; Cao et al., 2009). One of the first studies using proteomics to identify an avirulence gene was carried out by De Wit research group (1986 and 1988) on the pathogenic fungi *C. fulvum*. In these studies, the gene product encoded by *avr9* (Avr9) was characterized through protein purification of tomato apoplastic fluids during compatible/incompatible fungal-plant host interactions using polyacrylamide gel electrophoresis (PAGE), reverse-phase HPLC and EDMAN N-terminal sequencing (De Wit et al., 1986; Schottens-Toma and DeWit, 1988). In *F. oxysporum* f. sp *lycopersici* (FOL), one of the first effectors, termed SIX1 (Avr3), was identified and sequenced using mass spectrometry (MS) (Rep et al., 2004). More specifically, the peptide sequence SIX1 was obtained from purification of xylem sap from infected tomato plants, and a gene deletion was able to further confirm that the absence of the *SIX1* gene resulted in the loss of resistance on an I-3 (corresponding R gene) tomato line. Other studies conducted in the FOL isolate followed suite, where identification of 14 additional SIX proteins were determined through the analysis of the xylem sap proteome of infected tomato plants using two-dimensional gel separation and mass spectrometry (Houterman et al., 2008, 2009; Lievens et al., 2009; Ma et al., 2010; Schmidt et al., 2013). A recent study using a large-scale proteomics approach determined the protein changes in the xylem sap proteome following FOL infection from analyzing plants inoculated with either the wild type or *SIX* knockout FOL isolates (Gawehns et al., 2015). The authors illustrated that the absence of a single effector altered the abundance of the sap xylem proteins, which corresponded to the altered virulence visualized on the plants inoculated with the SIX knockout strains (Gawehns et al., 2015). Clearly, proteomics is an extremely effective tool for finding effector proteins that are expressed during pathogen-host interactions. Even though there are some limitations associated with this type of technique, such as low protein abundance, sample complexity, sensitivity, resolution, and speed of data acquisition (Gonzalez-Fernandez et al., 2010), proteomics appears to be more direct, where proteins that are necessary for plant-pathogen interactions are more precisely identified. This is in contrast to computational prediction systems, which have a tendency to broadly select effector proteins that may or may not have an actual function in plant-pathogen interactions. Furthermore, a more global application can be used in proteomics where the overall contribution of a specific effector on the abundance of proteins within the proteome can be assessed. This type of global analysis will allow for a better understanding of plant-pathogen association and provide insight into the mechanism involved in effector recognition.

FACTORS THAT AFFECT EVOLUTIONARY CHANGES IN EFFECTORS

With the development of new computational tools for predicting effectors, researchers have focused their attention on how factors such as genome structure and gene transfer impact fungal effector composition. Fungal effector composition seems to be driven by the evolutionary arms race between effector recognition by plant R-proteins, however where they are located within the genome and how they are transferred from one isolate to another can significantly influence fungal effector repertoire.

Association of the pathogen with its host plant results in either an incompatible (resistant) or a compatible (susceptible) interaction. The fundamental principle explaining the outcome of this interaction is based on the gene-for-gene model (Flor, 1956). This model proposes that the product of the R-gene from the host plant distinguishes an Avr (Avirulent) effector produced by the pathogen leading to a resistant or incompatible reaction. If there is no detection, as a result of allelic variation or lack of one of two components, then a susceptible or compatible reaction is observed. Interestingly, plant-associated microbes continually evolve new effectors to retain or improve their ability to cause disease and to minimize detection by a plant host. Plants are also compelled to maintain disease resistance and improve their ability to detect pathogen effectors by developing new R-gene allelic varations. The antagonistic cycles of evolution between pathogen effectors and plant R-proteins is described as the co-evolutionary "arms-race" or zigzag model (Jones and Dangl, 2006; Dodds and Rathjen, 2010; Tyler and Rouxel, 2013). As a result of this arms race, the genotype for each of these proteins become highly polymorphic and include single-nucleotide polymorphisms (SNPs), insertions and deletions within the gene sequence (Guttman et al., 2014).

An ongoing change in gene alleles especially effector genes, provides plant pathogens with the ability to evade detection while optimizing or maintaining virulence. To ensure survival, pathogens may have to evolve new effectors to successfully colonize and infect a new host target. Adaptation to hosts is believed to be enabled by compartmentalization of effector genes within the genome. Plant pathogenic fungi and oomycetes contain gene-sparse genomic regions that are exceedingly enriched in repetitive elements and putative effector genes (Raffaele and Kamoun, 2012). The movement of transposable elements (TEs) within these genome compartments results in gene duplication, horizontal gene transfer and gene loss, all of which contribute to virulence factor evolution (Lo Presti et al., 2015). For example, the fungal pathogen *L. maculans* has an unusual genome structure compared to other pathogenic fungi in that it contains alternating GC and AT-rich blocks termed GC- and AT-isochores. The GC-isochore regions are enriched with housekeeping genes, whereas the AT-isochores are relatively gene-sparse, enriched in TEs which are truncated and further reduced by repeat-induced point mutations (RIP). The AT-isochore harbors ~122 genes encoding small secreted proteins (Rouxel et al., 2011), which also includes the conditionally dispensable chromosome (Soyer et al., 2014). In *F. oxysporum*, all known effector genes are found in one of its four dispensable chromosomes (Ma et al., 2010) and in *P. infestans* effector genes are located in highly plastic genomic regions, enriched in TEs (Haas et al., 2009).

Horizontal gene transfer (HGT) is another important mechanism by which effector diversification occurs in fungal pathogens. One particular study conducted on *M. oryzae* revealed that the Avr-Pita effector had been translocated several times via mobile elements. The authors suggested that multiple translocations implicate deletions and recoveries mediated via parasexual transfer among individual isolates (Chuma et al., 2011). In addition to intraspecies genetic transfer an HGT event was also reported between different species documented by the transfer of ToxA along with hAT transposase from the wheat blotch pathogen *Stagonospora nodorum* to *Pyrenophora tritci-repentis*, which is the causal agent of tan spot in wheat (Friesen et al., 2006). These studies highlight how selection can influence specific evolutionary changes leading toward diversification of virulence-promoting effectors.

FUNCTION OF CHARACTERIZED FUNGAL EFFECTORS IN BIOTROPHS AND HEMIBIOTROPHS

Unlike bacterial and oomycete effectors, a limited number of fungal effectors have been functionally characterized. The primary obstacles in fungal effector research derive from difficulties associated with manipulating fungi in a laboratory setting, in particular obligate biotrophs such as rust fungi. Adding to these challenges is the fact that many effector mutants display no associated phenotype, likely due to functional redundancy, poor assay systems or the inability to accurately measure small changes in phenotype. However, technologies such as bimolecular fluorescent complementation (BiFC) (Kerppola, 2008; Kodama and Hu, 2012), immunocolocalization (Dunn et al., 2011) yeast-two hybrid systems (Bruckner et al., 2009) and gene expression assays have significantly advanced our understanding of the functional roles of secreted fungal effectors. In this section, we will discuss the function of biotrophic and hemibiotrophic fungal effectors and emphasize which of the above mentioned techniques were used to further understand *in planta*effector localization, or protein-protein interactions.

EFFECTORS IDENTIFIED IN BIOTROPHS

To date, functional analysis of effectors from biotrophic pathogens including *U. maydis* and *C. fulvum* are among the most extensively studied. Much has been revealed about the effectors in the maize pathogen *U. maydis*. With a relatively small genome of 20.5 Mb, ~50 secreted proteins have been predicted, 50% of which are novel (Kamper et al., 2006; Koeck et al., 2011; Djamei and Kahmann, 2012). Many of these novel genes are localized within gene clusters, and appear to be upregulated during host colonization, and encode effectors that have virulence function (Kamper et al., 2006; Schirawski et al., 2010). As shown in Table 1, five effector proteins, Pep1, Pit2, Cmu1, Tin2, and See1 have been well-characterized.

Pep1 (protein essential for penetration 1) is a small (178 amino acids) secreted effector of *U. maydis*and related smut fungi that has been shown to accumulate in the apoplast. Doehlemann et al. (2009) confirmed the presence of Pep1 within the apoplastic spaces within the leaves, in addition to finding protein accumulation at sites of cell-to-cell passage using *in vivo* immunocolocalization (Doehlemann et al., 2009). Inoculation of leaves with *pep1* deletion mutants showed a failure to establish compatible interactions along with large necrotic patches. Interestingly, the Δ*pep1*mutants were defective for penetration of the initial epidermal cell and cell-to-cell spread. Pep1 has also been shown to be required to overcome extensive host resistance. Using bimolecular fluorescence complementation (BiFC) a direct interaction between the Pep1 protein and the maize defense peroxidase POX12 was established. It appears that Pep1 is able to suppress plant defense mechanisms through scavenging reactive oxygen species (ROS) (Hemetsberger et al., 2012). Interestingly, a recent report further supported the role of Pep1, as it was shown to be necessary for inducing hypersensitive response that displayed necrotic cell death features (Hof et al., 2014).

Pit2 (protein involved in tumors 2) is a secreted, apoplastic effector of *U. maydis* that is required for virulence; studies have shown that a mutation within the *pit2* gene results in attenuation of tumor formation (Doehlemann et al., 2011). A recent study illustrated through the utilization of a yeast-two hybrid system that Pit2 directly interacts with maize cysteine proteases. Using a complementary approach, these researchers illustrated that when the recombinant Pit2 protein was co-inoculated with cysteine protease (CP2), the protease activity was significantly reduced, indicating that Pit2 inhibits the proteases activity via direct interaction. It was also noted that the CP2 protease utilized in the study was recently identified in the leaf apoplasts as an important factor required for Salicylic acid (SA)-associated defense (Mueller et al., 2013), suggesting the importance of Pit2 in virulence.

Cmu1 is an effector that is translocated into the host cell. The *cmu1* gene of *U. maydis* is most strongly induced during plant colonization, leading to an abundance of Cmu1 protein accumulation detected within the apoplast (Djamei et al., 2011). Through complementation analysis in yeast, along with *in vitro* enzymatic assays, Cmu1 was shown to be a chorismate mutase. As the branching metabolite of the shikimate pathway, chorismate is the precursor for synthesis of aromatic amino acids and the plant defense hormone salicylic acid. Data obtained from utilizing a two hybrid system indicated that Cmu1 can form heterodimers with the plastidic as well as the cytosolic forms of plant chorismate mutase (Eberhard et al., 1996; Djamei et al., 2011). Immunocolocalization studies illustrated that Cmu1 was located within the cytosol after translocation into the host plant cell, where the flow of chorismate was redirected through potential cooperation between the cytosolic maize chorismate mutase along with Cmu1, leading to a reduction in available chorismate for salicylic acid biosynthesis (Eberhard et al., 1996). Interestingly, Cmu1 has the ability to spread locally to neighboring yet uninfected host cells (most likely via plasmodesmata), which the authors suggest is a form of metabolic priming leading to lower salicylic acid levels, allowing cells to prepare for impending colonization by *U. maydis* (Eberhard et al., 1996).

The effector Tin2 (Tumor inducing 2) is part of the largest cluster of effectors identified in *U. maydis* and was shown to be responsible for anthocyanin induction during biotrophic phase and have a role in virulence (Brefort et al., 2009). Tin2 localization was shown using confocal microscopy, which illustrated the accumulation of the protein around the fungal hyphae, which is secreted into the apoplast and functions within the cytosol (Tanaka et al., 2014). Inside the plant, the Tin2 protein appears to interact with the protein TTK1 (Tin2-targeting kinase1), a maize protein kinase that regulates the anthocyanin biosynthetic pathway (Tanaka et al., 2014). The interaction between Tin2 and TKK1 was observed using a yeast two-hyrbid screen, where the region of contact between the two proteins coincided with a variable region containing a phosphodegron motif, which is the target for the ubiquitin ligase complex that leads to protein degredation through the ubiquitin-proteasome system (Ravid and Hochstrasser, 2008; Spoel et al., 2009). (Tanaka et al., 2014) illustrated that the Tin2 protein protects the active kinase against ubiquitination as the full length TKK1 was only detectable on SDS-PAGE when co-expressed *in planta* with a functional Tin2 protein. The stabilization of TKK1 via Tin2 appears to positively stimulate the production of anthocyanin in infected tissue and suppress lignin biosynthesis, a common defense pathway (Tanaka et al., 2014).

The *U. maydis* effector protein See1 (Seedling efficient effector 1) is recognized as an organ-specific effector that induces tumor expansion specifically on maize leaves (Redkar et al., 2015). It appears that this protein is expressed in the nucleus and cytoplasm of maize as live cell imaging and immunolabeling using transmission electron microscopy were able to detect the See1-mCherry tagged protein within these two areas of the plant cell (Redkar et al., 2015). See1 is required for reactivation of DNA synthesis, which is an important step for tumor progression in maize leaf cells (Walbot and Skibbe, 2010). Through yeast-two hybrid analysis, it was shown that this protein interacts with maize homolog of SGT1 (Suppressor of G2 allele of *skp1*), a factor that has been shown to act in cell cycle progression in yeast (Dubacq et al., 2002) and an essential component of innate immunity in plant and animals (Shirasu, 2009; Zhang et al., 2010). Further analysis of this interaction illustrated that the See1 protein restricts MAPK-triggered phosphorylation of maize SGT1, which could potentially modulate the immune responses and DNA reactivation synthesis within leaf cells (Redkar et al., 2015).

C. fulvum produces at least 10 different effector proteins (Table 1), three of which have been functionally characterized; Avr4, Ecp6, and Avr2. The Avr4 and Ecp6 effectors are involved in suppression of chitin-induced PTI. Avr4 binds to chitin in the fungal cell wall, which protects the fungi against hydrolysis caused by plant chitinases (van den Burg et al., 2006). The Ecp6 effector appears to sequester chitin oligosaccharides, which are released by cell walls during hyphal invasion, allowing the suppression of host immunity (Bolton et al., 2008; de Jonge et al., 2010). A study focused on structural formation of the Ecp6 protein illustrated that two of the three LysM protein domains undergo ligand-induced dimerization, resulting in a high affinity chitin-binding pocket, while the third LysM domain binds to chitin with lower affinity. The authors suggest that binding of this protein out-competes the plant chitin receptor by interfering with its dimerization (Sanchez-Vallet et al., 2013).

The Avr2 effector produced by *C. fulvum* specifically targets secreted cysteine proteases, similar to Cmu1 in *U. maydis*. One particular study using heterologous expression of *Avr2* in *A. thaliana*illustrated an enhanced susceptibility *in planta* toward extracellular fungal pathogens, which included *B. cinerea* and *Verticillium dahlia*. Interestingly, microarray analysis demonstrated that*Avr2* expression generated global changes in the transcriptome reflecting pathogen challenge. Through monitoring protease activity, Avr2 was found to inhibit multiple extracellular Cys proteases, including Rcr3 and its close relative Pip1. In addition, infection by *C. fulvum* was significantly reduced when Avr2 was silenced. Collectively these findings illustrate that Avr2 not only inhibits several Cys proteases required for plant basal defense but it also functions as a virulence factor (van Esse et al., 2008).

EFFECTORS IDENTIFIED IN HEMIBIOTROPHS

Hemibiotrophic fungi have a distinct lifestyle wherein they combine a biotrophic phase with a nectrotrophic phase. In this section, we will focus on two hemibiotrophs, *M. oryzae* and *L. maculans*.

A whole genome draft sequence of the *M. oryzae* isolate 70-15 revealed a genome size of 37.8 Mb, with nucleotides encoding ~11,109 proteins (Dean et al., 2005). From the predicted proteome, 1309 genes were predicted to encode secreted proteins (Yoshida et al., 2009). As shown in Table 1, ~16 different effectors have been cloned or characterized to date. Recently, map-based cloning of AvrPib was reported (Zhang et al., 2015), adding another Avr-type effector to the repertoire. MC69 was also shown to be required for pathogenicity; however no evidence was provided to show that this protein suppresses or interferes with plant defense responses (Saitoh et al., 2012). The best characterized effectors to date include SLP1, AvrPiz-t, and ACE1.

SLP1 is a secreted LysM protein that has been shown to be required for *M. oryzae* virulence and is similar to the Ecp6 chitin binding protein found in *C. fulvum*. In a study conducted by Mentlak et al. (2012) live-cell imaging of the GFP-tagged SLP1 illustrated that the protein accumulated at the plant-fungus interface upon rice infection and accrued at the tips of the invading hypae when fungus moved into invading cells (Mentlak et al., 2012). Deletion in the *SLP1* gene reduced the ability of *M. oryzae* to cause disease, which was associated with the inability to proliferate within the host tissues, rather than reducing production of structures required for successful penetration. In a high stringency yeast two-hybrid analysis, SLP1 monomers appeared to associate with one another, suggesting that this protein forms oligomers (Mentlak et al., 2012). The protein was also shown to bind specifically to chitin as purified forms of the protein only co-precipitated with chitin beads or insoluble crab chitin, and not with any other tested polysaccharide (Mentlak et al., 2012). In addition to these findings, the authors determined that SLP1 was able to out-compete the rice PRR chitin elicitor binding protein (CEBiP) for chitin binding when challenged in a competition assay, and suppress chitin-induced oxidative burst,. These findings illustrate that SLP1 binds directly to chitin thereby suppressing chitin-triggered immune response in rice. However, it seems that this function is reliant on the presence of CEBiP, as rice plants silenced with CEBiP RNAi generated disease symptoms when inoculated with the Δ*slp1* mutant (Mentlak et al., 2012). Taken together, the key function of SLP1 as a suppressor of chitin-triggered defense response appears to be reliant on the interaction between SLP1 and CEBiP, which determines the progression of rice blast disease.

AvrPiz-t is a cytoplasmic (Avr-type) effector that has been shown to target the cytosolic rice R-gene (Zhou et al., 2006; Li et al., 2009). AvrPiz-t contributes to disease development in rice when it lacks the corresponding resistance protein Piz-t. Expression of AvrPiz-t in transgenic plant lines illustrated that PTI was suppressed, leading to a reduction in ROS and enhanced susceptibility to*M. oryzae* (Park et al., 2012). A yeast two-hybrid screen identified the interaction between AvrPiz-t with a rice RING E3 ubiquitin ligase APIP6. Ubiquitination of AvrPiz-t by APIP6 was visualized via immune-blotting, and it was shown that after ubiquitination AvrPiz-t-mediated suppression of APIP6 activity occurred (Park et al., 2012).

ACE1, another avirulence gene found in *M. oryzae* has been shown to be upregulated during initiation of infection. ACE1 is not predicted to be secreted or localized to the plant cell cytoplasm, suggesting that the protein itself is not the actual effector but that ACE1 is responsible for the synthesis of an effector that is recognized by the plant R-protein (Bohnert et al., 2004). However, the mechanism by which this occurs has yet to be elucidated.

In *L. maculans*, a total of 122 candidate effectors have been predicted (Rouxel et al., 2011). In terms of Avr effectors, 11 avirulence genes (Avr) termed *AvrLm1, 2, 3, 4, 5, 6, 7, 9, 11, AvrLepR1*, and*AvrLmJ1* have been identified (Balesdent et al., 2002, 2005; Ghanbarnia et al., 2012; Van de Wouw et al., 2014). All of the proteins produced by these genes are known as effectors; small, secreted, cysteine-rich proteins that are non-homologous to any other known *L. maculans* proteins. In plants lacking the corresponding R-protein, the above mentioned Avr effector proteins contribute to disease development; however very little is known about their mechanistic function and whether they are involved in plant immune suppression or virulence specifically. Recently a new study illustrated that the AvrLm4-7 effector was strongly required for virulence (Nováková et al., 2015). Through analysis of plant hormone concentrations, defense gene transcription and ROS accumulation, the authors were able to show that various components of the plant immune system were affected after inoculation with an *L. maculans* isolate lacking a functional AvrLm4-7 allele. Plants inoculated with the isolate carrying a functional AvrLm4-7 gene produced larger cotyledon lesion than the control group isolates that did not harbor an active allele. Interestingly, incompatible interactions due to recognition of AvrLm4-7 by Rlm4 revealed strong early induction of SA and Ethylene (ET)-dependent signaling pathways, which was also previously shown to occur upon recognition of AvrLm1 by Rlm1 (Šašek et al., 2012; Nováková et al., 2015). Through LC-MS/MS chromatography and RT-qPCR analysis (Nováková et al., 2015) determined that an active AvrLm4-7 allele causes suppression of SA-signaling, affecting not only its biosynthesis, but also reducing the expression of the SA-related marker gene *PR1*. The same situation appears to hold true in terms of ET-signaling, where AvrLm4-7 reduces the expression of *ACS2* and *HEL*expression, two ET-responsive genes. These findings together with a recent report illustrating that AvrLm4-7 is translocated into the plant cell (Blondeau et al., 2015) suggest that the plant SA and/or ET hormone signaling pathways, which are critical for induction of the plant defense response (Wang and Irving, 2011), are the primary targets of AvrLm4-7.

CRISPR/CAS9: A NEW TECHNIQUE TO ACCELERATE OUR UNDERSTANDING OF EFFECTOR FUNCTION

From the findings described above, it is apparent that a great deal of progress has been made toward understanding the role of fungal effectors. The most common approach utilized to elucidate the function of these genes involves reverse genetics, where the functional gene is replaced by an antibiotic selection marker, or deleted through homologous recombination. Effector genes in *U. maydis* and *M. oryzae*, as well as pathogenicity genes in *L. maculans* have been successfully deleted using this method (Hemetsberger et al., 2012; Mentlak et al., 2012; Mueller et al., 2013;Feng et al., 2014; Lo Presti et al., 2015). Gene silencing through RNAi has also been used in a number of fungi (Panwar et al., 2013; Petre et al., 2014; Soyer et al., 2014; Whigham et al., 2015;Yin et al., 2015). Although these methods have proven extremely useful, it has become apparent that there is an increasing need in effector research for techniques that can address effector gene redundancy and function within a large protein family. For studies such as these, we are currently limited in terms of fungal selection markers, where it becomes increasingly difficult to generate strains carrying multiple mutations within defined sites. One particular method that is used to recycle antibiotic markers is called the FLP-FRT recombination technique. FLP is a site-specific recombinase, which catalyzes recombination between two directly oriented FRT sites, leading to excision of the intervening DNA segment. Because the antibiotic selection marker is located on the intervening DNA segment, the end result is an unmarked gene disruption (Wirth et al., 2007). In *U. maydis*, this approach represents one of the best ways to introduce multiple mutations within the genome without utilizing all available selection markers; however a number of retro-transformation cycles followed by growth cycles to stimulate the loss of the marker are often required (Khrunyk et al., 2010).

The clustered regularly interspaced short palindromic repeat (CRISPR)-Cas9 system was originally identified in bacteria and archaea as a defense mechanism to prevent the invasion of foreign DNA from phage or plasmids (Barrangou et al., 2007). The CRISPR/Cas9 system has been adapted for use in a number of genome editing applications in both eukaryotes and prokaryotes (Jinek et al., 2012; Cong et al., 2013; Jiang et al., 2013; Doudna and Charpentier, 2014; Nissim et al., 2014;Sander and Joung, 2014). The adapted system consists of a Cas9 nuclease, which is guided to a specific target site by a single guide RNA (sgRNA). The nuclease then generates a double stranded break (DSB) at the desired target site within the genome. The DSB within the DNA initiates the repair mechanism (non-homologous end joining pathway) resulting in short deletions or substitutions, which may cause frameshifts or generate premature stop codons within the ORF of the target DNA (Doudna and Charpentier, 2014). Application of the CRISPR/Cas9 system depends on the expression of the *cas9* gene from *Streptococcus pyogenes* and the ability of the sgRNA to express and fold into the secondary structure necessary for interaction with Cas9 nuclease inside the nucleus.

CRISPR/Cas9-based technologies have emerged as an effective way to generate unmarked mutations in several organisms (Sander and Joung, 2014), including *Saccharomyces cereviseae*(DiCarlo et al., 2013), *Trichoderma reesei* (Liu et al., 2015), *M. oryzae* (Arazoe et al., 2015), and*Neurospora crassa* (Matsu-ura et al., 2015). One particular group was able to show that CRISPR/Cas9 could be modified to enable disruption and replacement of an effector gene in the oomycete *P. sojae* (Fang and Tyler, 2015). In each of the mutations the authors examined, short indels were located specifically at the Cas9 cleavage site, and deletions of one, three and six base-pairs were observed within the target regions providing evidence regarding the effectiveness of this system. Nodvig and colleagues developed a CRISPR/Cas9 system that allows RNA-guided mutagenesis by transforming a target fungus with a single plasmid. This approach was further modified allowing enhanced functionality in a broader range of filamentous fungi through the use of strong promoters from *Aspergillus niger* for *cas9* and sgRNA expression (Nodvig et al., 2015). More intriguingly, the CRISPR/Cas9 system has high efficiency in targeting multiple unrelated genes using several sgRNA, as shown in the yeast *Saccharomyces cerevisiae* (Cong et al., 2013), the filamentous fungus *T. reesei* (Liu et al., 2015) and two plant systems, namely rice and *Arabidopsis*(Endo et al., 2015; Zhang et al., 2015).

Clearly, the CRISPR/Cas9 "craze" has led to a plethora of useful tools that target not only single genes, but multiple gene targets, which will be extremely useful for effector research. The most valuable attribute of the CRISPR/Cas9 systems is its versatility, where the tools can be used to specifically target genes for mutagenesis in many organisms. Certainly, this system will transform and improve biological research, leading to many new discoveries in gene regulation and function. Nevertheless, this system is still in its infancy and as such several aspects need to be refined, including off-targeting and expression of the components within the desired host.

CONCLUSION: FUTURE CHALLENGES

We have gained a wealth of knowledge toward understanding plant-pathogen interactions over the past several decades. Research has shown that effector proteins play a significant role in the ability of fungal pathogens to establish a compatible interaction with the host plant, and in some cases have confirmed their involvement as virulence factors. We have working methods for developing mutations within single genes, and with the introduction of CRISPR/Cas9 technology, we now have the capability to target multiple genes without the use of markers. This system will significantly improve our ability to target redundant or related effector genes at one time, perhaps providing a definitive or visual phenotype that will provide insight into effector function. Despite our improved knowledge in this area, some important factors remain unresolved. Many fungal effectors have been shown to function primarily within the plant cytosol, yet lack translocation signals within the primary amino acid sequence. This brings into question how effectors are targeted to and taken across the plasma membrane of the plant cell? Although some host targeting sequence motifs have been identified in oomycetes, most fungi do not share this common feature. Moreover, the low sequence homologies among these fungal proteins make it difficult to predict similar features. However, the rapid development of next-generation sequencing tools and new bioinformatics algorithms and pipelines will provide new whole-genome information for both biotrophic and hemibiotrophic fungi. Given the large numbers of effector candidates, research will be a challenge. However, with the continuous development of new tools and techniques we will undoubtedly move toward exceptional opportunities for investigating the roles of fungal pathogen effectors in plant-pathogen interactions.

CONFLICT OF INTEREST STATEMENT

The authors declare that the research was conducted in the absence of any commercial or financial relationships that could be construed as a potential conflict of interest.

REFERENCES

1. Ali, S., Laurie, J. D., Linning, R., Cervantes-Chavez, J. A., Gaudet, D., and Bakkeren, G. (2014). An immunity-triggering effector from the Barley smut fungus *Ustilago hordei* resides in an Ustilaginaceae-specific cluster bearing signs of transposable element-assisted evolution. *PLoS Pathog.* 10:e1004223. doi: 10.1371/journal.ppat.1004223
2. Arazoe, T., Miyoshi, K., Yamato, T., Ogawa, T., Ohsato, S., Arie, T., et al. (2015). Tailor-made CRISPR/Cas system for highly efficient targeted gene replacement in the rice blast fungus. *Biotechnol. Bioeng.* 112, 2543–2549. doi: 10.1002/bit.25662
3. Armstrong, M. R., Whisson, S. C., Pritchard, L., Bos, J. I., Venter, E., Avrova, A. O., et al. (2005). An ancestral oomycete locus contains late blight avirulence gene Avr3a, encoding a protein that is recognized in the host cytoplasm. *Proc. Natl. Acad. Sci. U.S.A.* 102, 7766–7771. doi: 10.1073/pnas.0500113102
4. Asai, S., and Shirasu, K. (2015). Plant cells under siege: plant immune system versus pathogen effectors. *Curr. Opin. Plant Biol.*28, 1–8. doi: 10.1016/j.pbi.2015.08.008
5. Balesdent, M. H., Attard, A., Kuhn, M. L., and Rouxel, T. (2002). New avirulence genes in the phytopathogenic fungus*Leptosphaeria maculans*. *Phytopathology* 92, 1122–1133. doi: 10.1094/PHYTO.2002.92.10.1122
6. Balesdent, M. H., Barbetti, M. J., Li, H., Sivasithamparam, K., Gout, L., and Rouxel, T. (2005). Analysis of *Leptosphaeria maculans* race structure in a worldwide collection of isolates. *Phytopathology* 95, 1061–1071. doi: 10.1094/PHYTO-95-1061
7. Balesdent, M. H., Fudal, I., Ollivier, B., Bally, P., Granaubert, J., Chevre, A., et al. (2013). The dispensable chromosome of*Leptosphaeria maculans* shelters an effector gene conferring avirulence towards *Brassica rapa*. *New Phytol.* 198, 889–898. doi: 10.1111/nph.12178
8. Barrangou, R., Fremaux, C., Deveau, H., Richards, M., Boyaval, P., Moineau, S., et al. (2007). *CRISPR-Cas9 provides acquired resistance against viruses in prokaryotes*. *Science* (New York, N.Y.) 315, 1709–1712. doi: 10.1126/science.1138140
9. Blondeau, K., Blaise, F., Graille, M., Kale, S. D., Linglin, J., Ollivier, B., et al. (2015). Crystal structure of the effector AvrLm4-7 of *Leptosphaeria maculans* reveals insights into its translocation into plant cells and recognition by resistance proteins. *Plant J.* 83, 610–624. doi: 10.1111/tpj.12913
10. Bohnert, H. U., Fudal, I., Dioh, W., Tharreau, D., Notteghem, J. L., and Lebrun, M. H. (2004). A putative polyketide synthase/peptide synthetase from *Magnaporthe grisea* signals pathogen attack to resistant rice. *Plant Cell* 16, 2499–2513. doi: 10.1105/tpc.104.022715

11. Bolton, M. D., van Esse, H. P., Vossen, J. H., de Jonge, R., Stergiopoulos, I., Stulemeijer, I. J., et al. (2008). The novel*Cladosporium fulvum* lysin motif effector Ecp6 is a virulence factor with orthologues in other fungal species. *Mol. Microbiol.*69, 119–136. doi: 10.1111/j.1365-2958.2008.06270.x

12. Bos, J. I., Armstrong, M. R., Gilroy, E. M., Boevink, P. C., Hein, I., Taylor, R. M., et al. (2010). *Phytophthora infestans* effector AVR3a is essential for virulence and manipulates plant immunity by stabilizing host E3 ligase CMPG1. *Proc. Natl. Acad. Sci. U.S.A.* 107, 9909–9914. doi: 10.1073/pnas.0914408107

13. Boutemy, L. S., King, S. R., Win, J., Hughes, R. K., Clarke, T. A., Blumenschein, T. M., et al. (2011). Structures of Phytophthora RXLR effector proteins: a conserved but adaptable fold underpins functional diversity. *J. Biol. Chem.* 286, 35834–35842. doi: 10.1074/jbc.M111.262303

14. Bozkurt, T. O., Schornack, S., Win, J., Shindo, T., Ilyas, M., Oliva, R., et al. (2011). *Phytophthora infestans* effector AVRblb2 prevents secretion of a plant immune protease at the haustorial interface. *Proc. Natl. Acad. Sci. U.S.A.* 108, 20832–20837. doi: 10.1073/pnas.1112708109

15. Brefort, T., Doehlemann, G., Mendoza-Mendoza, A., Reissmann, S., Djamei, A., and Kahmann, R. (2009). *Ustilago maydis* as a pathogen. *Annu. Rev. Phytopath.* 47, 423–445. doi: 10.1146/annurev-phyto-080508-081923

16. Bruckner, A., Polge, C., Lentze, N., Auerbach, D., and Schlattner, U. (2009). Yeast-two hybrid system, a powerful tool for systems biology. *Int. J. Mol. Sci.* 10, 2763–2788. doi: 10.3390/ijms10062763

17. Cao, T., Kim, Y. M., Kav, N. N. V., and Strelkov, S. E. (2009). A proteomic evaluation of Pyrenophora tritici-repentis, causal agent of tan spot of wheat, reveals major differences between virulent and avirulent isolates. *Proteomics* 9, 1177–1196. doi: 10.1002/pmic.200800475

18. Cao, Y., Liang, Y., Tanaka, K., Nguyen, C. T., Jedrezjczak, R. P., Joachimiak, A., et al. (2014). The kinase LYK5 is a major chitin receptor in *Arabidopsis* and forms a chitin-induced complex with related kinase CERK1. *eLife* 3:e03766. doi: 10.7554/eLife.03766

19. Catanzariti, A. M., Dodds, P. N., Lawrence, G. J., Ayliffe, M. A., and Ellis, J. G. (2006). Haustorially expressed secreted proteins from flax rust are highly enriched for avirulence elicitors. *Plant Cell* 18, 243–256. doi: 10.1105/tpc.105.035980

20. Cesari, S., Thilliez, G., Ribot, C., Chalvon, V., Michel, C., Jauneau, A., et al. (2013). The rice resistance protein pair RGA4/RGA5 recognizes the *Magnaporthe oryzae* effectors Avr-Pia and Avr1-CO39 by direct binding. *Plant Cell* 25, 1463–1481. doi: 10.1105/tpc.112.107201

21. Chuma, I., Isobe, C., Hotta, Y., Ibaragi, K., Futamata, N., Kusaba, M., et al. (2011). Multiple translocation of the AVR-Pita effector gene among chromosomes of the rice blast fungus *Magnaporthe oryzae* and related species. *PLoS Pathog.* 7:e1002147. doi: 10.1371/journal.ppat.1002147

22. Cong, L., Ran, F. A., Cox, D., Lin, S., Barretto, R., Habib, N., et al. (2013). *Multiplex genome engineering using CRISPR/Cas systems. Science* (New York, N.Y.) 339, 819–823. doi: 10.1126/science.1231143

23. Cui, H., Tsuda, K., and Parker, J. E. (2015). Effector-triggered immunity: from pathogen perception to robust defense. *Annu. Rev. Plant Biol.* 66, 487–511. doi: 10.1146/annurev-arplant-050213-040012

24. de Jonge, R., Bolton, M. D., and Thomma, B. P. (2011). How filamentous pathogens co-opt plants: the ins and outs of fungal effectors. *Curr. Opin. Plant Biol.* 14, 400–406. doi: 10.1016/j.pbi.2011.03.005

25. de Jonge, R., van Esse, H. P., Kombrink, A., Shinya, T., Desaki, Y., Bours, R., et al. (2010). Conserved fungal LysM effector Ecp6 prevents chitin-triggered immunity in plants. *Science* (New York, N.Y.) 329, 953–955. doi: 10.1126/science.1190859

26. De Wit, P. J., Brandwagt, B. F., van den Burg, H. A., Cai, X., van der Hoorn, R. A., de Jong, C. F., et al. (2002). The molecular basis of co-evolution between *Cladosporium fulvum* and tomato. *Antonie Van Leeuwenhoek* 81, 409–412. doi: 10.1023/A:1020553120889

27. De Wit, P. J., Buurlage, M. B., and Hammond, K. E. (1986). The occurrence of host-, pathogen- and interaction-specific proteins in the apoplast of *Cladosporium fulvum* (syn. *Fulvia fulva)* infected tomato leaves. *Physiol. Mol. Plant Pathol.* 29, 159–172. doi: 10.1016/S0048-4059(86)80018-2

28. Dean, R. A., Talbot, N. J., Ebbole, D. J., Farman, M. L., Mitchell, T. K., Orbach, M. J., et al. (2005). The genome sequence of the rice blast fungus *Magnaporthe grisea. Nature* 434, 980–986. doi: 10.1038/nature03449

29. DiCarlo, J. E., Norville, J. E., Mali, P., Rios, X., Aach, J., and Church, G. M. (2013). Genome engineering in *Saccharomyces cerevisiae* using CRISPR-Cas systems. *Nucleic Acids Res.* 41, 4336–4343. doi: 10.1093/nar/gkt135

30. Dixon, M. S., Jones, D. A., Keddie, J. S., Thomas, C. M., Harrison, K., and Jones, J. D. (1996). The tomato Cf-2 disease resistance locus comprises two functional genes encoding leucine-rich repeat proteins. *Cell* 84, 451–459. doi: 10.1016/S0092-8674(00)81290-8

31. Djamei, A., and Kahmann, R. (2012). *Ustilago maydis*: dissecting the molecular interface between pathogen and plant. *PLoS Pathog.* 8:e1002955. doi: 10.1371/journal.ppat.1002955

32. Djamei, A., Schipper, K., Rabe, F., Ghosh, A., Vincon, V., Kahnt, J., et al. (2011). Metabolic priming by a secreted fungal effector.*Nature* 478, 395–398. doi: 10.1038/nature10454

33. Dodds, P. N., and Rathjen, J. P. (2010). Plant immunity: towards an integrated view of plant-pathogen interactions. *Nat. Rev.*11, 539–548. doi: 10.1038/nrg2812

34. Dodds, P. N., Lawrence, G. J., Catanzariti, A. M., Ayliffe, M. A., and Ellis, J. G. (2004). The *Melampsora lini* AvrL567 avirulence genes are expressed in haustoria

and their products are recognized inside plant cells. *Plant Cell* 16, 755–768. doi: 10.1105/tpc.020040

35. Doehlemann, G., Reissmann, S., Assmann, D., Fleckenstein, M., and Kahmann, R. (2011). Two linked genes encoding a secreted effector and a membrane protein are essential for *Ustilago maydis* -induced tumour formation. *Mol. Microbiol.* 81, 751–766 doi: 10.1111/j.1365-2958.2011.07728.x

36. Doehlemann, G., van der Linde, K., Assmann, D., Schwammbach, D., Hof, A., Mohanty, A., et al. (2009). Pep1, a secreted effector protein of *Ustilago maydis*, is required for successful invasion of plant cells. *PLoS Pathog.* 5, e1000290. doi: 10.1371/journal.ppat.1000290

37. Dong, S., Qutob, D., Tedman-Jones, J., Kuflu, K., Wang, Y., Tyler, B. M., et al. (2009). The *Phytophthora sojae* avirulence locus Avr3c encodes a multi-copy RXLR effector with sequence polymorphisms among pathogen strains. *PLoS One* 4:e5556. doi: 10.1371/journal.pone.0005556

38. Dong, S., Yin, W., Kong, G., Yang, X., Qutob, D., Chen, Q., et al. (2011). *Phytophthora sojae* avirulence effector Avr3b is a secreted NADH and ADP-ribose pyrophosphorylase that modulates plant immunity. *PLoS Pathog.* 7:e1002353. doi: 10.1371/journal.ppat.1002353

39. Dou, D., Kale, S. D., Liu, T., Tang, Q., Wang, X., Arredondo, F. D., et al. (2010). Different domains of *Phytophthora sojae*effector Avr4/6 are recognized by soybean resistance genes Rps4 and Rps6. *Mol. Plant Microbe Interact.* 23, 425–435. doi: 10.1094/MPMI-23-4-0425

40. Dou, D., Kale, S. D., Wang, X., Jiang, R. H. Y., Bruce, N. A., Arredondo, F. D., et al. (2008a). RXLR-mediated entry of*Phytophthora sojae* effector Avr1b into soybean cells does not require pathogen-encoded machinery. *Plant Cell* 20, 1930–1947. doi: 10.1105/tpc.107.056093

41. Dou, D., Kale, S. D., Wang, X., Chen, Y., Wang, Q., Wang, X., et al. (2008b). Carboxy-terminal motifs common to many oomycete RXLR effectors are required for avirulence and suppression of BAX-mediated programmed cell death by*Phytophthora sojae* effector Avr1b. *Plant Cell* 20, 1118–1133. doi: 10.1105/tpc.107.057067

42. Doudna, J. A., and Charpentier, E. (2014). *Genome editing. The new frontier of genome engineering with CRISPR-Cas9.Science* (New York, N.Y.) 346, 1258096. doi: 10.1126/science.1258096

43. Dubacq, C., Guerois, R., Courbeyrette, R., Kitagawa, K., and Mann, C. (2002). Sgt1p contributes to cyclic AMP pathway activity and physically interacts with the adenylyl cyclase Cyr1p/Cdc35p in budding yeast. *Eukaryotic Cell* 1, 568–582. doi: 10.1128/EC.1.4.568-582.2002

44. Dunn, K. W., Kamocka, M. M., and McDonald, J. H. (2011). A practical guide to evaluating colocalization in biological microscopy. *Am. J. Physiol. Cell Physiol.* 300, C723–C742. doi: 10.1152/ajpcell.00462.2010

45. Duplessis, S., Cuomo, C. A., Lin, Y. C., Aerts, A., Tisserant, E., Veneault-Fourrey, C., et al. (2011). Obligate biotrophy features unraveled by the genomic analysis of rust fungi. *Proc. Natl. Acad. Sci. U.S.A.* 108, 9166–9171. doi: 10.1073/pnas.1019315108

46. Eberhard, J., Ehrler, T. T., Epple, P., Felix, G., Raesecke, H. R., Amrhein, N., et al. (1996). Cytosolic and plastidic chorismate mutase isozymes from *Arabidopsis thaliana*: molecular characterization and enzymatic properties. *Plant J.* 10, 815–821. doi: 10.1046/j.1365-313X.1996.10050815.x

47. Endo, M., Mikami, M., and Toki, S. (2015). Multigene knockout utilizing off-target mutations of the CRISPR/CAS9 system in rice. *Plant Cell Physiol.* 56, 41–47. doi: 10.1093/pcp/pcu154

48. Fang, Y., and Tyler, B. M. (2015). Efficient disruption and replacement of an effector gene in the oomycete *Phytophthora sojae*using CRISPR/CAS9. *Mol. Plant Pathol.* 17, 127–139. doi: 10.1111/mpp.12318

49. Farman, M. L., and Leong, S. A. (1998). Chromosome walking to the AVR1-CO39 avirulence gene of *Magnaporthe grisea*: discrepancy between the physical and genetic maps. *Genetics* 150, 1049–1058.

50. Feng, J., Zhang, H., Strelkov, S. E., and Hwang, S. F. (2014). The LmSNF1 gene is required for pathogenicity in the canola blackleg pathogen *Leptosphaeria maculans*. *PLoS One* 9:e92503. doi: 10.1371/journal.pone.0092503

51. Flor, H. H. (1956). The complementary systems in flax and flax rust. *Adv. Genet.* 8, 29–54. doi: 10.1016/S0065-2660(08)60498-8

52. Friesen, T. L., Stukenbrock, E. H., Liu, Z., Meinhardt, S., Ling, H., Faris, J. D., et al. (2006). Emergence of a new disease as a result of interspecific virulence gene transfer. *Nat. Genet.* 38, 953–956. doi: 10.1038/ng1839

53. Fudal, I., Bohnert, H. U., Tharreau, D., and Lebrun, M. H. (2005). Transposition of MINE, a composite retrotransposon, in the avirulence gene ACE1 of the rice blast fungus *Magnaporthe grisea*. *Fungal Genet. Biol.* 42, 761–772. doi: 10.1016/j.fgb.2005.05.001

54. Fudal, I., Ross, S., Brun, H., Besnard, A. L., Ermel, M., Kuhn, M. L., et al. (2009). Repeat-induced point mutation (RIP) as an alternative mechanism of evolution towards virulence in *Leptosphaeria maculans*. *Mol. Plant Microbe Interact.* 22, 932–941. doi: 10.1094/MPMI-22-8-0932

55. Fudal, I., Ross, S., Gout, L., Blaise, F., Kuhn, M. L., Eckert, M. R., et al. (2007). Heterochromatin-like regions as ecological niches for avirulence genes in the *Leptosphaeria maculans* genome: map-based cloning of AvrLm6. *Mol. Plant Microbe Interact.*20, 459–470. doi: 10.1094/MPMI-20-4-0459

56. Garnica, D. P., Nemri, A., Upadhyaya, N. M., Rathen, J. P., and Dodds, P. N. (2014). The ins and outs of rust haustoria. *PLoS Pathog.* 10:e1004329. doi: 10.1371/journal.ppat.1004329

57. Gawehns, F., Houterman, P. M., Ichou, F. A., Michielse, C. B., Hijdra, M., Cornelissen, B. J. C., et al. (2014). The *Fusarium oxysporum* effector Six6 contributes to virulence and suppresses I-2-mediated cell death. *Mol.*

Plant Microbe Interact. 27, 336–348. doi: 10.1094/MPMI-11-13-0330-R

58. Gawehns, F., Ma, L. S., Bruning, O., Houterman, P. M., Boeren, S., Cornelissen, B. J. C., et al. (2015). The effector repertoire of*Fusarium oxysporum* determines the tomato xylem proteome composition following infection. *Front. Plant Sci.* 6:967. doi: 10.3389/fpls.2015.00967

59. Ghanbarnia, K., Fudal, I., Larkan, N. J., Links, M. G., Balesdent, M.-H., Profotova, B., et al. (2015). Rapid identification of the*Leptosphaeria maculans avirulence* gene *AvrLm2* using an intraspecific comparative genomics approach. *Mol. Plant Pathol.* 16, 699–709. doi: 10.1111/mpp.12228

60. Ghanbarnia, K., Lydiate, D. J., Rimmer, S. R., Li, G., Kutcher, H. R., Larkan, N. J., et al. (2012). Genetic mapping of the*Leptosphaeria maculans* avirulence gene corresponding to the LepR1 resistance gene of Brassica napus. *TAG. Theor. Appl. Genet.* 124, 505–513. doi: 10.1007/s00122-011-1724-3

61. Gilroy, E. M., Breen, S., Whisson, S. C., Squires, J., Hein, I., Kaczmarek, M., et al. (2011b). Presence/absence, differential expression and sequence polymorphisms between PiAVR2 and PiAVR2-like in *Phytophthora infestans* determine virulence on R2 plants. *New Phytol.* 191, 763–776. doi: 10.1111/j.1469-8137.2011.03736.x

62. Gilroy, E. M., Taylor, R. M., Hein, I., Boevink, P., Sadanandom, A., and Birch, P. R. J. (2011a). CMPG1-dependent cell death follows perception of diverse pathogen elicitors at the host plasma membrane and is suppressed by *Phytophthora infestans*RXLR effector AVR3a. *New Phytol.* 190, 653–666. doi: 10.1111/j.1469-8137.2011.03643.x

63. Giraldo, M. C., and Valent, B. (2013). Filamentous plant pathogen effectors in action. *Nat. Rev. Microbiol.* 11, 800–814. doi: 10.1038/nrmicro3119

64. Giraldo, M. C., Dagdas, Y. F., Gupta, Y. K., Mentlak, T. A., Yi, M., Martinez-Rocha, A. L., et al. (2013). Two distinct secretion systems facilitate tissue invasion by the rice blast fungus *Magnaporthe oryzae*. *Nat. Commun.* 4, 1996. doi: 10.1038/ncomms2996

65. Godfrey, D., Bohlenius, H., Pedersen, C., Zhang, Z., Emmersen, J., and Thordal-Christensen, H. (2010). Powdery mildew fungal effector candidates share N-terminal Y/F/WxC-motif. *BMC Genomics* 11:317. doi: 10.1186/1471-2164-11-317

66. Gomez-Gomez, L., and Boller, T. (2002). Flagellin perception: a paradigm for innate immunity. *Trends Plant Sci.* 7, 251–256. doi: 10.1016/S1360-1385(02)02261-6

67. Gonzalez-Fernandez, R. Q., Prats, E., and Jorrin-Novo, J. V. (2010). Proteomics of plant pathogenic fungi. *J. Biomed. Biotech.*2010, 1–36. doi: 10.1155/2010/932527

68. Goritschnig, S., Steinbrenner, A. D., Grunwald, D. J., and Staskawicz, B. J. (2016). Structurally distinct*Arabidopsis thaliana*NLR immune receptors recognize tandem WY domains of an oomycete effector. *New Phytol.* 210, 984–996. doi: 10.1111/nph.13823

69. Gout, L., Fudal, I., Kuhn, M. L., Blaise, F., Eckert, M., Cattolico, L., et al. (2006). Lost in the middle of nowhere: the AvrLm1 avirulence gene of the Dothideomycete *Leptosphaeria maculans*. *Mol. Microbiol.* 60, 67–80. doi: 10.1111/j.1365-2958.2006.05076.x

70. Gout, L., Kuhn, M. L., Vincenot, L., Bernard-Samain, S., Cattolico, L., Barbetti, M., et al. (2007). Genome structure impacts molecular evolution at the AvrLm1 avirulence locus of the plant pathogen *Leptosphaeria maculans*. *Environ. Microbiol.* 9, 2978–2992. doi: 10.1111/j.1462-2920.2007.01408.x

71. Grewal, T. S., Rossnagel, B. G., Bakkeren, G., and Scoles, G. J. (2008). Identification of resistance genes to barley covered smut and mapping of the Ruh1 gene using *Ustilago hordei* strains with defined avirulence genes. *Can. J. Plant Pathol.* 30, 277–284. doi: 10.1080/07060661.2008.10540543

72. Gururania, M. A., Venkatesh, J., Upadhyaya, C. P., Nookaraju, A., Pandey, S. K., and Park, S. W. (2012). Plant disease resistance genes: current status and future directions. *Physiol. Mol. Plant Pathol.* 78, 51–65. doi: 10.1016/j.pmpp.2012.01.002

73. Guttman, D. S., McHardy, A. C., and Schulze-Lefert, P. (2014). Microbial genome-enabled insights into plant-microorganism interactions. *Nat. Rev.* 15, 797–813. doi: 10.1038/nrg3748

74. Guyon, K., Balague, C., Roby, D., and Raffaele, S. (2014). Secretome analysis reveals effector candidates associated with broad host range necrotrophy in the fungal plant pathogen *Sclerotinia sclerotiorum*. *BMC genomics* 15:336. doi: 10.1186/1471-2164-15-336

75. Haas, B. J., Kamoun, S., Zody, M. C., Jiang, R. H., Handsaker, R. E., Cano, L. M., et al. (2009). Genome sequence and analysis of the Irish potato famine pathogen *Phytophthora infestans*. *Nature* 461, 393–398. doi: 10.1038/nature08358

76. Harrison, M. J. (1998). Biotrophic interface and nutrient transport in plant/fungal symbioses. *J. Exp. Biol.* 50, 1013–1022.

77. Hemetsberger, C., Herrberger, C., Zechmann, B., Hillmer, M., and Doehlemann, G. (2012). The *Ustilago maydis* effector Pep1 suppresses plant immunity by inhibition of host peroxidase activity. *PLoS Pathog.* 8:e1002684. doi: 10.1371/journal.ppat.1002684

78. Hof, A., Zechmann, B., Schwammbach, D., Huckelhoven, R., and Doehlemann, G. (2014). Alternative cell death mechanisms determine epidermal resistance in incompatible barley–Ustilago interactions. *Mol. Plant Microbe Interact.* 27, 403–414. doi: 10.1094/MPMI-10-13-0317-R

79. Houterman, P. M., Cornelissen, B. J., and Rep, M. (2008). Suppression of plant resistance gene-based immunity by a fungal effector. *PLoS Pathog.* 4:e1000061. doi: 10.1371/journal.ppat.1000061

80. Houterman, P. M., Ma, L., van Ooijen, G., de Vroomen, M. J., Cornelissen, B. J., Takken, F. L., et al. (2009). The effector protein Avr2 of the xylem-colonizing fungus *Fusarium oxysporum* activates the tomato resistance protein I-2 intracellularly. *Plant J.* 58, 970–978. doi: 10.1111/j.1365-313X.2009.03838.x

81. Jiang, R. H., Tripathy, S., Govers, F., and Tyler, B. M. (2008). RXLR effector reservoir in two Phytophthora species is dominated by a single rapidly evolving superfamily with more than 700 members. *Proc. Natl. Acad. Sci. U.S.A.* 105, 4874–4879. doi: 10.1073/pnas.0709303105

82. Jiang, W., Bikard, D., Cox, D., Zhang, F., and Marraffini, L. A. (2013). RNA-guided editing of bacterial genomes using CRISPR-Cas systems. *Nat. Biotechnol.* 31, 233–239. doi: 10.1038/nbt.2508

83. Jinek, M., Chylinski, K., Fonfara, I., Hauer, M., Doudna, J. A., and Charpentier, E. (2012). *A programmable dual-RNA-guided DNA endonuclease in adaptive bacterial immunity. Science* (New York, N.Y.) 337, 816–821. doi: 10.1126/science.1225829

84. Jones, J. D., and Dangl, J. L. (2006). The plant immune system. *Nature* 444, 323–329. doi: 10.1038/nature05286

85. Joosten, M. H. A. J., Cozijnsen, T. J., and de Wit, P. J. G. M. (1994). Host resistance to a fungal tomato pathogen lost by a single base-pair change in an avirulence gene. *Nature* 367, 384–386. doi: 10.1038/367384a0

86. Kaku, H., Nishizawa, Y., Ishii-Minami, N., Akimoto-Tomiyama, C., Dohmae, N., Takio, K., et al. (2006). Plant cells recognize chitin fragments for defense signaling through a plasma membrane receptor. *Proc. Natl. Acad. Sci. U.S.A.* 103, 11086–11091. doi: 10.1073/pnas.0508882103

87. Kale, S. D., Gu, B., Capelluto, D. G. S., Dou, D., Feldman, E., Rumore, A., et al. (2010). External Lipid PI3P mediates entry of eukaryotic pathogen effectors into plant and animal host cells. *Cell* 142, 284–295. doi: 10.1016/j.cell.2010.06.008

88. Kamoun, S. (2006). A catalogue of the effector secretome of plant pathogenic oomycetes. *Ann. Rev. Phytopathol.* 44, 41–60. doi: 10.1146/annurev.phyto.44.070505.143436

89. Kamoun, S. (2007). Groovy times: filamentous pathogen effectors revealed. *Curr. Opin. in Plant Biol.* 10, 358–365. doi: 10.1016/j.pbi.2007.04.017

90. Kamoun, S., Klucher, K. M., Coffey, M. D., and Tyler, B. M. (1993). A gene encoding a host-specific elicitor protein of*Phytophthora parasitica. Mol. Plant Microb. Interact.* 6, 573–581. doi: 10.1094/MPMI-6-573

91. Kamper, J., Kahmann, R., Bolker, M., Ma, L. J., Brefort, T., Saville, B. J., et al. (2006). Insights from the genome of the biotrophic fungal plant pathogen *Ustilago maydis*. *Nature* 444, 97–101. doi: 10.1038/nature05248

92. Kelley, B. S., Lee, S. J., Damasceno, C. M., Chakravarthy, S., Kim, B. D., Martin, G. B., et al. (2010). A secreted effector protein (SNE1) from *Phytophthora infestans* is a broadly acting suppressor of programmed cell death. *Plant J.* 62, 357–366. doi: 10.1111/j.1365-313X.2010.04160.x

93. Kemen, E., Kemen, A. C., Rafiqi, M., Hempel, U., Mendgen, K., Hahn, M., et al. (2005). Identification of a protein from rust fungi transferred from haustoria into infected plant cells. *Mol. Plant Microbe Interact.* 18, 1130–1139. doi: 10.1094/MPMI-18-1130

94. Kerppola, T. K. (2008). Bimolecular fluorescence complementation (BiFC) analysis as a probe of protein interactions in living cells. *Annu. Rev. Biophys.* 37, 465–487. doi: 10.1146/annurev.biophys.37.032807.125842

95. Khang, C. H., Berruyer, R., Giraldo, M. C., Kankanala, P., Park, S.-Y., Czymmek, K., et al. (2010). Translocation of *Magnaporthe oryzae* effectors into rice cells and their subsequent cell-to-cell movement. *Plant Cell* 22, 1388–1403. doi: 10.1105/tpc.109.069666

96. Khang, C. H., Sweigard, J. A., and Valent, B. (1995). The PWL host specificity gene family in the blast fungus *Magnaporthe oryzae*. *Mol. Plant Microbe Interact.* 8, 939–948. doi: 10.1094/MPMI-8-0939

97. Khrunyk, Y., Munch, K., Schipper, K., Lupas, A. N., and Kahmann, R. (2010). The use of FLP-mediated recombination for the functional analysis of an effector gene family in the biotrophic smut fungus *Ustilago maydis*. *New Phytol* 187, 957–968. doi: 10.1111/j.1469-8137.2010.03413.x

98. Kodama, Y., and Hu, C.-D. (2012). Bimolecular fluorescence complementation (BiFC): a 5-year update and future perspectives.*Biotechniques* 53, 285–298. doi: 10.2144/000113943

99. Koeck, M., Hardham, A. R., and Dodds, P. N. (2011). The role of effectors of biotrophic and hemibiotrophic fungi in infection.*Cell. Microbiol.* 13, 1849–1857. doi: 10.1111/j.1462-5822.2011.01665.x

100. Larkan, N. J., Lydiate, D. J., Parkin, I. A. P., Nelson, M. N., Epp, D. J., Cowling, W. A., et al. (2013). The Brassica napus blackleg resistance LepR3 enclodes a receptor-like protein triggered by the *Leptosphaeria maculans* effector AVRLM1. *New Phytol.* 197, 595–607. doi: 10.1111/nph.12043

101. Lauge, R., Goodwin, P. H., de Wit, P. J., and Joosten, M. H. (2000). Specific HR-associated recognition of secreted proteins from *Cladosporium fulvum* occurs in both host and non-host plants. *Plant J.* 23, 735–745. doi: 10.1046/j.1365-313x.2000.00843.x

102. Lauge, R., Joosten, M. H., Van Den Ackerveken, G. F., Van Den Broek, H. W., and De Wit, P. J. (1997). The in planta produced extracellular proteins ECP1 and ECP2 of *Cladosporium fulvum* are virulence factors. *Mol. Plant Microbe Interact.* 10, 725–734. doi: 10.1094/MPMI.1997.10.6.725

103. Lawrence, G. J., Dodds, P. N., and Ellis, J. G. (2010). Transformation of the flax rust fungus, *Melampsora lini*: selection via silencing of an avirulence gene. *Plant J.* 61, 364–369. doi: 10.1111/j.1365-313X.2009.04052.x

104. Li, W., Wang, B., Wu, J., Lu, G., Hu, Y., Zhang, X., et al. (2009). The *Magnaporthe oryzae* avirulence gene AvrPiz t encodes a predicted secreted protein that triggers the immunity in rice mediated by the blast resistance gene Piz-t. *Mol. Plant Microbe Interact.* 22, 411–420. doi: 10.1094/MPMI-22-4-0411

105. Lievens, B., Houterman, P. M., and Rep, M. (2009). Effectorgenescreeningallowsunambiguousidentification of *Fusarium oxysporum* f.sp. *lycopersici* races and discrimination from other formae speciales. *FEMS*

Microbiol. Lett. 300, 201–215. doi: 10.1111/j.1574-6968.2009.01783.x

106. Liu, R., Chen, L., Jiang, Y., Zhou, Z., and Zou, G. (2015). Efficient genome editing in filamentous fungus *Trichoderma reesei*using the CRISPR/CAS9 system. *Cell Discov.* 1, 1–11. doi: 10.1038/celldisc.2015.7

107. Liu, T. L., Ye, W. W., Ru, Y. Y., Yang, X. Y., Gu, B., Tao, K., et al. (2011). Two host cytoplasmic effectors are required for pathogenesis of *Phytophthora sojae* by suppression of host defenses. *Plant Physiol.* 155, 490–501. doi: 10.1104/pp.110.166470

108. Liu, T., Liu, Z., Song, C., Hu, Y., Han, Z., She, J., et al. (2012). *Chitin-induced dimerization activates a plant immune receptor.Science* (New York, N.Y.) 336, 1160–1164. doi: 10.1126/science.1218867

109. Lo Presti, L., Lanver, D., Schweizer, G., Tanaka, S., Liang, L., Tollot, M., et al. (2015). Fungal effectors and plant susceptibility.*Annu. Rev. Plant Biol.* 66, 513–545. doi: 10.1146/annurev-arplant-043014-114623

110. Lokossou, A. A., Park, T.- H., Arkel, G. V., Arens, M., Ruyter-Spira, C., Morales, J., et al. (2009). Exploiting knowledge of R/Avr genes to rapidly clone a new LZ-NBS-LRR family of late blight resistance genes from potato linkage group IV. *Mol. Plant Microbe Interact.* 22, 630–641. doi: 10.1094/MPMI-22-6-0630

111. Ma, L. J., van der Does, H. C., Borkovich, K. A., Coleman, J. J., Daboussi, M. J., Di Pietro, A., et al. (2010). Comparative genomics reveals mobile pathogenicity chromosomes in Fusarium. *Nature* 464, 367–373. doi: 10.1038/nature08850

112. Ma, L., Houterman, P. M., Gawehns, F., Cao, L., Sillo, F., Richter, H., et al. (2015). The AVR2-SIX5 gene pair is required to activate I-2-mediated immunity in tomato. *New Phytol.* 208, 507–518. doi: 10.1111/nph.13455

113. Macho, A. P., and Zipfel, C. (2014). Plant PRRs and the activation of innate immune signaling. *Mol. Cell* 54, 263–272. doi: 10.1016/j.molcel.2014.03.028

114. Mackie, A. J., Roberts, A. M., Callow, J. A., and Green, J. R. (1991). Molecular differentiation in pea powdery-mildew haustoria.*Planta.* 183, 399–408. doi: 10.1007/BF00197739

115. Manners, J. M., and Gay, J. L. (1983). "The host-parasite interface and nutrient transfer in biotrophic parasitism," in*Biochemical Plant Pathoglogy*, ed J. A. Callow (Chichester: John Wiley and Sons Ltd.), 163–168.

116. Manning, V. A., Hamilton, S. M., Karplus, P. A., and Ciuffetti, L. M. (2008). The Arg-Gly-Asp-containing, solvent-exposed loop of Ptr ToxA is required for internalization. *Mol. Plant Microbe Interact.* 21, 315–325. doi: 10.1094/MPMI-21-3-0315

117. Matsu-ura, T., Baek, M., Kwon, J., and Hong, C. (2015). Efficient gene editing in *Neurospora crassa* with CRISPR/CAS9 technology. *Fungal Biol. Biotechnol.* 2, 1–7. doi: 10.1186/s40694-015-0015-1

118. Mentlak, T. A., Kombrink, A., Shinya, T., Ryder, L. S., Otomo, I., Saitoh, H., et al. (2012). Effector-mediated suppression of chitin-triggered immunity by *Magnaporthe oryzae* is necessary for rice blast disease. *Plant Cell* 24, 322–335. doi: 10.1105/tpc.111.092957

119. Miya, A., Albert, P., Shinya, T., Desaki, Y., Ichimura, K., Shirasu, K., et al. (2007). CERK1, a LysM receptor kinase, is essential for chitin elicitor signaling in Arabidopsis. *Proc. Natl. Acad. Sci. U.S.A.* 104, 19613–19618. doi: 10.1073/pnas.0705147104

120. Mosquera, G., Giraldo, M. C., Khang, C. H., Coughlan, S., and Valent, B. (2009). Interaction transcriptome analysis identifies*Magnaporthe oryzae* BAS1-4 as biotrophy-associated secreted proteins in rice blast disease. *Plant Cell* 21, 1273–1290. doi: 10.1105/tpc.107.055228

121. Mueller, A. N., Ziemann, S., Treitschke, S., Assmann, D., and Doehlemann, G. (2013). Compatibility in the *Ustilago maydis* -maize interaction requires inhibition of host cysteine proteases by the fungal effector Pit2. *PLoS Pathog.* 9:e1003177. doi: 10.1371/journal.ppat.1003177

122. Na, R., Yu, D., Qutob, D., Zhao, J., and Gijzen, M. (2013). Deletion of the *Phytophthora sojae* avirulence gene Avr1d causes gain of viruence on Rps1d. *Mol. Plant Microbe Interact.* 16, 969–976. doi: 10.1094/MPMI-02-13-0036-R

123. Nemri, A., Saunders, D. G., Anderson, C., Upadhyaya, N. M., Win, J., Lawrence, G. J., et al. (2014). The genome sequence and effector complement of the flax rust pathogen *Melampsora lini*. *Front. Plant Sci.* 5:98. doi: 10.3389/fpls.2014.00098

124. Newman, M.-A., Sundelin, T., Nielsen, J. T., and Erbs, G. (2013). MAMP (microbe-associated molecular pattern) triggered immunity in plants. *Front. Plant Sci.* 4:139. doi: 10.3389/fpls.2013.00139

125. Nirmala, J., Drader, T., Lawrence, P. K., Yin, C., Hulbert, S., Steber, C. M., et al. (2011). Concerted action of two avirulent spore effectors activates reaction to *Puccinia graminis* 1 (Rpg1)-mediated cereal stem rust resistance. *Proc. Natl. Acad. Sci. U.S.A.* 108, 14676–14681. doi: 10.1073/pnas.1111771108

126. Nissim, L., Perli, S. D., Fridkin, A., Perez-Pinera, P., and Lu, T. K. (2014). Multiplexed and programmable regulation of gene networks with an integrated RNA and CRISPR/Cas toolkit in human cells. *Mol. Cell* 54, 698–710. doi: 10.1016/j.molcel.2014.04.022

127. Nodvig, C. S., Nielsen, J. B., Kogle, M. E., and Mortensen, U. H. (2015). A CRISPR-Cas9 system for genetic engineering of filamentous fungi. *PLoS One* 10:e0133085. doi: 10.1371/journal.pone.0133085

128. Nováková, M., Šašek, V., Trdá, L., Krutinová, H., Mongin, T., Valentová, O., et al. (2015). *Leptosphaeria maculans* effector AvrLm4-7 affects SA- and ET-signalling and H_2O_2 accumulation in *Brassica napus*. *Mol. Plant Pathol.* doi: 10.1111/mpp.12332. [Epub ahead of print].

129. Oerke, E. C. (2006). Crop losses to pests. *J. Agric. Sci.* 144, 31–43. doi: 10.1017/S0021859605005708

130. Oh, S. K., Young, C., Lee, M., Oliva, R., Bozkurt, T. O., Cano, L. M., et al. (2009). In planta expression screens of *Phytophthora infestans* RXLR effectors reveal

diverse phenotypes, including activation of the Solanum bulbocastanum disease resistance protein Rpi-blb2. *Plant Cell* 21, 2928–2947. doi: 10.1105/tpc.109.068247

131. Orbach, M. J., Farrall, L., Sweigard, J. A., Chumley, F. G., and Valent, B. (2000). A telomeric avirulence gene determines efficacy for the rice blast resistance gene Pi-ta. *Plant Cell* 12, 2019–2032. doi: 10.1105/tpc.12.11.2019
132. Panwar, V., McCallum, B., and Bakkeren, G. (2013). Endogenous silencing of Puccinia triticina pathogenicity genes through in planta-expressed sequences leads to the suppression of rust diseases on wheat. *Plant J.* 73, 521–532. doi: 10.1111/tpj.12047
133. Park, C. H., Chen, S., Shirsekar, G., Zhou, B., Khang, C. H., Songkumarn, P., et al. (2012). The *Magnaporthe oryzae* effector AvrPiz-t targets the RING E3 ubiquitin ligase APIP6 to suppress pathogen-associated molecular pattern-triggered immunity in rice. *Plant Cell* 24, 4748–4762. doi: 10.1105/tpc.112.105429
134. Parlange, F., Daverdin, G., Fudal, I., Kuhn, M. L., Balesdent, M. H., Blaise, F., et al. (2009). *Leptosphaeria maculans* avirulence gene AvrLm4-7 confers a dual recognition specificity by the Rlm4 and Rlm7 resistance genes of oilseed rape, and circumvents Rlm4-mediated recognition through a single amino acid change. *Mol. Microbiol.* 71, 851–863. doi: 10.1111/j.1365-2958.2008.06547.x
135. Petre, B., and Kamoun, S. (2014). How do filamentous pathogens deliver effector proteins into plant cells? *PLoS Biol.*12:e1001801. doi: 10.1371/journal.pbio.1001801
136. Petre, B., Joly, D. L., and Duplessis, S. (2014). Effector proteins of rust fungi. *Front. Plant Sci.* 5:416. doi: 10.3389/fpls.2014.00416
137. Pieterse, C. M. J., Van West, P., Verbakel, H. M., Brasse, P. W. H. M., Van Den Berg-Velthuis, G. C. M., and Govers, F. (1994). Structure and genomic organization of the ipiB and ipiO gene clusters of *Phytophthora infestans*. *Gene* 138, 67–77. doi: 10.1016/0378-1119(94)90784-6
138. Plissonneau, C., Daverdin, G., Ollivier, B., Blaise, B., Degrave, A., Fudal, I., et al. (2016). A game of hide and seek between avirulence genes AvrLm4-7 and AvrLm3 in *Leptosphaeria maculans*. *New Phytol.* 209, 1613–1624. doi: 10.1111/nph.13736
139. Qutob, D., Kemmerling, B., Brunner, F., Kufner, I., Engelhardt, S., Gust, A. A., et al. (2006). Phytotoxicity and innate immune responses induced by Nep1-like proteins. *Plant Cell* 18, 3721–3724. doi: 10.1105/tpc.106.044180
140. Qutob, D., Tedman-Jones, J., Dong, S., Kuflu, K., Pham, H., Wang, Y., et al. (2009). Copy number variation and transcriptional polymorphisms of *Phytophthora sojae* RXLR Effector Genes Avr1a and Avr3a. *PLoS One* 4:e5066. doi: 10.1371/annotation/2a2adcf8-afbc-4d46-92c6-d543d6b29182
141. Raffaele, S., and Kamoun, S. (2012). Genome evolution in filamentous plant pathogens: why bigger can be better. *Nat. Rev. Microbiol.* 10, 417–430. doi: 10.1038/nrmicro2790
142. Rafiqi, M., Gan, P. H., Ravensdale, M., Lawrence, G. J., Ellis, J. G., Jones, D. A., et al. (2010). Internalization of flax rust avirulence proteins into flax and tobacco cells can occur in the absence of the pathogen. *Plant Cell* 22, 2017–2032. doi: 10.1105/tpc.109.072983
143. Ravid, T., and Hochstrasser, M. (2008). Diversity of degradation signals in the ubiquitin-proteasome system. *Nat. Rev. Mol. Cell Biol.* 9, 679–690. doi: 10.1038/nrm2468
144. Redkar, A., Hoser, R., Schilling, L., Zechmann, B., Krzymowska, M., Walbot, V., et al. (2015). Secreted effector protein of*Ustilago maydis* guides maize leaf cells to form tumors. *Plant Cell* 27, 1332–1351. doi: 10.1105/tpc.114.131086
145. Rep, M., van der Does, H. C., Meijer, M., van Wijk, R., Houterman, P. M., Dekker, H. L., et al. (2004). A small, cysteine-rich protein secreted by *Fusarium oxysporum* during colonization of xylem vessels is required for I-3-mediated resistance in tomato. *Mol. Microbiol.* 53, 1373–1383. doi: 10.1111/j.1365-2958.2004.04177.x
146. Ricci, P., Bonnet, P., Huet, J.-C., Sallantin, M., Beauvais-Cante, F., Bruneteau, M., et al. (1989). Structure and activity of proteins from pathogenic fungi Phytophthora eliciting necrosis and acquired resistance in tobacco. *Eur. J. Biochem.* 183, 555–563. doi: 10.1111/j.1432-1033.1989.tb21084.x
147. Ridout, C. J., Skamnioti, P., Porritt, O., Sacristan, S., Jones, J. D., and Brown, J. K. (2006). Multiple avirulence paralogues in cereal powdery mildew fungi may contribute to parasite fitness and defeat of plant resistance. *Plant Cell* 18, 2402–2414. doi: 10.1105/tpc.106.043307
148. Rose, J. K., Ham, K. S., Darvill, A. G., and Albersheim, P. (2002). Molecular cloning and characterization of glucanase inhibitor proteins: coevolution of a counterdefense mechanism by plant pathogens. *Plant Cell* 14, 1329–1345. doi: 10.1105/tpc.002253
149. Rouxel, T., Grandaubert, J., Hane, J. K., Hoede, C., van de Wouw, A. P., Couloux, A., et al. (2011). Effector diversification within compartments of the *Leptosphaeria maculans* genome affected by Repeat-Induced Point mutations. *Nat. Commun.* 2, 202. doi: 10.1038/ncomms1189
150. Sanchez-Vallet, A., Saleem-Batcha, R., Kombrink, A., Hansen, G., Valkenburg, D. J., Thomma, B. P., et al. (2013). Fungal effector Ecp6 outcompetes host immune receptor for chitin binding through intrachain LysM dimerization. *Elife* 2:e00790. doi: 10.7554/eLife.00790
151. Sander, J. D., and Joung, J. K. (2014). CRISPR-Cas systems for editing, regulating and targeting genomes. *Nat. Biotechnol.* 32, 347–355. doi: 10.1038/nbt.2842
152. Šašek, V., Nováková, M., Jindřichová, B., Bóka, K., Valentová, O., and Burketová, L. (2012). Recognition of avirulence gene AvrLm1 from hemibiotrophic ascomycete *Leptosphaeria maculans* triggers salicylic acid and ethylene signaling in Brassica napus. *Mol.*

Plant Microbe Interact. 25, 1238–1250. doi: 10.1094/MPMI-02-12-0033-R

153. Saunders, D. G., Win, J., Cano, L. M., Szabo, L. J., Kamoun, S., and Raffaele, S. (2012). Using hierarchical clustering of secreted protein families to classify and rank candidate effectors of rust fungi. *PLoS One* 7:e29847. doi: 10.1371/journal.pone.0029847

154. Schirawski, J., Mannhaupt, G., Munch, K., Brefort, T., Schipper, K., Doehlemann, G., et al. (2010). *Pathogenicity determinants in smut fungi revealed by genome comparison. Science* (New York, N.Y.) 330, 1546–1548. doi: 10.1126/science.1195330

155. Schmidt, S. M., Houterman, P. M., Schreiver, I., Ma, L., Amyotte, S., Chellappan, B., et al. (2013). MITEs in the promoters of effector genes allow prediction of nove lvirulence genes in Fusariumoxysporum. *BMC Genomics* 14:119. doi: 10.1186/1471-2164-14-119

156. Schornack, S., van Damme, M., Bozkurt, T. O., Cano, L. M., Smoker, M., Thines, M., et al. (2010). Ancient class of translocated oomycete effectors targets the host nucleus. *Proc. Natl. Acad. Sci. U.S.A.* 107, 17421–17426. doi: 10.1073/pnas.1008491107

157. Schottens-Toma, I. M. J., and DeWit, P. J. (1988). Purification and primary structure of a necrosis-inducing peptide from the apoplastic fluids of tomato infected with *Cladosporium fulvum* (syn. *Fulvia fulva). Physiol. Mol. Plant Pathol.* 33, 59–67. doi: 10.1016/0885-5765(88)90043-4

158. Schwessinger, B., and Ronald, P. C. (2012). Plant Innate immunity: perception of conserved microbial signatures. *Ann. Rev. Plant Biol.* 63, 451–482. doi: 10.1146/annurev-arplant-042811-105518

159. Schwessinger, B., Bart, R., Krasileva, K. V., and Coaker, G. (2015). Focus issue on plant immunity: from model systems to crop species. *Front. Plant Sci.* 6:195. doi: 10.3389/fpls.2015.00195

160. Senchou, V., Weide, R., Carrasco, A., Bouyssou, H., Pont-Lezica, R., Govers, F., et al. (2004). High affinity recognition of aPhytophthora protein by *Arabidopsis* via an RGD motif. *Cell. Mol. Life Sci.* 61, 502–509. doi: 10.1007/s00018-003-3394-z

161. Shan, W., Cao, M., Leung, D., and Tyler, B. M. (2004). The Avr1b locus of *Phytophthora sojae* encodes an elicitor and a regulator required for avirulence on soybean plants carrying resistance gene Rps1b. *Mol. Plant Microbe Interact.* 17, 394–403. doi: 10.1094/MPMI.2004.17.4.394

162. Shen, Q. H., Saijo, Y., Mauch, S., Biskup, C., Bieri, S., Keller, B., et al. (2007). Nuclear activity of MLA immune receptors links isolate-specific and basal disease-resistance responses. *Science* 315, 1098–1103. doi: 10.1126/science.1136372

163. Shimizu, T., Nakano, T., Takamizawa, D., Desaki, Y., Ishii-Minami, N., Nishizawa, Y., et al. (2010). Two LysM receptor molecules, CEBiP and OsCERK1, cooperatively regulate chitin elicitor signaling in rice. *Plant J.* 64, 204–214. doi: 10.1111/j.1365-313X.2010.04324.x

164. Shinya, T., Motoyama, N., Ikeda, A., Wada, M., Kamiya, K., Hayafune, M., et al. (2012). Functional characterization of CEBiP and CERK1 homologs in *Arabidopsis* and rice reveals the presence of different chitin receptor systems in plants. *Plant Cell Physiol.* 53, 1696–1706. doi: 10.1093/pcp/pcs113

165. Shirasu, K. (2009). The HSP90-SGT1 chaperone complex for NLR immune sensors. *Annu. Rev. Plant Biol.* 60, 139–164. doi: 10.1146/annurev.arplant.59.032607.092906

166. Saitoh, H., Fujisawa, S., Mitsuoka, C., Ito, A., Hirabuchi, A., Ikeda, K., et al. (2012). Large-scale gene disruption in *Magnaporthe oryzae* identifies MC69, a secreted protein required for infection by monocot and dicot fungal pathogens. *PLoS Pathog.*8:e1002711. doi: 10.1371/journal.ppat.1002711

167. Song, J., Win, J., Tian, M., Schornack, S., Kaschani, F., Ilyas, M., et al. (2009). Apoplastic effectors secreted by two unrelated eukaryotic plant pathogens target the tomato defense protease Rcr3. *Proc. Natl. Acad. Sci. U.S.A.* 106, 1654–1659. doi: 10.1073/pnas.0809201106

168. Soyer, J. L., El Ghalid, M., Glaser, N., Ollivier, B., Linglin, J., Grandaubert, J., et al. (2014). Epigenetic control of effector gene expression in the plant pathogenic fungus *Leptosphaeria maculans. PLoS Genet.* 10:e1004227. doi: 10.1371/journal.pgen.1004227

169. Sperschneider, J., Dodds, P. N., Gardiner, D. M., Manners, J. M., Singh, K. B., and Taylor, J. M. (2015a). Advances and challenges in computational prediction of effectors from plant pathogenic fungi. *PLoS Pathog.* 11:e1004806. doi: 10.1371/journal.ppat.1004806

170. Sperschneider, J., Gardiner, D. M., Dodds, P. N., Tini, F., Covarelli, L., Singh, K. B., et al. (2015b). EffectorP: predicting fungal effector proteins from secretomes using machine learning. *New Phytol.* 210, 743–761. doi: 10.1111/nph.13794

171. Spoel, S. H., Mou, Z., Tada, Y., Spivey, N. W., Genschik, P., and Dong, X. (2009). Proteasome-mediated turnover of the transcription coactivator NPR1 plays dual roles in regulating plant immunity. *Cell* 137, 860–872. doi: 10.1016/j.cell.2009.03.038

172. Stergiopoulos, I., and de Wit, P. J. (2009). Fungal effector proteins. *Ann. Rev. Phytopathol.* 47, 233–263. doi: 10.1146/annurev.phyto.112408.132637

173. Stotz, H. U., Mitrousia, G. K., de Wit, P. J., and Fitt, B. D. (2014). Effector-triggered defence against apoplastic fungal pathogens. *Trends Plant Sci.* 19, 491–500. doi: 10.1016/j.tplants.2014.04.009

174. Sweigard, J. A., Carroll, A. M., Kang, S., Farral, L., Chumely, F. G., and Valent, B. (1995). Identification, cloning and characterization of PWL2, a gene for host-species specificity in the rice blast fungus. *Plant Cell* 7, 1221–1233. doi: 10.1105/tpc.7.8.1221

175. Tanaka, S., Brefort, T., Neidig, N., Djamei, A., Kahnt, J., Vermerris, W., et al. (2014). A secreted *Ustilago maydis* effector promotes virulence by targeting anthocyanin biosynthesis in maize. *eLife* 3:e01355. doi: 10.7554/eLife.01355

176. Tian, M., Benedetti, B., and Kamoun, S. (2005). A second Kazal-like protease inhibitor from *Phytophthora infestans* inhibits and interacts with the apoplastic

pathogenesis-related protease P69B of tomato. *Plant Physiol.* 138, 1785–1793. doi: 10.1104/pp.105.061226

177. Tian, M., Huitema, E., Da Cunha, L., Torto-Alalibo, T., and Kamoun, S. (2004). A Kazal-like extracellular serine protease inhibitor from *Phytophthora infestans* targets the tomato pathogenesis-related protease P69B. *J. Biol. Chem.* 279, 26370–26377. doi: 10.1074/jbc.M400941200
178. Tian, M., Win, J., Song, J., van der Hoorn, R., van der Knaap, E., and Kamoun, S. (2007). A *Phytophthora infestans* cystatin-like protein targets a novel tomato papain-like apoplastic protease. *Plant Physiol.* 143, 364–377. doi: 10.1104/pp.106.090050
179. Tyler, B. M., and Rouxel, T. (2013). *Effectors of Fungi and Oomycetes: Their Virulence and Avirulence Functions and Translocation from Pathogen to HOST Cells.* Ames, IA: John Wiley and Sons, Inc.
180. Tyler, B. M., Tripathy, S., Zhang, X., Dehal, P., Jiang, R. H., Aerts, A., et al. (2006). *Phytophthora genome sequences uncover evolutionary origins and mechanisms of pathogenesis. Science* (New York, N.Y.) 313, 1261–1266. doi: 10.1126/science.1128796
181. Upadhyaya, N. M., Mago, R., Staskawicz, B. J., Ayliffe, M. A., Ellis, J. G., and Dodds, P. N. (2014). A bacterial type III secretion assay for delivery of fungal effector proteins into wheat. *Mol. Plant Microbe Interact.* 27 255–264. doi: 10.1094/mpmi-07-13-0187-fi
182. Van de Wouw, A. P., Lowe, R. G., Elliott, C. E., Dubois, D. J., and Howlett, B. J. (2014). An avirulence gene, AvrLmJ1, from the blackleg fungus, *Leptosphaeria maculans*, confers avirulence to *Brassica juncea* cultivars. *Mol. Plant Pathol.* 15, 523–530. doi: 10.1111/mpp.12105
183. Van de Wouw, A. P., Stonard, J. F., Howlett, B. J., West, J. S., Fitt, B. D. L., and Atkins, S. D. (2010). Determining frequencies of avirulent alleles in airborne *Leptosphaeria maculans* inoculum using quantitative PCR. *Plant Pathol.* 59, 809–818. doi: 10.1111/j.1365-3059.2010.02311.x
184. van den Burg, H. A., Harrison, S. J., Joosten, M. H., Vervoort, J., and de Wit, P. J. (2006). *Cladosporium fulvum* Avr4 protects fungal cell walls against hydrolysis by plant chitinases accumulating during infection. *Mol. Plant Microbe Interact.* 19, 1420–1430. doi: 10.1094/MPMI-19-1420
185. van den Hooven, H. W., Van Den Burg, H. A., Vossen, P., Boeren, S., DeWit, P. J. G. M., and Vervoort, J. (2001). Disulfide bond structure of the AVR9 elicitor of the fungal tomato pathogen *Cladosporium fulvum*: evidence for a cystine knot. *Biochemistry*40, 3458–3466. doi: 10.1021/bi0023089
186. van Esse, H. P., Van't Klooster, J. W., Bolton, M. D., Yadeta, K. A., van Baarlen, P., Boeren, S., et al. (2008). The *Cladosporium fulvum* virulence protein Avr2 inhibits host proteases required for basal defense. *Plant Cell* 20, 1948–1963. doi: 10.1105/tpc.108.059394
187. van Kan, J. A. L., van den Ackerveken, G. F. J. M., and de Wit, P. J. (1991). Cloning and characterization of cDNA of avirulence gene avr9 of the fungal pathogen *Cladosporium fulvum*, causal agent of tomato leaf mold. *Mol. Plant Microbe Interact.* 4, 52–59. doi: 10.1094/MPMI-4-052
188. van Poppel, P. M. J. A., Guo, J., van de Vondervoort, P. J. I., Jung, M. W. M., Birch, P. R. J., Whisson, S. C., et al. (2008). The*Phytophthora infestans* avirulence gene Avr4 encodes an RXLR-DEER effector. *Mol. Plant Microbe Interact.* 21, 1460–1470. doi: 10.1094/MPMI-21-11-1460
189. van Poppel, P. M., Jiang, R. H., Sliwka, J., and Govers, F. (2009). Recognition of *Phytophthora infestans* Avr4 by potato R4 is triggered by C-terminal domains comprising W motifs. *Mol. Plant Pathol.* 10, 611–620. doi: 10.1111/j.1364-3703.2009.00556.x
190. Vleeshouwers, V. G., Rietman, H., Krenek, P., Champouret, N., Young, C., Oh, S. K., et al. (2008). Effector genomics accelerates discovery and functional profiling of potato disease resistance and *Phytophthora infestans* avirulence genes. *PLoS One* 3:e2875. doi: 10.1371/journal.pone.0002875
191. Walbot, V., and Skibbe, D. S. (2010). Maize host requirements for *Ustilago maydis* tumor induction. *Sex. Plant Reprod.* 2, 1–13. doi: 10.1007/s00497-009-0109-0
192. Wang, C. I., Guncar, G., Forwood, J. K., Teh, T., Catanzariti, A. M., Lawrence, G. J., et al. (2007). Crystal structures of flax rust avirulence proteins AvrL567-A and -D reveal details of the structural basis for flax disease resistance specificity. *Plant Cell* 19, 2898–2912. doi: 10.1105/tpc.107.053611
193. Wang, Q., Han, C., Ferreira, A. O., Yu, X., Ye, W., Tripathy, S., et al. (2011). Transcriptional programming and functional interactions within the *Phytophthora sojae* RXLR effector repertoire. *Plant Cell* 23, 2064–2086. doi: 10.1105/tpc.111.086082
194. Wang, Y. H., and Irving, H. R. (2011). Developing a model for plant hormone interaction. *Plant Signal. Behav.* 6, 494–500. doi: 10.4161/psb.6.4.14558
195. Westerink, N., Brandwagt, B. F., de Wit, P. J., and Joosten, M. H. (2004). *Cladosporium fulvum* circumvents the second functional resistance gene homologue at the Cf-4 locus (Hcr9-4E) by secretion of a stable avr4E isoform. *Mol. Microbiol.* 54, 533–545. doi: 10.1111/j.1365-2958.2004.04288.x
196. Whigham, E., Qi, S., Mistry, D., Surana, P., Xu, R., Fuerst, G., et al. (2015). Broadly conserved fungal effector BEC1019 suppresses host cell death and enhances pathogen virulence in powdery mildew of Barley (*Hordeum vulgare* L.). *Mol. Plant Microbe Interact.* 28, 968–983. doi: 10.1094/MPMI-02-15-0027-FI
197. Whisson, S. C., Boevink, P. C., Moleleki, L., Avrova, A. O., Morales, J. G., Gilroy, E. M., et al. (2007). A translocation signal for delivery of oomycete effector proteins into host plant cells. *Nature* 450, 115–118. doi: 10.1038/nature06203
198. Win, J., Chaparro-Garcia, A., Belhaj, K., Saunders, D. G., Yoshida, K., Dong, S., et al. (2012). Effector biology of plant-associated organisms: concepts and

perspectives. *Cold Spring Harb. Symp. Quant. Biol.* 77, 235–247. doi: 10.1101/sqb.2012.77.015933

199. Wirth, D., Gama-Norton, L., Riemer, P., Sandhu, U., Schucht, R., and Hauser, H. (2007). Road to precision: recombinase-based targeting technologies for genome engineering. *Curr. Opin. Biotechnol.* 18, 411–419. doi: 10.1016/j.copbio.2007.07.013
200. Xu, S., Chen, J., Liu, L., Wang, X., Huang, X., and Zhai, Y. (2007). Proteomics associated with virulence differentiation of*Curvularia lunata* in maize in China. *J. Integr. Plant Biol.* 49, 487–496. doi: 10.1111/j.1744-7909.2007.00469.x
201. Yaeno, T., Li, H., Chaparro-Garcia, A., Schornack, S., Koshiba, S., Watanabe, S., et al. (2011). Phosphatidylinositol monophosphate-binding interface in the oomycete RXLR effector AVR3a is required for its stability in host cells to modulate plant immunity. *Proc. Natl. Acad. Sci. U.S.A.* 108, 14682–14687. doi: 10.1073/pnas.1106002108
202. Yin, C., Downey, S. I., Klages-Mundt, N. L., Ramachandran, S., Chen, X., Szabo, L. J., et al. (2015). Identification of promising host-induced silencing targets among genes preferentially transcribed in haustoria of Puccinia. *BMC Genomics* 16:1791. doi: 10.1186/s12864-015-1791-y
203. Yin, W., Dong, S., Zhai, L., Lin, Y., Zheng, X., and Wang, Y. (2013). *Phytophthora sojae* Avr1d genes encode an RxLR-DEER effector with presence and absence polymorphisms among pathogen strains. *Mol. Plant Microbe Interact.* 26, 958–968. doi: 10.1094/MPMI-02-13-0035-R
204. Yoshida, K., Saitoh, H., Fujisawa, S., Kanzaki, H., Matsumura, H., Yoshida, K., et al. (2009). Association genetics reveals three novel avirulence genes from the rice blast fungal pathogen *Magnaporthe oryzae. Plant Cell* 21, 1573–1591. doi: 10.1105/tpc.109.066324
205. Zhang, M., Kadota, Y., Prodromou, C., Shirasu, K., and Pearl, L. H. (2010). Structural basis for assembly of Hsp90-Sgt1-CHORD protein complexes: implications for chaperoning of NLR innate immunity receptors. *Mol. Cell* 39, 269–281. doi: 10.1016/j.molcel.2010.05.010
206. Zhang, S., and Xu, J. R. (2014). Effectors and effector delivery in *Magnaporthe oryzae. PLoS Pathog.* 10, e1003826. doi: 10.1371/journal.ppat.1003826
207. Zhang, Z. M., Zhang, X., Zhou, Z. R., Hu, H. Y., Liu, M., Zhou, B., et al. (2013). Solution structure of the *Magnaporthe oryzae*avirulence protein AvrPiz-t. *J. Biomol. NMR* 55, 219–223. doi: 10.1007/s10858-012-9695-5
208. Zhang, Z., Mao, Y., Ha, S., Liu, W., Botella, J. R., and Zhu, J. K. (2015). A multiplex CRISPR/CAS9 platform for fast and efficient editing of multiple genes in *Arabidopsis. Plant Cell Rep.* doi: 10.1007/s00299-015-1900-z. [Epub ahead of print].
209. Zhou, B., Qu, S., Liu, G., Dolan, M., Sakai, H., Lu, G., et al. (2006). The eight amino-acid differences within three leucine-rich repeats between Pi2 and Piz-t resistance proteins determine the resistance specificity to *Magnaporthe grisea. Mol. Plant Microbe Interact.* 19, 1216–1228. doi: 10.1094/MPMI-19-1216
210. Zhu, S., Li, Y., Vossen, J. H., Visser, R. G., and Jacobsen, E. (2012). Functional stacking of three resistance genes against*Phytophthora infestans* in potato. *Transgenic Res.* 21, 89–99. doi: 10.1007/s11248-011-9510-1

CHAPTER 5 GENETIC COMPLEXITY OF PATHOGEN PERCEPTION BY PLANTS

Mark S. Dixon, Catherine Golstein, Colwyn M. Thomas, Erik A. van der Biezen, and Jonathan D. G. Jones

VIRULENCE AND AVIRULENCE

It would be surprising if pathogens were to carry *Avr* genes that had no function other than to enable recognition by plants that carry the matching *R* genes. During infection, pathogens make an array of virulence factors. If plant *R* genes can evolve to recognize one of these molecules to trigger a defense response, any such virulence factor could become an avirulence determinant. *Avr* genes of the bacterial plant pathogens *Xanthomonas campestris* and *Pseudomonas syringae* encode hydrophilic proteins that are delivered inside the plant cell by a specialized type III secretion mechanism (2). In the absence of recognition by corresponding *R* genes, some of these *Avr* genes have been shown to confer enhanced virulence (3, 4). Viral pathogens also present potential ligands intracellularly, and for some *R* genes, the corresponding viral Avr protein has been defined as the replicase domain (5) or the coat protein (6).

We study resistance to two biotrophic pathogens, the tomato leaf mold fungus *Cladosporium fulvum* and the oomycete *Peronospora parasitica*, which causes downy mildew on cruciferous plants, including *Arabidopsis*. Studies on *C. fulvum,* which colonizes the intercellular spaces of infected leaves, led to the cloning of *Avr4*and *Avr9* that confer avirulence on tomato plants carrying the corresponding *Cf-4* and *Cf-9 R* genes (see below). *Avr4* and *Avr9* encode small secreted peptides that trigger *Cf* gene-dependent resistance (7, 8). No evidence to support a role for Avr4 or Avr9 as virulence determinants has been demonstrated. In contrast, ECP2 is a secreted peptide that has a virulence function in all *C. fulvum* strains analyzed (9) and is recognized as an avirulence factor in certain tomato lines (10). *P. parasitica*, like many other parasites, forms hyphae that produce feeding structures inside the plant cell wall, termed haustoria, that provide intimate association with the plant plasma membrane (11). Avr effectors may be delivered through haustoria inside the plant cell. This is plausible because their cognate R proteins are predicted to be intracellular (see next section).

R PROTEINS

Tomato *Cf-2*, *Cf-4*, *Cf-5*, and *Cf-9* genes confer recognition of different *C. fulvum Avr* genes. They encode an extracellular leucine-rich repeat (eLRR) domain, a transmembrane domain, and a short cytoplasmic domain with no sequence similarity to known signaling domains (Fig. 1*A*) (12). Cf-9 has recently been shown to be a plasma membrane-localized glycoprotein (13). *Xa21* is a rice gene that confers resistance to bacterial blight caused by *X. campestris* pv. *oryzae* and also encodes

Citation: Mark S. Dixon, Catherine Golstein, Colwyn M. Thomas, Erik A. van der Biezen, and Jonathan D. G. Jones, Genetic complexity of pathogen perception by plants, doi: 10.1073/pnas.97.16.8807

eLRRs and a transmembrane domain, but in addition has a cytoplasmic Ser/Thr protein kinase domain (Fig. 1*B*) (14). The presence of eLRRs in these R proteins is consistent with a presumed role in the recognition of extracellular ligands. Interestingly,*Pto*, a tomato gene that confers resistance to *P. syringae* pv. *tomato* strains expressing the *AvrPto* gene, also encodes a Ser/Thr protein kinase (Fig. 1*C*) (15). Pto lacks a signal peptide (but carries a putative myristoylation site), and its interaction with AvrPto in yeast suggests a cytoplasmic recognition capacity (16,17).

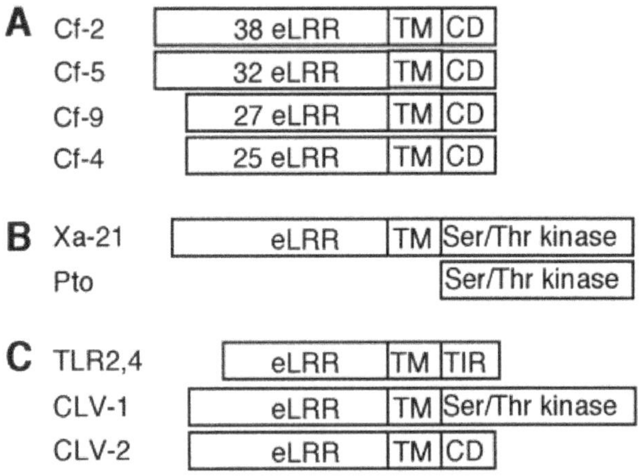

Figure 1: Tomato Cf proteins and structurally related proteins. (*A*) Cf proteins contain different numbers of N-terminal eLRRs, a transmembrane (TM) domain, and a short cytoplasmic domain (CD). (*B*) Rice Xa-21 and tomato Pto bacterial R proteins. (*C*) eLRR-TM domain-containing proteins with different cytoplasmic domains (CD). See text for descriptions.

Both the Cf class and Xa21 resemble the *Drosophila* Toll receptor and its human homologs, the Toll-like receptors (TLRs) (18). TLR2 and TLR4 (Fig. 1*C*) function in innate immunity and activate microbial defense pathways to trigger inflammatory responses after recognition of conserved molecular structures of distinct microbial pathogen classes (19). TLRs have therefore been designated "pattern recognition receptors" (20) and serve an analogous role to plant R proteins in detecting pathogen molecules and activating defense pathways. Interestingly, TLR-mediated inflammatory responses resemble plant defense responses, including production of antimicrobial active oxygen species and nitric oxide and ultimately cell death (21,22). The significance of these homologies for plant R protein mechanisms has yet to be established, but it is tempting to suggest that, whereas vertebrates use the TLRs to recognize conserved structures of different pathogen classes, plants have evolved these to detect strain-specific pathogen Avr products.

Arabidopsis RPP1 and *RPP5* genes confer resistance to *P. parasitica* and belong to the largest class of *R*genes that encode nucleotide-binding leucine-rich repeat (NB-LRR) proteins (Fig. 2*A*). Genome sequencing has shown that approximately 200 NB-LRR-encoding genes are present in *Arabidopsis* (23). NB-LRR proteins can be divided into two subclasses on the basis of their N-terminal domain. The leucine zipper (LZ)-NB-LRR is a broad class of NB-LRR proteins (with perhaps two subclasses) that contains an N-terminal putative heptad LZ or coiled-coil domain (23). Members of this class have been identified for resistance to bacteria, viruses, fungi, oomycetes, and even nematodes and aphids. Examples include*Arabidopsis* RPS2, RPM1, and RPS5 for resistance to *P. syringae* (24), RPP8 for resistance to *P. parasitica*(25), and maize Rp1 for resistance to *Puccinia sorghi* (26).

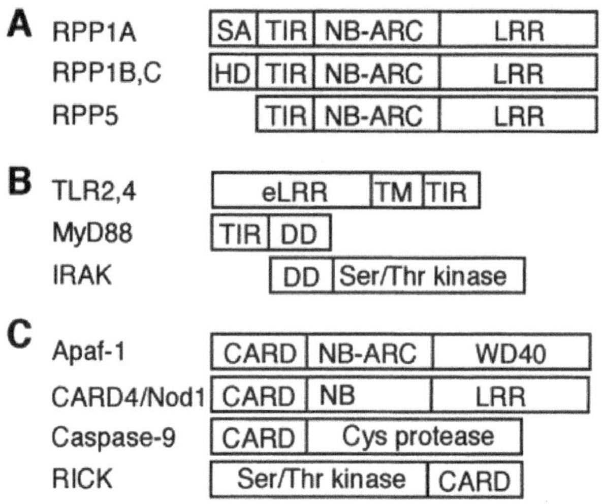

Figure 2: *Arabidopsis* RPP1, RPP5, and structurally related proteins. (*A*) RPP1 and RPP5 have an N-terminal TIR domain, a nucleotide-binding Apaf-1, R proteins and CED4 homology (NB-ARC) domain, and a LRR domain. The RPP1 family differ by their N-terminal domains, which are absent in RPP5: RPP1A has a putative signal anchor (SA) domain, RPP1B,C have hydrophobic domains (HD). (*B*) Human proteins that function in the NF-κB pathway. eLRR, extracellular LRR domain; TM, transmembrane domain. (*C*) Human proteins with CARDs that function in NF-κB and/or apoptosis pathways. See text for descriptions.

Members of the TIR-NB-LRR class carry N-terminal homology with the cytoplasmic domain of the Toll and interkeukin-1 receptors (Fig. 2*B*) involved in vertebrate innate immunity (27, 28). Besides RPP1 and RPP5, this class also includes tobacco N for resistance to tobacco mosaic virus, flax L6 for resistance to flax rust, and *Arabidopsis* RPS4 for resistance to *P. syringae* (24, 29). After activation, the TIR domain of human TLR2 and TLR4 binds, through homophilic interactions, to the TIR domain of the MyD88 adaptor protein (Fig. 2*B*) (30). MyD88 also carries a death domain (DD) that then binds to the DD of the Ser/Thr protein kinase IRAK (the human ortholog of *Drosophila* Pelle, and interestingly, also homologous to tomato Pto; Fig.1*B*), leading to the translocation of the transcription factor NF-κB and induction of the inflammatory response (27). These homologies may suggest a role for the TIR domain of the plant R proteins in homophilic TIR–TIR interactions with TIR domain-containing plant adaptor proteins, but these have yet to be reported in plants.

An intriguing homology has been noted between plant NB-LRR proteins and the apoptotic adaptor CED4 from *Caenorhabditis elegans* (31). This region of homology also extends to its human ortholog Apaf-1 (Fig.2*C*) (32) and has been designated the NB-ARC domain (Fig. 2*C*) (33) or Ap-ATPase domain (34). The NB-ARC domain is an ancient motif, because it is also present in proteins of Gram-positive bacteria (34). Another protein that resembles plant NB-LRR proteins is human caspase recruitment domain (CARD)4/Nod1 (35, 36), which contains LRRs and a NB site (the homology to the rest of the NB-ARC domain is less convincing). Instead of TIR or LZ domains at their N termini, Apaf-1 and CARD4/Nod1 carry CARDs (Fig. 2*C*). Induction of apoptosis by both Apaf-1 and CARD4/Nod1 occurs through homophilic CARD–CARD interactions with caspase-9, a CARD-containing cysteine protease (36, 37). CARD4/Nod1, in addition, plays a role in activation of the NF-κB pathway through homophilic CARD–CARD interactions with the Ser/Thr kinase RICK (Fig. 2*C*) (35, 36). Caspases and CARDs have not yet been reported in plants; however, the use of peptide inhibitors specific to caspases in animal cells suggests plants may possess caspase activity (38). Although the significance of these homologies is not clear, they provide interesting paradigms for NB-LRR protein mechanisms. For example, the WD40 repeat domain of Apaf-1 negatively regulates its NB-ARC domain, which is alleviated after binding to cytochrome *c*, then allowing ATP-dependent oligomerization and caspase-9 activation (37). By analogy, pathogen Avr signals perceived by the LRR domain of plant NB-LRR proteins might alleviate negative regulation of their NB-ARC domain such that ATP hydrolysis and oligomerization can be initiated, possibly leading to homo- or heterodimerization (through homophilic interactions) of the N-terminal TIR or LZ domains.

R GENE EVOLUTION

Plants can activate both localized and systemic defense mechanisms in response to pathogen infection. This activation is often a consequence of R protein perception of a pathogen-encoded Avr determinant in cells targeted by the pathogen (24). This contrasts with the situation in mammals where circulating defender cells migrate to the sites of pathogen ingress. Furthermore, in mammals, the elaboration of recognition capacity and the consequent resistance to an array of potential pathogens occurs somatically through the adaptive immune system and germinally in genes encoding the MHC (39). In plants, *R* gene sequence evolution takes place exclusively through the germline. Population geneticists have proposed that in natural populations, where plants and their pathogens coevolve, disease is kept in check by "balancing polymorphisms" at *R* gene loci (40). In this model, no single *R* gene allele is present at high frequency because, as a result of selection pressure, it may be overcome by a novel pathogen variant (40). For example, *R* gene monocultures in crops eventually succumb to pathogen infection, and breeding durable disease resistance remains a major challenge. Overdominance, or heterozygote advantage, has been proposed to explain the polymorphism at the human MHC locus. However, overdominance cannot explain polymorphism at *R* gene loci in inbreeding species. We and others have proposed that frequency-dependent selection may account for the maintenance of sequence polymorphism at the *Arabidopsis RPP5*,*RPM1*, and other *R* gene loci (41, 42).

Recent analyses have revealed the molecular organization of a number of *R* gene loci and provided insight to the molecular mechanisms that generate sequence diversity. We have analyzed two *Cf* loci in tomato and two *RPP* loci in *Arabidopsis*. The *Cf-0*, *Cf-9*, and *Cf-4* haplotypes of tomato contain distinct complements of tandemly duplicated *Cf-9* homologs (*Hcr9*s; refs. 43 and 44). Several *Hcr9*s have been shown to confer resistance to *C. fulvum* through recognition of distinct Avr determinants (43, 45). A similar comparison has revealed the polymorphic nature of *Cf-2* homologs (*Hcr2*s) at the tomato *Cf-2*/*Cf-5* locus (46). The sequence polymorphism at *Cf* loci is because of interspecific variation (12). However, in *Arabidopsis* the*RPP1* and *RPP5* loci (41, 47), which are also comprised of linked multigene families, show pronounced intraspecific polymorphism when different landraces are compared by DNA gel blot analysis (Fig. 3).

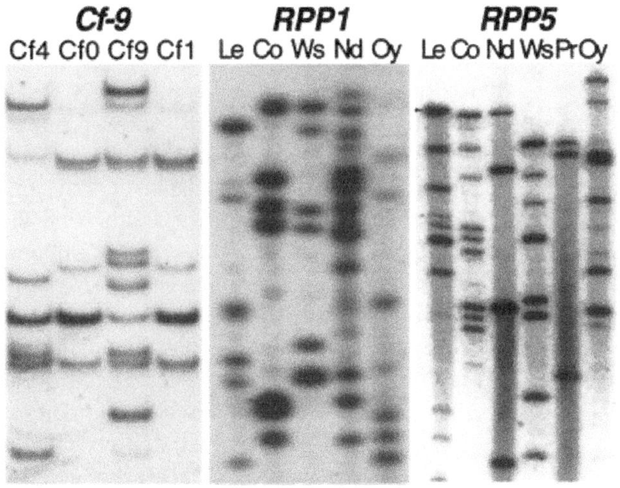

Figure 3: Polymorphic *R* gene families revealed by DNA gel blot analysis. Comparisons of tomato lines with introgressed chromosome segments containing different *Cf* genes show interspecific polymorphisms (44). Comparisons of different *Arabidopsis* landraces show intraspecific polymorphisms at the *RPP1*(47) and *RPP5* loci (48).

DNA sequence analysis of tomato *Hcr9*s and *Hcr2*s (43, 46), the *Arabidopsis RPP1*, *RPP5*, and *RPP8*families (25, 41, 47, 48) and NB-LRR homologs at the lettuce *Dm3* locus (49) have recently been reported. Like *RPP5* and *RPP1* homologs, *Hcr2*s contain variable numbers of LRRs, suggesting this is an important mechanism to generate novel recognition specificities (12, 46). *Hcr9*s, however, contain similar numbers of LRRs, and most sequence variation occurs in nucleotides that encode the putative solvent-exposed amino acids of a conserved β-strand/β-turn structural motif of LRR proteins. These amino acids are thought to be critical for the recognition specificity of ligand binding in LRR proteins (50). Sequence comparisons of NB-LRR multigene families such as *RPP1*, *RPP5*, and *Dm3* have also shown high ratios of nonsynonymous to synonymous substitutions in sequences encoding the corresponding amino acids of their LRRs. These sequences within the LRRs have clearly undergone diversifying selection, whereas sequences encoding putative signaling domains have undergone purifying selection (23, 41, 47).

As in studies on the evolution of the mammalian MHC locus (39), the relative contribution of interallelic recombination, gene conversion, unequal crossing over, and the accumulation of point mutations in generating *R* gene diversity is controversial. Classical genetic experiments suggested that interallelic sequence exchange at the flax *L* locus (51) or chromosomal mispairing and unequal crossing over between homologs at the maize *Rp1* locus (52) could generate *R* gene variation. This was subsequently verified by molecular analyses of the *L* and *Rp1* loci (26, 51). At some *R* gene loci, a "birth-and-death" model of *R* gene evolution has been proposed (53), based on a model for evolution of the mammalian MHC locus (54). In this model, expansion or contraction of members of a multigene family is caused by unequal crossing over, and evolution of alleles proceeds by sequence exchange between gene orthologs, random mutation and selection.

A comparison of the *Cf-0*, *Cf-4* and *Cf-9* haplotypes in tomato showed that the copy number of *Hcr9*s at this locus can vary. In the *Cf-4* and *Cf-9* haplotypes (each containing five *Hcr9*s), evidence for extensive intergenic sequence exchange between *Hcr9* paralogs was shown that generated recombinant *Hcr9*s. This sequence exchange could have contributed to the creation of novel recognition specificities. Also, the order of putative gene orthologs differed in the two haplotypes (43) and is inconsistent with the proposal that negligible sequence exchange occurs between paralogs. *Hcr9*s have also accumulated multiple point mutations that generate sequence diversity in the solvent-exposed residues of their LRRs (see above), and this appears to be another important source of sequence variation. The comparison of *Col-0* and *La-er*haplotypes of the *Arabidopsis RPP5* locus also revealed extensive sequence exchange between *RPP5*homologs (41). With the exception of homologs at the ends of the cluster, orthologous genes were difficult to discern. Furthermore, many "mutated" homologs were identified. It is possible that these sequences provide a reservoir for intergenic sequence exchange and the generation of sequence diversity.

It is likely that the rate and molecular mechanism for generating *R* gene novelty will depend on a number of factors. Further insight into these mechanisms will come from comparative analysis of natural *R* gene variants such as the numerous recombinant *Rp1* alleles in maize (26, 52). These analyses should reveal the range of molecular mechanisms that generate *R* gene diversity in different plants, at different *R* gene loci within plants, and the molecular basis for R protein-mediated pathogen perception.

RESULTS AND DISCUSSION

Mutational Analysis Identifies *Rcr3*, a Gene Required for *Cf-2* Function.

Mutational analysis is a powerful tool to dissect signaling pathways and has been used to identify genes required for *R* gene function. Mutational analysis of genes required for *Cf-9* function has revealed two loci required for full *C. fulvum* resistance, *Rcr1* and *Rcr2* (55). Here, we report the results of mutagenesis of the tomato Cf2 line. Twenty-five M_2 seeds from 3,200 diepoxybutane- or ethyl methanesulfonate-treated Cf2 M_1 plants were screened for their response to *C. fulvum* race 4 β-glucuronidase (GUS) infection, and four M_2 families that segregated for resistance or sensitivity to infection were identified (Table 1). PCR analysis of disease-sensitive plants confirmed these plants contained *Cf-2*. To determine the genetic basis of disease sensitivity in these mutants, all four were crossed to the Cf0 near-isogenic line that lacks known resistance genes to *C. fulvum*. In crosses to Cf0, mutants at *Cf-2* should give susceptible F_1 progeny as neither parent would have contained a functional gene for resistance to *C. fulvum* infection. Crosses between the four mutant lines and Cf0 gave resistant F_1 progeny, confirming the mutations were not in *Cf-2*. Progeny from intercrosses between all four mutants were sensitive to *C. fulvum* race 4 GUS infection, demonstrating that these mutations are allelic. This gene has been designated *Rcr3* (required for *C. fulvum* resistance *3*; Table 1).

Table 1: *C. fulvum*-sensitive Cf2 mutants

Mutant allele	Mutagen	Frequency¶
rcr3-1*	DEB‡	5/23
rcr3-2*	EMS§	2/22
rcr3-3†	EMS	5/23
rcr3-4†	EMS	4/23

*Identified with fluorometric GUS assay.
†Identified by visual inspection.
‡Diepoxybutane.
§Ethyl methanesulfonate.
¶Number of susceptible individuals in total M_2 family.

Strong and Weak Mutant Alleles of Rcr3.

Inoculation with *C. fulvum* race 5 revealed that rcr3–2 (the line carrying the *rcr3–2* mutation) and rcr3–3 allowed as much fungal growth as Cf0, whereas rcr3–1 appeared significantly less susceptible (Fig. 4). From 2 weeks after inoculation, *C. fulvum* growth on rcr3–1 was intermediate to Cf0 and Cf2 (Fig. 4*B*). Fungal development was barely detectable on rcr3–1 at 1 week after inoculation, whereas hyphae had already invaded the mesophyll layers and were proliferating around the vascular tissue of rcr3–2, rcr3–3, and Cf0 cotyledons (Fig. 5). Microscopic inspection of 5-bromo-4-chloro-3-indolyl-β-D-glucuronic acid-stained tissue confirmed that the fungus was able to sporulate on all three mutants (not shown).

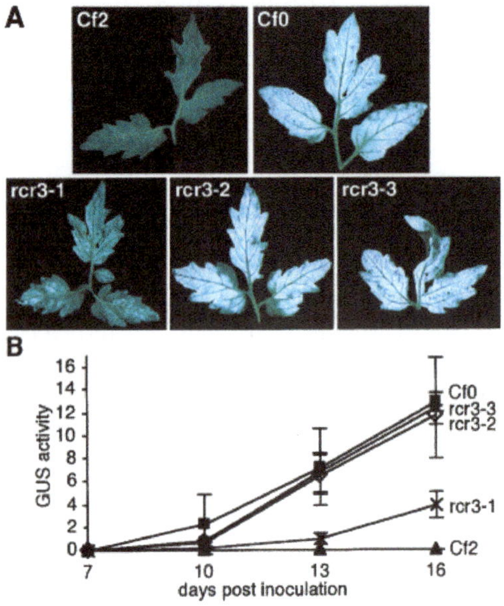

Figure 4: Infection phenotypes of rcr3 mutants. Seedlings were immersed in a solution containing 3.5 × 10^5 ml^{-1}spores of the appropriate *C. fulvum* race. Inoculated plants were kept for 3 to 6 days in a phytochamber at saturated humidity and subsequently at 70% relative humidity, 24°C, during the 16-hr day–light, 18°C overnight. (*A*) Comparison of infection phenotype of the rcr3 mutants and the Cf0-sensitive and Cf2-resistant control lines. Plants were inoculated with *C. fulvum* race 5 15 days after sowing, and leaves were photographed 17 days after inoculation. Fungal colonization is visible on the abaxial surface of the leaves as white mycelium. (*B*) Comparison of fungal biomass accumulation in rcr3–1 (X), rcr3–2 (∘), rcr3–3 (•), Cf0 (■), and Cf2 (▲) lines by using a fluorometric GUS assay (56) over a time course after inoculation. Twelve-day-old plants were inoculated with *C. fulvum* race 4 GUS (57). No GUS activity (4-methyl umbelliferone mg^{-1} protein min^{-1}) was detected on Cf2 controls in contrast to Cf0 controls. Three cotyledons were harvested and analyzed separately for each sample at each time point (the mean value for each set of triplicate samples is shown and error bars indicate the standard deviation).

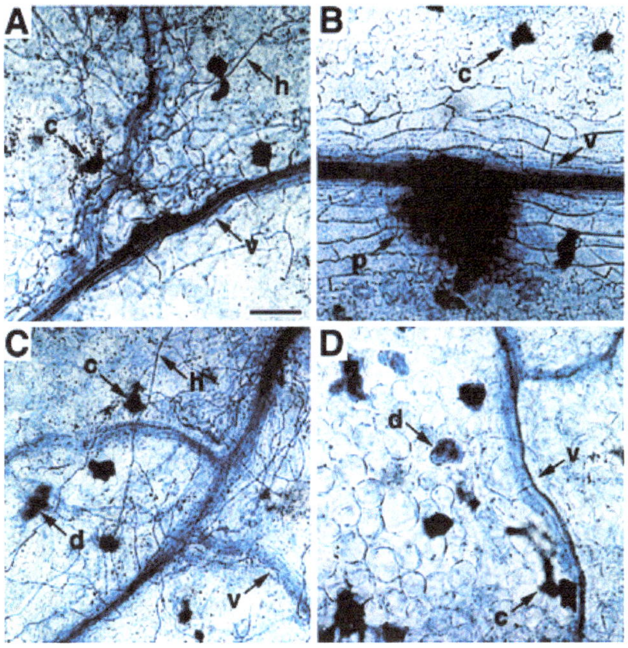

Figure 5: Comparison of fungal development and plant cell death response during the *C. fulvum* infection of rcr3 mutants and control lines. At least three cotyledons and two leaves were sampled every 3 days from each genotype during the experiment described in Fig. 4*A* and were stained with lactophenol-trypan blue (58). Photomicrographs of cotyledons were taken 7 days after inoculation at the same magnification. Scale bar = 100 µm. (*A*) Fungal growth in the Cf0 line: *C. fulvum* hyphae (h) tend to localize close to plant veins (v) where they start swelling. (*B*) Characteristic early resistance response in the Cf2 line: patches of dead mesophyll cells (p) develop at the vicinity of the plant veins, whereas the fungus is undetectable. (*C*) Infection phenotype of rcr3–3, and rcr3–2 (not shown): once the fungus has penetrated the leaf apoplast, the fungus develops as in Cf0. However, these fungal foci appear to occur at a lower frequency on rcr3–3 and rcr3–2 than on Cf0. (*D*) rcr3–1 intermediate phenotype: although rcr3–1 allows sparse fungal growth in some areas; the fungus remains undetectable on the major part of the cotyledon. Discrete dead cells (d) were scattered around the upper mesophyll of the mutants but not in wild type, irrespective of *C. fulvum* infection (noninoculated material not shown). This cell death phenotype was poorly reproducible. Calcium oxalate crystals (c) in some mesophyll cells are common in tomato leaves. These phenotypes were observed in three independent infections with *C. fulvum* race 5 and in an infection with *C. fulvum* race 4 GUS, where over 300 infection sites were analyzed in cotyledon and leaf samples.

The rcr3–1, rcr3–2, and rcr3–3 mutants were each back crossed to Cf2 and selfed to produce F_2 seed. Approximately one-quarter of the F_2 progeny from crosses between Cf2 and rcr3–2 or rcr3–3 appeared to be fully susceptible to *C. fulvum* race 4 GUS infection. One-quarter of the F_2 progeny from the Cf2 × rcr3–1 cross showed an intermediate level of disease sensitivity as observed in the original M_2 seedlings, suggesting this phenotype is caused by allelic variation at *rcr3* and not by another unlinked mutation (data not shown). Therefore, *rcr3–1* is a weak, and *rcr3–2* and *rcr3–3* are strong, suppressors of *Cf-2* function.

Rcr3 Maps to Chromosome 2.

To generate a mapping population where *rcr3* and molecular markers were segregating, rcr3–1 was crossed to *Lycopersicon pennellii* (59). F_1 progeny from this cross were resistant to *C. fulvum* race 4 GUS infection, demonstrating that the *L. pennellii* *Rcr3* allele can complement the *rcr3* mutation. A single F_1 plant was back crossed to rcr3–1 to give back cross (BC_1) individuals that segregated 1:1 for resistance or susceptibility to *C. fulvum* race 4 GUS infection. DNA gel blot analysis was used to screen resistant BC_1 plants for *Cf-2* homozygotes, and one was used as male parent in a back cross to rcr3–1. In the resulting BC_2 population, all plants were homozygous for *Cf-2* and, as expected, segregated 1:1 for resistance or susceptibility to *C. fulvum* race 4 GUS infection.

The chromosomal location of *Rcr3* was determined by amplified restriction fragment polymorphism (AFLP) analysis (60) of bulked segregant pools. Fifty-eight BC_2 individuals were inoculated with *C. fulvum* race 4 GUS and screened for resistance or susceptibility to infection. DNA was extracted from pools of 24 resistant and 34 susceptible plants and subjected to AFLP analysis. Several AFLP markers linked to the *L. pennellii Rcr3* allele were identified (Fig. 6*A*). AFLP analysis of *L. esculentum* × *L. pennellii* introgression lines (61) showed these markers were located on tomato chromosome 2 (Fig. 6*B*).

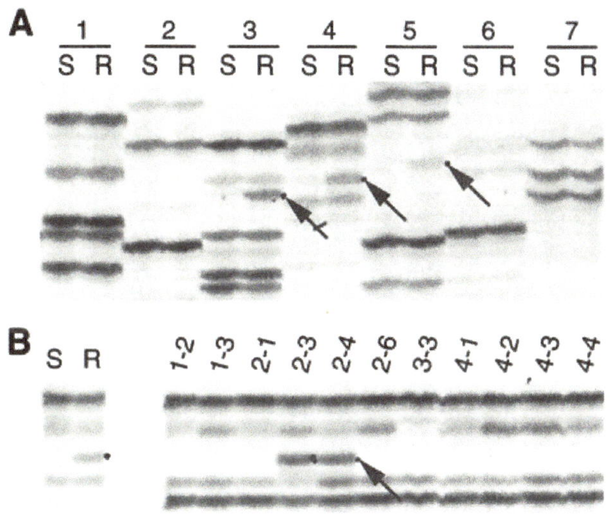

Figure 6: Rcr3 maps to chromosome 2. AFLP analysis was carried out as previously described (59). *(A)*Identification of AFLP markers linked to *Rcr3* (arrows). Susceptible (S) and resistant (R) pools of the BC_2 mapping populations were subjected to AFLP analysis. Pairs of lanes (1 to 7) represent analysis with different AFLP primer combinations. *(B)* Mapping of a linked AFLP marker by using *L. pennellii*introgression lines (1–2 to 4–4) with primer combination 4. Susceptible and resistant pools of the BC_2mapping populations were analyzed as controls. The arrows indicate the AFLP fragment linked to *L. pennellii Rcr3* present in introgression lines 2–3 and 2–4 that localizes that marker to chromosome 2.

Rcr3 Is Required Specifically for *Cf-2* Function.

To determine whether *Rcr3* is required for the function of other *Cf* genes, rcr3 mutants were crossed to Cf5 and Cf9 near isogenic lines. *Rcr3* will segregate independently of *Cf-5* and *Cf-9*, because they map to different chromosomes (12). In a cross between Cf9 and rcr3 lines, F_2 progeny should segregate 3 resistant to 1 susceptible when inoculated with*C. fulvum* race 2,4 if *Rcr3* is not required for *Cf-9* function (*C. fulvum* race 2,4 lacks *Avr2* and *Avr4* and is virulent on tomato lines carrying either *Cf-2* or *Cf-4*, so in a cross between Cf9 and rcr3, *Cf-2* confers no effective resistance to *C. fulvum* race

2,4). If *Rcr3* is required for *Cf-9* function, the progeny should segregate 9 resistant to 7 susceptible when inoculated with *C. fulvum* race 2,4. Progeny from this cross segregated 3 resistant to 1 susceptible, demonstrating *Rcr3* is not required for *Cf-9*-mediated resistance (Table 2). Further confirmation of this result was obtained from analysis of back-cross progeny. As predicted, the progeny segregated 1:1 for resistance and susceptibility to *C. fulvum* race 2,4 infection (Table2), confirming that *Rcr3* is not required for *Cf-9* function. DNA gel blot analysis confirmed that none of the susceptible progeny contained *Cf-9* (data not shown).

Table 2: Effect of *rcr3* mutations on *Cf*-5 and *Cf*-9 functions

		Segregation	Goodness of fit (χ^2)¶		
Cf line	Cross	R:S	1:1	3:1	9:7
Cf9*	F_2‡	41:17	—	0.57	4.9
Cf9*	BC_1‡	16:10	1.38	18.5	—
Cf5†	F_2¶	91:32	—	0.068	15.7
Cf5†	F_2§	61:52	—	25	0.24

S, susceptible; R, resistant.
*Crosses with rcr3-1.
†Crosses with rcr3-2.
‡Screened with *C. fulvum* race 2,4.
§Screened with *C. fulvum* race 5.
¶Significance level at $P = 0.05$ for a χ^2 with one degree of freedom is 3.84.

A similar analysis was used to determine whether *Rcr3* is required for *Cf-5*-mediated resistance. *Cf-5* and*Cf-2* are allelic, and their products are more than 90% identical (46). F_2 progeny from a cross between rcr3–2 and Cf5 were screened by inoculation with *C. fulvum* race 2,4 and segregated at a ratio of 3 resistant to 1 susceptible (Table 2), demonstrating *Rcr3* is not required for *Cf-5*-mediated resistance. PCR analysis confirmed that all *C. fulvum* race 2,4 susceptible plants lacked *Cf-5* (data not shown). To confirm the authenticity of the cross, the same F_2 plants were reinoculated with *C. fulvum* race 5 (a race that lacks Avr5 and is virulent on lines carrying only *Cf-5*, so only *Cf-2* will confer resistance to infection in this test). The segregation ratio of resistant to sensitive progeny was close to a 9:7 ratio predicted on the basis of independent segregation of *Cf-2* and *Rcr3* (Table 2).

Cf Protein Function.

One approach to understanding Cf protein function is to analyze the biochemical and physiological responses that are activated on Avr-dependent elicitation. R proteins appear to activate similar cellular responses that are effective against a broad range of plant pathogens. For example, in transgenic tobacco, *Cf-9* can confer resistance to Potato Virus X expressing *Avr9* (62). We have defined five main Avr9-dependent responses that ensue on elicitation in tobacco *Cf-9* plants and cell cultures. These include stimulation of ion fluxes (63), production of active oxygen species (64), activation of two mitogen-activated protein kinases (65), and a calcium-dependent protein kinase (66). We have also defined a large set of rapidly induced genes (67) and are currently trying to identify plasma membrane proteins that interact with epitope-tagged Cf-9 in tobacco (13).

Another approach is to dissect the signaling pathway of *Cf*-mediated resistance by using mutational analysis. Previously we identified two genes (*Rcr1* and *Rcr2*) that are required for full *Cf-9* function (55). In the present study, mutations in *Rcr3,* a gene required for *Cf-2* function, were identified. No *Cf-2* mutants were recovered because the Cf2 line carries two functional copies of the gene (68). Four mutations in *Rcr3*were identified, and no mutations were identified in other genes, indicating that the genetic screen for suppressors of *Cf-2* function may be saturated. The lack of additional mutants in the *Cf-2*-mediated signaling pathway may reflect functional redundancy or lethality. Genetic analysis demonstrated that *Rcr3* is not required for *Cf-9* function, therefore Rcr3 is not a component of a conserved Cf signal transduction pathway. We suggested previously that Cf-2 and Cf-5 may interact with a common partner to activate an Avr-dependent plant defense response (12, 46). Because *Rcr3* is not required for *Cf-5* function, *Rcr3*cannot be a Cf-2/Cf-5 common interacting partner.

In *Arabidopsis*, several genes appear to be required for the function of either of the two classes of NB-LRR genes. *EDS1* is required for the function of the TIR-NB-LRR genes *RPP5, RPP1*, and *RPS4*, whereas *NDR1*and *Pbs2* are required for the function of the LZ-NB-LRR genes *RPS2, RPS5* and *RPM1* (29, 69, 70). These studies appeared to identify at least two parallel signaling pathways for *R* genes from the two different NB-LRR classes. *Pbs3* is required for the function of *RPS2, RPS5* and *RPM1* (all LZ-NB-LRR class) and also*RPS4* (TIR-NB-LRR class), thus establishing a possible convergence point for these pathways (70). Another gene, *Pbs1*, has been identified that is specifically required for *RPS5* (70). *Rcr3* and *Pbs1* are therefore unique because they appear to suppress the function of single *R* genes. These analyses have revealed additional genetic complexity in the pathogen perception mechanism of plants.

There are several possible roles for Rcr3 in Cf-2/Avr2-mediated resistance to *C. fulvum.* Rcr3 could be specifically required for the expression, stability, or modification of Cf-2. Rcr3 may also play a role in the perception of Avr2 or in its proteolytic processing/modification to generate a mature ligand that can be recognized by Cf-2. It is known that maturation of Avr9 and Avr4 requires the activity of both fungal and plant proteases (71, 72). Proteolytic processing of ligand molecules is common in other systems, e.g., processing of the Spätzle ligand of the *Drosophila* Toll receptor (73). However, if Rcr3 functions solely as an extracellular protease or modifying enzyme, infiltrating cotyledons of *rcr3* mutants with intercellular fluids isolated from a compatible tomato–*C. fulvum* race 5 interaction should restore the characteristic chlorotic response observed in Cf2 plants (74). No complementation of the mutant phenotype was observed, suggesting Rcr3 does not function as an extracellular modifying enzyme (data not shown).

It is possible that Cf-2 and Cf-5 utilize independent signaling pathways to activate the plant defense response, and that Rcr3 is a component of the Cf-2-dependent signaling pathway, downstream of Avr2 perception. An analogous situation may occur in barley, where at least two distinct classes of *Mla* alleles can be identified on the basis of their requirement for the *Rar1* gene (75). Alternatively, Rcr3 may interact directly with Cf-2 to function as a signaling partner. A paradigm for Cf protein function was suggested from studies on three genes (*CLV1, CLV2*, and *CLV3*) that condition the *Arabidopsis* "clavata" mutant phenotype (76–78). CLV1 is an Xa21-like extracellular receptor kinase, and CLV2 is an extracellular membrane anchored Cf-like protein (Fig. 1*C*).

CLV1 and CLV2 are proposed to form a heterodimer that may be activated on binding of a peptide ligand (CLV3) resulting in phosphorylation of the kinase domain of CLV1 and association with downstream factors such as a kinase-associated protein phosphatase, KAPP, and the small GTPase, Rop (77, 79). Likewise, the truncated rice Xa21D protein for resistance to *X. campestris* pv.*oryzae* may signal through a full-length Xa21 homolog (80). A similar model has been postulated for the sporophytic self-incompatibility mechanism in *Brassica* species to explain the requirement of SLG, a secreted *S*-locus glycoprotein, and SRK, an *S*-locus receptor kinase that contains an extracellular SLG domain.

We with others have investigated whether Cf-9 and Avr9 directly interact with each other. These experiments have failed to reveal a direct interaction between Avr9 and Cf-9, suggesting that other factors are required for Avr9 perception (J.D.G.J., P. J. G. M. De Wit & T. Nürnberger, unpublished work). These factors and other proteins required for *Cf-* function are conserved in solanaceous plants because *Cf-9*confers responsiveness to Avr9 in tobacco and potato (81) and *Cf-2* to Avr2 in tobacco (M.S.D., unpublished work). Furthermore, studies using labeled Avr9 have identified a high-affinity binding site on plasma membranes of tomato and tobacco that is present irrespective of their *Cf-9* genotype (82). Similarly in soybean, a 34-kDa protein has been identified that binds syringolide elicitors, avirulence factors recognized by Rpg4, irrespective of the *Rpg4* genotype (83). By analogy, Avr2 may not interact directly or exclusively with Cf-2 but may interact with Rcr3. Another example where accessory proteins are required for full function of recognition proteins is provided by the mammalian CD14 surface protein. CD14 contributes to the perception of microbial products and acts in concert with specific TLRs that discriminate between pathogens and initiate transmembrane signaling (84).

Pathogen-encoded (a)virulence factors could interact with host proteins to modify their functions to access nutrients or to suppress defense mechanisms. For example, the human pathogen *Yersinia* uses a type III secretion system, in common with many plant pathogenic bacteria (2), to deliver YopJ into host cells to suppress host defense mechanisms (85). R proteins may have evolved to specifically recognize the physical association of pathogen-encoded virulence factors with their plant cellular targets to subsequently activate defense mechanisms, i.e., R proteins may "guard" host proteins and initiate an Avr-dependent defense response (86, 87). The physical interaction of Pto with AvrPto provides the only published evidence for an Avr product binding to its corresponding R protein (16, 17). We have proposed that Prf, a NB-LRR protein required for Pto function (88), might best be thought of as an R protein that recognizes the AvrPto/Pto pathogenicity complex to activate a resistance response (86). A similar model has been proposed to account for yeast two-hybrid interactions between RPP5 and an *Arabidopsis* RelA/SpoT homolog (87). Accordingly, Rcr3 may be the pathogenicity target of AVR2, and Cf-2 could specifically "guard" Rcr3. The tendency for a lower disease sensitivity observed 10 days after *C. fulvum* infection in rcr3–2 and rcr3–3 mutants compared with the Cf0 susceptible control (Fig. 4*B*) could be caused by absence of the Rcr3 pathogenicity target. The characterization of a Cf0/*rcr3* mutant will reveal whether *Rcr3* is necessary for full pathogenicity of *C. fulvum*.

The Guard hypothesis predicts that for each R protein there is both a corresponding pathogen Avr product and a host target. Such a model would explain the dual recognition capacity of some NB-LRR proteins such as RPM1 and Mi-1 if they "guard" the same host component targeted by unrelated Avr products. Evolutionary mechanisms sustaining *R* gene diversity are essential for the plant to be able to detect distinct pathogen (a)virulence products that target R protein-"guarded" host components.

ACKNOWLEDGMENTS

All members of the Jones lab are thanked for useful discussion. We thank Sara Perkins, Margaret Shailer, and Justine Campling for their excellent horticultural service. This work was supported in part by the United Kingdom Biotechnology and Biological Sciences Research Council (M.S.D.) and by a Gatsby Ph.D. studentship (C.G.). The Sainsbury Laboratory is supported by the Gatsby Charitable Foundation.

REFERENCES

1. Flor H H(1971) Annu Rev Phytopathol 9:275–296.
2. Galan E J, Collmer A(1999) Science 284:1322–1328.
3. Kearney B, Staskawicz B J(1990) Nature (London) 346:385–386.
4. Ritter C, Dangl J L(1995) Mol Plant-Microbe Interact 8:444–453.
5. Erickson F L, Holzberg S, Calderon-Urrea A, Handley V, Axtell M, Corr C, Baker B(1999) Plant J 18:67–75.
6. Bendahmane A, Kohm B A, Dedi C, Baulcombe D C(1995) Plant J 8:933–941.
7. Van den Ackerveken G F J M, Van Kan J A L, De Wit P J G M(1992) Plant J 2:359–366.
8. Joosten M H A J, Cozijnsen T J, De Wit P J G M(1994) Nature (London) 367:384–386.
9. Lauge R, Joosten M H A J, Van den Ackerveken G F J M, Van den Broek H W J, De Wit P J G M(1997) Mol Plant-Microbe Interact 10:725–734.
10. Lauge R, Joosten M H A J, Haanstra J P W, Goodwin P H, Lindhout P, De Wit P J G M(1998) Proc Natl Acad Sci USA 95:9014–9018.
11. Koch E S, Slusarenko A(1990) Plant Cell 2:437–445.
12. Thomas C M, Dixon M S, Parniske M, Golstein C, Jones J D G(1998) Philos Trans R Soc London 353:1413–1424.
13. Piedras P, Rivas S, Droge S, Hillmer S, Jones J D G(2000) Plant J 21:529–536.
14. Song W-Y, Wang G-L, Chen L-L, Kim H-S, Pi L-Y, Holsten T, Gardner J, Wang B, Zhai W-X, Zhu L-H, et al.(1995) Science 270:1804–1806.
15. Bogdanove A J, Martin G B(2000) Proc Natl Acad Sci USA 97:8836–8840.
16. Scofield S R, Tobias C M, Rathjen J P, Chang J H, Lavelle D T, Michelmore R W, Staskawicz B J(1996) Science 274:2063–2065.
17. Tang X, Frederick R D, Zhou J, Halterman D A, Jia Y, Martin G B(1996) Science 274:2060–2063.

18. Rock F L, Hardiman G, Timans J C, Kastelein R A, Bazan J F(1998) Proc Natl Acad Sci USA 95:588–593.

19. Janeway C A, Medzhitov R(1999) Curr Biol 9:R879–R882.

20. Medzhitov R, Janeway C A(1997) Curr Opin Immunol 9:4–9.

21. Brightbill H D, Libraty D H, Krutzik S R, Yang R B, Belisle J T, Bleharski J R, Maitland M, Norgard M V, Plevy S E, Smale S T, et al.(1999) Science 285:732–736.

22. Aliprantis A O, Yang R B, Mark M R, Suggett S, Devaux B, Radolf J D, Klimpel G R, Godowski P, Zychlinsky A(1999) Science 285:736–739.

23. Meyers B C, Dickerman A W, Michelmore R W, Sivaramakrishnan S, Sobral B W, Young N D(1999) Plant J 20:317–332.

24. Hammond-Kosack K E, Jones J D G(1997) Annu Rev Plant Physiol Plant Mol Biol 48:575–607.

25. McDowell J M, Dhandaydham M, Long T A, Aarts M G M, Goff S, Holub E B, Dangl J L(1998) Plant Cell 10:1861–1874.

26. Collins N, Drake J, Ayliffe M, Sun Q, Ellis J, Hulbert S, Pryor T(1999) Plant Cell 11:1365–1376.

27. O'Neill L A J, Greene C(1998) J Leukocyte Biol 63:650–657. Abstract

28. Hoffmann J A, Kafatos F C, Janeway C A, Ezekowitz R A B(1999) Science 284:1313–1318.

29. Gassmann W, Hinsch M E, Staskawicz B J(1999) Plant J 20:265–277.

30. Medzhitov R, Preston-Hurlburt P, Kopp E, Stadlen A, Chen C Q, Ghosh S, Janeway C A(1998) Mol Cell 2:253–258.

31. Chinnaiyan A M, Chaudhary D, O'Rourke K, Koonin E V, Dixit V M(1997) Nature (London) 388:728–729.

32. Zou H, Henzel W J, Liu X S, Lutschg A, Wang X D(1997) Cell 90:405–413.

33. Van der Biezen E A, Jones J D G(1998) Curr Biol 8:226–227.

34. Aravind L, Dixit V M, Koonin E V(1999) Trends Biochem Sci 24:47–53.

35. Bertin J, Nir W J, Fischer C M, Tayber O V, Errada P R, Grant J R, Keilty J J, Gosselin M L, Robison K E, Wong G H W, et al.(1999) J Biol Chem 274:12955–12958.

36. Inohara N, Koseki T, Del Peso L, Hu Y M, Yee C, Chen S, Carrio R, Merino J, Liu D, Ni J, Nunez G(1999) J Biol Chem 274:14560–14567.

37. Hu Y M, Benedict M A, Ding L Y, Nunez G(1999) EMBO J 18:3586–3595.

38. Del Pozo O, Lam E(1998) Curr Biol 8:1129–1132.

39. Hughes A L, Yeager M(1998) Annu Rev Genet 32:415–435.

40. Leonard K J(1997) in The Gene-for-Gene Relationship in Plant-Parasite Interactions, eds Crute I R, Holub E B, Burdon J J(CAB, Wallingford, U. K.) pp 211–230.

41. Noel L, Moores T L, Van der Biezen E A, Parniske M, Daniels M J, Parker J E, Jones J D G(1999) Plant Cell 11:2099–2012.

42. Stahl E A, Dwyer G, Mauricio R, Kreitman M, Bergelson J(1999) Nature (London) 400:667–671.

43. Parniske M, Hammond-Kosack K E, Golstein C, Thomas C M, Jones D A, Harrison K, Wulff B B H, Jones J D G(1997) Cell 91:821–832.

44. Thomas C M, Jones D A, Parniske M, Harrison K, Balint-Kurti P J, Hatzixanthis K, Jones J D G(1997) Plant Cell 9:2209–2224.

45. Takken F L W, Thomas C M, Joosten M, Golstein C, Westerink N, Hille J, Nijkamp H J J, De Wit P J G M, Jones J D G(1999) Plant J 20:279–288.

46. Dixon M S, Hatzixanthis K, Jones D A, Harrison K, Jones J D G(1998) Plant Cell 10:1915–1925.

47. Botella M A, Parker J E, Frost L N, BittnerEddy P D, Beynon J L, Daniels M J, Holub E B, Jones J D G(1998) Plant Cell 10:1847–1860.

48. Parker J E, Coleman M J, Szabo V, Frost L N, Schmidt R, Van der Biezen E A, Moores T, Dean C, Daniels M J, Jones J D G(1997) Plant Cell 9:879–894.

49. Meyers B C, Shen K A, Rohani P, Gaut B S, Michelmore R W(1998) Plant Cell 10:1833–1846.

50. Kobe B, Deisenhofer J(1994) Trends Biochem Sci 19:415–421.

51. Ellis J, Lawrence G, Ayliffe M, Anderson P, Collins N, Finnegan J, Frost D, Luck J, Pryor T(1997) Annu Rev Phytopathol 35:271–291.

52. Hulbert S H(1997) Annu Rev Phytopathol 35:293–310.

53. Michelmore R W, Meyers B C(1998) Genome Res 8:1113–1130.

54. Nei M, Gu X, Sitnikova T(1997) Proc Natl Acad Sci USA 94:7799–7806.

55. Hammond-Kosack K E, Jones D A, Jones J D G(1994) Plant Cell 6:361–374.

56. Jefferson R A, Kavanagh T A, Bevan M W(1987) EMBO J 6:3901–3907.

57. Oliver R P, Farman M L, Jones J D G, Hammond-Kosack K E(1993) Mol Plant-Microbe Interact 6:521–525.

58. Keogh R C, Deverall B J, Mcleod S(1980) Trans Brit Mycol Soc 74:329–333.

59. Thomas C M, Vos P, Zabeau M, Jones D A, Norcott K A, Chadwick B P, Jones J D G(1995) Plant J 8:785–794.

60. Vos P, Hogers R, Bleeker M, Reijans M, Van de Lee T, Hornes M, Frijters A, Pot J, Peleman J, Kuiper M, Zabeau M(1995) Nucleic Acids Res 23:4407–4414.

61. Eshed Y, Zamir D(1994) Euphytica 79:175–179.

62. Thomas C M, Tang S, Hammond-Kosack K E, Jones J D G(2000) Mol Plant-Microbe Interact 13:465–469.

63. Blatt M R, Grabov A, Brearley J, Hammond-Kosack K E, Jones J D G(1999) Plant J 19:453–462.

64. Piedras P, Hammond-Kosack K E, Harrison K, Jones J D G(1998) Mol Plant-Microbe Interact 11:1155–1166.

65. Romeis T, Piedras P, Zhang S Q, Klessig D F, Hirt H, Jones J D G(1999) Plant Cell 11:273–287.

66. Romeis T, Piedras P, Jones J D G(2000) Plant Cell 12:803–815.

67. Durrant W, Rowland O, Piedras P, Jones J D G(2000) Plant Cell 12:963–977.
68. Dixon M S, Jones D A, Keddie J S, Thomas C M, Harrison K, Jones J D G(1996) Cell 84:451–459.
69. Aarts N, Metz M, Holub E, Staskawicz B J, Daniels M J, Parker J E(1998) Proc Natl Acad Sci USA 95:10306–10311.
70. Warren R F, Merritt P M, Holub E, Innes R W(1999) Genetics 152:401–412.
71. Van den Ackerveken G F J M, Vossen J P M J, De Wit P J G M(1993) Plant Physiol 103:91–96. Abstract
72. Joosten M H A J, Vogelsang R, Cozijnsen T J, Verberne M C, De Wit P J G M(1997) Plant Cell 9:367–379.
73. Van Eeden F, St Johnston D(1999) Curr Opin Genet Dev 9:396–404.
74. Hammond-Kosack K E, Jones J D G(1994) Mol Plant-Microbe Interact 7:58–70.
75. Jorgensen J H(1996) Genome 39:492–498.
76. Clark S E, Williams R W, Meyerowitz E M(1997) Cell 89:575–585.
77. Jeong S, Trotochaud A E, Clark S E(1999) Plant Cell 11:1925–1933.
78. Fletcher L C, Brand U, Running M P, Simon R, Meyerowitz E M(1999) Science 283:1911–1914.
79. Trotochaud A E, Hao T, Wu G, Yang Z B, Clark S E(1999) Plant Cell 11:393–405.
80. Wang G L, Ruan D L, Song W Y, Sideris S, Chen L L, Pi L Y, Zhang S P, Zhang Z, Fauquet C, Gaut B S, et al.(1998) Plant Cell 10:765–779.
81. Hammond-Kosack K E, Tang S J, Harrison K, Jones J D G(1998) Plant Cell 10:1251–1266.
82. Kooman-Gersmann M, Honee G, Bonnema G, De Wit P J G M(1996) Plant Cell 8:929–938.
83. Ji C, Boyd C, Slaymaker D, Okinaka Y, Takeuchi Y, Midland S L, Sims J J, Herman E, Keen N(1998) Proc Natl Acad Sci USA 95:3306–3311.
84. Ulevitch R J(1999) Nature (London) 401:755–756.
85. Orth K, Palmer L E, Bao Z Q, Stewart S, Rudolph A E, Bliska J B, Dixon J E(1999) Science 285:1920–1923.
86. Van der Biezen E A, Jones J D G(1998) Trends Biochem Sci 23:454–456.
87. Van der Biezen E A, Sun J, Coleman M J, Bibb M J, Jones J D G(2000) Proc Natl Acad Sci USA 97:3747–3752.
88. Salmeron J M, Oldroyd G E D, Rommens C M T, Scofield S R, Kim H-S, Lavelle D T, Dahlbeck D, Staskawicz B J(1996) Cell 86:123–133.

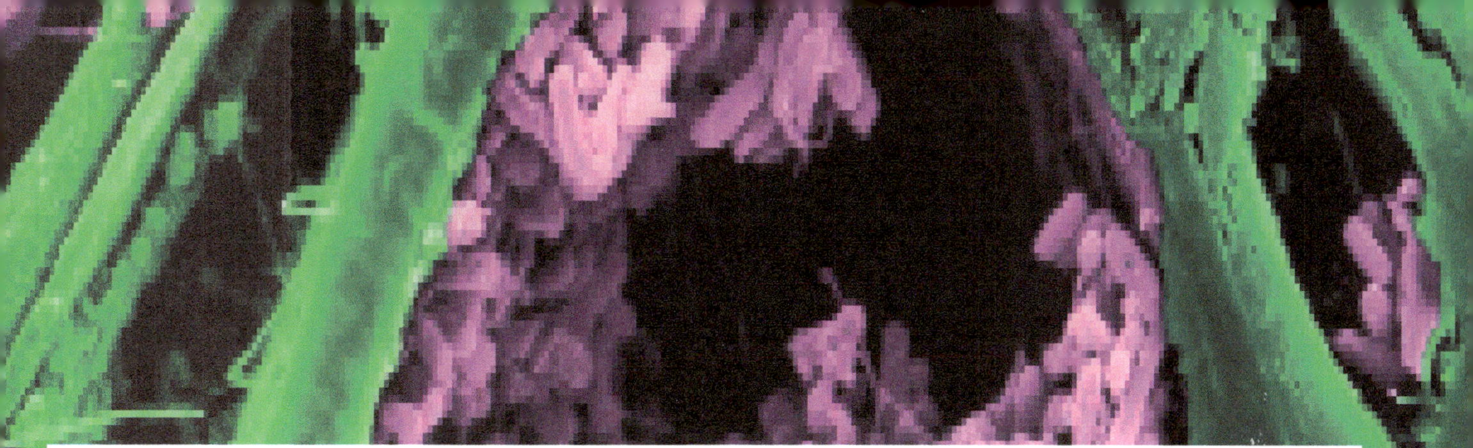

CHAPTER 6

HOST VERSUS NONHOST RESISTANCE: DISTINCT WARS WITH SIMILAR ARSENALS

Upinder S. Gill, Seonghee Lee, and Kirankumar S. Mysore

Keywords: Gene-for-gene resistance, plant disease resistance, *R* gene resistance.

Plants are important components of any food chain. During their life cycle, plants are often challenged by numerous abiotic and biotic stresses. Among biotic stresses, diseases caused by fungi, oomycete, bacteria, and viruses have significant economic significance. Fungal and oomycete diseases such as wheat stem rust (caused by *Puccinia graminis*), Asian soybean rust (caused by *Phakopsora pachyrhizi*), potato blight (caused by*Phytophthora infestans*), banana black sigatoka (caused by *Mycosphaerella fijiensis*), and rice blast (caused by *Magnaporthe oryzae*) are considered major threats to world food security (Pennisi 2010). In general, plants are resistant to most microorganisms, and there are only a few plant pathogens that have the ability to cause disease (Lipka et al. 2008). In response to pathogen infection, plants have a dynamic immune system comprised of several mechanisms to defend against plant pathogens as previously explained in a four-phased zig-zag model (Jones and Dangl 2006).

Depending upon the nature of the plant–pathogen interactions, plants generally have two types of disease resistance—host resistance and nonhost resistance. Host resistance is mostly cultivar- or accession-specific and less durable (Mysore and Ryu 2004). So far, most of the breeding efforts have focused on host resistance where a single or multiple resistance (*R*) genes were introduced into susceptible cultivars. Unfortunately, resistance imparted by most *R* genes is less durable because of monocultures, which puts tremendous selection pressure on pathogens to lose or mutate their corresponding effectors to evade detection by the host (Dangl et al. 2013). However, there are some core effectors that contribute toward pathogen virulence (Dangl et al. 2013). Loss/mutation of such effectors sometimes has negative implications on pathogen virulence and results in improved durability of corresponding host *R* genes (Vera Cruz et al. 2000). Another form of resistance is nonhost resistance; unlike host resistance, nonhost resistance can act against all races of a particular pathogen and can occur in all cultivars of a host plant species (Heath 2000). Thus, nonhost resistance is more durable and is the common form of plant defense mechanism exhibited by plants toward a vast majority of potential pathogens (Heath 2000; Mysore and Ryu 2004; Niks and Marcel 2009; Senthil-Kumar and Mysore 2013). For a long time, nonhost resistance has been considered to protect crops from many potential pathogens because of its durability (Niks and Marcel 2009). Complete understanding of nonhost resistance mechanisms is imperative to develop cultivars by the transfer of genes involved in nonhost resistance across species (Ayliffe and Lagudah 2004;Zhao et al. 2005). Moreover, nonhost resistance is usually more complicated due to the involvement of multiple pathways.

Citation: Upinder S. Gill, Seonghee Lee, and Kirankumar S. Mysore, Host Versus Nonhost Resistance: Distinct Wars with Similar Arsenals, http://dx.doi.org/10.1094/PHYTO-11-14-0298-RVW

In this review, we briefly describe the mechanisms of host and nonhost resistance against fungal, oomycete, and bacterial plant pathogens. In addition, we compare host and nonhost resistance mechanisms, and highlight the similarities and differences between them. We also compare host and nonhost resistance in terms of their practical application in crop improvement.

MECHANISMS OF HOST AND NONHOST RESISTANCE

Host resistance and nonhost resistance are most commonly differentiated based on pathogen adaptation to a particular species (host) and lack of adaptation to other species (nonhost). Both host and nonhost resistances are the outcomes of the plant immune response. Plant innate immunity has been well described in previous reviews (Chisholm et al. 2006; Cohn et al. 2001; Dangl and Jones 2001; Dodds and Rathjen 2010; Jones and Dangl 2006). Among several components of the plant immune response, basal defense is the first line of defense and is initiated during the early phases of pathogen detection. Basal defense in plants is initiated with the perception of evolutionarily conserved microbial- or pathogen-associated molecular patterns (MAMPs or PAMPs) such as flagellin and EF-Tu by plant extracellular pathogen recognition receptors (PRRs) such as leucine-rich repeat kinases (Dangl et al. 2013; Zipfel 2008). Such responses are referred to as PAMP-triggered immunity (PTI). There are also some overlaps between basal defense and nonhost resistance because it is possible that both host and nonhost plants may recognize similar factors to initiate a defense response. Initiation of plant defense responses and the counter attack of pathogen are well explained by the widely accepted zig-zag model proposed by Jones and Dangl (2006). According to this model, there are numerous PRRs in plants to recognize PAMPs and to initiate basal defense responses, but some well-adapted pathogens secrete effectors to evade recognition by plant PRRs and to promote pathogen growth and virulence (Bent and Mackay 2007; Dangl et al. 2013; Zipfel 2008). Suppression of PTI by pathogen effectors leads to effector-triggered susceptibility (ETS). In turn, plants have evolved to recognize these effectors or avirulence (Avr) proteins with the help of R proteins (Nimchuk et al. 2003). Therefore, in response to ETS, host plants trigger R proteins to interact directly or indirectly with pathogen effectors and induce a stronger defense response, referred to as effector-triggered immunity (ETI). There are several examples of direct recognition of pathogen Avr proteins by plant R proteins (Dodds et al. 2006), but in some cases indirect recognition is also possible in which plant decoy proteins interact with Avr proteins and subsequent activation of R proteins (van der Hoorn and Kamoun 2008). This mechanism is also referred to as guard hypothesis (Dangl and Jones 2001; van der Biezen and Jones 1998). Direct and indirect recognition of Avr proteins by R proteins is followed by a strong defense signaling response that is associated with localized programmed cell death, called hypersensitive response (HR), to limit the pathogen growth.

Both pathogen and host undergo natural selection to diversify effectors and R proteins to achieve superiority in this molecular arms race (Boller and He 2009). ETI, which is determined by the recognition of a pathogen's effector or Avr protein by corresponding plant R protein, is also the basis of classical gene-for-gene resistance (Flor 1971). Most R proteins contain a nucleotide-binding site and leucine rich repeats (NBS-LRRs), and are involved, upon recognition of Avr protein, in the development of HR through the generation of reactive oxygen species (ROS). The well-studied host resistance is often considered gene-for-gene resistance and is usually very specific to a particular plant genotype or cultivar and against a particular race of the pathogen (Flor 1971). In contrast, nonhost resistance is not pathogen-race-specific and is a broad spectrum resistance exhibited by the whole plant species against a particular pathogen (Heath 2000). Nonhost resistance is often multi-tiered, with several obstacles depending upon a particular host to stop the colonization by a potential pathogen (Thordal-Christensen 2003). These obstacles include, but are not limited to, the presence/absence of signals from plants, such as surface topology features that are required to initiate pathogen growth; preformed barriers such as cell wall, cuticle, phytoanticipins, etc.; and induced defense responses such as lignin accumulation, production of antimicrobials like phytoalexins, HR response, induction of pathogenesis-related (PR) proteins, etc. (Senthil-Kumar and Mysore 2013; Thordal-Christensen 2003). Depending upon the presence or absence of visual symptoms, nonhost resistance is again divided into type I and type II nonhost resistance (Mysore and Ryu 2004). Type I nonhost resistance does not produce any visual symptoms, whereas type II nonhost resistance is associated with visual necrosis/cell death due to HR (Mysore and Ryu 2004). Type I nonhost resistance typically involves passive or preformed barriers and/or active defense mechanisms that are induced in response to general elicitors of pathogens such as PAMPs (Mysore and Ryu 2004; Thordal-Christensen 2003). Type I nonhost resistance resembles PTI, whereas type II nonhost resistance involves HR and is triggered in response to pathogen elicitor/effector recognition and is similar to ETI (Senthil-Kumar and Mysore 2013). For example, recognition of pathogen effectors by nonhost plant species has been reported previously when *AvrA* and *AvrD* from *Pseudomonas syringae* pv.*tomato* were recognized by *R* genes *Rpg2* and *Rpg4*, respectively, of the nonhost plant soybean (Ashfield et al. 1995; Keen et al. 1991). In addition, *AvrRxo1* from *Xanthomonas oryzae* pv. *oryzae* was recognized by the *R* gene *Rxo1* of the nonhost plant maize (Zhao et al. 2004). Despite the fact that there are some overlaps between host and nonhost resistance, nonhost resistance is more complex and the mechanism of resistance may vary, depending upon the nature of pathogen (virus, bacteria, and fungi) and the plant species.

HOST VERSUS NONHOST RESISTANCE AGAINST FUNGAL AND OOMYCETE PATHOGENS IN PLANTS

Although nonhost resistance is considered to be a different type than host resistance, there are significant overlaps between them. Fungal pathogens recognize specific plant cues in the form of surface topology or composition of plant waxes (Hoch et al. 1987). Preformed barriers such as surface waxes have been shown to be important for growth and differentiation of nonhost fungal pathogens such as *Blumeria graminis* f. sp. *hordei* on cabbage and tobacco, and *Phakopsora pachyrhizi* on *Medicago truncatula* (Tsuba et al. 2002; Uppalapati et al. 2012). A mutation in the *IRG1* (*INHIBITOR OF RUST GERM TUBE DIFFERENTIATION 1*) gene in *M. truncatula* that causes loss of epicuticular wax accumulation on the abaxial side of the leaf was found to inhibit growth and development of germ tubes of both host and nonhost pathogens (Uppalapati et al. 2012). In contrary, epicuticular waxes were shown to be important for host resistance in cocoa against black pod disease caused by *Phytophthora palmivora* and *Phytophthora megakarya* (Nyadanu et al. 2012). In addition, a role of waxes in host resistance was also reported in *Lolium* spp. against powdery mildew pathogen, *Erysiphe graminis* (Carver et al. 1990). In both cases, epicuticular wax was directly correlated with resistance, suggesting a positive role of epicuticular waxes in host resistance.

Actin microfilaments are part of the plant cytoskeleton, and the role of actin microfilaments in plant defense is well established (Day et al. 2011). Involvement of*ACTIN DEPOLYMERIZATION FACTORS* (*Adf3* and *TaAdf7*, a *Triticum aestivum* homolog of*Adf7*) in host resistance against rust pathogens has been shown (Fu et al. 2014; Moscou et al. 2011; Wang et al. 2013). Apart from host resistance, nonhost resistance against several nonhost plant–pathogen combinations was compromised when actin polymerization was disrupted with specific inhibitors, indicating the importance of actin microfilaments in nonhost resistance as well (Kobayashi et al. 1997).

Another level of resistance against nonhost pathogens is triggered at the time of pathogen penetration through the plasma membrane. Three *PENETRATION* genes, *PEN1*, *PEN2*, and*PEN3*, have been identified in Arabidopsis and are involved in penetration resistance against the nonhost powdery mildew pathogen *Blumeria graminis* f. sp. *hordei* and Asian soybean rust pathogen *Phakopsora pachyrhizi* (Collins et al. 2003; Lipka et al. 2005;Loehrer et al. 2008; Stein et al. 2006). In Arabidopsis, *PEN1* encodes a membrane-associated syntaxin containing a SNARE (soluble N-ethylmaleimide sensitive factor attachment protein receptor) domain which forms a complex with SNAP33 (soluble N-ethylmaleimide-sensitive factor adaptor protein 33) and VAMP721/VAMP722 (vesicle-associated membrane proteins), and is involved in vesicle transport to cell periphery in response to nonhost pathogen *B. graminis* f. sp. *hordei* (Assaad et al. 2004; Collins et al. 2003; Kwon et al. 2008). Similarly, *Required for mlo-specific resistance 1* (*Ror1*) and *Ror2*, an ortholog of *PEN1*, genes are also involved in penetration resistance in barley against nonhost rust pathogen *Phakopsora pachyrhizi* (Hoefle et al. 2009). *Ror2* is also involved in pre-invasive resistance against host pathogen *B. graminis* f. sp. *hordei* (Collins et al. 2003). In Arabidopsis, *PEN2*, *ARABIDOPSIS G PROTEIN BETA SUBUNIT 1* (*AGB1*),*POWDERY MILDEW RESISTANT 5* (*PMR5*), and *MILDEW RESISTANCE LOCUS O* (*MLO2*) genes are involved in both pre- and postinvasive nonhost resistance against *B. graminis* f. sp. *hordei* (Nakao et al. 2011). *PMR5* restricts fungal hyphae elongation whereas *AGB1* and*MLO2* restrict both hyphae elongation and branching of nonhost pathogen *B. graminis* f. sp.*hordei* (Nakao et al. 2011). The *Mlo* family gene was first identified in barley, but unlike Arabidopsis *MLO2*, mutation in the *Mlo* locus of barley conferred durable resistance against biotrophic host pathogen *B. graminis* f. sp. *hordei* and hyper-susceptibility against hemi-biotrophic nonhost pathogen *Magnaporthe oryzae*, indicating the involvement of yet another gene in both host and nonhost resistance, even though the outcome was antagonistic (Buschges et al. 1997; Jarosch et al. 1999). However, the *Mlo* gene product is considered to act as a negative regulator of cell death due to the association of a mutation in *Mlo* with resistance against biotrophic fungal pathogens and susceptibility against necrotrophic fungal pathogens (Király et al. 2002). Barley *mlo* alleles cause enhanced susceptibility against host pathogens *Bipolaris sorokiniana*, a necrotroph, and *Ramularia collo-cygni*, which lives in host as an endophyte but later changes to necrotrophic mode (Kumar et al. 2001; McGrann et al. 2014).

Fungal pathogens that successfully overcome penetration or pre-invasion resistance produce infection hyphae and/or haustoria to absorb nutrients from plant cells. At this point, pathogens encounter postinvasion resistance from plants. Plant genes such as*ENHANCED DISEASE SUSCEPTIBILITY 1* (*EDS1*), *PHYTOALEXIN DEFICIENT 4* (*PAD4*), and*SENESCENCE ASSOCIATED GENE 101* (*SAG101*), which are also involved in basal defense during host–pathogen interactions, are also implicated in postinvasion defense (Lipka et al. 2005; Stein et al. 2006). Individual mutations in these genes have an effect on postinvasion defense in Arabidopsis against nonhost pathogen *B. graminis* f. sp. *hordei*. In addition, a triple mutant of *pen2/pad4/eds1* completely broke down the nonhost resistance barrier against *B. graminis* f. sp. *hordei* and pea powdery mildew pathogen *Erysiphe pisi*(Lipka et al. 2005). *EDS1* was initially identified in Arabidopsis as a component of basal defense against host pathogen *Hyaloperonospora parasitica* (Parker et al. 1996). *EDS1* and*PAD4* are also required for several Arabidopsis *R* genes belonging to TIR-NB-LRR class-mediated resistance (Wiermer et al. 2005). Apart from a few plant genes that are commonly required for host and nonhost resistance, some effectors are also commonly recognized by host and nonhost plant species. For example, *Avr3a* of *Phytophthora infestans* and *Avr1b-1*, a homolog of *Avr3a* in *Phytophthora sojae*, are recognized by their corresponding host *R* genes, *R3a* of potato and *Rps1b* of soybean, respectively (Armstrong et al. 2005; Shan et al. 2004). In addition, *PcAvr3a1* of *Phytophthora capsici*, which is also a homolog of *Avr3a* of *Phytophthora infestans*, is recognized by a dominant *R* gene of nonhost *Nicotiana* species, indicating parallel mechanisms among host and nonhost resistance (Vega-Arreguín et al. 2014).

Additionally, there are several examples where potential *R* genes were identified in nonhost plant species, and after transferring to host plant species, those *R* genes have successfully conferred resistance against host pathogens (Sumit 2013; Zhao et al. 2005). Examples of successful transgenic deployment of *R* genes between plant species have been reviewed previously (Wulff et al. 2011). These studies indicate a conserved mechanism involved downstream of recognition among various plant species, irrespective of their classification as host or nonhost (Stam et al. 2014).

At the cellular level, both nonhost and host resistance are associated with similar cellular responses such as cytoplasmic aggregation involving relocation of organelles, production of ROS, and HR (Hückelhoven et al. 2001; Hückelhoven and Kogel 2003; Takemoto et al. 2003; Vleeshouwers et al. 2000). However, apart from overlaps or similarities in host and nonhost resistance mechanisms, there is evidence that also suggests distinct or independent mechanisms. Distinct cellular processes and timing of HR have also been reported among host and nonhost interactions against penetration of rust and powdery mildew fungi in cowpea and French beans (Christopher-Kozjan and Heath 2003). In nonhost interactions, the cell death process was rapid (less than one hour) and the plant cell nucleus was seen associated with fungal haustorium. During host resistance in contrast to nonhost resistance, the cell death response took longer (two or more hours) and no consistent association of plant cell nucleus with the intracellular fungi was found (Christopher-Kozjan and Heath 2003). Nonetheless, no differences were observed in cellular signaling during HR, such as influx of calcium and protein kinase activity between host and nonhost resistance (Christopher-Kozjan and Heath 2003). However, there are also few examples where plant nucleus has been shown to be associated with the pathogen structures, such as haustoria, during host resistance (Griffis et al. 2014). For example, during host resistance of a potato cultivar (*Solanum tuberosum* L. cv. Datura) against oomycete pathogen *Phytophthora infestans* (race 4) the plant cell nucleus conglomerate around the intracellular oomycete haustorium which is followed by cell death response (Freytag et al. 1994). Based on our current understanding, host and nonhost mechanisms against fungal and oomycete pathogens have some similarities and overlaps by the involvement of common genes in both types of resistance but are yet distinct in recognition and initiation of defense response.

HOST VERSUS NONHOST RESISTANCE AGAINST BACTERIAL PATHOGENS IN PLANTS

Similar to the way plants have evolved to combat infections by fungal and oomycete pathogens, plants have also evolved defense mechanisms to prevent attacks from bacterial pathogens. Most bacterial pathogens get access to the plant milieu through natural openings (hydathodes and stomata) and wounds (Hirano and Upper, 1983). The most studied plant pathogenic bacteria belong to the genus *Pseudomonas* (type III secretion), *Xanthomonas* (type III secretion), *Erwinia* (type II secretion), *Ralstonia* (type III secretion), and *Agrobacterium* (type IV secretion). Species within these genera of bacterial pathogens use different types of secretion systems to secrete effector proteins into a host cell for the suppression of plant defense systems. Many gram-negative plant bacterial pathogens employ a type III secretion system (T3SS) to colonize host plants by suppressing plant defense (Buttner and He 2009). For example, AvrPto from *Pseudomonas syringae* pv. *tomato* blocks callose deposition in tomato plants (Hauck et al. 2003) and suppresses HR induced in nonhost plants (Kang et al. 2004). In Arabidopsis, AvrRpm1 (from *Pseudomonas syringae* pv. *maculicola*) and AvrRpt2 (from *Pseudomonas syringae*pv. *tomato*) effectors inhibit PAMP-mediated defense signaling pathways (Kim et al. 2009). Plants, on the other hand, have evolved to express R proteins that can detect the presence of T3SS effectors inside the plant cell and initiate defense responses such as HR, ROS production, and expression of *PR* genes. As described above, this phenomenon is also referred to as ETI. In contrast to R protein-mediated host resistance, early recognition of bacterial PAMPs is also important for initiating the downstream signaling pathway for nonhost resistance. PRRs recognize bacterial PAMPs, induce defense signaling and defend against nonadapted bacterial pathogens. In Arabidopsis, flagellin (a PAMP)-induced defense plays an important role in nonhost resistance against bacterial pathogens (Li et al. 2005). The flg22 peptide of flagellin from *Pseudomonas* bacteria is perceived by FLS2, a receptor-like kinase that activates downstream events. It has been reported that ROS and defense phytohormones, such as salicylic acid (SA), jasmonic acid (JA), and abscisic acid (ABA), play a role in the defense signaling pathway for nonhost resistance (Li et al. 2012; Mishina and Zeier 2007; van Wees and Glazebrook 2003). Since SA, JA, and ABA are the major hormones for plant defense and development, genes involved in their signaling pathway are not only involved in nonhost resistance but also closely associated with plant growth and development.

Several plant genes and proteins have been identified that play a role in both host and nonhost resistance. For example, in *Nicotiana benthamiana*, the suppressor of the G2 allele of S-phase kinase-associated protein 1 (SGT1) has been known for the production of ROS and HR cell death against nonhost bacterial pathogens (Shibata et al. 2011; Wang et al. 2010). Also, SGT1 together with Rar1 (Required for Mla12 Resistance 1) is required for resistance responses mediated by NBS-LRR and other R proteins (Azevedo et al. 2002;Peart et al. 2002). *GLYCOLATE OXIDASE* (*GOX*) and *NONHOST 1* (*NHO1*) genes have also been demonstrated to be required for both host and nonhost resistance in Arabidopsis (Lu et al. 2001; Rojas et al. 2012). Another important gene for the SA signaling pathway,*EDS1*, also contributes to host, nonhost, and basal resistance against bacterial pathogens (Aarts et al. 1998; Glazebrook 2001; Moreau et al. 2012). Apart from plant genes, several pathogen genes also determine host/nonhost status of a plant. In *Pseudomonas*, several*survival in Arabidopsis extracts* (*sax*) genes in host bacterial strains are implicated in overcoming Arabidopsis defense. Depending upon the presence or absence of *sax* genes, Arabidopsis host or nonhost status against certain bacterial pathogens is decided (Fan et al. 2011).

EVOLUTIONARY PERSPECTIVE OF HOST AND NONHOST RESISTANCE

The concept of host and nonhost resistance is derived from the fact that not all pathogens can infect all plant species. Host and nonhost interactions between plants and pathogens are dynamic and greatly influenced by evolution. At the molecular level, depending upon the genetic distance from the host species, host resistance is predominantly regulated by ETI, whereas nonhost resistance is regulated by PTI and/or ETI (Schulze-Lefert and Panstruga 2011; Senthil-Kumar and Mysore 2013) (Fig. 1). Within host species, the variation of resistance among cultivars and accessions is mainly defined by a large repertoire of *R* genes that continuously evolve due to gene duplication/deletion, recombination, mutations, and selection pressure for resistance against pathogens (Ellis et al. 2000). On the other hand, pathogens also evolve their effector proteins to maintain virulence on host species. Co-evolution of host and pathogen is also the basis of the zig-zag model that has been developed for plant immunity against pathogens (Jones and Dangl 2006). In some instances, the host plant evolves and acquires a nonhost status by one-sided evolution of the plant while the pathogen is specialized to a particular host and does not counter evolve (Antonovics et al. 2013). Generally, durability of nonhost resistance increases against a specialized pathogen when the nonhost species is evolutionarily divergent from the source host species (Fig. 1). On the contrary, highly evolving pathogens have the potential to overcome the defense exhibited by nonhost species evolutionarily divergent from their source host (Fig. 1). In general, an increase in genetic distance of nonhost from adapted host increases the durability of nonhost resistance because of the presence of several mechanisms to block the pathogen infection (Antonovics et al. 2013). Additionally, pathogens which keep evolving generally have a large host range. Several factors which influence genetic structure and evolutionary trajectory of a pathogen population include the host range, mode of reproduction, transmission and dispersal, life-cycle complexity and epidemiology, and host longevity (Barrett et al. 2008). A good example of a pathogen with a large host range, long-distance transmission, sexual and asexual modes of reproduction, and a complex life cycle including an alternate host is *Puccinia* spp. The *Puccinia* spp. cause rust diseases in monocots and cause economically significant damage on several crop plants in the United States (Roelfs 1978). The host range of *Puccinia graminis* alone is very large and covers 365 species in 54 genera (Anikster 1984). Another example is *Puccinia coronata* var.*hordei*, also has a broad host range and infects *Hordeum*, *Bromus*, *Agropyron*, *Aegilops*,*Phalaris*, and *Secale* genera (Jin and Steffenson 1999). It can be presumed that such pathogens have a large repertoire of effectors compared to a specialized pathogen (Niks and Marcel 2009). Recently, a new race (Ug99) of *Puccinia graminis* f. sp. *tritici* emerged in Uganda in 1998 which is virulent on most wheat cultivars grown all over the world (Pretorius et al. 2000; Singh et al. 2011). Sexual reproduction and long distance dispersal of windborne spores are the major contributing factors for emergence of new races and rust epidemics. Intercontinental long-distance dispersal of *Puccinia graminis* f. sp. *tritici*has been reported previously (Watson 1981). *Puccinia graminis* f. sp. *tritici* overwinters and completes its sexual cycle on the alternate host barberry (*Berberis vulgaris*), and the presence of barberries increases genetic diversity and infectivity of rust (Burdon and Roelfs 1985). Therefore, a barberry eradication program was conducted from 1918 to 1980 in the United States to reduce wheat stem rust epidemics (Roelfs 1982).

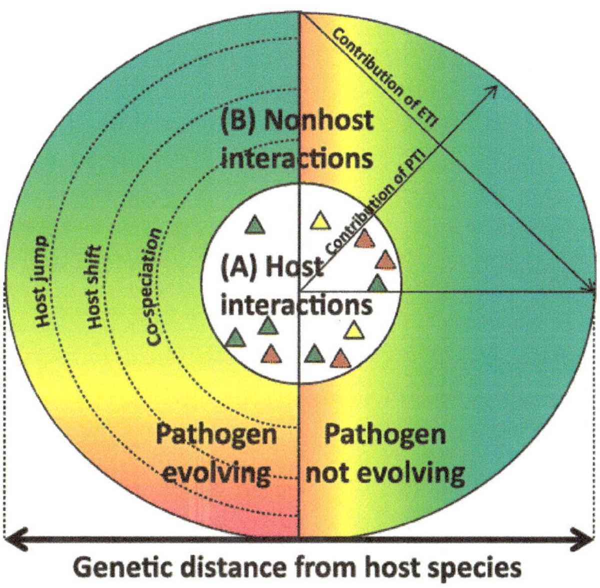

Figure 1: Evolution of host and nonhost resistance. **A,** Host resistance is primarily controlled by R-AVR recognition. A micro-evolution creates diversity within host species for resistance/susceptibility and also within pathogen to develop new races with diverse suite of effectors. Red, yellow, and green indicate host variations for susceptibility, partial resistance/susceptibility, and resistance, respectively.**B,** Outcomes of nonhost interactions vary with genetic distance from host species and the pathogen's ability to evolve. A rapidly evolving pathogen due to co-speciation, host shift, and host jump has better capability to adapt to new nonhost species by breaking the nonhost barriers. Durability of nonhost resistance with respect to genetic distance is depicted as a gradient from red to green, where red is less durable and green is more durable. The contribution of PAMP-triggered immunity and effector-triggered immunity also increases and decreases, respectively, with the increase in genetic distance from host to nonhost (based on information from Schulze-Lefert and Panstruga 2011).

Apart from co-evolution, which largely determines host–pathogen interactions, co-speciation, host-shift speciation, and host jump play a role in the adaptation of pathogens to new plant species (Fig. 1) (Schulze-Lefert and Panstruga 2011; de Vienne et al. 2013). Co-speciation happens when both host and pathogen undergo speciation simultaneously to maintain their interactions. The presence of co-speciation has been experimentally proven by comparing phylogenetic trees of rodents and their ectoparasites, which showed a high degree of concordance (Hafner and Nadler 1988). Host-shift happens when pathogens usually shift from their natural host to related nonhost species. For example, *Blumeria graminis* that causes powdery mildew disease on barley underwent host-shift from barley to the related nonhost species, wheat (Inuma et al. 2007). Similarly, rice blast pathogen*Magnaporthe oryzae* also shifted host to cause wheat blast in the mid-1980s (Couch et al. 2005). On the other hand, host jump involves adaption of the pathogen to a distantly related plant species from their source host or sometimes even across kingdom (Raffaele et al. 2010; van Baarlen et al. 2007). For example, many plant bacterial and fungal pathogens such as *Pseudomonas aeruginosa*, *Agrobacterium radiobacter*, *Alternaria* spp., and *Fusarium* spp. have a wide plant host range but are also capable of infecting mammalian hosts indicating a cross kingdom jump in the past (van Baarlen et al. 2007). A very good example of host jump is related to *Phytophthora infestans* that was responsible for the famous 19th century Irish potato famine. *Phytophthora infestans* that infect several*Solanum* species evolved through host jump from *Phytophthora mirabilis* that infect the four o'clock plant (*Mirabilis jalapa*) (Grünwald and Flier 2005; Raffaele et al. 2010). Recent findings suggest that the host specialization followed by host jump is dependent on reciprocal amino acid changes in effector proteins such as a single amino acid change in one of the effector protein that binds and inhibits a plant's protease defense (Dong et al. 2014; Raffaele et al. 2010). Horizontal gene transfer of pathogenicity genes between pathogen species beyond reproductive barriers also helps bacterial and fungal pathogens to shift their hosts and infect nonhosts (de la Cruz and Davies 2000; Rosewich and Kistler 2000). Interspecific horizontal gene transfer has been reported for the *ToxA* gene that was transferred from *Phaeosphaeria nodorum* to *Pyrenophora tritici-repentis*, which causes tan spot disease of wheat (Friesen et al. 2006).

Based on the current understanding, plant pathogen interactions are highly influenced by evolutionary forces that act on both plants and pathogens. Host and nonhost status is decided by both plant and pathogen, which is again affected by speciation, biology of plant, and pathogen, including mode of reproduction, geographical proximity, epidemiology, and host range of the pathogen.

HOST VERSUS NONHOST RESISTANCE: PRACTICAL APPLICATION FOR CROP IMPROVEMENT

Compared with host resistance, nonhost resistance is considered more durable; but at the same time, nonhost resistance is more complicated and difficult to apply in crop breeding. Availability of both susceptible and resistant individuals in host populations against a particular pathogen makes it easier to identify underlying genetic factors responsible for resistance by employing genetic and quantitative trait loci mapping. Contrarily, a nonhost plant population generally lacks susceptible individuals, especially if the nonhost is distantly related to the susceptible host. Under these circumstances, inheritance of nonhost resistance can be achieved by employing three different strategies: crossing host × nonhost where cross species hybridization is possible; crossing nonhost × nonhost if there are variations in resistance; and crossing rare susceptible and common resistant individuals in marginal host species (Niks and Marcel 2009). Marginal host species are those which are mostly nonhost to a pathogen species, but rarely individuals with lower susceptibility than host species still exist (Niks and Marcel 2009). Lack of natural variation for resistance in host and nonhost species can be overcome by generating variations using mutagenesis and gene silencing. Several genes involved in nonhost resistance of *M. truncatula* against *Phakopsora pachyrhizi* were identified by screening a large *Tnt1*insertion mutagenized population of *M. truncatula* (Tadege et al. 2008; Uppalapati et al. 2012). In addition, *PEN* genes were identified and characterized in nonhost Arabidopsis by screening a mutant population upon infection with *Blumeria graminis* f. sp. *hordei* (Collins et al. 2003; Lipka et al. 2005; Stein et al. 2006). In another approach, virus-induced gene silencing (VIGS)-based forward genetics screening was utilized to identify the role of genes such as *SQUALENE SYNTHASE* (*SQS*), *GOX*, *ORNITHINE DELTA-AMINOTRANSFERASE*(*OAT*), and *PROLINE DEHYDROGENASE* (*ProDH*) in nonhost resistance of *N. benthamiana*against several bacterial pathogens (Rojas et al. 2012; Senthil-Kumar and Mysore 2012;Wang et al. 2012).

The next step after identification of genes involved in nonhost resistance is to successfully transfer them to susceptible host species. Genes involved in nonhost resistance are often difficult to transfer compared with host resistance genes, since they need either wide hybridization or transgenic approaches. Wide hybridization has been successfully employed in the past to transfer sources of resistance against multiple diseases in wheat by transferring individual rye chromosomes to wheat (Riley and Macer 1966). More recently, a barley leaf rust resistance gene, *Rph22*, was successfully transferred from nonhost species *Hordeum bulbosum* to *Hordeum vulgare* (cultivated barley) (Johnston et al. 2013). Similarly, a transgenic approach has also been implemented successfully in the past in which a nonhost resistance gene, *Rxo1*, from maize was transferred to rice and imparted resistance against the bacterial streak pathogen *Xanthomonas oryzae* pv. *oryzicola* (Zhao et al. 2005). Recently, another Arabidopsis gene involved in nonhost resistance,*PHYTOPHTHORA SOJAE SUSCEPTIBLE 1* (*PSS1*), was identified in the Arabidopsis *pen1-1*mutant background which is required for both pre- and post-penetration nonhost resistance against the hemibiotrophic oomycete fungal pathogen *Phytophthora sojae* and necrotrophic fungal pathogen *Fusarium virguliforme* (Sumit et al. 2012). *PSS1* encodes a glycine-rich protein (GRP), and expression of *AtGRP1* in soybean cultivar Williams-82 led to enhanced resistance against *Phytophthora sojae* and *F. virguliforme* (Sumit 2013).

CONCLUDING REMARKS

Host resistance has been studied extensively, but nonhost resistance is only beginning to be understood. Host resistance is mainly controlled by gene-for-gene resistance, whereas nonhost resistance is usually more complicated due to the involvement of multiple pathways. Due to the durability of nonhost resistance over host resistance, nonhost resistance holds great promise for agriculture. Based on the available information on host and nonhost resistance, we compared these two forms of resistance. Despite their classification into two classes, host and nonhost resistance share more similarities than differences in their mechanisms and resistance process. Due to similarities in downstream mechanisms, genes involved in nonhost resistance have been successfully transferred from nonhost to host species (Johnston et al. 2013; Sumit 2013; Zhao et al. 2005). But identification and utilization of sources of nonhost resistance in crop improvement are still challenging. We speculate that with the aid of the latest molecular biology and genomics tools, such as improved transformation methods, advanced tissue culture techniques, mutagenesis, genome editing, VIGS, and genome sequencing, more novel sources of nonhost resistance will be identified and effectively utilized in future plant breeding.

REFERENCES

1. Aarts, N., Metz, M., Holub, E., Staskawicz, B. J., Daniels, M. J., and Parker, J. E. 1998. Different requirements for EDS1 and NDR1 by disease resistance genes define at least two R gene-mediated signaling pathways in Arabidopsis. Proc. Natl. Acad. Sci. USA 95:10306-10311. 10.1073/pnas.95.17.10306
2. Anikster, Y. 1984. The formae speciales. Pages 115-130 in: The Cereal Rusts. Vol. I. W. R. Bushnell and A. P. Roelfs, eds. Academic Press, Orlando.
3. Antonovics, J., Boots, M., Ebert, D., Koskella, B., Poss, M., and Sadd, B. M. 2013. The origin of specificity by means of natural selection: Evolved and nonhost resistance in host–pathogen interactions. Evolution 67:1-9. 10.1111/j.1558-5646.2012.01793.x
4. Armstrong, M. R., Whisson, S. C., Pritchard, L., Bos, J. I., Venter, E., Avrova, A. O., Rehmany, A. P., Bohme, U., Brooks, K., Cherevach, I., Hamlin, N., White, B., Fraser, A., Lord, A., Quail, M. A., Churcher, C., Hall, N., Berriman, M., Huang, S., Kamoun, S., Beynon, J. L., and Birch, P. R. 2005. An ancestral oomycete locus contains late blight avirulence gene Avr3a, encoding a protein that is recognized in the host cytoplasm. Proc. Natl. Acad. Sci. USA 102:7766-7771. 10.1073/pnas.0500113102
5. Ashfield, T., Keen, N. T., Buzzell, R. I., and Innes, R. W. 1995. Soybean resistance genes specific for different

Pseudomonas syringae avirulence genes are allelic, or closely linked, at the RPG1 locus. Genetics 141:1597-1604.

6. Assaad, F. F., Qiu, J. L., Youngs, H., Ehrhardt, D., Zimmerli, L., Kalde, M., Wanner, G., Peck, S. C., Edwards, H., Ramonell, K., Somerville, C. R., and Thordal-Christensen, H. 2004. The PEN1 syntaxin defines a novel cellular compartment upon fungal attack and is required for the timely assembly of papillae. Mol. Biol. Cell 15:5118-5129. 10.1091/mbc.E04-02-0140
7. Ayliffe, M. A., and Lagudah, E. S. 2004. Molecular genetics of disease resistance in cereals. Ann. Bot. (Lond.) 94:765-773. 10.1093/aob/mch207
8. Azevedo, C., Sadanandom, A., Kitagawa, K., Freialdenhoven, A., Shirasu, K., and Schulze-Lefert, P. 2002. The RAR1 interactor SGT1, an essential component of R gene-triggered disease resistance. Science 295:2073-2076. 10.1126/science.1067554
9. Barrett, L. G., Thrall, P. H., Burdon, J. J., Nicotra, A. B., and Linde, C. C. 2008. Population structure and diversity in sexual and asexual populations of the pathogenic fungus Melampsora lini. Mol. Ecol. 17:3401-3415. 10.1111/j.1365-294X.2007.03503.x
10. Bent, A. F., and Mackey, D. 2007. Elicitors, effectors, and R genes: The new paradigm and a lifetime supply of questions. Annu. Rev. Phytopathol. 45:399-436. 10.1146/annurev.phyto.45.062806.094427
11. Boller, T., and He, S. Y. 2009. Innate immunity in plants: An arms race between pattern recognition receptors in plants and effectors in microbial pathogens. Science 324:742-744. 10.1126/science.1171647
12. Burdon, J. J., and Roelfs, A. P. 1985. The Effect of Sexual and Asexual Reproduction on the Isozyme Structure of Populations of Puccinia graminis. American Phytopathological Society, St. Paul, MN.
13. Buschges, R., Hollricher, K., Panstruga, R., Simons, G., Wolter, M., Frijters, A., van Daelen, R., van der Lee, T., Diergaarde, P., Groenendijk, J., Topsch, S., Vos, P., Salamini, F., and Schulze-Lefert, P. 1997. The barley Mlo gene: A novel control element of plant pathogen resistance. Cell 88:695-705. 10.1016/S0092-8674(00)81912-1
14. Buttner, D., and He, S. Y. 2009. Type III protein secretion in plant pathogenic bacteria. Plant Physiol. 150:1656-1664. 10.1104/pp.109.139089
15. Carver, T. W. L., Thomas, B. J., Ingerson-Morris, S. M., and Roderick, H. W. 1990. The role of abaxial leaf surface waxes of Lolium spp. in resistance to Erysiphe graminis. Plant Pathol. 39:573-583. 10.1111/j.1365-3059.1990.tb02536.x
16. Chisholm, S. T., Coaker, G., Day, B., and Staskawicz, B. J. 2006. Host–microbe interactions: Shaping the evolution of the plant immune response. Cell 124:803-814. 10.1016/j.cell.2006.02.008
17. Christopher-Kozjan, R., and Heath, M. C. 2003. Cytological and pharmacological evidence that biotrophic fungi trigger different cell death execution processes in host and nonhost cells during the hypersensitive response. Physiol. Mol. Plant Pathol. 62:265-275. 10.1016/S0885-5765(03)00088-2
18. Cohn, J., Sessa, G., and Martin, G. B. 2001. Innate immunity in plants. Curr. Opin. Immunol. 13:55-62. 10.1016/S0952-7915(00)00182-5
19. Collins, N. C., Thordal-Christensen, H., Lipka, V., Bau, S., Kombrink, E., Qiu, J. L., Huckelhoven, R., Stein, M., Freialdenhoven, A., Somerville, S. C., and Schulze-Lefert, P. 2003. SNARE-protein-mediated disease resistance at the plant cell wall. Nature 425:973-977. 10.1038/nature02076
20. Couch, B. C., Fudal, I., Lebrun, M. H., Tharreau, D., Valent, B., van Kim, P., Notteghem, J. L., and Kohn, L. M. 2005. Origins of host-specific populations of the blast pathogen Magnaporthe oryzae in crop domestication with subsequent expansion of pandemic clones on rice and weeds of rice. Genetics 170:613-630. 10.1534/genetics.105.041780
21. Dangl, J. L., Horvath, D. M., and Staskawicz, B. J. 2013. Pivoting the plant immune system from dissection to deployment. Science 341:746-751. 10.1126/science.1236011
22. Dangl, J. L., and Jones, J. D. G. 2001. Plant pathogens and integrated defense responses to infection. Nature 411:826-833. 10.1038/35081161
23. Day, B., Henty, J. L., Porter, K. J., and Staiger, C. J. 2011. The pathogen-actin connection: A platform for defense signaling in plants. Annu. Rev. Phytopathol. 49:483-506. 10.1146/annurev-phyto-072910-095426
24. de la Cruz, F., and Davies, J. 2000. Horizontal gene transfer and the origin of species: Lessons from bacteria. Trends Microbiol. 8:128-133. 10.1016/S0966-842X(00)01703-0
25. de Vienne, D. M., Refregier, G., Lopez-Villavicencio, M., Tellier, A., Hood, M. E., and Giraud, T. 2013. Cospeciation vs. host-shift speciation: Methods for testing, evidence from natural associations and relation to coevolution. New Phytol. 198:347-385. 10.1111/nph.12150
26. Dodds, P. N., Lawrence, G. J., Catanzariti, A. M., Teh, T., Wang, C. I. A., Ayliffe, M. A., Kobe, B., and Ellis, J. G. 2006. Direct protein interaction underlies gene-for-gene specificity and coevolution of the flax resistance genes and flax rust avirulence genes. Proc. Natl. Acad. Sci. USA 103:8888-8893. 10.1073/pnas.0602577103
27. Dodds, P. N., and Rathjen, J. P. 2010. Plant immunity: Towards an integrated view of plant–pathogen interactions. Nat. Rev. Genet. 11:539-548. 10.1038/nrg2812
28. Dong, S., Stam, R., Cano, L. M., Song, J., Sklenar, J., Yoshida, K., Bozkurt, T. O., Oliva, R., Liu, Z., Tian, M., Win, J., Banfield, M. J., Jones, A. M., van der Hoorn, R. A., and Kamoun, S. 2014. Effector specialization in a lineage of the Irish potato famine pathogen. Science 343:552-555. 10.1126/science.1246300
29. Ellis, J., Dodds, P., and Pryor, T. 2000. Structure, function and evolution of plant disease resistance genes. Curr. Opin. Plant Biol. 3:278-284. 10.1016/S1369-5266(00)00080-7

30. Fan, J., Crooks, C., Creissen, G., Hill, L., Fairhurst, S., Doerner, P., and Lamb, C. 2011. Pseudomonas sax genes overcome aliphatic isothiocyanate–mediated non-host resistance in Arabidopsis. Science 331:1185-1188. 10.1126/science.1199707

31. Flor, H. H. 1971. Current status of the gene-for-gene concept. Annu. Rev. Phytopathol. 9:275-296. 10.1146/annurev.py.09.090171.001423

32. Freytag, S., Arabatzis, N., Hahlbrock, K., and Schmelzer, E. 1994. Reversible cytoplasmic rearrangements precede wall apposition, hypersensitive cell death and defense-related gene activation in potato/Phytophthora infestans interactions. Planta 194:123-135. 10.1007/BF00201043

33. Friesen, T. L., Stukenbrock, E. H., Liu, Z., Meinhardt, S., Ling, H., Faris, J. D., Rasmussen, J. B., Solomon, P. S., McDonald, B. A., and Oliver, R. P. 2006. Emergence of a new disease as a result of interspecific virulence gene transfer. Nat. Genet. 38:953-956. 10.1038/ng1839

34. Fu, Y., Duan, X., Tang, C., Li, X., Voegele, R. T., Wang, X., Wei, G., and Kang, Z. 2014. TaADF7, an actin-depolymerizing factor, contributes to wheat resistance against Puccinia striiformis f. sp. tritici. Plant J. 78:16-30. 10.1111/tpj.12457

35. Glazebrook, J. 2001. Genes controlling expression of defense responses in Arabidopsis-2001 status. Curr. Opin. Plant Biol. 4:301-308. 10.1016/S1369-5266(00)00177-1

36. Griffis, A. H., Groves, N. R., Zhou, X., and Meier, I. 2014. Nuclei in motion: Movement and positioning of plant nuclei in development, signaling, symbiosis, and disease. Front. Plant Sci. 5:129. 10.3389/fpls.2014.00129

37. Grünwald, N. J., and Flier, W. G. 2005. The biology of Phytophthora infestans at its center of origin. Annu. Rev. Phytopathol. 43:171-190. 10.1146/annurev.phyto.43.040204.135906

38. Hafner, M. S., and Nadler, S. A. 1988. Phylogenetic trees support the coevolution of parasites and their hosts. Nature 332:258-259. 10.1038/332258a0

39. Hauck, P., Thilmony, R., and He, S. Y. 2003. A Pseudomonas syringae type III effector suppresses cell wall-based extracellular defense in susceptible Arabidopsis plants. Proc. Natl. Acad. Sci. USA 100:8577-8582. 10.1073/pnas.1431173100

40. Heath, M. C. 2000. Nonhost resistance and nonspecific plant defenses. Curr. Opin. Plant Biol. 3:315-319. 10.1016/S1369-5266(00)00087-X

41. Hirano, S. S., and Upper, C. D. 1983. Ecology and epidemiology of foliar bacterial plant pathogens. Annu. Rev. Phytopathol. 21:243-270. 10.1146/annurev.py.21.090183.001331

42. Hoch, H. C., Staples, R. C., Whitehead, B., Comeau, J., and Wolf, E. D. 1987. Signaling for growth orientation and cell differentiation by surface topography in uromyces. Science 235:1659-1662. 10.1126/science.235.4796.1659

43. Hoefle, C., Loehrer, M., Schaffrath, U., Frank, M., Schultheiss, H., and Huckelhoven, R. 2009. Transgenic suppression of cell death limits penetration success of the soybean rust fungus Phakopsora pachyrhizi into epidermal cells of barley. Phytopathology 99:220-226. 10.1094/PHYTO-99-3-0220 [Abstract]

44. Hückelhoven, R., Dechert, C., and Kogel, K. H. 2001. Non-host resistance of barley is associated with a hydrogen peroxide burst at sites of attempted penetration by wheat powdery mildew fungus. Mol. Plant Pathol. 2:199-205. 10.1046/j.1464-6722.2001.00067.x

45. Hückelhoven, R., and Kogel, K. H. 2003. Reactive oxygen intermediates in plant-microbe interactions: Who is who in powdery mildew resistance? Planta 216:891-902.

46. Inuma, T., Khodaparast, S. A., and Takamatsu, S. 2007. Multilocus phylogenetic analyses within Blumeria graminis, a powdery mildew fungus of cereals. Mol. Phylogenet. Evol. 44:741-751. 10.1016/j.ympev.2007.01.007

47. Jarosch, B., Kogel, K. H., and Schaffrath, U. 1999. The ambivalence of the barley Mlo locus: Mutations conferring resistance against powdery mildew (Blumeria graminis f. sp. hordei) enhance susceptibility to the rice blast fungus Magnaporthe grisea. Mol. Plant-Microbe Interact. 12:508-514. 10.1094/MPMI.1999.12.6.508 [Abstract]

48. Jin, Y., and Steffenson, B. J. 1999. Puccinia coronata var. hordei var. nov.: Morphology and pathogenicity. Mycologia 91:877-884. 10.2307/3761541

49. Johnston, P. A., Niks, R. E., Meiyalaghan, V., Blanchet, E., and Pickering, R. 2013. Rph22: Mapping of a novel leaf rust resistance gene introgressed from the non-host Hordeum bulbosum L. into cultivated barley (Hordeum vulgare L.). Theor. Appl. Genet. 126:1613-1625. 10.1007/s00122-013-2078-9

50. Jones, J. D., and Dangl, J. L. 2006. The plant immune system. Nature 444:323-329. 10.1038/nature05286

51. Kang, L., Tang, X., and Mysore, K. S. 2004. Pseudomonas type III effector AvrPto suppresses the programmed cell death induced by two nonhost pathogens in Nicotiana benthamiana and tomato. Mol. Plant-Microbe Interact. 17:1328-1336. 10.1094/MPMI.2004.17.12.1328 [Abstract]

52. Keen, N., Kobayashi, D., Tamaki, S., Shen, H., Stayton, M., Lawrence, D., Sharma, A., Midland, S., Smith, M., and Sims, J. 1991. Avirulence gene D from Pseudomonas Syringae pv. tomato and its interaction with resistance gene Rpg4 in soybean. Pages 37-44 in: Advances in Molecular Genetics of Plant–Microbe Interactions. Vol. 1. H. Hennecke and D. Verma, eds. Springer, The Netherlands.

53. Kim, M. G., Geng, X., Lee, S. Y., and Mackey, D. 2009. The Pseudomonas syringae type III effector AvrRpm1 induces significant defenses by activating the Arabidopsis nucleotide-binding leucine-rich repeat protein RPS2. Plant J. 57:645-653. 10.1111/j.1365-313X.2008.03716.x

54. Király, L., Kumar, J., Hückelhoven, R., and Kogel, K. H. 2002. mlo5, a resistance gene effective against a biotrophic pathogen (Blumeria graminis f. sp. hordei) confers enhanced susceptibility of barley to the necrotrophic fungus Bipolaris sorokiniana (teleomorph: Cochliobolus sativus). Acta. Biol. Szeged. 46:135-136.

55. Kobayashi, Y., Yamada, M., Kobayashi, I., and Kunoh, H. 1997. Actin microfilaments are required for the expression of nonhost resistance in higher plants. Plant Cell Physiol. 38:725-733. 10.1093/oxfordjournals.pcp.a029226

56. Kumar, J., Huckelhoven, R., Beckhove, U., Nagarajan, S., and Kogel, K. H. 2001. A compromised Mlo pathway affects the response of barley to the necrotrophic fungus Bipolaris sorokiniana (teleomorph: Cochliobolus sativus) and its toxins. Phytopathology 91:127-133. 10.1094/PHYTO.2001.91.2.127 [Abstract]

57. Kwon, C., Bednarek, P., and Schulze-Lefert, P. 2008. Secretory pathways in plant immune responses. Plant Physiol. 147:1575-1583. 10.1104/pp.108.121566

58. Li, X., Lin, H., Zhang, W., Zou, Y., Zhang, J., Tang, X., and Zhou, J. M. 2005. Flagellin induces innate immunity in nonhost interactions that is suppressed by Pseudomonas syringae effectors. Proc. Natl. Acad. Sci. USA 102:12990-12995. 10.1073/pnas.0502425102

59. Li, W., Xu, Y. P., Zhang, Z. X., Cao, W. Y., Li, F., Zhou, X., Chen, G. Y., and Cai, X. Z. 2012. Identification of genes required for nonhost resistance to Xanthomonas oryzae pv. oryzae reveals novel signaling components. PLoS ONE 7:e42796. 10.1371/journal.pone.0042796

60. Lipka, U., Fuchs, R., and Lipka, V. 2008. Arabidopsis non-host resistance to powdery mildews. Curr. Opin. Plant Biol. 11:404-411. 10.1016/j.pbi.2008.04.004

61. Lipka, V., Dittgen, J., Bednarek, P., Bhat, R., Wiermer, M., Stein, M., Landtag, J., Brandt, W., Rosahl, S., Scheel, D., Llorente, F., Molina, A., Parker, J., Somerville, S., and Schulze-Lefert, P. 2005. Pre- and postinvasion defenses both contribute to nonhost resistance in Arabidopsis. Science 310:1180-1183. 10.1126/science.1119409

62. Loehrer, M., Langenbach, C., Goellner, K., Conrath, U., and Schaffrath, U. 2008. Characterization of nonhost resistance of Arabidopsis to the Asian soybean rust. Mol. Plant-Microbe Interact. 21:1421-1430. 10.1094/MPMI-21-11-1421

63. Lu, M., Tang, X., and Zhou, J. M. 2001. Arabidopsis NHO1 is required for general resistance against Pseudomonas bacteria. Plant Cell 13:437-447. 10.1105/tpc.13.2.437

64. McGrann, G. R. D., Stavrinides, A., Russell, J., Corbitt, M. M., Booth, A., Chartrain, L., Thomas, W. T. B., and Brown, J. K. M. 2014. A trade off between mlo resistance to powdery mildew and increased susceptibility of barley to a newly important disease, Ramularia leaf spot. J. Exp. Bot. 65:1025-1037. 10.1093/jxb/ert452

65. Mishina, T. E., and Zeier, J. 2007. Bacterial non-host resistance: Interactions of Arabidopsis with non-adapted Pseudomonas syringae strains. Physiol. Plant. 131:448-461. 10.1111/j.1399-3054.2007.00977.x

66. Moreau, M., Degrave, A., Vedel, R., Bitton, F., Patrit, O., Renou, J. P., Barny, M. A., and Fagard, M. 2012. EDS1 contributes to nonhost resistance of Arabidopsis thaliana against Erwinia amylovora. Mol. Plant-Microbe Interact. 25:421-430. 10.1094/MPMI-05-11-0111

67. Moscou, M. J., Lauter, N., Steffenson, B., and Wise, R. P. 2011. Quantitative and qualitative stem rust resistance factors in barley are associated with transcriptional suppression of defense regulons. PLoS Genet. 7:e1002208. 10.1371/journal.pgen.1002208

68. Mysore, K. S., and Ryu, C. M. 2004. Nonhost resistance: How much do we know? Trends Plant Sci. 9:97-104. 10.1016/j.tplants.2003.12.005

69. Nakao, M., Nakamura, R., Kita, K., Inukai, R., and Ishikawa, A. 2011. Non-host resistance to penetration and hyphal growth of Magnaporthe oryzae in Arabidopsis. Sci. Rep. 1:171. 10.1038/srep00171

70. Niks, R. E., and Marcel, T. C. 2009. Nonhost and basal resistance: How to explain specificity? New Phytol. 182:817-828. 10.1111/j.1469-8137.2009.02849.x

71. Nimchuk, Z., Eulgem, T., Holt, B. F., 3rd, and Dangl, J. L. 2003. Recognition and response in the plant immune system. Annu. Rev. Genet. 37:579-609. 10.1146/annurev.genet.37.110801.142628

72. Nyadanu, D., Akromah, R., Adomako, B., Kwoseh, C., Dzahini-Obiatey, H., Lowor, S. T., Akrofi, A. Y., and Assuah, M. K. 2012. Host plant resistance to Phytophthora pod rot in cacao (Theobroma cacao L.): The role of epicuticular wax on pod and leaf surfaces. Int. J. Bot. 8:13-21. 10.3923/ijb.2012.13.21

73. Parker, J. E., Holub, E. B., Frost, L. N., Falk, A., Gunn, N. D., and Daniels, M. J. 1996. Characterization of eds1, a mutation in Arabidopsis suppressing resistance to Peronospora parasitica specified by several different RPP genes. Plant Cell 8:2033-2046. 10.1105/tpc.8.11.2033

74. Peart, J. R., Lu, R., Sadanandom, A., Malcuit, I., Moffett, P., Brice, D. C., Schauser, L., Jaggard, D. A. W., Xiao, S., Coleman, M. J., Dow, M., Jones, J. D. G., Shirasu, K., and Baulcombe, D. C. 2002. Ubiquitin ligase-associated protein SGT1 is required for host and nonhost disease resistance in plants. Proc. Natl. Acad. Sci. USA 99:10865-10869. 10.1073/pnas.152330599

75. Pennisi, E. 2010. Armed and dangerous. Science 327:804-805. 10.1126/science.327.5967.804

76. Pretorius, Z. A., Singh, R. P., Wagoire, W. W., and Payne, T. S. 2001. Detection of virulence to wheat stem rust resistance gene Sr31 in Puccinia graminis f. sp. tritici in Uganda. Plant Dis. 84:203. 10.1094/PDIS.2000.84.2.203B

77. Raffaele, S., Farrer, R. A., Cano, L. M., Studholme, D. J., MacLean, D., Thines, M., Jiang, R. H., Zody, M. C., Kunjeti, S. G., Donofrio, N. M., Meyers, B. C., Nusbaum, C., and Kamoun, S. 2010. Genome evolution following host jumps in the Irish potato famine pathogen lineage. Science 330:1540-1543. 10.1126/science.1193070

78. Riley, R., and Macer, R. C. F. 1966. The chromosomal distribution of the genetic resistance of rye to wheat pathogens. Can. J. Genet. Cytol. 8:640-653. 10.1139/g66-079

79. Roelfs, A. P. 1978. Estimated losses caused by rust in small grain cereals in the United States, 1918-76. U.S. Department of Agriculture, Agricultural Research Service, U.S. Govt. Print. Off., Washington.

80. Roelfs, A. P. 1982. Effects of barberry eradication on stem rust in the United States. Plant Dis. 66:177-181. 10.1094/PD-66-177

81. Rojas, C. M., Senthil-Kumar, M., Wang, K., Ryu, C. M., Kaundal, A., and Mysore, K. S. 2012. Glycolate oxidase modulates reactive oxygen species-mediated signal transduction during nonhost resistance in Nicotiana benthamiana and Arabidopsis. Plant Cell 24:336-352. 10.1105/tpc.111.093245

82. Rosewich, U. L., and Kistler, H. C. 2000. Role of horizontal gene transfer in the evolution of fungi. Annu. Rev. Phytopathol. 38:325-363. 10.1146/annurev.phyto.38.1.325

83. Schulze-Lefert, P., and Panstruga, R. 2011. A molecular evolutionary concept connecting nonhost resistance, pathogen host range, and pathogen speciation. Trends Plant Sci. 16:117-125. 10.1016/j.tplants.2011.01.001

84. Senthil-Kumar, M., and Mysore, K. S. 2012. Ornithine-delta-aminotransferase and proline dehydrogenase genes play a role in non-host disease resistance by regulating pyrroline-5-carboxylate metabolism-induced hypersensitive response. Plant Cell Environ. 35:1329-1343. 10.1111/j.1365-3040.2012.02492.x

85. Senthil-Kumar, M., and Mysore, K. S. 2013. Nonhost resistance against bacterial pathogens: Retrospectives and prospects. Annu. Rev. Phytopathol. 51:407-427. 10.1146/annurev-phyto-082712-102319

86. Shan, W., Cao, M., Leung, D., and Tyler, B. M. 2004. The Avr1b locus of Phytophthora sojae encodes an elicitor and a regulator required for avirulence on soybean plants carrying resistance gene Rps1b. Mol. Plant-Microbe Interact. 17:394-403. 10.1094/MPMI.2004.17.4.394

87. Shibata, Y., Kawakita, K., and Takemoto, D. 2011. SGT1 and HSP90 are essential for age-related resistance of Nicotiana benthamiana against the oomycete pathogen Phytophathora infestans. Physiol. Mol. Plant Pathol. 75:120-128. 10.1016/j.pmpp.2010.10.001

88. Singh, R. P., Hodson, D. P., Huerta-Espino, J., Jin, Y., Bhavani, S., Njau, P., Herrera-Foessel, S., Singh, P. K., Singh, S., and Govindan, V. 2011. The emergence of Ug99 races of the stem rust fungus is a threat to world wheat production. Annu. Rev. Phytopathol. 49:465-481. 10.1146/annurev-phyto-072910-095423

89. Stam, R., Mantelin, S., McLellan, H., and Thilliez, G. 2014. The role of effectors in nonhost resistance to filamentous plant pathogens. Front. Plant Sci. 5:582. 10.3389/fpls.2014.00582

90. Stein, M., Dittgen, J., Sanchez-Rodriguez, C., Hou, B. H., Molina, A., Schulze-Lefert, P., Lipka, V., and Somerville, S. 2006. Arabidopsis PEN3/PDR8, an ATP binding cassette transporter, contributes to nonhost resistance to inappropriate pathogens that enter by direct penetration. Plant Cell 18:731-746. 10.1105/tpc.105.038372

91. Sumit, R. 2013. Investigation of the Arabidopsis nonhost resistance mechanism against the soybean pathogen, Phytophthora sojae. Graduate Thesis and Dissertations, Iowa State University.

92. Sumit, R., Sahu, B. B., Xu, M., Sandhu, D., and Bhattacharyya, M. K. 2012. Arabidopsis nonhost resistance gene PSS1 confers immunity against an oomycete and a fungal pathogen but not a bacterial pathogen that cause diseases in soybean. BMC Plant Biol. 12:87. 10.1186/1471-2229-12-87

93. Tadege, M., Wen, J., He, J., Tu, H., Kwak, Y., Eschstruth, A., Endre, G., Zhao, P. X., Chabaud, M., Ratet, P., and Mysore, K. S. 2008. Large scale insertional mutagenesis using the Tnt1 retrotransposon in the model legume Medicago truncatula. Plant J. 54:335-347. 10.1111/j.1365-313X.2008.03418.x

94. Takemoto, D., Jones, D. A., and Hardham, A. R. 2003. GFP-tagging of cell components reveals the dynamics of subcellular re-organization in response to infection of Arabidopsis by oomycete pathogens. Plant J. 33:775-792. 10.1046/j.1365-313X.2003.01673.x

95. Thordal-Christensen, H. 2003. Fresh insights into processes of nonhost resistance. Curr. Opin. Plant Biol. 6:351-357. 10.1016/S1369-5266(03)00063-3

96. Tsuba, M., Katagiri, C., Takeuchi, Y., Takada, Y., and Yamaoka, N. 2002. Chemical factors of the leaf surface involved in the morphogenesis of Blumeria graminis. Physiol. Mol. Plant Pathol. 60:51-57. 10.1006/pmpp.2002.0376

97. Uppalapati, S. R., Ishiga, Y., Doraiswamy, V., Bedair, M., Mittal, S., Chen, J. H., Nakashima, J., Tang, Y. H., Tadege, M., Ratet, P., Chen, R. J., Schultheiss, H., and Mysore, K. S. 2012. Loss of abaxial leaf epicuticular wax in Medicago truncatula irg1/palm1 mutants results in reduced spore differentiation of anthracnose and nonhost rust pathogens. Plant Cell 24:353-370. 10.1105/tpc.111.093104

98. van Baarlen, P., van Belkum, A., Summerbell, R. C., Crous, P. W., and Thomma, B. P. 2007. Molecular mechanisms of pathogenicity: How do pathogenic microorganisms develop cross-kingdom host jumps? FEMS Microbiol. Rev. 31:239-277. 10.1111/j.1574-6976.2007.00065.x

99. van der Biezen, E. A., and Jones, J. D. G. 1998. The NB-ARC domain: A novel signaling motif shared by plant resistance gene products and regulators of cell death in animals. Curr. Biol. 8:R226-R228. 10.1016/S0960-9822(98)70145-9

100. van der Hoorn, R. A., and Kamoun, S. 2008. From guard to decoy: A new model for perception of plant pathogen effectors. Plant Cell 20:2009-2017. 10.1105/tpc.108.060194

101. van Wees, S. C., and Glazebrook, J. 2003. Loss of non-host resistance of Arabidopsis NahG to Pseudomonas syringae pv. phaseolicola is due to degradation products of salicylic acid. Plant J. 33:733-742. 10.1046/j.1365-313X.2003.01665.x

102. Vega-Arreguin, J. C., Jalloh, A., Bos, J. I., and Moffett, P. 2014. Recognition of an Avr3a homologue plays a major role in mediating nonhost resistance to Phytophthora capsici in Nicotiana species. Mol. Plant-Microbe Interact. 27:770-780. 10.1094/MPMI-01-14-0014-R

103. Vera Cruz, C. M., Bai, J., Oña, I., Leung, H., Nelson, R. J., Mew, T. W., and Leach, J. E. 2000. Predicting durability of a disease resistance gene based on an assessment of the fitness loss and epidemiological consequences of

avirulence gene mutation. Proc. Natl. Acad. Sci. USA 97:13500-13505. 10.1073/pnas.250271997

104. Vleeshouwers, V. G., van Dooijeweert, W., Govers, F., Kamoun, S., and Colon, L. T. 2000. The hypersensitive response is associated with host and nonhost resistance to Phytophthora infestans. Planta 210:853-864. 10.1007/s004250050690

105. Wang, K., Senthil-Kumar, M., Ryu, C. M., Kang, L., and Mysore, K. S. 2012. Phytosterols play a key role in plant innate immunity against bacterial pathogens by regulating nutrient efflux into the apoplast. Plant Physiol. 158:1789-1802. 10.1104/pp.111.189217

106. Wang, K., Uppalapati, S. R., Zhu, X., Dinesh-Kumar, S. P., and Mysore, K. S. 2010. SGT1 positively regulates the process of plant cell death during both compatible and incompatible plant-pathogen interactions. Mol. Plant Pathol. 11:597-611.

107. Wang, X., Richards, J., Gross, T., Druka, A., Kleinhofs, A., Steffenson, B., Acevedo, M., and Brueggeman, R. 2013. The rpg4-mediated resistance to wheat stem rust (Puccinia graminis) in barley (Hordeum vulgare) requires Rpg5, a second NBS-LRR gene, and an actin depolymerization factor. Mol. Plant-Microbe Interact. 26:407-418. 10.1094/MPMI-06-12-0146-R [Abstract]

108. Watson, I. A. 1981. Wheat and its rust parasites in Australia. Pages 129-147 in: Wheat Science—Today and Tomorrow. L. T. Evans and W. J. Peacock, eds. Cambridge University Press, Cambridge.

109. Wiermer, M., Feys, B. J., and Parker, J. E. 2005. Plant immunity: The EDS1 regulatory node. Curr. Opin. Plant Biol. 8:383-389. 10.1016/j.pbi.2005.05.010

110. Wulff, B. B. H., Horvath, D. M., and Ward, E. R. 2011. Improving immunity in crops: New tactics in an old game. Curr. Opin. Plant Biol. 14:468-476. 10.1016/j.pbi.2011.04.002

111. Zhao, B., Ardales, E. Y., Raymundo, A., Bai, J., Trick, H. N., Leach, J. E., and Hulbert, S. H. 2004. The avrRxo1 gene from the rice pathogen Xanthomonas oryzae pv. oryzicola confers a nonhost defense reaction on maize with resistance gene Rxo1. Mol. Plant-Microbe Interact. 17:771-779. 10.1094/MPMI.2004.17.7.771 [Abstract]

112. Zhao, B., Lin, X., Poland, J., Trick, H., Leach, J., and Hulbert, S. 2005. A maize resistance gene functions against bacterial streak disease in rice. Proc. Natl. Acad. Sci. USA 102:15383-15388. 10.1073/pnas.0503023102

113. Zipfel, C. 2008. Pattern-recognition receptors in plant innate immunity. Curr. Opin. Immunol. 20:10-16. 10.1016/j.coi.2007.11.003

CHAPTER 7

THE REACTIVE OXYGEN SPECIES NETWORK PATHWAYS: AN ESSENTIAL PREREQUISITE FOR PERCEPTION OF PATHOGEN ATTACK AND THE ACQUIRED DISEASE RESISTANCE IN PLANTS

Simeon O Kotchoni and Emma W Gachomo

Keywords: Disease resistancegene expressionpathogensROSsignal transduction

INTRODUCTION

The plant pathogens have developed various independent and well-elaborated mechanisms of penetrating and accessing plant cell contents. Stopping the penetration of pathogens during plant infection is generally dependent on the accurate time-course of the pathogen perception by the plant host cells and the activation of networking systems resulting in induction of secondary metabolites, reactive oxygen species (ROS) and pathogenesis related proteins, working often in combination to mount an adequate defence mechanism against the pathogen infection (Bolwell et al 2001). This mechanism includes modifications of the plant cell wall, deposition of callose containing papillae and production of hydroxyproline rich glycoproteins and phenolic compounds (Mazau and Esquerré-Tugayé 1986; Aist and Bushnell 1991). Being sessile organisms, the plants have evolved various other ways to hinder the pathogens attempting to enter the cells. Here we selectively review the molecular implication of ROS network mediating the perception and induction of defence mechanisms to hinder pathogen infection in plants.

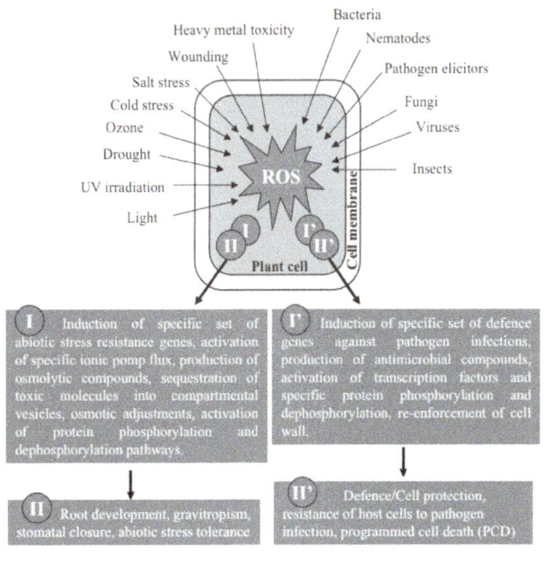

Figure 1: Involvement of ROS in cellular metabolic processes of plants response to both various environmental stresses. I

Citation: Simeon O. Kotchoni, and Emma W. Gachomo, The reactive oxygen species network pathways: an essential prerequisite for perception of pathogen attack and the acquired disease resistance in plants, doi:10.1007/BF02704112

and II indicate the subsequent downstream events mediating by ROS in plant cells exposed to abiotic stresses, while I' and II' indicate the subsequent downstream events mediating by ROS in plants cells exposed to pathogens and pathogen-elicitors. The resistance of the plant cells to the stress condition is dependent of the intensity and the speed of these downstream events.

ROS, which include superoxide radical, hydrogen peroxide and singlet oxygen are ubiquitous molecules produced as a consequence of normal cellular metabolism (Kotchoni 2004). Under normal conditions, ROS are rapidly metabolised with help of constitutive antioxidant enzymes and other metabolites via non-enzymatic pathways such as antioxidant vitamins, proteins and non-protein thiols (Kovtun et al 2000; Lim et al 1993; Scandalios et al 1997). However, when subjected to environmental stresses such as cold, high light, ozone (O_3), drought, salt stress, pathogens, and UV irradiation, excessive ROS is generated (figure 1). This excessive accumulation of ROS necessitates the activation of additional defences (Doke 1997; Scandalios et al 1997). Unless these ROS are effi ciently metabolized, they rapidly oxidize and damage lipids in cellular membranes, proteins, and other cellular components leading consequently to cellular dysfunction and ultimately to cell death, or appearance of necrotic lesions (Doke 1997; Foyer and Noctor 2005).

Recently, responses of plants to various environmental stresses have been analysed extensively at biochemical and molecular level (Gachomo et al 2003; Rizhsky et al 2004, 1996; Kotchoni et al 2005). Evidence is increasing that ROS are required for different aspects of the life cycle including adaptation processes to environmental stress conditions (Bartels 2001; Ramanjulu and Bartels 2002). In animals, ROS have been implicated in infl ammation, and cancer, but in plant-microbe interactions, ROS have just been recently suggested to participate in plant defence system in a number of ways including acting as signalling agents (Lamb and Dixon 1997; Gachomo and Kotchoni 2006) or causing reinforcement of cell wall through oxidative cross-linking (Brisson et al 1994; Brown et al 1998; McLusky et al 1999; Mellersh et al 2002). Various enzymatic mechanisms for generation of ROS, the type and source of ROS have been shown to vary depending on the plant-pathogen combination (Bolwell 1999). ROS have taken a centre stage in signal transduction network of stress-inducible genes in higher plants (Bartels 2001; Mittler et al 2004; Davletova et al 2005) and it is emerging that a balanced amount of ROS is crucial for many different metabolic processes in plants (Bartels 2001; Kotchoni 2004; Kotchoni et al 2006; Davletova et al 2005). Manipulating the ROS levels in plants has been proposed as a promising way to generate transgenic plants that can resist multiple environmental stresses (Bartels 2001). We emphasize on the accumulation of ROS in plants response to pathogen infections with the aim of elucidating the mechanisms of disease resistance acquisition in higher plants. This review describes new research trends and highlights several stress-inducible gene families of ROS network pathways that trigger a rapid pathogen attack perception and disease resistance in plants by using bioinformatic tools and software applications (Zimmermann et al 2004; Schmid et al 2005).

PRODUCTION SITES OF ROS IN PLANT CELLS

ROS are generally produced during aerobic phase of photosynthesis and photorespiration (Asada and Takahashi 1987; Mittler 2002; Kotchoni et al 2006). Accumulation of these molecules can also be detected in peroxisomes under abiotic stress (Ramanjulu and Bartels 2002; Mittler 2002) and biotic stress (Mittler 2002). During cellular metabolism, oxygen molecules are often converted into several intermediates such as $O_2^{\ -}$, H_2O_2, hydroxyl radical, which often leak out from electron transport chain (Banerjee et al 2003), and can therefore be detected as traces in various cell compartments. New sources of ROS production include cell-wall-bound peroxidase, chloroplasts, and mitochondria (Mittler 2002; Davletova et al 2005). The chloroplast is considered to be a focal point of ROS metabolism. It is a major producer of $O_2^{\ -}$ and H_2O_2 and contains also a large array of ROS-scavenging mechanisms that have been extensively studied (Davletova et al 2005). Most of the oxygen turnover in aerobic organism is utilized in the mitochondria for substrate metabolism and production of ATP. The production of ROS is related to environmental stresses, among which the most predominant are drought/ desiccation, salinity, heat shock, heavy metals, UV radiation, ozone, nutrient deprivation, and pathogen attack (Bartels 2001; Davletova et al 2005; Foyer and Noctor 2005; Gachomo 2005). Despite being part of a normal process in the life of aerobic organism, accumulation of ROS is the source of oxidative damage and has been considered also as part of a defensive mechanism of cells. The production of ROS is recently shown to be the underlying mechanism of a series of biochemical and physiological changes that occur under environmental stress conditions, which subsequently mediate the disease resistance in plants (Gachomo and Kotchoni 2006).

ROS NETWORKING BEYOND THE BOUNDARY OF TOXICITY

ROS are known to play a dual role depending on their accumulation levels. High intracellular concentration of ROS can cause extensive cell injury or death. The levels of ROS need therefore to be tightly regulated to avoid cell damage (Neill et al 2002; Kotchoni 2004; Mittler et al 2004). Here we particularly focus on the benefi cial steadystate level of ROS production in plants response to pathogen attacks. ROS are partially reduced or excited forms of atmospheric oxygen (O_2) are continuously produced in cells during aerobic metabolism (Halliwell and Gutteridge 1989). A stable and balanced intracellular redox is of vital importance to all organisms. A moderate accumulation of ROS function as key inducers for secondary programmed metabolisms, defence signal, cell wall differentiation and activation of mitogen-activated protein kinases (MAPKs) leading to the environmental stress tolerance (Conrath et al 2002; Gachomo et al 2003). Different members of gene families involved in the protective mechanism against pathogen attacks were up-regulated in plants under exogenous application of ROS (H_2O_2) (Rizhsky et al 2004). Mellersh et al (2002) demonstrated that localized generation of H_2O_2 is one of the earliest cytologically detectable defence responses to penetration of plant cell walls by various fungal pathogens. Rapid generation of H_2O_2 in response to cell wall penetration is one of the most important determinants of pathogen penetration failure in invading epidermal cells (Mellersh et al 2002). H_2O_2 generation was the only response detected to account for fungal penetration failure. Enzymatic removal of H_2O_2 resulted in increased penetration success of fungi in the host plant cells (Mellersh et al 2002). Although the chemical nature and reactivity of ROS prove them to be potentially harmful to cells, plants use them as secondary messengers in signal transduction cascades regulating diverse processes such as mitosis, tropisms, cell death and defence mechanisms (Pavet et al 2005). It is now well accepted that ROS accumulation is crucial to plant development as well as in defence mechanisms (Foyer and Noctor 2005). Exogenous application of H_2O_2 was found to be essential to activate different pathogenesis-related proteins and to provide adequate protection against the pathogenic fungus Diplocarpon rosae causing black spot disease of rose leaves (Gachomo and Kotchoni 2006).

Knock-out (KO) plants deficient in ROS-scavenging proteins are of particular interest in elucidating role of ROS in disease defence systems. They have been used to study implication of high ROS content in plants response to pathogens infections. These KO-plants maintain a high steady-state level of H_2O_2 in cells and activate ROS defence mechanisms when grown under control conditions (Pnueli et al 2003; Rizhsky et al 2004; Davletova et al 2005). These null mutant plants (KO) provide an ideal experimental system to study plant responses to ROS accumulation and the effect of ROS mediating the activation of environmental stress-related proteins (Davletova et al 2005; Kotchoni et al 2006). On the other hand, transgenic plants expressing H_2O_2 -generating enzymes have been reported to display increased protection against bacterial and fungal pathogens (Wu et al 1995; Schweizer et al 1999).

ROS NETWORK MEDIATING GENE EXPRESSION IN PLANTS RESPONSE TO PATHOGEN ATTACKS

Response of plants to pathogens depends on the regulatory gene expression at genomic level through complex genetic controls infl uenced by redox regulation (Pavet et al 2005). Control gene expression is one of the key-regulatory mechanisms used by living cells to execute biological functions. Elucidating the gene expression at genomic level in higher plants is still a subject of contemporary research. Microarray technology has provided new insight into the transcriptomes involved in mechanisms of environmental stress tolerance in higher plants. Transcriptomics and proteomics are among the current and most powerful strategies for quantitative, real-time analysis of gene expressions and protein functions in the whole plant. Several robust screening methods such as DNA microarry analysis, mass spectrometry-based proteomics and forward/reverse genetics (Tyres and Mann 2003; Pavet et al 2005) are being now used to identify stress-inducible genes at genomic level. Genomics-based approaches, and bioinformatic tools are now available and being used to identify specific set of genes that are up-regulated by ROS (fi gure 2a). Bioinformatic tools have facilitated integration of experimental data into a computational framework, thereby allowing structured and systematic processing of information (Zimmermann et al 2004). Structured databases and data querying tools provide nowadays the means to assign putative functional information to genes. These tools were used to retrieve ROSrelated genes in Arabidopsis thaliana which is a model plant with a complete genome sequence (fi gure 2 a, b), and to view the expression patterns of the selected ROS-related genes in A. thaliana exposed to different biotic stresses (table 1). It is now possible to analyse consequences of various stresses by monitoring expression of several genes simultaneously and record subsequent alteration of protein-protein or proteinDNA interactions in the whole plants. As a major tool, the hybridization with cDNA arrays allows for the analysis of thousands of gene expressions, which may be extended to the whole genome of plant. For example, Shimono et al (2003) have used the cDNA microarray analysis to elucidate various genes of disease resistance in rice plants. In addition, Scheideler et al (2002) have monitored, simultaneously global changes in the A. thaliana transcriptome (the whole set of transcripts present in a cell, tissue or organism) after infecting the plant with the incompatible bacteria pathogen Pseudomonas syringae pv. tomato by using cDNA arrays comprising 13000 unique expressed tags (ESTs).

Understanding regulatory mechanisms of gene expression under pathogen attack has been the subject of many research efforts. Evidence has shown that pathogen infection leads to oxidative damaging effect in plants (Smirnoff 1998). To prevent oxidative damage, especially in chloroplasts due to excessive ROS accumulation, the chloroplasts containing multiple ROS scavenging systems – such as ascorbateglutathione cycle (Asada and Takahashi 1987) superoxide dismutase (SOD), ascorbate peroxidase (APX), and catalase (Asada 1999; Mittler 2002; Apel and Hirt 2004) – must regulate the steady-state level of ROS production in that cell compartment. For example, four different isozymes of SOD – a CuZnSOD (CSD2), three FeSODs, and enzymes of the ascorbate-glutathione cycle capable of reducing oxidized ascorbic acid and glutathione (Asada and Takahashi 1987) – were identifi ed in chloroplasts. However, Davletova et al (2005) demonstrated that in the absence of the cytosolic H_2O_2 -scavenging enzyme APX1, the entire chloroplastic H_2O_2 -scavenging system of A. thaliana collapses; H_2O_2 levels increase, and protein oxidation occurs. This suggests that the accumulation of ROS in cytosol and in other cellular compartments must be well regulated to avoid oxidation of macromolecules.

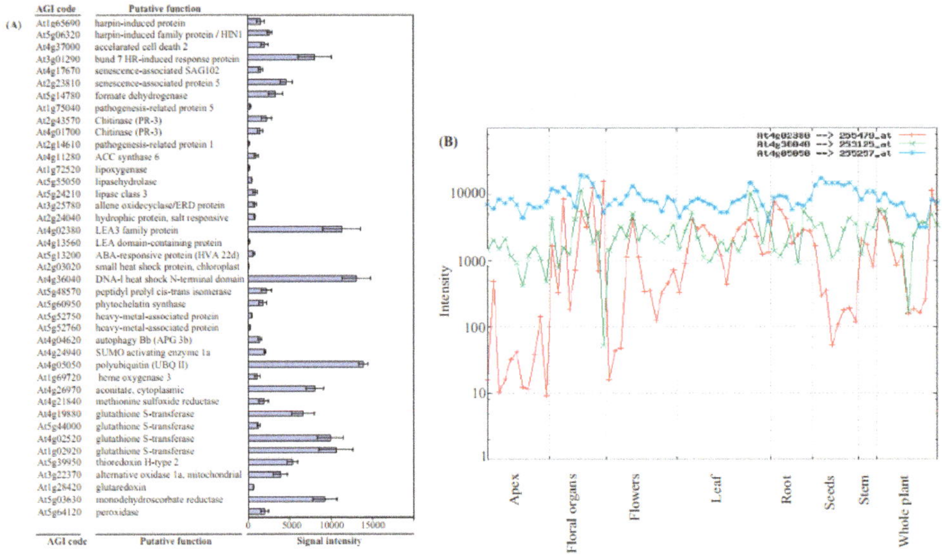

Figure 2: Expression level of ROS inducible genes in A. thaliana. Only the ROS-induced genes with an induction rate more than 100 (± SE) compared to non-treated plants are shown. (A) The Response Viewer of GENEVESTIGATOR database was used to retrieve the expression level of the genes compared to the expression profile (non-induced) in control plants. (B) Organ specific expression pattern of selected ROS-induced genes using Weigelworld's database web-browser. The 12 depicted points for each of the three genes in Apex (for example) indicate the time course considered for the probe set used (probe set code) within the life time of this specific organ for the three gene expression profiles addressed in this case.

Table 1: Expression of selected ROS-related genes and their putative functions in plant defence mechanism.

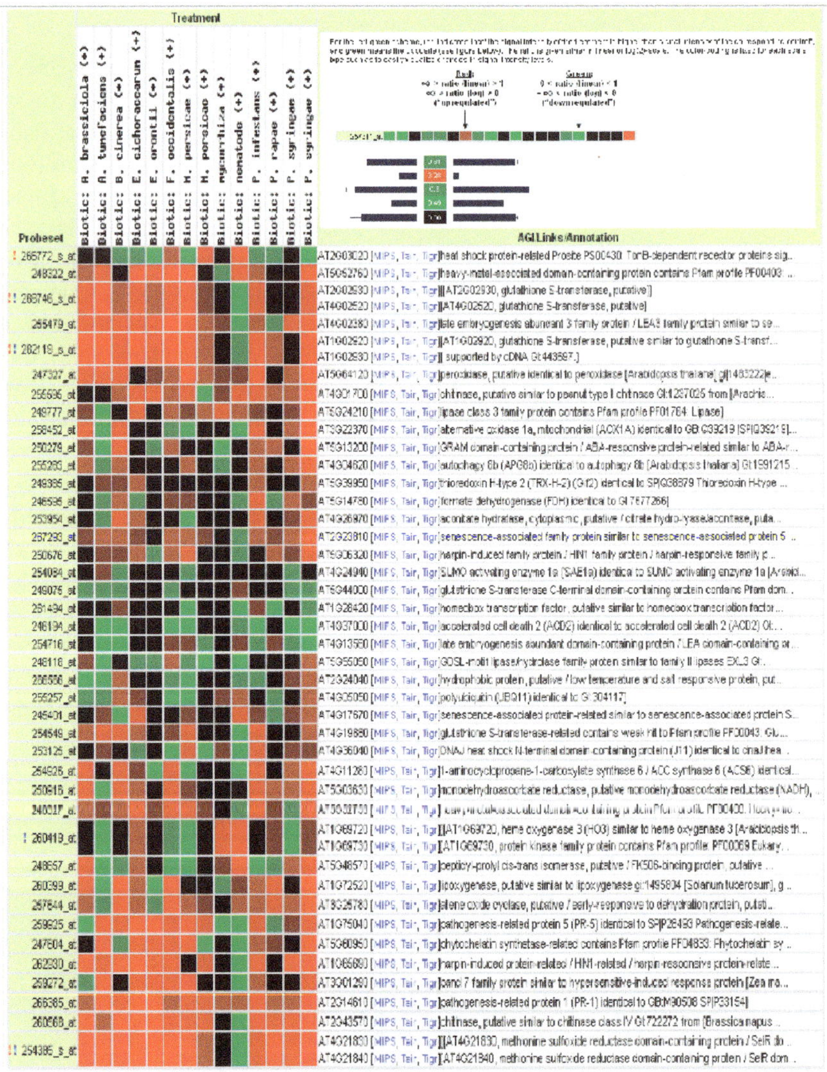

ROS MEDIATING STRESS-SENSING AND NETWORK TRANSDUCTION

Sensing and stopping pathogen attack/penetration is among the primary detecting mechanisms associated with disease resistance in plants. The detecting machinery of pathogen infection by the host cells include cross-linking of preexisting or else induced cell-wall proteins and phenolic compounds (Aist and Bushnell 1991; Brisson et al 1994; Thordal-Christensen et al 1997; McLusky et al 1999; Bolwell et al 2001; Davletova et al 2005); formation of calcium-pectate gels (Kieffer et al 2000); accumulation of glycoproteins (Mazau and Esquerré-Tugayé 1986); silica (Aist and Bushnell 1991); deposition of callose-containing pappille (Aist and Bushnell 1991) and the generation of ROS (Thordal-Christensen et al 1997; Gachomo and Kotchoni 2006). Generally, production of these compounds occurs simultaneously to mount a programmed and efficient infection arrest in plants (Perumalla and Hearth 1991). It is widely accepted that ROS may play a central role in many signalling pathways during stress perception, regulation of photosynthesis, pathogen response, hormonal action, and plant growth and development (Dat et al 2000; Mittler 2002; Mullineaux and Karpinski 2002; Neill et al 2002; Torres et al 2002; Foreman et al 2003; Kwak et al 2003; Apel and Hirt 2004). This assertion was supported by the fact that localized extracellular ROS have been detected during plant-pathogen interations (Doke 1983; ThordalChristensen et al 1997), and has been found to have an antimicrobial effect on phytopathogens (Bestwick et al 1997; Gachomo and Kotchoni 2006). H_2O_2 is an electronaccepting substrate for a wide range of plant peroxidases, the well characterized ROS-detoxifying enzymes (Yamasaki et al 1997). In this close relation of donor and acceptor of electron, various ion channels are activated, among which the Ca^{2+}-channel is the most predominant of all (Kawano et al 2000). These ion-signalling pathways lead to intracellular pH changes needed for subsequent activation of plant protective mechanisms (Kawano et al 2000). The regulatory function of the cytosolic pool upon ion fluxes has been well demonstrated by Britto and Kronzucker (2001). Their work brought insight in the pivotal role of ROS in modulating the transmenbrane ion flux regulation of plant protection against external challenges.

Under pathogen attack, ROS metabolism is regulated by a network involving at least 152 genes in A. thaliana (Mittler et al 2004). The regulatory network controls the rates of ROS production and ROS scavenging enzymes in the different cellular compartments and modulates the steady state level of ROS for signalling as well as defence purposes. The generation of H_2O_2 and the stress inducible genes act within the penetration-inhibiting pathway, possibly through the involvement of phenolic materials (Aist and

Brushnell 1991). ROS alone and especially H_2O_2 generation is necessary and suffi cient to account for fungal penetration failure in plants (Mellersh et al 2002). In addition, ROS have the potential to be directly toxic as exogenous H_2O_2 can inhibit pathogen growth in vitro (Lu and Higgins 1999) and can prevent fungal infection of detached leaves as well as leaf disk (Peng and Kuc 1992; Gachomo and Kotchoni 2006). H_2O_2 most likely acts within a pathway involving transcription/translation and the expression of wall-associated responses such as the accumulation of fungal inhibiting compounds (Aist and Brushnell 1991). Recent studies suggest that cross-talk between salicylic acid (SA)-, jasmonic acid (JA)-, and ethylene-dependent signalling pathways regulates plant responses to both biotic and abiotic stress factors. Treating cell cultures with H_2O_2 alone induces defence responses and cell death (Levine et al 1994; Solomon et al 1999). However, although sublethal H_2O_2 concentrations induce expression of defence genes, complete induction of defence genes and cell death requires additional signalling molecules such as SA at the whole-plant level (Chamnongpol et al 1998; Rao and Davis 1999). These studies have established that ROS probably require additional downstream components to transduce or amplify the signal (Van Camp et al 1998; Bolwell 1999). Among several molecules proposed to act downstream of ROS, SA, JA, and ethylene are considered to be the major regulators of plant defence responses (Penninckx et al 1998; Spoel et al 2003; Veronese et al 2006). SA is one of the most widely studied stress-signalling molecules, and its role in infl uencing plant resistance to pathogens and other stress factors is well documented (Draper 1997; Shirasu et al 1997; Surplus et al 1998; Rao and Davis 1999). Similar to SA, JA is also believed to play an important role in infl uencing plant resistance to pathogens and other stress factors (Penninckx et al 1998; Spoel et al 2003).

Production of ROS occurs generally by direct transfer of excitation energy from the chlorophyll to produce singlet oxygen, or by univalent oxygen reduction at photosystem I in the Mehler reaction (Foyer et al 1994; Allen 1995). They lead to oxidative injuries, which are mainly caused either by an imbalance of the cellular antioxidant capacity or by a deficiency in the antioxidant system controlling ROS levels (Halliwell and Gutteridge 2002). These damaging effects force the plants to develop complex redox homeostatic mechanisms to cope with the accumulation of ROS (Moon et al 2003). A slight alteration in the homeostatic level of intracellular ROS triggers therefore the expression of a specific set of genes generally encoding for antioxidant proteins, compatible molecules, oxidative scavengers, and protein phosporylation cascade via activated MAPK proteins (Borsani et al 2003).

On the other hand, accumulation of ROS inhibits mitochondrial electron transport leading to ATP depletion (Jones 2000). Both ROS accumulation and ATP depletion lead rapidly to Ca^{2+} influx, which triggers programmed cell death (PCD) in plants (Jones 2000). A moderate concentration of ROS activates the cellular defence response (Levine et al 1994). Tobacco plants inoculated with the tobacco mosaic virus (TMV) developed systemic acquired resistance (SAR) that was mediated by a burst of ROS (Lamb and Dixon 1997). Rapid production of ROS could also inhibit pathogen growth by restricting pathogen penetration via cross-linking of cell wall glycoproteins (Bradley et al 1992), by induction of phytoalexin accumulation (Gachomo et al 2003), or by induction of defence-related genes (Desikan et al 2001; Gachomo et al 2003; Gachomo and Kotchoni 2006). For example, incubation of potato tuber slices with an incompatible race of Phytophthora infestans stimulates the NADPH oxidase, which is strictly under the control of O_2^- accumulation (Doke and Miura 1995). This is evidence of potential role of ROS to mediate expression of pathogen defence mechanism in plants.

ROLE OF ROS IN MITOGEN-ACTIVATED PROTEIN KINASE (MAPK) CASCADE

The resistance of plants to pathogen infections involves an elaborate system of signal perception and transduction to activate cellular defence. After perception of pathogen attack, the cell host synthesizes various molecules including ROS, protein-kinases and transcription factors, relaying the pathogen-perception signal to activate downstream defence responses (Gachomo and Kotchoni 2006; Veronese et al 2006). The speed and intensity of activation of these downstream cascade reactions are crucial for plant resistance to pathogen infection. ROS are one of the most important components of the signalling pathways, which influence plant defence responses, including cell death (Jabs et al 1996; Alvarez et al 1998; Karpinski et al 1999; Solomon et al 1999). In plants, the signal transduction is controlled by protein phosphorylation and dephosphorylation involving MAPKs (Khokhlatchev et al 1998; Tena et al 2001). Induction of MAPK activity has been detected after exposure of plants to various stresses (Kovtun et al 2000; Xiong and Yang 2003). It was recently discovered that ROS mediate the activation of MAPK cascade and subsequent responses of plants to external stimuli (Moon et al 2003). The MAPK-signalling pathways involve three distinct components of the protein kinase family including (i) the MAPK, (ii) the MAPK kinase (MAPKK) and (iii) the MAPKK kinase (MAPKKK). During signal transduction, the MAPKKKs are phosphorylated and could then be able to activate particular MAPKKs, which become phosphorylated and then activate thereafter specifi c MAPKs. Activated MAPKs (phosphorylated MAPKs) are often imported into the nucleus, where they phosphorylate and activate specific downstream signalling components such as transcription factors (Khokhlatchev et al 1998). A connection between the activation of plant MAPK cascades and the mediating action of ROS ($H_2 O_2$) has been demonstrated using the Arabidopsis protoplast transient expression assay (Tena et al 2001). This interactive process underlines the pivotal role of ROS as the key activator of the MAPK cascade and the subsequent expression of the targeted genes. Kovtun et al (2000) showed that the ANP (Nicotinama protein kinase) class of MAPK kinase-kinases (MAPKKKs) from Arabidopsis is induced specifically by $H_2 O_2$, which subsequently induced the activation of other set of specifi c classes of MAPKs under environmental stresses. Moon et al (2003) demonstrated that increased accumulation of ROS in transgenic plants activates several MAPKs, which regulates the state of cellular redox and enhances tolerance to multiple stresses. Their results demonstrated that NDP kinase (NDPK) is associated with hydrogen peroxide-mediated MAPK signalling in plants. In addition, $H_2 O_2$ induces the expression of the NADPK2 gene in A. thaliana (AtNDPK2) and that of AtMAPK3 and AtMAPK6 (Moon et al 2003) and proteins from transgenic plants overexpressing AtNDPK2 showed high levels of autophosphorylation and NDPK activity than the-wild type plants (Moon et al 2003).

NITRIC OXIDE IN PLANTS: ANOTHER FASCINATING ROS MOLECULE

ROS ($H_2 O_2$, O_2^- and nitric oxide) have been shown to serve as diffusible intra and intercellular signals for activation of various defence genes in animals, bacteria and plants (Durner et al 1998). Nitric oxide (NO) in particular has recently emerged as one of the pivotal ROS mediators of various physiological, biochemical and stress responses such as infl ammation, acute phase responses, PCD and disease resistance in plants (Durner et al 1998; Zeidler et al 2004). The various activities of NO, including redox signalling have been explored principally in animals. Based on its molecular stability and reactivity, NO and its redox-activated derivates are regarded as the ROS that can fulfi ll the requirements of a true intra- and intercellular signalling molecule (Stamler 1994). A major advance in our understanding of the multiple functions of NO in plants has been the identifi cation of enzymes that catalyze NO synthesis. Nitrate reductase (NR) was the fi rst enzymatic source of NO to be identifi ed in plants (Yamasaki and Sakihama 2000). In addition to its role in nitrate reduction, NR catalyzes the reduction of nitrite to NO using NAD(P)H as co-factor (Meyer et al 2004). The observation that NO produced by NR is required for abscisic acid (ABA)-induced stomatal closure in A. thaliana guard cells (Desikan et al 2002), suggests the involvement of NO in signalling processes leading to abiotic stress tolerance such as drought and osmotic stress through the ABA-regulatory pathway. A similar result of drought and salt stress tolerance was also observed in A. thaliana after $H_2 O_2$ treatment (Kotchoni et al 2006). Another enzymatic source of NO in plants, the Nicotiana-nitric oxide reductase (Ni-NOR), has been identified (Stöhr et al 2001). Ni–NOR, a 310 kDa plasma membrane-bound enzyme from tobacco has been shown to catalyze the reduction of nitrite to NO using cytochrome C as an electron donor (Stöhr et al 2001). Recently, a protein displaying NO synthase (NOS) activity has also been identifi ed in A. thaliana. This protein, termed as 'AtNOS1' (Guo et al 2003), is unrelated to previously described mammalian NOS proteins, and belongs to a novel family of putative NOS which is conserved in eukaryotes and bacteria (Zemojtel et al 2004). These proteins contain a GTP-binding domain and possess a GTPase activity which is Ca^{2+}-CaM/NADPH-dependent, but FAD, FMN, BH_4, and heme-independent. The characterization of AtNOS1 and Atnos1 mutant mark an important step in elucidating the multiple functions of NO in plants (Guo et al 2003). In addition, the identifi cation of nox1, an A. thaliana mutant overproducing NO (He et al 2004), paved a way for the functional analysis of NO. AtNOS1 was therefore confi rmed as a source of NO in ABA-induced stomatal closure (Guo et al 2003), in the repression of fl owering (He et al 2004), and in lipopolysaccharides-induced defence responses (Zeidler et al 2004). In animals, NO is known to act as a redox transmitter in the regulation of a diverse array of physiological processes, and NOS plays a central role in host responses to pathogen infection. Particularly two key enzymes involved in mammalian macrophage action, NADPH oxidase and NOS, have been found to play a crucial role in plant defence system during plantmicrobe interactions (Durner et al 1998). The analysis of the phenotypic traits of Atnos1 mutants with an impaired NO synthesis indicated that AtNOS1 also plays a central role in plant development as Atnos1 plants showed reduced shoot, root, growth, and fertility compared to the wild type plants (Guo et al 2003). Several arguments suggest that AtNOS1 could also be involved in NO synthesis in the process of senescence of pea leaves (Lamotte et al 2005). In addition to AtNOS1, another NOS-like protein has been reported in plant peroxisomes (Corpas et al 2004). It has been observed that NO production occurs within the same time frame with that of H_2O_2, and a critical balance between the two ROS regulates cellular outcomes such as sensitivity or resistance to a given stress condition (Delledonne et al 2001), because H_2O_2 and NO have been shown to regulate in tandem the expression of a number of common genes whose products are involved in limiting pathogen growth, in cellular protection or in other signalling responses (Desikan et al 2001; Huang et al 2002). For example, a correlating accumulation of H_2O_2 or O_2^- and NO is often observed during the oxidative burst preceding the hypersensitive response (HR), cell death and disease resistance (Grün et al 2006). The evidence that NO, in combination with other ROS, is required for cell death and induced disease resistance was given by Delledonne et al (1998). NO was found to react with O_2^- and/or H_2O_2 to the non-HR-inducing peroxynitrite ($ONOO^-$). The importance of the fi ne-tuning of NO/H_2O_2 balance was demonstrated using thylakoidal ascorbate peroxidases (tAPX) mutants with enhanced or reduced expression of tAPX gene (Murgia et al 2004; Tarantino et al 2005). Antisense reduction of tAPX in A. thaliana enhances paraquat-induced H_2O_2 accumulation and NO-induced cell death and disease resistance (Tarantino et al 2005), confi rming therefore the correlation of H_2O_2 and NO acting in tandem to induce multiple signal transductions. Understanding the mechanisms by which NO-based signals are initiated, processed, and propagated in plant cells will require the elucidation of subcellular location of the enzymes implicated in NO synthesis and the identifi cation of their cofactors. NO modulates their activities by nitrosylation or, in certain cell types including endothelial and neuronal cells, via indirect signalling pathways involving cyclic GMP and/or cyclic ADP ribose (cADPR) (Stamler et al 2001; Willmott et al 1996). In plants, NO treatment has indeed been shown to increase cGMP levels both in tobacco and A. thaliana (Durner et al 1998; Clarke et al 2000). NO primary targets in plant cells include Ca^{2+} channels and MAPK, because artifi cially generated NO stimulated MAPK in both tobacco and A. thaliana leaves (Clarke et al 2000; Kumar et al 2000; Capone et al 2004). How is NO integrated in these various plant physiological processes? Microarrays analyses of NOresponsive transcripts in A. thaliana, show that NO governs the regulation of expression of numerous genes via Ca^{2+} channels, cGMP, cADPR and phytohormones. These genes encode proteins related to defense responses, metabolism, cellular detoxifi cation, transport, iron homeostasis, signalling, fl owering, and lignin biosynthesis (Durner et al 1998; Polverari et al 2003; Parani et al 2004). NO also modulates the synthesis of SA, JA, and ethylene involved in wound responses (Huang et al 2004; Parani et al 2004).

A full understanding of NO signalling in plants also requires information about mechanisms that switch off NO signals and/or protect cells from excess of NO. For example, Hausladen et al (1998) reported that the bacterial fl avohaemoglobin functions as a dioxygenase, metabolizing NO into nitrate, controlling therefore the excessive accumulation of NO. In plants, overexpression of class 1 non-symbiotic haemoglobin was shown to reduce NO levels and consequently protects alfalfa root cultures challenged by hypoxic conditions (Dordas et al 2003). Besides the haemoglobin, the involvement of S-nitrosoglutathione (GSNO) reductase (also named glutathione dependent formaldehyde dehydrogenase, or FALDH in plants) has also been reported (Sakamoto et al 2002; Diaz et al 2004). In humans, GSNO, which is one of the major metabolites of NO formed primarily by nitrosylation of GSH, serves as a reservoir of NO and is believed to mediate some of its action. By converting GSNO to GSSG and NH_3, GSNO reductase controls intracellular levels of GSNO and limits NO toxicity. Like its mammalian counterpart, FALDH from A. thaliana displays GSNO-reductase activity in vitro (Sakamoto et al 2002), suggesting its involvement in the cellular protection against nitrosative stress. The NO donors GSNO and Snitroso-N-acetyl-penicillamine (SNAP), as well as the membrane analog of the NO second messenger cGMP were found to induce phenylalanine ammonia lyase (PAL) expression in tobacco suspension cells. PAL is involved in the biosynthesis of SA (Durner et al 1998) confi rming that NO indirectly mediate the synthesis of SA. Administration of NO donors or recombinant mammalian NOS to tobacco plants triggered expression of the defence-related gene encoding pathogenesis-related 1 (PR-1) protein. PR-1 exhibits antifungal activity and considered as an excellent marker of plant disease resistance against pathogens.

One of the major questions to answer is how the correct specific response is evoked and how ROS particularly NO and H_2O_2, which shared common secondary messengers are coordinated in the plants to induce multiple signal transductions. Based on overwhelming research contributions, NO has been shown to trigger the expression of a wide range of genes and transcription factors through cGMP/ cADPR and S-nitosylation via direct regulation of ion channels and phytohormones such as ABA, gibbereline, SA and others (Stamler 1994). Several lines of evidence point to an interrelationship between NO, SA and wounding/JAsignalling pathway, because NO donors are found to affect both wounding-induced H_2O_2 synthesis and wounding- or JA-induced expression of defence genes (Grün et al 2006). The beneficial effects of this fascinating molecule, NO, are overwhelming and far from exhausted. Till now, our understanding of how NO participates in signalling function remains rudimentary. NO affects several aspects of the plant development, including seed germination, primary and lateral root growth, pollen tube growth regulation, fertility, circadian clock regulation, flowering, senescence, defence response and abiotic stress. Discussing these in detail will require a book chapter but because of the page limitation, we have chosen to focus on NO-mediating stress responses and signal transduction pathways of physiological importance in plants and selectively discussed here the molecular implication of NO in different metabolic pathways that fits the scope of this topic.

ROLE OF PEROXISOMAL ROS IN PLANT SIGNAL TRANSDUCTION

Peroxisomes are subcellular organelles like chloroplasts and mitochondria, with a potential of generation and integration of ROS into signal transduction regulation the expression of various genes (del Río et al 2002). Plants peroxisomes have been demonstrated as source of superoxide radicals, and H_2O_2 production (Corpas et al 2001; del Río et al 2002). At least, two sites of superoxide generation have been identifi ed and characterized in peroxisomal organelles: one in the organelle matrix; the generating system being xanthine oxidase, and the other site is located in the peroxisomal membrane and depends on NAD(P)H. In this second site, integrated peroxisomal membrane polypeptides (PMPs) are the generating sources of the ROS especially the superoxide radicals. Three PMPs with 18, 29 and 32 kDa have been characterized in peroxisomal membrane and proved to produced essential superoxide radicals necessary for the reducing power to transform NAD(P) H to NAD(P)+ initiating therefore the peroxisomal metabolism in the matrix of the organelle, which includes various cascade reactions such as ascorbate-glutathione cycle, catalase-peroxidase cycle, generation of GSNO, conversion of xanthine into ONOO- and production of NO via peroxisomal NOS using L-arginine. Identifi cation of NOS in plant peroxisomes (del Río et al 2002) confi rms that these organelles, with their ability to produce superoxide radicals and $H_2 O_2$ could function in plant cells as a source of signal molecules like NO and S-nitrisoglutathione (GSNO) to trigger various metabolic reactions including plants' response to external stimuli. Research on NO has gained considerable attention due to the function of NO in plant growth and response to the environment, and as a key-signalling molecule in different intracellular processes. For instance, the physiological function of NO in plants mainly involves the induction of different processes, including the expression of defence-related genes against pathogens and apoptosis/ PCD, maturation and senescence, stomatal closure, seed germination, root development and the induction of ethylene emission (Durner et al 1998; Guo et al 2003, He et al 2004). Different experimental evidences suggested that peroxisomes function as subcellular source of ROS production and particularly NO required as signal molecules to mediate plant disease resistance, plant growth, and resistance to multiple environmental stresses (Corpas et al 2001; del Río et al 2004). In addition to peroxisomal organelles, research contributions indicate that there are other potential enzymatic sources of ROS particularly NO in plants, including cytosolic xanthine oxidoreductase, peroxidase, cytochrome P450, and some hemeproteins, but the important role of plant peroxisomes in a variety of metabolic reactions is here underlined. NO and $H_2 O_2$ can permeate the peroxisomal membrane and superoxide radicals can be produced on the cytosolic side of the membrane. The generating signal molecules (NO, $H_2 O_2$ and superoxide-radicals) of peroxisomes have important implications for cellular metabolism, particularly under abiotic and biotic stress.

ROLE OF ROS AS POTENTIAL MEDIATORS IN APOPTOSIS AND ACQUIRED DISEASE RESISTANCE IN PLANTS

Eukaryotes such as plants, animals and yeast have evolved ways of cellular suicide that are known as PCD, a restricted zone of cell death to effectively contain pathogens at their site of entry, with the purpose of removing unwanted, damaged or infected cells. The term 'apoptosis' generally referred to a PCD observed in biological systems during their life cycle, and especially under environmental stress or pathogen attack (Kerr et al 1972). In other words, apoptosis/PCD is a general biological phenomenon of targeted self-cell suicide with the aim of overcoming the invading organism/system or maintaining homeostasis of the cellular metabolism at a normal biological status (Beers and McDowell 2001). The well-characterized model system for study of plantPCD/apoptosis is the HR, which is often observed during plant-microbe interactions (Wyllie et al 1980; Beers and McDowell 2001). The cytological network events that mediate cell-death activation in several plant-PCD model systems include various phytohormones such as SA, JA, ABA, ethylene, the ROS (NO, $H_2 O_2$, O_2^-), the lipid related signals (sphyncolipids) and the ion fl ux signal pathways (Guo et al 2003; Khurana et al 2005). With respect to the topic addressed in this review, we singularly focused our attention on the potential role of ROS to mediate apoptosis and subsequent disease resistance acquisition in plants. Khurana et al (2005) have elegantly discussed the molecular basis of apoptosis induction in various organisms including animals, yeast and plants. The involvement of ROS in the activation of apoptosis/PCD was demonstrated as a consequence of ROS accumulation detected during the early and late phases of plant-pathogen interaction (Durner et al 1998; Beers and McDowell 2001). Recent observations also support the role for ROS in cell-death signalling (Kachroo et al 2003). Recently, mutations of A. thaliana genes, rbohD and rbohF, that encode orthologues of the mammalian $GP91^{phox}$ NADPH oxidase catalytic subunit, have revealed insights into the role of ROS in PCD formation (Torres et al 2002). Loss-of-function of rbohD and rbohF in A. thaliana mutants resulted in decreased ROS production during the HR, with a corresponding decrease in cell death (Torres et al 2002). These results provided the first direct evidence of ROS implication in cell-death response via plant NADPH oxidases. The role of ROS as potential mediator of apoptosis and subsequent acquired disease resistance has been well documented (Khurana et al 2005). As discussed above, ROS and NO generally interfere in PAL activity and ascorbate, and glutathione (GSH) metabolisms. Based on current research contributions, the simultaneous implication of induced PAL, GSH-metabolisms and APX mediates the induction of PCD (Khurana et al 2005). It has been proposed that the balance between intracellular NO and hydrogen peroxide ($H_2 O_2$), but not superoxide, concentrations is the key determinant for the HR cell-death response (Delledonne et al 2001). In addition, NO cooperates with SA to induce HR cell death and activate defence, which is analogous to its role in animal systems (Durner et al 1998). Increased NO production is sufficient to induce cell death in an A. thaliana cell culture (Clarke et al 2000). However, Khurana et al (2005) recently reported that O_2^- rather than $H_2 O_2$ was identifi ed as the primary signal molecule for pathogen inducing GST, and the rates of production and dismutation of O_2^- generated during the oxidative burst is crucial for the modulation and integration of NO/$H_2 O_2$ signalling in HR inducing plant programmed, localized cell death at the site of infection limits pathogen spread (Levine et al 1994). $H_2 O_2$ is also involved in signal transduction leading to SAR, a phenomena in which localized infection confers systemic resistance to subsequent attacks by the same or unrelated pathogens (Conrath et al 2002). The ROS detected in plant pathogen interactions are. O_2^-(. O_2H), $H_2 O_2$ and. OH (Scheel 2002). Plants also respond to aggression of other parasites such as insects whose feeding causes wounding by producing ROS in the damaged tissue (Kessler and Baldwin 2002). In addition to direct wounding by herbivory insects, insects' enzymes have also been linked to increased production of ROS in plant tissues (Felton and Eichenseer 1999). These ROS are believed to work together with NO to induce a hypersensitive response cell death as a defence reaction against avirulent pathogens (Delledonne et al 1998). Successful pathogens have to overcome or suppress the host defence mechanisms for example, the secretion of superoxide dismutase and catalase converting the ROS into less reactive molecules.

CURRENT RESEARCH TRENDS IN PLANTS RESPONSE TO ENVIRONMENTAL STRESS

A major challenge in biology is the large-scale determination of gene function (Boyes et al 2001). Plants rarely experience stressful conditions caused by a single environmental constraint. Pathogen infection under field conditions is often accompanied by various additional abiotic stresses such as wound, water deficit, mineral stresses, and others (figure1), causing multistress interactions. Databases and web-browser data mining interface for Affymetrix GeneChips are now available to scientists. As shown in figure 2 users can query the database to retrieve the expression patterns of individual genes throughout chosen environmental conditions, growth stages, or organs (Zimmermann et al 2004; Schmid et al 2005). Different genes specifi cally expressed during selected stresses, growth stages, or in particular organs can now be identified at any growing time point of the plant development. Thousands of arrays have since been processed, of which a significant number are publicly available through services and repositories such as Nottingham Arabidopsis Stock Centre Transcriptomics Service (NASCArrays), ArrayExpress at the European Bioinformatics Institute (EBI), or Gene Expression Omnibus (GEO) at the National Center for Biotechnology Information (NCBI) (Edgar et al 2002; Brazma et al 2003; Craigon et al 2004). Public repositories such as GEO, ArrayExpress and GENEVESTIGATOR provide to scientists some potential storage tools and retrieval of heterogeneous data sets as well as coherent data from a single organism generated on a common hybridisation platform. Advantage of the designed bioinformatic tools is to retrieve biologically meaningful results from the highly diverse experiments represented in the database, which can be easily interpreted to generate new experimental hypotheses or more precise understanding of the behaviour of plants in their environment. Despite these major research contributions toward elucidating the molecular network pathways mediating environmental resistant mechanism in plants, we still have only a fragmented view of the relevant pathways. Some of the remaining questions include the identities of up and downstream receptors, substrates of various kinases, which loci interact directly or indirectly during ROS signal network and the identities of additional signalling elements linking the yet known and characterized elements to complete network pathway to mount the appropriate defence mechanism in plants. The interaction of pathogen infection with other stress factors must be investigated in field trials in order to really understand plant defence mechanism beyond the boundary of laboratory trials.

CONCLUSION

It is now clear that ROS is involved in physiological and biochemical mechanisms of disease resistance in plants. Plant responses to ROS are dose dependent. They interact both positively and negatively during plants responses to external challenges. High dosage of ROS results to hypersensitive reactions and oxidative damage in cells, whereas biologically balanced levels of ROS interact with other major signalling pathways including those regulated by ABA, JA, SA, ethylene, ion channels to trigger a complex metabolism leading to up regulation of stress inducible genes. Arabidopsis thaliana adopted as model plant (Meinke et al 1998) has broadened our understanding in the pivotal role of ROS to trigger the disease resistance in plants. Our current understanding of ROS signal transduction in plant responses to various pathogen attacks is still limited, but essential to successfully engineer plants with enhanced disease resistance. Complexity of plant responses to multiple stresses has shown a need to develop new research approaches to elucidate the overwhelming benefit of ROS in plant defence mechanisms. With the advance of biotechnology, ribosome-inactivating proteins (RIPs) can now be used as a potential tool to engineer plants resistant to various stresses. RIPs are a group of RNA N-glycosidases with an extremely high site-specific deadenylation activity towards ribosomal RNA (Nielson and Boston 2001). Due to the removal of a single adenine by RIP, the ribosomes are no longer capable of binding the elongation factors EF-1 and EF-2, and therefore loose their elongation capacity. Based on this property, an overexpression of specifi c RIP genes can arrest the synthesis of targeted proteins that enhance the excessive accumulation of ROS, and thereby regulate ROS levels in plants during multiple environmental challenges. Desmyter et al (2003) have used this approach to demonstrate that the expression of the type-1 ribosome-inactivating protein from iris bulbs (IRIP) in transgenic tobacco increases the resistance of the plants against infection with TMV. Introduction of mechanisms that keep the concentration of ROS at normal redox status in response to pathogen attacks is a promising way to engineer transgenic plants that can resist pathogen infections and even other environmental stresses (Bartels 2001). The notion of ROS accumulation mediating the up-regulation of specific set of stress inducible genes is a positive aspect of plant defence systems, which needs to be evaluated in various agricultural important crops and the ROS network pathways may be adopted as a highly beneficial pre-requisite for disease resistance in plants.

REFERENCES

1. Aist J R and Brushnell W R 1991 Invasion of plants by powdery mildew fungi, and cellular mechanisms of resistance; in:*The fungal spore and disease interaction in plants and animals*(eds) G T Cole and H C Hoch (New York: Plenum Press) pp 321–345
2. Allen R 1995 Dissection of oxidative stress tolerance using transgenic plants;*Plant Physiol.* **107**1049–1054
3. Alvarez M E, Pennell R I, Meijer P -J, Ishikawa A, Dixon R A and Lamb C 1998 Reactive oxygen intermediates mediate a systemic signal network in the establishment of plant immunity;*Cell***92** 773–784
4. Apel K and Hirt H 2004 Reactive oxygen species: Metabolism, oxidative stress, and signal transduction;*Annu. Rev. Plant Biol.* **55** 373–399
5. Asada K 1999 The water-water cycle in chloroplasts: Scavenging of active oxygen and dissipation of excess photons;*Annu. Rev. Plant Physiol. Plant Mol. Biol.* **50** 601–639
6. Asada K and Takahashi M 1987 Production and scavenging of active oxygen in photosynthesis;

in*Photoinhibition* (eds) D J Kyle, C B Osmond and C J Arntzen (Amsterdam: Elsevier) pp 227–287

7. Banerjee A K, Mandal A, Chanda D and Chakraborti S 2003 Oxidant, antioxidant and physical exercise;*Mol. Cell. Biochem.* **253** 307–312
8. Bartels D 2001 Targeting detoxification pathways: an efficient approach to obtain plants with multiple stress tolerence?;*Trends Plant Sci.* **6** 284–286
9. Beers E P and McDowell J M 2001 Regulation and execution of programmed cell death in response to pathogens, stress and developmental cues;*Curr. Opin. Plant Biol.* **4** 561–567
10. Bestwick C S, Brown I R, Bennett M H R and Mansfield J W 1997 Localization of hydrogen peroxide accumulation during the hypersensitive reaction of lettuce cells to*Pseudomonas syringae* pv.*phaseolicola*;*Plant Cell* **9** 209–221
11. Bolwell G P 1999 Role of active oxygen species and NO in plant defence responses;*Curr. Opin. Plant Biol.* **2** 287–294
12. Bolwell P P, Page A, Pislewska M and Wojtaszek P 2001 Pathogenic infection and the oxidative defenses in plants apoplast;*Protoplasma* **217** 20–32
13. Borsani O, Valpuesta V and Botella M A 2003 Developing salt tolerance plants in a new century: a molecular biology approach;*Plant Cell Tissue Organ Cult.* **73** 101–115
14. Boyes D C, Zayed A M, Ascenzi R, McCaskill A J, Hoffman N E, Davis K R and Gorlach J 2001 Growth stage-based phenotypic analysis of Arabidopsis: a model for high throughput functional genomics in plants;*Plant Cell* **13** 1499–1510
15. Bradley D J, Kjellbom P and Lamb C J 1992 Elicitor- and woundinduced oxidative cross-linking of a proline-rich plant cell wall protein: A novel, rapid defense response;*Cell* **70** 21–30
16. Brazma A, Parkinson H, Sarkans U, Shojatalab M, Vilo J, Abeygunawardena N, Holloway E, Kapushesky M, Kemmeren P, Lara G G,*et al* 2003 ArrayExpress: a public repository for microarray gene expression data at the EBI;*Nucleic Acids Res.* **31** 68–71
17. Brisson L F, Tenhaken R and Lamb C 1994 Function of oxidative crosslinking of cell wall structural proteins in plant disease resistance;*Plant Cell* **6** 1703–1712
18. Britto D T and Kronzucker H J 2001 Can unidirectional influx be measured in higher plants? A mathematical approach using parameters from efflux analysis;*New Phytol.* **150** 37–47
19. Brown I, Trethowan J, Kerry M, Mansfield J and Bolwell G P 1998 Localization of components of the oxidative cross-linking of glycoproteins and of callose synthesis in papillae formed during the interaction between non-pathogenic strains of*Xanthomonas campestris* and French bean mesophyll cells;*Plant J.* **15** 333–343
20. Capone R, Tiwari B S and Levine A 2004 Rapid transmission of oxidative and nitrosative stress signals from roots to shoots in*Arabidopsis*;*Plant Physiol. Biochem.* **42** 425–428
21. Chamnongpol S, Willekens H, Moeder W, Langebartels C, Sandermann H Jr*et al* 1998 Defense activation and enhanced pathogen tolerance induced by $H_{-2}O_2$ in transgenic tobacco;*Proc. Natl. Acad. Sci. USA* **95** 5818–5823
22. Clarke A, Desikan R, Hurst R D, Hancock J T and Neill S J 2000 NO way back: nitric oxide and programmed cell death in*Arabidopsis thaliana* suspension cultures;*Plant J.* **24** 667–677
23. Conrath U, Pieterse C M J and Mauch-Mani B 2002 Priming in plant-pathogen interactions;*Trends Plant Sci.* **7** 110–116
24. Corpas F J, Barroso J B, Carreras A, Quiros M, Leon A M, Romero Puertas M C, Esteban F J, Valderrama R, Palma J M, Sandalio L M, Gomez M and Del Rio L A 2004 Cellular and subcellular localization of endogenous nitric oxide in young and senescent pea plants;*Plant Physiol.* **136** 2722–2733
25. Corpas F J, Barroso J B and del Río L A 2001 Peroxisomes as a source of reactive oxygen species and nitric oxide signal molecules in plant cells;*Trend Plant Sci.* **6** 145–150
26. Craigon D J, James N, Okyere J, Higgins J, Jotham J and May S 2004 NASCArrays: a repository for microarray data generated by NASC's transcriptomics service;*Nucleic Acids Res. (Database issue)* **32** 575–577
27. Dat J, Vandenabeele S, Vranová E, Van MontaguM,. Inzé D and Van Breusegem F 2000 Dual action of active oxygen species during plant stress responses;*Cell. Mol. Life Sci.* **57** 779–795
28. Davletova S, Rizhsky L, Liang H, Shenggiang Z, Oliver D J, Coutu J, Shulaev V, Schlauch K and Mittler R 2005 Cytolosic ascorbate peroxidase 1 is a central component of the reactive oxygen gene network of Arabidopsis;*Plant Cell* **17** 268–281
29. Delledonne M, Zeier J, Marocco A and Lamb C2001 Signal interactions between nitric oxide and reactive oxygen intermediates in the plant hypersensitive disease resistance response;*Proc. Natl. Acad. Sci. USA* **98** 13454–13459
30. Delledone M, Xia Y, Dixon R A and Lamb C 1998 Nitric oxide functions as a signal in plant disease resistance;*Nature (London)* **394** 585–588
31. del Río LA, Corpas F J and Barroso J B 2004 Nitric oxide and nitric oxide synthase activity in plants;*Phytochemistry* **65** 783–792
32. del Río L A, Corpas F J, Sandalio L M, Palma José M, Gómez M and Barroso J B 2002 Reactive oxygen species, antioxidant systems and nitric oxide in peroxisomes;*J. Exp. Bot.* **53** 1255–1272
33. Desikan R, Mackerness A H S, Hancock J T and Neil S J 2001 Regulation of*Arabidopsis*transcriptome by oxidative stress;*Plant Physiol.* **127** 159–172
34. Desikan R, Griffiths R, Hancock J and Neill S 2002 A new role for an old enzyme: nitrate reductase-mediated nitric oxide generation is required for abscisic acid-induced stomatal closure in*Arabidopsis thaliana*;*Proc. Natl. Acad. Sci. USA* **99** 16314–16318

35. Desmyter S, Vandenbussche F, Hao Q, Proost P, Peumans J W, and Van Damme Els J M 2003 Type-1 ribosome-inactivating protein from iris bulbs: a useful agronomic tool to engineer virus resistance?;*Plant Mol. Biol.* **51** 567–576

36. Diaz M, Achkor H, Titarenko E and Martinez M C 2004 The gene encoding glutathione-dependent formaldehyde dehydrogenase/ GSNO reductase is responsive to wounding, jasmonic acid and salicylic acid;*FEBS Lett.* **543** 136–139

37. Doke N 1983 Involvement of superoxide anion generation in the hypersensitive response of potato tuber tissues to infection with an incompatible race of*Phytophthora infestans* and to the hyphal wall components;*Physiol. Plant Pathol.* **23** 345–357

38. Doke N 1997 The oxidative burst: roles in signal transduction and plant stress; in*Oxidative stress and the molecular biology of antioxidant defenses* (ed.) J G Scandalios (New York: Cold Spring Harbor Laboratory Press) pp 785–813

39. Doke N and Miura Y 1995 In vitro activation of NADPH-dependent O_2^- generating system in a plasma membrane-rich fraction of potato tuber tissues by treatment with an elicitor from*Phytophthora infestans* or with digitonin;*Physiol. Mol. Plant Pathol.* **46** 17–28

40. Dordas C, Hasinoff B B, Igamberdiev A U, Manac'h N, Rivoal J and Hill R D 2003 Expression of a stress-induced hemoglobin affects NO levels produced by alfalfa root cultures under hypoxic stress;*Plant J.* **35** 763–770

41. Draper J 1997 Salicylate, superoxide synthesis and cell suicide in plant defence;*Trends Plant Sci.* **2** 162–165

42. Durner J, Wendehenne D and Klessig D F 1998 Defense gene induction in tobacco by nitric oxide, cyclic GMP and cyclic ADP-ribose;*Proc. Natl. Acad. Sci. USA* **95** 10328–10333

43. Edgar R, Domrachev M and Lash A E 2002 Gene Expression Omnibus: NCBI gene expression and hybridisation array data repository;*Nucleic Acids Res.* **30** 207–210

44. Felton G W and Eichenseer H 1999 Herbivore saliva and its effect on plant defense against herbivores and pathogens: in*Induced plant defenses against pathogens and herbivores: Ecology and agriculture* (eds) A A Agrawal, S Tuzun and E Bent (St. Paul, Minnesota: American Phytopathologial Society Press) pp 19–36

45. Foreman J, Demidchik V, Bothwell J H F, Mylona P, Miedema H, Torres M A, Linstead P, Costa S, Brownlee C, Jones JDG et*al* 2003 Reactive oxygen species produced by NADPH oxidase regulate plant cell growth;*Nature (London)* **422** 422–446

46. Foyer C H and Noctor G 2005 Redox homeostasis and antioxidant signalling: a metabolic interface between stress perception and physiological responses;*Plant Cell* **17** 1866–1875

47. Foyer C H, Descourvieres P and Kunert K J 1994 Protection against oxygen radicals: an important defense mechanism studied in transgenic plants;*Plant Cell Environ.* **17** 507–523

48. Gachomo W E 2005 Studies of the life cycle of*Diplocarpon rosae* Wolf on roses and the effectiveness of fungicides on pathogenesis;*Cuvillier Verlag Goettingen*, ISBN 3-86537-460-3, pp 1–147

49. Gachomo W E and Kotchoni O S 2006 Functional analysis of hydrogen peroxide-activated protein expression mediating black spot disease resistance in rose leaves infected by two differentially aggressive field isolates of*Diplocarpon rosae*;*Belg. J. Bot.* (in press)

50. Gachomo W E, Shonukan O O and Kotchoni O S 2003 The molecular initiation and subsequent acquisition of disease resistance in plants;*Afr. J. Biotechnol.* **2** 26–32

51. Grün S, Lindermayr C, Sell S and Durner J 2006 Nitric oxide and gene regulation in plants;*J. Exp. Bot.* **57** 507–516

52. Guo F-Q, Okamoto M and Crawford N M 2003 Identification of a plant nitric oxide synthase gene involved in hormonal signalling;*Science* **302** 100–103

53. Halliwell B and Gutteridge J M C 1989*Free radicals in biology and medicine* (Oxford: Clarendon Press)

54. Halliwell B and Gutteridge J M C 2002*Free radicals in biology and medicine* 3rd edition (Oxford, UK: Oxford University Press)

55. Hausladen A, Gow A J and Stamler J S 1998 Nitrosative stress: metabolic pathway involving the flavohemoglobin;*Proc. Natl. Acad. Sci. USA* 95 14100–14105

56. He Y, Tang R H, Hao Y, Stevens R D, Cook C W, Ahn S M, Jing L, Yang Z, Chen L, Guo F, Fiorani F, Jackson R B, Crawford N M and Pei Z M 2004 Nitric oxide represses the*Arabidopsis*floral transition;*Science* **305** 1968–1971

57. Huang X, Stettmaier K, Michel C, Hutzler P, Mueller M J and Durner J 2004 Nitric oxide is induced by wounding and influences jasmonic acid signaling in*Arabidopsis thaliana*;*Planta* 218 938–946

58. Huang X, Von Rad U and Durner J 2002 Nitric oxide induces the nitric oxide-tolerant alternative oxidase in Arabidopsis suspension cells;*Planta* **215** 914–923

59. Jabs T, Dietrich R A and Dangl J L 1996 Initiation of runaway cell death in an*Arabidopsis*mutant by extracellular superoxide;*Science* **273** 1853–1856

60. Jones A 2000 Does the plant mitochondrion integrate cellular stress and regulate programmed cell death?;*Trends Plant Sci.* **5** 225–230

61. Kachroo A, He Z, Patkar R and Zhu Q 2003 Induction of H_2O_2 in transgenic rice leads to cell death and enhanced resistance to both bacterial and fungal pathogens;*Transgenic Res.* **12** 577–586

62. Karpinski S, Reynolds H, Karpinska B, Wingsle G, Creissen G and Mullineaux P 1999 Systemic signaling and acclimation in response to excess excitation energy in Arabidopsis;*Science* **284**654–657

63. Kawano T, Pinontoan R, Uozumi N, Miyake C, Asada K, Kolattukudy P E and Muto S 2000 Aromatic monoamine-induced immediate oxidative burst leading to an increase in cytosolic Ca^{2+} concentration in tobacco suspension culture;*Plant Cell Physiol.* **41** 1251–1258

64. Kerr J F R, Wyllie A H and Currie A R 1972 Apoptosis: A basic biological phenomenon with wide-ranging implications in tissue kinetics;*Br. J. Cancer* **26** 239–257

65. Kessler A and Baldwin I T 2002 Plant response to herbivory: the emerging molecular analysis;*Annu. Plant Biol.* **53** 299–328

66. Khokhlatchev A V, Canagarajah B, Wilsbacher J, Robinson M, Atkinson M, Goldsmith E and Cobb M H 1998 Phosphorylation of the MAP kinase ERK2 promotes its homodimerization and nuclear translocation;*Cell* 93 605–615

67. Khurana S M P, Pandey S K, Sarkar D and Chanemougasoundharam A 2005 Apoptosis in plant disease response: a close encounter of the pathogen*kind;Curr. Sci.* **88** 740–752

68. Kieffer F, Jherminier J, Simon-Plas F, Nicole M, Paynot M, Elmayan T and Blein J 2000 The fungal elicitor cryptogein induces cell wall modifications on tobacco cell suspension;*J. Exp. Bot.***51** 1799–1811

69. Kotchoni O S 2004*Molecular and physiological characterization of transgenic Arabidopsis plants expressing different aldehyde dehydrogenase (ALDH) genes*, Ph.D. thesis, University of Bonn, Germany

70. Kotchoni O S, Kuhns C, Ditzer A, Kirch H-H and Bartels D 2006 Over-expression of different aldehyde dehydrogenase genes in*Arabidopsis thaliana* confers tolerance to abiotic stress and protects plants against lipid peroxidation and oxidative stress;*Plant Cell Environ.* **29** 1033–1048

71. Kovtun Y, Chiu W -L, Tena G and Sheen J 2000 Functional analysis of oxidative stress-activated mitogen-activated protein kinase cascade in plants;*Proc. Natl. Acad. Sci. USA* **97**2940–2945

72. Kumar D and Klessig D F 2000 Differential induction of tobacco MAP kinases by the defense signals nitric oxide, salicylic acid, ethylene, and jasmonic acid;*Mol. Plant Microbe Interact.* **13**347–351

73. Kwak J M, Mori I C, Pei Z M, Leonhardt N, Torres M A, Dangl J L, Bloom R E, Bodde S, Jones J D and Schroeder J I 2003 NADPH oxidase AtrbohD and AtrbohF genes function in ROS-dependent ABA signaling in Arabidopsis,*EMBO J.* **22** 2623–2633

74. Lamb C and Dixon R A 1997 The oxidative burst in plant disease resistance response;*Annu. Rev. Plant Physiol. Plant Mol. Biol.* **48** 251–275

75. Lamotte O, Courtois C and Barnavon L 2005 Nitric oxide in plants: the biosynthesis and cell signalling properties of a fascinating molecule;*Planta* **221** 1–4

76. Levine A, Tenhaken R, Dixon R A and Lamb C J 1994 H_2O_2 from the oxidative burst orchestrates the plant hypersensitive disease resistance response;*Cell* **79** 583–593

77. Lim Y S, Cha M -K, Kim H K, Uhm T B, Park J W, Kim K and Kim I H 1993 Removal of hydrogen peroxide and hydroxyl radical by thiol-specific antioxidant protein as a possible role*in vitro*;*Biochem. Biophys. Res. Commun.* **192** 273–280

78. Lu H and Higgins V J 1999 The effect of hydrogen peroxide on the viability of tomato cells and of the fungal pathogen*Cladosporium fulvum*;*Physiol. Mol. Plant Pathol.* **54** 131–143

79. Mazau D and Esquerré-Tugayé M T 1986 Hydroxyprolinerich glycoprotein accumulation in the cell walls of plants infected by various pathogens;*Physiol. Mol. Plant Pathol.* **29** 147–157

80. McLusky S R, Bennett M H, Beale M H, Lewis M J, Gaskin P and Mansfield J W 1999 Cell wall alteration and localized accumulation of feruloyl-3'-methoxytyramine in onion epidermis at site of attempted penetration by*Botrytis allii* are associated with actin polarisation, peroxidase activity and suppression of flavonoid biosynthesis;*Plant J.* **17** 523–534

81. Meinke D W, Cherry J M, Dean C D, Rounsley S and Koonneef M 1998*Arabidopsis thaliana*: a model plant for genome analysis;*Science* **282** 662–682

82. Mellersh D G, Foulds I V, Higgins V J and Heath C M 2002 H_2O_2 plays different roles in determining penetration failure in three diverse plant-fungal interactions;*Plant J.* **29** 257–268

83. Meyer C, Lea U S, Provan F, Kaiser W M and Lillo C 2004 Is nitrate reductase a major player in the plant NO (nitric oxide) game?;*Photosynth Res.* (in press)

84. Mittler R 2002 Oxidative stress, antioxidants, and stress tolerance;*Trends Plant Sci.* **7** 405–410

85. Mittler R, Vanderauwera S, Gollery M and Van Breusegem F 2004 The reactive oxygen gene network of plants;*Trends Plant Sci.* **9** 490–498

86. Moon H, Lee B, Choi G, Shin D, Prasad D T, Lee O, Kwak S -S, Kim D H, Nam J, Bahk J, Hong J C, Lee S Y, Cho M Je and Lim C Oh 2003 NDP kinase 2 interacts with two oxidative stress-activated MAPKs to regulate cellular redox state and enhances multiple stress tolerance in transgenic plants;*Proc. Natl. Acad. Sci. USA* **100** 358–363

87. Mullineaux P and Karpinski S 2002 Signal transduction in response to excess light: getting out of the chloroplast;*Curr. Opin. Plant Biol.* **5** 43–48

88. Murgia I, Tarantino D, Vannini C, Bracale M, Carravieri S and Soave C 2004*Arabidopsis thaliana* plants overexpressing thylakoidal ascorbate peroxidase show increased resistance to paraquat-induced photo-oxidative stress and to nitric oxideinduced cell death;*Plant J.* **38** 940–953

89. Neill S, Desikan R and Hancock J 2002 Hydrogen peroxide signalling,*Curr. Opin. Plant Biol.* **5**388–395

90. Nielsen K and Boston R S 2001 Ribosome-inactivating proteins: a plant perspective;*Annu. Rev. Plant Physiol. Plant Mol. Biol.* **52** 785–816

91. Parani M, Rudrabhatla S, Myers R, Weirich H, Smith B, Leaman D W and Goldman S L 2004 Microarray analysis of nitric oxide responsive transcripts in*Arabidopsis*;*Plant Biotech. J.* **2**359–366

92. Pavet V, Olmos E, Kiddle G, Mowla S, Kumar S, Antoniw J, Alvarez M E and Foyer C H 2005 Ascorbic acid deficiency activates cell death and disease resistance

responses in*Arabidopsis*;*Plant Physiol.* **139** 1291–1303

93. Peng M and Kuc J 1992 Peroxidase-generated hydrogen peroxide as source of antifungal activity in vitro and on tobacco leaf disks;*Phytopathology* **82** 696–699

94. Penninckx I A, Thomma B P, Buchala A, Metraux J P and Broekaert W F 1998 Concomitant activation of jamonate and ethylene response pathways is required for induction of a plant defensin gene in Arabidopsis;*Plant Cell* **10** 2103–2113

95. Perumalla C J and Heath MC 1991 The effect of inhibitors of various cellular processes on the wall modifications induced in bean leaves by the cowpea rust fungus;*Physiol. Mol. Plant Pathol.***38** 293–300

96. Pnueli L, Hongjian L and Mittler R 2003 Growth suppression, abnormal guard cell response, and augmented induction of heat shock proteins in cytosolic ascorbate peroxidase (Apx1)-deficient Arabidopsis plants;*Plant J.* **34** 187–203

97. Polverari A, Molesini B, Pezzotti M, Buonaurio R, Marte M and Delledonne M 2003 Nitric oxide-mediated transcriptional changes in*Arabidopsis thaliana*;*Mol. Plant. Microbe Interact.***16** 1094–1105

98. Ramanjulu S and Bartels D 2002 Drought- and desiccation-induced modulation of gene expression in plants;*Plant Cell Environ.* **25** 141–151

99. Rao M V and Davis R D 1999 Ozone-induced cell death occurs via two distinct mechanisms in*Arabidopsis*: the role of salicylic acid;*Plant J.* **17** 603–614

100. Rizhsky L, Davletova S, Liang H and Mittler R 2004 The zinc finger protein Zat12 is required for cytosolic ascorbate peroxidase 1 expression during oxidative stress in*Arabidopsis*;*J. Biol. Chem.***279** 11736–11743

101. Sakamoto A, Ueda M and Morikawa H 2002*Arabidopsis* glutathione-dependent formaldehyde dehydrogenase is an S-nitrosoglutathione reductase;*FEBSLett.* **515** 20–24

102. Scandalios J G, Guan L and Polidoros A N 1997 Catalases in plants: gene structure, properties, regulation, and expression; in*Oxidative stress and the molecular biology of antioxidant defenses* (ed.) J G Scandalios (New York: Cold Spring Harbor Laboratory Press) pp 343–406

103. Scheel D 2002 Oxidative burst and the role of reactive oxygen species in plant-pathogen interactions; in*Oxidative stress in plants.* (eds) D Inzè and M Van Montagu (London: Taylor and Francis) pp 137–153

104. Scheideler M, Schlaich N L, Fellenberg K, Beissbarth T, Hauser N C, Vingron M, Slusarenko A J and Hoheisel J D 2002 Monitoring the switch from housekeeping to pathogen defense metabolism in*Arabidopsis thaliana* using cDNA arrays;*J. Biol. Chem.* **277** 10555–10561

105. Schmid M, Davison T S, Henz S R, Pape U J, Demar M, Vingron M, Schölkopf B Weigel D and Lohmann J 2005 A gene expression map of*Arabidopsis thaliana* development;*Nat. Genet.* **37**501–506

106. Schweizer P, Christoffel A and Dudler R 1999 Transient expression of members of the germin-like gene family in epidermal cells of wheat confers disease resistance;*Plant J.* **20** 541–552

107. Shimono M, Yazaki J, Nakamura K, Kishimoto N, Kikuchi S, Iwano M, Yamamoto K, Sakata K, Sasaki T and Nishiguchi M 2003 cDNA microarray analysis of gene expression in rice plants treated with probenazole, a chemical inducer of disease resistance;*J. Gen. Plant Pathol.* **69** 76–82

108. Shirasu K, Nakajima H, Rajasekhar V K, Dixon R A and Lamb C 1997 Salicylic acid potentiates an agonist-dependent gain control that amplifies pathogen signals in the activation of defense mechanisms;*Plant Cell* **9** 261–270

109. Smirnoff N 1998 Plant resistance to environmental stress;*Curr. Opin. Biotechnol.* **9** 214–219

110. Solomon M, Belenghi B, Delledonne M, Menachem E and Levine A 1999 The involvement of cysteine proteases and protease inhibitor genes in the regulation of programmed cell death in plants;*Plant Cell* **11** 431–443

111. Spoel S H Koornnef A, Claessens SMC, Korzelius J P*et al* 2003 NPR1 modulates cross-talk between salicylate- and jasmonate- dependent defense pathways through a novel function in the cytosol;*Plant Cell* **15** 760–770

112. Stamler J S 1994 Redox signalling: nitrosylation and related target interactions of nitric oxide;*Cell* **78** 931–936

113. Stamler J S, Lamas S and Fang F C 2001 Nitrosylation: the prototypic redox-based signaling mechanism;*Cell* **106** 675–683

114. Stöhr C, Strube F, Marx G, Ullrich W R and Rockel P 2001 A plasma membrane-bound enzyme of tobacco roots catalyses the formation of nitric oxide from nitrite;*Planta* **21** 835–841

115. Surplus S L, Jordan B R, Murphy A M, Carr J P, Thomas B and Mackerness S A -H 1998 Ultraviolet-B-induced responses in*Arabidopsis thaliana*: role of salicylic acid and reactive oxygen species in the regulation of transcripts encoding photosynthetic and acidic pathogenesis-related proteins;*Plant Cell Environ.* **21** 685–694

116. Tarantino D, Vannini C, Brascale M, Campa M, Soave C and Murgia I 2005 Antisense reduction of thylakoidal ascorbate peroxidase in Arabidopsis enhances paraquat-induced photooxidative stress and nitric oxide-induced cell death;*Planta* 10.1007/s00425-0051485-9

117. Tena G, Asai T, Chiu W -L, and Sheen J 2001 Plant mitogenactivated protein kinase signaling cascades;*Curr. Opin. Plant Biol.* **4** 392–400

118. Thordal-Christensen H, Zhang Z, Wei Y and Collinge D B 1997 Subcellular localization of H_2O_2in plants. H_2O_2 accumulation in papillae and hypersensitive response during the barley-powdery mildew interaction;*Plant J.* **11** 1187–1194

119. Torres M A, Dangl J L and Jones J D 2002 Arabidopsis gp91phox homologues AtrbohD and AtrbohF are required for accumulation of reactive oxygen intermediates in the plant defense response,*Proc. Natl. Acad. Sci. USA* **99** 517–522

120. Tyres M and Mann M 2003 From genomics to proteomics;*Nature(London)* **422** 193–197

121. Van Camp W, Van MontaguM and Inzé D 1998 H_2O_2 and NO: redox signals in disease resistance;*Trends Plant Sci.* **3** 330–334

122. Veronese P, Nakagami H, Bluhm B, AbuQamar S, Chen X, Salmeron J, Dietrich R A, Hirt H and Mengiste T 2006 The membrane-anchored BOTRYTIS-INDUCED KINASE 1 plays distinct roles in*Arabidopsis* resistance to necrotrophic and biotrophic pathogens;*Plant Cell* **18** 257–273

123. Willmott N, Sethi J K, Walseth T F, Lee H C, White A M and Galione A 1996 Nitric oxide-induced mobilization of intracellular calcium via cyclic ADP-ribose signalling pathway;*J. Biol. Chem.* **271** 3699–3705

124. Wu G, Shortt B J, Lawrence E B, Levine E B, Fitzsimmons K C and Shah D M 1995 Disease resistance conferred by expression of a gene encoding H_2O_2-generating glucose oxidase in transgenic potato plants;*Plant Cell* **7** 1357–1368

125. Wyllie A H, Kerr J F and Currie A R 1980 Cell death: the significance of apoptosis;*Int. Rev. Cytol.* **68** 251–306

126. Xiong L and Yang Y 2003 Disease resistance and abiotic stress tolerance in rice are inversely modulated by an abscisic acidinducible mitogen-activated protein kinase;*Plant cell* **15** 745–759

127. Yamasaki H and Sakihama Y 2000 Simultaneous production of nitric oxide and peroxynitrite by plant nitrate reductase: in vitro evidence for the NR-dependent formation of active nitrogen species;*FEBSLett.* **468** 89–92

128. Yamasaki H, Sakihama Y and Ikehara N 1997 Flavonoid-peroxidase reaction as detoxification mechanism of plant cells against H_2O_2;*Plant Physiol.* **115** 1405–1412

129. Zeidler D, Zähringer U, Gerber I, Dubery I, Hartung T, Bors W, Hutzler P and Durner J 2004 Innate immunity in*Arabidopsis thaliana*: lipopolysaccharides activate nitric oxide synthase (NOS) and induce defense gene;*Proc. Natl. Acad. Sci. USA* (in press)

130. Zemojtel T, Penzkofer T, Dandekar T and Schultz J 2004 A novel conserved family of nitric oxide synthase?;*Trends Biochem. Sci.* **29** 224–226

131. Zimmermann P, Hirsch-Hoffmann M, Hennig L and Gruissem W 2004 GENEVESTIGATOR: Arabidopsis microarray database and analysis toolbox;*Plant Physiol.* **136** 2621–2632

CHAPTER 8 SECONDARY METABOLITES IN FUNGUS-PLANT INTERACTIONS

Tünde Pusztahelyi, Imre J. Holb, and István Pócsi

Keywords: Host-pathogen interaction, phytotoxin, phytoalexin, secondary metabolite, mycotoxin

INTRODUCTION

Phytopathogenic fungi that are basically classified as necrotrophs, hemibiotrophs and biotrophs constitute one of the main infectious agents in plants, causing alterations during developmental stages including post-harvest, gaining nutrients from the plants they invade and, therefore, resulting in huge economic damage. Plants and fungi are rich sources of thousands of secondary metabolites (SMs), which consist of low-molecular weight compounds (the number of the described compounds exceeds 100,000) that are usually regarded as not essential for life while their role are quite versatile (Perez-Nadales et al., 2014; Scharf et al., 2014). Here, our primary aims were to overview the fungal-plant interactions and summarize special SM productions (e.g., phytotoxins and phytoalexins) in context of these interactions. Furthermore, the review also considers data from new fungal genome and transcriptome analyses. These data have hypothesized the biosynthesis of a much wider spectrum of SMs than we have actually isolated and characterized, and which could have strong effect on crop quality. In addition, these data proposed more possible physiological activities for the SMs than we thought before.

Citation: Pusztahelyi T, Holb IJ and Pócsi I (2015) Secondary metabolites in fungus-plant interactions. Front. Plant Sci. 6:573. doi: 10.3389/fpls.2015.00573

PHYTOPATHOGENIC FUNGI

Lifestyle of Phytopathogenic Fungi

While the initial phases of pathogenesis do not differ fundamentally between necrotrophs, hemibiotrophs and obligate biotrophic fungi, different strategies are used to acquire nutrients. Necrotrophic fungi have broader host ranges than biotrophs and often enlist cell-wall-degrading enzymes and toxins, which can be small peptides or SMs (Howlett, 2006). In contrast to necrotrophic and hemibiotrophic fungal pathogens, obligate biotrophs are entirely dependent on living plant tissue and characterized by a number of sophisticated infection structures including appressoria, penetration hyphae and infection hyphae allowing the invader to suppress plant defense responses and to gain excess to host nutrients (reviewed by Mendgen and Hahn, 2002;Schulze-Lefert and Panstruga, 2003). Biotrophs establish haustoria for nutrient uptake (Panstruga, 2003), suppress induction of host defense and reprogram metabolism (Biemelt and Sonnewald, 2006). Biotrophic fungi and their metabolism has been studied on nonobligate biotrophs, such as*Cladosporium fulvum* (Thomma et al., 2005), *Magnaporthe grisea* (Talbot, 2003) and*Mycosphaerella graminicola* (Palmer and Skinner, 2002; Deller et al., 2011). Much less is known about the obligate biotrophs, such as powdery mildews or rust fungi. However, it appears that biotrophy is associated with a convergent loss of secondary metabolic enzymes and reduction in genes encoding specific transporters of toxin secretion and extrusion of host defense compounds usual in necrotrophic fungi. Nevertheless, the infection strategy of necrotrophic fungi is less complex than that of obligate biotrophs. Appressoria formed by typical necrotrophs such as*Cercospora, Ramularia, Rhynchosporium, Alternaria, Fusarium, Botrytis, Helminthosporium, Sclerotinia*, or *Verticillium* species, are inconspicuous, and infection hyphae formed within the host are quite uniform (reviewed by Horbach et al., 2011). Condon et al. (2013) suggested that, while necrotrophs and hemibiotrophs employ fundamentally contrasting mechanisms of promoting disease, the tools they utilize e.g., host-selective toxins (HST) and protein effectors basically overlap.

It cannot be forgot that there are numerous examples of fungi associated with plants as symptomless endophytes (e.g., black Aspergilli, Penicillia). However, in association with host plants, the symptomless endophytes have the capacity to either develop as pathogens or saprophytes, and in either state can become producers of mycotoxins (Palencia et al., 2010), rich sources of effector molecules.

Fungal Secondary Metabolites

Fungal SMs can be divided into four main chemical classes: polyketides, terpenoids, shikimic acid derived compounds, and non-ribosomal peptides. Moreover, hybrid metabolites composed of moieties from different classes are common, as in the meroterpenoids, which are fusions between terpenes and polyketides. Analysis of available fungal genomes revealed that ascomycetes have more genes of secondary metabolism than basidiomycetes, archeo-ascomycetes, and chytridiomycetes, whereas hemi-ascomycetes and zygomycetes have none (Collemare et al., 2008). Ascomycete genomes code for on average 16 polyketide synthases (PKS), 10 non-ribosomal protein synthases (NRPS), two tryptophan synthetases (TS), and two dimethylallyl tryptophan synthetases (DMATS) with crucial importance in SM synthesis. These types of SM genes encode signature enzymes that can be enriched in secondary metabolism gene clusters and responsible for main synthesis steps of metabolites. PKS–NRPSs have been identified only in ascomycetes, with an average of three genes per species. *Neurospora crassa* as well as human pathogens *Coccidioides*spp. and *Histoplasma capsulatum* have a lower number of PKSs (1–9 genes), NRPS (3–6 genes) and PKS-NRPSs (0–2 genes) than other ascomycetes. High number of fungal species have more than 40 genes encoding PKS, NRPS, hybrids, TS, and DMATS in their genome, including *M. grisea*(45 genes) (Collemare et al., 2008) (Table 1). Synthesis of siderophores, a class of SMs for iron uptake also involves a NRPS that is also very important for the virulence of several fungi (e.g.,*Cochliobolus heterostrophus, C. miyabeanus, F. graminearum*, and *A. brassicicola*) (Oide et al., 2006).

Table 1: Distribution of secondary metabolite gene families in selected pathogenic fungi.

Species	PKS[a]	PKS-like	NRPS[b]	NRPS-like	Hybrid[c]	DMAT[d]	Total	References
A. alternata	10	n.d.[e]	n.d.	n.d.	n.d.	n.d.	~10	Saha et al., 2012
A. arborescens	29	n.d.	5	n.d.	2	n.d.	~36	Hu et al., 2012
C. fulvum	10	n.d.	10	n.d.	2	1	~23	de Wit et al., 2012
C. lunatus CX-3	16	1	6	10	2	1	36	Gao et al., 2014
C. lunatus m118	14	1	5	9	2	2	33	Gao et al., 2014
C. heterostrophus C5	22	3	9	7	0	3	44	Gao et al., 2014
C. zea-maydis	11	2	7	8	1	1	30	Gao et al., 2014
P. nodorum	12	9	9	5	1	2	38	Gao et al., 2014
P. tritici-repentis	14	6	12	6	1	0	39	Gao et al., 2014
P. teres f. teres	18	1	27	n.d.	2	1	~49	Amselem et al., 2011
B. cinerea	16	6	6	8	0	1	37	Islam et al., 2012
S. sclerotiorum	16	2	5	5	0	1	29	Islam et al., 2012
M. grisea	12	3	5	6	3	3	32	Islam et al., 2012
M. oryzae	23	2	8	6	5	3	47	Gao et al., 2014
A. flavus	25	3	18	14	2	8	70	Gao et al., 2014
A. niger	15	1	12	2	5	0	35	Amselem et al., 2011
F. graminearum	12	2	10	10	0	0	34	Gao et al., 2014
S. turcica	23	3	9	7	2	2	46	Gao et al., 2014
M. phaseolina	19	16	15	13	12	0	75	Islam et al., 2012

Listed organisms: Alternaria alternata; Alternaria arborescens; Cladosporium fulvum; Cochliobolus lunatus CX-3; Cochliobolus lunatus m118; Cochliobolus heterostrophus C5; Cercospora zea-maydis; Phaeosphaeria nodorum; Pyrenophora tritici-repentis; Pyrenophora teres f. teres; Botrytis cinerea; Sclerotinia sclerotiorum; Magnaporthe grisea; Magnaporthe oryzae; Aspergillus flavus; Aspergillus niger; Fusarium graminearum; Setosphaeria turcica; Macrophomina phaseolina.

[a]*polyketide synthase.*

[b]*non-ribosomal peptide synthase.*

[c]*PKS-NRPS hybrid.*

[d]*dimethylallyl tryptophan synthetase.*

[e]*not determined.*

Whole-genomic analysis have identified 12–15 PKS genes in *F. graminearum* (Kroken et al., 2003;Gaffoor and Trail, 2006; Gao et al., 2014; Sieber et al., 2014), where six have been linked to metabolites. The remaining PKSs have no assigned products yet even though they were expressed under tested conditions. In *F. graminearum*, the genes with known functions (13 SM genes) cover only a minor fraction of the 51 predicted SM genes: 15 PKSs, 19 NPSs and 17 TSs were identified (Sieber et al., 2014). Besides the classical SM genes (TS, NPS, and PKS) the 114 predicted genes encoding cytochrome P450 enzymes are also suitable candidates for searching SM gene clusters. Cytochrome P450s play an essential role in many known biosynthetic pathways of fungal compounds, for instance in the biosynthesis of trichothecene mycotoxins (Tokai et al., 2007) and gibberellins (Hedden et al., 2001) (Figure 1).

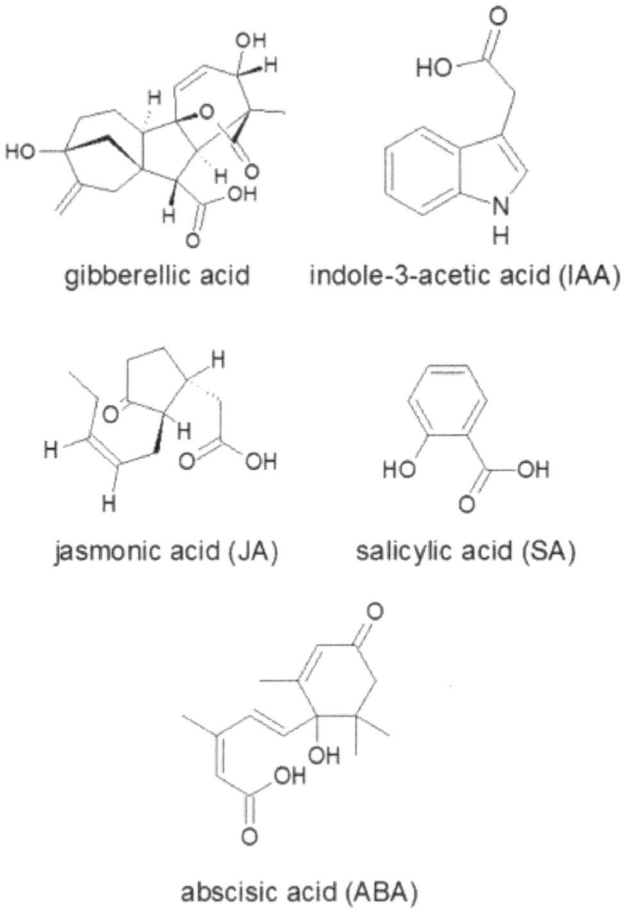

Figure 1: Chemical structures of some plant hormones. Source: National Center for Biotechnology Information. PubChem Compound Database (accessed Jun. 6, 2015) (Bolton et al., 2008).

In the *Macrophomina phaseolina* anamorphic fungus from the ascomycete family*Botryosphaeriaceae*, Islam et al. (2012) identified 75 putative SM genes compared with 32 in *M. grisea*, 37 in *B. cinerea*, 29 in *S. sclerotiorum*, and 37 in *F. graminearum*. A high number of NRPSs which catalyze the production of cyclic peptides including numerous toxins were also found (Table1). In *M. phaseolina* an NRPS, which showed 46% identity to *Cochliobolus carbonum* HST1, the key enzyme responsible for the biosynthesis of the maize HST cyclic tetrapeptide HC-toxin (Figure2) (Panaccione, 1993; Walton, 2006). In 10 different *Fusarium* species including *F. graminearum, F. verticillioides, F. solani, F. culmorum, F. pseudograminearum, F. fujikuroi, F. acuminatum, F. avenaceum, F. equiseti*, and *F. oxysporum* comparative analyses of PKSs and NRPSs led to identification of 52 NRPSs and 52 PKSs orthology groups, respectively (Hansen et al., 2015). A core collection of eight NRPSs (NRPS2–4, 6, 10–13) and two PKSs (PKS3 and PKS7) were only conserved in the investigated strains. The genome of the saprophytic model organism *A. nidulans*contained 56 putative SM core genes including 27 PKS, two PKS-like, 11 NRPS, 15 NRPS-like genes, and one hybrid NRPS-PKS gene (Yaegashi et al., 2014).

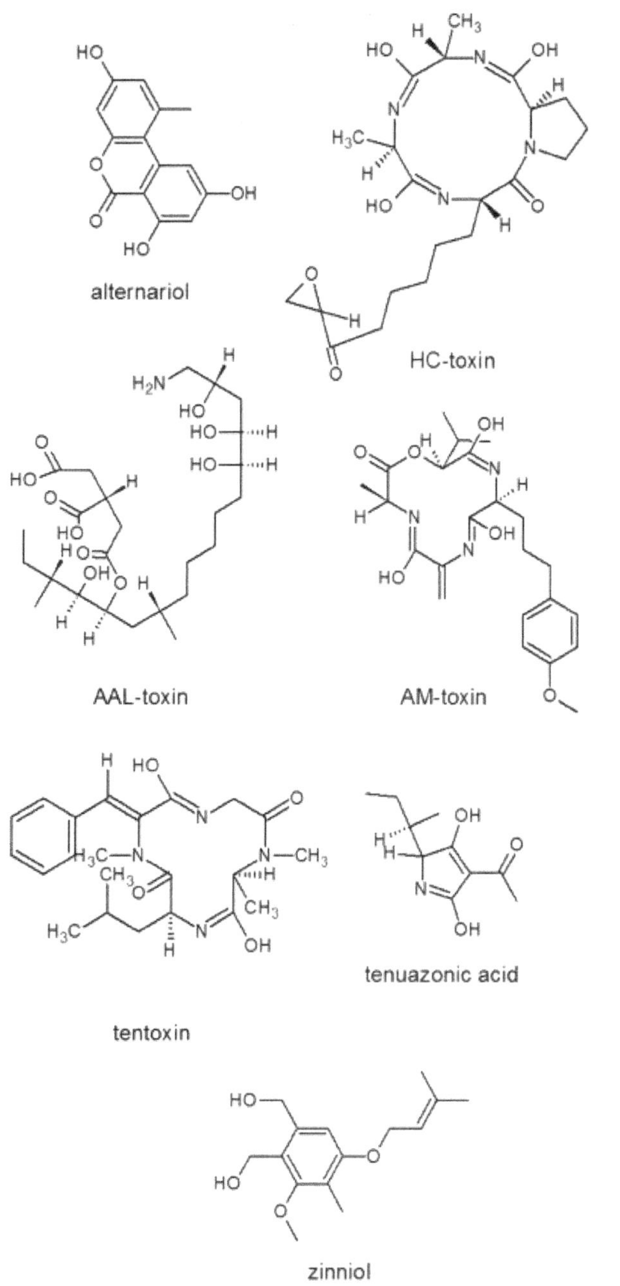

Figure 2: Structures of representatives of *Alternaria* **SMs**. Source: National Center for Biotechnology Information. PubChem Compound Database (accessed Jun. 6, 2015) (Bolton et al., 2008).

The genome sequences of *B. cinerea* and *Sclerotinia sclerotiorum* were determined by Amselem et al. (2011). The *B. cinerea* genome showed high sequence identity and a similar arrangement of genes to *S. sclerotiorum*. The genomes contained a significant number of genes encoding key SM enzymes (Islam et al., 2012), however, the two fungi differed strikingly in the number and diversity of SM gene clusters, which may be involved in the adaptation to different ecological niches (Islam et al., 2012). These fungi had the potential to produce ~26 and 40 main SMs, respectively, as some SM pathways have more than one key enzyme (Amselem et al., 2011).

Stimuli in Fungal SM Production

A high degree of environmental interaction, particularly sources of abiotic stress for either the host or the fungus such as drought or heat stress, also affect on the interactions (e.g., Fountain et al., 2014). Fungal genes involved in stress related responses, especially to oxidative stress, are highly represented in phytopathogenic fungi (see e.g., FSRD: Fungal Stress Response Database; Karányi et al., 2013) and fungal SM toxins often play a role in triggering these responses. Some fungal SMs, such as pigments, polyols and mycosporines, are associated with pathogenicity and/or fungal tolerance to several stress-inducing environmental factors, including temperature and UV light (Sinha et al., 2007). Moreover, environmental factors (e.g., light, temperature, pH, calcium, and nutrients) regulate SM production in a concerted way.

Light is a requirement for deoxynivalenol (DON) toxin (Figure 3) to exert its deleterious effect similarly to the induction of programmed cell death (PCD) during *Botrytis* infections (Govrin and Levine, 2002). This might reflect the plant›s need for light to produce reactive oxygen during the oxidative burst (Howlett, 2006). Meanwhile, regulation of toxin production is also light-dependent (Avalos and Estrada, 2010) through one of the most important light-regulatory protein complex, the velvet complex, comprising at least *Fg*Vel and *Fg*VeB in *Fusarium* with homologous components in other fungi (Yang et al., 2013; Amare and Keller, 2014). *Fg*Vel homolog VeA has been demonstrated to regulate trichothecene production at the level of the biosynthetic genes *Tri4*and *Tri5* and the transcriptional regulator genes *Tri6* and *Tri10* (Jiang et al., 2011; Merhej et al., 2012). Disruption of VeB gene led to several phenotypic defects, including suppression of aerial hyphae formation, reduced hyphal hydrophobicity, highly increased conidiation and reduced DON biosynthesis through the regulation of *Tri5* and *Tri6* (Jiang et al., 2012). Deletion of LaeA (a nuclear regulator from the velvet complex) homolog Lae1 in *F. verticillioides* resulted in reduced expression of gene clusters responsible for synthesis of the SMs bikaverin, fumonisins (Figure 3), fusaric acid and fusarins (Figure 3). Analysis of SMs in the *F. verticillioides* LAE1 mutant revealed differences of regulation from that of in *F. fujikuroi* LAE1 mutant (Wiemann et al., 2010) as bikaverin production was reduced, but the amount of fumonisin B1 (FB1) (Figure 3) remained unchanged (Butchko et al., 2012).

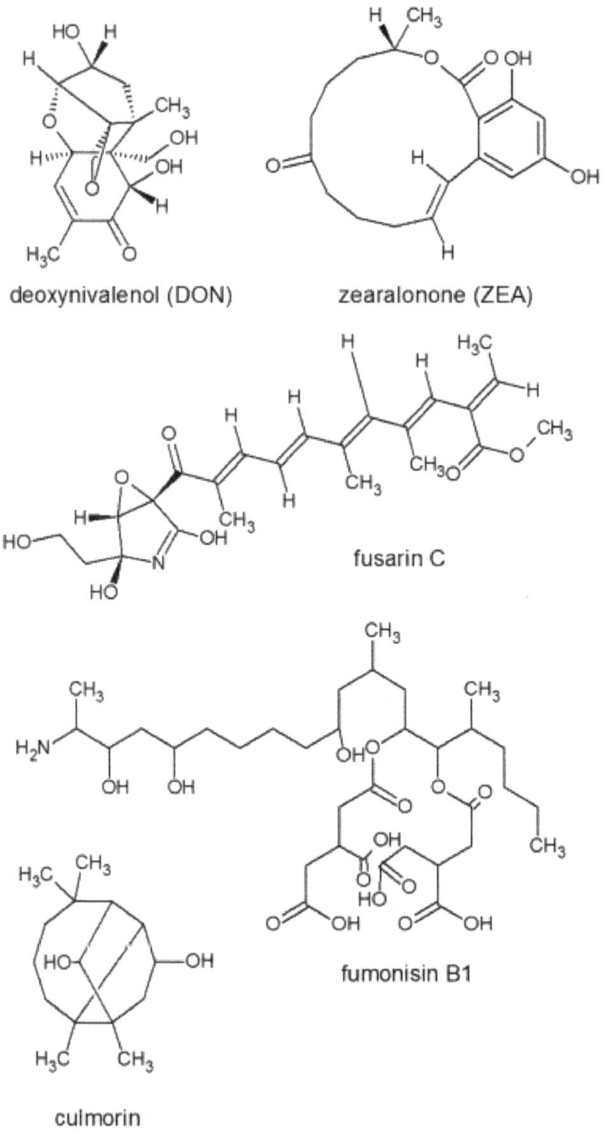

Figure 3: Structures of representatives of *Fusarium* **SMs**. Source: National Center for Biotechnology Information. PubChem Compound Database (accessed Jun. 6, 2015) (Bolton et al., 2008).

Nitrogen limitation have appeared to be an essential stimulus for the activation of virulence functions in phytopathogenic fungi. The ability to metabolize a wide variety of nitrogen sources enables fungi to colonize different environmental niches and survive nutrient limitations (Tudzynski, 2014). Amino acids are required for SM biosynthesis, especially for the NRPS. Amino acid limitation in fungi results in the induction of a genetic network that induces genes for enzymes of multiple amino acid biosynthetic pathways as well as for aminoacyl-tRNA synthases. Inorganic N sources are also affect SM production. Ammonium activated the expression of aflatoxin (AF) (Figure 4) genes (Feng and Leonard, 1998), while nitrate served as an inhibitor of AF biosynthesis of *Aspergillus parasiticus* (Bagheri-Gavkosh et al., 2009). In all fungal species studied, the major GATA transcription factor AreA and its co-repressor Nmr were central players of the nitrogen regulatory network (Tudzynski, 2014). The importance of global nitrogen regulators for the development of pathogenicity was shown for *M. grisea* (Talbot et al., 1997) and many other fungal plant pathogens, e.g., *Colletotrichum lindemuthianum, C. acutatum*, and *F. oxysporum* (Kroll et al., 2014). In *F. graminearum*, which causes crop disease, nitrogen starvation activated the trichothecene pathway and induced the biosynthesis of the DON toxin (Figure 3) that was identified as a virulence factor (Desjardins et al., 1993; Audenaert et al., 2014), similar to the host selective T-toxin from *Cochliobolus heterostrophus* (*Bipolaris maydis*) (Turgeon and Baker, 2007) and the cyclic peptide AM-toxin (Figure 2) from *Alternaria alternata* (Markham and Hille, 2001).

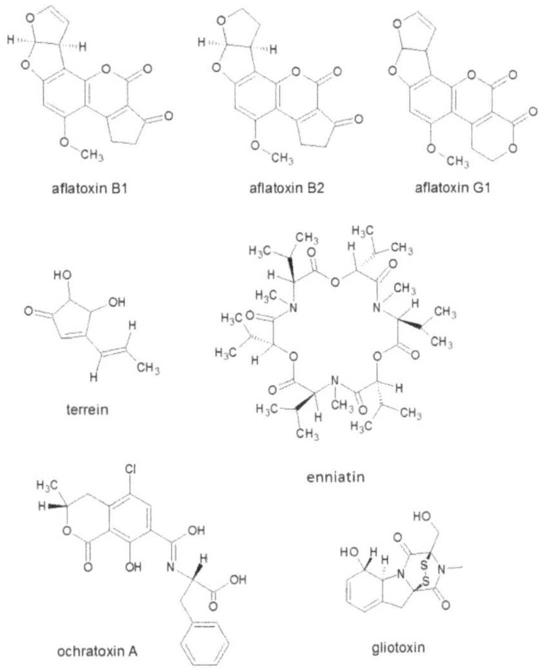

Figure 4: Structures of representatives of *Aspergillus* **SMs**. Source: National Center for Biotechnology Information. PubChem Compound Database (accessed Jun. 6, 2015) (Bolton et al., 2008).

Fungal toxin production is also regulated by signals or even substrates from plant. The well-characterized oxylipins (a group of diverse oxygenated polyunsaturated fatty acids) such as jasmonic acid (JA) (Figure 1) and its immediate precursor 12-oxo-phytodienoic acid are formed enzymatically in plants and accumulate in response to various stresses, in particular wounding and pathogen infection (Wasternack, 2007). These compounds are also formed non-enzymatically *via*the action of reactive oxygen species (ROS) (Wu and Ge, 2004), which also accumulate in response to pathogen infection, heavy metal uptake, or other stresses.

Fungal species have been shown to harbor or secrete JA and its derivatives (Miersch et al., 1999). Fungal oxylipins are able to mimic plant oxylipins; therefore, a reciprocal crosstalk was proposed between plant and fungus (Brodhagen et al., 2008), and several examples have proven this theory. The tomato-infecting *F. oxysporum* produced JAs using a lipoxygenase enzyme related to those found in plants, suggesting that JA biosynthesis in pathogenic fungi occurs *via* a pathway similar to that in plants (Brodhun et al., 2013). In *Aspergillus flavus*, oxylipins are molecules of quorum sensing. At low extracellular oxylipin concentration the cultures were characterized by increased sexual reproduction (sclerotia production), reduced conidiation (Horowitz Brown et al., 2008) and increased AF biosynthesis (Horowitz Brown et al., 2008; Affeldt et al., 2012; Amare and Keller, 2014). Moreover, deletion of oxylipin-encoding dioxygenase genes (*ppo* genes) of *A. flavus* resulted in decreased pathogenicity on host seeds. Exposure to the exogenous plant oxylipins 9(S)-hydroperoxyoctadecadienoic (9(S)-HpODE) acid and 13(S)-hydroperoxyoctadecadienoic acid (13(S)-HpODE) influenced positively the sporulation and effected precursor sterigmatocystin and AF synthesis in *A. flavus* as well as in *A. nidulans* and *A. parasiticus* (Calvo et al., 1999). In a lipidomic approach, Scarpari et al. (2014) have proven the important role of maize oxylipins in driving SM production in *A. flavus*; however, the mechanism of the action has been remained unsolved.

Effects of Phytotoxins on Host Plant

Fungal phytotoxins are usually divided into host-selective toxins (HSTs) and non-host selective (NHSTs) toxins.

Typically, HSTs are active only toward host plants, have unique modes of action and toxicity to the host (Otani et al., 1995); moreover, the production of the HSTs is crucial for the virulence of these fungi (Walton, 1996; Horbach et al., 2011; Tsuge et al., 2013). Nearly all HSTs identified so far are produced by necrotrophic pathogens of the order of *Pleosporales* within the class of *Dothideomycetes* and especially in *Alternaria* and *Cochliobolus* species (Friesen et al., 2008; Stergiopoulos et al., 2013). These HST toxins are diverse chemically ranging from low-molecular-weight compounds to cyclic peptides. Genes encoding polypeptides for biosynthesis of these HSTs have been shown to reside on a conditionally dispensable chromosome that controls host-specific pathogenicity (Hatta et al., 2002). The mechanism of host-selective pathogenesis, through the HSTs, is well understood and about 20 HSTs have been documented (Otani et al., 1995; Walton, 1996). In some cases, host sensitivity was mediated by gene-for-gene interactions, and the toxin sensitivity was mandatory for disease development (Wolpert et al., 2002). Contrarily, NHSTs are not primary determinants of host range and not essential for pathogenicity, although they may contribute to virulence. These toxins have a broader range of activity, causing symptoms not only on hosts of the pathogenic fungus but also on other plant species (Walton, 1996).

Several microbial phytotoxic compounds either inhibited an amino transferase or appeared to have such a mode of action, like cornexistin (Figure 5) from *Paecilomyces variotii* (Amagasa et al., 1994), which was patented as an herbicide; or tentoxin (Figure 2), a cyclic tetrapeptide from *A. alternata*, which indirectly inhibited the chloroplast development(Halloin et al., 1970). A series of structurally related fungal metabolites specifically inhibited ceramide synthase (sphinganine-*N*-acyltransferase) in plants, e.g., several analogs of AAL-toxin (*A. alternata*) (Figure 2) and FB1 (Figure 3) (*Fusarium* spp.) (e.g., Abbas et al., 1994). Fusicoccin (Figure 5) [*Fusicoccum*(*Phomopsis*) *amygdali*] irreversibly activated the plant plasma membrane H^+-ATPase (Paiardini et al., 2014). Alternariol (Figure 2) and monomethyl alternariol are natural phytotoxins, produced by*Nimbya* and *Alternaria*, inhibited the electron transport chain (Demuner et al., 2013). Cerulenin (Figure 5) (*Cephalosporium cerulens*) inhibited *de novo* fatty acid synthesis in plastids (Laskay et al., 1985). T-toxin (a family of C35 to C49 polyketides) from *C. heterostrophus* (Levings et al., 1995;Inderbitzin et al., 2010), which is a HST trichothecene phytotoxin, inhibited mitochondrial respiration by binding to an inner mitochondrial membrane protein in sensitive plants, resulting in pore formation, leakage of NAD^+, and other ions, as well as subsequent mitochondrial swelling (reviewed by Rocha et al., 2005). Zinniol (Figure 2) (*Alternaria* species and one *Phoma* species) bound plant protoplasts and stimulated Ca^{2+} entry into cells (Thuleau et al., 1988). The availability of fungal genome sequences, the knowledge of the biosynthesis of these toxins and gene disruption techniques, allows the development of tools for discovering the role of more and more toxins in plant cell death and disease.

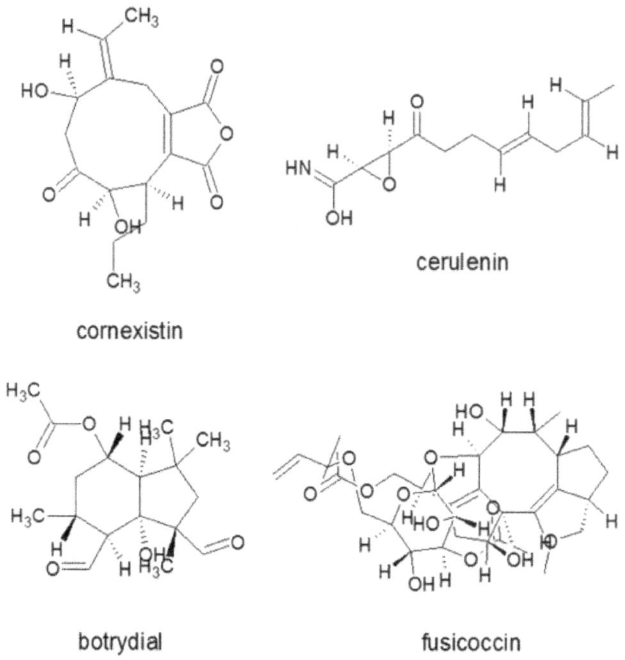

Figure 5: Phytotoxic SM molecules from diverse fungi. Cornexistin from*Paecilomyces variotii*, fusicoccin from *Fusicoccum* (*Phomopsis*) *amygdali*, cerulenin*Cephalosporium caerulens* and botrydial from *Botrytis cinerea*.Source: National Center for Biotechnology Information. PubChem Compound Database (accessed Jun. 6, 2015) (Bolton et al., 2008).

SECONDARY METABOLITE PRODUCTION ON THE HOST'S SIDE

Based on their biosynthetic origins, plant SMs can be divided into three major groups, (i) flavonoids and allied phenolic and polyphenolic compounds, (ii) terpenoids, (iii) nitrogen-containing alkaloids and sulfur-containing compounds, while other researchers have classified plant SMs into more specific groups (Wink, 2003) (Table 2). Plant SMs functions as defense molecules against microbes, viruses or other competing plants or as signal molecules like hormones and attracting molecules for pollinators or seed dispersal animals. Therefore, these compounds have importance for survival and fitness (Wink, 2003).

Table 2: Classification of plant secondary metabolites.

Groups	Chemical structures	Examples
Phenolics with one aromatic ring	C6	Phenol, hydroquinone, pyrogallol acid
	C6-C1	Gallic acid, salicylic acid, methyl syringate, vanillic acid
	C6-C2	Acetophenones, apocynin
	C6-C3	Hydroxycinnamic acid, ferulic acid, sinapic acid, coumaric acid, eugenol, zosteric acid
Phenolics with two aromatic rings	C6-C1-C6 Xanthones	Mangostin
	C6-C2-C6 Stylbenes	Reservatrol, chlorophorin
	C6-C3-C6 Flavonoids	Quercetin, glyceollin, sakuranetin
Quinones	Naphthoquinones Antraquinones Benzoquinones	Alizarin, emodin
Flavonoid polymers and non flavonoid polymers		Tannins
Terpenoids	C5 Hemiterpene	Isoprene, prenol, isovaleric acid
	C10 Monoterpene	Limonene, cineol, pinene, thymol, camphor, turpentin, carvacrol, citral, γ-terpinene, myrcene
	C15 Sesquiterpene	Abscisic acid, humulanes, culmorin, gossypol, zealexin
	C20 Diterpene	Gibberellin, taxol, oryzalexins, phytocassanes, momilactone, kauralexin
	C30 Triterpene	Brassinosteroids, squalen, lanosterol, avenacin
	C40 Tetraterpene	Carotenoids, lycopen
	C > 40 Polyterpenes	Rubber, glisoprenin
	Mixed origin (meroterpenes)	Cytokines, vitamine E
Nitrogen-containing	Alkaloides	Tomatin, solanin, nicotine
	Glucosinolates	Sinigrin, glucobrassicin
	Non protein amino acids	L-canavanine
	Amines	Phenylethylamine, tyramine, morphin
	Cyanogenic glycosides	Amygdalin, sambunigrin, linamarin

Hormone Production and Plant Resistance

Hormone biosynthetic pathways are typically involved in the regulation of plant resistance to pathogens and pests. Endogenous signaling molecules, e.g., ethylene (ET) (Ton et al., 2002), SA (Figure 1; Janda and Ruelland, 2014), JA (Figure 1; Wasternack, 2007; Van der Ent et al., 2009) and abscisic acid (ABA; Figure 1) (Hauser et al., 2011) have been associated with plant defense signaling against biotic stress. Generally, SA signaling induces defense against biotrophic pathogens, whereas JA against necrotrophic pathogens (Glazebrook, 2005).

SA synthesis is a crucial way a plant responds to a biotic attack and involved in both local and systemic resistance (Janda and Ruelland, 2014). Systemic acquired resistance (SAR) is a plant immune response (Shah et al., 2014) that is induced after a local infection and confers immunity throughout the plant to a broad spectrum of pathogens. The onset of SAR (Durrant and Dong, 2004) is usually associated not only with increased levels of SA but additional small metabolites (Figure 6) have also been involved as effectors. Some of these metabolites have been implicated in the rapid activation of defenses in SAR in response to subsequent exposure to the pathogen that called priming (Shah et al., 2014).

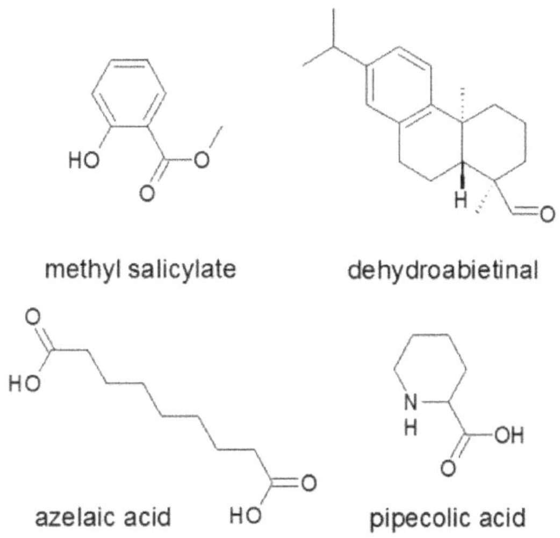

Figure 6: Small metabolites as effectors in SAR signaling. Methyl salicylate, the abietane diterpenoid dehydroabietinal, pipecolic acid from lysine catabolism, and the dicarboxylic acid azelaic acid are effectors in SAR signaling additionally to SA. Source: National Center for Biotechnology Information. PubChem Compound Database (accessed Jun. 6, 2015) (Bolton et al., 2008).

The induced systemic resistance (ISR) pathway is stimulated during necrotrophic bacterial attack but was shown to protect *Arabidopsis* against the necrotrophic fungal pathogens *Alternaria brassicicola* (Ton et al., 2002), *Botrytis cinerea* (Van der Ent et al., 2008) and also*Plectosphaerella cucumerina* (Segarra et al., 2009), where SAR was ineffective (Van der Ent et al., 2009). Investigations of the regulation of ISR revealed the role of JA and ET (Ton et al., 2002; Yan et al., 2002; Kazan and Lyons, 2014). SAR and ISR were characterized by the coordinated activation of pathogenesis-related (PR) genes, many of which encode PR proteins with antimicrobial activity such as chitinases (Van Loon et al., 2006). Soluble chitin fragments released from fungal cell wall through the action of plant chitinases were found to serve as biotic elicitors of defense-related responses like phytoalexin synthesis in plants (Ren and West, 1992; Walker et al., 2003). ISR-related effect of methyl JA and SA was shown to activate also some defense enzymes (Derksen et al., 2013), which play a role to save plant cell wall and also raise the antioxidant capacity in plant cells (Yao and Tian, 2005).

The main auxin in higher plants, indole-3-acetic acid (IAA) (Figure 1), has profound effects on plant growth and development (Zhao, 2010). Only the free form of IAA and related compounds are considered to be active. The majority of produced auxin, however, is conjugated mainly to amino acids and sugars and thereby inactivated. IAA induces e.g., the production of expansins, the proteins whose function is to loosen the cell wall. But, the loose cell wall is more vulnerable to the invasion of different types of pathogens (Ludwig-Müller, 2011). Similarly to bacterial pathogens, hemibiotrophic or necrotrophic fungi produced IAA, manipulated plant growth and subverted plant defense responses such as PCD to provide nutrients for their growth and colonization (Ludwig-Müller, 2015). *Magnaporthe oryzae* secreted IAA in its biotrophic phase especially in the area of the infection hyphae (Tanaka et al., 2011) and, in turn, provoked rice to synthesize its own IAA at the infection sites (Li et al., 2013). However, it has not been elucidated yet whether IAA production is for the manipulation of the host plant or also for the fungus›s own benefit. The activation of an auxin-inducible promoter by fungal IAA indicated that the host plant responds transcriptionally to the secreted auxin. The molecular processes that lead to plant disease and also the prospects for sustainable control were reviewed by Wilson and Talbot (2009). Treatment of *F. culmorum* infected barley with IAA resulted in a reduction of symptoms and yield losses, even though IAA did not inhibit the growth of the fungus *in vitro*. The results indicated increases in the gene regulation for defense-associated genes (Petti et al., 2012).

Plant gibberellins are important phytohormones promoting plant growth and fungi also synthesize gibberellins among other several important terpenes (Keller et al., 2005; Khan et al., 2011). However, higher plants and fungi have evolved their complex gibberellic acid (Figure 1) biosynthetic pathways convergently as it was indicated by the amino acid sequence homology analysis of the proteins in their biosythetic pathways (Hedden et al., 2001). Nevertheless, gibberellic acids produced as SMs in the rice-infecting *F. fujikuroi* were good examples of phytohormone mimics (Bömke and Tudzynski, 2009). Fungal gibberellins were involved in plant infection, e.g., as growth modulators like IAA, cytokinins, and ABA (Figure 1). Interestingly, other*Fusarium* species seem to have lost the ability to synthesize gibberellic acid, suggesting that this is an advantage for *F. fujikuroi* over other pathogens (Wiemann et al., 2013). *Aspergillus fumigatus*also produced gibberellins, and the role of this fungal species was also rectified by its regulatory effect on other phytohormones (ABA, SA, and JA) under stress condition (Khan et al., 2011).

Plant Secondary Metabolites–Antifungal Compounds

Most of the SMs like phytocassanes (Koga et al., 1995) have been reported to have antifungal properties at least *in vitro*. The flavonoids and allied phenolics, e.g., coumarins, lignans, and polyphenolic compounds, including tannins and derived polyphenols form one major group of phytochemicals (reviewed by Crozier et al., 2008). These compounds or their precursors are present in high concentrations in leaves and the skin of fruits and are involved in important defense processes such as UV resistance, pigmentation, disease resistance, stimulation of nitrogen-fixing nodules (Pierpoint, 2000). Phenolic compounds (reviewed by Balasundram et al., 2006) are derivatives of the pentose phosphate, shikimate, and phenylpropanoid pathways in plants. These are known to alter microbial cell permeability and to interact with membrane proteins, which cause deformation in the structure and functionality of these proteins. These disadvantageous changes may lead to dysfunction and subsequent disruption of the membranes including the following events: (i) dissipation of the pH gradient and electrical potential components of the proton motive force, (ii) interference with the energy (ATP) generating and conservation system of the cell; (iii) inhibition of membrane-bound enzymes, and (iv) prevention of substrate utilization for energy production (De Oliveira et al., 2011; El-Mogy and Alsanius, 2012).

Antimicrobial compounds such as the steroidal glycoalkaloid saponins, e.g., avenacin (Figure 7) and α-tomatine, restrict the growth of pathogens in the apoplast. Saponins have strong antifungal activity; the tomato saponin α-tomatine activates phosphotyrosine kinase and monomeric G-protein signaling pathways leading to Ca^{2+} elevation and ROS burst by binding to cell membranes followed by leakage of cell components in *F. oxysporum* cells (Ito et al., 2007). Different plant species produce different types of saponins, which are effective against a wide range of pathogenic fungi (Osbourn, 1996). Terpenes are composed of several isoprene units, and can be linear or cyclic, and even saturated or unsaturated. The best-known terpenes are odoriferous plant metabolites like camphor and turpentine. The industrial and medical significances of plant terpenes, e.g., those of taxol, are reviewed by Bohlmann and Keeling (2008). In maize, sesquiterpenoid phytoalexins, zealexins (Figure 7), were discovered through characterization of physiological responses to the toxinogenic pathogen *F. graminearum*. Importantly, zealexins exhibited antifungal activity against several phytopathogenic fungi (*F. graminearum, A. flavus, Rhizopus microsporus*) at physiologically relevant concentrations (Huffaker et al., 2011).

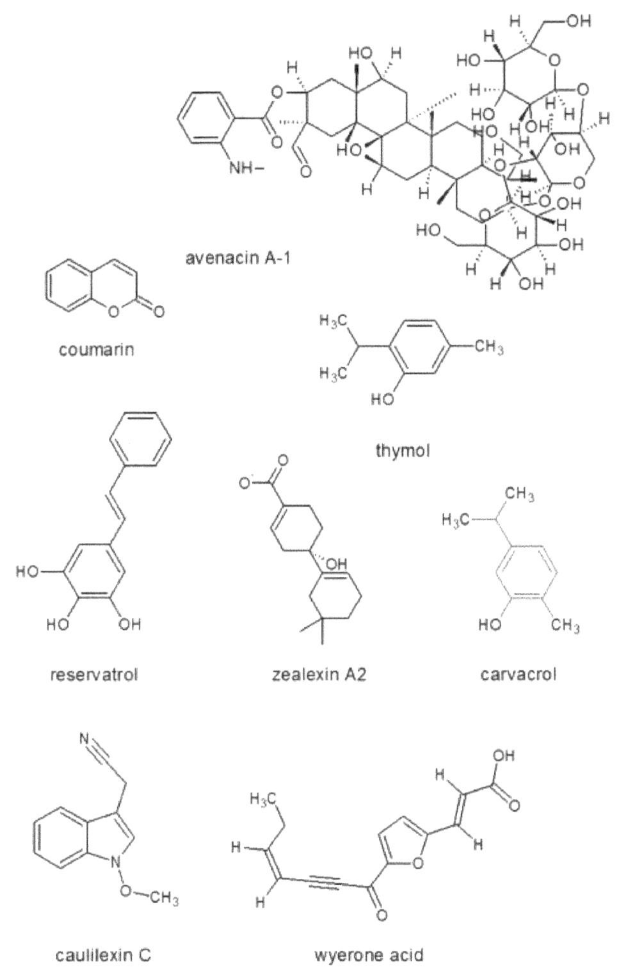

Figure 7: Antimicrobials from plants. Source: National Center for Biotechnology Information. PubChem Compound Database (accessed Jun. 6, 2015) (Bolton et al., 2008).

Plant antifungal metabolites are preformed inhibitors that constitutively produced in healthy plants (phytoanticipins), or they may be synthesized *de novo* in response to pathogen attack or various non-biological stress factors such as short-wavelength UV light, treatment with heavy metal ions (e.g., copper or mercury salts). The letter pathogen and environmental stress elicited compounds are called phytoalexins. These groups cannot be separated strictly as the same compound may be a preformed antifungal substance in one species and can be phytoalexin in another. For example, flavanone sakuranetin (Figure 8) was found to be a phytoanticipin in *Ribes nigra* (Atkinson and Blakeman, 1982) and in *Hebe cupressoides* (Perry and Foster, 1994) but was induced in the leaves of rice *Oryza sativa* (Kodama et al., 1992). These metabolites can be constitutively present in one organ and can be induced in another. Plant SMs usually accumulate in smaller quantities than the primary metabolites (e.g., Dewick, 2002); however, they can accumulate in particular tissues (e.g.,Takanashi et al., 2012) at a higher concentration. This accumulation is regulated in a highly sophisticated manner in appropriate compartments because some plant SMs are even toxic to the plants themselves if they are mislocalized. In the compartmentation and translocation processes, both primary and secondary transporters are involved and many transporter genes, especially genes belonging to the multidrug and toxin extrusion type transporter family, have been identified as responsible for the membrane transport of SMs (Yazaki, 2006; Yazaki et al., 2008). High number of SMs are well characterized in the families Fabaceae, Solanaceae and Labiaceae (Wink, 2003) as well as in cereals (reviewed by Du Fall and Solomon, 2011). Phytoalexins in families Fabaceae and Rosaceae and in rice were reviewed by Grayer and Kokubun (2001); while, SMs in a range of crop plants from families Cruciferae, Fabaceae, Solanaceae (Pedras and Ahiahonu, 2005), Brassicaceae, Vitaceae, and Poaceae (reviewed by Ahuja et al., 2012) have also been described recently.

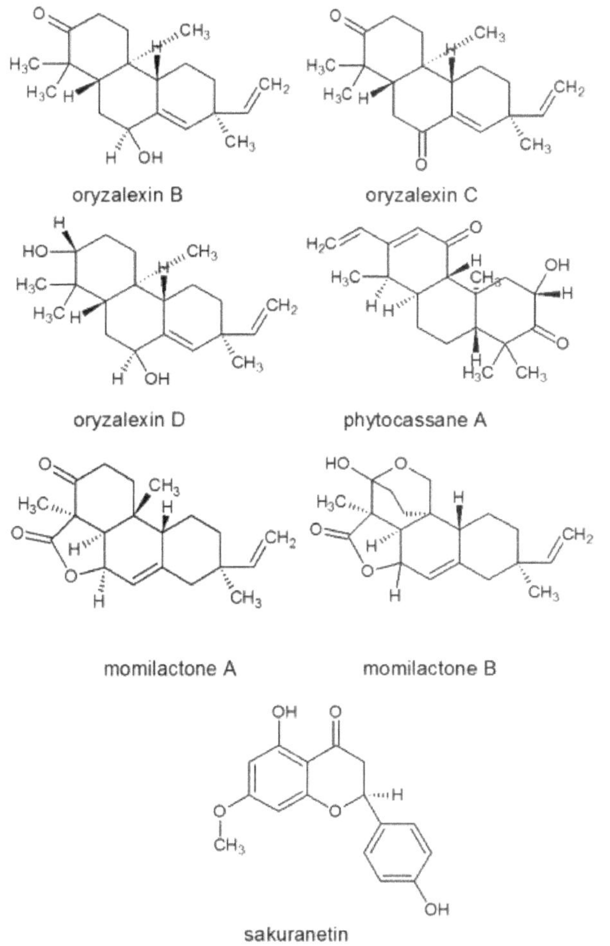

Figure 8: Structures of some phytoalexins isolated from rice. Source: National Center for Biotechnology Information. PubChem Compound Database (accessed Jun. 6, 2015) (Bolton et al., 2008).

Phyllosphere and rhizosphere microorganisms can live in a close mutualistic association with plants or even colonize plant tissues (endophytes). Plant growth-promoting non-pathogenic microorganisms like rhizobacteria and fungi are soil-borne microbes with beneficial effects on plant performance in the rhizosphere. They can stimulate plant growth by increasing tolerance to abiotic stress or by suppressing plant diseases (Van der Ent et al., 2009). Plants may actively shape microbial communities either inhabiting their outer surface or colonizing their interior (Bednarek et al., 2010). The growing plant secretes a wide range of chemicals, e.g., in root exudates, to communicate with rhizosphere microbes (Kolattukudy et al., 1995; Suryanarayanan et al., 2009;Baetz and Martinoia, 2014) such as arbuscular mycorrhiza. Altered exudation patterns, putative direct arbuscular mycorrhiza effects, different root size and architecture, altered physiology may contribute to quantitative and qualitative microbial community changes in the mycorrhizosphere caused by arbuscular mycorrhiza fungi (Wehner et al., 2010). Organic acids, amino acids and phenolic compounds present in root exudates play an active role in root-microbe communications (Dakora and Phillips, 2002; Tanimoto, 2005; Crowley and Kraemer, 2007; Li et al., 2013) and stimuli such as microbial elicitors trigger compositional changes in root exudates (Baetz and Martinoia, 2014). Walker et al. (2003) analyzed root exudates of *Arabidopsis thaliana* elicited by SA, JA and chitosan as well as by two fungal cell-wall elicitors and profiled the SMs subsequently secreted. Among the several compounds detected butanoic acid, trans-cinnamic acid, *o*-coumaric acid, *p*-coumaric acid, ferulic acid, *p*-hydroxybenzamide, methyl *p*-hydroxybenzoate, 3-indolepropanoic acid, gallic acid, and vanillic acid were successfully inhibited the growth of *F. oxysporum, Phytophthora drechsleri* and *Rhizoctonia solani* phytopathogenic fungi.

Manipulation of Programmed Cell Death

Different fungal strategies mediate killing of the plant host cells such as secretion of low molecular weight or peptide toxins or eliciting PCD in the host by secretion of ROS (Horbach et al., 2011;Barna et al., 2012). From the host's side, chloroplasts have a critical role in plant defense as these organelles are not only sites for the biosynthesis of the plant signaling compounds: SA, JA and nitric oxide but for ROS production as well (e.g., Lee et al., 2015). Therefore, chloroplasts are regarded as important players in the induction and regulation of PCD in response to both abiotic stresses and pathogen attack. Moreover, toxin effectors from necrotrophic fungi can target one of the host's central signaling/regulatory pathway to trigger resistance (R) gene-mediated resistance or to down-regulate defense enzymes, and, as a consequence, to increase thereby host susceptibility to fungal attack (Wang et al., 2014). *Aspergillus* mycotoxin ochratoxin A (Figure 4) induced necrotic lesions in detached leaves through oxidative burst induction with increased ROS level and concomitant down-regulation of plant antioxidant defense enzymes (Peng et al., 2010). SMs fusarenon, nivalenol, DON, T-2, HT-2, diacetoxyscirpenol, beauvericine and neosolaniol (Figure 3) from *Fusaria* caused complete inhibition of seed germination and induced PCD and alteration to ascorbate metabolism in tomato protoplasts (Paciolla et al., 2004). T-2 trichothecene toxin, produced by e.g., *F. sporotrichioides*, also induced cell death, callose deposition, generation of hydrogen peroxide, and accumulation of SA, while DON toxin inhibited translation without induction of the elicitor-like signaling pathway in the non-host plant *A. thaliana* (Nishiuchi et al., 2006). The ascomycete *Cochliobolus victoriae* is a necrotrophic fungal pathogen of *Arabidopsis*and oats with HST victorine, which induced defense related responses such as phytoalexin synthesis, extracellular alkalization and PCD causing Victoria blight (Tada et al., 2005). Zhang et al. (2011) proposed that both JA and ET promote the *A. alternata* AAL toxin-induced PCD in detached*Solanum lycopersicum* leaves by disruption of sphingolipid metabolism (Spassieva et al., 2002). In*Arabidopsis*, free sphingoid bases were again shown to be involved in the control of PCD, presumably through the regulation of the ROS level upon receiving different developmental or environmental cues (Raffaele et al., 2009).

PHYTOTOXINS, PHYTOALEXINS AND SPECIAL SMS

Magnaporthe Grisea Species Complex

The *M. grisea* species complex comprises many phylogenetic species (Couch and Kohn, 2002) that cause disease to some 50 grass and sedge species. These include rice, wheat, barley, maize, oats, rye, finger millet, perennial ryegrass, weed and ornamental grasses. Within this species complex,*M. oryzae* (previously known as *M. grisea*) isolates form the pathotype *Oryza*, which causes rice blast disease. Approximately 10–30% of the annual rice harvest is lost due to the infection. The fungus infects all aerial parts of rice, leading to leaf blast, neck and panicle rot, collar rot and node blast (reviewed by Skamnioti and Gurr, 2009).

Chemical signals are responsible for appressorium formation in *M. grisea*. The appressorial glue of*M. grisea* contains glycoproteins, neutral lipids and glycolipids (Ebata et al., 1998). The non-toxic plant metabolite zosteric acid (Figure 9) (Todd et al., 1993) binds water and enhances the hydrophilicity of the surface, thereby weakening the binding capacity of the appressorial glue, which is highest with hydrophobic surfaces. Therefore, zosteric acid inhibits spore adhesion and infection by *M. grisea* and also by *Colletotrichum lindemuthianum* on artificial hydrophobic surfaces as well as on plant leaves (Stanley et al., 2002).

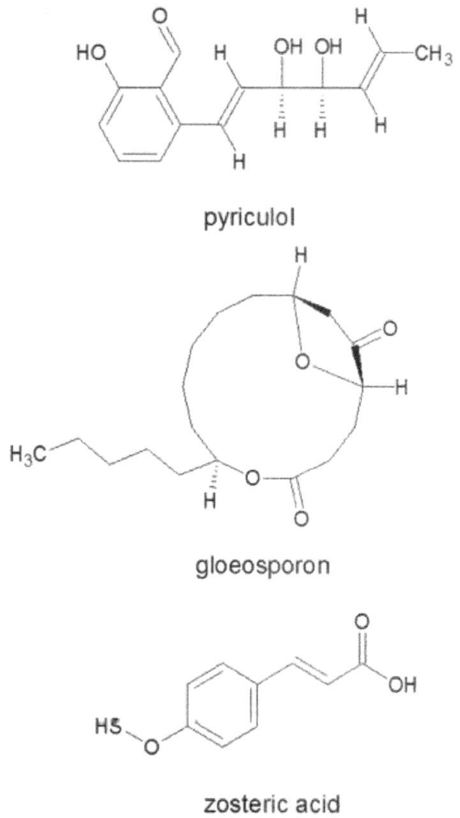

Figure 9: Autoinhibitor signal molecules responsible for inhibition of conidia germination. Phytotoxin pyriculol and gloeosporone, and/or appressorium formation: zosteric acid. Source: National Center for Biotechnology Information. PubChem Compound Database (accessed Jun. 6, 2015) (Bolton et al., 2008).

The two most effective inducers of the germination and appressorium formation were found to be 1,16-hexadecanedial and 1,16-hexadecanediol from cutin monomers in *M. grisea* (Gilbert et al., 1996). Besides cutin monomers, surface waxes also activated development processes in fungi (Liu et al., 2011). Appressorium formation was induced by leaf wax of rice or other plants or synthetic n-C22 fatty acid, fatty alcohol or alkane (Hegde and Kolattukudy, 1997). Self-inhibitors of the germination are related to phytotoxin pyriculol (Figure 9) (Kono et al., 1991). Moreover, other fungi can also produce specific, non-toxic inhibitors of conidial germination and appressorium formation of *M. grisea* like flaviolin, tenuazonic acid (Figure 2), and glisoprenins (Thines et al., 2004).

In the appressorium, several key biochemical and morphogenetic events take place under the generation of turgor pressure which, in *M. grisea*, is the highest pressure recorded in any living cell at up to 8 MPa to penetrate the tough rice cuticle (Howard et al., 1991). This exceptionally high pressure and mechanical penetration question the role of secreted cell wall-degrading enzymes in the first steps of invasion at least on the natural host (Howard and Valent, 1996). In order to generate the high turgor pressure, a thick melanin layer is deposited outside the primary cell wall. Several natural products inhibit melanin biosynthesis in a similarly specific and non-toxic manner, presumably hitting the same targets (Thines et al., 1995). Typical examples are coumarin (Figure7), a common SM from plants (Wheeler and Bell, 1988), scytalol D from fungus *Scytalidium* sp. (Thines et al., 1998) and lipid biosynthesis inhibitor cerulenin (Figure 5) originally obtained from an isolate called *Cephalosporium caerulens* (Ohtake et al., 1999) that conspecific to *Sarocladium oryzae* phytopathogenic fungus of rice (Bills et al., 2004).

The first identified rice phytoalexins (reviewed by Peters, 2006) were the momilactones A and B (Figure 8) (Cartwright et al., 1981). Momilactones exhibit antifungal activity against *M. grisea* and only appear in rice leaves after infection (Kodama et al., 1988). These compounds were originally isolated and identified as plant growth inhibitors from rice seed (Kato-Noguchi et al., 2002). Another group of diterpenoid phytoalexins are oryzalexins (A-F) also isolated from rice (Akatsuka et al., 1985; Kato et al., 1993, 1994). Oryzalexins B, C and D (Figure 8) were identified as *ent*-pimarane diterpenoids, and found in *M. grisea* infected, but not healthy rice leaves (Akatsuka et al., 1985). Oryzalexin S (Tamogani et al., 1993), and phytocassanes A (Figure 8) to E (Koga et al., 1995,1997) are also labdane-related diterpenoid phytoalexins in rice. Overexpression of another rice flavonon phytoalexin sakuranetin (Figure 8) resulted in an increased resistance to *M. grisea*(Kodama et al., 1992; Kim et al., 2009).

Genus Colletotrichum

Colletotrichum usually shares similar lifestyles and infection strategies with *M. grisea*, particularly during the early stages of pathogenesis. However, in hemibiotrophic *C. gloeosporioides* and in other *Colletotrichum* species unlike in the case of *M. grisea*, blocking the cell cycle did not prevent spore germination and appressoria formation (Nesher et al., 2008). The differentiation of infection structures including appressoria preceded mitosis and could proceed without nuclear division. Spore cell death did not occur during plant infection and the fungus primary infection structures remained viable throughout the infection cycle (Nesher et al., 2008).

The spores of many phytopathogenic fungi contain potent autoinhibitors, which prevent germination until they have been washed or diluted out of the spore. *Colletotrichum* spp. appeared to be a rich source of germination autoinhibitors. The first self-inhibitor to be isolated from conidia of *C. gloeosporioides* was gloeosporone (Figure 9) (Meyer et al., 1983) followed by (Z)- (E)-ethylidene-1,3-dihydroindole-2-one that were active at lower concentrations than gloeosporone (Tsurushima et al., 1995). At higher concentrations, the latter two compounds also inhibited the germination of conidia of other *Colletotrichum* spp. and of *F. oxysporum* (Tsurushima et al., 1995). Mycosporin-alanine is also a potent autoinhibitor of conidial germination in *C. graminicola* that was synthesized during the development of conidia in the pycnidium, and it was quite effective in preventing germination of the spores until they have become dispersed (Leite and Nicholson, 1992). Interestingly, the synthesis of mycosporines and mycosporine-like amino acids occurs in fungi, bacteria, cyanobacteria, phytoplankton and macroalgae but not in animals because it needs the shikimate pathway (Sinha et al., 2007). Conidia germination is induced by ET in ripened fruits (Flaishman and Kolattukudy, 1994), additionally, by fatty alcohols from cuticular waxes (Podila et al., 1993; Hwang and Kolattukudy, 1995).

Botrytis Cinerea

The gray mold fungus *B. cinerea* is a typical necrotrophic phytopathogenic fungus with a very wide host range. It causes vast economic damage pre- and postharvesting (Amselem et al., 2011). Two groups of its phytotoxic metabolites have been characterized, the sesquiterpene botrydial (Figure 5) and related compounds (Colmenares et al., 2002) and botcinic acid and its derivatives (Tani et al., 2006). The sesquiterpene-derived phytotoxin botrydial has been implicated in virulence, as it can be detected *in planta* and its addition facilitates fungal penetration and colonization of plants (Deighton et al., 2001). In addition to SM toxins, ROS play in important role in *B. cinerea* as the fungus actively contributes to the elevated levels of ROS detected at infection sites and causing an oxidative burst during cuticle penetration and lesion formation (Tiedemann, 1997; Tudzynski and Kokkelink, 2009).

After fungal attack of grapevine and berries, leaves produced phytoalexins such as resveratrol (trans-3,5,4′-trihydroxystilbene) (Figure 7) (Langcake and Pryce, 1976) and related compounds, which have antifungal activity toward *B. cinerea* and also a number of other fungal pathogens including *Rhizopus stolonifer* and *Plasmopara viticola* (Jeandet et al., 2002). Strong antifungal activity of carvacrol (Figure 7) and thymol (Figure 7) was also confirmed against *B. cinerea* and *R. solani, Fusarium moniliforme* and *S. sclerotiorum* (Mueller-Riebau et al., 1995; Tsao and Zhou, 2000; Camele et al., 2012). High inhibitory activity was detected against *B. cinerea* by monoterpene γ-terpinene (Espinosa-García and Langenheim, 1991), while monoterpene citral has been reported as a potent antimicrobial compound against *B. cinerea* (Tsao and Zhou, 2000) and*Penicillium italicum* (Saddiq and Khayyat, 2010). Essential plant oils (e.g., D-limonene, cineole, β-myrcene, α-pinene, β-pinene, and camphor) showed remarkably high antifungal activity against *B. cinerea* (Wilson et al., 1997). In *Vicia faba* tissues, low-molecular-weight phytoalexins such as wyerone acid (Figure 7) and wyerone furanoacetylenic were produced as part of the post-infection defense response against fungal pathogens. Wyerone acid accumulated in *B. cinerea* lesions, whereas in *Botrytis fabae* lesions the phytoalexin started to accumulate but later tended to decrease. The enhanced ability of *B. fabae* to colonize e.g., broad bean tissues seemed to be related to its capacity to detoxify broad bean phytoalexins (Buzi et al., 2003).

Fusaria

The filamentous fungus *F. graminearum* (teleomorph: *Gibberella zeae*) is a worldwide pathogen of maize and small grains such as wheat, barley and oats. In infected grains, *F. graminearum* can produce several mycotoxins, including trichothecene derivatives (e.g., DON), polyketide zearalenone (ZEA), fusarin C (Figure 3) (Desjardins et al., 1993; Kimura et al., 2007) among which trichothecenes were related to the pathogenicity of *F. graminearum* (Gaffoor and Trail, 2006;Foroud and Eudes, 2009) reducing crop yield and quality. In plants, trichothecenes produced by*Fusarium* spp. cause necrosis, chlorosis, and mortality enabling them to mediate a wide variety of plant diseases, including wilts, stalk rot, root rot, and leaf rot in many important crop and ornamental plants (Abbas et al., 2013).

F. verticillioides (teleomorph: *Gibberella moniliformis*) is a ubiquitous pathogen of maize, attacking stalks, kernels, and seedlings. Considering the maize developmental stages, silking (R1), blister (R2), milk (R3), dough (R4), dent (R5), and physiological maturity (R6), infecting the seed at stages R2–R5 with *F. verticillioides* revealed that the pathogen colonized seeds equally well (Bluhm and Woloshuk, 2005). Nevertheless, significant sphingoid-derived fumonisin B1 (FB1) mycotoxin (Figure 3) production (Abbas et al., 1994; Bluhm and Woloshuk, 2005; Picot et al., 2011) occurred only in the R5 (dent)-stage kernels where the R5 kernel acidic state also induced more FB1 production (Picot et al., 2011). Expression of FUM8 and FUM12 fumonisin biosynthetic genes as well as low amounts of FB1 was detected in the R3 (milk) and R4 (dough) stages. In contrast, no FB1 or FUM gene expression was detectable in the R2 (blister) stage. Other experiments revealed that the fungus produced fourfold more FB1 on maize polysaccharide amylopectin than on glucose carbon source (Bluhm and Woloshuk, 2005; Picot et al., 2011).

Usually extracellular ATP functions as an endogenous external metabolite regulating plant cell viability. FB1 toxin could trigger the depletion of extracellular ATP, which altered the abundance of particular intracellular plant proteins and ended in cell death, which process was reversible by exogenous ATP (Chivasa et al., 2005). Nevertheless, FB1 did not appear to be a primary virulence factor, while DON (Figure 3) was considered to have a key role as a virulence factor at least in *F. graminearum*, and their induction is quite different. The production of DON and the spread of the fungus in the spikes correlated well with the presence of several polyamine compounds that accumulate as the infection progresses through the spike (Gardiner et al., 2010).

SM culmorins are tricyclic sesquiterpene diols that have been reported from *F. culmorum, F. graminearum* and *F. venenatum* (Langseth et al., 2001). Culmorin had weak phytotoxicity to wheat coleoptile tissue (Wang and Miller, 1988), but its role in wheat head scab was not reported. Contaminated grain samples are usually not screened for culmorins, because there are no limits for these SMs. However, culmorin and hydroxyculmorins were detected at relatively high levels in naturally contaminated Norwegian wheat, barley, and oat samples co-occurring with high DON concentrations (Ghebremeskel and Langseth, 2001).

Inhibition of toxigenesis in *Fusaria* has also been studied. Velluti et al. (2004) explored the efficacy of cinnamon, clove, lemongrass, oregano and palmarosa essential oils in order to prevent ZEA and DON accumulation when inoculated with *F. graminearum*; however, it should be noted that this assay was based on non-sterilized, naturally contaminated maize grain. Dambolena et al. (2011)studied the capacity of 10 natural phenolic compounds to inhibit FB1 synthesis by *F. verticillioides*and revealed that thymol, carvacrol (Figure 7), isoeugenol as well as eugenol were the most active. The plant phenol chlorophorin was also effective in reducing FB1 toxin production (94% reduction), followed by caffeic acid (hydroxycinnamic acid), ferulic acid, vanillic acid and iroko (Beekrum et al., 2003). In *F. proliferatum*, aqueous extracts of host plants inhibited the fungal growth in dose dependent manner, resulting in growth induction at low doses. While, pea extract inhibited the FB production in most of the strains (St□pień et al., 2015).

Modulation of spore germination is also often based on low-molecular weight substances produced by the plant host. For example, flavonoids stimulated the germination of conidia of *F. solani* on the leaves of vegetables (Ruan et al., 1995). Moreover, Garcia et al. (2012) found that production of FB1 and FB2 by *F. verticillioides*, and ZEA and DON by *F. graminearum* was stimulated or similar to the controls in most of the conditions tested using *Equisetum arvense* and *Stevia rebaudiana*extracts.

Aspergilli

Aspergillus species can be saprophytic, or symptomless endophytes or weak and opportunistic plant pathogen. *A. flavus* from yellow Aspergilli is a weak and opportunistic plant pathogen. It lacks host specificity (St Leger et al., 2000) as it can attack seeds of both monocots and dicots such as maize, cotton, groundnuts (peanuts) and other nuts like tree nuts such as Brazil nuts, pecans, pistachio nuts, and walnuts. *A. flavus* can cause ear rot on maize and preharvest contamination of these crops with SM AFs is common, but *A. flavus* also causes the spoilage of post-harvest grains during storage resulting in significant economic losses to farmers (Figure 10).

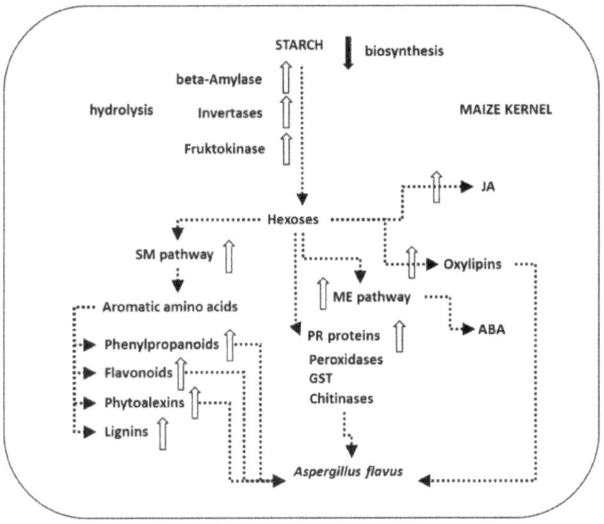

Figure 10: Schematic presentation of main plant defense processes in*Aspergillus flavus*-maize interactions. *A. flavus* can attack kernels during all the six stages of their development. However, infection in non-injured kernels takes place later in the field, during the dent (R5) developmental stage just prior to physiological maturity (R6) (Marsh and Payne, 1984). As soon as 4 days after inoculation *A. flavus* mycelium reaches the aleurone, endosperm and germ tissue (Dolezal et al., 2013). Transcriptional analysis of the maize—*A. flavus* pathogen interaction revealed down-regulated (black arrow) starch biosynthesis and up-regulated genes (white arrows) of plant starch hydrolytic enzymes like β-amylase as well as downstream invertases and fructokinase. The produced hexoses flow through the up-regulated shikimate (SM) pathway, the methylerithryole (ME) pathway and toward up-regulated jasmonic acid (JA) and oxylipin biosynthesis, and feed pathogenesis related (PR) protein synthesis, e.g., peroxidases, glutathione S-transferase (GST) or chitinases that were also found up-regulated during infection. Oxylipins up-regulate aflatoxin (AF) biosynthesis and sexual reproduction in *A. flavus* and down-regulate fungal growth. Up-regulation of the SM pathway leads to the production of antifungal compounds flavonoids, phenylpropanoids, phytoalexins, and up-regulated lignin production in maize. Up-regulated plant hormone JA and abscisic acid (ABA) production is crucial in these defense mechanisms (Dolezal et al., 2014).

A. flavus, A. parasiticus and *A. nidulans* are proposed to derive acetyl CoA for the biosynthesis of SM toxins (i.e., sterigmatocystin and AF) (Figure 5) from fatty acids present in the kernel of maize (Howlett, 2006). *In vitro* supplemented oleic acid induced the biogenesis of fungal peroxisomes, as well as catalase activity and β-oxidation. Concomitantly with the increased expression of biosynthetic genes for precursor sterigmatocystin and AF in hyphae, colonizing the embryo and aleuronic layer, where most seed lipids are stored, AF precursor norsolorinic acid accumulated in peroxisomes (Maggio-Hall et al., 2005).

Bagheri-Gavkosh et al. (2009) showed that AF B1 production by *A. parasiticus* was inhibited by methanolic extracts of *Ephedra major* aerial parts and roots, whereas the essential oil of the plant aerial parts did not exhibit any effect on AF B1 biosynthesis. The authors attributed the inhibition of *A. parasiticus* growth and AF B1 production to the presence of flavonoid compounds such as *p*-coumaric acid and quercetin in plant extracts. Dos Santos and Furlong (2008) noted that AF B1 and AF B2 production by *A. flavus* was inhibited in the presence of methanolic extracts from banana pulp and peel, orange, eggplant and potato pulp. However, these authors found that in the presence of banana pulp and potato pulp extracts, *A. flavus* produced AF B2, which was not detected in the control. Crude essential oil of *Betula alba* also inhibited both AF production and fungal growth in parallel. Jermnak et al. (2012) found that after roughly purifying the oil by silica gel column chromatography an active fraction was obtained that was identified as methyl syringate. This compound strongly inhibited norsolorinic acid production, an early step of the AF biosynthetic pathway: it inhibited the AF B1 and AF G1 production of *A. parasiticus* in liquid medium in a dose-dependent manner and also inhibited AF B1 production by *A. flavus* on raw peanuts.

Black aspergilli are common soil organisms decomposing dead plant residues, some of them are capable of a biotrophic endophytic existence with maize and onion. *A. niger var. niger* and *A. carbonarius* black Aspergilli are the two major producers of ochratoxin A (Figure 4) that is nephrotoxic, teratogenic, carcinogenic, and immunosuppressive in animals, and of FB1 mycotoxin (Palencia et al., 2010).

In *A. terreus* infection, crops such as wheat, ryegrass and potatoes were shown to acquire disease. A number of SMs and mycotoxins, including territrem A, citreoviridin, citrinin, gliotoxin (Figure 4), patulin, terrein, terreic acid, and terretonin are coded in *A. terreus* (Guo and Wang, 2014). The phytotoxic SM terrein possessed ecological, antimicrobial, antiproliferative, and antioxidative activities was also highly induced in plant-derived media and in induced lesions on fruit surfaces (Zaehle et al., 2014).

Rhizoctonia Solani Species Complex

The soil-borne fungus *Rhizoctonia solani* (teleomorph *Thanatephorus cucumeris*), belonging to the phylum *Basidiomycota*, is an economically important plant pathogen. *R. solani*, as a non-obligate necrotrophic pathogen, causes diseases in many crops including species in the families*Solanaceae, Fabaceae, Asteraceae, Poaceae*, and *Brassicaceae* as well as ornamental plants and forest trees throughout the world (Gonzalez Garcia et al., 2006). The anamorph *R. solani* is a species complex and consists species of at least 14 different, genetically isolated populations [=anastomosis groups (AG)] that differ in their ecology and host range (Carling et al., 2002;Stodart et al., 2007).

Recently it was considered that *R. solani* synthesizes HSTs and NHSTs (Vidhyasekaran et al., 1997). HSTs from *R. solani* could increase the virulence of the pathogen (e.g., HC-toxin on maize) (Brooks, 2007), and were often pathogenicity determinants required for a pathogen to incite disease (reviewed in Wolpert et al., 2002). For example, significantly increased five new superoxide dismutase (SOD) activities were detected in plant under treatment of rice leaf sheaths with *R. solani*-toxin, which eliminated the antifungal oxidative burst (Paranidharan et al., 2005).

Previous research has shown that *R. solani* also produces phenyl acetic acid and its derivatives (Bartz et al., 2013), a phenolic compound, and a carbohydrate as phytotoxins that supported the broad host range and diversity within the *R. solani* species. Xu et al. (2015) identified eight compounds from fermentation broth of *R. solani*, from which *m*-hydroxymethylphenyl pentanoate, (Z)-3-methylpent-

2-en-1,5-dioic acid and 3-methoxyfuran-2-carboxylic acid showed phytotoxicity*in vitro*.

Interestingly, in the *R. solani* AG1-IA genome project, genes homologous to reported mycotoxin biosynthesis genes could not be identified (Zheng et al., 2013). However, genes for a putative phytotoxin enniatin (Figure 4) and sequences featuring homology to putative trichothecene citrinin, and AF (Figure 4) and terpene biosynthesis genes (e.g., genes encoding sesquiterpene synthases) were identified. In addition, three volvatoxin genes that homologous to volvatoxin from*Volvariella volvacea* (*Basidiomycota*) were also detected (Wibberg et al., 2014). In proteome analysis, a trichothecene 3-O-acetyltransferase that is required for trichothecene biosynthesis and is involved in reducing the toxicity of trichothecene mycotoxin DON of *Fusaria* (Audenaert et al., 2014), was differentially expressed during the development stage in *R. solani* AG1 (Kwon et al., 2014).

In the last decade, newly related chemical structures have been reported to have significant antifungal activity against *R. solani*: arvelexin isolated from *Thlaspi arvense* (stinkweed) (Pedras et al., 2003), isalexin, brassicanate A and rutalexin from *Brassica napus*, ssp. *rapifera* (Pedras et al., 2004). Cauliflower (*Brassica oleracea* var. *botrytis*) produced other phytoalexins caulilexins A, B, and C (Figure 7), which were also active against the economically important pathogenic fungi*Leptosphaeria maculans* and *S. sclerotiorum* (Pedras et al., 2006). Pedras and Ahiahonu (2005)reviewed the detoxification metabolism of phytoalexins in phytopathogenic fungi. Indole-3-acetaldoxime is an intermediate in the biosynthesis of diverse plant SMs such as indole-3-acetonitrile, brassinin, and brassilexin, as well as the indole glucosinolate (glucobrassicin) and the plant hormone IAA (Figure 1) in *Cruciferae*. Metabolism of indole-3-acetaldoxime to IAA *via*indole-3-acetonitrile by fungi could support the development of plant diseases in crucifers (Pedras and Montaut, 2003).

Genus Cochliobolus

The filamentous ascomycete genus *Cochliobolus* (anamorph *Bipolaris*/*Curvularia*; Manamgoda et al., 2014) is composed of more than 40 closely related pathogenic species with particular specificity to their host plants (Condon et al., 2013). Gao et al. (2014) reported the genome sequence of a highly virulent *C. lunatus* strain, and phylogenomic analysis indicated that *C. lunatus* was evolved from *C. heterostrophus*. *C. lunatus* CX-3 strain was capable of producing diverse SMs (Table 1) such as NHSTs and melanin that could aid in niche exploitation and pathogenicity.

It was known previously that the ability to produce HST T-toxin requires three genes encoded at two unlinked loci (Baker et al., 2006). However, Inderbitzin et al. (2010) reported further six genes including two PKSs, one decarboxylase, five dehydrogenases, and one unknown protein that were involved in T-toxin production and high virulence of *C. heterostrophus* to maize. HST1, one NPRS of *C. carbonum* (*Bipolaris zeicola*) played a key role in the cyclic tetrapeptide HC-toxin (Figure 2) biosynthesis (Walton, 2006), which was also produced by*Alternaria jesenskae* (Wight et al., 2013) and was also encoded in another maize pathogen *Setosphaeria turcica* (Condon et al., 2013). Six other known PKSs were found to be involved in different kinds of toxin biosynthesis such as *A. alternata* ACT-toxin, *F. graminearum* ZEA, *F. verticilloides* fumonisin, *A. ochraceus* OTA and *C. heterostrophus* T-toxin. Phylogenetic and modular analyses suggested that the protein structures of*C. lunatus* CX-3 NRPSs were obviously different from other known NRPSs being involved in the biosynthesis of mycotoxins such as HC-toxin (Figure 2) of *C. carbonum*, similarly to AM-toxin (Figure 2) of *A. alternata*, gliotoxin (Figure 4) of *A. fumigatus* and enniatin of *F. equiseti* (Gao et al., 2014).

Genus Alternaria

Alternaria species have different lifestyles ranging from saprophytes to endophytes and to pathogens. Phylogenetic relations of the *Alternaria* complex was revisited and delineated byWoudenberg et al. (2013) within *Alternaria* and related genera based on nucleotide sequence data.*A. alternata* has the ability to produce more than 60 SMs from which at least 10 PKS products can be found (Saha et al., 2012) (Table 1). *Alternaria* species have been reported to cause diseases in nearly 400 plant species including a wide variety of economically important crops and cause severe economic problems. *A. alternata* alone can infect more than 100 plant species (Thomma, 2003). The production of diverse phytotoxins and HSTs can be considered as a key reason for the success of these pathogens (Nishimura and Kohmoto, 1983). From about 20 HSTs that have been documented (Otani et al., 1995; Walton, 1996), at least seven are from *A. alternata* pathotypes (Otani et al., 1995). For HST ACR-toxin production and pathogenicity, PKS gene ACRTS2 was found to be essential of the rough lemon pathotype of *A. alternata* (Izumi et al., 2012). Several NHSTs are also produced in *Alternaria* such as brefeldin A, altertoxin, and tentoxin (Figure 2) and also other mycotoxins. Alternariol (Figure 2) and alternariol-9-methyl ether are major NHSTs that are common contaminants of food such as cereals, fruits and fruit juices (Scott, 2001). A PKS involved in melanin biosynthesis was also characterized and named ALM (albino) (Kimura and Tsuge, 1993). For production of SM siderophores and virulence, the *A. alternata* gene AaNPS6, encoding a polypeptide analogous to fungal NRPS was demonstrated by Chen et al. (2013). The*Alternaria* toxins provide prospect for biocontrol of weeds due to high phytotoxic effect against weeds but low mammalian toxicity (Abbas et al., 1995; Chen et al., 2005; Evidente et al., 2009;Yang et al., 2012).

From the plant side, high concentrations of alkaloid phytoalexin camalexin (Tsuji et al., 1992) have been observed at the infection site of*A. alternata* (Schuhegger et al., 2007) and also in the proximity to the lesions induced by *Botrytis* species (Kliebenstein et al., 2005). In *A. thaliana*leaves both biotrophic and necrotrophic plant pathogens induced camalexin formation (Thomma et al., 1999). *A. brassicicola* could detoxify camalexin but at much slower rate than phytoalexin brassinin from *Brassicaceae* (Pedras et al., 2014).

Secondary Metabolite Production of Biotrophs

Growth and reproduction of obligate biotrophic phytopathogens that are very poor in SM production like powdery mildews are entirely dependent on living plant cells. Spanu et al. (2010)hypothesized that *Blumeria* synthesized only one iron siderophore and one simple polyketide pigment of the cleistothecia. Similar trends have been observed in other biotrophs, such as the basidiomycete corn smut fungus *Ustilago maydis* and the plant symbiotic fungus *Tuber melanosporum*.

Cladosporium fulvum (*Passalora fulva*) is also a biotrophic fungus that infecting tomato, grows extracellularly in close contact with host mesophyll cells. The only known SMs produced by *C. fulvum* is cladofulvin (de Wit et al., 2012; Collemare et al., 2014) anthraquinone pigment. However, cladofulvin has not been detected to cause necrosis on *Solanaceae* plants or to show any antimicrobial activity (Collemare et al., 2014). *C. fulvum* has also the potential to produce elsinochrome and cercosporin toxins, but the corresponding core genes were not expressed during infection of tomato (Collemare et al., 2014). It has been suggested that loss of SM biosynthetic pathways is associated with biotrophy (Spanu et al., 2010); nevertheless, the biotrophic *C. fulvum*has twice the number of key SM genes compared to the closely related hemibiotrophic *Dothistroma septosporum* (teleomorph *Mycosphaerella pini*), of which 14 and 9, respectively, are organized into gene clusters along with other SM-related genes (de Wit et al., 2012). The numbers of its SM enzyme-encoding genes were comparable to those of *M. graminicola*, but were lower than those in most other sequenced Dothideomycete (Table 1). It could be concluded that, in contrast to reduced SM production capacity, down-regulation of high number of SM biosynthetic pathways might represent another mechanism associated with a biotrophic lifestyle (Collemare et al., 2014).

CONCLUSIONS AND FUTURE ASPECTS

Fungal-plant host interactions represent biochemically complex and challenging scenarios that are being investigated also by metabolomic approaches (Allwood et al., 2008). It is noteworthy that although SMs play important roles in the virulence and lifestyle of fungal plant pathogens only about 25% of the fungal SM gene clusters have already been characterized functionally and this number is much lower at plant side.

Concomitantly, comparative genomics and transcriptomics are employed to obtain insights into the genetic features that enable fungal pathogens to adapt successfully to various ecological niches and to adopt different pathogenic lifestyles. The suites of fungal SM genes reflect astounding diversity among species, hinting that gene products, particularly those associated with unique genomic regions, are candidates for pathogenic lifestyle differences. Furthermore, horizontal gene and chromosome transfers provide a means for pathogens to broaden their host range (Mehrabi et al., 2011; Fitzpatrick, 2012). The increasing availability of fungal pathogen genome sequences and next-generation genomic tools allow us to survey the SM gene clusters in individual fungi. The recent availability of next-generation RNA-Seq technologies has revolutionized transcriptomic profiling and are used to probe the expression of SM gene clusters during various stages of infection. Unlike microarray, RNA-Seq allows the simultaneous quantification of transcripts from more than one organism and is thus perfectly suited for the study of plant-pathogen interactions (Chooi and Solomon, 2014). Moreover, manipulations of strain-unique SM genes associated with host-specific virulence provide possibility to investigate fungal-plant interaction.

The great structural diversity of phytotoxins, the high potency and exclusive mechanisms of action (compared to synthetic herbicides) make fungal toxins highly attractive for discovering herbicidal activity. Even if natural phytotoxins are not necessarily suitable for direct use as a commercial herbicide, the identification of mechanisms are very important for new herbicide developments. Newly developed herbicides with environmentally friendly component could be used more safely in integrated pest management systems. On the other side, plant SMs can be used against plant pathogens (especially in sprayable forms) as natural plant extracts for example in organic agricultural production systems. Deeper knowledge of the fungus-plant interaction may help resistance breeding of new plant cultivars/hybrids against stresses such as abiotic (e.g., heat) stress or against fungal diseases.

CONFLICT OF INTEREST STATEMENT

The authors declare that the research was conducted in the absence of any commercial or financial relationships that could be construed as a potential conflict of interest.

ACKNOWLEDGMENTS

This work was supported by the Hungarian Scientific Research Fund (OTKA K100464 and OTKA K108333).

REFERENCES

1. Abbas, H. K., Tanaka, T., Duke, S. O., Porter, J. K., Wray, E. M., Hodges, L., et al. (1994). Fumonisin- and AAL-toxin-induced disruption of sphingolipid metabolism with accumulation of free sphingoid bases. *Plant Physiol.* 106, 1085–1093.
2. Abbas, H. K., Tanaka, T., and Shier, W. T. (1995). Biological activities of synthetic analogues of *Alternaria alternata* toxin (AAL-toxin) and fumonisin in plant and mammalian cell cultures. *Phytochemistry* 40, 1681–1689.
3. Abbas, H. K., Yoshizawa, T., and Shier, W. T. (2013). Cytotoxicity and phytotoxicity of trichothecene mycotoxins produced by*Fusarium* spp. *Toxicon* 74, 68–75. doi: 10.1016/j.toxicon.2013.07.026
4. Affeldt, K. J., Brodhagen, M., and Keller, N. P. (2012). *Aspergillus* oxylipin signaling and quorum sensing pathways depend on G protein-coupled receptors. *Toxins (Basel.)* 4, 695–717. doi: 10.3390/toxins4090695
5. Ahuja, I., Kissen, R., and Bones, A. M. (2012). Phytoalexins in defense against pathogens. *Trends Plant Sci.* 17, 73–90. doi: 10.1016/j.tplants.2011.11.002
6. Akatsuka, T., Kodama, O., Sekido, H., Kono, Y., and Takeuchi, S. (1985). Novel phytoalexins (oryzalexins A, B and C) isolated from rice blast leaves infected with *Pyricularia oryzae*. *Agric. Biol. Chem.* 49, 1689–1701. doi: 10.1080/00021369.1985.10866951
7. Allwood, J. W., Ellis, D. I., and Goodacre, R. (2008). Metabolomic technologies and their application to the study of plants and plant-host interactions. *Physiol. Plant.* 132, 117–1135. doi: 10.1111/j.1399-3054.2007.01001.x

8. Amagasa, T., Paul, R. N., Heitholt, J. J., and Duke, S. O. (1994). Physiological effects of cornexistin on *Lemna pausicostata.Pestic. Biochem. Physiol.* 49, 37–52. doi: 10.1006/pest.1994.1032

9. Amare, M. G., and Keller, N. P. (2014). Molecular mechanisms of *Aspergillus flavus* secondary metabolism and development.*Fungal Genet. Biol.* 66, 11–18. doi: 10.1016/j.fgb.2014.02.008

10. Amselem, J., Cuomo, C. A., van Kan, J. A. L., Viaud, M., Benito, E. P., Couloux, A., et al. (2011). Genomic analysis of the necrotrophic fungal pathogens *Sclerotinia sclerotiorum* and *Botrytis cinerea. PLoS Genet.* 7:e1002230. doi: 10.1371/journal.pgen.1002230

11. Atkinson, P., and Blakeman, J. P. (1982). Seasonal occurrence of an antimicrobial flavanone, sakuranetin, associated with glands on leaves of *Ribes nigrum. New Phytol.* 92, 63–74. doi: 10.1111/j.1469-8137.1982.tb03363.x

12. Audenaert, K., Vanheule, A., Höfte, M., and Haesaert, G. (2014). Deoxynivalenol: a major player in the multifaceted response of*Fusarium* to its environment. *Toxins (Basel.)* 6, 1–19. doi: 10.3390/toxins6010001

13. Avalos, J., and Estrada, A. F. (2010). Regulation by light in *Fusarium. Fungal Genet. Biol.* 47, 930–938. doi: 10.1016/j.fgb.2010.05.001

14. Baetz, U., and Martinoia, E. (2014). Root exudates: the hidden part of plant defense. *Trends Plant Sci.* 19, 90–98. doi: 10.1016/j.tplants.2013.11.006

15. Bagheri-Gavkosh, S., Bigdeli, M., Shams-Ghahfarokhi, M., and Razzaghi-Abyaneh, M. (2009). Inhibitory effects of Ephedra major host on *Aspergillus parasiticus* growth and aflatoxin production. *Mycopathologia* 168, 249–255. doi: 10.1007/s11046-009-9220-x

16. Baker, S. E., Kroken, S., Inderbitzin, P., Asvarak, T., Li, B. Y., Shi, L., et al. (2006). Two polyketide synthase-encoding genes are required for biosynthesis of the polyketide virulence factor, T-toxin, by *Cochliobolus heterostrophus. Mol. Plant Microbe Interact.* 19, 139–149. doi: 10.1094/MPMI-19-0139

17. Balasundram, N., Sundram, K., and Samman, S. (2006). Phenolic compounds in plants and agri-industrial by-products: antioxidant activity, occurrence, and potential uses. *Food Chem.* 99, 191–203. doi: 10.1016/j.foodchem.2005.07.042

18. Barna, B., Fodor, J., Harrach, B. D., Pogány, M., and Király, Z. (2012). The Janus face of reactive oxygen species in resistance and susceptibility of plants to necrotrophic and biotrophic pathogens. *Plant Physiol. Biochem.* 59, 37–43. doi: 10.1016/j.plaphy.2012.01.014

19. Bartz, F. E., Glassbrook, N. J., Danehower, D. A., and Cubeta, M. A. (2013). Modulation of the phenylacetic acid metabolic complex by quinic acid alters the disease-causing activity of *Rhizoctonia solani* on tomato. *Phytochemistry* 89, 47–52. doi: 10.1016/j.phytochem.2012.09.018

20. Bednarek, P., Kwon, C., and Schulze-Lefert, P. (2010). Not a peripheral issue: secretion in plant-microbe interactions. *Curr. Opin. Plant Biol.* 13, 378–387. doi: 10.1016/j.pbi.2010.05.002

21. Beekrum, S., Govinden, R., Padayachee, T., and Odhav, B. (2003). Naturally occurring phenols: a detoxification strategy for fumonisin B1. *Food Addit. Contam.* 20, 490–493. doi: 10.1080/0265203031000098678

22. Biemelt, S., and Sonnewald, U. (2006). Plant-microbe interactions to probe regulation of plant carbon metabolism. *J. Plant Physiol.* 163, 307–318. doi: 10.1016/j.jplph.2005.10.011

23. Bills, G. F., Platas, G., and Gams, W. (2004). Conspecificity of the cerulenin and helvolic acid producing "*Cephalosporium caerulens*," and the hypocrealean fungus *Sarocladium oryzae. Mycol. Res.* 108, 1291–1300. doi: 10.1017/S0953756204001297

24. Bluhm, B. H., and Woloshuk, C. P. (2005). Amylopectin induces fumonisin B1 production by *Fusarium verticillioides* during colonization of maize kernels. *Mol. Plant Microbe Interact.* 18, 1333–1339. doi: 10.1094/MPMI-18-1333

25. Bohlmann, J., and Keeling, C. I. (2008). Terpenoid biomaterials. *Plant J.* 54, 656–669. doi: 10.1111/j.1365-313X.2008.03449.x

26. Bolton, E. E., Wang, Y., Thiessen, P. A., and Bryant, S. H. (2008). "PubChem: integrated platform of small molecules and biological activities," in *Annual Reports in Computational Chemistry*, eds R. A. Wheeler and D. C. Spellmeyer (Oxford: Elsevier), 217–241. doi: 10.1016/S1574-1400(08)00012-1

27. Bömke, C., and Tudzynski, B. (2009). Diversity, regulation, and evolution of the gibberellin biosynthetic pathway in fungi compared to plants and bacteria. *Phytochemistry* 70, 1876–1893. doi: 10.1016/j.phytochem.2009.05.020

28. Brodhagen, M., Tsitsigiannis, D. I., Hornung, E., Goebel, C., Feussner, I., and Keller, N. P. (2008). Reciprocal oxylipin-mediated cross-talk in the *Aspergillus*-seed pathosystem. *Mol. Microbiol.* 67, 378–391. doi: 10.1111/j.1365-2958.2007.06045.x

29. Brodhun, F., Cristobal-Sarramian, A., Zabel, S., Newie, J., Hamberg, M., and Feussner, I. (2013). An iron 13S-lipoxygenase with an α-linolenic acid specific hydroperoxidase activity from *Fusarium oxysporum. PLoS ONE* 8:e64919. doi: 10.1371/journal.pone.0064919

30. Brooks, S. A. (2007). Sensitivity to a phytotoxin from *Rhizoctonia solani* correlates with sheath blight susceptibility in rice.*Phytopathology* 97, 1207–1212. doi: 10.1094/PHYTO-97-10-1207

31. Butchko, R. A. E., Brown, D. W., Busman, M., Tudzynski, B., and Wiemann, P. (2012). Lae1 regulates expression of multiple secondary metabolite gene clusters in *Fusarium verticillioides. Fungal Genet. Biol.* 49, 602–612. doi: 10.1016/j.fgb.2012.06.003

32. Buzi, A., Chilosi, G., Timperio, A. M., Zolla, L., Rossall, S., and Magro, P. (2003). Polygalacturonase produced by *Botrytis fabae*as elicitor of two furanoacetylenic phytoalexins in *Vicia faba* pods. *J. Plant Pathol.* 85, 111–116. doi: 10.4454/jpp.v85i2.1018

33. Calvo, A. M., Hinze, L. L., Gardner, H. W., and Keller, N. P. (1999). Sporogenic effect of polyunsaturated fatty

acids on development of *Aspergillus spp*. *Appl. Environ. Microbiol.* 65, 3668–3673.

34. Camele, I., Altieri, L., De Martino, L., De Feo, V., Mancini, E., and Rana, G. L. (2012). *In vitro* control of post-harvest fruit rot fungi by some plant essential oil components. *Int. J. Mol. Sci.* 13, 2290–2300. doi: 10.3390/ijms13022290
35. Carling, D. E., Baird, R. E., Gitaitis, R. D., Brainard, K. A., and Kuninaga, S. (2002). Characterization of AG-13, a newly reported anastomosis group of *Rhizoctonia solani*. *Phytopathology* 92, 893–899. doi: 10.1094/PHYTO.2002.92.8.893
36. Cartwright, D. W., Langcake, P., Pryce, R. J., Leworthy, D. P., and Ride, J. P. (1981). Isolation and characterization of two phytoalexins from rice as momilactones A and B. *Phytochemistry* 20, 535–537. doi: 10.1016/S0031-9422(00)84189-8
37. Chen, L. H., Lin, C. H., and Chung, K. R. (2013). A nonribosomal peptide synthetase mediates siderophore production and virulence in the citrus fungal pathogen *Alternaria alternata*. *Mol. Plant Pathol.* 14, 497–505. doi: 10.1111/mpp.12021
38. Chen, S., Dai, X., Qiang, S., and Tang, Y. (2005). Effect of a nonhost-selective toxin from *Alternaria alternata* on chloroplast-electron transfer activity in *Eupatorium adenophorum*. *Plant Pathol.* 54, 671–677. doi: 10.1111/j.1365-3059.2005.01249.x
39. Chivasa, S., Ndimba, B. K., Simon, W. J., Lindsey, K., and Slabas, A. R. (2005). Extracellular ATP functions as an endogenous external metabolite regulating plant cell viability. *Plant Cell* 17, 3019–3034. doi: 10.1105/tpc.105.036806
40. Chooi, Y. H., and Solomon, P. S. (2014). A chemical ecogenomics approach to understand the roles of secondary metabolites in fungal cereal pathogens. *Front. Microbiol.* 5:640. doi: 10.3389/fmicb.2014.00640
41. Collemare, J., Billard, A., Böhnert, H. U., and Lebrun, M.-H. (2008). Biosynthesis of secondary metabolites in the rice blast fungus *Magnaporthe grisea*: the role of hybrid PKS-NRPS in pathogenicity. *Mycol. Res.* 112, 207–215. doi: 10.1016/j.mycres.2007.08.003
42. Collemare, J., Griffiths, S., Iida, Y., Karimi Jashni, M., Battaglia, E., Cox, R. J., et al. (2014). Secondary metabolism and biotrophic lifestyle in the tomato pathogen *Cladosporium fulvum*. *PLoS ONE* 9:e85877. doi: 10.1371/journal.pone.0085877
43. Colmenares, A., Aleu, J., Durán-Patrón, R., Collado, I. G., and Hernández-Galán, R. (2002). The putative role of botrydial and related metabolites in the infection mechanism of *Botrytis cinerea*. *J. Chem. Ecol.* 28, 997–1005. doi: 10.1023/A:1015209817830
44. Condon, B. J., Leng, Y., Wu, D., Bushley, K. E., Ohm, R. A., Otillar, R., et al. (2013). Comparative genome structure, secondary metabolite, and effector coding capacity across *Cochliobolus* pathogens. *PLoS Genet.* 9:e1003233. doi: 10.1371/journal.pgen.1003233
45. Couch, B. C., and Kohn, L. M. (2002). A multilocus gene genealogy concordant with host preference indicates segregation of a new species, *Magnaporthe oryzae*, from *M. grisea*. *Mycologia* 94, 683–693. doi: 10.2307/3761719
46. Crowley, D. E., and Kraemer, S. M. (2007). "Function of siderophores in the plant rhizosphere," in *The Rhizosphere, Biochemistry and Organic Substances at the Soil-Plant Interface*, eds R. Pinton, Z. Varanini, and P. Nannipieri (New York, NY: CRC Press), 73–109.
47. Crozier, A., Jaganath, I. B., and Clifford, M. N. (2008). "Phenols, poliphenols and tannins: an overview," in *Plant Secondary Metabolites: Occurrence, Structure and Role in the Human Diet*, eds A. Crozier, M. N. Clifford, and H. Ashihara (Oxford, UK: Blackwell Publishing Ltd), 1–24.
48. Dakora, F., and Phillips, D. (2002). Root exudates as mediators of mineral acquisition in low-nutrient environments. *Plant Soil*245, 35–47.
49. Dambolena, J. S., Zygadlo, J. A., and Rubinstein, H. R. (2011). Antifumonisin activity of natural phenolic compounds A structure-property-activity relationship study. *Int. J. Food Microbiol.* 145, 140–146. doi: 10.1016/j.ijfoodmicro.2010.12.001
50. Deighton, N., Muckenschnabel, I., Colmenares, A. J., Collado, I. G., and Williamson, B. (2001). Botrydial is produced in plant tissues infected by *Botrytis cinerea*. *Phytochemistry* 57, 689–692. doi: 10.1016/S0031-9422(01)00088-7
51. Deller, S., Hammond-Kosack, K. E., and Rudd, J. J. (2011). The complex interactions between host immunity and non-biotrophic fungal pathogens of wheat leaves. *J. Plant Physiol.* 168, 63–71. doi: 10.1016/j.jplph.2010.05.024
52. Demuner, A. J., Barbosa, L. C. A., Miranda, A. C. M., Geraldo, G. C., da Silva, C. M., Giberti, S., et al. (2013). The fungal phytotoxin alternariol 9-methyl ether and some of its synthetic analogues inhibit the photosynthetic electron transport chain. *J. Nat. Prod.* 76, 2234–2345. doi: 10.1021/np4005882
53. De Oliveira, T. L. C., de Araújo Soares, R., Ramos, E. M., das Graças Cardoso, M., Alves, E., and Piccoli, R. H. (2011). Antimicrobial activity of *Satureja montana* L. essential oil against *Clostridium perfringens* type A inoculated in mortadella-type sausages formulated with different levels of sodium nitrite. *Int. J. Food Microbiol.* 144, 546–555. doi: 10.1016/j.ijfoodmicro.2010.11.022
54. Derksen, H., Rampitsch, C., and Daayf, F. (2013). Signaling cross-talk in plant disease resistance. *Plant Sci.* 207, 79–87. doi: 10.1016/j.plantsci.2013.03.004
55. Desjardins, A. E., Hohn, T. M., and McCormick, S. P. (1993). Trichothecene biosynthesis in *Fusarium* species: chemistry, genetics, and significance. *Microbiol. Rev.* 57, 595–604.
56. Dewick, P. M. (2002). The biosynthesis of C5-C25 terpenoid compounds. *Nat. Prod. Rep.* 19, 181–222. doi: 10.1039/b002685i
57. de Wit, P. J. G. M., van der Burgt, A., Ökmen, B., Stergiopoulos, I., Abd-Elsalam, K. A., Aerts, A. L., et al. (2012). The genomes of the fungal plant pathogens *Cladosporium fulvum* and *Dothistroma*

septosporum reveal adaptation to different hosts and lifestyles but also signatures of common ancestry. *PLoS Genet.* 8:e1003088. doi: 10.1371/journal.pgen.1003088

58. Dolezal, A. L., Obrian, G. R., Nielsen, D. M., Woloshuk, C. P., Boston, R. S., and Payne, G. A. (2013). Localization, morphology and transcriptional profile of *Aspergillus flavus* during seed colonization. *Mol. Plant Pathol.* 14, 898–909. doi: 10.1111/mpp.12056

59. Dolezal, A. L., Shu, X., OBrian, G. R., Nielsen, D. M., Woloshuk, C. P., Boston, R. S., et al. (2014). *Aspergillus flavus* infection induces transcriptional and physical changes in developing maize kernels. *Front. Microbiol.* 5:384. doi: 10.3389/fmicb.2014.00384

60. Dos Santos, M. O., and Furlong, E. B. (2008). Screening of antifungal and antimycotoxigenic activity of plant phenolic extracts.*World Mycotoxin J.* 1, 139–146. doi: 10.3920/WMJ2008.1006

61. Du Fall, L. A., and Solomon, P. S. (2011). Role of cereal secondary metabolites involved in mediating the outcome of plant-pathogen interactions. *Metabolites* 1, 64–78. doi: 10.3390/metabo1010064

62. Durrant, W. E., and Dong, X. (2004). Systemic acquired resistance. *Annu. Rev. Phytopathol.* 42, 185–209. doi: 10.1146/annurev.phyto.42.040803.140421

63. Ebata, Y., Yamamoto, H., and Uchiyama, T. (1998). Chemical composition of the glue from apressoria of *Magnaporthe grisea.Biosci. Biotechnol. Biochem.* 62, 672–674.

64. El-Mogy, M. M., and Alsanius, B. W. (2012). Cassia oil for controlling plant and human pathogens on fresh strawberries. *Food Control* 28, 157–162. doi: 10.1016/j.foodcont.2012.04.036

65. Espinosa-García, F. J., and Langenheim, J. H. (1991). Effects of sabinene and γ-terpinene from coastal redwood leaves acting singly or in mixtures on the growth of some of their fungus endophytes. *Biochem. Syst. Ecol.* 19, 643–650. doi: 10.1016/0305-1978(91)90080-J

66. Evidente, A., Punzo, B., Andolfi, A., Berestetskiy, A., and Motta, A. (2009). Alternethanoxins A and B, polycyclic ethanones produced by *Alternaria sonchi*, potential mycoherbicides for *Sonchus arvensis* biocontrol. *J. Agric. Food Chem.* 57, 6656–6660. doi: 10.1021/jf9014944

67. Feng, G. H., and Leonard, T. J. (1998). Culture conditions control expression of the genes for aflatoxin and sterigmatocystin biosynthesis in *Aspergillus parasiticus* and *A. nidulans. Appl. Environ. Microbiol.* 64, 2275–2277.

68. Fitzpatrick, D. A. (2012). Horizontal gene transfer in fungi. *FEMS Microbiol. Lett.* 329, 1–8. doi: 10.1111/j.1574-6968.2011.02465.x

69. Flaishman, M. A., and Kolattukudy, P. E. (1994). Timing of fungal invasion using host›s ripening hormone as a signal. *Proc. Natl. Acad. Sci. U.S.A.* 91, 6579–6583.

70. Foroud, N. A., and Eudes, F. (2009). Trichothecenes in cereal grains. *Int. J. Mol. Sci.* 10, 147–173. doi: 10.3390/ijms10010147

71. Fountain, J. C., Scully, B. T., Ni, X., Kemerait, R. C., Lee, R. D., Chen, Z. Y., et al. (2014). Environmental influences on maize-*Aspergillus flavus* interactions and aflatoxin production. *Front. Microbiol.* 5:40. doi: 10.3389/fmicb.2014.00040

72. Friesen, T. L., Faris, J. D., Solomon, P. S., and Oliver, R. P. (2008). Host-specific toxins: effectors of necrotrophic pathogenicity.*Cell. Microbiol.* 10, 1421–1428. doi: 10.1111/j.1462-5822.2008.01153.x

73. Gaffoor, I., and Trail, F. (2006). Characterization of two polyketide synthase genes involved in zearalenone biosynthesis in*Gibberella zeae. Appl. Environ. Microbiol.* 72, 1793–1799. doi: 10.1128/AEM.72.3.1793-1799.2006

74. Gao, S., Li, Y., Gao, J., Suo, Y., Fu, K., Li, Y., et al. (2014). Genome sequence and virulence variation-related transcriptome profiles of *Curvularia lunata*, an important maize pathogenic fungus. *BMC Genomics* 15:627. doi: 10.1186/1471-2164-15-627

75. Garcia, D., Ramos, A. J., Sanchis, V., and Marín, S. (2012). Effect of *Equisetum arvense* and *Stevia rebaudiana* extracts on growth and mycotoxin production by *Aspergillus flavus* and Fusarium verticillioides in maize seeds as affected by water activity.*Int. J. Food Microbiol.* 153, 21–27. doi: 10.1016/j.ijfoodmicro.2011.10.010

76. Gardiner, D. M., Kazan, K., Praud, S., Torney, F. J., Rusu, A., and Manners, J. M. (2010). Early activation of wheat polyamine biosynthesis during Fusarium head blight implicates putrescine as an inducer of trichothecene mycotoxin production. *BMC Plant Biol.* 10:289. doi: 10.1186/1471-2229-10-289

77. Ghebremeskel, M., and Langseth, W. (2001). The occurrence of culmorin and hydroxy-culmorins in cereals. *Mycopathologia*152, 103–108. doi: 10.1023/A:1012479823193

78. Gilbert, R. D., Johnson, A. M., and Dean, R. A. (1996). Chemical signals responsible for appressorium formation in the rice blast fungus *Magnaporthe grisea. Physiol. Mol. Plant Pathol.* 48, 335–346. doi: 10.1006/pmpp.1996.0027

79. Glazebrook, J. (2005). Contrasting mechanisms of defense against biotrophic and necrotrophic pathogens. *Annu. Rev. Phytopathol.* 43, 205–227. doi: 10.1146/annurev.phyto.43.040204.135923

80. Gonzalez Garcia, V., Portal Onco, M. A., and Rubio Susan, V. (2006). Review. Biology and systematics of the form genus*Rhizoctonia. Spanish J. Agric. Res.* 4, 55. doi: 10.5424/sjar/2006041-178

81. Govrin, E. M., and Levine, A. (2002). Infection of *Arabidopsis* with a necrotrophic pathogen, *Botrytis cinerea*, elicits various defense responses but does not induce systemic acquired resistance (SAR). *Plant Mol. Biol.* 48, 267–276. doi: 10.1023/A:1013323222095

82. Grayer, R. J., and Kokubun, T. (2001). Plant–fungal interactions: the search for phytoalexins and other antifungal compounds from higher plants. *Phytochemistry* 56, 253–263. doi: 10.1016/S0031-9422(00)00450-7

83. Guo, C. J., and Wang, C. C. C. (2014). Recent advances in genome mining of secondary metabolites in *Aspergillus terreus*. *Front. Microbiol.* 5:717. doi: 10.3389/fmicb.2014.00717

84. Halloin, J. M., De Zoeten, G. A., and Walker, J. C. (1970). The effects of tentoxin on chlorophyll synthesis and plastid structure in cucumber and cabbage. *Plant Physiol.* 45, 310–314.

85. Hansen, F. T., Gardiner, D. M., Lysøe, E., Fuertes, P. R., Tudzynski, B., Wiemann, P., et al. (2015). An update to polyketide synthase and non-ribosomal synthetase genes and nomenclature in *Fusarium*. *Fungal Genet. Biol.* 75, 20–29. doi: 10.1016/j.fgb.2014.12.004

86. Hatta, R., Ito, K., Hosaki, Y., Tanaka, T., Tanaka, A., Yamamoto, M., et al. (2002). A conditionally dispensable chromosome controls host-specific pathogenicity in the fungal plant pathogen *Alternaria alternata*. *Genetics* 161, 59–70.

87. Hauser, F., Waadt, R., and Schroeder, J. I. (2011). Evolution of abscisic acid synthesis and signaling mechanisms. *Curr. Biol.* 21, R346–R355. doi: 10.1016/j.cub.2011.03.015

88. Hedden, P., Phillips, A. L., Rojas, M. C., Carrera, E., and Tudzynski, B. (2001). Gibberellin biosynthesis in plants and fungi: a case of convergent evolution? *J. Plant Growth Regul.* 20, 319–331. doi: 10.1007/s003440010037

89. Hegde, Y., and Kolattukudy, P. E. (1997). Cuticular waxes relieve self-inhibition of germination and appressorium formation by the conidia of *Magnaporthe grisea*. *Physiol. Mol. Plant Pathol.* 51, 75–84. doi: 10.1006/pmpp.1997.0105

90. Horbach, R., Navarro-Quesada, A. R., Knogge, W., and Deising, H. B. (2011). When and how to kill a plant cell: infection strategies of plant pathogenic fungi. *J. Plant Physiol.* 168, 51–62. doi: 10.1016/j.jplph.2010.06.014

91. Horowitz Brown, S., Zarnowski, R., Sharpee, W. C., and Keller, N. P. (2008). Morphological transitions governed by density dependence and lipoxygenase activity in *Aspergillus flavus*. *Appl. Environ. Microbiol.* 74, 5674–5685. doi: 10.1128/AEM.00565-08

92. Howard, R. J., Ferrari, M. A., Roach, D. H., and Money, N. P. (1991). Penetration of hard substrates by a fungus employing enormous turgor pressures. *Proc. Natl. Acad. Sci. U.S.A.* 88, 11281–11284. doi: 10.1073/pnas.88.24.11281

93. Howard, R. J., and Valent, B. (1996). Breaking and entering: host penetration by the fungal rice blast pathogen *Magnaporthe grisea*. *Annu. Rev. Microbiol.* 50, 491–512. doi: 10.1146/annurev.micro.50.1.491

94. Howlett, B. J. (2006). Secondary metabolite toxins and nutrition of plant pathogenic fungi. *Curr. Opin. Plant Biol.* 9, 371–375. doi: 10.1016/j.pbi.2006.05.004

95. Hu, J., Chen, C., Peever, T., Dang, H., Lawrence, C., and Mitchell, T. (2012). Genomic characterization of the conditionally dispensable chromosome in *Alternaria arborescens* provides evidence for horizontal gene transfer. *BMC Genomics* 13:171. doi: 10.1186/1471-2164-13-171

96. Huffaker, A., Kaplan, F., Vaughan, M. M., Dafoe, N. J., Ni, X., Rocca, J. R., et al. (2011). Novel acidic sesquiterpenoids constitute a dominant class of pathogen-induced phytoalexins in maize. *Plant Physiol.* 156, 2082–2097. doi: 10.1104/pp.111.179457

97. Hwang, C. S., and Kolattukudy, P. E. (1995). Isolation and characterization of genes expressed uniquely during appressorium formation by *Colletotrichum gloeosporioides* conidia induced by the host surface wax. *Mol. Gen. Genet.* 247, 282–294.

98. Inderbitzin, P., Asvarak, T., and Turgeon, B. G. (2010). Six new genes required for production of T-toxin, a polyketide determinant of high virulence of *Cochliobolus heterostrophus* to maize. *Mol. Plant. Microbe. Interact.* 23, 458–472. doi: 10.1094/MPMI-23-4-0458

99. Islam, M. S., Haque, M. S., Islam, M. M., Emdad, E. M., Halim, A., Hossen, Q. M. M., et al. (2012). Tools to kill: genome of one of the most destructive plant pathogenic fungi *Macrophomina phaseolina*. *BMC Genomics* 13:493. doi: 10.1186/1471-2164-13-493

100. Ito, S. I., Ihara, T., Tamura, H., Tanaka, S., Ikeda, T., Kajihara, H., et al. (2007). alpha-Tomatine, the major saponin in tomato, induces programmed cell death mediated by reactive oxygen species in the fungal pathogen *Fusarium oxysporum*. *FEBS Lett.*581, 3217–3222. doi: 10.1016/j.febslet.2007.06.010

101. Izumi, Y., Ohtani, K., Miyamoto, Y., Masunaka, A., Fukumoto, T., Gomi, K., et al. (2012). A polyketide synthase gene, ACRTS2, is responsible for biosynthesis of host-selective ACR-toxin in the rough lemon pathotype of *Alternaria alternata*. *Mol. Plant Microbe Interact.* 25, 1419–1429. doi: 10.1094/MPMI-06-12-0155-R

102. Janda, M., and Ruelland, E. (2014). Magical mystery tour: salicylic acid signalling. *Environ. Exp. Bot.* 114, 117–128. doi: 10.1016/j.envexpbot.2014.07.003

103. Jeandet, P., Douillet-Breuil, A. C., Bessis, R., Debord, S., Sbaghi, M., and Adrian, M. (2002). Phytoalexins from the Vitaceae: biosynthesis, phytoalexin gene expression in transgenic plants, antifungal activity, and metabolism. *J. Agric. Food Chem.* 50, 2731–2741. doi: 10.1021/jf011429s

104. Jermnak, U., Yoshinari, T., Sugiyama, Y., Tsuyuki, R., Nagasawa, H., and Sakuda, S. (2012). Isolation of methyl syringate as a specific aflatoxin production inhibitor from the essential oil of *Betula alba* and aflatoxin production inhibitory activities of its related compounds. *Int. J. Food Microbiol.* 153, 339–344. doi: 10.1016/j.ijfoodmicro.2011.11.023

105. Jiang, J., Liu, X., Yin, Y., and Ma, Z. (2011). Involvement of a velvet protein FgVeA in the regulation of asexual development, lipid and secondary metabolisms and virulence in *Fusarium graminearum*. *PLoS ONE* 6:e28291. doi: 10.1371/journal.pone.0028291

106. Jiang, J., Yun, Y., Liu, Y., and Ma, Z. (2012). FgVELB is associated with vegetative differentiation, secondary metabolism and virulence in *Fusarium graminearum*. *Fungal Genet. Biol.* 49, 653–662. doi: 10.1016/j.fgb.2012.06.005

107. Karányi, Z., Holb, I., Hornok, L., Pócsi, I., and Miskei, M. (2013). FSRD: fungal stress response database. *Database*2013:bat037. doi: 10.1093/database/bat037

108. Kato, H., Kodama, O., and Akatsuka, T. (1993). Oryzalexin E, A diterpene phytoalexin from UV-irradiated rice leaves.*Phytochemistry* 33, 79–81. doi: 10.1016/0031-9422(93)85399-C

109. Kato, H., Kodama, O., and Akatsuka, T. (1994). Oryzalexin F, a diterpene phytoalexin from UV-irradiated rice leaves.*Phytochemistry* 36, 299–301. doi: 10.1016/S0031-9422(00)97064-X

110. Kato-Noguchi, H., Ino, T., Sata, N., and Yamamura, S. (2002). Isolation and identification of a potent allelopathic substance in rice root exudates. *Physiol. Plant.* 115, 401–405. doi: 10.1034/j.1399-3054.2002.1150310.x

111. Kazan, K., and Lyons, R. (2014). Intervention of phytohormone pathways by pathogen effectors. *Plant Cell* 26, 2285–2309. doi: 10.1105/tpc.114.125419

112. Keller, N. P., Turner, G., and Bennett, J. W. (2005). Fungal secondary metabolism - from biochemistry to genomics. *Nat. Rev. Microbiol.* 3, 937–947. doi: 10.1038/nrmicro1286

113. Khan, A. L., Hamayun, M., Kim, Y.-H., Kang, S. M., Lee, J.-H., and Lee, I.-J. (2011). Gibberellins producing endophytic*Aspergillus fumigatus* sp. LH02 influenced endogenous phytohormonal levels, isoflavonoids production and plant growth in salinity stress. *Process Biochem.* 46, 440–447. doi: 10.1016/j.procbio.2010.09.013

114. Kim, J.-A., Cho, K., Singh, R., Jung, Y.-H., Jeong, S.-H., Kim, S.-H., et al. (2009). Rice *OsACDR1* (*Oryza sativa* accelerated cell death and resistance 1) is a potential positive regulator of fungal disease resistance. *Mol. Cells* 28, 431–439. doi: 10.1007/s10059-009-0161-5

115. Kimura, M., Tokai, T., Takahashi-Ando, N., Ohsato, S., and Fujimura, M. (2007). Molecular and genetic studies of *Fusarium*trichothecene biosynthesis: pathways, genes, and evolution. *Biosci. Biotechnol. Biochem.* 71, 2105–2123. doi: 10.1271/bbb.70183

116. Kimura, N., and Tsuge, T. (1993). Gene cluster involved in melanin biosynthesis of the filamentous fungus *Alternaria alternata*.*J. Bacteriol.* 175, 4427–4435.

117. Kliebenstein, D. J., Rowe, H. C., and Denby, K. J. (2005). Secondary metabolites influence *Arabidopsis/Botrytis* interactions: variation in host production and pathogen sensitivity. *Plant J.* 44, 25–36. doi: 10.1111/j.1365-313X.2005.02508.x

118. Kodama, O., Miyakawa, J., Akatsuka, T., and Kiyosawa, S. (1992). Sakuranetin, a flavanone phytoalexin from ultraviolet-irradiated rice leaves. *Phytochemistry* 31, 3807–3809. doi: 10.1016/S0031-9422(00)97532-0

119. Kodama, O., Suzuki, T., Miyakawa, J., and Akatsuka, T. (1988). Ultraviolet-induced accumulation of phytoalexins in rice leaves.*Agric. Biol. Chem.* 52, 2469–2473.

120. Koga, J., Ogawa, N., Yamauchi, T., Kikuchi, M., Ogasawara, N., and Shimura, M. (1997). Functional moiety for the antifungal activity of phytocassane E, a diterpene phytoalexin from rice. *Phytochemistry* 44, 249–253. doi: 10.1016/S0031-9422(96)00534-1

121. Koga, J., Shimura, M., Oshima, K., Ogawa, N., Yamauchi, T., and Ogasawara, N. (1995). Phytocassanes A, B, C and D, novel diterpene phytoalexins from rice, *Oryza sativa* L. *Tetrahedron* 51, 7907–7918. doi: 10.1016/0040-4020(95)00423-6

122. Kolattukudy, P. E., Rogers, L. M., Li, D., Hwang, C. S., and Flaishman, M. A. (1995). Surface signaling in pathogenesis. *Proc. Natl. Acad. Sci. U.S.A.* 92, 4080–4087. doi: 10.1073/pnas.92.10.4080

123. Kono, Y., Sekido, S., Yamaguchi, I., Kondo, H., Suzuki, Y., Neto, G. C., et al. (1991). Structures of two novel pyriculol-related compounds and identification of naturally produced epipyriculol from *Pyricularia oryzae*. *Agric. Biol. Chem.* 55, 2785–2791.

124. Kroken, S., Glass, N. L., Taylor, J. W., Yoder, O. C., and Turgeon, B. G. (2003). Phylogenomic analysis of type I polyketide synthase genes in pathogenic and saprobic ascomycetes. *Proc. Natl. Acad. Sci. U.S.A.* 100, 15670–15675. doi: 10.1073/pnas.2532165100

125. Kroll, K., Pähtz, V., and Kniemeyer, O. (2014). Elucidating the fungal stress response by proteomics. *J. Proteomics* 97, 151–163. doi: 10.1016/j.jprot.2013.06.001

126. Kwon, Y. S., Kim, S. G., Chung, W. S., Bae, H., Jeong, S. W., Shin, S. C., et al. (2014). Proteomic analysis of *Rhizoctonia solani*AG-1 sclerotia maturation. *Fungal Biol.* 118, 433–443. doi: 10.1016/j.funbio.2014.02.001

127. Langcake, P., and Pryce, R. J. (1976). The production of resveratrol by *Vitis vinifera* and other members of the Vitaceae as a response to infection or injury. *Physiol. Plant Pathol.* 9, 77–86. doi: 10.1016/0048-4059(76)90077-1

128. Langseth, W., Ghebremeskel, M., Kosiak, B., Kolsaker, P., and Miller, D. (2001). Production of culmorin compounds and other secondary metabolites by *Fusarium culmorum* and *F. graminearum* strains isolated from Norwegian cereals. *Mycopathologia*152, 23–34. doi: 10.1023/A:1011964306510

129. Laskay, G., Farkas, T., and Lehoczki, E. (1985). Cerulenin-induced changes in lipid and fatty acid content of chloroplasts in detached greening barley leaves. *J. Plant Physiol.* 118, 267–275. doi: 10.1016/S0176-1617(85)80228-5

130. Lee, W. S., Devonshire, J. B., Hammond-Kosack, K. E., Rudd, J. J., and Kanyuka, K. K. (2015). Deregulation of plant cell death through disruption of chloroplast functionality affects asexual sporulation of *Zymoseptoria tritici* on wheat. *Mol. Plant Microbe Interact.* 28, 590–604. doi: 10.1094/MPMI-10-14-0346-R

131. Leite, B., and Nicholson, R. L. (1992). Mycosporine-alanine: a self-inhibitor of germination from the conidial mucilage of*Colletotrichum graminicola*. *Exp. Mycol.* 16, 76–86. doi: 10.1016/0147-5975(92)90043-Q

132. Levings, I. I. I. C. S., Rhoads, D. M., and Siedow, J. N. (1995). Molecular interactions of *Bipolaris*

maydis T-toxin and maize.*Can. J. Bot.* 73, 483–489. doi: 10.1139/b95-286

133. Li, Y., Huang, F., Lu, Y., Shi, Y., Zhang, M., Fan, J., et al. (2013). Mechanism of plant–microbe interaction and its utilization in disease-resistance breeding for modern agriculture. *Physiol. Mol. Plant Pathol.* 83, 51–58. doi: 10.1016/j.pmpp.2013.05.001

134. Liu, W., Zhou, X., Li, G., Li, L., Kong, L., Wang, C., et al. (2011). Multiple plant surface signals are sensed by different mechanisms in the rice blast fungus for appressorium formation. *PLoS Pathog.* 7:e1001261. doi: 10.1371/journal.ppat.1001261

135. Ludwig-Müller, J. (2011). Auxin conjugates: their role for plant development and in the evolution of land plants. *J. Exp. Bot.* 62, 1757–1773. doi: 10.1093/jxb/erq412

136. Ludwig-Müller, J. (2015). Bacteria and fungi controlling plant growth by manipulating auxin: balance between development and defense. *J. Plant Physiol.* 172, 4–12. doi: 10.1016/j.jplph.2014.01.002

137. Maggio-Hall, L. A., Wilson, R. A., and Keller, N. P. (2005). Fundamental contribution of beta-oxidation to polyketide mycotoxin production in planta. *Mol. Plant Microbe Interact.* 18, 783–793. doi: 10.1094/MPMI-18-0783

138. Manamgoda, D. S., Rossman, A. Y., Castlebury, L. A., Crous, P. W., Madrid, H., Chukeatirote, E., et al. (2014). The genus*Bipolaris*. *Stud. Mycol.* 79, 221–288. doi: 10.1016/j.simyco.2014.10.002

139. Markham, J. E., and Hille, J. (2001). Host-selective toxins as agents of cell death in plant-fungus interactions. *Mol. Plant Pathol.*2, 229–239. doi: 10.1046/j.1464-6722.2001.00066.x

140. Marsh, S. F., and Payne, G. A. (1984). Preharvest infection of corn silks and kernels by *Aspergillus flavus*. *Phytopathology* 74, 1284–1289.

141. Mehrabi, R., Bahkali, A. H., Abd-Elsalam, K. A., Moslem, M., Ben M›barek, S., Gohari, A. M., et al. (2011). Horizontal gene and chromosome transfer in plant pathogenic fungi affecting host range. *FEMS Microbiol. Rev.* 35, 542–554. doi: 10.1111/j.1574-6976.2010.00263.x

142. Mendgen, K., and Hahn, M. (2002). Plant infection and the establishment of fungal biotrophy. *Trends Plant Sci.* 7, 352–356. doi: 10.1016/S1360-1385(02)02297-5

143. Merhej, J., Urban, M., Dufresne, M., Hammond-Kosack, K. E., Richard-Forget, F., and Barreau, C. (2012). The velvet gene, FgVe1, affects fungal development and positively regulates trichothecene biosynthesis and pathogenicity in *Fusarium graminearum*. *Mol. Plant Pathol.* 13, 363–374. doi: 10.1111/j.1364-3703.2011.00755.x

144. Meyer, W., Lax, A., Templeton, G., and Brannon, M. (1983). The structure of gloeosporone, a novel germination self-inhibitor from conidia of*Colletotrichum gloeosporioides*. *Tetrahedron Lett.* 24, 5059–5062.

145. Miersch, O., Bohlmann, H., and Wasternack, C. (1999). Jasmonates and related compounds from *Fusarium oxysporum*.*Phytochemistry* 50, 517–523. doi: 10.1016/S0031-9422(98)00596-2

146. Mueller-Riebau, F., Berger, B., and Yegen, O. (1995). Chemical composition and fungitoxic properties to phytopathogenic fungi of essential oils of selected aromatic plants growing wild in turkey. *J. Agric. Food Chem.* 43, 2262–2266. doi: 10.1021/jf00056a055

147. Nesher, I., Barhoom, S., and Sharon, A. (2008). Cell cycle and cell death are not necessary for appressorium formation and plant infection in the fungal plant pathogen *Colletotrichum gloeosporioides*. *BMC Biol.* 6:9. doi: 10.1186/1741-7007-6-9

148. Nishimura, S., and Kohmoto, K. (1983). Host-specific toxins and chemical structures from *Alternaria* species. *Annu. Rev. Phytopathol.* 21, 87–116. doi: 10.1146/annurev.py.21.090183.000511

149. Nishiuchi, T., Masuda, D., Nakashita, H., Ichimura, K., Shinozaki, K., Yoshida, S., et al. (2006). *Fusarium* phytotoxin trichothecenes have an elicitor-like activity in *Arabidopsis thaliana*, but the activity differed significantly among their molecular species. *Mol. Plant Microbe Interact.* 19, 512–520. doi: 10.1094/MPMI-19-0512

150. Ohtake, M., Yamamoto, H., and Uchiyama, T. (1999). Influences of metabolic inhibitors and hydrolytic enzymes on the adhesion of appressoria of *Pyricularia oryzae* to wax-coated cover-glasses. *Biosci. Biotechnol. Biochem.* 63, 978–982.

151. Oide, S., Moeder, W., Krasnoff, S., Gibson, D., Haas, H., Yoshioka, K., et al. (2006). NPS6, encoding a nonribosomal peptide synthetase involved in siderophore-mediated iron metabolism, is a conserved virulence determinant of plant pathogenic ascomycetes. *Plant Cell* 18, 2836–2853. doi: 10.1105/tpc.106.045633

152. Osbourn, A. (1996). Saponins and plant defence—a soap story. *Trends Plant Sci.* 1, 4–9. doi: 10.1016/S1360-1385(96)80016-1

153. Otani, H., Kohmoto, K., and Kodama, M. (1995). *Alternaria* toxins and their effects on host plants. *Can. J. Bot.* 73, 453–458. doi: 10.1139/b95-282

154. Paciolla, C., Dipierro, N., Mulè, G., Logrieco, A., and Dipierro, S. (2004). The mycotoxins beauvericin and T-2 induce cell death and alteration to the ascorbate metabolism in tomato protoplasts. *Physiol. Mol. Plant Pathol.* 65, 49–56. doi: 10.1016/j.pmpp.2004.07.006

155. Paiardini, A., Aducci, P., Cervoni, L., Cutruzzolà, F., Di Lucente, C., Janson, G., et al. (2014). The phytotoxin fusicoccin differently regulates 14-3-3 proteins association to mode III targets. *IUBMB Life* 66, 52–62. doi: 10.1002/iub.1239

156. Palencia, E. R., Hinton, D. M., and Bacon, C. W. (2010). The black *Aspergillus* species of maize and peanuts and their potential for mycotoxin production. *Toxins (Basel)* 2, 399–416. doi: 10.3390/toxins2040399

157. Palmer, C. L., and Skinner, W. (2002). *Mycosphaerella graminicola*: latent infection, crop devastation and genomics. *Mol. Plant Pathol.* 3, 63–70. doi: 10.1046/j.1464-6722.2002.00100.x

158. Panaccione, D. G. (1993). The fungal genus *Cochliobolus* and toxin-mediated plant disease. *Trends Microbiol.* 1, 14–20.

159. Panstruga, R. (2003). Establishing compatibility between plants and obligate biotrophic pathogens. *Curr. Opin. Plant Biol.* 6, 320–326. doi: 10.1016/S1369-5266(03)00043-8

160. Paranidharan, V., Palaniswami, A., Vidhyasekaran, P., and Velazhahan, R. (2005). A host-specific toxin of *Rhizoctonia solani*triggers superoxide dismutase (SOD) activity in rice. *Arch. Phytopathol. Plant Prot.* 38, 151–157. doi: 10.1080/03235400500094159

161. Pedras, M. S. C., and Ahiahonu, P. W. K. (2005). Metabolism and detoxification of phytoalexins and analogs by phytopathogenic fungi. *Phytochemistry* 66, 391–411. doi: 10.1016/j.phytochem.2004.12.032

162. Pedras, M. S. C., Chumala, P. B., and Suchy, M. (2003). Phytoalexins from *Thlaspi arvense*, a wild crucifer resistant to virulent*Leptosphaeria maculans*: structures, syntheses and antifungal activity. *Phytochemistry* 64, 949–956. doi: 10.1016/S0031-9422(03)00441-2

163. Pedras, M. S. C., Minic, Z., and Abdoli, A. (2014). The phytoalexin camalexin induces fundamental changes in the proteome of*Alternaria brassicicola* different from those caused by brassinin. *Fungal Biol.* 118, 83–93. doi: 10.1016/j.funbio.2013.11.005

164. Pedras, M. S. C., and Montaut, S. (2003). Probing crucial metabolic pathways in fungal pathogens of crucifers: biotransformation of indole-3-acetaldoxime, 4-hydroxyphenylacetaldoxime, and their metabolites. *Bioorg. Med. Chem.* 11, 3115–3120. doi: 10.1016/S0968-0896(03)00241-4

165. Pedras, M. S. C., Montaut, S., and Suchy, M. (2004). Phytoalexins from the crucifer rutabaga: structures, syntheses, biosyntheses, and antifungal activity. *J. Org. Chem.* 69, 4471–4476. doi: 10.1021/jo049648a

166. Pedras, M. S. C., Sarwar, M. G., Suchy, M., and Adio, A. M. (2006). The phytoalexins from cauliflower, caulilexins A, B and C: isolation, structure determination, syntheses and antifungal activity. *Phytochemistry* 67, 1503–1509. doi: 10.1016/j.phytochem.2006.05.020

167. Peng, X. L., Xu, W. T., Wang, Y., Huang, K. L., Liang, Z. H., Zhao, W. W., et al. (2010). Mycotoxin Ochratoxin A-induced cell death and changes in oxidative metabolism of *Arabidopsis thaliana*. *Plant Cell Rep.* 29, 153–161. doi: 10.1007/s00299-009-0808-x

168. Perez-Nadales, E., Almeida Nogueira, M. F., Baldin, C., Castanheira, S., El Ghalid, M., Grund, E., et al. (2014). Fungal model systems and the elucidation of pathogenicity determinants. *Fungal Genet. Biol.* 70, 42–67. doi: 10.1016/j.fgb.2014.06.011

169. Perry, N. B., and Foster, L. M. (1994). Antiviral and antifungal flavonoids, plus a triterpene from *Hebe cupressoides*. *Planta Medica* 60, 491–492.

170. Peters, R. J. (2006). Uncovering the complex metabolic network underlying diterpenoid phytoalexin biosynthesis in rice and other cereal crop plants. *Phytochemistry* 67, 2307–2317. doi: 10.1016/j.phytochem.2006.08.009

171. Petti, C., Reiber, K., Ali, S. S., Berney, M., and Doohan, F. M. (2012). Auxin as a player in the biocontrol of Fusarium head blight disease of barley and its potential as a disease control agent. *BMC Plant Biol.* 12:224. doi: 10.1186/1471-2229-12-224

172. Picot, A., Barreau, C., Pinson-Gadais, L., Piraux, F., Caron, D., Lannou, C., et al. (2011). The dent stage of maize kernels is the most conducive for fumonisin biosynthesis under field conditions. *Appl. Environ. Microbiol.* 77, 8382–8390. doi: 10.1128/AEM.05216-11

173. Pierpoint, W. S. (2000). Why should plants make medicine – don›t they do enough for mankind already? *Biochemistry (Lond.)*22, 37–40.

174. Podila, G. K., Rogers, L. M., and Kolattukudy, P. E. (1993). Chemical signals from avocado surface wax trigger germination and appressorium formation in *Colletotrichum gloeosporioides*. *Plant Physiol.* 103, 267–272.

175. Raffaele, S., Leger, A., and Roby, D. (2009). Very long chain fatty acid and lipid signaling in the response of plants to pathogens.*Plant Signal. Behav.* 4, 94–99. doi: 10.4161/psb.4.2.7580

176. Ren, Y. Y., and West, C. A. (1992). Elicitation of diterpene biosynthesis in rice (*Oryza sativa* L.) by chitin. *Plant Physiol.* 99, 1169–1178.

177. Rocha, O., Ansari, K., and Doohan, F. M. (2005). Effects of trichothecene mycotoxins on eukaryotic cells: a review. *Food Addit. Contam.* 22, 369–378. doi: 10.1080/02652030500058403

178. Ruan, Y., Kotraiah, V., and Straney, D. C. (1995). Flavonoids stimulate spore germination in *Fusarium solani* pathogenic on legumes in a manner sensitive to inhibitors of cAMP-dependent protein kinase. *MPMI* 8, 929–938.

179. Saddiq, A. A., and Khayyat, S. A. (2010). Chemical and antimicrobial studies of monoterpene: Citral. *Pestic. Biochem. Physiol.*98, 89–93. doi: 10.1016/j.pestbp.2010.05.004

180. Saha, D., Fetzner, R., Burkhardt, B., Podlech, J., Metzler, M., Dang, H., et al. (2012). Identification of a polyketide synthase required for alternariol (AOH) and alternariol-9-methyl ether (AME) formation in *Alternaria alternata*. *PLoS ONE* 7:e40564. doi: 10.1371/journal.pone.0040564

181. Scarpari, M., Punelli, M., Scala, V., Zaccaria, M., Nobili, C., Ludovici, M., et al. (2014). Lipids in *Aspergillus flavus*-maize interaction. *Front. Microbiol.* 5:74. doi: 10.3389/fmicb.2014.00074

182. Scharf, D. H., Heinekamp, T., and Brakhage, A. A. (2014). Human and plant fungal pathogens: the role of secondary metabolites.*PLoS Pathog.* 10:e1003859. doi: 10.1371/journal.ppat.1003859

183. Schuhegger, R., Rauhut, T., and Glawischnig, E. (2007). Regulatory variability of camalexin biosynthesis. *J. Plant Physiol.* 164, 636–644. doi: 10.1016/j.jplph.2006.04.012

184. Schulze-Lefert, P., and Panstruga, R. (2003). Establishment of biotrophy by parasitic fungi and reprogramming of host cells for disease resistance. *Annu.*

Rev. Phytopathol. 41, 641–667. doi: 10.1146/annurev.phyto.41.061002.083300

185. Scott, P. M. (2001). Analysis of agricultural commodities and foods for *Alternaria* mycotoxins. *J. AOAC Int.* 84, 1809–1817.

186. Segarra, G., Van der Ent, S., Trillas, I., and Pieterse, C. M. J. (2009). MYB72, a node of convergence in induced systemic resistance triggered by a fungal and a bacterial beneficial microbe. *Plant Biol. (Stuttg.)* 11, 90–96. doi: 10.1111/j.1438-8677.2008.00162.x

187. Shah, J., Chaturvedi, R., Chowdhury, Z., Venables, B., and Petros, R. A. (2014). Signaling by small metabolites in systemic acquired resistance. *Plant J.* 79, 645–658. doi: 10.1111/tpj.12464

188. Sieber, C. M. K., Lee, W., Wong, P., Münsterkötter, M., Mewes, H. W., Schmeitzl, C., et al. (2014). The *Fusarium graminearum*genome reveals more secondary metabolite gene clusters and hints of horizontal gene transfer. *PLoS ONE* 9:e110311. doi: 10.1371/journal.pone.0110311

189. Sinha, R. P., Singh, S. P., and Häder, D. P. (2007). Database on mycosporines and mycosporine-like amino acids (MAAs) in fungi, cyanobacteria, macroalgae, phytoplankton and animals. *J. Photochem. Photobiol. B.* 89, 29–35. doi: 10.1016/j.jphotobiol.2007.07.006

190. Skamnioti, P., and Gurr, S. J. (2009). Against the grain: safeguarding rice from rice blast disease. *Trends Biotechnol.* 27, 141–150. doi: 10.1016/j.tibtech.2008.12.002

191. Spanu, P. D., Abbott, J. C., Amselem, J., Burgis, T. A., Soanes, D. M., Stüber, K., et al. (2010). Genome expansion and gene loss in powdery mildew fungi reveal tradeoffs in extreme parasitism. *Science* 330, 1543–1546. doi: 10.1126/science.1194573

192. Spassieva, S. D., Markham, J. E., and Hille, J. (2002). The plant disease resistance gene Asc-1 prevents disruption of sphingolipid metabolism during AAL-toxin-induced programmed cell death. *Plant J.* 32, 561–572. doi: 10.1046/j.1365-313X.2002.01444.x

193. Stanley, M. S., Callow, M. E., Perry, R., Alberte, R. S., Smith, R., and Callow, J. A. (2002). Inhibition of fungal spore adhesion by zosteric acid as the basis for a novel, nontoxic crop protection technology. *Phytopathology* 92, 378–383. doi: 10.1094/PHYTO.2002.92.4.378

194. St□pień, Ł., Waśkiewicz, A., and Wilman, K. (2015). Host extract modulates metabolism and fumonisin biosynthesis by the plant-pathogenic fungus *Fusarium proliferatum*. *Int. J. Food Microbiol.* 193, 74–81. doi: 10.1016/j.ijfoodmicro.2014.10.020

195. Stergiopoulos, I., Collemare, J., Mehrabi, R., and De Wit, P. J. G. M. (2013). Phytotoxic secondary metabolites and peptides produced by plant pathogenic Dothideomycete fungi. *FEMS Microbiol. Rev.* 37, 67–93. doi: 10.1111/j.1574-6976.2012.00349.x

196. St Leger, R. J., Screen, S. E., and Shams-Pirzadeh, B. (2000). Lack of host specialization in *Aspergillus flavus*. *Appl. Environ. Microbiol.* 66, 320–324. doi: 10.1128/AEM.66.1.320-324.2000

197. Stodart, B. J., Harvey, P. R., Neate, S. M., Melanson, D. L., and Scott, E. S. (2007). Genetic variation and pathogenicity of anastomosis group 2 isolates of *Rhizoctonia solani* in Australia. *Mycol. Res.* 111, 891–900. doi: 10.1016/j.mycres.2007.05.008

198. Suryanarayanan, T. S., Thirunavukkarasu, N., Govindarajulu, M. B., Sasse, F., Jansen, R., and Murali, T. S. (2009). Fungal endophytes and bioprospecting. *Fungal Biol. Rev.* 23, 9–19. doi: 10.1016/j.fbr.2009.07.001

199. Tada, Y., Kusaka, K., Betsuyaku, S., Shinogi, T., Sakamoto, M., Ohura, Y., et al. (2005). Victorin triggers programmed cell death and the defense response via interaction with a cell surface mediator. *Plant Cell Physiol.* 46, 1787–1798. doi: 10.1093/pcp/pci193

200. Takanashi, K., Takahashi, H., Sakurai, N., Sugiyama, A., Suzuki, H., Shibata, D., et al. (2012). Tissue-specific transcriptome analysis in nodules of *Lotus japonicus*. *Mol. Plant Microbe Interact.* 25, 869–876. doi: 10.1094/MPMI-01-12-0011-R

201. Talbot, N. J. (2003). On the trail of a cereal killer: exploring the biology of *Magnaporthe grisea*. *Annu. Rev. Microbiol.* 57, 177–202. doi: 10.1146/annurev.micro.57.030502.090957

202. Talbot, N. J., McCafferty, H. R. K., Ma, M., Moore, K., and Hamer, J. E. (1997). Nitrogen starvation of the rice blast fungus*Magnaporthe grisea* may act as an environmental cue for disease symptom expression. *Physiol. Mol. Plant Pathol.* 50, 179–195. doi: 10.1006/pmpp.1997.0081

203. Tamogani, S., Mitani, M., Kodama, O., and Akatsuka, T. (1993). Oryzalexin S structure: a new stemarane-type rice plant phytoalexin and its biogenesis. *Tetrahedron* 49, 2025–2032. doi: 10.1016/S0040-4020(01)86302-X

204. Tanaka, E., Koga, H., and Mori, M. M. (2011). Auxin production by the rice blast fungus and its localization in host tissue. *J. Phytopathol.* 159, 522–530. doi: 10.1111/j.1439-0434.2011.01799.x

205. Tani, H., Koshino, H., Sakuno, E., Cutler, H. G., and Nakajima, H. (2006). Botcinins E and F and botcinolide from *Botrytis cinerea* and structural revision of botcinolides. *J. Nat. Prod.* 69, 722–725. doi: 10.1021/np060071x

206. Tanimoto, E. (2005). Regulation of root growth by plant hormones – roles for auxin and gibberellins. *Crit. Rev. Plant. Sci.* 24, 249–265. doi: 10.1080/07352680500196108

207. Thines, E., Anke, H., and Sterner, O. (1998). Scytalols A, B, C, and D and other modulators of melanin biosynthesis from*Scytalidium sp.* 36–93. *J. Antibiot.* (Tokyo) 51, 387–393.

208. Thines, E., Anke, H., and Weber, R. W. S. (2004). Fungal secondary metabolites as inhibitors of infection-related morphogenesis in phytopathogenic fungi. *Mycol. Res.* 108, 14–25. doi: 10.1017/S0953756203008943

209. Thines, E., Daußmann, T., Semar, M., Sterner, O., and Anke, A. (1995). Fungal melanin biosynthesis inhibitors: introduction of a test system based on the production of dihydroxynaphthalene (DHN) melanin in agar cultures. *Z. Naturforsch.* 50, 813–819.

210. Thomma, B. P. (2003). *Alternaria spp.*: from general saprophyte to specific parasite. *Mol. Plant Pathol.* 4, 225–236. doi: 10.1046/j.1364-3703.2003.00173.x

211. Thomma, B. P., Nelissen, I., Eggermont, K., and Broekaert, W. F. (1999). Deficiency in phytoalexin production causes enhanced susceptibility of *Arabidopsis thaliana* to the fungus *Alternaria brassicicola*. *Plant J.* 19, 163–171. doi: 10.1046/j.1365-313X.1999.00513.x

212. Thomma, B. P., Van Esse, H. P., Crous, P. W., and De Wit, P. J. G. M. (2005). *Cladosporium fulvum* (syn. *Passalora fulva*), a highly specialized plant pathogen as a model for functional studies on plant pathogenic Mycosphaerellaceae. *Mol. Plant Pathol.* 6, 379–393. doi: 10.1111/j.1364-3703.2005.00292.x

213. Thuleau, P., Graziana, A., Rossignol, M., Kauss, H., Auriol, P., and Ranjeva, R. (1988). Binding of the phytotoxin zinniol stimulates the entry of calcium into plant protoplasts. *Proc. Natl. Acad. Sci. U.S.A.* 85, 5932–5935.

214. Tiedemann, A. V. (1997). Evidence for a primary role of active oxygen species in induction of host cell death during infection of bean leaves with *Botrytis cinerea*. *Physiol. Mol. Plant Pathol.* 50, 151–166. doi: 10.1006/pmpp.1996.0076

215. Todd, J. S., Zimmerman, R. C., Crews, P., and Alberte, R. S. (1993). The antifouling activity of natural and synthetic phenol acid sulphate esters. *Phytochemistry* 34, 401–404. doi: 10.1016/0031-9422(93)80017-M

216. Tokai, T., Koshino, H., Takahashi-Ando, N., Sato, M., Fujimura, M., and Kimura, M. (2007). *Fusarium* Tri4 encodes a key multifunctional cytochrome P450 monooxygenase for four consecutive oxygenation steps in trichothecene biosynthesis.*Biochem. Biophys. Res. Commun.* 353, 412–417. doi: 10.1016/j.bbrc.2006.12.033

217. Ton, J., Van Pelt, J. A., Van Loon, L. C., and Pieterse, C. M. J. (2002). Differential effectiveness of salicylate-dependent and jasmonate/ethylene-dependent induced resistance in *Arabidopsis*. *Mol. Plant Microbe Interact.* 15, 27–34. doi: 10.1094/MPMI.2002.15.1.27

218. Tsao, R., and Zhou, T. (2000). Antifungal activity of monoterpenoids against postharvest pathogens *Botrytis cinerea* and*Monilinia fructicola*. *J. Essent. Oil Res.* 12, 113–121. doi: 10.1080/10412905.2000.9712057

219. Tsuge, T., Harimoto, Y., Akimitsu, K., Ohtani, K., Kodama, M., Akagi, Y., et al. (2013). Host-selective toxins produced by the plant pathogenic fungus *Alternaria alternata*. *FEMS Microbiol. Rev.* 37, 44–66. doi: 10.1111/j.1574-6976.2012.00350.x

220. Tsuji, J., Jackson, E. P., Gage, D. A., Hammerschmidt, R., and Somerville, S. C. (1992). Phytoalexin accumulation in *Arabidopsis thaliana* during the hypersensitive reaction to *Pseudomonas syringae* pv syringae. *Plant Physiol.* 98, 1304–1309.

221. Tsurushima, T., Ueno, T., Fukami, H., Irie, H., and Inoue, M. (1995). Germination self-inhibitors from *Colletotrichum gloeosporioides f. sp. jussiaea*. *MPMI* 8, 652–657.

222. Tudzynski, B. (2014). Nitrogen regulation of fungal secondary metabolism in fungi. *Front. Microbiol.* 5:656. doi: 10.3389/fmicb.2014.00656

223. Tudzynski, P., and Kokkelink, L. (2009). "*Botrytis cinerea*: molecular aspects of a necrotrophic life style," in *Plant Relationships*, ed H. B. Deising (Berlin; Heidelberg: Springer-Verlag), 29–50.

224. Turgeon, B. G., and Baker, S. E. (2007). Genetic and genomic dissection of the *Cochliobolus heterostrophus* Tox1 locus controlling biosynthesis of the polyketide virulence factor T-toxin. *Adv. Genet.* 57, 219–261. doi: 10.1016/S0065-2660(06)57006-3

225. Van der Ent, S., Van Wees, S. C. M., and Pieterse, C. M. J. (2009). Jasmonate signaling in plant interactions with resistance-inducing beneficial microbes. *Phytochemistry* 70, 1581–1588. doi: 10.1016/j.phytochem.2009.06.009

226. Van der Ent, S., Verhagen, B. W. M., Van Doorn, R., Bakker, D., Verlaan, M. G., Pel, M. J. C., et al. (2008). MYB72 is required in early signaling steps of rhizobacteria-induced systemic resistance in *Arabidopsis*. *Plant Physiol.* 146, 1293–1304. doi: 10.1104/pp.107.113829

227. Van Loon, L. C., Rep, M., and Pieterse, C. M. J. (2006). Significance of inducible defense-related proteins in infected plants.*Annu. Rev. Phytopathol.* 44, 135–162. doi: 10.1146/annurev.phyto.44.070505.143425

228. Velluti, A., Marın, S., Gonzalez, P., Ramos, A. J., and Sanchis, V. (2004). Initial screening for inhibitory activity of essential oils on growth of *Fusarium verticillioides, F. proliferatum* and *F. graminearum* on maize-based agar media. *Food Microbiol.* 21, 649–656. doi: 10.1016/j.fm.2004.03.009

229. Vidhyasekaran, P., Ponmalar, T. R., Samiyappan, R., Velazhahan, R., Vimala, R., Ramanathan, A., et al. (1997). Host-specific toxin production by *Rhizoctonia solani*, the rice sheath blight pathogen. *Phytopathology* 87, 1258–1263. doi: 10.1094/PHYTO.1997.87.12.1258

230. Walker, T. S., Bais, H. P., Halligan, K. M., Stermitz, F. R., and Vivanco, J. M. (2003). Metabolic profiling of root exudates of*Arabidopsis thaliana*. *J. Agric. Food Chem.* 51, 2548–2554. doi: 10.1021/jf021166h

231. Walton, J. D. (1996). Host-selective toxins: agents of compatibility. *Plant Cell* 8, 1723–1733. doi: 10.1105/tpc.8.10.1723

232. Walton, J. D. (2006). HC-toxin. *Phytochemistry* 67, 1406–1413. doi: 10.1016/j.phytochem.2006.05.033

233. Wang, X., Jiang, N., Liu, J., Liu, W., and Wang, G. L. (2014). The role of effectors and host immunity in plant-necrotrophic fungal interactions. *Virulence* 5. doi: 10.4161/viru.29798

234. Wang, Y. Z., and Miller, J. D. (1988). Effects of *Fusarium graminearum* metabolites on wheat tissue in relation to Fusarium Head Blight resistance. *J. Phytopathol.* 122, 118–125. doi: 10.1111/j.1439-0434.1988.tb00998.x

235. Wasternack, C. (2007). Jasmonates: an update on biosynthesis, signal transduction and action in plant stress response, growth and development. *Ann. Bot.* 100, 681–697. doi: 10.1093/aob/mcm079

236. Wehner, J., Antunes, P. M., Powell, J. R., Mazukatow, J., and Rillig, M. C. (2010). Plant pathogen protection

by arbuscular mycorrhizas: a role for fungal diversity? *Pedobiologia (Jena)* 53, 197–201. doi: 10.1016/j.pedobi.2009.10.002

237. Wheeler, M. H., and Bell, A. A. (1988). "Melanins and their importance in pathogenic fungi," in *Current Topics in Medical Mycology 2nd Edn.,* ed M. R. McGinnis (New York, NY: Springer), 338–387.

238. Wibberg, D., Jelonek, L., Rupp, O., Kröber, M., Goesmann, A., Grosch, R., et al. (2014). Transcriptome analysis of the phytopathogenic fungus *Rhizoctonia solani* AG1-IB 7/3/14 applying high-throughput sequencing of expressed sequence tags (ESTs). *Fungal Biol.* 118, 800–813. doi: 10.1016/j.funbio.2014.06.007

239. Wiemann, P., Brown, D. W., Kleigrewe, K., Bok, J. W., Keller, N. P., Humpf, H. U., et al. (2010). FfVel1 and FfLae1, components of a velvet-like complex in *Fusarium fujikuroi*, affect differentiation, secondary metabolism and virulence. *Mol. Microbiol.* 77, 972–994. doi: 10.1111/j.1365-2958.2010.07263

240. Wiemann, P., Sieber, C. M. K., von Bargen, K. W., Studt, L., Niehaus, E. M., Espino, J. J., et al. (2013). Deciphering the cryptic genome: genome-wide analyses of the rice pathogen *Fusarium fujikuroi* reveal complex regulation of secondary metabolism and novel metabolites. *PLoS Pathog.* 9:e1003475. doi: 10.1371/journal.ppat.1003475

241. Wight, W. D., Labuda, R., and Walton, J. D. (2013). Conservation of the genes for HC-toxin biosynthesis in *Alternaria jesenskae*. *BMC Microbiol.* 13:165. doi: 10.1186/1471-2180-13-165

242. Wilson, C. L., Solar, J. M., El Ghaouth, A., and Wisniewski, M. E. (1997). Rapid evaluation of plant extracts and essential oils for antifungal activity against *Botrytis cinerea*. *Plant Dis.* 81, 204–210.

243. Wilson, R. A., and Talbot, N. J. (2009). Under pressure: investigating the biology of plant infection by *Magnaporthe oryzae*.*Nat. Rev. Microbiol.* 7, 185–195. doi: 10.1038/nrmicro2032

244. Wink, M. (2003). Evolution of secondary metabolites from an ecological and molecular phylogenetic perspective.*Phytochemistry* 64, 3–19. doi: 10.1016/S0031-9422(03)00300-5

245. Wolpert, T. J., Dunkle, L. D., and Ciuffetti, L. M. (2002). Host-selective toxins and avirulence determinants: what›s in a name?*Annu. Rev. Phytopathol.* 40, 251–285. doi: 10.1146/annurev.phyto.40.011402.114210

246. Woudenberg, J. H. C., Groenewald, J. Z., Binder, M., and Crous, P. W. (2013). *Alternaria* redefined. *Stud. Mycol.* 75, 171–212. doi: 10.3114/sim0015

247. Wu, J., and Ge, X. (2004). Oxidative burst, jasmonic acid biosynthesis, and taxol production induced by low-energy ultrasound in *Taxus chinensis* cell suspension cultures. *Biotechnol. Bioeng.* 85, 714–721. doi: 10.1002/bit.10911

248. Xu, L., Wang, X., Luo, R., Lu, S., Guo, Z., Wang, M., et al. (2015). Secondary metabolites of rice sheath blight pathogen*Rhizoctonia solani* Kühn and their biological activities. *J. Integr. Agric.* 14, 80–87. doi: 10.1016/S2095-3119(14)60905-9

249. Yaegashi, J., Oakley, B. R., and Wang, C. C. C. (2014). Recent advances in genome mining of secondary metabolite biosynthetic gene clusters and the development of heterologous expression systems in *Aspergillus nidulans*. *J. Ind. Microbiol. Biotechnol.* 41, 433–442. doi: 10.1007/s10295-013-1386-z

250. Yan, Z., Reddy, M. S., Ryu, C. M., McInroy, J. A., Wilson, M., and Kloepper, J. W. (2002). Induced systemic protection against tomato late blight elicited by plant growth-promoting rhizobacteria. *Phytopathology* 92, 1329–1333. doi: 10.1094/PHYTO.2002.92.12.1329

251. Yang, F. Z., Li, L., and Yang, B. (2012). *Alternaria* toxin-induced resistance against rose aphids and olfactory response of aphids to toxin-induced volatiles of rose plants. *J. Zhejiang Univ. Sci. B* 13, 126–135. doi: 10.1631/jzus.B1100087

252. Yang, Q., Chen, Y., and Ma, Z. (2013). Involvement of BcVeA and BcVelB in regulating conidiation, pigmentation and virulence in*Botrytis cinerea*. *Fungal Genet. Biol.* 50, 63–71. doi: 10.1016/j.fgb.2012.10.003

253. Yao, H., and Tian, S. (2005). Effects of pre- and post-harvest application of salicylic acid or methyl jasmonate on inducing disease resistance of sweet cherry fruit in storage. *Postharvest Biol. Technol.* 35, 253–262. doi: 10.1016/j.postharvbio.2004.09.001

254. Yazaki, K. (2006). ABC transporters involved in the transport of plant secondary metabolites. *FEBS Lett.* 580, 1183–1191. doi: 10.1016/j.febslet.2005.12.009

255. Yazaki, K., Sugiyama, A., Morita, M., and Shitan, N. (2008). Secondary transport as an efficient membrane transport mechanism for plant secondary metabolites. *Phytochem. Rev.* 7, 513–524. doi: 10.1007/s11101-007-9079-8

256. Zaehle, C., Gressler, M., Shelest, E., Geib, E., Hertweck, C., and Brock, M. (2014). Terrein biosynthesis in *Aspergillus terreus*and its impact on phytotoxicity. *Chem. Biol.* 21, 719–731. doi: 10.1016/j.chembiol.2014.03.010

257. Zhang, L., Jia, C., Liu, L., Zhang, Z., Li, C., and Wang, Q. (2011). The involvement of jasmonates and ethylene in *Alternaria alternata* f. sp. lycopersici toxin-induced tomato cell death. *J. Exp. Bot.* 62, 5405–5418. doi: 10.1093/jxb/err217

258. Zhao, Y. (2010). Auxin biosynthesis and its role in plant development. *Annu. Rev. Plant Biol.* 61, 49–64. doi: 10.1146/annurev-arplant-042809-112308

259. Zheng, A., Lin, R., Zhang, D., Qin, P., Xu, L., Ai, P., et al. (2013). The evolution and pathogenic mechanisms of the rice sheath blight pathogen. *Nat. Commun.* 4, 1424. doi: 10.1038/ncomms2427

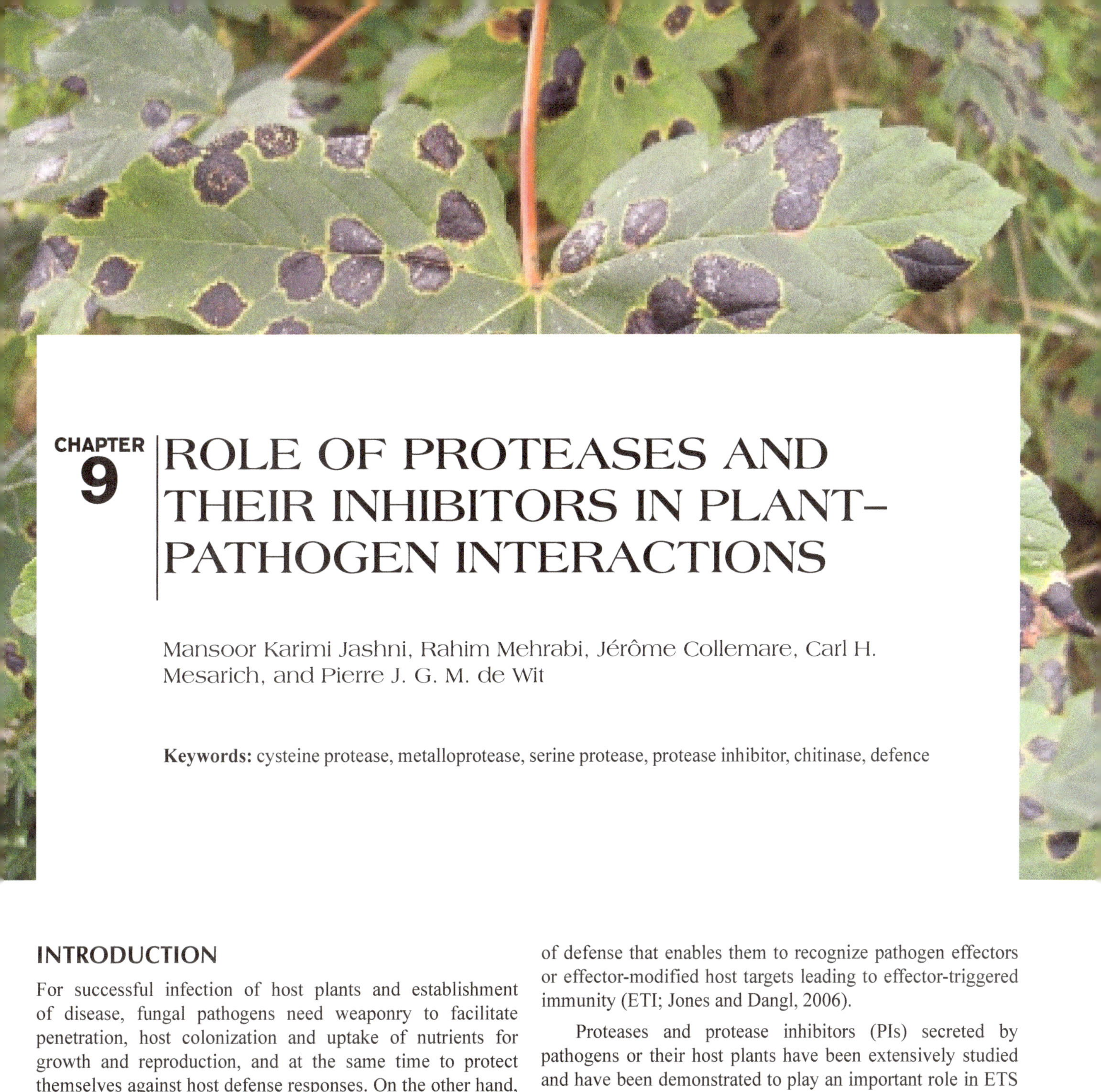

CHAPTER 9

ROLE OF PROTEASES AND THEIR INHIBITORS IN PLANT–PATHOGEN INTERACTIONS

Mansoor Karimi Jashni, Rahim Mehrabi, Jérôme Collemare, Carl H. Mesarich, and Pierre J. G. M. de Wit

Keywords: cysteine protease, metalloprotease, serine protease, protease inhibitor, chitinase, defence

INTRODUCTION

For successful infection of host plants and establishment of disease, fungal pathogens need weaponry to facilitate penetration, host colonization and uptake of nutrients for growth and reproduction, and at the same time to protect themselves against host defense responses. On the other hand, plants have developed surveillance systems to recognize and defend themselves against invading pathogens. Plant immune receptors recognize conserved microbe-associated molecular patterns (MAMPs) like chitin oligomers released from fungal cell walls during infection. This recognition leads to MAMP-triggered immunity (MTI) and initiates basal defense responses including the activation of structural and (bio)chemical barriers (Jones and Dangl, 2006; Spoel and Dong, 2012). However, adapted plant pathogens have gained the ability to overcome MTI by producing effector molecules that suppress or compromise MTI responses, thereby facilitating effector-triggered susceptibility (ETS; Stergiopoulos and de Wit, 2009). In response, plants have developed an additional layer of defense that enables them to recognize pathogen effectors or effector-modified host targets leading to effector-triggered immunity (ETI; Jones and Dangl, 2006).

Proteases and protease inhibitors (PIs) secreted by pathogens or their host plants have been extensively studied and have been demonstrated to play an important role in ETS and ETI (van der Hoorn, 2008). However, little is known about their role in MTI and related plant basal defense responses. Plant basal defense responses include the induction of pathogenesis-related proteins (PRs) such as antimicrobial chitinases, β-1,3-glucanases and proteases that hydrolyse the fungal cell wall components chitin, glucans, and polypeptides, respectively. The induction of these PR proteins upon plant infection, their antifungal activity, as well as their exploitation in engineering resistance in transgenic plants are very well documented (Wubben et al., 1996; Sels et al., 2008;Balasubramanian et al., 2012; Cletus et al., 2013). An early report in the literature suggested that pathogens might overcome the deleterious effects of plant chitinases by secreting proteases that modified them (Lange et al., 1996; Sela-Buurlage, 1996). This was further supported by recent studies, which indicate that chitinases are targeted by pathogen proteases and protected by PIs (Naumann et al., 2011; Slavokhotova et al., 2014). This encouraged us to review the recent advances on proteases and PIs that play a role in the arms race between plants and their fungal and oomycete pathogens.

Citation: Karimi Jashni M, Mehrabi R, Collemare J, Mesarich CH, and de Wit PJGM (2015) role of proteases and their inhibitors in plant–pathogen interactions. Front. Plant Sci. 6:584. doi: 10.3389/fpls.2015.00584

PLANT PROTEASES AND PROTEASE INHIBITORS INVOLVED IN BASAL DEFENSE

Most PR proteins exhibit direct antimicrobial activities, such as chitinases that degrade chitin present in fungal cell walls. PR proteins play a role in both constitutive and induced basal defense responses (Avrova et al., 2004; Shabab et al., 2008; van Esse et al., 2008). For example, tomato and potato contain basal levels of proteases in their apoplast, including serine proteases like P69, and papain-like cysteine proteases (PLCPs) like Rcr3, which are required for resistance of tomato against *Cladosporium fulvum* (Song et al., 2009), as well as Pip1 (*Phytophthora* inhibited protease 1; Tian et al., 2007; Shabab et al., 2008) and C14, which play a role in the resistance of potato against *Phytophthora infestans* (Kaschani et al., 2010; Bozkurt et al., 2011). After being challenged by pathogens, proteases are induced both locally (Tian et al., 2005) and systemically in the apoplast (Tian et al., 2007; Shabab et al., 2008; Song et al., 2009), which suggests that their activity affects pathogen growth directly or indirectly. Deletion or silencing of genes encoding these proteases enhanced the susceptibility of plants to pathogens, supporting their role in defense responses. Deletion of *Rcr3* increased the susceptibility of tomato to the late blight pathogen *P. infestans*(Song et al., 2009), to the leaf mold pathogen *C. fulvum* (Dixon et al., 2000), and also to the potato cyst nematode *Globodera rostochiensis* (Lozano-Torres et al., 2012). Likewise, silencing of *C14* in*Nicotiana benthamiana* significantly increased their susceptibility to *P. infestans* (Kaschani et al., 2010). These findings suggest that proteases have a determinative role in the execution of defense against plant pathogens.

Plant PIs have also been reported to play a role in plant immunity, through the inhibition of pathogen proteases, or the regulation of endogenous plant proteases (Ryan, 1990; Mosolov et al., 2001; Valueva and Mosolov, 2004; Kim et al., 2009). This has been shown for PIs from barley (*Hordeum vulgare*) against proteases from *Fusarium culmorum* (Pekkarinen et al., 2007), as well as for PIs from broad bean (*Vicia faba*), which inhibited the mycelial growth of several pathogens (Ye et al., 2001). The *A. thaliana* unusual serine protease inhibitor (UPI) was shown to play a role in defense against the necrotrophic fungi *Botrytis cinerea* and *Alternaria brassicicola* (Laluk and Mengiste, 2011). The UPI protein strongly inhibited the serine protease chymotrypsin but also affected the cysteine protease papain (Laluk and Mengiste, 2011). Plants harboring a loss-of-function *UPI* allele displayed enhanced susceptibility to *B. cinerea* and *A. brassicicola*, but not to the bacterium *Pseudomonas syringae*. Also, hevein-like antimicrobial peptides from wheat (WAMPs) were shown to inhibit class IV chitinase degradation by fungalysin, a metalloprotease secreted by *Fusarium verticillioides* (Slavokhotova et al., 2014). WAMPs bind to fungalysin, but are not cleaved by the enzyme due to the presence of a Ser residue between the Gly and Cys residues where cleavage of class IV chitinase by fungalysin normally takes place (Naumann et al., 2011;Slavokhotova et al., 2014). Adding equal molar quantities of WAMP and chitinase to fungalysin was sufficient to completely inhibit fungalysin activity suggesting a higher affinity of the protease to the WAMP than to the chitinase.

Interestingly, some pathogens can also manipulate the transcription of plant PIs to inhibit deleterious effects of plant proteases in their favor. For example, production of maize cysteine proteases is induced during infection by *Ustilago maydis*, but at the same time the fungus induces the production of maize cystatin CC9 that inhibits cysteine proteases to facilitate infection (van der Linde et al., 2012b; Mueller et al., 2013). This suggests an evolutionary arms race in which the infection strategy of the pathogen benefits from the host's antimicrobial defense to suppress its defense responses.

FUNGAL PROTEASES TARGETING HOST DEFENSE PROTEINS

The arms race between pathogens and their hosts is often explained by recognition of MAMPs or effectors through pattern recognition receptors or resistance proteins, which results in MTI or ETI (Jones and Dangl, 2006). However, several components of basal defense are both constitutive and induced after interaction between MAMPs/effectors and immune receptors. PR proteins provide an excellent example of this. PR proteins are generally stable proteins that often exhibit a basal level of expression, but are also strongly induced after infection (Sels et al., 2008). PR proteins and their antifungal activity have been exploited to improve broad-spectrum resistance in plants. Plants such as tobacco, tomato, potato, peanut, and cacao have been engineered to over-express chitinases alone (Schickler and Chet, 1997; de las Mercedes Dana et al., 2006; Maximova et al., 2006; Iqbal et al., 2012; Cletus et al., 2013) or in combination with other PR proteins in pea and rice (Sridevi et al., 2008; Amian et al., 2011), and showed enhanced resistance to fungal pathogens.

Plant chitinases and especially chitin-binding domain (CBD)-containing chitinases play an important role in defense against pathogenic fungi (Iseli et al., 1993; Suarez et al., 2001). Some fungal pathogens such as *C. fulvum* secrete chitin-binding effector proteins like CfAvr4 into the colonized extracellular space of tomato leaves to protect themselves against the antifungal activity of apoplastic plant chitinases (van den Burg et al., 2006). Indeed, CfAvr4 binds to chitin of fungal cell walls, making chitin inaccessible to plant chitinases, thereby preventing hydrolysis by these enzymes (van den Burg et al., 2006). Functional homologs of CfAvr4 have been identified in other Dothideomycete plant pathogens, in which they likely also protect the fungal cell wall against plant chitinases (Stergiopoulos et al., 2010; de Wit et al., 2012; Mesarich et al., 2015). However, many fungal pathogens do not carry homologs of the *CfAvr4* gene in their genome. It appears that some fungi secrete proteases that cleave CBD-chitinases. For example, *F. solani* f. sp. *phaseoli* is able to modify chitinases during infection of bean to facilitate host colonization (Lange et al., 1996). Also an extracellular subtilisin protease from *F. solani* f. sp. *eumartii* was reported to modify chitinases and β-1,3-glucanases present in intercellular washing fluids of potato (Olivieri et al., 2002). More recently, it was shown that *F. verticillioides* and other maize pathogens, including *Bipolaris zeicola*and *Stenocarpella maydis*, secrete two types of proteases that truncate maize class IV CBD-chitinases (Naumann, 2011). A fungalysin metalloprotease of *F. verticillioides* was found to cleave within the CBD domain between conserved Gly and Cys residues (Naumann et al., 2011), while a novel polyglycine hydrolase present in many fungi belonging to the family of *Pleosporineae* cleaved within the polyglycine linker present in the hinge domain of class IV chitinases (Naumann et al., 2014, 2015). In another recent study it was shown that the fungal tomato pathogens *B. cinerea*, *V. dahliae*, and *F. oxysporum* f. sp. *lycopersici* secrete proteases that modify tomato CBD-chitinases (Karimi Jashni et al., 2015). For *F. oxysporum* f. sp. *lycopersici*, the synergistic action of a serine protease, FoSep1, and a metalloprotease, FoMep1 (the ortholog of fungalysin from *F. verticillioides*), was required for cleavage and removal of the CBD from two tomato CBD-chitinases (Karimi Jashni et al., 2015). Removal of the CBD from two tomato CBD-chitinases by these two enzymes led to a reduction of their chitinase and antifungal activity. In addition, mutants of *F. oxysporum* f. sp. *lycopersici* lacking both *FoSep1* and *FoMep1* exhibited reduced virulence on tomato, confirming that secreted fungal proteases are important virulence factors by targeting CDB-chitinases to compromise an important component of plant basal defense (Karimi Jashni et al., 2015).

Collectively, the activity of fungal proteases might explain why overexpression of plant chitinases in transgenic plants has not become an effective strategy to obtain durable resistance against fungal pathogens. Secretion of proteases and PIs by pathogens to modify, degrade, or inhibit basal defense proteins might have played an important role during co-evolution with their host plants (Hörger and van der Hoorn, 2013). Therefore, overexpression of chitinases from a heterologous source in transgenic plants might be a more efficient approach to obtain durable resistance against pathogens, as they have not co-evolved with these “foreign” defense proteins.

FUNGAL PROTEASE INHIBITORS TARGETING HOST PROTEASES

Plant pathogens also secrete PI effectors to inhibit plant defense proteases and promote disease development. These effectors are targeted to various host compartments (Tian et al., 2009). One such effector, Avr2, secreted by *C. fulvum* during infection, is required for full virulence of this fungus on tomato (Rooney et al., 2005). Avr2 inhibits the tomato apoplastic PLCPs Rcr3 and Pip1 to support growth of *C. fulvum* in the apoplast. Also, plants expressing Avr2 showed increased susceptibility to other pathogenic fungi, including *B. cinerea* and *V. dahliae* (van Esse et al., 2008). Moreover, *A. thaliana* plants expressing Avr2 triggered global transcriptional reprogramming, reflecting a typical host response to pathogen attack (van Esse et al., 2008). Two other PI effectors are the cystatin-like proteins EPIC1 (extracellular proteinase inhibitor C1) and EPIC2B (extracellular proteinase inhibitor C2B), whose expression is strongly induced in the oomycete *P. infestans* during biotrophic growth on tomato leaves (Tian et al., 2007; Song et al., 2009). These PIs selectively target the plant PLCPs Rcr3, Pip1, and C14 in the apoplast of potato and tomato. The EPICs inhibit C14 and possibly other PLCPs over a wider pH range than that observed for Avr2, which only inhibits Pip1 and Rcr3 at pH values occurring in the apoplast where the pathogen grows. In addition, *P. infestans* secretes two serine PIs (EPI1 and EPI10) that target and inhibit the major apoplastic serine protease P69B, likely to decrease its role in defense (Tian et al., 2004, 2005). It was proposed that EPI1 protects EPIC1 and EPIC2B proteins from degradation by P69B (Tian, 2005).

Furthermore, the maize pathogen *U. maydis* secretes the cysteine PI Pit2 that strongly inhibits three abundant defense-related maize cysteine proteases (CP2 and its two isoforms CP1A and CP1B; Van der Linde et al., 2012a; van der Linde et al., 2012b; Mueller et al., 2013). These findings indicate that cysteine and serine PIs secreted by different groups of filamentous fungal and oomycete pathogens, as well as their activity against plant proteases, can compromise plant basal defense responses. A schematic overview of different types of interactions between pathogen and host proteases and PIs at the plant–pathogen interface is presented in Figure 1.

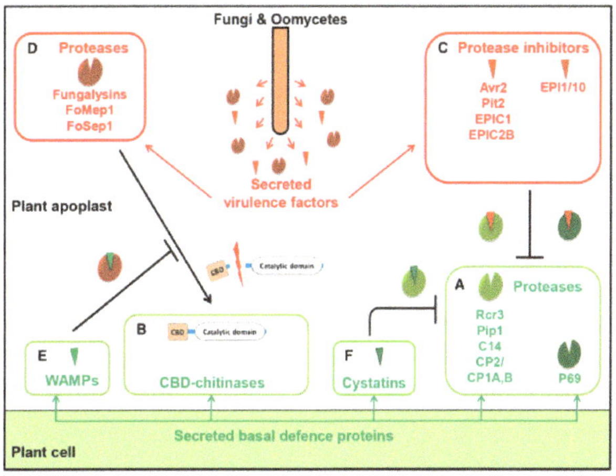

Figure 1: Proteases and protease inhibitors at the plant–pathogen interface. As part of their basal defense response, plants secrete deleterious enzymes such as proteases(A) and chitin-binding domain (CBD)-containing chitinases (B) that target pathogen components. In response, filamentous pathogens secrete protease inhibitors (C) that inhibit plant cysteine or serine proteases. Filamentous pathogens also secrete fungalysin metallo- or serine proteases (D) that process antifungal CBD-chitinases of plants. In response, plants secrete antimicrobial peptides such as hevein-like antimicrobial peptides from wheat (WAMPs) **(E)** that inhibit fungalysin metalloproteases or cystatins **(F)** that inhibit endogenous plant cysteine proteases. Examples shown in this figure are discussed in the text.

PROTEASES, PI EFFECTORS, AND THEIR ROLE IN RECEPTOR-MEDIATED HOST DEFENSE RESPONSES

The plant immune system is able to recognize pathogen effectors to mount receptor-mediated defense responses. Although the intrinsic function of protease and PI effectors secreted by some pathogenic fungi promote disease through manipulation of host defense, proteases and PI effectors can also be recognized by host immune receptors mediating defense responses. This adaptation and counter-adaptation reflects the arms race between pathogens and their host plants. A clear example of such an evolutionary arms race are the cysteine PIs Avr2 from *C. fulvum* and Gr-VAP1 (*Globodera rostochiensis* Venom Allergen-like Protein) from *G. rostochiensis* that bind and inhibit the tomato cysteine protease Rcr3^{pim}. The tomato immune receptor protein Cf-2 senses this interaction and mediates the induction of defense responses (Song et al., 2009; Lozano-Torres et al., 2012). Most likely, the interaction causes a conformational change in Rcr3, which is recognized by the Cf-2 receptor (Krüger et al., 2002; Rooney et al., 2005). This hypothesis is supported by the finding that a natural variant of Rcr3 is recognized by Cf-2 in an Avr2-independent manner (Dixon et al., 2000). Moreover, in tomato plants lacking the Cf-2 receptor, targeting of Rcr3 is not sensed and plants are more susceptible to *G. rostochiensis* (Lozano-Torres et al., 2012).

CO-EVOLUTION BETWEEN PLANTS AND THEIR PATHOGENS IS REFLECTED BY THE NUMEROUS VARIANT PROTEASES AND PIS IN THE GENOMES OF BOTH ORGANISMS

The genomes of fungal plant pathogens encode predicted proteases belonging to various subfamilies that vary in number between pathogens with different lifestyles. Generally, hemi-biotrophs and saprotrophs contain higher numbers of secreted proteases than biotrophs (Ohm et al., 2012). However, these predictions are based on gene numbers and may not be supported by their transcription and translation profiles. For example, *C. fulvum*, which is a biotrophic fungus, has numbers of proteases that are comparable to the phylogenetically closely related hemi-biotroph*Dothistroma septosporum* (de Wit et al., 2012). However, likely due to its adaptation to a different host and lifestyle, many *C. fulvum* protease genes are not expressed *in planta* and some have undergone pseudogenization (van der Burgt et al., 2014). Deletion and duplication of protease genes were reported to occur in the genome of the grass endophytic fungus Epichloë festucae (Bryant et al., 2009) but their biological implications have not yet been studied.

Adaptation of PI effectors from pathogens to inhibit different host proteases has been observed in several cases. The Avr2 PI of *C. fulvum*, for example, has a high affinity for the host proteases Rcr3 and Pip1 and a low affinity for C14 (Shabab et al., 2008; Hörger et al., 2012). *P. infestans* EPICs have a high affinity for C14 and a low affinity for Rcr3 and Pip1 (Kaschani et al., 2010). Furthermore, *U. maydis* Pit2 inhibits the maize cysteine proteases CP1, CP2, and XCP2, but does not inhibit cathepsin CatB (Mueller et al., 2013). Different types of selection pressure may lead to the circumvention of protease inhibition by PIs. For example, purifying or diversifying selection has been reported for the proteases Rcr3, C14, and Pip1, and has been shown to act at their PI binding sites. Sequencing of the tomato proteases Rcr3 and Pip1 across different wild tomato species has shown that these proteins are under strong diversifying selection imposed by Avr2. For instance, one of the variant residues in the binding site of Rcr3 prevented inhibition by Avr2, indicating selection for evasion from recognition by this inhibitor (Shabab et al., 2008). C14 from solanaceous plants is also the target of EPICs secreted by *P. infestans* and is under diversifying selection in potato and under conservative selection in tomato. This demonstrates that C14 plays an active role in host immunity against this pathogen and variations in the sequence of C14 in natural hosts of *P. infestans* highlight the co-evolutionary arms race at the plant–pathogen interface (Kaschani et al., 2010).

Evolutionary diversification may vary from point mutation to gene deletion or insertion. EPIC1 and EPIC2 are PIs present in *P. infestans*, however their orthologs were lost in *P. sojae* and *P. ramorum*(Tian et al., 2007). *P. mirabilis*, a species closely related to *P. infestans*, is a pathogen of *Mirabilis jalapa*, and secretes the PI PmEPIC1, an ortholog of EPIC1 that inhibits C14 but not Rcr3 (Dong et al., 2014). However, *M. jalapa* secretes MRP2, a PLCP homolog of Rcr3, that is more effectively inhibited by PmEPIC1 than by EPIC1 (Dong et al., 2014). Substitution of one amino acid residue in PmEPIC1 and EPIC1 restored the inhibitory function of PmEPIC1 on Rcr3 and of EPIC1 on MRP2, respectively. These results show that proteases and PIs have played important roles in adaptation of the two *Phytophthora* species to their respective host plants, although the two species diverged only a 1000 years ago (Dong et al., 2014). This is an excellent example for a role of a protease and PI in the arms race between a plant and its pathogen and exemplifies how diversification and adaptation of a protease-PI complex may work at the molecular level.

CONCLUSION AND PERSPECTIVE

The recent advances reviewed here exemplify determinative roles of proteases and PIs in shaping plant–pathogen interactions. Analyses of genome databases of both plants and pathogens show that these organisms encode numerous proteases and PIs, of which we are just beginning to understand some of their roles. Advanced transcriptome and proteome tools such as RNA sequencing and protease profiling will facilitate identification of important proteases and PIs for further functional analysis. The redundancy of proteases in pathogens is a technical challenge that has so far hampered defining their biological functions. Targeted deletion of one or even two protease genes failed to change virulence of the plant pathogenic fungi *Glomerella cingulata*(Plummer et al., 2004) and *B. cinerea* (ten Have et al., 2010), respectively. Karimi Jashni et al. (2015) could only show decreased virulence of a double protease mutant of the tomato pathogen *F. oxysporum* by a combined biochemical and genetic approach, and using a defined plant enzyme (CBD-chitinase) as a substrate that was presumed to be involved in plant defense. This indicates that multi-gene targeting of protease and PI genes to identify their role in virulence or avirulence remains a challenge in filamentous fungi. Targeting multiple protease and PI genes might also be hampered by lack of sufficient numbers of selection markers for targeted gene replacement. In the latter case multiple protease and PI genes might be targeted by targeted gene silencing.

CONFLICT OF INTEREST STATEMENT

The authors declare that the research was conducted in the absence of any commercial or financial relationships that could be construed as a potential conflict of interest.

ACKNOWLEDGMENTS

MJ was financed by the Ministry of Science, Research and Technology (MSRT) of Iran; PdW, JC, RM, and CM were financed by Wageningen University and the Royal Netherlands Academy of Arts and Sciences.

REFERENCES

1. Amian, A. A., Papenbrock, J., Jacobsen, H. J., and Hassan, F. (2011). Enhancing transgenic pea (*Pisum sativum* L.) resistance against fungal diseases through stacking of two antifungal genes (chitinase and glucanase). *GM Crops* 2, 104–109. doi: 10.4161/gmcr.2.2.16125

2. Avrova, A. O., Taleb, N., Rokka, V. M., Heilbronn, J., Campbell, E., Hein, I., et al. (2004). Potato oxysterol binding protein and cathepsin B are rapidly up-regulated in independent defence pathways that distinguish R gene-mediated and field resistances to*Phytophthora infestans*. *Mol. Plant Pathol.* 5, 45–56. doi: 10.1111/j.1364-3703.2004.00205.x

3. Balasubramanian, V., Vashisht, D., Cletus, J., and Sakthivel, N. (2012). Plant β-1,3-glucanases: their biological functions and transgenic expression against phytopathogenic fungi. *Biotechnol. Lett.* 34, 1983–1990. doi: 10.1007/s10529-012-1012-6

4. Bozkurt, T. O., Schornack, S., Win, J., Shindo, T., Ilyas, M., Oliva, R., et al. (2011). *Phytophthora infestans* effector Avrblb2 prevents secretion of a plant immune protease at the haustorial interface. *Proc. Natl. Acad. Sci. U.S.A.* 108, 20832–20837. doi: 10.1073/pnas.1112708109

5. Bryant, M. K., Schardl, C. L., Hesse, U., and Scott, B. (2009). Evolution of a subtilisin-like protease gene family in the grass endophytic fungus Epichloë festucae. *BMC Evol. Biol.* 9:168. doi: 10.1186/1471-2148-9-168

6. Cletus, J., Balasubramanian, V., Vashisht, D., and Sakthivel, N. (2013). Transgenic expression of plant chitinases to enhance disease resistance. *Biotechnol. Lett.* 35, 1719–1732. doi: 10.1007/s10529-013-1269-4

7. de las Mercedes Dana, M., Pintor-Toro, J. A., and Cubero, B. (2006). Transgenic tobacco plants overexpressing chitinases of fungal origin show enhanced resistance to biotic and abiotic stress agents. *Plant Physiol.* 142, 722–730. doi: 10.1104/pp.106.086140

8. de Wit, P. J. G. M, van der Burgt, A., Ökmen, B., Stergiopoulos, I., Abd-Elsalam, K. A., Aerts, A. L., et al. (2012). The genomes of the fungal plant pathogens *Cladosporium fulvum* and *Dothistroma septosporum* reveal adaptation to different hosts and lifestyles but also signatures of common ancestry. *PLoS Genet.* 8:e1003088. doi: 10.1371/journal.pgen.1003088

9. Dixon, M. S., Golstein, C., Thomas, C. M., van der Biezen, E. A., and Jones, J. D. G. (2000). Genetic complexity of pathogen perception by plants: the example of *Rcr3*, a tomato gene required specifically by Cf-2. *Proc. Natl. Acad. Sci. U.S.A.* 97, 8807–8814. doi: 10.1073/pnas.97.16.8807

10. Dong, S., Stam, R., Cano, L. M., Song, J., Sklenar, J., Yoshida, K., et al. (2014). Effector specialization in a lineage of the Irish potato famine pathogen. *Science* 343, 552–555. doi: 10.1126/science.1246300

11. Hörger, A. C., Ilyas, M., Stephan, W., Tellier, A., van der Hoorn, R. A. L., and Rose, L. E. (2012). Balancing selection at the tomato *Rcr3* guardee gene family maintains variation in strength of pathogen defense. *PLoS Genet.* 8:e1002813. doi: 10.1371/journal.pgen.1002813

12. Hörger, A. C., and van der Hoorn, R. A. L. (2013). The structural basis of specific protease–inhibitor interactions at the plant–pathogen interface. *Curr. Opin. Struct. Biol.* 23, 842–850. doi: 10.1016/j.sbi.2013.07.013

13. Iqbal, M., Nazir, F., Ali, S., Asif, M. A., Zafar, Y., Iqbal, J., et al. (2012). Overexpression of rice chitinase gene in transgenic peanut (*Arachis hypogaea* L.) improves resistance against leaf spot. *Mol. Biotechnol.* 50, 129–136. doi: 10.1007/s12033-011-9426-2

14. Iseli, B., Boller, T., and Neuhaus, J. M. (1993). The N-terminal cysteine-rich domain of tobacco class I chitinase is essential for chitin binding but not for catalytic or antifungal activity. *Plant Physiol.* 103, 221–226. doi: 10.1104/pp.103.1.221

15. Jones, J. D. G., and Dangl, J. L. (2006). The plant immune system. *Nature* 444, 323–329. doi: 10.1038/nature05286

16. Karimi Jashni, M., Dols, H. M. I., Iida, Y., Boeren, S., Beenen, H. G., Mehrabi, R., et al. (2015). Synergistic action of a metalloprotease and a serine protease from *Fusarium oxysporum* f. sp. *lycopersici* cleaves chitin-binding tomato chitinases, reduces their antifungal activity and enhances fungal virulence. *Mol. Plant Microbe Interact.* doi: 10.1094/MPMI-04-15-0074-R [Epub ahead of print].

17. Kaschani, F., Shabab, M., Bozkurt, T., Shindo, T., Schornack, S., Gu, C., et al. (2010). An effector-targeted protease contributes to defense against *Phytophthora infestans* and is under diversifying selection in natural hosts. *Plant Physiol.* 154, 1794–1804. doi: 10.1104/pp.110.158030

18. Kim, J. Y., Park, S. C., Hwang, I., Cheong, H., Nah, J. W., Hahm, K. S., et al. (2009). Protease inhibitors from plants with antimicrobial activity. *Int. J. Mol. Sci.* 10, 2860–2872. doi: 10.3390/ijms10062860

19. Krüger, J., Thomas, C. M., Golstein, C., Dixon, M. S., Smoker, M., Tang, S., et al. (2002). A tomato cysteine protease required for Cf-2-dependent disease resistance and suppression of autonecrosis. *Science* 296, 744–747. doi: 10.1126/science.1069288

20. Laluk, K., and Mengiste, T. (2011). The *Arabidopsis* extracellular unusual serine protease inhibitor functions in resistance to necrotrophic fungi and insect herbivory. *Plant J.* 68, 480–494. doi: 10.1111/j.1365-313X.2011.04702.x

21. Lange, J., Mohr, U., Wiemken, A., Boller, T., and Vögeli-Lange, R. (1996). Proteolytic processing of class IV chitinase in the compatible interaction of bean roots with *Fusarium solani*. *Plant Physiol.* 111, 1135–1144. doi: 10.1104/pp.111.4.1135

22. Lozano-Torres, J. L., Wilbers, R. H. P., Gawronski, P., Boshoven, J. C., Finkers-Tomczak, A., Cordewener, J. H. G., et al. (2012). Dual disease resistance mediated by the immune receptor Cf-2 in tomato requires a common virulence target of a fungus and a nematode. *Proc. Natl. Acad. Sci. U.S.A.* 109, 10119–10124. doi: 10.1073/pnas.1202867109

23. Maximova, S., Marelli, J. P., Young, A., Pishak, S., Verica, J., and Guiltinan, M. (2006). Over-expression of a cacao class I chitinase gene in *Theobroma cacao* L. Enhances resistance against the pathogen, *Colletotrichum gloeosporioides*. *Planta* 224, 740–749. doi: 10.1007/s00425-005-0188-6

24. Mesarich, C. H., Stergiopoulos, I., Beenen, H. G., Cordovez, V., Guo, Y., Karimi Jashni, M., et al. (2015). A conserved proline residue in Dothideomycete Avr4

effector proteins is required to trigger a Cf-4-dependent hypersensitive response *Mol. Plant Pathol.* doi: 10.1111/mpp.12265 [Epub ahead of print].

25. Mosolov, V. V., Grigor'eva, L. I., and Valueva, T. A. (2001). Involvement of proteolytic enzymes and their inhibitors in plant protection (review). *Appl. Biochem. Microbiol.* 37, 115–123. doi: 10.1023/A:1002824527463

26. Mueller, A. N., Ziemann, S., Treitschke, S., Aßmann, D., and Doehlemann, G. (2013). Compatibility in the *Ustilago maydis*–maize interaction requires inhibition of host cysteine proteases by the fungal effector Pit2. *PLoS Pathog.* 9:e1003177. doi: 10.1371/journal.ppat.1003177

27. Naumann, T. A. (2011). Modification of recombinant maize ChitA chitinase by fungal chitinase-modifying proteins. *Mol. Plant Pathol.* 12, 365–372. doi: 10.1111/j.1364-3703.2010.00677.x

28. Naumann, T. A., Naldrett, M. J., Ward, T. J., and Price, N. P. J. (2015). Polyglycine hydrolases: fungal β-lactamase-like endoproteases that cleave polyglycine regions within plant class IV chitinases. *Protein Sci.* 24, 1147–1157. doi: 10.1002/pro.2705

29. Naumann, T. A., Wicklow, D. T., and Price, N. P. J. (2011). Identification of a chitinase-modifying protein from *Fusarium verticillioides*. *J. Biol. Chem.* 286, 35358–35366. doi: 10.1074/jbc.M111.279646

30. Naumann, T. A., Wicklow, D. T., and Price, N. P. J. (2014). Polyglycine hydrolases secreted by Pleosporineae fungi that target the linker region of plant class IV chitinases. *Biochem. J.* 460, 187–198. doi: 10.1042/BJ20140268

31. Ohm, R. A., Feau, N., Henrissat, B., Schoch, C. L., Horwitz, B. A., Barry, K. W., et al. (2012). Diverse lifestyles and strategies of plant pathogenesis encoded in the genomes of eighteen *Dothideomycetes* fungi. *PLoS Pathog.* 8:e1003037. doi: 10.1371/journal.ppat.1003037

32. Olivieri, F., Eugenia Zanetti, M., Oliva, C. R., Covarrubias, A. A., and Casalongué, C. A. (2002). Characterization of an extracellular serine protease of *Fusarium eumartii* and its action on pathogenesis related proteins. *Eur. J. Plant Pathol.* 108, 63–72. doi: 10.1023/A:1013920929965

33. Pekkarinen, A. I., Longstaff, C., and Jones, B. L. (2007). Kinetics of the inhibition of *Fusarium* serine proteinases by barley (*Hordeum vulgare* L.) inhibitors. *J. Agric. Food Chem.* 55, 2736–2742. doi: 10.1021/jf0631777

34. Plummer, K. M., Clark, S. J., Ellis, L. M., Loganathan, A., Al-Samarrai, T. H., Rikkerink, E. H. A., et al. (2004). Analysis of a secreted aspartic peptidase disruption mutant of *Glomerella cingulata*. *Eur. J. Plant Pathol.* 110, 265–274. doi: 10.1023/b:cjpp.0000019796.78598.8c

35. Rooney, H. C. E., van't Klooster, J. W., van der Hoorn, R. A. L., Joosten, M. H. A. J., Jones, J. D. G., and de Wit, P. J. (2005). Cladosporium Avr2 inhibits tomato Rcr3 protease required for Cf-2-dependent disease resistance. *Science* 308, 1783–1786. doi: 10.1126/science.1111404

36. Ryan, C. A. (1990). Protease inhibitors in plants: genes for improving defenses against insects and pathogens. *Annu. Rev. Phytopathol.* 28, 425–449. doi: 10.1146/annurev.py.28.090190.002233

37. Schickler, H., and Chet, I. (1997). Heterologous chitinase gene expression to improve plant defense against phytopathogenic fungi. *J. Ind. Microbiol. Biotechnol.* 19, 196–201. doi: 10.1038/sj.jim.2900447

38. Sela-Buurlage, M. B. (1996). *In vitro Sensitivity and Tolerance of Fusarium solani Towards Chitinases and -1,3-glucanases*. Ph.D. thesis, Wageningen University, Wageningen.

39. Sels, J., Mathys, J., De Coninck, B. M. A., Cammue, B. P. A., and De Bolle, M. F. C. (2008). Plant pathogenesis-related (PR) proteins: a focus on PR peptides. *Plant Physiol. Biochem.* 46, 941–950. doi: 10.1016/j.plaphy.2008.06.011

40. Shabab, M., Shindo, T., Gu, C., Kaschani, F., Pansuriya, T., Chintha, R., et al. (2008). Fungal effector protein AVR2 targets diversifying defense-related Cys proteases of tomato. *Plant Cell* 20, 1169–1183. doi: 10.1105/tpc.107.056325

41. Slavokhotova, A. A., Naumann, T. A., Price, N. P. J., Rogozhin, E. A., Andreev, Y. A., Vassilevski, A. A., et al. (2014). Novel mode of action of plant defense peptides-hevein-like antimicrobial peptides from wheat inhibit fungal metallo-proteases. *FEBS J.* 281, 4754–4764. doi: 10.1111/febs.13015

42. Song, J., Win, J., Tian, M., Schornack, S., Kaschani, F., Ilyas, M., et al. (2009). Apoplastic effectors secreted by two unrelated eukaryotic plant pathogens target the tomato defense protease Rcr3. *Proc. Natl. Acad. Sci. U.S.A.* 106, 1654–1659. doi: 10.1073/pnas.0809201106

43. Spoel, S. H., and Dong, X. (2012). How do plants achieve immunity? Defence without specialized immune cells. *Nat. Rev. Immunol.* 12, 89–100. doi: 10.1038/nri3141

44. Sridevi, G., Parameswari, C., Sabapathi, N., Raghupathy, V., and Veluthambi, K. (2008). Combined expression of chitinase and β-1, 3-glucanase genes in indica rice (*Oryza sativa* L.) enhances resistance against *Rhizoctonia solani*. *Plant Sci.* 175, 283–290. doi: 10.1016/j.plantsci.2008.04.011

45. Stergiopoulos, I., and de Wit, P. J. (2009). Fungal effector proteins. *Annu. Rev. Phytopathol.* 47, 233–263. doi: 10.1146/annurev.phyto.112408.132637

46. Stergiopoulos, I., van den Burg, H. A., Ökmen, B., Beenen, H. G., van Liere, S., Kema, G. H. J., et al. (2010). Tomato Cf resistance proteins mediate recognition of cognate homologous effectors from fungi pathogenic on dicots and monocots. *Proc. Natl. Acad. Sci. U.S.A.* 107, 7610–7615. doi: 10.1073/pnas.1002910107

47. Suarez, V., Staehelin, C., Arango, R., Holtorf, H., Hofsteenge, J., and Meins, F. Jr. (2001). Substrate specificity and antifungal activity of recombinant tobacco class I chitinases. *Plant Mol. Biol.* 45, 609–618. doi: 10.1023/A:1010619421524

48. ten Have, A., Espino, J. J., Dekkers, E., van Sluyter, S. C., Brito, N., Kay, J., et al. (2010). The *Botrytis*

cinerea aspartic proteinase family. *Fungal. Genet. Biol.* 47, 53–65. doi: 10.1016/j.fgb.2009.10.008

49. Tian, M. (2005). *Functional Characterization of Extracellular Protease Inhibitors of Phytophthora infestans*. Ph.D. thesis, The Ohio State University, Columbus.

50. Tian, M., Benedetti, B., and Kamoun, S. (2005). A second Kazal-like protease inhibitor from *Phytophthora infestans* inhibits and interacts with the apoplastic pathogenesis-related protease P69b of tomato. *Plant Physiol.* 138, 1785–1793. doi: 10.1104/pp.105.061226

51. Tian, M., Huitema, E., da Cunha, L., Torto-Alalibo, T., and Kamoun, S. (2004). A Kazal-like extracellular serine protease inhibitor from *Phytophthora infestans* targets the tomato pathogenesis-related protease P69b. *J. Biol. Chem.* 279, 26370–26377. doi: 10.1074/jbc.m400941200

52. Tian, M., Chaudhry, F., Ruzicka, D. R., Meagher, R. B., Staiger, C. J., and Day, B. (2009). *Arabidopsis* actin-depolymerizing factor AtADF4 mediates defense signal transduction triggered by the *Pseudomonas syringae* effector AvrPphB. *Plant Physiol.* 150, 815–824. doi: 10.1104/pp.109.137604

53. Tian, M., Win, J., Song, J., van der Hoorn, R., van der Knaap, E., and Kamoun, S. (2007). A *Phytophthora infestans* cystatin-like protein targets a novel tomato papain-like apoplastic protease. *Plant Physiol.* 143, 364–377. doi: 10.1104/pp.106.090050

54. Valueva, T., and Mosolov, V. (2004). Role of inhibitors of proteolytic enzymes in plant defense against phytopathogenic microorganisms. *Biochemistry* 69, 1305–1309. doi: 10.1007/s10541-005-0015-5

55. van den Burg, H. A., Harrison, S. J., Joosten, M. H. A. J., Vervoort, J., and de Wit, P. J. (2006). *Cladosporium fulvum* Avr4 protects fungal cell walls against hydrolysis by plant chitinases accumulating during infection. *Mol. Plant Microbe Interact.* 19, 1420–1430. doi: 10.1094/MPMI-19-1420

56. van der Burgt, A., Karimi Jashni, M., Bahkali, A. H., and de Wit, P. J. (2014). Pseudogenization in pathogenic fungi with different host plants and lifestyles might reflect their evolutionary past. *Mol. Plant Pathol.* 15, 133–144. doi: 10.1111/mpp.12072

57. van der Hoorn, R. A. L. (2008). Plant proteases: from phenotypes to molecular mechanisms. *Annu. Rev. Plant Biol.* 59, 191–223. doi: 10.1146/annurev.arplant.59.032607.092835

58. van der Linde, K., Mueller, A. N., Hemetsberger, C., Kashani, F., van der Hoorn, R. A. L., and Doehlemann, G. (2012a). The maize cystatin CC9 interacts with apoplastic cysteine proteases. *Plant Signal. Behav.* 7, 1397–1401. doi: 10.4161/psb.21902

59. van der Linde, K., Hemetsberger, C., Kastner, C., Kaschani, F., van der Hoorn, R. A. L., Kumlehn, J., et al. (2012b). A maize cystatin suppresses host immunity by inhibiting apoplastic cysteine proteases. *Plant Cell* 24, 1285–1300. doi: 10.1105/tpc.111.093732

60. van Esse, H. P., van't Klooster, J. W., Bolton, M. D., Yadeta, K. A., van Baarlen, P., Boeren, S., et al. (2008). The *Cladosporium fulvum* virulence protein Avr2 inhibits host proteases required for basal defense. *Plant Cell* 20, 1948–1963. doi: 10.1105/tpc.108.059394

61. Wubben, J. P., Lawrence, C. B., and de Wit, P. J. (1996). Differential induction of chitinase and 1,3-β-glucanase gene expression in tomato by *Cladosporium fulvum* and its race-specific elicitors. *Physiol. Mol. Plant Pathol.* 48, 105–116. doi: 10.1006/pmpp.1996.0010

62. Ye, X. Y., Ng, T. B., and Rao, P. F. (2001). A Bowman–Birk-type trypsin-chymotrypsin inhibitor from broad beans. *Biochem. Biophys. Res. Commun.* 289, 91–96. doi: 10.1006/bbrc.2001.5965

CHAPTER 10 ROLE OF PHYTOALEXINS IN PLANT-MICROORGANISM INTERACTIONS

Philippe Jeandet, Claire Hébrard, Marie-Alice Deville, Sylvain Cordelier, Stéphan Dorey, Aziz Aziz, and Jérôme Crouzet

Keywords: Phytoalexins; plants; defense mechanisms; microorganisms; biological activity

PHYTOALEXINS: A GLOBAL SURVEY

Phytoalexins take part in an intricate defense system used by plants against pests and pathogens [1,2]. These are low molecular weight antimicrobial compounds both synthesized by and accumulated in plants as a response to biotic and abiotic stresses. The concept of phytoalexins was first introduced over 70 years ago by Müller and Börger [3] after observing that infection of potato tubers with a strain of *Phytophthora infestans* capable of initiating hypersensitive reactions, significantly inhibited the effect of a subsequent infection with another strain of *P. infestans*. This inhibition was linked to a "principle" produced by the plant cells reacting hypersensitively that they named *phytoalexin* [4].

Most of what is known about phytoalexins derives from extensive work on a limited number of plant families: Leguminosae or Fabaceae and Solanaceae [5,6], on one hand, and investigations on one or a few species within other plant families, namely Amaryllidaceae, Euphorbiaceae, Orchidaceae, Chenopodiaceae, Compositae, Convolvulaceae, Ginkgoaceae, Poaceae, Linaceae, Moraceae, Orchidaceae, Piperaceae, Rosaceae, Rutaceae and Umbelliferae on the other hand [7]. More intensive studies recently focused on phytoalexins from plant families of significant economic importance: Poaceae (maize and rice) [8], Vitaceae [9,10] and Malvaceae (cotton) [11]. Camalexin, the main phytoalexin from Brassicaceae (Cruciferae) has also been the subject of numerous studies focusing on its biosynthetic pathway and the regulatory networks involved in its production in the model plant *Arabidopsis thaliana* [1,12]. However, the question of the ubiquity of phytoalexins throughout the plant kingdom still remains.

Phytoalexins are restricted to compounds produced from remote precursors, through *de novo* synthesis of enzymes. This peculiarity makes deciphering their biosynthesis and regulation mechanisms very complex [1,2]. Phosphorylation cascades, defense-related marker genes, calcium sensors and elicitors as well as hormone signaling are potentially important regulators for the modulation of phytoalexin production and pathogen resistance. As a corollary, knowledge of the control mechanisms of phytoalexin accumulation has served as the basis for the genetic manipulation of those compounds in engineered plants for enhanced disease resistance [1,13,14].

The question as to whether phytoalexins are active *in vivo* and play a significant role in plant defense mechanisms has long been debated addressing both the actual antimicrobial

Citation: Philippe Jeandet, Claire Hébrard, Marie-Alice Deville, Sylvain Cordelier, Stéphan Dorey, Aziz Aziz, and Jérôme Crouzet, Role of Phytoalexins in Plant-Microorganism Interactions, Molecules 2014, 19(11), 18033-18056; doi:10.3390/molecules191118033

activity of phytoalexins under the conditions found within plant tissues and their localization around invading organisms [15,16]. These intractable interrogations are indeed crucial to their proposed role as microbial growth regulators in infected plant tissues. Nonetheless there is considerable evidence that these compounds exhibit *in vitro* toxicity across much of the biological spectrum, prokaryotic and eukaryotic.

The nature of the interaction between plants and pathogens largely depends on the ability of the latter to metabolize the phytoalexins to which they are exposed. Engineering of fungal genes responsible for detoxification of phytoalexins in plants has pointed out their role in the interactions between plants and pathogens [17]. In phytopathogenic fungi, ATP-Binding Cassette (ABC) transporters may also extrude plant defense products as well as fungicides. These transporters act as virulence factors providing protection against phytoalexins produced by the host. Many factors thus interplay to affect the outcome of the interaction between plants and pathogens.

It has recently been demonstrated that phytoalexins may also display health-promoting effects in humans. For instance, resveratrol produced by Vitaceae has been acclaimed for its wondrous effects and its wide range of purported healing and preventive powers as a cardioprotective, antitumor, neuroprotective and antioxidant agent as well as an antifungal and antibacterial compound [14] (see Section 8).

Work on phytoalexins has been prolific and the production of these compounds in infected tissues has become one of the most intensively studied mechanisms of disease resistance in plants. This review will focus on some of the main features of phytoalexins:

- Chemical diversity
- Main biosynthetic pathways and regulation networks
- Biological activity against microorganisms
- Molecular engineering for disease resistance in plants
- Metabolism/Transport in fungi
- Role in human health

CHEMICAL DIVERSITY OF PHYTOALEXINS

Most phytoalexins produced by the Leguminosae belong to six isoflavonoid classes: isoflavones, isoflavanones, pterocarpans, pterocarpenes, isoflavans and coumestans (Table 1) ([1] and references therein). Some pterocarpan phytoalexins are especially well known: pisatin, phaseollin, glyceollin, medicarpin and maackiain. Pisatin was the first phytoalexin to be isolated and characterized from garden pea, *Pisum sativum* [18]. Besides these compounds, a small number of legumes also produce non-isoflavonoid phytoalexins such as furanoacetylenes and stilbenes (Table 1).

Table 1: Phytoalexins from different plant families.

Plant Families (in Alphabetical Order)	Types of Phytoalexins/Examples	References
Amaryllidaceae	Flavans	[19]
Brassicaceae (Cruciferae)	Indole phytoalexins/camalexin	[20]
	Sulfur-containing phytoalexins/brassinin	[21]
Chenopodiaceae	Flavanones/betagarin Isoflavones/betavulgarin	[22]
Compositae	Polyacetylenes/safynol	[23]
Convolvulaceae	Furanosesquiterpenes/Ipomeamarone	[24]
Euphorbiaceae	Diterpenes/casbene	[25]
Poaceae	Diterpenoids:Momilactones; Oryzalexins; Zealexins; Phytocassanes; Kauralexins	[8,26]
	Deoxyanthocyanidins/luteolinidin and apigeninidin	[26,27]
	Flavanones/sakuranetin	[1]
	Phenylamides	[28]
Leguminosae	Isoflavones Isoflavanones Isoflavans Coumestans Pterocarpans/pisatin, phaseollin, glyceollin and maiackiain Furanoacetylenes/wyerone Stilbenes/resveratrol Pterocarpens	[1] and references therein
Linaceae	Phenylpropanoids/coniferyl alcohol	[29]
Malvaceae	Terpenoids naphtaldehydes/gossypol	[11]
Moraceae	Furanopterocarpans/moracins A-H	[30]
Orchidaceae	Dihydrophenanthrenes/loroglossol	[31]
Rutaceae	Methylated phenolic compounds/xanthoxylin	[32]
Umbelliferae	Polyacetylenes/falcarinol	[33]
	Phenolics: xanthotoxin	[34]
	6-methoxymellein	[35]
Vitaceae	Stilbenes/resveratrol	[9]
Rosaceae	Biphenyls/auarperin	[36]
	Dibenzofurans/cotonefurans	
Solanaceae	Phenylpropanoid related compounds	[1] and references therein
	Steroid glycoalkaloids	
	Norsequi and sesquiterpenoids	
	Coumarins	
	Polyacetylenic derivatives	

Chen *et al.*, describe a series of compounds produced by the genus *Tephrosia*, which belongs to the Leguminosae family, and possesses phytoalexin-like activities [37]. Five main classes of phytoalexins have been reported in Solanaceae: phenylpropanoid-related compounds, steroid glycolalkaloids, norsesqui- and sesquiterpenoids, coumarins and polyacetylenic derivatives (Table 1) ([1] and references therein).

Although considerable work has been done on phytoalexins from the Leguminosae and Solanaceae families, it has been recently overshadowed by the discovery of two new phytoalexin classes from the Poaceae [8,26] and Brassicaceae families [20,21]. The main phytoalexins of Poaceae (rice, maize and sorghum) are represented by members of the labdane-related diterpenoid superfamily (zealexins, kauralexins, momilactones, oryzalexins and phytocassanes) [8,26], flavanones, an unusual group of flavonoid phytoalexins, the 3-deoxyanthocyanidins [27] and phenylamides [28] (Table 1 and references therein). The current knowledge on phytoalexins produced by sorghum (3-deoxy-anthocyanidins like luteolinidin and apigeninidin) and maize (zealexins, kauralexins) has been reviewed previously [26]. Indole compounds such as camalexin and brassinin represent the major phytoalexins from the Brassicaceae family (Table 1 and references therein).

Not unexpectedly, phytoalexins from very diverse plant families are represented by many different chemical classes. Naphtaldehyde compounds such as gossypol and its derivatives constitute the main Malvaceae phytoalexins [11] (Table 1). Antifungal polyacetylenes have been isolated as phytoalexins from the Compositae and Umbelliferae families [23,33]. Furanosesquiterpenes and diterpenes constitute the phytoalexins from the Convolvulaceae and Euphorbiaceae families [24,25]. The majority of phytoalexins found in the following plant families are phenolic compounds: flavans in Amaryllidaceae [19], flavanones and isoflavones from Chenopodiaceae [22], Linaceae phenylpropanoids [29], furanopterocarpans in Moraceae [30], dihydrophenanthrenes from Orchidaceae [31], Rutaceae methylated phenolics [32], biphenyls and dibenzofurans in Rosaceae [36], xanthotoxin and 6-methoxymellein in Umbellifereae [34,35] and finally hydroxystilbenes from Vitaceae (Table 1) [1,9,10].

MAIN BIOSYNTHETIC PATHWAYS

Various pathways are utilized for producing different phytoalexins. As it is not our goal to describe each of these biosynthetic routes in details, we will simply outline the three most characteristic ones:

- The phenylpropanoic-polymalonic acid route
- The methylerythritol phosphate and geranyl-geranyl diphosphate pathway
- The indole phytoalexin pathway

Phytoalexins Deriving from the Phenylpropanoic-Polymalonic Acid Route

All flavonoid phytoalexins (isoflavonoids, isoflavones, pterocarpans, isoflavans, coumestans and arylbenzofurans) as well as stilbene phytoalexins and derivatives (dihydrophenanthrenes) are formed through the universal phenylpropanoic-polymalonic acid pathway. It begins with phenylalanine and the phenylalanine ammonia lyase (PAL) or to a lesser extent with tyrosine and the tyrosine ammonia lyase (TAL). The obtained *para*-coumaric acid is activated in *para*-coumaroyl-CoA by ligation to a coenzyme A by 4-coumaroyl:CoA ligase (C4L). Subsequently, chalcone synthase (CHS) on the one hand and stilbene synthase (STS) on the other hand use this same substrate and condense it with three successive units of malonyl-CoA, leading respectively to the production of naringenin chalcone, the first C15 intermediate in the flavonoid pathway and resveratrol, the precursor of all stilbenes. The possible biosynthetic routes to the main flavonoid and stilbene-like phytoalexins from the Leguminosae family are illustrated in Figure 1 [10,38,39].

Mevalonoid-Derived Phytoalexins

These phytoalexins are represented by members of the monoterpene, sesquiterpene, carboxylic sesquiterpene and diterpene families. Specific attention will be given to the diterpene phytoalexin class [8]. This assumption has been confirmed by the observed synchronous accumulation of seven MEP pathway gene transcripts (OsDXS3, OsDXR, OsCMS, OsCMK, OsMCS, OsHDS and OsHDR) in elicitor-induced rice (*Oryza sativa*) cells and the next steps of this biosynthesis are predicted to occur in plastids. Diterpenoids result from the subsequent action of diverse enzymes using GGDP as the starting block. Class II diterpene cyclases named copalyldiphosphate synthases (CPS) are the first to act on GGDP catalyzing the initial cyclization of the latter to copalyldiphosphate (CDP). CDP is the required substrate for class I diterpene synthases named kaurene synthase like (KSL). Sequential action of CPS and KSL produces the olefin precursors of the main diterpene phytoalexin families [8]. Stereochemically differentiated isomers are used subsequently by KSL: the *ent*-CDP in the biosynthesis of phytocassanes A-E and oryzalexins A-F and the *syn*-CDP in the construction of momilactones A and B (Figure 2). Further additions of oxygen in the formation of oryzalexins, momilactones and phytocassanes require a series of cytochrome P450 (CYPs) (Figure 2).

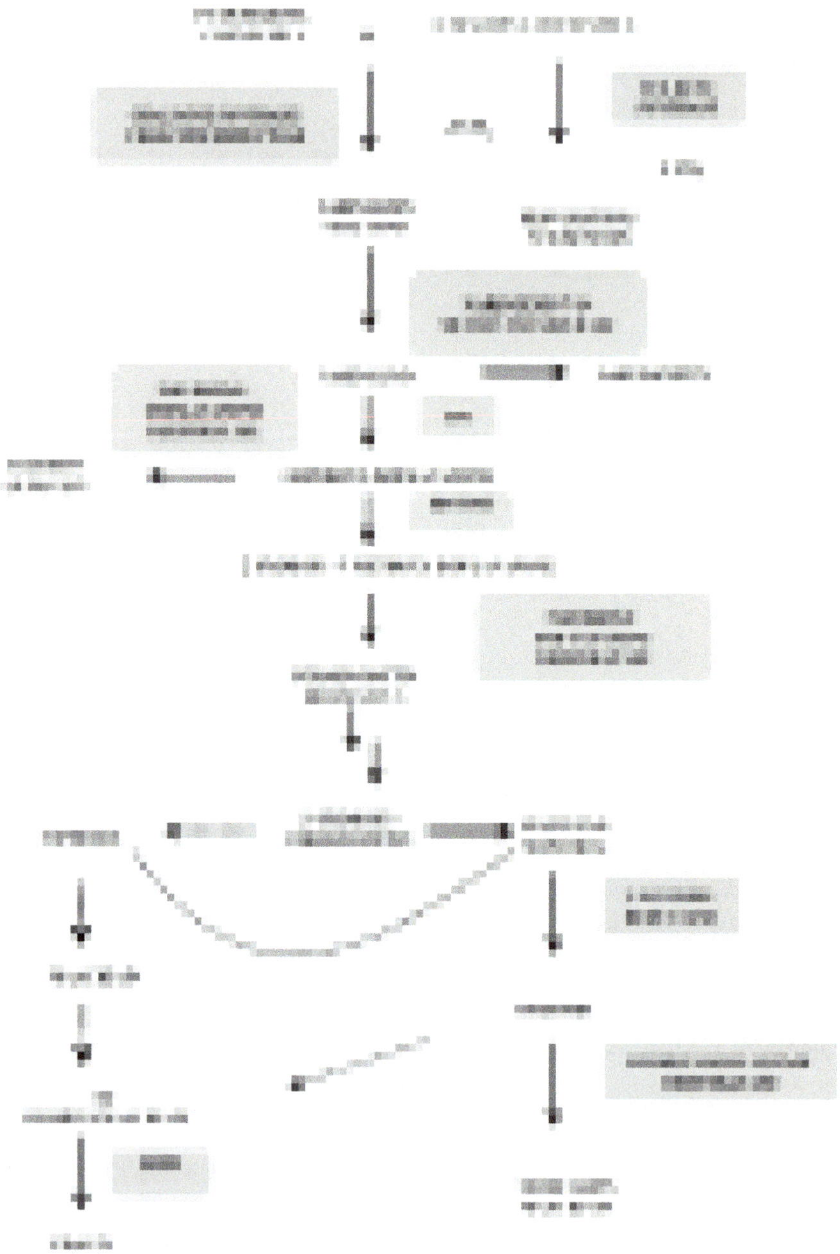

Figure 1: Biosynthetic pathways to the main flavonoid and stilbenoid phytoalexins from the Leguminosae family. (adapted from [10,38,39]). The dashed arrows represent hypothetical steps and the solid arrows denote reactions for which the catalyzing enzymes have been cloned.

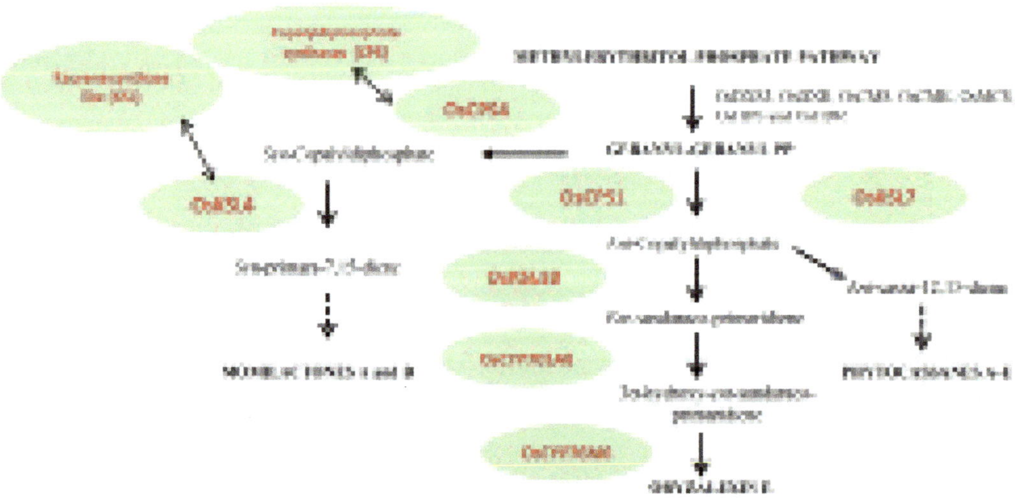

Figure 2: Biosynthetic pathway of diterpenoid phytoalexins.

Indole Phytoalexins

Specific attention will be paid in this section to camalexin, the major phytoalexin of Arabidopsis. The indolic ring of camalexin is derived from tryptophan (Trp) which in turn arises from chorismate (Figure 3). The first step in the route from Trp to camalexin is under the control of two cytochrome P450 homologues CYP79B2 and CYP79B3, leading to indole-3-acetaldoxime. The latter is then transformed into indole-3-acetonitrile (IAN) via the cytochrome P450, CYP71A13. Subsequent conjugation of IAN with glutathione is performed by the combined action of a glutathione-S-transferase and most likely a cytochrome P450. The IAN glutathionyl derivative is then converted into IAN cysteinyl-glycine via a phytochelatin synthase or into γ-glutamyl-cysteine IAN through the action of two γ-glutamyltranspeptidases 1 and 3 [2]. Both intermediates lead to the IAN cysteine conjugate. The last steps of this biosynthesis pathway are under the control of a*CYP71B15* (*PHYTOALEXIN DEFICIENT 3*, *PAD 3*) gene encoding a multifunctional enzyme which forms camalexin via dihydrocamalexic acid (Figure 3).

REGULATION NETWORKS

Phytoalexin biosynthesis is up- or downregulated by expression of many endogenous molecules such as phytohormones (jasmonic acid, salicylic acid, ethylene, auxins, abscisic acid, cytokinins and to a lesser extent gibberellins), transcriptional regulators, defense-related genes, phosphorylation relays and cascades [1,2].

Regulatory mechanisms of phytoalexin biosynthesis also depend on the nature of the infecting pathogen as well as the nature of the induced phytoalexin itself. For example, in the Arabidopsis-*Alternaria brassicicola* interaction, accumulation of camalexin was reported to be independent from jasmonic acid (JA) [40,41] though JA was involved in the regulatory signaling pathways of this phytoalexin in Arabidopsis plants challenged with the fungal pathogen *Botrytis cinerea* [42]. Similarly, existence of JA-dependent and independent pathways in the modulation of diterpenoid phytoalexins in the interaction between rice and the fungal agent *Magnaporthe oryzae* was clearly evidenced by the use of rice mutants [43]. These mutants lacking a functional allene oxide cyclase required for JA production were impaired in momilactone accumulation upon fungal infection whereas phytocassane production was not altered.

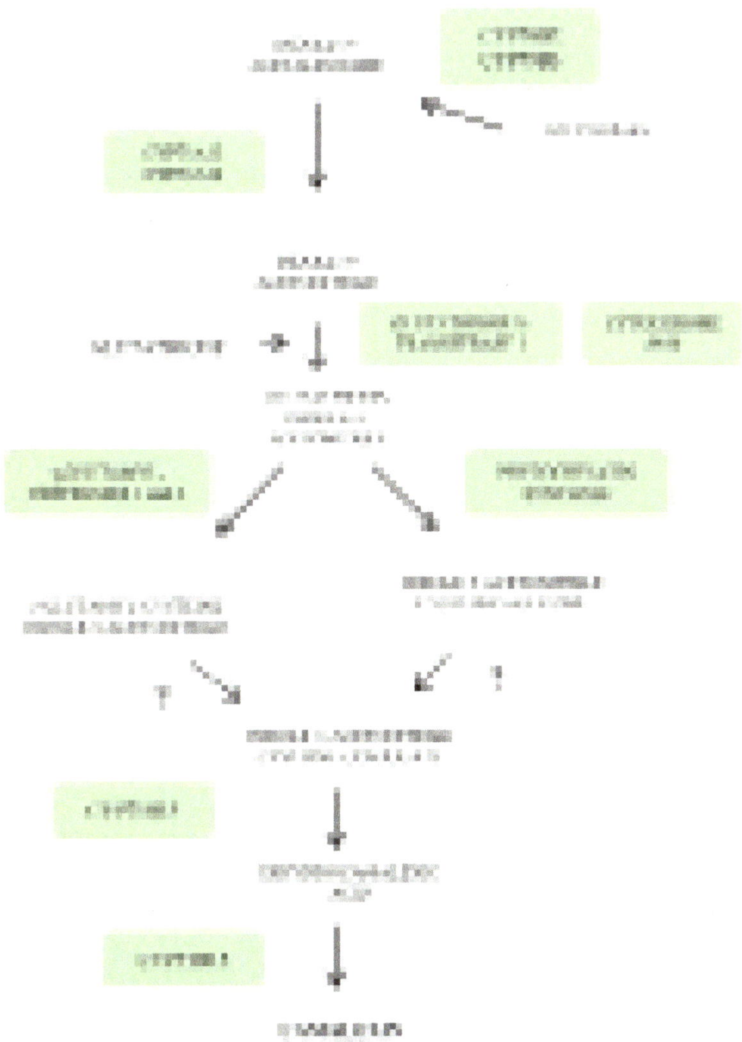

Figure 3: Biosynthetic pathway from tryptophan to camalexin (adapted from [2]).

Besides, regulation of camalexin production in *Arabidopsis* is controlled by either salicylic acid (SA)-independent [44,45] or SA-dependent signaling pathways [46]. Indeed, biosynthesis of this phytoalexin was also found to be lower in SA-induction deficient mutants of *Arabidopsis* with impaired production of ethylene upon bacterial infection by *Pseudomonas syringae* [47].

Other phytohormones have been involved in the regulatory mechanisms of phytoalexin biosynthesis. Auxins and abscisic acid (ABA) generally appear to negatively regulate phytoalexin production [1]. Suppression of auxin signaling has recently been shown to increase the resistance of Arabidopsis to biotrophic pathogens and to redirect phytoalexin metabolism [48]. The biosynthesis of numerous phytoalexins is downregulated by ABA. For example, synthesis of kievitone in bean [49], synthesis of glyceollin in soybean [50,51] and production of rishitin and lubimin in potato [52] are all decreased by ABA. Tobacco mutants deficient in ABA exhibit twice as much capsidiol as wild-type plants [53]. In contrast, cytokinin overexpression was shown to enhance resistance of tobacco to *P. syringae* [54]. This increased resistance correlated well with the up-regulated synthesis of two phytoalexins, capsidiol and scopoletin.

Mitogen-Activated Protein Kinases (MAPKs) have been involved in the induction of camalexin accumulation in Arabidopsis plants upon treatment with Microbe-Associated Molecular Patterns (MAMPs) [12]. Specifically two MAP kinases, MPK3 and MPK6 take part in the up-regulation of numerous enzymes of the camalexin biosynthetic route. For example, expression of the CYP71B15 gene, which encodes the multifunctional enzyme acting at the end of the pathway showed a 400-fold increase upon overexpression of these two MAPKs [12]. In Arabidopsis *mpk3/mpk6* double mutants, camalexin production was completely abolished concomitantly with an increased susceptibility to *B. cinerea*.

Protein phosphorylation-induced phytoalexin production is also under the control of cell calcium transfers which in turn are decoded and transmitted by a toolkit of calcium binding proteins [1]. Several families of calcium sensors are indeed involved in the phytoalexin regulation networks. For instance, overexpression of two genes encoding a calcineurin B-like protein-interacting protein kinase in rice was found to induce two phytoalexin classes, phytocassanes and momilactones upon MAMP treatment [55].

Other regulators of phytoalexin biosynthesis have been identified [1]. Overexpression of Rac proteins in rice induced disease resistance to bacterial blight together with a 19- to 180-fold increase in the accumulation of the rice phytoalexin momilactone A [56]. Production of this phytoalexin is also controlled by selenium-binding protein homologues as shown in the interaction between rice and both the rice blast fungus and the rice bacterial blight [57]. Overexpression of microbial virulence factors belonging to the Nep1-like protein family in *Arabidopsis* was associated with a strong transcriptional activation of genes involved in the camalexin route [58]. Various sugars (sucrose, glucose and fructose) acting as endogenous signals, have been reported to regulate the biosynthesis and accumulation of some phytoalexins [59]. Finally, overexpression of non-expressor of pathogenesis-related genes-1 which play a critical role in the systemic acquired resistance was reported to induce the biosynthesis of the cotton phytoalexin gossypol [60]. Knowledge of the regulatory mechanisms of phytoalexin biosynthesis thus paves the way for metabolic engineering of plants for disease resistance (seeSection 6).

BIOLOGICAL ACTIVITY AGAINST MICROORGANISMS

Are phytoalexins biologically active compounds? Do phytoalexins show antibacterial activities? Over 70 years after their discovery, the actual role of phytoalexins in plant defense mechanisms is still debated. Phytoalexins are considerably less toxic than chemical fungicides. Lack of activity of isoflavonoid phytoalexins was indeed reported in comparison to classic fungicides like benomyl and mancozeb [61]. Effective doses of phytoalexins generally fall within orders of magnitude 10^{-5} to 10^{-4} M [62,63]. Phytoalexin fungitoxicity is clearly evidenced by the inhibition of germ-tube elongation, radial mycelial growth and/or mycelia dry weight increase, as best illustrated by the action of resveratrol on *B. cinerea*, the causal agent for gray mold in grapevine [63,64]. Phytoalexin antifungal activity can considerably vary from one compound to another. For example, Hasegawa *et al.*, show that the rice phytoalexin sakuranetin displays a higher activity against the blast fungus than does another rice phytoalexin, momilactone A, both *in vivo* and *in vitro* [65].

Phytoalexins may also exert some effects on the cytological, morphological and physiological characteristics of fungal cells. The activity of four phytoalexins from the Solanaceae family (rishitin, phytuberin, anhydro-β-rotunol and solavetivone) on three *Phytophthora* species resulted in loss of motility of the zoospores, rounding-up of the cells associated with some level of swelling, cytoplasmic granulation and bursting of the cell membrane [66]. The two latter are very general features of the action of phytoalexins on fungal cells ([63,64,67] and references therein). The extensive membrane damage occurring after fungal exposure to phytoalexins is reflected in substantial leakage of electrolytes and metabolites [68]. However, it has been observed that despite the presence of wyerone acid or resveratrol, surviving *B. cinerea* fungal cells could produce secondary and to a lesser extent tertiary germ tubes suggesting that some sort of escape from phytoalexin damage could take place [63,69]. Asymetric growth of the germ tube resulting in the production of "curved-germ tubes" has also been observed in *B. cinerea* conidia treated with sub-lethal doses of resveratrol [63]. This cytological abnormality suggests that stilbenic compounds may interact with tubulin polymerization, the mode of action of many synthetic fungicides and anticancer agents [70]. Moreover, phytoalexins may affect glucose uptake by fungal cells as reported in the interactions between phaseollin or kievitone/and *Rhizoctonia solani* [68]. Observations of *B. cinerea* conidia showed a complete disorganization of mitochondria and disruption of the plasma membrane upon treatment with the stilbene phytoalexins, resveratrol and pterostilbene [63,64,67]. Pterostilbene especially led to a rapid and complete cessation of respiration in *B. cinerea* conidia which can be explained by its activity as an uncoupling agent of electron transport and phosphorylation [67]. Camalexin has recently been involved in the induction of fungal apoptotic programmed cell death in*B. cinerea* [71].

The efficaciousness *in vivo* of some phytoalexins, namely the coumarin phytoalexin, scopoletin on the reduction of green mold symptoms caused by *Penicillium digitatum* on oranges was shown [72]. In the same way, phenolic phytoalexins (resveratrol, scopoletin, scoparone and umbelliferone) were shown to significantly inhibit the growth of*Penicillium expansum* and patulin accumulation in apples [73]. To increase the fungitoxicity of phytoalexins, design and synthesis of more active phytoalexin derivatives is needed [74,75].

Beside their antifungal activity, phytoalexins possess some antibacterial activity. Rishitin for instance decreased the viability of cells of *Erwinia atroseptica* by around 100% at a dose of 360 µg/L [76]. Resveratrol also exerts some activity against numerous bacteria affecting humans: *Chlamydia*, *Helicobacter*, *Staphylococcus*, *Enterococcus*, *Pseudomonas* and *Neisseria* ([14] and references therein). It is thus clear that phytoalexins exhibit toxicity across much of the biological spectrum, prokaryotic and eukaryotic.

ENGINEERING OF PHYTOALEXINS AND ROLE IN PLANT DEFENSE MECHANISMS

Gain- or loss-of-function genetic approaches addressing phytoalexin production for disease resistance have provided direct and indirect proofs of their implication in plant/microorganism interactions. Relatively simple genetic constructs involving the introduction of a single gene in plants are required in the case of the grapevine phytoalexin resveratrol, synthesis of which is controlled by the stilbene synthase (STS). The first report of increased disease resistance resulting from foreign phytoalexin expression in a novel plant was brought by the group of Kindl with the transfer of two grapevine STS genes (*Vst 1* and *Vst 2*) into tobacco [77]. Introduction of these two genes was shown to confer higher resistance to *B. cinerea*. From that point, a number of transformations were then completed in alfalfa, rice, barley, wheat, tomato, papaya and Arabidopsis using the same STS genes or STS genes from other plant origins, conferring resistance to various pathogens [1]. All these results clearly showed that phytoalexins could act as determinant factors in the expression of the defense mechanisms of plants against phytopathogenic microorganisms though there are rare examples of STS overexpression not being associated with disease resistance [1].

Following the works on stilbene phytoalexins, other genetic transformations were achieved with other phytoalexin genes. Surprisingly, engineering phytoalexins seems to have been limited to exploiting only a few phytoalexin biosynthetic genes. This has mainly concerned the genetic manipulation of phytoalexin glycosylation by the use of a tobacco glucosyltransferase acting on scopoletin [78]. Overexpression of the isoflavonoid-7-*O-methyltransferase* in alfalfa, an enzyme with a crucial role in the biosynthesis of the phytoalexin maiackiain, was also linked to an increased resistance of that plant to *Phoma medicaginis*[79]. Transformation of soybean hairy roots with both the peanut resveratrol synthase 3 *AhRS3* gene and resveratrol-*O-methyltransferase ROMT* gene catalyzing the transformation of resveratrol to pterostilbene [80] resulted in the resistance of that plant to *Rhizoctonia solani* [81]. In many cases, engineering the entire phytoalexin biosynthetic pathway is not feasible and the problem researchers are facing is to choose the right enzyme catalyzing the limitant step of this pathway.

Loss-of-function genetic approaches clearly evidenced the role played by phytoalexins in plant-microorganism interactions. In almost all experiments, mutants impaired in phytoalexin production showed increased susceptibility to pathogens. Reduced amounts of pisatin in hairy roots of pea transformed with antisense 6-α-hydroxymaiackiain-3-*O*-methyltransferase were associated with a decreased resistance to the fungal pathogen *Nectria haematococca* [17]. RNAi silencing of isoflavone synthase or chalcone reductase in soybean suppressed by 90% the accumulation of daidzein and glyceollin as well as disease resistance to *P. sojae* [82]. Loss-of-function alleles of the *yellow seed1* gene encoding CHS, chalcone isomerase, dihydroflavonol reductase and flavonoid-3'-hydroxylase induced deficiency in the accumulation of 3-deoxyanthocyanidin associated with severe symptoms of the anthracnose disease in sorghum [83]. Effect of the*PHYTOALEXIN DEFICIENT* mutation on camalexin in Arabidopsis was found to be dependent on the infecting pathogen. This mutation was not associated with increased susceptibility to *P. syringae*, *Perenospora parasitica*, *Erysiphae oronti* and *B. cinerea* though it markedly affected its susceptibilty to *A. brassicicola* ([1] and references therein). Finally loss-of-function genetic approaches have underlined the role of phytoalexin glycosylation in plant-pathogen interactions. Transgenic tobacco leaves downregulated for a tobacco specific phenylpropanoid glucosyltransferase saw their scopolin content decreased by 70% to 75% associated with a 63% increase in TMV lesion surfaces [84].

Indirect modulation of phytoalexin levels through manipulation of hormone signaling, phosphorylation cascades or defense-related marker genes also demonstrated the role of phytoalexins in plant defense mechanisms. For instance cytokinin overexpression in tobacco led to increased resistance to *P. syringae* which strongly correlated with up-regulated synthesis of two phytoalexins, capsidiol and scopoletin [54]. Mutations in two MAP kinases MPK3 and MPK6 impaired camalexin production and disease resistance to *B. cinerea* in Arabidopsis [12].

Though phytoalexin engineering seems to have been limited to exploiting only a few genes mainly stilbene and isoflavonoid ones, indirect modulation of phytoalexin accumulation employing transcriptional regulators or components of upstream regulatory pathways becomes a useful approach to improve plant disease resistance [1].

FUNGAL METABOLISM AND TRANSPORTERS

The interaction between a plant and its pathogen can be envisaged as a balance between phytoalexin production by the host and phytoalexin metabolism or inactivation via transporters (mainly ATP Binding Cassette, ABC transporters) by the second actor of this interaction. Modification of any of the factors contributing to this balance could modify the outcome of the interaction. We will further see that in phytopathogenic fungi, ABC transporters act as virulence factors, conferring protection against defense compounds produced by the host. In several plant-fungus interactions, it has become evident that the ability to weaken or neutralize the effects of phytoalexins is one of the essential determinants of fungal/host coupling.

Detoxification processes of phytoalexins by fungi are far from being clearly understood [85].

It is rather difficult to derive any comprehensive generalizations from the existing data on phytoalexin metabolism *per se*. The known catabolic pathways of phytoalexins by fungi may involve monoxygenation, reduction, hydration, oxidation, oxidative dimerization, glycosylation and demethylation reactions. Since most phytoalexins are lipophilic compounds that efficiently penetrate cell membrane structures, phytoalexin metabolism usually involves their conversion to more polar products. Creation of new hydroxyl groups by oxygenation, demethylation, reduction of aldehydes and ketones or hydration of double bonds as well as glycosylation increase the degree of polarity of phytoalexins. Detoxification of the cruciferous phytoalexins, brassinin, 1-methoxybrassinin and cyclobrassinin, by the stem rot fungus *Sclerotinia sclerotiorum*indeed requires a brassinin glucosyltransferase [86]. Methylated phytoalexins are well known to be more fungitoxic than the non-methylated ones owing to the fact that phytoalexin methylation enhances their lipophilic character. Moreover the presence of methylated groups or any other electron-attracting groups on the aromatic ring of some phytoalexins plays an important role in the formation of charge transfer complexes, favoring contact and affinity with (membrane) proteins and acting as uncoupling agents of electron transport and photophosphorylation. A cytochrome P450 pisatin demethylase transforming pisatin into 6-α-hydroxymaiackiain and 3-hydroxymaackiain-isoflavan was characterized from the fungal pathogen *Nectria haematococca* [87]. Interestingly, fungal isolates with the highest pisatin demethylating activity were shown to be the most virulent on pea [87]. In addition, overexpression of this pisatin demethylating activity in hairy roots of pea resulted in reduced amounts of this phytoalexin in the plant tissues upon infection by *N. haematococca* with a correlated decreased resistance to this pathogen [17]. Two hydratases, a kievitone hydratase from *Fusarium solani* [88] and one inducible hydrolase of *Leptosphaeria maculans* acting on brassinin [89] were implicated in the detoxification process of these two phytoalexins.

Oxidation and oxidative dimerization processes can also take place in the metabolism of phytoalexins. A brassinin detoxifying oxidase with a molecular mass of 57 kDa has been characterized and purified from the blackleg fungus *L. maculans*. This oxidase transforms the cruciferous phytoalexin brassinin into the less fungitoxic compound indol-3-carboxaldehyde [90]. Interestingly, this pathogen was unable to metabolize camalexin, another major phytoalexin from crucifers, conferring this plant family protection against *L. maculans*. Simple stilbenes produced by members of the Vitaceae family may undergo oxidative dimerization by a laccase-like stilbene oxidase from *B. cinerea* with a molecular mass of 32 kDa [91]. This process involves the 4›-hydroxyphenyl group of one resveratrol unit leading to a dehydrodimer with a dihydrobenzofuran structure named δ-viniferin [92,93]. Importantly, *B. cinerea* isolates possessing the highest oxidative activity were found to be the most virulent on grapevine [94]. Very recently, the stilbene-type phytoalexin astringin produced by Norway spruce in the interaction with the bark beetle (*Ips* spp.) and its fungal associate, *Ceratocystis polonica*was shown to undergo metabolism by the latter [95]. *C. polonica* converted astringin to ring-opened lactones, aglycones and dehydrodimers *in vitro*. In this study, the virulence of the fungal pathogen on Norway spruce correlated well with differential usage of the various pathways for stilbene biotransformation.

Phytopathogenic fungi evolved mechanisms of insensitivity or resistance to protect themselves against phytoalexins. One of them involves extruding toxic compounds out of the cell through transporters, conferring them protection against plant defense products. In fungi, the role of ATP-binding cassette (ABC) transporters in the efflux of natural and synthetic toxicants is well known [96,97,98]. In addition, several genes encoding ABC transporters have been shown to be involved in fungal virulence on plant hosts [99,100,101,102,103,104]. A number of phytoalexins and other toxic compounds induced expression of these fungal transporters [99,100,105,106,107]. However, only a few studies have demonstrated the ability of these transporters to confer tolerance to a known phytoalexin. In *B. cinerea*, the BcatrB ABC transporter has been shown to be a virulence factor that increases tolerance of the pathogen towards phytoalexins. Indeed, *BcatrB* replacement mutants revealed increased sensitivity to resveratrol and reduced virulence on grapevine leaves [100]. Moreover, a *B. cinerea* strain lacking functional BcatrB was more sensitive to camalexin *in vitro* and less virulent on *A. thaliana* wild-type plants, but was fully virulent on camalexin-deficient *A. thaliana* mutants [103]. In the same way, ABC transporters related to *BcatrB*, such as GpABC1 from *Gibberella pulicaris*, act as virulence factors on potato. GpABC1 provides tolerance to rishitin, while a *GpABC1*mutant is essentially non-pathogenic [101]. In *N. haematococca*, the *NhABC1* gene is induced after treatment with pisatin *in vitro* and during infection of pea plants. Mutation in *NhABC1* gene rendered the fungus even more sensitive to pisatin and led to lower pathogenicity on pea, indicating that *NhABC1* contributes to the tolerance to pisatin and acts as a virulence factor [102]. The substrate range of ABC transporters can vary from a single compound to a wide spectrum of molecules with no identified common feature. BcatrB from *B. cinerea* has a wide substrate range, comprising mainly aromatic compounds [100,103,107,108,109] such as the phytoalexins eugenol, resveratrol and camalexin, the fungicides fenpiclonil and fludioxonil, as well as the antibiotics phenazine-1-carboxylic acid and phenazine-1-carboxamide. The closest identified homologue of BcatrB, AtrB from *Aspergillus nidulans*, shows a similar function in multidrug resistance [110]. Two other homologues with similar substrate ranges are MgAtr5 from *Mycosphaerella graminicola* [111] and PMR5 from *Penicillium digitatum* [112], indicating that other homologues of BcatrB may function also in multidrug resistance and pathogenesis. Taken together, it becomes evident that ABC transporters can be essential for the development of phytopathogenic fungi, providing protection against phytoalexins produced by the host plant and acting as virulence factors.

ROLE OF PHYTOALEXINS IN HUMAN HEALTH

Phytoalexins may display health-promoting effects in humans. A few of them have been reported to exert antioxidant, anticarcinogenic and cardiovascular protective activities.

Maslinic acid, a natural phytoalexin-type triterpene from olives exerts a wide range of biological activities as an antitumor, antidiabetic, neuroprotective, cardioprotective, antiparasitic and growth-stimulating agent, providing evidence of the potential of this molecule as a nutraceutical [113]. Health benefit properties of Brassicaceae were attributed in part to their phytoalexins, camalexin and related indolic compounds [114]. Camalexin namely is able to induce apoptosis in prostate cancer cells [115]. 3-deoxyanthocyanidins, flavonoid phytoalexins produced by members of the Poaceae family, are helpful in reducing the incidence of gastro-intestinal cancer [116]. Moreover, other indolic phytoalexins such as brassinin and its derivative, homobrassinin, show marked antiproliferative activities in human colorectal cancer cells *in vitro* [117]. The most promising results in this area were obtained with resveratrol, the phytoalexin from grapevine. This compound is indeed considered to be an antiproliferative agent exerting antitumor activity either as a cytostatic or a cytotoxic agent in various cancers [118]. The first report of the cancer chemoprotective activity of a phytoalexin is the study of the group of Pezzuto [119]. This pioneering work was then confirmed on many other human cancer models. The most frequently described mode of antitumor action for phytoalexins concerns apoptosis which may be via the inhibition of antiapoptotic molecules such as survivin [120] or, for instance as reported in this issue, alterations of expression and activity of lysosomal protease cathepsin D [115].

Resveratrol exerts antitumor activities *in vivo*, namely in skin cancers by topical applications. It does not seem to be very effective in inhibiting leukemia despite displaying antileukemic activity *in vitro* and shows anticancer effects in experimentally-induced breast cancers only at high doses. Resveratrol presents some anticancer activities in hepatoma, lung carcinoma and intestinal tumors ([14] and references therein). However lack of efficacy of natural phytoalexins in reducing tumors has led to a number of investigations regarding the design and synthesis of more potent anticancer derivatives of known phytoalexins such as brassinin [121], methoxybrassinol [122] and resveratrol [123].

There are also several studies providing evidence of the cardioprotective activity of phytoalexins such as indoles and stilbenes [14,114]. Resveratrol namely was proven to inhibit LDL peroxidation in *ex vivo* rat heart studies, to have a potent role in preventing atherosclerosis and to block platelet aggregation from high-cholesterol-fed rabbits ([14] and references therein). Besides, this compound has an effect in neurological diseases such as cerebral ischemia, Parkinson's disease, pain and cognitive impairment in rats, spinal cord lesion in rabbits and finally brain edema and tumors in human cells ([14] and references therein). Interestingly, additional studies have also demonstrated that resveratrol increases lifespan in lower organisms (yeast, metazoans) and higher organisms through the activation of the sirtuin proteins [124,125]. Resveratrol's mechanisms of action are likely to be pleitropic and mediated by the interaction of this compound with key signaling proteins controlling cellular calcium homeostasis [126]. Interestingly, quercetin and umbelliferone were also reported to reduce mycotoxin accumulation in apple fruits by *P. expansum* by down-regulating relative expression of genes encoding patulin biosynthesis [127].

Some other phytoalexins like the steroid glycoalkaloids from potato or the dimeric sesquiterpene gossypol from cotton display a certain level of toxicity for humans explaining the crucial interest in engineering those plants for abolishing production of these undesirable compounds [11,128].

CONCLUDING REMARKS

Works on phytoalexins from diverse chemical families have generated a lot of data regarding basic aspects of plant defenses and their regulatory mechanisms. As a result, engineering of phytoalexins has arisen as a new area in the development of useful approaches to disease control. Nonetheless, while a variety of genetic transfers were carried out in order to investigate the potential of stilbene and flavonoid phytoalexin biosynthetic genes in conferring disease resistance, strategies focusing on the other phytoalexin chemical families did not [1].

Some studies have attempted to determine the actual concentration and the nature of phytoalexins directly in plant tissues in response to invading microorganisms using spectroscopic methods [129,130]. However, our general knowledge remains limited by the difficulty to analyse the events occurring under natural conditions between the plant and the pathogen.

On the other hand, the potential value of several phytoalexins on a therapeutic point of view has made their large-scale production a necessity. Engineering yeast and bacteria, may represent valuable means for the production of phytoalexins at an industrial scale [14,131]. However, their tailoring is needed as they do not possess the genes encoding phytoalexin biosynthesis. Another approach is large-scale production of phytoalexins using plant cell suspensions in bioreactors. Some experiments are underway to optimize stilbene phytoalexin production in bioreactors [132,133].

Although considerable work has already been done on phytoalexins, the ways in which they act against microorganisms and the mechanisms the latter have developed to counteract their action are still poorly understood keeping this subject an active field of research even after over 70 years.

REFERENCES

1. Jeandet, P.; Clément, C.; Courot, E.; Cordelier, S. Modulation of phytoalexin biosynthesis in engineered plants for disease resistance. *Int. J. Mol. Sci.* **2013**, *14*, 14136–14170.
2. Ahuja, I.; Kissen, R.; Bones, A.M. Phytoalexins in defense against pathogens. *Trends Plant Sci.* **2012**, *17*, 73–90.
3. Müller, K.O.; Börger, H. Experimentelle Untersuchungen über die *Phytophthora* Resistenz der Kartoffel. *Arb. Biol. Reichsanst. Land Forstwirtsch.* **1940**, *23*, 189–231.
4. Deverall, B.J. Introduction. In *Phytoalexins*; Bailey, J.A., Mansfield, J.W., Eds.; Blackie: Glasgow/London, UK, 1982; pp. 1–20.
5. Ingham, J.L. Phytoalexins from the Leguminosae. In *Phytoalexins*; Bailey, J.A., Mansfield, J.W., Eds.; Blackie: Glasgow/London, UK, 1982; pp. 21–80.

6. Kuc, J. Phytoalexins from the Solanaceae. In *Phytoalexins*; Bailey, J.A., Mansfield, J.W., Eds.; Blackie: Glasgow/London, UK, 1982; pp. 81–105.
7. Coxon, D.T. Phytoalexins from other plant families. In *Phytoalexins*; Bailey, J.A., Mansfield, J.W., Eds.; Blackie: Glasgow/London, UK, 1982; pp. 106–132.
8. Schmelz, E.A.; Huffaker, A.; Sims, J.W.; Christensen, S.A.; Lu, X.; Okada, K.; Peters, R.J. Biosynthesis, elicitation and roles of monocot terpenoid phytoalexins. *Plant J.* **2014**, *79*, 659–678.
9. Langcake, P.; Pryce, R.J. The production of resveratrol by *Vitis vinifera* and other members of the *Vitaceae* as a response to infection or injury. *Physiol. Plant Pathol.* **1976**, *9*, 77–86.
10. Jeandet, P.; Delaunois, B.; Conreux, A.; Donnez, D.; Nuzzo, V.; Cordelier, S.; Clément, C.; Courot, E. Biosynthesis, metabolism, molecular engineering and biological functions of stilbene phytoalexins in plants. *BioFactors* **2010**, *36*, 331–341.
11. Sunilkumar, G.; Campbell, L.M.; Pukhaber, L.; Stipanovic, R.D.; Rathore, K.S. Engineering cottonseed for use in human nutrition by tissue-specific reduction of toxic gossypol. *Proc. Natl. Acad. Sci. USA* **2006**, *103*, 18054–18059.
12. Ren, D.; Liu, Y.; Yang, K.Y.; Han, L.; Mao, G.; Glazebrook, J.; Zhang, S. A fungal-responsive MAPK cascade regulates phytoalexin biosynthesis in *Arabidopsis*. *Proc. Natl. Acad. Sci. USA* **2008**, *105*, 5638–5643.
13. Delaunois, B.; Cordelier, S.; Conreux, A.; Clément, C.; Jeandet, P. Molecular engineering of resveratrol in plants. *Plant Biotechnol. J.* **2009**, *7*, 2–12.
14. Jeandet, P.; Delaunois, B.; Aziz, A.; Donnez, D.; Vasserot, Y.; Cordelier, S.; Courot, E. Metabolic engineering of yeast and plants for the production of the biologically active hydroxystilbene, resveratrol. *J. Biomed. Biotechnol.* **2012**.
15. Mansfield, J.W.; Bailey, J.A. Phytoalexins: Current problems and future prospects. In *Phytoalexins*; Bailey, J.A., Mansfield, J.W., Eds.; Blackie: Glasgow/London, UK, 1982; pp. 319–323.
16. Hammerschmidt, R. Phytoalexins: What have we learned after 60 years? *Annu. Rev. Phytopathol.* **1999**, *37*, 285–306.
17. Wu, Q.; VanEtten, H.D. Introduction of plant and fungal genes into pea (*Pisum sativum* L.) hairy roots reduces their ability to produce pisatin and affects their response to a fungal pathogen. *Mol. Plant-Microbe Interact.* **2004**, *17*, 798–804.
18. Cruickshank, I.A.M.; Perrin, D.R. Isolation of a phytoalexin from *Pisum sativum* L. *Nature* **1960**, *187*, 799–800.
19. Coxon, D.T.; O'Neill, T.M.; Mansfield, J.W.; Porter, A.E.A. Identification of three hydroxyflavan phytoalexins from daffodil bulbs. *Phytochemistry* **1980**, *19*, 889–891.
20. Browne, L.M.; Conn, K.L.; Ayert, W.A.; Tewari, J.P. The camalexins: New phytoalexins produced in the leaves ofCamelia sativa (Cruciferae). *Tetrahedron* **1991**, *47*, 3909–3914.
21. Pedras, M.S.C.; Okanga, F.I.; Zaharia, I.L.; Khan, A.G. Phytoalexins from crucifers: Synthesis, biosynthesis and biotransformation. *Phytochemistry* **2000**, *53*, 161–176.
22. Geigert, J.; Stermitz, F.R.; Johnson, G.; Maag, D.D.; Johnson, D.K. Two phytoalexins from sugarbeet (*Beta vulgaris*) leaves. *Tetrahedron* **1973**, *29*, 2703–2706.
23. Allen, E.H.; Thomas, C.A. Trans-trans-3,11-tridecadiene-5,7,9-triyne-1,2-diol, an antifungal polyacetylene from diseased safflower (*Carthamus tinctorius*). *Phytochemistry* **1971**, *10*, 1579–1582.
24. Uritani, I.; Uritani, M.; Yamada, H. Similar metabolic alterations induced in sweet potato by poisonous chemicals and by *Ceratostomella fimbriata*. *Phytopathology* **1960**, *50*, 30–34.
25. Sitton, D.; West, C.A. Casbene: An antifungal diterpene produced in cell-free extracts of *Ricinus communis* seedlings.*Phytochemistry* **1975**, *14*, 1921–1925.
26. Poloni, A.; Schirawski, J. Red card for pathogens: Phytoalexins in sorghum and maize. *Molecules* **2014**, *19*, 9114–9133.
27. Lo, S.C.; de Verdier, K.; Nicholson, R. Accumulation of 3-deoxyanthocyanidin phytoalexins and resistance to*Colletotrichum sublineolum* in sorghum. *Physiol. Mol. Plant Pathol.* **1999**, *55*, 263–273.
28. Lin Park, H.; Lee, S.W.; Jung, K.H.; Hahn, T.R.; Cho, M.H. Transcriptomic analysis of UV-treated rice leaves reveals UV-induced phytoalexin biosynthetic pathways and their regulatory networks in rice. *Phytochemistry* **2013**, *96*, 57–71.
29. Keen, N.T.; Littlefield, L.J. The possible association of phytoalexins with resistant gene expression in flax to*Melamspora lini*. *Physiol. Plant Pathol.* **1975**, *14*, 275–280.
30. Takasugi, M.; Nagao, S.; Masamune, T.; Shirata, A.; Takahashi, K. Structures of moracins E, F, G and H, new phytoalexins from diseased mulberry. *Tetrahedron Lett.* **1979**, *28*, 4675–4678.
31. Ward, E.W.B.; Unwin, C.H.; Stoessel, A. Loroglossol: An orchid phytoalexin. *Phytopathology* **1975**, *65*, 632–633.
32. Hartmann, G.; Nienhaus, F. The isolation of xanthoxylin from the bark of *Phytophthora-* and *Hendersonula-* infected*Citrus lemon* and its fungitoxic effect. *Phytopathol. Z.* **1974**, *81*, 97–113.
33. Harding, V.K.; Heale, J.B. The accumulation of inhibitory compounds in the induced resistance response of carrot root slices to *Botrytis cinerea*. *Physiol. Plant Pathol.* **1981**, *18*, 7–15.
34. Johnson, C.; Brannon, D.R.; Kuc, J. Xanthotoxin: A phytoalexin of *Pastinaca sativa* root. *Phytochemistry* **1973**, *12*, 2961–2962.
35. Condon, P.; Kuc, J.; Draudt, H.N. Production of 3-methyl-6-methoxy-8-hydroxy-3,4-dihydroisocoumarin by carrot root tissue. *Phytopathology* **1963**, *53*, 1244–1250.
36. Kokubun, T.; Harborne, J.B. Phytoalexin induction in the sapwood of plants of the Maloideae (Rosaceae):

Biphenyls or dibenzofurans. *Phytochemistry* **1995**, *40*, 1649–1654.

37. Chen, Y.; Tao, Y.; Gao, C.; Cao, W.; Huang, R. Natural products from the genus *Tephrosia*. *Molecules* **2014**, *19*, 1432–1458.
38. Deavours, B.E.; Dixon, R.A. Metabolic engineering of isoflavonoid biosynthesis in alfalfa. *Plant Physiol.* **2005**, *138*, 2245–2259.
39. Kaimoyo, E.; VanEtten, H.D. Inactivation of pea genes by RNAi supports the involvement of two similar *O*-methyltransferases in the biosynthesis of (+)-pisatin and of chiral intermediates with a configuration opposite that found in (+)-pisatin. *Phytochemistry* **2008**, *69*, 76–87.
40. Thomma, B.P.H.J.; Nelissen, I.; Eggermont, K.; Broekaert, W.F. Deficiency in phytoalexin production causes enhanced susceptibilty of *Arabidopsis thaliana* to the fungus *Alternaria brassicicola*. *Plant J.* **1999**, *19*, 163–117.
41. Van Wees, S.C.; Chang, H.S.; Zhu, T.; Glazebrook, J. Characterization of the early response of Arabidopsis to *Alternaria brassicicola* infection using expression profiling. *Plant Physiol.* **2003**, *132*, 606–617.
42. Rowe, H.C.; Walley, J.W.; Corwin, J.; Chan, E.K.F.; Dehesh, K.; Kliebenstein, D.J. Deficiencies in jasmonate-mediated plant defense reveal quantitative variation in *Botrytis cinerea* pathogenesis. *PLoS Pathog.* **2010**, *6*, e1000861.
43. Riemann, M.; Haga, K.; Shimizu, T.; Okada, K.; Ando, S.; Mochizuki, S.; Nishizawa, Y.; Yamanouchi, U.; Nick, P.; Yano, M.; *et al.* Identification of rice *Allene Oxide Cyclase* mutants and the function of jasmonate for defence against*Magnaporthe oryzae*. *Plant J.* **2013**, *74*, 226–238.
44. Nawrath, C.; Métraux, J.P. Salicylic acid induction-deficient mutants of *Arabidopsis* express *PR-2* and *PR-5* and accumulate high levels of camalexin after pathogen inoculation. *Plant Cell* **1999**, *11*, 1393–1404.
45. Roetschi, A.; Si-Ammour, A.; Belbahri, L.; Mauch, F.; Mauch-Mani, B. Characterization of an *Arabidopsis-Phytophthora*pathosystem: Resistance requires a functional PAD2 gene and is independent of salicylic acid, ethylene and jasmonic acid signalling. *Plant J.* **2001**, *28*, 293–305.
46. Denby, K.J.; Jason, L.J.M.; Murray, S.L.; Last, R.L. *ups1*, an *Arabidopsis thaliana* camalexin accumulation mutant defective in multiple defence signalling pathways. *Plant J.* **2005**, *41*, 673–684.
47. Heck, S.; Grau, T.; Buchala, A.; Métraux, J.P.; Nawrath, C. Genetic evidence that expression of NahG modifies defence pathways independent of salicylic acid biosynthesis in the *Arabidopsis-Pseudomonas syringae* pv. tomato interaction.*Plant J.* **2003**, *36*, 342–352.
48. Robert-Seilaniantz, A.; MacLean, D.; Jikumaru, Y.; Hill, L.; Yamaguchi, S.; Kamiya, Y.; Jones, J.D.G. The microRNA miR393 re-directs secondary metabolite biosynthesis away from camalexin and towards glucosinates. *Plant J.* **2011**, *67*, 218–231.
49. Goosens, J.F.V.; Vendrig, J.C. Effects of abscissic acid, cytokinins, and light on isoflavonoid phytoalexin accumulation in *Phaseolus vulgaris*. *Planta* **1982**, *154*, 441–446.
50. Ward, E.W.; Cahill, D.M.; Bhattacharyya, M.K. Abscisic acid suppression of phenylalanine ammonia-lyase activity and mRNA, and resistance of soybeans to *Phytophthora megasperma* f.s.p. *glycinea*. *Plant Physiol.* **1989**, *91*, 23–27.
51. Mohr, P.; Cahill, D.M. Relative roles of glyceollin, lignin and the hypersensitive response and the influence of ABA in compatible and incompatible interactions of soybeans with *Phytophthora sojae*. *Physiol. Mol. Plant Pathol.* **2001**, *58*, 31–41.
52. Henfling, J.W.D.M.; Bostock, R.M.; Kuc, J. Effect of abscisic acid on rishitin and lubimin accumulation and resistance to *Phytophthora infestans* and *Cladosporium cucumerinum* in potato tuber tissue slices. *Phytopathology* **1980**, *70*, 1074–1078.
53. Mialoundama, A.S.; Heintz, D.; Debayle, D.; Rahier, A.; Camara, B.; Bouvier, F. Abscisic acid negatively regulates elicitor-induced synthesis of capsidiol in wild tobacco. *Plant Physiol.* **2009**, *150*, 1556–1566.
54. Grosskinsky, D.K.; Naseem, M.; Abdelmoshem, U.A.; Plickert, N.; Engelke, T.; Griebel, T.; Zeier, J.; Novak, O.; Strand, M.; Pfeifhofer, H.; *et al.* Cytokinins mediate resistance against *Pseudomonas syringae* in tobacco through increased antimicrobial phytoalexin synthesis independent of salicylic acid signaling. *Plant Physiol.* **2011**, *157*, 815–830.
55. Kurusu, T.; Hamada, J.; Nokajima, H.; Kitagawa, Y.; Kiyoduka, M.; Takahashi, A.; Hanamata, S.; Ohno, R.; Hayashi, T.; Okada, K.; *et al.* Regulation of microbe-associated molecular pattern-induced hypersensitive cell death, phytoalexin production, and defense gene expression by calcineurin B-like protein-interacting protein kinases, OsCIPK14/15, in rice cultured cells. *Plant Physiol.* **2010**, *153*, 678–692.
56. Ono, E.; Wong, H.L.; Kawasaki, T.; Hasegawa, M.; Kodama, O.; Shimamoto, K. Essential role of the small GTPase Rac in disease resistance of rice. *Proc. Natl. Acad. Sci. USA* **2001**, *98*, 759–764.
57. Sadawa, K.; Hasegawa, M.; Tokuda, L.; Kameyama, J.; Kodama, O.; Kohchi, T.; Yoshida, K.; Shinmyo, A. Enhanced resistance to blast fungus and bacterial blight in transgenic rice constitutively expressing *OsSBP*, a rice homologue of mamalian selenium-binding proteins. *Biosci. Biotechnol. Biochem.* **2004**, *68*, 873–880.
58. Rauhut, T.; Luberacki, B.; Seitz, H.U.; Glawischnig, E. Inducible expression of a Nep1-like protein serves as a model trigger system of camalexin biosynthesis. *Phytochemistry* **2009**, *70*, 185–189.
59. Formela, M.; Samardakiewicz, S.; Marczak, L.; Nowak, W.; Narozna, D.; Waldemar, B.; Kasprowicz-Maluski, A.; Morkunas, I. Effects of endogenous signals and *Fusarium oxysporum* on the mechanism regulating genistein synthesis and accumulation in yellow lupine and their impact on plant cell cytoskeleton. *Molecules* **2014**, *19*, 13392–13421.

60. Parkhi, V.; Kumar, V.; Campbell, L.M.; Bell, A.A.; Shah, J.; Rathore, K.S. Resistance against various fungal pathogens and reniform nematode in transgenic cotton plants expressing Arabidopsis *NRP1*. *Transgenic Res.* **2010**, *19*, 959–975.
61. Rathmell, W.G.; Smith, D.A. Lack of activity of selected isoflavonoid phytoalexins as protectant fungicides. *Pestic. Sci.* **1980**, *11*, 568–572.
62. Ingham, J.L.; Deverall, B.J.; Kuc, J.; Coxon, D.T.; Stoessl, A.; VanEtten, H.D.; Matthews, D.E.; Smith, D.A. Toxicity of phytoalexins. In *Phytoalexins*; Bailey, J.A., Mansfield, J.W., Eds.; Blackie: Glasgow/London, UK, 1982; pp. 218–252.
63. Adrian, M.; Jeandet, P.; Veneau, J.; Weston, L.A.; Bessis, R. Biological activity of resveratrol, a stilbenic compound from grapevines, against *Botrytis cinerea*, the causal agent for gray mold. *J. Chem. Ecol.* **1997**, *23*, 1689–1702.
64. Adrian, M.; Jeandet, P. Effects of resveratrol on the ultrastructure of *Botrytis cinerea* conidia and biological significance in plant/pathogen interactions. *Fitoterapia* **2012**, *83*, 1345–1350.
65. Hasegawa, M.; Mitsuhara, I.; Seo, S.; Okada, K.; Yamane, H.; Iwai, T.; Ohashi, Y. Analysis on blast fungus-responsive characters of a flavonoid phytoalexin sakuranetin; Accumulation in infected rice leaves, antifungal activity and detoxification by fungus. *Molecules* **2014**, *19*, 11404–11418.
66. Harris, J.E.; Dennis, C. The effect of post-infectional potato tuber metabolites and surfactants on zoospores of Oomycetes. *Physiol. Plant Pathol.* **1977**, *11*, 163–169.
67. Pezet, R.; Pont, V. Ultrastructural observations of pterostilbene fungitoxicity in dormant conidia of*Botrytis cinerea*Pers. *J. Phytopathol.* **1990**, *129*, 29–30.
68. VanEtten, H.D.; Bateman, D.F. Studies on the mode of action of the phytoalexin phaseollin. *Phytopathology* **1971**, *61*, 1363–1372.
69. Rossall, S.; Mansfield, J.W.; Huston, R.A. Death of *Botrytis cinerea* and *B. fabae* following exposure to wyerone derivatives *in vitro* and during infection development in broad bean leaves. *Physiol. Plant Pathol.* **1980**, *16*, 135–146.
70. Woods, J.A.; Hafield, J.A.; Pettit, G.R.; Fox, B.W.; McGown, A.T. The interaction with tubulin of a series of stilbenes based on combretastatin A-4. *Br. J. Cancer* **1995**, *71*, 705–711.
71. Shlezinger, N.; Minz, A.; Gur, Y.; Hatam, I.; Dagdas, Y.F.; Talbot, N.J.; Sharon, A. Anti-apoptotic machinery protects the necrotrophic fungus *Botrytis cinerea* from host-induced apoptotic-like cell death during plant infection. *PLoS Pathog.* **2011**, *7*, e1002185.
72. Sanzani, S.; Schena, L.; Ippolito, A. Effectiveness of phenolic compounds against citrus green mould. *Molecules* **2014**,*19*, 12500–12508.
73. Sanzani, S.M.; de Girolamo, A.; Schena, L.; Solfrizzo, M.; Ippolito, A.; Visconti, A. Control of *Penicillium expansum* and patulin accumulation on apples by quercetin and umbelliferone. *Eur. Food Res. Technol.* **2009**, *228*, 381–389.
74. Pont, V.; Pezet, R. Relation between the chemical structure and biological activity of hydroxystilbenes against *Botrytis cinerea*. *J. Phytopathol.* **1990**, *130*, 1–8.
75. Chalal, M.; Klinguer, A.; Echairi, A.; Meunier, P.; Vervandier-Fasseur, D.; Adrian, M. Antimicrobial activity of resveratrol analogues. *Molecules* **2014**, *19*, 7679–7688.
76. Lyon, G.D.; Bayliss, C.E. The effect of rishitin on *Erwinia carotovora* var. *atroseptica* and other bacteria. *Physiol. Plant Pathol.* **1975**, *6*, 177–186.
77. Hain, R.; Reif, H.J.; Krause, E.; Langebartels, R.; Kindl, H.; Vornam, B.; Wiese, W.; Schmelzer, E.; Schreier, P.; Stöcker, R.; *et al.* Disease resistance results from foreign phytoalexin expression in a novel plant. *Nature* **1993**, *361*, 153–156.
78. Gachon, C.; Baltz, R.; Saindrenan, P. Over-expression of a scopoletin glucosyltransferase in *Nicotiana tabacum* leads to precocious lesion formation during the hypersensitive response to tobacco mosaic virus but does not affect virus resistance. *Plant Mol. Biol.* **2004**, *54*, 137–146.
79. He, X.Z.; Dixon, R.A. Genetic manipulation of isoflavone 7-*O-methyltransferase* enhances biosynthesis of 4'-*O*-methylated isoflavonoid phytoalexins and disease resistance in alfalfa. *Plant Cell* **2000**, *12*, 1689–1702.
80. Schmidlin, L.; Poutaraud, A.; Claudel, P.; Mestre, P.; Prado, E.; Santos-Rosa, M.; Wiedemann-Merdinoglu, S.; Karst, F.; Merdinoglu, D.; Hugueney, P. A stress-inducible resveratrol *O-methyltransferase* involved in the biosynthesis of pterostilbene in grapevine. *Plant Physiol.* **2008**, *148*, 1630–1639.
81. Zernova, O.V.; Lygin, A.V.; Pawlowski, M.L.; Hill, C.B.; Hartman, G.L.; Widholm, J.M.; Lozovaya, V.V. Regulation of plant immunity through modulation of phytoalexin synthesis. *Molecules* **2014**, *19*, 7480–7496.
82. Graham, T.L.; Graham, M.Y.; Subramanian, S.; Yu, O. RNAi silencing of genes for elicitation or biosynthesis of 5-deoxyisoflavonoids suppresses race-specific resistance and hypersensitive cell death in *Phytophthora sojae* infected tissues. *Plant Physiol.* **2007**, *144*, 728–740.
83. Ibraheem, F.; Gaffoor, I.; Chopra, S. Flavonoid phytoalexin-dependent resistance to anthracnose leaf blight requires a functional *yellow seed1* in *Sorghum bicolor*. *Genetics* **2010**, *184*, 915–926.
84. Chong, J.; Baltz, R.; Schmitt, C.; Beffa, R.; Fritig, B.; Saindrenan, P. Downregulation of a pathogen-responsive tobacco UDP-Glc:phenylpropanoid glucosyltransferase reduces scopoletin glucoside accumulation, enhances oxidative stress, and weakens virus resistance. *Plant Cell* **2002**, *14*, 1093–1107.
85. Cho, Y.; Ohm, R.A.; Devappa, R.; Lee, H.B.; Grigoriev, I.V.; Kim, B.Y.; Ahn, J.S. Transcriptional responses of the *bdtf1*-deletion mutant to the phytoalexin brassinin in the necrotrophic fungus *Alternaria brassicicola*. *Molecules* **2014**, *19*, 10717–10732.
86. Pedras, M.S.C.; Ahiahonu, P.W.K.; Hossain, M. Detoxification of the cruciferous phytoalexin brassinin in *Sclerotinia sclerotiorum* requires an inducible

glucosyltransferase. *Phytochemistry* **2004**, *65*, 2685–2694.

87. Delserone, L.M.; McCluskey, K.; Matthews, D.E.; VanEtten, H.D. Pisatin demethylation by fungal pathogens and non pathogens on pea: Association with pisatin tolerance and virulence. *Physiol. Mol. Plant Pathol.* **1999**, *55*, 317–326.
88. Li, D.; Chung, K.R.; Smith, D.A.; Schardl, C.L. The *Fusarium solani* gene encoding kievitone hydratase, a secreted enzyme that catalyzes detoxification of a bean phytoalexin. *Mol. Plant-Microbe Interact.* **1995**, *8*, 388–397.
89. Pedras, M.S.C.; Gadagi, R.S.; Jha, M.; Sarma-Mamillapalle, V.K. Detoxification of the phytoalexin brassinin by isolates of *Leptosphaeria maculans* pathogenic on brown mustard involves an inducible hydrolase. *Phytochemistry* **2007**, *68*, 1572–1578.
90. Pedras, M.S.C.; Minic, Z.; Jha, M. Brassinin oxidase, a fungal detoxifying enzyme to overcome a plant defense—purification, characterization and inhibition. *FEBS J.* **2008**, *275*, 3691–3705.
91. Pezet, R. Purification and characterization of a 32-kDa laccase-like stilbene oxidase produced by *Botrytis cinerea* Pers.: Fr. *FEMS Microbiol. Lett.* **1998**, *167*, 203–208.
92. Breuil, A.C.; Adrian, M.; Pirio, N.; Meunier, P.; Bessis, R.; Jeandet, P. Metabolism of stilbene phytoalexins by *Botrytis cinerea*: Characterization of a resveratrol dehydrodimer. *Tetrahedron Lett.* **1998**, *39*, 537–540.
93. Breuil, A.C.; Jeandet, P.; Chopin, F.; Adrian, M.; Pirio, N.; Meunier, P.; Bessis, R. Characterization of a pterostilbene dehydrodimer produced by laccase of *Botrytis cinerea*. *Phytopathology* **1999**, *89*, 298–302.
94. Sbaghi, M.; Jeandet, P.; Bessis, R.; Leroux, P. Metabolism of stilbene type-phytoalexins in relation to the pathogenicity of *Botrytis cinerea* to grapevines. *Plant Pathol.* **1996**, *45*, 139–144.
95. Hammerbacher, A.; Schmidt, A.; Wadke, N.; Wright, L.P.; Schneider, B.; Bohlmann, J.; Brand, W.A.; Fenning, T.M.; Gershenzon, J.; Paetz, C. A common fungal associate of the spruce bark beetle metabolizes the stilbene defenses of Norway spruce. *Plant Physiol.* **2013**, *162*, 1324–1336.
96. Del Sorbo, G.; Schoonbeek, H.J.; de Waard, M.A. Fungal transporters involved in efflux of natural toxic compounds and fungicides. *Fungal Genet. Biol.* **2000**, *30*, 1–15.
97. De Ward, M.A.; Andrade, A.C.; Hayashi, K.; Schoonbeek, H.J.; Stergiopoulos, I.; Zwiers, L.H. Impact of fungal drug transporters on fungicide sensitivity, multidrug resistance and virulence. *Pest Manag. Sci.* **2006**, *62*, 195–207.
98. Coleman, J.J.; Mylonakis, E. Efflux in fungi: La pièce de résistance. *PLoS Pathog.* **2009**, *5*, e1000486.
99. Urban, M.; Bhargava, T.; Hamer, J.E. An ATP-driven efflux pump is a novel pathogenicity factor in rice blast disease.*EMBO J.* **1999**, *18*, 512–521.
100. Schoonbeek, H.J.; del Sorbo, G.; de Waard, M.A. The ABC transporter BcatrB affects the sensitivity of *Botrytis cinerea* to the phytoalexin resveratrol and the fungicide fenpiclonil. *Mol. Plant-Microbe Interact.* **2001**, *14*, 562–571.
101. Fleissner, A.; Sopalla, C.; Weltring, K.M. An ATP-binding cassette multidrug-resistance transporter is necessary for tolerance of *Gibberella pulicaris* to phytoalexins and virulence on potato tubers. *Mol. Plant-Microbe Interact.* **2002**, *15*, 102–108.
102. Del Sorbo, G.; Ruocco, M.; Schoonbeek, H.J.; Scala, F.; Pane, C.; Vinale, F.; de Waard, M.A. Cloning and functional characterization of BcatrA, a gene encoding an ABC transporter of the plant pathogenic fungus *Botryotinia fuckeliana*(*Botrytis cinerea*). *Mycol. Res.* **2008**, *112*, 737–746.
103. Stefanato, F.L.; Abou-Mansour, E.; Buchala, A.; Kretschmer, M.; Mosbach, A.; Hahn, M.; Bochet, C.G.; Métraux, J.P.; Schoonbeek, H.J. The ABC transporter BcatrB from *Botrytis cinerea* exports camalexin and is a virulence factor on*Arabidopsis thaliana*. *Plant J.* **2009**, *58*, 499–510.
104. Coleman, J.J.; White, G.J.; Rodriguez-Carres, M.; VanEtten, H.D. An ABC transporter and a cytochrome P450 of *Nectria haematococca* MPVI are virulence factors on pea and are the major tolerance mechanisms to the phytoalexin pisatin.*Mol. Plant-Microbe Interact.* **2010**, *24*, 368–376.
105. Zwiers, L.H.; de Waard, M.A. Characterization of the ABC transporter genes *MgAtr1* and *MgAtr2* from the wheat pathogen *Mycosphaerella graminicola*. *Fungal Genet. Biol.* **2000**, *30*, 115–125.
106. Stergiopoulos, I.; van Nistelrooy, J.G.M.; Kema, G.H.J.; de Waard, M.A. Multiple mechanisms account for variation in base-line sensitivity to azole fungicides in field isolates of *Mycosphaerella graminicola*. *Pest Manag. Sci.* **2003**, *59*, 1333–1343.
107. Schoonbeek, H.J.; van Nistelrooy, J.G.M.; de Waard, M.A. Functional analysis of ABC transporter genes from *Botrytis cinerea* identifies BcatrB as a transporter of eugenol. *Eur. J. Plant Pathol.* **2003**, *109*, 1003–1011.
108. Schoonbeek, H.J.; Raaijmakers, J.M.; de Waard, M.A. Fungal ABC transporters and microbial interactions in natural environments. *Mol. Plant-Microbe Interact.* **2002**, *15*, 1165–1172.
109. Vermeulen, T.; Schoonbeek, H.J.; de Waard, M.A. The ABC transporter BcatrB from *Botrytis cinereais* a determinant of the activity of the phenylpyrrole fungicide fludioxonil. *Pest Manag. Sci.* **2001**, *57*, 393–402.
110. Andrade, A.C.; del Sorbo, G.; van Nistelrooy, J.G.M.; de Waard, M.A. The ABC transporter AtrB from *Aspergillus nidulans* mediates resistance to all major classes of fungicides and some natural toxic compounds. *Microbiology* **2000**,*146*, 1987–1997.
111. Zwiers, L.H.; Stergiopoulos, I.; Gielkens, M.M.C.; Goodall, S.D.; de Waard, M.A. ABC transporters of the wheat pathogen *Mycosphaerella graminicola* function as protectants against biotic and xenobiotic toxic compounds. *Mol. Genet. Genomics* **2003**, *269*, 499–507.

112. Nakaune, R.; Hamamoto, H.; Imada, J.; Akutsu, K.; Hibi, T. A novel ABC transporter gene, *PMR5*, is involved in multidrug resistance in the phytopathogenic fungus *Penicillium digitatum*. *Mol. Genet. Genomics* **2002**, *267*, 179–185.

113. Lozano-Mena, G.; Sanchez-Gonzalez, M.; Juan, M.E.; Planas, J.M. Maslinic acid, a natural phytoalexin-type triterpene from olives—A promising nutraceutical? *Molecules* **2014**, *19*, 11538–11559.

114. Jahangir, M.; Kim, H.K.; Choi, Y.H.; Verpoorte, R. Health-affecting compounds in Brassicaceae. *Compr. Rev. Food Sci. Food Saf.* **2009**, *8*, 31–43.

115. Smith, B.; Randle, D.; Mezencev, R.; Thomas, L.; Hinton, C.; Odero-Marah, V. Camalexin-induced apoptosis in prostate cancer cells involves alterations of expression and activity of lysosomal protease cathepsin D. *Molecules* **2014**,*19*, 3988–4005.

116. Yang, L.; Browning, J.D.; Awika, J.M. Sorghum 3-deoxyanthocyanidins possess strong phase II enzyme inducer activity and cancer cell growth inhibition properties. *J. Agric. Food Chem.* **2009**, *57*, 1797–1804.

117. Kello, M.; Drutovic, D.; Chripkova, M.; Pilatova, M.; Budovska, L.; Kulikova, L.; Urdzik, P.; Mojzis, J. ROS-dependent antiproliferative effect of brassinin derivative homobrassinin in human colorectal cancer Caco2 cells. *Molecules* **2014**,*19*, 10877–10897.

118. Athar, M.; Back, J.H.; Tang, X.; Kim, K.H.; Kopelovich, L.; Bickers, D.R.; Kim, A.L. Resveratrol: A review of preclinical studies for human cancer protection. *Toxicol. Appl. Pharmacol.* **2007**, *224*, 274–283.

119. Jang, M.; Cai, L.; Udeani, G.O.; Slowing, K.V.; Thomas, C.F.; Beecher, C.W.; Fong, H.H.; Farnsworth, N.R.; Kinghorn, A.D.; Metha, R.G.; *et al.* Cancer chemopreventive activity of resveratrol, a natural product derived from grapes.*Science* **1997**, *275*, 218–220.

120. Fuda, S.; Debatin, K.M. Sensitization for tumor necrosis factor-related apotosis-inducing ligand-induced apoptosis by the chemopreventive agent resveratrol. *Cancer Res.* **2004**, *64*, 337–346.

121. Budovska, M.; Pilatova, M.; Varinska, L.; Mojzis, J.; Mezncev, R. The synthesis and anticancer activity of analogs of the indole phytoalexins brassinin, 1-methoxyspirobrassinol methyl eher and cyclobrassinin. *Bioorg. Med. Chem.* **2013**, *21*, 6623–6633.

122. Mezencev, R.; Kutschy, P.; Salayova, A.; Updegrove, T.; McDonald, J.F. The design, synthesis and anticancer activity of new nitrogen mustard derivatives of natural indole phytoalexin 1-methoxybrassinol. *Neoplasma* **2009**, *56*, 321–330.

123. Chalal, M.; Delmas, D.; Meunier, P.; Latruffe, N.; Vervandier-Fasseur, D. Inhibition of cancer derived cell lines proliferation by synthesized hydroxylated stilbenes and new ferrocenyl-stilbene analogs. Comparison with resveratrol. *Molecules* **2014**, *19*, 7850–7868.

124. Wood, J.G.; Rogina, B.; Lavu, S.; Howitz, K.; Helfand, S.L.; Tatar, M.; Sinclair, D. Sirtuin activators mimic caloric restriction and delay ageing in metazoans. *Nature* **2004**, *430*, 686–689.

125. Barger, J.L.; Kayo, T.; Vann, J.M.; Arias, E.B.; Wang, J.; Hacker, T.A.; Wang, Y.; Raederstorff, D.; Morrow, J.D.; Leeuwenburgh, C.; *et al.* A Low dose of dietary resveratrol partially mimics caloric restriction and retards aging parameters in mice. *PLoS One* **2008**, *3*, e2264.

126. McCalley, A.E.; Kaja, S.; Payne, A.J.; Koulen, P. resveratrol and calcium signaling: Molecular mechanisms and clinical relevance. *Molecules* **2014**, *19*, 7327–7340.

127. Sanzani, S.M.; Schena, L.; Nigro, F.; de Girolamo, A.; Ippolito, A. Effect of quercetin and umbelliferone on the transcript level of *Penicillium expansum* genes involved in patulin biosynthesis. *Eur. J. Plant Pathol.* **2009**, *125*, 223–233.

128. Matthews, D.; Jones, H.; Gans, P.; Coates, S.; Smith, L.M.J. Toxic secondary metabolite production in genetically modified potatoes in response to stress. *J. Agric. Food Chem.* **2005**, *53*, 7766–7776.

129. Becker, L.; Carré, V.; Poutaraud, A.; Merdinoglu, D.; Chaimbault, P. MALDI mass spectrometry imaging for the simultaneous location of resveratrol, pterostilbene and viniferins on grapevine leaves. *Molecules* **2014**, *19*, 10587–10600.

130. Marti, G.; Schnee, S.; Andrey, Y.; Simoes-Pires, C.; Carrupt, P.A.; Wolfender, J.L.; Gindro, K. Study of leaf metabolome modifications induced by UV-C radiations in representative *Vitis*, *Cissus* and *Cannabis* species by LC-MS based metabolomics and antioxidant assays. *Molecules* **2014**, *19*, 14004–14021.

131. Donnez, D.; Jeandet, P.; Clément, C.; Courot, E. Bioproduction of resveratrol and stilbene derivatives by plant cells and microorganisms. *Trends Biotechnol.* **2009**, *27*, 706–713.

132. Jeandet, P.; Vasserot, Y.; Chastang, T.; Courot, E. Engineering microbial cells for the biosynthesis of natural compounds of pharmaceutical significance. *BioMed Res. Int.* 2013.

133. Jeandet, P.; Clément, C.; Courot, E. Resveratrol production at large scale using plant cell suspensions. *Eng. Life Sci.***2014**.

CHAPTER 11

PHYTOHORMONES ROLE IN INTERACTIONS BETWEEN PLANTS AND THE SOIL-BORNE FUNGUS FUSARIUM OXYSPORUM

Xiaotang Di, Frank L. W. Takken and Nico Tintor

Keywords: Plant immunity, root pathogen, vascular wilt disease, effectors, endophyte

INTRODUCTION

Fusarium oxysporum (*Fo*), one of the most relevant plant pathogens in global agriculture, is a widespread soil-borne fungus that invades roots and causes vascular wilt disease through colonization of xylem tissue (Tjamos and Beckman, 1989). Pathogenic *Fo* strains have been classified in more than 120 *formae speciales* (ff.spp.), which refers to a specific plant host, as a particular isolate typically produces disease only within a limited range of host species (Armstrong and Armstrong, 1981; Katan and Di Primo, 1999; Table 1). The infection process occurs following attachment to the root surface and subsequent penetration and colonization of the plant root and proliferation within the xylem vessels, leading to both local and systemic induction of a broad spectrum of plant defense responses (Berrocal-Lobo and Molina, 2008). Vascular browning, stunting, progressive wilting, and eventually plant death are typical disease symptoms in infected plants (Pietro et al., 2003; Agrios, 2005). In contrast to the potential of pathogenic *Fo* isolates to cause destructive plant diseases, many *Fo* strains are non-pathogenic and survive either saprophytically in the soil, as non-invasive colonizer of the rhizosphere, or as endophyte inside plant tissues (Kuldau and Yates, 2000; Edel-Hermann et al., 2015; Imazaki and Kadota, 2015). There is relatively little known about the lifestyle strategies of these inconspicuous endophytic strains, but some of them have been successfully employed in biocontrol strategies to combat plant diseases (Alabouvette et al., 2009; Vos et al., 2014).

Based on their lifestyle plant pathogenic fungi have been classified as biotrophs and necrotrophs. Biotrophic pathogens derive nutrients from living cells and deploy complex manipulation strategies to exploit their hosts while keeping them alive. In contrast, necrotrophic pathogens generally kill host cells and feed on their contents, resulting in extensive necrosis, tissue maceration, and plant rot (Glazebrook, 2005). A third type, termed hemi-biotrophs, displays both forms of nutrient acquisition, shifting from a biotrophic phase early in infection to necrotrophy at later stages. These pathogens typically produce toxins only at later stages of disease development in order to kill the host cells and to complete their life cycle on dead tissues (Horbach et al., 2011). The strategy of different pathogenic *Fo* strains can vary, but is usually best described by a hemi-biotrophic lifestyle (Michielse and Rep, 2009). Consistently, the *Fo* genomes show an expansion of genes that encode small, secreted proteins as well as cell-wall degrading enzymes, a feature shared by many hemi-biotrophic fungi (Lo Presti et al., 2015). Analysis of the xylem sap proteome from *Fol*-infected tomato plants identified numerous fungal

---

Citation: X, Takken FLW and Tintor N (2016), Phytohormones role in Interactions between Plants and the Soil-Borne Fungus Fusarium oxysporum, Front. Plant Sci. 7:170. doi: 10.3389/fpls.2016.00170

proteins, termed Secreted in Xylem (Six) protein. For several Six proteins a contribution to virulence has been demonstrated, designating them as *bona fide* effectors (Takken and Rep, 2010; de Sain and Rep, 2015). However, their molecular mode of action and putative virulence targets remain unknown.

Table 1: *Fo* strains and their host described in the manuscript.

Abbreviations	*Formae speciales*	Host
Foal	*Fusarium oxysporum* f.sp. *albedinis*	Date palm
Focn	*Fusarium oxysporum* f.sp. *conglutinans*	Cabbage, Arabidopsis
Focb	*Fusarium oxysporum* f.sp. *cubense*	Banana
Fol	*Fusarium oxysporum* f.sp. *lycopersici*	Tomato
Fomt	*Fusarium oxysporum* f.sp. *matthioli*	Garden stock, Arabidopsis
Foph	*Fusarium oxysporum* f.sp. *phaseoli*	Common bean
Forp	*Fusarium oxysporum* f.sp. *raphani*	Radish, Arabidopsis
Forl	*Fusarium oxysporum* f.sp. *radicis-lycopersici*	Tomato

Phytohormones such as SA, JA, and ET, are known to play major roles in regulating plant defense responses against various pathogens. Generally, SA signaling triggers resistance against biotrophic and hemibiotrophic pathogens, whereas a combination of JA and ET signaling activates resistance against necrotrophic pathogens (Glazebrook, 2005). All these hormones are part of a larger signaling network that integrates environmental inputs and provides robustness against microbial manipulations (Katagiri and Tsuda, 2010; Pieterse et al., 2012). Additional hormones such as auxins, abscisic acid (ABA), gibberellic acids (GAs), and brassinosteroids (BRs) have also been reported to be involved in plant immunity and to fine-tune immunity and growth/development in plants (Table 2; Robert-Seilaniantz et al., 2011). In this review, we summarize the current knowledge on the role of phytohormones in plant disease and resistance triggered by different *Fo*ff.spp. to uncover how they shape the outcome of this widespread plant-fungal interaction.

Table 2: Phytohormone mutants involved in the defense response against *Fo*infection

Hormones	Mutants and transgenic lines	Process affected	Plant species	Susceptibility	References
SA	*NahG*	SA accumulation	Arabidopsis	Increased to *Focn* and *Fol*	Berrocal-Lobo and Molina, 2004; Diener and Ausubel, 2005; Thatcher et al., 2009; Trusov et al., 2009
	sid2-1	SA biosynthesis	Arabidopsis	Increased to *Focn* and *Fol*	Berrocal-Lobo and Molina, 2004; Diener and Ausubel, 2005
	eds1-1, eds1-22	SA signaling	Arabidopsis	Unaltered to *Focn* and *Fol*	Berrocal-Lobo and Molina, 2004; Trusov et al., 2009
	eds3, eds4, eds10	SA signaling	Arabidopsis	Increased to *Focn*	Diener and Ausubel, 2005
	eds5-1	SA biosynthesis	Arabidopsis	Increased to *Focn* and *Fol*	Berrocal-Lobo and Molina, 2004; Diener and Ausubel, 2005; Thatcher et al., 2009; Trusov et al., 2009
	pad4-1	SA signaling	Arabidopsis	Unaltered to *Focn* and *Fol*	Berrocal-Lobo and Molina, 2004
	pad4			Increased to *Focn*	Diener and Ausubel, 2005
	npr1-1	SA perception	Arabidopsis	Increased to *Focn* and *Fol*	Berrocal-Lobo and Molina, 2004
	npr1-1, npr1-2, npr1-3, npr1-4	SA perception	Arabidopsis	Unaltered to *Focn*	Diener and Ausubel, 2005; Trusov et al., 2009
	35S::NPR1	SA perception	Tomato	Reduced to *Fol*	Lin et al., 2004
	hpSAMT	SA metabolism	Tomato	Reduced to *Fol*	Ament et al., 2010
JA	*aos, fad3-2, opr3 fad7-1 fad8*	JA biosynthesis	Arabidopsis	Unaltered to *Focn*	Thatcher et al., 2009
	coi1, coi1-21	JA perception	Arabidopsis	Reduced to *Focn* and *Fomt*,	Thatcher et al., 2009; Trusov et al., 2009; Cole et al., 2014
	jar1-1	JA-Ile biosynthesis	Arabidopsis	Increased to *Focn* and *Fol*	Berrocal-Lobo and Molina, 2004; Trusov et al., 2009
	jar1-1			Unaltered to *Focn*	Thatcher et al., 2009
	jin1-9(atmyc2-3), jin1-9/myc2	JA signaling	Arabidopsis	Reduced to *Focn*	Anderson et al., 2004; Trusov et al., 2009
	35S::AtERF2	Positive regulator of MeJA response	Arabidopsis	Reduced to *Focn*	McGrath et al., 2005
	35S::AtERF4	Negative regulator of MeJA response	Arabidopsis	Increased to *Focn*	McGrath et al., 2005
	pft1-1, med8	JA signaling	Arabidopsis	Reduced to *Focn*	Kidd et al., 2009
	def1	JA biosynthesis	Tomato	Increased to *Forl* and *Fol*	Thaler et al., 2004; Kavroulakis et al., 2007
	jai1	JA perception (Coi1 homolog)	Tomato	Unaltered to *Fol*	Cole et al., 2014
ET	*ein2-1*	ET signaling	Arabidopsis	Reduced to *Focn* and *Forp*	Trusov et al., 2009; Cole et al., 2014
	ein2, etr1	ET signaling	Arabidopsis	Unaltered to *Focn*	Thatcher et al., 2009
	ein2-5	ET signaling	Arabidopsis	Increased to *Focn* and *Fol*	Berrocal-Lobo and Molina, 2004
	etr1-1	ET perception	Arabidopsis	Reduced to *Forp*	Pantelides et al., 2013
	35S::ERF1	ET signaling	Arabidopsis	Reduced to *Focn* and *Fol*	Berrocal-Lobo and Molina, 2004
	Never ripe	ET perception	Tomato	Reduced to *Fol*	Lund et al., 1998; Francis et al., 2007
	Never ripe, epinastic (epi1)	ET signaling	Tomato	Unaltered to *Forl*	Kavroulakis et al., 2007
ABA	*aba1-6, aba2-1*	ABA biosynthesis	Arabidopsis	Reduced to *Focn*	Anderson et al., 2004; Trusov et al., 2009
	aba2	ABA biosynthesis	Arabidopsis	Reduced to *Focn*	Cole et al., 2014
Auxin	*cyp79b2 cyp79b3, atr4/sur2, myb51/hig1, atr1, atr2d, pad3*	auxin biosynthesis	Arabidopsis	Unaltered to *Focn*	Kidd et al., 2011
	35S:ATR1/MYB34, atr1d, 35S:ATR2				
	axr1, axr2, axr3, sgt1b	auxin signaling	Arabidopsis	Reduced to *Focn*	Kidd et al., 2011
	tir1	auxin perception	Arabidopsis	Unaltered to *Focn*	Kidd et al., 2011
	arf1, arf2, arf1arf2	auxin signaling	Arabidopsis	Reduced to *Focn*	Lyons et al., 2015

SA PROMOTES RESISTANCE TO F_O

Defense to biotrophic or hemibiotrophic pathogens is frequently mediated via SA signaling (Glazebrook, 2005). Arabidopsis plants with impaired SA accumulation showed increased susceptibility to *Fo* f.sp. *conglutinans* (*Focn*), but not to *Fo* f.sp. *raphani* (*Forp*) pointing to an isolate-specific role of SA-dependent defense responses (Table 2; Berrocal-Lobo and Molina, 2004;Diener and Ausubel, 2005; Trusov et al., 2009; Cole et al., 2014). Interestingly, mutants of the SA master-signaling regulator NPR1 showed wildtype (WT)-like susceptibility to *Focn* when 2–3-week-old soil grown plants were examined (Diener and Ausubel, 2005; Trusov et al., 2009). In contrast, when seedlings were infected on sterile agar plates *npr1* mutants displayed clearly enhanced susceptibility in comparison to WT (Berrocal-Lobo and Molina, 2004). SA accumulation and signaling is also influenced by the nucleo-cytoplasmic proteins PAD4 and EDS1. Soil-grown*pad4*, but not *eds1*, Arabidopsis showed increased susceptibility, whereas sterile grown *pad4*seedlings behaved like WT (Berrocal-Lobo and Molina, 2004; Diener and Ausubel, 2005; Trusov et al., 2009). Thus, the influence of SA signaling regulators depends on growth and inoculation conditions, plant age and the *Fo* isolate used. It appears that at least in soil-grown plants, SA enhances immunity to *Fo* via NPR1- and EDS1-independent pathways.

Consistent with its defense promoting role in Arabidopsis, exogenous application of SA, or synthetic SA analogs reduced *Fo* disease symptoms in a broad range of tested plants including tomato, common bean, date palm, and Arabidopsis (Edgar et al., 2006; Mandal et al., 2009; Dihazi et al., 2011; Xue et al., 2014). Furthermore, stable overexpression of Arabidopsis *NPR1* in tomato reduced disease symptoms as well as *Fo* f.sp. *lycopersici* (*Fol*) colonization of the stem (Lin et al., 2004). Similarly, preventing SA volatilization by silencing of a *Salicylic Acid Methyltransferase*reduced tomato susceptibility to *Fol*, however without significantly changing overall SA levels (Ament et al., 2010). How exactly Methyl-SA levels influence tomato defense to *Fol* remains to be addressed.

Despite a clear effect of SA on disease severity, global transcriptome profiling of Arabidopsis plants inoculated with *Focn* revealed relatively mild changes in the expression of known SA marker genes (Kidd et al., 2011; Zhu et al., 2013; Chen et al., 2014; Lyons et al., 2015). In fact, expression of *PR1*was even slightly down regulated both in roots and shoots of inoculated plants (Kidd et al., 2011). It is possible that activation of SA signaling occurs at rather late stages, which would be missed by the present studies that focus on the time points 1–6 days-post-inoculation (dpi). Alternatively, SA signaling could activate previously uncharacterized defense mechanisms, in line with the observed NPR1- and EDS1-independency, especially in roots that are still little explored in terms of plant immunity (De Coninck et al., 2015). A third possibility has been suggested by Cole et al.: SA signaling could serve to dampen activation of JA responses that are promoting Arabidopsis infection by *Fo* (see below; Cole et al., 2014). Similar to Arabidopsis, transcriptome profiling of banana saplings infected with virulent and avirulent *Fo* f. sp. *cubense* (*Focb*) strains also failed to detect activation of typical SA marker genes, at least at the relatively early time points analyzed (Li et al., 2013). This led to the suggestion that in banana defense against *Fo* is mainly mediated via ET/JA signaling (Swarupa et al., 2014).

Taken together, SA signaling positively regulates defense to *Fo* in most tested plant species, which is in line with a predominantly hemi-biotrophic lifestyle of this pathogen (Figure 1). However, the exact mechanisms by which SA reduces susceptibility to *Fo* are not understood and yet unknown SA targets possibly play a role during defense to this root-infecting pathogen.

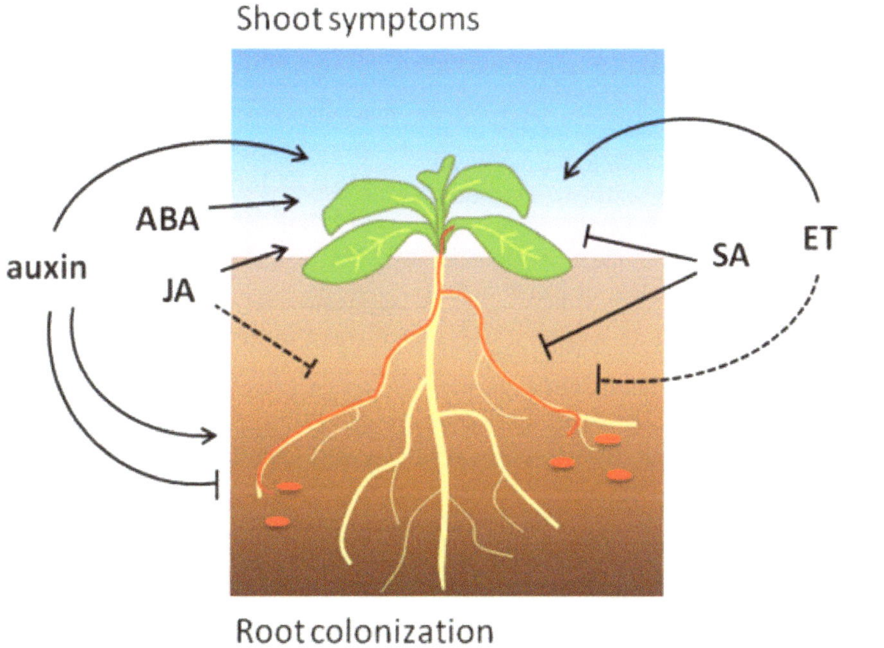

Figure 1: The effects of phytohormone signaling on Arabidopsis root colonization and shoot symptoms induced by *Focn* **(shown in red)**. Arrows and blunt ends indicate promotion and reduction, respectively. The dashed line indicates a presumed positive role of ET-signaling on (root) defense.

JA SIGNALING CAN PROMOTE EITHER RESISTANCE OR SUSCEPTIBILITY IN DIFFERENT HOST–F_O INTERACTIONS

JA signaling generally mediates resistance to necrotrophic pathogens and insect herbivores. These two functions are exerted by two separate and often mutually antagonistic branches: the former is regulated by ERF transcription factors and is associated with ET signaling, whereas the latter requires MYC transcription factors and often involves ABA signaling (Pieterse et al., 2012).

In Arabidopsis JA biosynthesis is not critical for defense to *Focn* as a whole series of mutants with impaired JA accumulation behaved like WT plants (Table 2; Thatcher et al., 2009). Similarly, *jar1*mutants that are defective in synthesis of the bioactive JA–Isoleucine conjugate showed WT-like or slightly increased susceptibility (Thatcher et al., 2009; Trusov et al., 2009). Considering these results it was unexpected that mutations in *CORONATINE INSENSITIVE1* (*COI1*), an essential component of JA perception, strongly increased resistance to *Focn* (Thatcher et al., 2009; Trusov et al., 2009). Importantly, loss-of-function mutations in additional regulators of JA signaling that are involved in pathogen defense, such as the transcriptional regulators MYC2, PFT1, and LBD20, also resulted in increased resistance to *Focn* (Anderson et al., 2004; Kidd et al., 2009; Thatcher et al., 2012b). These findings indicate that JA signaling-capacity of the host is critical for *Fo* pathogenesis in Arabidopsis.

Hijacking the host JA signaling machinery is a common strategy employed by many (hemi-)biotrophic pathogens and often involves suppression of SA-dependent defense responses (Pieterse et al., 2012; Kazan and Lyons, 2014). However, COI1 promotes *Fo* infection in an SA-independent manner as *coi1 NahG* plants remain as resistant as *coi1* plants (Thatcher et al., 2009). This indicates that JA signaling supports *Fo* pathogenesis by mechanisms other than antagonizing SA responses. Grafting experiments demonstrated that expression of *COI1* in roots, but not in shoots, is required for *Fo* infection (Thatcher et al., 2009). Cole and co-workers observed reduced colonization of the vasculature in *coi1* plants pointing to a role of JA signaling at relatively early stages (Cole et al., 2014). Thatcher et al. detected similar levels of fungal biomass in WT and *coi1*shoots before the switch to necrotrophic growth, and hence concluded that *COI1* is predominantly required for triggering plant decay (Thatcher et al., 2009). These differences might be explained by the inoculation methods used: uprooting of plants before inoculation could have created additional entry sites leading to stronger vascular colonization (as observed in Thatcher et al., 2009). Nevertheless, the studies indicate that COI1 influences the interaction with *Fo* at several stages. Moreover, the finding that even strongly colonized *coi1* plants remain essentially symptomless reveals an uncoupling of colonization and plant disease symptoms (Thatcher et al., 2009).

Despite its strong effects on the interaction with *Fo*, it is not well-understood how the host JA signaling machinery is re-wired by the pathogen to promote disease. Cole and co-workers found that two Arabidopsis-infecting strains, *Focn* and f.sp. *matthioli* (*Fomt*) produce JA-isoleucin and JA-leucin conjugates in culture filtrates that induce senescence-like symptoms on Arabidopsis leaves in a COI1-dependent manner (Cole et al., 2014). However, it is not yet known if these fungal derived hormones are also generated during infection, and to what extent they contribute to virulence. Alternatively, *Fo* effectors could play a role in JA signaling manipulation as has been suggested for the effector SIX4 (Thatcher et al., 2012a). It will be interesting to explore which JA signaling regulators are targeted by *Fo* and how their activity is modulated.

The dependency on host JA signaling for successful colonization however, is not conserved among all *Fo* ff.spp. For instance, *Forp* colonizes WT and *coi1* Arabidopsis plants to a similar extent and this isolate does not produce bioactive JA-conjugates *in vitro*. Similarly, *Fol* seems to infect tomato without JA-signaling manipulation as this f.sp. does not produce JAs and is not dependent on the tomato *COI1* homolog to cause disease (Cole et al., 2014). Thus, different *Fo* isolates have developed distinct infection strategies that either strongly rely on host JA signaling manipulation or involve alternative virulence mechanisms that are JA-independent.

In addition, several lines of evidence point to a role of JA in promoting resistance rather than susceptibility in plant species other than Arabidopsis. Tomato *def1* mutants that are defective in JA synthesis show enhanced susceptibility to *Fol*, which can be suppressed by exogenous JA treatment (Thaler et al., 2004). Similarly, *def1* tomato plants were more susceptible to root rot caused by *Fo*f.sp. *radicis-lycopersici* (*Forl*; Kavroulakis et al., 2007). Consistently, Sun and co-workers found that spraying of banana plants with Methyl-JA reduced disease incidence and severity caused by*Focb* Tropical Race 4 (Sun et al., 2013). In addition, resistant cultivars of strawberry and watermelon showed strong induction of the JA biosynthesis enzyme AOC upon inoculation with *Fo*f.sp. *fragariae* (Lu et al., 2011; Fang et al., 2013). A positive role of JA for defense activation was also found in date palms inoculated with *Fo* f. sp. *albenidis* (Jaiti et al., 2009).

In summary, JA signaling promotes defense to *Fo* in different plant species, but can also be hijacked to induce pathogenicity in at least Arabidopsis (Figure 1). Further research will be necessary to reveal via which mechanisms JA contributes to disease reduction or induction, and which function is predominant among different plant species.

DUAL ROLE OF ET IN ACTIVATION OF BOTH DEFENSE RESPONSES AND DISEASE SYMPTOMS

Generally, ET together with JA mediates the resistance response to necrotrophic pathogens. However, ET can also positively influence defense responses to hemi-biotrophs and induced systemic resistance, which is triggered by beneficial root-associated microbes (Pieterse et al., 2014;Broekgaarden et al., 2015).

Pre-treatment of Arabidopsis seedlings with either MeJA or the ET precursor 1-aminocyclopropane-1-carboxylic acid (ACC) resulted in enhanced disease symptoms upon *Focn*inoculation, indicating that both these hormones promote disease development (Trusov et al., 2009). Accordingly, the ET-insensitive Arabidopsis *ein2-1* and *etr1-1* mutants showed a reduction of disease symptoms compared to WT Col plants when inoculated with *Focn* or *Forp*, respectively (Table 2; Trusov et al., 2009; Pantelides et al.,

2013). It is worth mentioning that enhanced disease upon MeJA treatment, as well as reduced disease in ET-insensitive plants was not observed in similar studies, indicating that these effects are either weak or depend on yet unknown factors (Edgar et al., 2006; Thatcher et al., 2009). Moreover, a different *ein2* allele (*ein2-5*) even showed markedly enhanced susceptibility to *Focn* under sterile conditions (Berrocal-Lobo and Molina, 2004). These findings point to an age- and/or condition-dependent role of ET in Arabidopsis interaction with *Fo*.

Global transcriptome profiling in Arabidopsis and banana plants inoculated with virulent *Fo* strains revealed a massive induction of ET/JA-responsive genes such as *Plant Defensins* (*PDFs*) and*Pathogenesis-Related* (*PR*) genes as well as genes encoding ethylene biosynthesis enzymes (McGrath et al., 2005; Kidd et al., 2011; Li et al., 2013; Zhu et al., 2013; Lyons et al., 2015). Furthermore, the transcriptome profiles indicated that initial activation of ET-dependent genes precedes the activation of JA, SA and ABA signaling (Li et al., 2013; Zhu et al., 2013). Altogether, these findings suggest a model in which initial ET/JA-associated defenses are mounted in response to *Fo* infection, but these are typically insufficient to stop the pathogen. At later stages however, ET signaling can rather enhance disease symptoms and possibly also pathogen proliferation. This hypothesis is further supported by the observation that Arabidopsis plants overexpressing certain*ERF* transcription factors, thereby constitutively activating ET/JA-dependent defense responses, become less susceptible to *Focn* (Berrocal-Lobo and Molina, 2004; McGrath et al., 2005). However, whether ET signaling is actively suppressed and/or at later stages co-opted by *Fo* to promote pathogenesis remains to be addressed.

The exact role of ET in other plant species is relatively little understood. The tomato *Never ripe*(*Nr*) mutant is impaired in ethylene perception and shows reduced disease symptoms upon *Fol*inoculation (Lund et al., 1998; Francia et al., 2007). Interestingly, previous work revealed a role for ET in mediating xylem occlusion through formation of gels in castor bean (Vandermolen et al., 1983). Xylem occlusion is thought to limit pathogen spread, but also to contribute to the typical wilting symptoms (Yadeta and Thomma, 2013). Thus, it is an interesting question whether *Nr*tomato plants allow systemic fungal spread and how this would correspond to the observed reduction in disease symptoms. Resistance to *Forl*, a pathogen which adopts a necrotroph-like lifestyle, was largely unaffected in two tested ET-insensitive tomato lines, *Nr* and *epinastic* (*epi*;Kavroulakis et al., 2007). However, protection mediated by an endophytic *Fusarium solani* strain was greatly reduced in *Nr* and *epi* tomato plants and hence required intact ET signaling (Kavroulakis et al., 2007).

In conclusion, the present studies underline a multifaceted role of ET signaling that strongly depends on the interaction stage, the host plant and environmental conditions (Figure 1).

ABA PROMOTES SHOOT DISEASE SYMPTOMS BUT NOT ROOT COLONIZATION IN ARABIDOPSIS

Besides its well-described role in development and abiotic stress responses, ABA has been increasingly recognized as a critical regulator of biotic interactions. ABA can either positively or negatively influence resistance largely depending on the encountered pathogen (Ton et al., 2009;Robert-Seilaniantz et al., 2011).

Similarly to Methyl-JA and ACC, exogenous treatment with ABA increased Arabidopsis susceptibility to *Focn* (Trusov et al., 2009). Consistently, Arabidopsis mutants in which either ABA biosynthesis or signaling is disrupted showed fewer symptoms (Table 2; Anderson et al., 2004;Trusov et al., 2009). The reduced susceptibility of ABA mutants was associated with hyper-activation of ET/JA-dependent defense genes, likely due to antagonistic interactions between ABA and ET signaling (Anderson et al., 2004). ABA could also antagonize SA-dependent responses (Yasuda et al., 2008), but currently it is unknown if the interaction with other hormones explains reduced *Fo* symptoms in ABA-deficient mutants.

Interestingly, *Fo* successfully colonized the roots of ABA-deficient mutants to a similar extent as those from WT plants (Cole et al., 2014). This would point to a role of ABA during the switch to the necrotrophic phase. However, transcriptome profiling revealed activation of numerous ABA responsive genes in the roots of *Fo*-inoculated plants (Lyons et al., 2015). Previous studies indicated that ABA mediates root-to-shoot defense signaling in plants (Balmer et al., 2013). This raises the possibility that *Fo* co-opts systemic ABA signaling to manipulate root-shoot signaling. Moreover, it has been shown that, for example during defense against herbivorous insects, ABA signaling can serve to activate or enhance the MYC2-regulated branch of JA signaling (Kazan and Manners, 2013; Vos et al., 2013). However, if and how exactly *Fo* manipulates ABA signaling, is currently unknown.

AUXINS AFFECTS BOTH ROOT COLONIZATION AND SHOOT SYMPTOM DEVELOPMENT

Auxins are major regulators of plant growth and development, but have also profound effects on interactions with both pathogenic and mutualistic microbes (Robert-Seilaniantz et al., 2011;Zamioudis et al., 2013).

Exogenous application of auxin or auxin biosynthesis inhibitors did not affect disease development in *Focn*-inoculated Arabidopsis (Kidd et al., 2011). Similarly, mutants with either reduced or increased auxin levels behaved like WT plants. In contrast, *Focn*-inoculated auxin-signaling mutants showed markedly reduced symptoms relative to WT plants (Table 2). Additionally, alteration of polar auxin transport, either by chemical inhibitors or in mutants, resulted in increased resistance to *Focn* (Kidd et al., 2011). These data indicate that local changes of auxin levels and/or distribution are important for disease susceptibility. Indeed, histological visualization of *DR5* expression, a well-known auxin reporter gene, revealed activation of auxin signaling at root tips and lateral root initials, two preferred *Fo* entry sites in Arabidopsis (Czymmek et al., 2007;Kidd et al., 2011; Diener, 2012). Additionally, Diener revealed that fewer root tips are colonized in plants mutated in the auxin efflux carrier *PIN2/EIR1* (Diener, 2012). In contrast, *tir3* mutants which are defective in polar auxin transport show 2–3 fold higher *Fo* biomass in roots, but disease symptoms of the shoot remained strongly reduced (Kidd et al., 2011; Diener, 2012). These findings suggest that auxin signaling and transport affect several stages of the *Fo*–Arabidopsis interaction from initial root tip colonization to disease symptom expression in the shoot. However, the mechanisms by which auxin promotes colonization and symptom development are still enigmatic. Previous studies describe an antagonistic relationship between auxin and SA signaling, however the*Fo* disease phenotypes of auxin signaling mutants were not SA-dependent (Kidd et al., 2011).

Auxin accumulation, transport, and signaling are modulated by numerous different symbiotic and pathogenic organisms including bacteria, fungi, nematodes, and even parasitic plants during their interaction with roots (Grunewald et al., 2009; Kazan and Manners, 2009; Zamioudis and Pieterse, 2012). This suggests that manipulation of the host auxin signaling pathway represents a common strategy employed by diverse root colonizers resulting in either detrimental or beneficial effects for plants.

ARE PHYTOHORMONES DETERMINANTS OF F_O LIFESTYLE?

Interestingly, changes in the phytohormone network can uncouple colonization by *Fo* from plant disease development. For instance, specific mutants with impaired JA, ABA, and auxin signaling still allow extensive root (and sometimes shoot) colonization but have greatly reduced disease symptoms (Thatcher et al., 2009; Diener, 2012; Cole et al., 2014). Similarly, resistant tomato plants that are completely free of symptoms can have their shoots and stems extensively colonized by *Fol*(Mes et al., 1999). Furthermore, a *Fol* knockout strain lacking the Six6 effector triggered vascular browning in a susceptible tomato cultivar, indicative of successful xylem colonization, but exerted almost no negative effects on plant growth and development (Gawehns et al., 2014). Altogether, these findings suggest that manipulation of plant hormone signaling rather than colonization triggers disease symptom development.

How does *Fo* manipulate the plant hormone network? One mechanism could be the production by the fungus of hormone-like secondary metabolites, including JAs, auxins, gibberellic acids, and ethylene (Hasan, 2002; Cole et al., 2014; Bitas et al., 2015). *Fo* also secretes numerous small proteins during plant infection, which might be another means to manipulate the host. For several of these proteins a virulence-promoting function has been shown designating them as effectors*sensu strictu* (Takken and Rep, 2010; de Sain and Rep, 2015). Among these, SIX4 was found to enhance JA signaling during infection of Arabidopsis (Thatcher et al., 2012a). Infection of tomato plants with *Fol* knockout strains lacking specific effectors revealed common and unique effects on the xylem proteome composition raising the possibility that each effector targets a distinct hormone signaling pathway (Gawehns et al., 2015). However, for the vast majority of *Fo* effectors their working mechanism remains unknown. It will be interesting to explore if plant hormone-synthesis or -signaling represents a recurrent virulence target of *Fo* strains on various hosts. Furthermore, it is tempting to speculate that at least a subset of the proteinacious effectors mediate immune suppression enabling (endophytic) colonization during the biotrophic phase of infection, whereas secondary metabolites with hormonal- or toxic-activity trigger plant damage/death during necrotrophic growth.

A growing body of evidence suggests that the majority of *Fo* strains survive in soil, in the rhizosphere or within plant tissues without causing disease symptoms, and some strains even confer extensive beneficial effects (Alabouvette et al., 2009; Edel-Hermann et al., 2015; Imazaki and Kadota, 2015). The existence of such a widespread and intimate co-habitation points to a finely balanced interaction between plant and fungus. The finding that also pathogenic strains can reside inside plant tissues without damaging the host indicate that one *Fo* isolate can employ diverse interaction/colonization strategies whose outcome possibly depends on the "compatibility" of a putative host plant. Thus, the "infection tools" of a *Fo* strain, likely comprised of a combination of effectors, enzymes, and secondary metabolites, determine the outcome of an interaction: either endophytic with potential beneficial effects for the host or pathogenic with various levels of disease and in extreme cases plant death. This idea is supported by the observation that transfer of one specific *Fol* chromosome, which contains most of its effector genes plus a secondary metabolite cluster, conferred pathogenicity to an endophytic strain (Ma et al., 2010).

Clearly more research in the areas of genomics and effector biology is required to understand how*Fo* manages to trick the hormonal network of its hosts, and how these interactions can have opposite outcomes ranging from pathogenesis to mutualism.

CONCLUSIONS AND OUTLOOK

Phytohormones determine colonization and disease symptom development during interactions with pathogenic *Fo* strains. However, their roles vary depending on the host plant and the fungal strain involved, suggesting that the manipulation of the host hormonal network differs between individual *Fo* strains. Genetic interference with hormone regulators mostly reduced disease symptoms, as seen for JA, ABA, and auxin, indicating that the ability to hijack plant hormone pathways is a requirement for pathogenesis. This scenario implies a strong adaptation to a particular host plant, potentially leading to the narrow host range observed in the *Fo* species complex. How exactly the manipulation of phytohormone signaling differs between *Fo* strains—and if this is indeed the key difference between pathogenic and endophytic interactions—remains an intriguing question for future research. Comparison of the respective effector repertoires as well as a better understanding of their mode of action will help answering these questions and may furthermore reveal novel approaches for plant protection, either by breeding or by optimizing *Fo*strains for biocontrol.

CONFLICT OF INTEREST STATEMENT

The authors declare that the research was conducted in the absence of any commercial or financial relationships that could be construed as a potential conflict of interest.

ACKNOWLEDGMENTS

We apologize to our colleagues whose work is not cited due to space limitation. The authors wish to thank Prof. Ben Cornelissen and Dr. Martijn Rep for critically reading and commenting on the manuscript. XD was supported by funding from the program of China Scholarship Council. FT received funding from the European Union's Horizon 2020 research and innovation programme under the Marie Skodowska-Curie grant agreement No. 676480 (Bestpass) and from the NWO-Earth and Life Sciences funded VICI project No. 865.14.003. NT was financed from a NWO VICI grant (project No. 865.10.002) granted to Dr. Martijn Rep.

REFERENCES

1. Agrios, G. N. (2005). *Plant Pathology*. St. Louis, MO: Academic Press.
2. Alabouvette, C., Olivain, C., Migheli, Q., and Steinberg, C. (2009). Microbiological control of soil-borne phytopathogenic fungi with special emphasis on wilt-inducing *Fusarium oxysporum. New phytol.* 184, 529–544. doi: 10.1111/j.1469-8137.2009.03014.x
3. Ament, K., Krasikov, V., Allmann, S., Rep, M., Takken, F. L., and Schuurink, R. C. (2010). Methyl salicylate production in tomato affects biotic interactions. *Plant J.* 62, 124–134. doi: 10.1111/j.1365-313X.2010.04132.x
4. Anderson, J. P., Badruzsaufari, E., Schenk, P. M., Manners, J. M., Desmond, O. J., Ehlert, C., et al. (2004). Antagonistic interaction between abscisic acid and jasmonate-ethylene signaling pathways modulates defense gene expression and disease resistance in Arabidopsis. *Plant Cell* 16, 3460–3479. doi: 10.1105/tpc.104.025833
5. Armstrong, G. M., and Armstrong, J. K. (1981). "Formae speciales and races of *Fusarium oxysporum* causing wilt diseases," in*Fusarium, Diseases, Biology and Taxonomy*, eds P. E. Nelson, T. A. Toussoun, and R. J. Cook (University park, TX: Pennsylvania State University Press), 391–399.
6. Balmer, D., De Papajewski, D. V., Planchamp, C., Glauser, G., and Mauch-Mani, B. (2013). Induced resistance in maize is based on organ-specific defence responses. *Plant J.* 74, 213–225. doi: 10.1111/tpj.12114
7. Berrocal-Lobo, M., and Molina, A. (2004). Ethylene response factor 1 mediates Arabidopsis resistance to the soilborne fungus*Fusarium oxysporum. Mol. Plant Microb. Interact.* 17, 763–770. doi: 10.1094/MPMI.2004.17.7.763
8. Berrocal-Lobo, M., and Molina, A. (2008). Arabidopsis defense response against *Fusarium oxysporum. Trends PlantSci.* 13, 145–150. doi: 10.1016/j.tplants.2007.12.004
9. Bitas, V., McCartney, N., Li, N., Demers, J., Kim, J. E., Kim, H. S., et al. (2015). Volatiles enhance plant growth via affecting Auxin transport and signaling. *Front. Microbiol.* 6:1248. doi: 10.3389/fmicb.2015.01248
10. Broekgaarden, C., Caarls, L., Vos, I. A., Pieterse, C. M., and Van Wees, S. C. (2015). Ethylene: traffic controller on hormonal crossroads to defense. *Plant Physiol.* 169, 2371–2379. doi: 10.1104/pp.15.01020
11. Chen, Y. C., Wong, C. L., Muzzi, F., Vlaardingerbroek, I., Kidd, B. N., and Schenk, P. M. (2014). Root defense analysis against*Fusarium oxysporum* reveals new regulators to confer resistance. *Sci. Rep.* 4:5584. doi: 10.1038/srep05584
12. Cole, S. J., Yoon, A. J., Faull, K. F., and Diener, A. C. (2014). Host perception of jasmonates promotes infection by *Fusarium oxysporum* formae speciales that produce isoleucine- and leucine-conjugated jasmonates. *Mol. Plant Pathol.* 15, 589–600. doi: 10.1111/mpp.12117
13. Czymmek, K. J., Fogg, M., Powell, D. H., Sweigard, J., Park, S. Y., and Kang, S. (2007). *In vivo* time-lapse documentation using confocal and multi-photon microscopy reveals the mechanisms of invasion into the Arabidopsis root vascular system by*Fusarium oxysporum. Fungal Genet. Biol.* 44, 1011–1023. doi: 10.1016/j.fgb.2007.01.012
14. De Coninck, B., Timmermans, P., Vos, C., Cammue, B. P. A., and Kazan, K. (2015). What lies beneath: belowground defense strategies in plants. *Trends Plant Sci.* 20, 91–101. doi: 10.1016/j.tplants.2014.09.007
15. de Sain, M., and Rep, M. (2015). The role of pathogen-secreted proteins in fungal vascular wilt diseases. *Int. J. Mol. Sci.* 16, 23970–23993. doi: 10.3390/ijms161023970
16. Diener, A. (2012). Visualizing and quantifying *Fusarium oxysporum* in the plant host. *Mol. Plant Microbe Interact.* 25, 1531–1541. doi: 10.1094/MPMI-02-12-0042-TA
17. Diener, A. C., and Ausubel, F. M. (2005). Resistance to *Fusarium oxysporum* 1, a dominant Arabidopsis disease-resistance gene, is not race specific. *Genetics* 171, 305–321. doi: 10.1534/genetics.105.042218
18. Dihazi, A., Serghini, M. A., Jaiti, F., Daayf, F., Driouich, A., Dihazi, H., et al. (2011). Structural and biochemical changes in salicylic-acid-treated date palm roots challenged with *Fusarium oxysporum* f. sp. *albedinis. J. Pathog.* 2011:280481. doi: 10.4061/2011/280481
19. Edel-Hermann, V., Gautheron, N., Mounier, A., and Steinberg, C. (2015). *Fusarium* diversity in soil using a specific molecular approach and a cultural approach. *J. Microbiol. Methods* 111, 64–71. doi: 10.1016/j.mimet.2015.01.026
20. Edgar, C. I., McGrath, K. C., Dombrecht, B., Manners, J. M., Maclean, D. C., Schenk, P. M., et al. (2006). Salicylic acid mediates resistance to the vascular wilt pathogen *Fusarium oxysporum* in the model host *Arabidopsis thaliana. Australas. Plant Pathol.*35, 581–591. doi: 10.1071/AP06060
21. Fang, X. L., Jost, R., Finnegan, P. M., and Barbetti, M. J. (2013). Comparative proteome analysis of the strawberry-*Fusarium oxysporum* f. sp *fragariae* pathosystem reveals early activation of defense responses as a crucial determinant of host resistance.*J. Proteome Res.* 12, 1772–1788. doi: 10.1021/pr301117a
22. Francia, D., Demaria, D., Calderini, O., Ferraris, L., Valentino, D., Arcioni, S., et al. (2007). Wounding induces resistance to pathogens with different lifestyles in tomato: role of ethylene in cross-protection. *Plant Cell Environ.* 30, 1357–1365. doi: 10.1111/j.1365-3040.2007.01709.x
23. Gawehns, F., Houterman, P. M., Ichou, F. A., Michielse, C. B., Hijdra, M., Cornelissen, B. J. C., et al. (2014). The *Fusarium oxysporum* effector Six6 contributes to virulence and suppresses I-2-mediated cell death. *Mol.*

Plant Microb. Interact. 27, 336–348. doi: 10.1094/MPMI-11-13-0330-R

24. Gawehns, F., Ma, L. S., Bruning, O., Houterman, P. M., Boeren, S., Cornelissen, B. J. C., et al. (2015). The effector repertoire of*Fusarium oxysporum* determines the tomato xylem proteome composition following infection. *Front. Plant Sci.* 6:967. doi: 10.3389/fpls.2015.00967

25. Glazebrook, J. (2005). Contrasting mechanisms of defense against biotrophic and necrotrophic pathogens. *Annu. Rev. Phytopathol.* 43, 205–227. doi: 10.1146/annurev.phyto.43.040204.135923

26. Grunewald, W., Van Noorden, G., Van Isterdael, G., Beeckman, T., Gheysen, G., and Mathesius, U. (2009). Manipulation of Auxin transport in plant roots during rhizobium symbiosis and nematode parasitism. *Plant Cell* 21, 2553–2562. doi: 10.1105/tpc.109.069617

27. Hasan, H. A. H. (2002). Gibberellin and auxin production by plant root-fungi and their biosynthesis under salinity-calcium interaction. *Rost. Vyroba* 48, 101–106. doi: 10.1556/amicr.49.2002.1.11

28. Horbach, R., Navarro-Quesada, A. R., Knogge, W., and Deising, H. B. (2011). When and how to kill a plant cell: infection strategies of plant pathogenic fungi. *J. Plant Physiol.* 168, 51–62. doi: 10.1016/j.jplph.2010.06.014

29. Imazaki, I., and Kadota, I. (2015). Molecular phylogeny and diversity of *Fusarium* endophytes isolated from tomato stems.*FEMS Microbiol. Ecol.* 91:fiv098. doi: 10.1093/femsec/fiv098

30. Jaiti, F., Verdeil, J. L., and El Hadrami, I. (2009). Effect of jasmonic acid on the induction of polyphenoloxidase and peroxidase activities in relation to date palm resistance against *Fusarium oxysporum* f. sp *albedinis*. *Physiol. Mol. Plant Pathol.* 74, 84–90. doi: 10.1016/j.pmpp.2009.09.005

31. Katagiri, F., and Tsuda, K. (2010). Understanding the plant immune system. *Mol. Plant Microb. Interact.* 23, 1531–1536. doi: 10.1094/MPMI-04-10-0099

32. Katan, T., and Di Primo, P. (1999). Current status of vegetative compatibility groups in *Fusarium oxysporum. Phytoparasitica*27(Suppl.), 273–277. doi: 10.1007/BF02981483

33. Kavroulakis, N., Ntougias, S., Zervakis, G. I., Ehaliotis, C., Haralampidis, K., and Papadopoulou, K. K. (2007). Role of ethylene in the protection of tomato plants against soil-borne fungal pathogens conferred by an endophytic *Fusarium solani* strain. *J. Exp. Bot.* 58, 3853–3864. doi: 10.1093/jxb/erm230

34. Kazan, K., and Lyons, R. (2014). Intervention of phytohormone pathways by pathogen effectors. *Plant Cell* 26, 2285–2309. doi: 10.1105/tpc.114.125419

35. Kazan, K., and Manners, J. M. (2009). Linking development to defense: auxin in plant-pathogen interactions. *Trends Plant Sci.*14, 373–382. doi: 10.1016/j.tplants.2009.04.005

36. Kazan, K., and Manners, J. M. (2013). Myc2: the master in action. *Mol. Plant* 6, 686–703. doi: 10.1093/mp/sss128

37. Kidd, B. N., Edgar, C. I., Kumar, K. K., Aitken, E. A., Schenk, P. M., Manners, J. M., et al. (2009). The mediator complex subunit PFT1 is a key regulator of jasmonate-dependent defense in Arabidopsis. *Plant Cell* 21, 2237–2252. doi: 10.1105/tpc.109.066910

38. Kidd, B. N., Kadoo, N. Y., Dombrecht, B., Tekeoglu, M., Gardiner, D. M., Thatcher, L. F., et al. (2011). Auxin signaling and transport promote susceptibility to the root-infecting fungal pathogen *Fusarium oxysporum* in Arabidopsis. *Mol. Plant Microb. Interact.* 24, 733–748. doi: 10.1094/MPMI-08-10-0194

39. Kuldau, G., and Yates, I. (2000). "Evidence for *Fusarium* endophytes in cultivated and wild plants,"in *Microbial Endophytes*, eds C. W. Bacon and J. F. White (New York, NY: Marcel Dekker), 85–117.

40. Li, C. Q., Shao, J. F., Wang, Y. J., Li, W. B., Guo, D. J., Yan, B., et al. (2013). Analysis of banana transcriptome and global gene expression profiles in banana roots in response to infection by race 1 and tropical race 4 of *Fusarium oxysporum* f. sp *cubense.BMC Genomics* 14:851. doi: 10.1186/1471-2164-14-851

41. Lin, W. C., Lu, C. F., Wu, J. W., Cheng, M. L., Lin, Y. M., Yang, N. S., et al. (2004). Transgenic tomato plants expressing the Arabidopsis NPR1 gene display enhanced resistance to a spectrum of fungal and bacterial diseases. *Transgenic Res.* 13, 567–581. doi: 10.1007/s11248-004-2375-9

42. Lo Presti, L., Lanver, D., Schweizer, G., Tanaka, S., Liang, L., Tollot, M., et al. (2015). Fungal effectors and plant susceptibility.*Annu. Rev. Plant. Biol.* 66, 513–545. doi: 10.1146/annurev-arplant-043014-114623

43. Lu, G. Y., Guo, S. G., Zhang, H. Y., Geng, L. H., Song, F. M., Fei, Z. J., et al. (2011). Transcriptional profiling of watermelon during its incompatible interaction with *Fusarium oxysporum* f. sp *niveum*. *Eur. J. Plant Pathol.* 131, 585–601. doi: 10.1007/s10658-011-9833-z

44. Lund, S. T., Stall, R. E., and Klee, H. J. (1998). Ethylene regulates the susceptible response to pathogen infection in tomato.*Plant Cell* 10, 371–382. doi: 10.1105/tpc.10.3.371

45. Lyons, R., Stiller, J., Powell, J., Rusu, A., Manners, J. M., and Kazan, K. (2015). *Fusarium oxysporum* triggers tissue-specific transcriptional reprogramming in *Arabidopsis thaliana*. *PLoS ONE* 10:e0121902. doi: 10.1371/journal.pone.0121902

46. Ma, L. J., Van Der Does, H. C., Borkovich, K. A., Coleman, J. J., Daboussi, M. J., Di Pietro, A., et al. (2010). Comparative genomics reveals mobile pathogenicity chromosomes in *Fusarium*. *Nature* 464, 367–373. doi: 10.1038/nature08850

47. Mandal, S., Mallick, N., and Mitra, A. (2009). Salicylic acid-induced resistance to *Fusarium oxysporum* f. sp *lycopersici* in tomato. *Plant Physiol. Biochem.* 47, 642–649. doi: 10.1016/j.plaphy.2009.03.001

48. McGrath, K. C., Dombrecht, B., Manners, J. M., Schenk, P. M., Edgar, C. I., Maclean, D. J., et al. (2005). Repressor- and activator-type ethylene response factors functioning in jasmonate signaling and disease resistance identified via a genome-wide screen of Arabidopsis transcription

factor gene expression. *Plant Physiol.* 139, 949–959. doi: 10.1104/pp.105.068544

49. Mes, J. J., Weststeijn, E. A., Herlaar, F., Lambalk, J. J., Wijbrandi, J., Haring, M. A., et al. (1999). Biological and molecular characterization of *Fusarium oxysporum* f. sp *lycopersici* divides race 1 isolates into separate virulence groups. *Phytopathology*89, 156–160. doi: 10.1094/PHYTO.1999.89.2.156

50. Michielse, C. B., and Rep, M. (2009). Pathogen profile update: *Fusarium oxysporum*. *Mol. Plant Pathol.* 10, 311–324. doi: 10.1111/j.1364-3703.2009.00538.x

51. Pantelides, I. S., Tjamos, S. E., Pappa, S., Kargakis, M., and Paplomatas, E. J. (2013). The ethylene receptor ETR1 is required for*Fusarium oxysporum* pathogenicity. *Plant Pathol.* 62, 1302–1309. doi: 10.1111/ppa.12042

52. Pieterse, C. M., Van der Does, D., Zamioudis, C., Leon-Reyes, A., and Van Wees, S. C. (2012). Hormonal modulation of plant immunity. *Annu. Rev. Cell Dev. Biol.* 28, 489–521. doi: 10.1146/annurev-cellbio-092910-154055

53. Pieterse, C. M., Zamioudis, C., Berendsen, R. L., Weller, D. M., Van Wees, S. C., and Bakker, P. A. (2014). Induced systemic resistance by beneficial microbes. *Annu. Rev. Phytopathol.* 52, 347–375. doi: 10.1146/annurev-phyto-082712-102340

54. Pietro, A. D., Madrid, M. P., Caracuel, Z., Delgado-Jarana, J., and Roncero, M. I. (2003). *Fusarium oxysporum*: exploring the molecular arsenal of a vascular wilt fungus. *Mol. Plant Pathol.* 4, 315–325. doi: 10.1046/j.1364-3703.2003.00180.x

55. Robert-Seilaniantz, A., Grant, M., and Jones, J. D. (2011). Hormone crosstalk in plant disease and defense: more than just jasmonate-salicylate antagonism. *Annu. Rev. Phytopathol.* 49, 317–343. doi: 10.1146/annurev-phyto-073009-114447

56. Sun, D. Q., Lu, X. H., Hu, Y. L., Li, W. M., Hong, K. Q., Mo, Y. W., et al. (2013). Methyl jasmonate induced defense responses increase resistance to *Fusarium oxysporum* f. sp *cubense* race 4 in banana. *Sci. Hortic.* 164, 484–491. doi: 10.1016/j.scienta.2013.10.011

57. Swarupa, V., Ravishankar, K. V., and Rekha, A. (2014). Plant defense response against *Fusarium oxysporum* and strategies to develop tolerant genotypes in banana. *Planta* 239, 735–751. doi: 10.1007/s00425-013-2024-8

58. Takken, F., and Rep, M. (2010). The arms race between tomato and *Fusarium oxysporum*. *Mol. Plant Pathol.* 11, 309–314. doi: 10.1111/j.1364-3703.2009.00605.x

59. Thaler, J. S., Owen, B., and Higgins, V. J. (2004). The role of the jasmonate response in plant susceptibility to diverse pathogens with a range of lifestyles. *Plant Physiol.* 135, 530–538. doi: 10.1104/pp.104.041566

60. Thatcher, L. F., Gardiner, D. M., Kazan, K., and Manners, J. M. (2012a). A highly conserved effector in *Fusarium oxysporum* is required for full virulence on Arabidopsis. *Mol. Plant Microb. Interact.* 25, 180–190. doi: 10.1094/MPMI-08-11-0212

61. Thatcher, L. F., Manners, J. M., and Kazan, K. (2009). *Fusarium oxysporum* hijacks COI1-mediated jasmonate signaling to promote disease development in Arabidopsis. *Plant J.* 58, 927–939. doi: 10.1111/j.1365-313X.2009.03831.x

62. Thatcher, L. F., Powell, J. J., Aitken, E. A., Kazan, K., and Manners, J. M. (2012b). The lateral organ boundaries domain transcription factor LBD20 functions in *Fusarium* wilt susceptibility and jasmonate signaling in Arabidopsis. *Plant Physiol.* 160, 407–418. doi: 10.1104/pp.112.199067

63. Tjamos, E. C., and Beckman, C. H. (1989). *Vascular Wilt Diseases of Plants*. Berlin: Springer-Verlag.

64. Ton, J., Flors, V., and Mauch-Mani, B. (2009). The multifaceted role of ABA in disease resistance. *Trends PlantSci.* 14,310–317.doi:10.1016/j.tplants.2009.03.006

65. Trusov, Y., Sewelam, N., Rookes, J. E., Kunkel, M., Nowak, E., Schenk, P. M., et al. (2009). Heterotrimeric G proteins-mediated resistance to necrotrophic pathogens includes mechanisms independent of salicylic acid-, jasmonic acid/ethylene- and abscisic acid-mediated defense signaling.*Plant J. Cell Mol. Biol.* 58, 69–81. doi: 10.1111/j.1365-313X.2008.03755.x

66. Vandermolen, G. E., Labavitch, J. M., Strand, L. L., and Devay, J. E. (1983). Pathogen-induced vascular gels - ethylene as a host intermediate. *Physiol. Plantarum* 59, 573–580. doi: 10.1111/j.1399-3054.1983.tb06282.x

67. Vos, C. M., Yang, Y., De Coninck, B., and Cammue, B. P. A. (2014). Fungal (-like) biocontrol organisms in tomato disease control.*Biol. Control* 74, 65–81. doi: 10.1016/j.biocontrol.2014.04.004

68. Vos, I. A., Verhage, A., Schuurink, R. C., Watt, L. G., Pieterse, C. M., and Van Wees, S. C. (2013). Onset of herbivore-induced resistance in systemic tissue primed for jasmonate-dependent defenses is activated by abscisic acid. *Front. Plant Sci.* 4:539. doi: 10.3389/fpls.2013.00539

69. Xue, R. F., Wu, J., Wang, L. F., Blair, M. W., Wang, X. M., Ge, W. D., et al. (2014). Salicylic acid enhances resistance to *Fusarium oxysporum* f. sp. *phaseoli* in common beans (*Phaseolus vulgaris* L.). *J. Plant Growth Regul.* 33, 470–476. doi: 10.1007/s00344-013-9376-y

70. Yadeta, K., and Thomma, B. P. (2013). The xylem as battleground for plant hosts and vascular wilt pathogens. *Front. Plant Sci.*4:97. doi: 10.3389/fpls.2013.00097

71. Yasuda, M., Ishikawa, A., Jikumaru, Y., Seki, M., Umezawa, T., Asami, T., et al. (2008). Antagonistic interaction between systemic acquired resistance and the abscisic acid-mediated abiotic stress response in Arabidopsis. *Plant Cell* 20, 1678–1692. doi: 10.1105/tpc.107.054296

72. Zamioudis, C., Mastranesti, P., Dhonukshe, P., Blilou, I., and Pieterse, C. M. (2013). Unraveling root developmental programs initiated by beneficial *Pseudomonas* spp. bacteria. *Plant Physiol.* 162, 304–318. doi: 10.1104/pp.112.212597

73. Zamioudis, C., and Pieterse, C. M. (2012). Modulation of host immunity by beneficial microbes. *Mol. Plant Microb. Interact.* 25, 139–150. doi: 10.1094/MPMI-06-11-0179

74. Zhu, Q. H., Stephen, S., Kazan, K., Jin, G. L., Fan, L. J., Taylor, J., et al. (2013). Characterization of the defense transcriptome responsive to *Fusarium oxysporum*-infection in Arabidopsis using RNA-seq. *Gene* 512, 259–266. doi: 10.1016/j.gene.2012.10.036

CHAPTER 12

NUTRITION ACQUISITION STRATEGIES DURING FUNGAL INFECTION OF PLANTS

Hege H. Divon and Robert Fluhr

Keywords: Plant pathogenfungal nutritionnitrogen and carbon regulation

INTRODUCTION

Parasitism defines the state in which one organism benefits at the expense of another. An inherent part of the success of the parasite in this misbalanced relationship is the securing of its nourishment. Much research has focused towards understanding pathogenicity and virulence factors. These include study of avirulence factors, toxins, cell wall-degrading enzymes (CWDE) and other mechanisms directly linked to the ability of the pathogen to obtain access to its host. A less-studied aspect of this interaction is the appreciation of how nutrients are acquired and how the parasite adapts to changing nutritional environments. Ultimately, a pathogenic lifestyle is characterized by metabolic dependency on the host. This is a fine-tuned association where host physiology is exploited and modified to optimize production of nutrients for the pathogen, and reciprocally pathogen metabolism is adapted to utilize what is made available. The final fitness of the fungus in this metabolic 'tug-of-war' will determine the success of colonization.

In this review, we examine *in planta* nutrition by plant–pathogenic fungi as established by biochemical and molecular tools. Many of the general concepts have been adopted from studies in saprophytes such as *Saccharomyces cerevisiae*, *Aspergillus nidulans* and *Neurospora crassa*. The central questions are as follows: what nutrients are acquired by the pathogen from the host, what metabolic adjustments have to be made during the pathogen's life cycle and what are the regulatory mechanisms that facilitate this adaptation?

DYNAMIC NUTRIENT REQUIREMENTS THROUGHOUT THE INFECTION CYCLE

Fungal plant pathogens have been traditionally classified by their lifestyle into biotrophic and necrotrophic types of parasitism. Biotrophic pathogens are defined by a total dependency on the host to complete their life cycle, deriving nutrients from living host cells by differentiation of specialized infection structures called haustoria (Mendgen & Hahn, 2002). In contrast, necrotrophs do not possess specialized infection structures and derive nutrients from sacrificed cells (Lewis, 1973). To accommodate further diversification of lifestyles, the term 'hemibiotrophic' has been coined to consider situations in which the pathogen renders its host largely alive while establishing itself within the host tissue. Only after this brief biotrophic-like phase does it switch to a necrotrophic lifestyle (Perfect & Green, 2001). The infection process itself can be

Citation: Hege H. Divon and Robert Fluhr, Nutrition acquisition strategies during fungal infection of plants, DOI:10.1111/j.1574-6968.2006.00504.x

separated into germination, proliferation and sporulation (Solomon *et al.*, 2003). The utility of this division is used here by following the dynamic nutritional challenges facing the pathogen in the context of its respective infectious phases (Fig. 1).

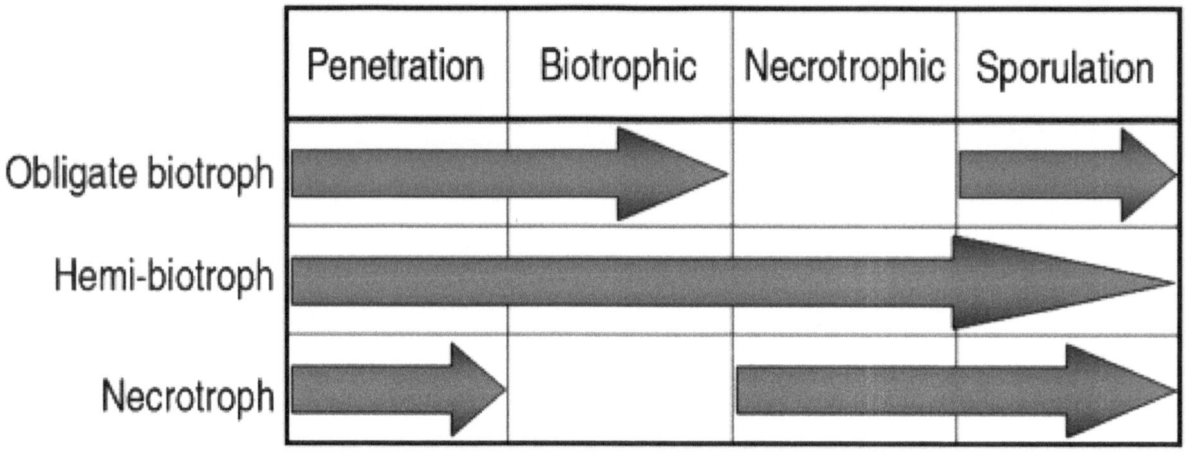

Figure 1: Infectious phases in phytopathogenic fungi that are relevant to nutrition acquisition strategies during fungal infection of plants. The penetration and sporulation phases are common to all fungi. Hemibiotrophs are distinguished by going through successive biotrophic and necrotrophic phases. Necrotrophs do not have a biotrophic phase, whereas obligate biotrophs complete their life cycle in this phase, without proceeding to necrotrophic stage.

Spore germination and penetration

Pathogen penetration of the host is facilitated by direct local mechanical pressure during appressorium development, enzymatic digestion of plant barriers or by the more serendipitous availability of stomata, wounds and cracks at the plant environment interface. It is generally assumed that during spore germination and the penetration phase, the phytopathogenic fungi are in a state of starvation and are completely reliant on nutrients derived from internal stores. The major components of fungal spore stores include glycogen, trehalose, polyols such as mannitol and lipids (Thevelein, 1984; Thines *et al.*, 2000; Voegele *et al.*, 2005). Recent evidence from several plant–pathogen systems has focused on the role of fatty acid mobilization during the penetration process (Table 1). Mobilization of lipid stores occurs through lipolysis and β-oxidation cycles to form acetyl-CoA, which is further assimilated into the tricarboxylic acid cycle (TCA cycle) via the glyoxylate cycle. The glyoxisomal enzyme isocitrate lyase (ICL), a marker for lipolytic activity, is induced in the absence of glucose and during growth on gluconeogenic carbon sources such as acetate and ethanol, and on fatty acids (Bowyer *et al.*, 2000). During infection, ICL promoter activity was detected predominantly in the prepenetration stage on the plant surface and declined after appressorial penetration, indicating a switch from gluconeogenic to glycolytic metabolism at this stage. Several mutants in which virulence is reduced were found to be defective in peroxisomal development (Kimura *et al.*, 2001), or disrupted in the glyoxylate cycle enzymes ICL (Idnurm & Howlett, 2002; Wang *et al.*, 2003) and malate synthase (MLS; Solomon *et al.*, 2004a). The findings indicate a universal role for fatty acid metabolism and glyoxylate cycle during plant penetration, regardless of pathogenic lifestyle. Interestingly, in some studies, glucose was able to complement the loss of pathogenesis (Kimura *et al.*, 2001; Idnurm & Howlett, 2002; Solomon *et al.*, 2004a). The implications are that a critical role for the fatty acid metabolic activity is the support of the TCA cycle through acetyl-CoA production in order to generate ATP under the poor nutrient conditions that exist at the plant surface. An additional role for fatty acid metabolism in glycerol production was demonstrated in the ICL mutant in *Magnaporthe grisea*, which shows delayed appressorium turgor generation (Wang *et al.*, 2003).

Table 1: Fungal nutrition-related genes and ESTs expressed during plant infection. Signalling components are not included

Gene/EST name	Acc. No.	Fungal species	Similarity	Function	EP*	References
Biotrophic fungi						
Aat1 (Pig27)	AJ308252	*U. fabae*	Amino acid transporter	Histidine uptake	ND	Struck *et al.* (2002)
Aat2 (Pig2)	U81794	*U. fabae*	Amino acid transporter	Amino acid transport	ND	Hahn *et al.* (1997)
Ard1	AJ809335	*U. fabae*	D-arabitol dehydrogenase 1	D-arabitol biosynthesis	ND	Link *et al.* (2005)
Bgl1	AJ575269	*U. fabae*	β-glucosidase	Cellobiose degradation	ND	Haerter & Voegele (2004)
Gat1	AF271266	*C. fulvum*	GABA transaminase	GABA metabolism	ND	Solomon & Oliver (2002)
Gltrn1	AF376000	*B. graminis*	Glucose transporter	Glucose uptake	ND	Zhang *et al.* (2005)
GS-1	AF395112	*B. graminis*	Glutamine synthetase	Glutamine synthesis	ND	Zhang *et al.* (2005)
Hxt1	AJ310209	*U. fabae*	Hexose transporter	Uptake of D-glc, D-frc	ND	Voegele *et al.* (2001)
Met6	AF226997	*C. fulvum*	Methionine synthase	Methionine biosynthesis	Yes	Solomon *et al.* (2000)
Mad1 (Pig8)	O00058	*U. fabae*	Mannitol dehydrogenase	Mannitol metabolism	ND	Voegele *et al.* (2005)
Pma	AF406807	*B. graminis*	Plasma membrane H+ ATPase	Nutrient uptake	ND	Zhang *et al.* (2005)
pSI-9	Y14555	*C. fulvum*	Aldehyde dehydrogenase	Carbon metabolism	ND	Coleman *et al.* (1997)
pSI-10	Y14556	*C. fulvum*	Alcohol dehydrogenase	Carbon metabolism	ND	Coleman *et al.* (1997)
Thi1 (Pig1)	AJ250426	*U. fabae*	Pyrimidine biosynthetic enzyme	Vitamin B1 biosynthesis	ND	Sohn *et al.* (2000), Hahn & Mendgen (1997)
Thi2 (Pig4)	AJ250427	*U. fabae*	Thiazole biosynthetic enzyme	Vitamin B1 biosynthesis	ND	Sohn *et al.* (2000), Hahn & Mendgen (1997)
Tr4	BU672643	*P. triticina*	Pyrimidine biosynthetic enzyme	Vitamin B1 biosynthesis	ND	Thara *et al.* (2003)
Tr21	BU672660	*P. triticina*	Thiazole biosynthetic enzyme	Vitamin B1 biosynthesis	ND	Thara *et al.* (2003)
UfAat3	AJ628939	*U. fabae*	Amino acid transporter	leu, met, cys uptake	ND	Struck *et al.* (2004)
UfInv1	AJ640083	*U. fabae*	Invertase	Sugar hydrolysis	ND	Voegele *et al.* (2006)
Uf-Pma1	AJ003067	*U. fabae*	Plasma membrane H+ ATPase	Proton-driven nutrient uptake	ND	Struck *et al.* (1996), Struck *et al.* (1998)
Hemi-biotrophic fungi						
Arg1	AB045736	*F. oxysporum*	Argininosuccinate lyase	Arginine biosynthesis	Yes	Namiki *et al.* (2001)
Gap1 (Fol5B10)	CK615491	*F. oxysporum*	General amino acid permease	Amino acid uptake	ND	Divon *et al.* (2005)
Gpdh	AY331190	*C. gloeosporioides*	Glycerol-3-phosphate dehydrogenase	Carbon utilization	No	Wei *et al.* (2004)
Icl1	AY118108	*L. maculans*	Isocitrate lyase	Glyoxylate cycle	Yes	Idnurm & Howlett, 2002
Mtd1 (Fol8D6)	CK615494	*F. oxysporum*	Oligopeptide transporter	Peptide uptake	ND	Divon *et al.* (2005)
pCgGS	L78067	*C. gloeosporioides*	Glutamine synthetase	Glutamine synthesis	ND	Stephenson *et al.* (1997)
Uricase (Fol6D12)	CK615495	*F. oxysporum*	Uricase	Uric acid anabolism	ND	Divon *et al.* (2005)
Necrotrophic fungi						
Als1	DQ167577	*S. nodorum*	Delta-aminolaevulinic acid synthase	Heme biosynthesis	Yes	Solomon *et al.* (2006)
Cbl1	EAA68828	*F. graminearum*	Cystathionine betalyase	Methionine biosynthesis	Yes	Seong *et al.* (2005)
Glo1	AY576607	*S. nodorum*	Glyoxylase I	Glycolytic bypass pathway	No	Solomon & Oliver (2004)
Icl1	AF540383	*M. grisea*	Isocitrate lyase	Glyoxylate cycle	Yes	Wang *et al.* (2003)
Mls1	AY508881	*S. nodorum*	Malate synthase	Glyoxylate cycle	Yes	Solomon *et al.* (2004a)
Mpd1	AY587541	*S. nodorum*	Mannitol 1-phosphate dehydrogenase	Mannitol metabolism	Yes	Solomon *et al.* (2005)
Msy1	EAA75179	*F. graminearum*	Methionine synthase	Methionine biosynthesis	Yes	Seong *et al.* (2005)
Odc1	AJ249387	*S. nodorum*	Ornithine decarboxylase	Polyamine biosynthesis	Yes	Bailey *et al.* (2000)
Pth2	AF027979	*M. grisea*	Carnitine acetyltranferase	Fatty acid transport	Yes	Sweigard *et al.* (1998)
Pth3	O42621	*M. grisea*	Imidazoleglycerol-P dehydratase	Histidine biosynthesis	Yes	Sweigard *et al.* (1998)
Ptr2	AY187281	*S. nodorum*	di/tripeptide transporter	di/tri peptide uptake	No	Solomon *et al.* (2004b)
	BM135467	*F. graminearum*	NADP-specific glutamate dehydrogenase	Glutamate synthesis	ND	Kruger *et al.* (2002)
	BM137184	*F. graminearum*	Hexose transporter protein	Hexose uptake	ND	Kruger *et al.* (2002)
	MG03533	*M. grisea*	Formamidase	trp catabolism	ND	Donofrio *et al.* (2006)
	MG04385	*M. grisea*	Urea transporter	Urea uptake	ND	Donofrio *et al.* (2006)
	MG05526	*M. grisea*	Methylammonium permease	Nitrogen sensing and uptake	ND	Donofrio *et al.* (2006)

*EP, effects of mutation on pathogenicity.
The reader is referred to Guldener *et al.* (2006) and Jakupovic *et al.* (2006), for additional genes expressed *in planta* in *F. graminearum* and *U. fabae*, respectively.

Nitrogen constitutes a smaller part of the storage compounds in the spore and might represent a rate-limiting factor during penetration. Nevertheless, elevated expression of a di/tri-peptide transporter (Ptr2) at the prepenetration stage in *Stagonospora nodorum* is not essential for pathogenicity (Solomon *et al.*, 2004b), suggesting that recycling of internal stores can serve as the primary source of nitrogen essentials at this stage. Indeed, a recent transcriptome survey of barley powdery mildew revealed a large number of genes involved in amino acid metabolism, protein turnover and amino acid recycling, which were expressed during spore germination and appressorium formation (Thomas *et al.*, 2001).

Biotrophic growth phase and the entrainment of plant metabolism

Once inside the host, and with internal stores likely exhausted, the pathogen needs to establish itself rapidly by mobilizing mechanisms that will ensure adequate nutrient uptake from the host. For the obligate biotrophs, this phase is initiated by the formation of a haustorium (Voegele, 2006, and references therein). For hemibiotrophs, deficient in such feeding structures, some recent evidence suggests that they have developed mechanisms that entrain host metabolism to establish new sink tissues to suit the needs of the pathogen. Invertases and hexose sugar transporters play a key role in defining a leaf as a 'source' or as a 'sink' of carbohydrates. The up-regulation of plant extracellular invertase appears to be a common response to various biotic and abiotic stress-related stimuli, serving to fuel plant defence responses in the infected tissue (reviewed in Biemelt & Sonnewald, 2006). This defence strategy is used by *Arabidopsis*; however, during powdery mildew infection, this physiological response of the plant is exploited by the fungus to ensure a ready supply of carbohydrates for its own use (Fig. 2; Fotopoulos *et al.*, 2003).

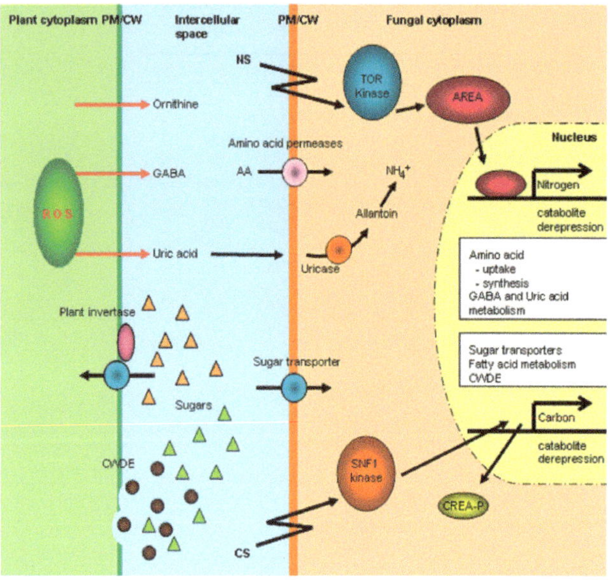

Figure 2: Examples of the entrainment of plant metabolism by phytopathogenic fungi. Sensing of nitrogen or carbon availability (NS, CS; flash arrow), or the absence of preferable nutrient sources, triggers signal transduction pathways that initiate transcription of transporters and pathway-specific metabolic enzymes. The de-repression of these pathways is regulated by de-phosphorylation and binding of AREA to nitrogen catabolite-regulated gene promoters (Ravagnani *et al.*, 1997; Beck & Hall, 1999), and by SNF1-dependent phosphorylation and release of CREA repressor from carbon catabolite-regulated gene promoters (DeVit *et al.*, 1997; Gancedo, 1998). Plant resistance responses (red arrows), involving ornithine or ROS scavengers such as GABA, are taken up through amino acid (AA) permeases and metabolized by the fungus (Oliver & Solomon, 2004; Guldener *et al.*, 2006). Uric acid, the product of disease-induced enhanced catabolic processes in the plant, can serve as a fungal nutrient by conversion to allantoin by uricase and subsequent recovery of ammonia (Divon *et al.*, 2005). Plant invertases that are induced by the plant to fuel its cellular defence needs, as well as CWDE secreted by the pathogen, will produce depolymerized sugars that the fungus can utilize (Ospina-Giraldo *et al.*, 2003; Biemelt & Sonnewald, 2006). PM, plasma membrane; CW, cell wall.

In a manner similar to the utilization of plant carbohydrates during infection, nitrogen acquisition can also be subverted for fungal benefit as illustrated by the metabolism of γ-amino butyric acid (GABA) during*Cladosporium fulvum* infection of tomato (Solomon & Oliver, 2001). GABA, normally a major amino-acid constituent of the tomato apoplast, accumulates to even higher (millimolar) levels during the compatible interaction. The accumulation was concomitant with the induction of tomato glutamate decarboxylase (GAD) responsible for GABA synthesis, and the fungal GABA transaminase (*Gat1*; Table 1), metabolizing GABA into succinic semialdehyde (Solomon & Oliver, 2002). The utilization of GABA as a nutrient has so far only been demonstrated for *Cladosporium fulvum*, and may be part of the adaptation to a biotrophic lifestyle without haustoria, which occurs exclusively in the apoplastic spaces of the leaf. Interestingly, the ramifications of modulation of GABA metabolism may be more subtle. The GABA shunt can also serve as part of a cellular scavenger system for oxidative stress signals. The role of reactive oxygen species (ROS) in plant defence signalling is well established, and one can speculate that as in the case of carbohydrate acquisition successful colonization will subvert plant signals for defence to provide a source of nutrients for pathogenic fungi (Fig. 2; Oliver & Solomon, 2004). A similar scenario may occur in the *Fusarium oxysporum* interaction with tomato, where uricase, an enzyme catalysing the conversion of uric acid to allantoin, was induced during the early nonnecrotic stages of infection (Table 1; Divon *et al.*, 2005). Uric acid can act as a scavenger of ROS (Becker *et al.*, 1989) and like GABA may serve the plant in that role. Hence, the fungus may utilize uric acid derived from host catabolism and defence reactions (Fig. 2). Additional evidence for the role of metabolites with scavenging capacity comes from a study of the rust fungus *Uromyces fabae*. Voegele (2005) suggested that the pathogen-derived compound mannitol, which the host, *Vicia faba*, can neither produce nor metabolize, accumulates in infected leaves to serve as a ROS scavenger to protect the pathogen in the infection process.

While great strides have been made in understanding how bacterial pathogen effectors are delivered through the type III secretion systems found in plant–pathogenic bacteria interactions, less is known at the molecular level of how fungi entrain host metabolism. For example, in the *Cladosporium fulvum*interaction with tomato, an extracellular fungal protein was detected that modifies a plant cysteine protease and subsequent plant defence responses (Rooney *et al.*, 2005). Interestingly, proteins from the rust fungi *Melamspora lini* and *U. fabae* are localized to the host cytoplasm (Dodds *et al.*, 2004;Kemen *et al.*, 2005). This may indicate that fungal effector proteins can gain intracellular access and thus target cellular metabolic crossroads.

Necrotrophic growth phase and entrainment of plant metabolism

Shortly after penetration, necrotrophic plant pathogenic fungi, aided by production of toxins, hydrolytic enzymes and necrosis-related proteins, cause cell lysis and death of host cells. One may surmise that the lysis of host cells would lead to an increase in accessible nutrients; however, as evidenced from examination of auxotrophic mutants, certain amino acids such as methionine and histidine are not supplied in sufficient quantity by the host tissue and their biosynthesis is carried out by the fungus (Sweigard *et al.*, 1998; Balhadere *et al.*, 1999; Seong *et al.*, 2005). The coordinate expression of nutrient acquisition genes during infection that includes possible sources for amino acid backbones was revealed in the transcriptome analysis of *Fusarium graminearum* infection (Guldener *et al.*, 2006). In this case, transporters of sugars and nitrogenous compounds were expressed in early infection, including an ornithine transporter expressed from day 2 postinfection. Ornithine is a nonprotein amino acid and an important intermediate in metabolic pathways such as arginine and proline biosynthesis, as well as the biosynthesis of putrescine, a precursor for other polyamines and pyrimidines. Ornithine uptake was followed by expression of arginine, proline and putrescine biosynthetic genes from day 3 to 6 postinfection (arginine: ornithine carbamoyltransferase, carbamoyl-P synthase and acetylglutamate kinase; proline: ornithine aminotransferase; putrescine: ornithine decarboxylase). In line with this observation, a putrescine auxotrophic mutant from the wheat pathogen *Stagonospora nodorum*, which was disrupted in the gene encoding ornithine decarboxylase, *odc1*, was found to have greatly reduced virulence on wheat (Table 1; Bailey *et al.*, 2000). Ornithine, as a precursor in polyamine synthesis, is involved in plant stress responses (Walters, 2003). Hence, we encounter, as in the case of carbohydrates, GABA and uric acid metabolism, the concept that successful pathogens hijack potential plant defences for their nutritional benefit during colonization (Fig. 2).

Sporulation

Sporulation marks the completion of the pathogenic life cycle. True biotrophic fungi complete their life cycle by maintaining a biotrophic state, whereas hemibiotrophs and necrotrophs complete their life cycle in a necrotrophic state (Fig. 1). *In vitro* studies have shown that sporulation can be triggered by exhaustion of nitrogen and carbon sources (Griffin, 1994). Thus, in a manner analogous to the finding that fungi undergo nutrient limitation during the initial penetration, it may be anticipated that nutrient limitation triggers sporulation. However, experiments describing the *in planta* sporulation, from the hemibiotrophic interaction of *Mycosphaerella graminicola* with wheat, points to a different situation. cDNA microarray analysis during *in planta* growth revealed fungal expression of a subset of genes that resemble transcripts associated with *in vitro* growth in rich medium rather than poor medium (Keon *et al.*, 2005). The results suggest an alternative trigger for the sporulation event such as, for example, a quorum-sensing-like mechanism.

A particularly illuminating example of dependence on nutrient transport in the late growth stages was revealed in the glycerol-3-phosphate dehydrogenase (*Gpdh*) disruption mutant of the hemibiotrophic plant pathogen, *Colletotrichum gloeosporioides* f.sp. *malvae* (Table 1; Wei *et al.*, 2004). The mutant contains reduced amounts of glycerol, and during heterotrophic growth on defined media, the mutant strain exhibited severe defects in carbon utilization and failed to conidiate. This defect could be complemented by exogenous addition of glycerol. Surprisingly, in spite of these profound growth defects during heterotrophic growth, the *gpdh* mutant strain displayed normal pathogenicity *in planta*, progressing normally through infection and conidiation. Importantly, the analysis of plant tissues showed significant depletion of glycerol in the infection zone, indicating glycerol transfer from the plant to the fungal pathogen.

REGULATION OF FUNGAL NUTRITION GENE EXPRESSION DURING INFECTION

Nutrient availability in the external environment dictates expression of fungal nutrition genes via general metabolic processes known as catabolite repression. This process implies that, as long as sufficient amounts of a favourable nutrient source are available, transcription of alternative uptake systems and catabolic enzymes is repressed. The preferred source for the carbon-regulatory system is glucose, and it is ammonia and glutamine for the nitrogen-regulatory system. Catabolite repression has been extensively studied in saprobes like *Saccharomyces cerevisiae, N. crassa* and *A. nidulans* (Marzluf, 1997; Carlson, 1998; Gancedo, 1998; Magasanik & Kaiser, 2002), and given the high degree of conservation in this process, these saprobes have served as models to guide similar research in phytopathogenic fungi.

Carbon catabolite repression (CCR)

CCR signifies the transcriptional regulation of genes involved in utilization of alternative carbon sources. It is driven by glucose repression and substrate induction. In *Saccharomyces cerevisiae*, the *SNF1*(sucrose nonfermenting 1) gene is required for expression of catabolite-repressed genes (Fig. 2;Gancedo, 1998, and references therein). It encodes a protein kinase whose major function is to phosphorylate Mig1, a DNA-binding transcriptional repressor. When glucose is limiting, phosphorylation of Mig1 causes its dissociation from the promoters of repressed genes (DeVit *et al.*, 1997), thus motivating the release from CCR. *SNF1*-orthologues have been cloned from the maize pathogen*Cochliobolus carbonum* (CcSnf1; Tonukari *et al.*, 2000) and the vascular wilt pathogen *F. oxysporum*(*FoSnf1*; Ospina-Giraldo *et al.*, 2003). Disruption mutants were compromised for growth on secondary carbon sources, and production of several secreted CWDE was reduced with a concomitant reduction in virulence. This phenomenon was observed at the penetration stage (Tonukari *et al.*, 2000), as well as in later stages of plant colonization (Ospina-Giraldo *et al.*, 2003). Thus, CCR is directly linked to pathogenicity. In addition, SNF1 might facilitate de-repression of hexose transporters for uptake and utilization of sugars released from degraded cell wall polymers (Fig. 2; Carlson, 1998). The orthologue of *MIG1* in filamentous fungi is called *CreA*, encoding a repressor with two zinc fingers of the C2H2 class (Fig. 2; Ronne, 1995). *CreA* genes were isolated from the plant pathogenic fungi *Gibberella fujikuroi*and *Botrytis cinerea* (Tudzynski *et al.*, 2000); however, their contribution to nutrient acquisition during infection was not determined. In another approach, random insertional mutagenesis in the hemibiotroph*Colletotrichum lindemuthianum* identified a mutant disrupted in the metabolic switch between the biotrophic and the necrotrophic phases (Dufresne *et al.*, 2000). The corresponding gene, encoding a putative zinc finger GAL4-like transcriptional activator, was designated *Clta1*.

Nitrogen catabolite repression (NCR) and the 'nitrogen starvation hypothesis'

The hypothesis that plant–pathogenic fungi experience nitrogen nutrient limitation during *in planta*growth (Snoeijers *et al.*, 2000) is based on observations that showed the induction of starvation-regulated genes in *Cladosporium fulvum*, *Magnaporthe grisea* and others during the infection process (Talbot *et al.*, 1993; Van den Ackerveken *et al.*, 1994; Coleman *et al.*, 1997; Stephenson *et al.*, 1997; Snoeijers *et al.*, 2000). NCR embodies the transcriptional regulation of permeases and catabolic enzymes needed for the utilization of secondary nitrogen sources. In yeast, the gene *GLN3* encodes a global activator of the GATA-type zinc finger family, recognizing the core consensus sequence 5′-(A/T/C)GATA(A/G)-3′ found in NCR-regulated promoters (Ravagnani *et al.*, 1997; Starich *et al.*, 1998). In the presence of favourable nitrogen sources such as ammonia and glutamine, Gln3, phosphorylated by the TOR (target of rapamycin) kinase (Beck & Hall, 1999; Fitzgibbon *et al.*, 2005), is sequestered to the cytosol by inhibitory binding to the negative regulator Nil2. During nitrogen starvation, due to the lack of favourable nitrogen sources, de-phosphorylation of Gln3 relieves it from Nil2 repression, and it relocates to the nucleus where it activates transcription of NCR-controlled genes, either alone or in combination with pathway-specific activators (Marzluf, 1997; Magasanik & Kaiser, 2002). Once identified as a key regulator of nitrogen metabolism, orthologues of *GLN3* have been extensively studied in saprobes like *A. nidulans* and *N. crassa* (*AreA* and *nit-2*, respectively; Marzluf, 1997), and in several phytopathogenic fungi. They act as central control devices aiding the fungus in adapting to different nutrient environments (Fig. 2).

As the nutrient environment varies with the pathogenic lifestyle, the dependency on AREA-like regulators during infection is expected to vary accordingly. Disruption of *Clnr1* in the hemibiotrophic fungus*Colletotrichum lindemuthianum* severely compromised the infection cycle at the stage of secondary hyphae formation and almost completely eliminated the occurrence of anthracnose symptoms (Pellier*et al.*, 2003). Disruption of *Fnr1* in *F. oxysporum* caused a delay in disease symptom development, attributed in part to the abrogated expression of nutrition genes normally up-regulated during *in planta*growth (Divon *et al.*, 2006). Conversely, disruption of*Nut1* in the rice blast fungus, *Magnaporthe grisea*, had only a minor effect on virulence (Froeliger & Carpenter, 1996). In *Cladosporium fulvum*, recent evidence shows that deletion of the *Nrf1* gene significantly reduces *Cladosporium fulvum* virulence on tomato; however, this reduction is not connected to the abolished expression of *Avr9* (Thomma *et al.*, 2006). Interestingly, GABA uptake via the GABA permease in *A. nidulans* is subject to NCR (Hutchings*et al.*, 1999). Under the assumption that GABA constitutes a significant nutrient for *Cladosporium fulvum*during pathogenicity (Solomon & Oliver, 2002), it is likely that loss of virulence in the *nrf1* mutant is, at least in part, caused by nutrient deficiency due to the inability to take up and metabolize GABA.

Based on the existing evidence, it is clear that the main impact of AREA on pathogenic ability is through the regulation of nutrition. Interestingly, in hemibiotrophs, the impact of AREA on regulation seems to somewhat correlate with the 'degree' of biotrophy, i.e. the time spent in the plant apoplast. Thus, in*Magnaporthe grisea*, which grows necrotrophically almost immediately after penetration, the effect of*AreA* disruption is minor (Froeliger & Carpenter, 1996). In contrast, in the hemibiotrophs *F. oxysporum, Colletotrichum lindemuthianum* and *Cladosporium Fulvum*, such a disruption shows a clear effect (Pellier *et al.*, 2003; Divon *et al.*, 2006; Thomma *et al.*, 2006). Apparently, hemibiotrophs that have an extended growth period in the nutrient-poor apoplast will tend to rely on AREA-regulatory nutrient control. In comparison, this regulatory control will be less important in the obligate biotrophic pathogens that feed through specialized infection structures or necrotrophs to whom ample nutrients are available. Finally, during *in planta* growth, nitrogen starvation might be seen as a stage-specific or recurring condition, rather than a continuous condition. Transition periods from one infectious phase to the next, i.e. germination, sporulation and transition between the biotrophic and necrotrophic phase, are likely to increase the demand for energy and metabolite precursors and require precise regulation.

Linkage of nutritional regulation and development

Apart from the influence of the external environment, stage-specific requirements for nutrients are also linked to the development of specialized infection structures. This is illustrated by the fact that, in obligate biotrophic fungi, nutrition genes are insensitive to the exogenous nutrient availability, and instead their transcriptional control is linked to the development of haustoria. For example, thiamine biosynthetic genes, *thi1* and *thi2*, in the rust fungus *U. fabae* do not respond with down-regulation to exogenously applied thiamine as do saprophytic fungi (Sohn *et al.*, 2000). Similarly, appressorium turgor generation depends on accumulation of large quantities of glycerol, which is acquired from breakdown of internal stores of lipids and glycogen (Thines *et al.*, 2000). Such processes are usually under CCR (Hardie *et al.*, 1998); however, impairment of the CCR regulatory pathway in *Cochliobolus carbonum* by disruption of *Snf1* did not affect appressorium formation and turgor generation (Tonukari*et al.*, 2000; see below), indicating that other mechanisms override SNF1 regulation of fatty acid breakdown and peroxisomal function.

The complexity of metabolic and developmental interaction was investigated in the process of turgor generation in developmental, nonpathogenic mutants in *Magnaporthe grisea* (Thines *et al.*, 2000). In the MAP kinase mutant, Δ*pmk1*, appressorium formation is impaired and glycogen and lipid mobilization do not occur during germination. The same processes were also retarded markedly in a cAMP-dependent protein kinase A (PKA) mutant (Δ*cpkA)*. Furthermore, in a Δ*mac1 sum1-99* mutant, carrying a mutation in the regulatory subunit of PKA that renders it cAMP-independent and nearly constitutively active, glycogen and lipid degradation were rapid and preceded appressorium morphogenesis (Thines*et al.*, 2000). In another example, appressorium formation was also impaired in the *nir1* mutant in*Colletotrichum acutatum* (Horowitz *et al.*, 2006). *Nir1*, an orthologue of *NirA* in *A. nidulans*, is the pathway-specific activator of nitrate utilization genes, and the *Nir1* mutation abolished activation of nitrate and nitrite reductase and was assumed to account partly for the penetration-defect phenotype. However, the *nir1* mutant displayed additional appressorium developmental defects not present in nitrate reductase mutants (Horowitz *et al.*, 2006). Together, these results indicate that considerable crosstalk occurs during development of the functional appressorium, involving both traditional catabolite-repressed regulatory pathways, MAP kinase pathways and cAMP-dependent PKA (**Fig. 3**).

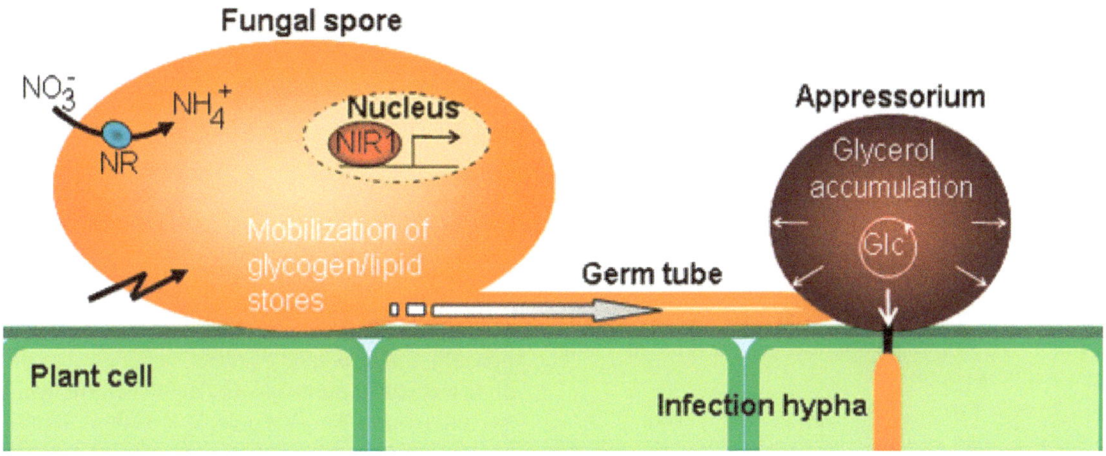

Figure 3: Nutrition-related processes that take place during appressorium-mediated penetration. MAP kinase- and cAMP-dependent PKA signalling pathways are involved in mobilization of glycogen and lipid stores from the spore. Turgor generation is accomplished by compartmentalization and rapid degradation of lipid and glycogen reserves in the appressorium (Thines *et al.*, 2000). The glyoxylate cycle (Glc) is involved in turgor generation, but also in replenishing intermediates to the TCA cycle from acetyl-CoA generated in fatty acid oxidation (Kimura *et al.*, 2001; Idnurm & Howlett, 2002; Wang *et al.*, 2003; Solomon *et al.*, 2004a). An *NirA*-like transcription factor, *Nir1*, is required for transcription of nitrate reductase (NR), necessary for nitrate assimilation in *Colletotrichum acutatum*, but is also involved in additional processes related to appressorium formation (Horowitz *et al.*, 2006). The grey/white arrow in the germ tube indicates transport of glycogen and lipid metabolic products; the flash arrow indicates sensing mechanism(s); and the small white arrows in the appressorium signify turgor generation.

CONCLUSIONS

An understanding of fungal metabolism and adaptation during host colonization is of critical importance to evaluate fully the complex orchestration of the activities that rule host–pathogen interactions. This knowledge will provide a rational basis for the development of new disease control strategies. Studies of gene expression and auxotrophic mutants have provided considerable information regarding those nutrients available from the plant and those that are not. Elaborate mechanisms have been unravelled that indicate how host and pathogen compete for nutrient supplies, and how plant defence mechanisms can be exploited by fungi for nutritional purposes. Besides amino acids, the nonprotein amino acids GABA and ornithine stand out as potential providers of nitrogen, and as precursors in biosynthetic pathways for compounds not adequately provided by the plant. More case studies will be needed to determine whether these represent specific or global mechanisms. Key regulatory factors involved in nutrient signalling and gene expression have been characterized. Although their impact on pathogenicity appears to be species specific, commonalties between lifestyle preference and regulatory control are now emerging. While important advances have been made in our ability to describe the metabolic events surrounding fungal ingress, little is known at the molecular level of how fungi entrain host metabolism. This accentuates the future challenge in understanding metabolite requisition by fungal parasites.

REFERENCES

1. Bailey A., Mueller E., Bowyer P. (2000) Ornithine decarboxylase of Stagonospora (Septoria) nodorum is required for virulence toward wheat. J Biol Chem 275: 14242–14247.
2. Balhadere P.V., Foster A.J., Talbot N.J. (1999) Identification of pathogenicity mutants in the rice blast fungus Magnaporthe grisea by insertional mutagenesis. Mol Plant-Microbe Interact 12: 129–142.CrossRefWeb of Science
3. Beck T., Hall M.N. (1999) The TOR signalling pathway controls nuclear localization of nutrient-regulated transcription factors. Nature 402: 689–692.
4. Becker B.F., Reinholz N., Özcelik T., Leipert B., Gerlach E. (1989) Uric acid as radical scavenger and antioxidant in the heart. Pflueg Arch Eur J Physiol 415: 127–135.
5. Biemelt S., Sonnewald U. (2006) Plant-microbe interactions to probe regulation of plant carbon metabolism. J Plant Physiol 163: 307–318.
6. Bowyer P., Mueller E., Lucas J. (2000) Use of an isocitrate lyase promoter — GFP fusion to monitor carbon metabolism of the plant pathogen Tapesia yallundae during infection of wheat. Mol Plant Pathol 1: 253–262.
7. Carlson M. (1998) Regulation of glucose utilization in yeast. Curr Opin Genet Dev 8: 560–564.
8. Coleman M., Henricot B., Arnau J., Oliver R.P. (1997) Starvation-induced genes of the tomato pathogen Cladosporium fulvum are also induced during growth in planta. Mol Plant-Microbe Interact 10: 1106–1109.
9. DeVit M.J., Waddle J.A., Johnston M. (1997) Regulated nuclear translocation of the Mig1 glucose repressor. Mol Biol Cell 8: 1603–1618.
10. Divon H.H., Rothan-Denoyes B., Davydov O., Di Pietro A., Fluhr R. (2005) Nitrogen responsive genes are differentially regulated in planta during Fusarium oxysporum f. sp. lycopersici infection. Mol Plant Pathol 6: 459–470.

11. Divon H.H., Ziv C., Davydov O., Yarden O., Fluhr R. (2006) The global nitrogen regulator, FNR1, regulates fungal nutrition-genes and fitness during Fusarium oxysporum pathogenesis. Mol Plant Pathol. 7: 485–497.

12. Dodds P.N., Lawrence G.J., Catanzariti A.M., Ayliffe M.A., Ellis J.G. (2004) The Melampsora lini AvrL567 avirulence genes are expressed in haustoria and their products are recognized inside plant cells. Plant Cell 16: 755–768.

13. Donofrio N.M., Oh Y., Lundy R., Pan H., Brown D.E., Jeong J.S., Coughlan S., Mitchell T.K., Dean R.A. (2006) Global gene expression during nitrogen starvation in the rice blast fungus, Magnaporthe grisea. Fungal Genet Biol 43: 605–617.

14. Dufresne M., Perfect S., Pellier A-L., Bailey J.A., Langin T. (2000) A GAL4-like protein is involved in the switch between biotrophic and necrotrophic phases of the infection process of Colletotrichum lindemuthianum on Common Bean. Plant Cell 12: 1579–1589.

15. Fitzgibbon G.J., Morozov I.Y., Jones M.G., Caddick M.X. (2005) Genetic analysis of the TOR pathway in Aspergillus nidulans. Eukaryot Cell 4: 1595–1598.

16. Fotopoulos V., Gilbert M.J., Pittman J.K., et al. (2003) The monosaccharide transporter gene, AtSTP4, and the cell-wall invertase, Atβfruct1, are induced in Arabidopsis during infection with the fungal biotroph Erysiphe cichoracearum. Plant Physiol 132: 821–829.

17. Froeliger E.H., Carpenter B.E. (1996) NUT1, a major nitrogen regulatory gene in Magnaporthe grisea, is dispensable for pathogenicity. Mol Gen Genet 251: 647–656.

18. Gancedo J.M. (1998) Yeast carbon catabolite repression. Microbiol Mol Biol Rev 62: 334–361.

19. Griffin D.H. (1994) Spore development. Fungal Physiology, 2nd edn. pp. 338–342. Wiley-Liss, Inc, NY.

20. Guldener U., Seong K.Y., Boddu J., et al. (2006) Development of a Fusarium graminearum Affymetrix GeneChip for profiling fungal gene expression in vitro and in planta. Fungal Genet Biol 43: 316–325.

21. Haerter A.C., Voegele R.T. (2004) A novel β-glucosidase in Uromyces fabae: feast or fight? Curr Genet 45: 96–103.

22. Hahn M., Mendgen K. (1997) Characterization of in planta-induced rust genes isolated from a haustorium-specific cDNA library. Mol Plant-Microbe Interact 10: 427–437.

23. Hahn M., Neef U., Struck C., Gottfert M., Mendgen K. (1997) A putative amino acid transporter is specifically expressed in haustoria of the rust fungus Uromyces fabae. Mol Plant-Microbe Interact 10: 438–445.

24. Hardie D.G., Carling D., Carlson M. (1998) The AMP-activated/SNF1 protein kinase subfamily: metabolic sensors of the eukaryotic cell? Annu Rev Biochem 67: 821–855.

25. Horowitz S., Freeman S., Zveibil A., Yarden O. (2006) A defect in nir1, a nirA-like transcription factor, confers morphological abnormalities and loss of pathogenicity in Colletotrichum acutatum. Mol Plant Pathol 7: 341–354.

26. Hutchings H., Stahmann K-P., Roels S., Espeso E.A., Timberlake W.E., Arst H.N., Tilburn J. (1999) The multiply-regulated gabA gene encoding the GABA permease of Aspergillus nidulans: a score of exons. Mol Microbiol 32: 557–568.

27. Idnurm A., Howlett B.J. (2002) Isocitrate lyase is essential for pathogenicity of the fungus Leptosphaeria maculans to canola (Brassica napus). Eukaryot Cell 1: 719–724.

28. Jakupovic M., Heintz M., Reichmann P., Mendgen K., Hahn M. (2006) Microarray analysis of expressed sequence tags from haustoria of the rust fungus Uromyces fabae. Fungal Genet Biol 43: 8–19.

29. Kemen E., Kemen A.C., Rafiqi M., Hempel U., Mendgen K., Hahn M., Voegele R.T. (2005) Identification of a protein from rust fungi transferred from haustoria into infected plant cells. Mol Plant-Microbe Interact 18: 1130–1139.

30. Keon J., Antoniw J., Rudd J., Skinner W., Hargreaves J., Hammond-Kosack K. (2005) Analysis of expressed sequence tags from the wheat leaf blotch pathogen Mycosphaerella graminicola (anamorph Septoria tritici). Fungal Genet Biol 42: 376–389.

31. Kimura A., Takano Y., Furusawa I., Okuno T. (2001) Peroxisomal metabolic function is required for appressorium-mediated plant infection by Colletotrichum lagenarium. Plant Cell 13: 1945–1957.

32. Kruger W.M., Pritsch C., Chao S., Muehlbauer G.J. (2002) Functional and comparative bioinformatic analysis of expressed genes from wheat spikes infected with Fusarium graminearum. Mol Plant-Microbe Interact 15: 445–455.

33. Lewis D.H. (1973) Concepts in fungal nutrition and the origin of biotrophy. Biol Rev 48: 261–278.CrossRefWeb of Science

34. Link T., Lohaus G., Heiser I., Mendgen K., Hahn M., Voegele R.T. (2005) Characterization of a novel NADP+-dependent d-arabitol dehydrogenase from the plant pathogen Uromyces fabae. Biochem J 389: 289–295.

35. Magasanik B., Kaiser C.A. (2002) Nitrogen regulation in Saccharomyces cerevisiae. Gene 290: 1–18.

36. Marzluf G.A. (1997) Genetic regulation of nitrogen metabolism in the fungi. Microbiol Mol Biol Rev 61: 17–32.

37. Mendgen K., Hahn M. (2002) Plant infection and the establishment of fungal biotrophy. Trends Plant Sci 7: 352–356.

38. Namiki F., Matsunaga M., Okuda M., Inoue I., Nishi K., Fujita Y., Tsuge T. (2001) Mutation of an arginine biosynthesis gene causes reduced pathogenicity in Fusarium oxysporum f. sp.melonis. Mol Plant-Microbe Interact 14: 580–584.

39. Oliver R.P., Solomon P.S. (2004) Does the oxidative stress used by plants for defence provide a source of nutrients for pathogenic fungi? Trends Plant Sci 9: 472–473.

40. Ospina-Giraldo M.D., Mullins E., Kang S. (2003) Loss of function of the Fusarium oxysporum SNF1 gene

reduces virulence on cabbage and Arabidopsis. Curr Genet 44: 49–57.

41. Pellier A-L., Lauge R., Venault-Fourrey C., Langin T. (2003) CLNR1, the AREA/NIT2-like global nitrogen regulator of the plant fungal pathogen Colletotrichum lindemuthianum is required for the infection cycle. Mol Microbiol 48: 639–655.
42. Perfect E.S., Green J.R. (2001) Infection structures of biotrophic and hemibiotrophic fungal plant pathogens. Mol Plant Pathol 2: 101–108.
43. Ravagnani A., Gorfinkiel L., Langdon T., et al. (1997) Subtle hydrophobic interactions between the seventh residue of the zinc finger loop and the first base of an HGATAR sequence determine promoter-specific recognition by the Aspergillus nidulans GATA factor AREA. EMBO J 16: 3974–3986.
44. Ronne H. (1995) Glucose repression in fungi. Trends Genet 11: 12–17.
45. Rooney H.C., Van›t Klooster J.W., Van Der Hoorn R.A., Joosten M.H., Jones J.D., De Wit P.J. (2005) Cladosporium Avr2 inhibits tomato Rcr3 protease required for Cf-2-dependent disease resistance. Science 308: 1783–1786.
46. Seong K., Hou Z., Tracy M., Kistler H.C., Xu J.R. (2005) Random insertional mutagenesis identifies genes associated with virulence in the wheat scab fungus Fusarium graminearum. Phytopathology 95: 744–750.
47. Snoeijers S.S., Perez-Garcia A., Joosten MHAJ, De Wit PJGM (2000) The effect of nitrogen on disease development and gene expression in bacterial and fungal plant pathogens. Eur J Plant Pathol 106: 493–506. CrossRefWeb of Science
48. Sohn J., Voegele R.T., Mendgen K., Hahn M. (2000) High level activation of vitamin B1 biosynthesis genes in haustoria of the rust fungus Uromyces fabae. Mol Plant-Microbe Interact 13: 629–636.
49. Solomon P.S., Oliver R.P. (2001) The nitrogen content of the tomato leaf apoplast increases during infection by Cladosporium fulvum. Planta 213: 241–249.
50. Solomon P.S., Oliver R.P. (2002) Evidence that γ-aminobutyric acid is a major nitrogen source during Cladosporium fulvum infection of tomato. Planta 214: 414–420.
51. Solomon P.S., Oliver R.P. (2004) Functional characterisation of glyoxalase I from the fungal wheat pathogen Stagonospora nodorum. Curr Genet 46: 115–121.Medline
52. Solomon P.S., Nielsen P.S., Clark A.J., Oliver R.P. (2000) Methionine synthase, a gene required for methionine synthesis, is expressed in planta by Cladosporium fulvum. Mol Plant Pathol 1: 315–323.
53. Solomon P.S., Tan K.C., Oliver R.P. (2003) The nutrient supply of phytopathogenic fungi; a fertile field for study. Mol Plant Pathol 4: 203–210.
54. Solomon P.S., Lee R.C., Wilson T.J., Oliver R.P. (2004a) Pathogenicity of Stagonospora nodorum requires malate synthase. Mol Microbiol 53: 1065–1073.
55. Solomon P.S., Thomas S.W., Spanu P., Oliver R.P. (2004b) The utilisation of di/tripeptides by Stagonospora nodorum is dispensable for wheat infection. Physiol Mol Plant Pathol 63: 191–199.
56. Solomon P.S., Tan K.C., Oliver R.P. (2005) Mannitol 1-phosphate metabolism is required for sporulation in planta of the wheat pathogen Stagonospora nodorum. Mol Plant-Microbe Interact 18: 110–115.
57. Solomon P.S., Jorgens C.I., Oliver R.P. (2006) Delta-aminolaevulinic acid synthesis is required for virulence of the wheat pathogen Stagonospora nodorum. Microbiol 152: 1533–1538.
58. Starich M.R., Wikstrom M., Arst H.N., Clore G.M., Gronenborn A.M. (1998) The solution structure of a fungal AREA protein-DNA complex: an alternative binding mode for the basic carboxyl tail of GATA factors. J Mol Biol 277: 605–620.
59. Stephenson S.A., Green J.R., Manners J.M., Maclean D.J. (1997) Cloning and characterisation of glutamine synthetase from Colletotrichum gloeosporioides and demonstration of elevated expression during pathogenesis on Stylosanthes guianensis. Curr Genet 31: 447–454.
60. Struck C., Hahn M., Mendgen K. (1996) Plasma Membrane H+-ATPase activity in spores, germ tubes, and haustoria of the rust fungus Uromyces viciae-fabae. Fungal Genet Biol 20: 30–35.
61. Struck C., Siebels C., Rommel O., Wernitz M., Hahn M. (1998) The plasma membrane H(+)-ATPase from the biotrophic rust fungus Uromyces fabae: molecular characterization of the gene (PMA1) and functional expression of the enzyme in yeast. Mol Plant-Microbe Interact 11: 458–465.
62. Struck C., Ernst M., Hahn M. (2002) Characterization of a developmentally regulated amino acid transporter (AAT1p) of the rust fungus Uromyces fabae. Mol Plant Pathol 3: 23–30.
63. Struck C., Mueller E., Martin H., Lohaus G. (2004) The Uromyces fabae UfAAT3 gene encodes a general amino acid permease that prefers uptake of in planta scarce amino acids. Mol Plant Pathol 5: 183–190.
64. Sweigard J.A., Carroll A.M., Farrall L., Chumley F.G., Valent B. (1998) Magnaporthe grisea pathogenicity genes obtained through insertional mutagenesis. Mol Plant-Microbe Interact 11: 404–412.
65. Talbot N., Ebbole D., Hamer J. (1993) Identification and Characterization of MPG1, a gene involved in pathogenicity from the rice blast fungus Magnaporthe grisea. Plant Cell 5: 1575–1590.
66. Thara V.K., Fellers J.P., Zhou J-M. (2003) In planta induced genes of Puccinia triticina. Mol Plant Pathol 4: 51–56.
67. Thevelein J.M. (1984) Regulation of trehalose mobilization in fungi. Microbiol Rev 48: 42–59.FREE Full Text
68. Thines E., Weber R.W., Talbot N.J. (2000) MAP kinase and protein kinase A-dependent mobilization of triacylglycerol and glycogen during appressorium turgor

generation by Magnaporthe grisea. Plant Cell 12: 1703–1718.

69. Thomas S.W., Rasmussen S.W., Glaring M.A., Rouster J.A., Christiansen S.K., Oliver R.P. (2001) Gene identification in the obligate fungal pathogen Blumeria graminis by expressed sequence tag analysis. Fungal Genet Biol 33: 195–211.

70. Thomma BPHJ, Bolton M.D., Clergeot P-H., DeWit P.J. (2006) Nitrogen controls in planta expression of Cladosporium fulvum Avr9 but no other effector genes. Mol Plant Pathol 7: 125–130.

71. Tonukari N.J., Scott-Craig J.S., Walton J.D. (2000) The Cochliobolus carbonum SNF1 gene is required for cell wall — degrading enzyme expression and virulence on maize. Plant Cell 12: 237–247.

72. Tudzynski B., Liu S., Kelly J.M. (2000) Carbon catabolite repression in plant pathogenic fungi: isolation and characterization of the Gibberella fujikuroi and Botrytis cinereacreA genes. FEMS Microbiol Lett 184: 9–15.

73. Van den Ackerveken GFJM, Dunn R.M., Cozijnsen T.J., Vossen P., Van den Broek HWJ, De Wit PJGM (1994) Nitrogen limitation induces expression of the avirulence gene avr9 in Cladosporum fulvum. Mol Gen Genet 243: 277–285.

74. Voegele R.T. (2006) Uromyces fabae: development, metabolism, and interactions with its host Vicia faba. FEMS Microbiol Lett 259: 165–173.

75. Voegele R.T., Struck C., Hahn M., Mendgen K. (2001) The role of haustoria in sugar supply during infection of broad bean by the rust fungus Uromyces fabae. Proc Natl Acad Sci USA 98: 8133–8138.

76. Voegele R.T., Hahn M., Lohaus G., Link T., Heiser I., Mendgen K. (2005) Possible roles for mannitol and mannitol dehydrogenase in the biotrophic plant pathogen Uromyces fabae. Plant Physiol 137: 190–198.

77. Voegele R.T., Wirsel S., Möll U., Lechner M., Mendgen K. (2006) Cloning and characterization of a novel invertase from the obligate biotroph Uromyces fabae and analysis of expression patterns of host and pathogen invertases in the course of infection. Mol Plant-Microbe Interact 19: 625–634.

78. Walters D. (2003) Resistance to plant pathogens: possible roles for free polyamines and polyamine catabolism. New Phytologist 159: 109–115.CrossRefWeb of Science

79. Wang Z.Y., Thornton C.R., Kershaw M.J., Debao L., Talbot N.J. (2003) The glyoxylate cycle is required for temporal regulation of virulence by the plant pathogenic fungus Magnaporthe grisea. Mol Microbiol 47: 1601–1612.

80. Wei Y., Shen W., Dauk M., Wang F., Selvaraj G., Zou J. (2004) Targeted gene disruption of glycerol-3-phosphate dehydrogenase in Colletotrichum gloeosporioides reveals evidence that glycerol is a significant transferred nutrient from host plant to fungal pathogen. J Biol Chem 279: 429–435.

81. Zhang Z., Henderson C., Perfect E., Carver TLW, Skamnioti P., Gurr S.J. (2005) Of genes and genomes, needles and haystacks: Blumeria graminis and functionality. Mol Plant Pathol 6: 561–575.

CHAPTER 13 CELL DEATH CONTROL IN PLANT FUNGAL INTERACTIONS

Martin B. Dickman and Paul de Figueiredo

While the concept of programmed cell death (PCD) or its morphological equivalent, apoptosis, was recognized in pockets of research prior to the 1970s, it was not until 1972 that Kerr, Wyllie, and Currie [1] first promulgated the phenomenon. It took nearly 20 years for the original idea to gain acceptance, thanks in large part to seminal studies conducted with *Caenorhabditus elegans*, which provided a solid genetic basis for these observations [2]. These studies also brought forth the idea that cell suicide is central to the life and well-being of multicellular organisms, and is neither uncommon nor normally detrimental to the organism.

Despite the simplicity and finality of the concept of death, the manifold processes by which cells die are not necessarily equivalent. While the role of cell death is firmly established in mammalian disease, it has not been as deeply characterized in other systems. In this Pearl article, we highlight several major ways in which cells die in the context of the high-stakes arms race between fungal pathogens and their plant hosts (Fig. 1).

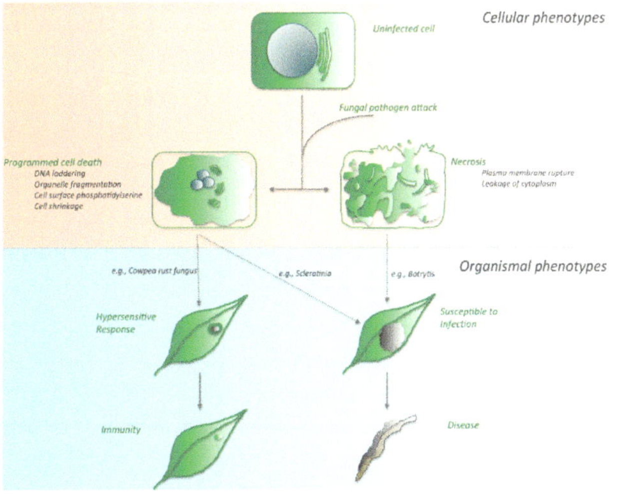

Figure 1: Cell death outcomes following fungal pathogen attack. Depending on the genotypes of both the plant host and the fungal invader, any one of several cell death pathways can be activated. Although these pathways intersect in cell death, they culminate in disparate outcomes, immunity, or disease as detailed in the text. During the recognition of fungal challenge by the plant, host-controlled HR-PCD leads to a restricted cell death phenotype and ultimately immunity. Conversely, pathogen-mediated PCD suppresses this host recognition, leading to unrestricted spread of the pathogen accompanied by PCD, susceptibility, and disease.

Citation: Dickman MB, de Figueiredo P (2013) Cell Death Control in Plant Fungal Interactions. PLoS Pathog 9(9): e1003542. doi:10.1371/journal.ppat.1003542

SOME HAVE CALLED THEE MIGHTY AND DREADFUL [3]

The material reality of death can sometimes involve the transition of a mighty living system into dreadful chaos. The process of necrotic cell death exemplifies this notion, and is associated with organelle swelling, a complete disorganization and de-compartmentalization of cellular contents, cell lysis via rupture of the plasma membrane, and leakage of cellular debris into the environment. During necrosis, the dying cell has little to no genetic control of these events. In contrast, apoptotic PCD events unfold in a genetically controlled, highly orchestrated process that includes cell shrinkage, fragmentation and "laddering" of DNA, plasma membrane blebbing, and phosphatidylserine externalization. A hybrid phenomenon, termed "programmed necrosis" (necroptosis) has also been recently described [4]. In this physiologically and pathologically relevant process, first documented in mammalian systems [5], the observed necrotic cell death functions as a back-up system that is activated when the PCD machinery is impaired. In many cases, necroptosis is induced following the initial activation of the extrinsic (receptor-mediated) PCD pathway [6]. Elevations in reactive oxygen species (ROS) are typical of necroptotic cells [7]. As functional parallels exist between plant and animal PCD (see below), it is plausible that necroptosis may constitute a novel hub of plant PCD systems.

Pathways for apoptotic programmed cell death are relatively well characterized in mammals. This is in contrast to plants and phytopathogenic fungi where core regulatory elements of apoptosis pathways have not been found [8]. However, expression of mammalian apoptosis regulators confers phenotypes in plants and fungi that are consistent with their known functions[9]. In addition, numerous studies with baker's yeast (*Saccharomyces cerevisiae*) have shown that this organism exhibits several, but not all, of the hallmarks of apoptosis [10]. Production of ROS, chromatin condensation, DNA fragmentation, and externalization of phosphatidylserine have been shown in yeast, but there is not complete agreement as to whether "true" apoptosis occurs. As is the case in plants, yeast do not possess *bona fide* caspases that possess the precise structural characteristics that define these proteases. However, both yeast and plants harbor distantly related meta-caspases, identified computationally (for a lively discussion of the pros and cons of yeast PCD vs. mammalian PCD, see [10], [11]).

Finally, there have been several comparative studies that examined plant apoptotic-like PCD in the context of biotic and abiotic stress responses, senescence, or other aspects of plant growth and development [12]. Plant and animal cell death regimes clearly displayed differences. For example, plant cells have a rigid cell wall and lack caspases or phagocytic machinery. However, plant and animal PCD share important similarities, including chromatin condensation, DNA laddering, the generation of ROS, and the externalization of phosphatidylserine. Importantly, the underlying conceptual framework for PCD, ranging from development to pathogen attack to abiotic stress, is remarkably conserved for all eukaryotes. Thus, an analysis of cell suicide in plant and animal systems suggests that these processes are observed across kingdoms. For a more detailed discussion of this topic, see [13].

WITH POISON, WAR, AND SICKNESS DWELL [3]

Many necrotrophic plant pathogenic fungi produce phytotoxic metabolites and peptides that play a central role in their pathogenic programs. These "poisons" are used by pathogens to attack their susceptible plant hosts. Phytotoxins can be non-host-specific and thus target a broad range of host plants, or host-specific and thus target a single plant species or even a particular cultivar within a given species [14]. However, while fungal toxins can be simply toxic, increasing evidence suggests that some toxins induce signaling that directs host pathways towards PCD, which exclusively benefits the fungus. For example, the filamentous ascomycete*Cochliobolus victoriae* is a necrotrophic fungal pathogen of *Arabidopsis* and oats and the causative agent of Victoria blight [7], a disease which decimated U.S. oat production in the 1940s. The fungal host–selective toxin victorin is a chlorinated cyclic pentapeptide that elicits several defense responses, PCD, and disease. In *Arabidopsis*, victorin targets a Nod-like NB-LRR-type resistance (R) protein, designated LOV1 [15]. This R protein must be present for fungal susceptibility. The fungus exploits R gene– mediated resistance to incite disease by inducing plant defense responses. Thus, the host R gene, which triggers cell death as a presumable defense, can function as both a resistance and a susceptibility factor. Taken together, the evidence indicates that death-inducing "poisons" can display subtlety in controlling the outcome of fungal pathogen–host interactions.

THOU ART SLAVE TO FATE [3]

Host-controlled PCD is often required for defense against pathogens. Indeed, the well-studied plant PCD commonly known as the hypersensitive response (HR) is correlated with host defenses that serve to restrict pathogen growth. In the plant host, HR-PCD limits the spread of the pathogen, which is particularly relevant in the case of interactions between biotrophic pathogens and their plant hosts, where host-controlled cell death is correlated with resistance. Conversely, several fungal pathogens appear to promote host cell death, thereby subverting plant cell death pathways for fungal nutrient acquisition. Thus, programmed cell death is a common readout observed during both susceptible and resistant interactions. The eventual victor is decided by which side is in control of cell death, cellular context, and the activities of the combatants. Therefore, the way host cells die provides insight into the eventual fate of the fungal host interaction.

POPPIES AND CHARMS CAN MAKE US SLEEP AS WELL [3]

The cellular mechanisms that can drive cell death are manifold. Two such "poppies and charms" that can lead to cell death are pyroptosis and autophagy. In animal systems, apoptosis can provide for the "clean" removal of dying cells, with limited induction of inflammatory responses. However, in pyroptosis, cell death is accompanied by the activation of inflammatory signaling. First noted in macrophages [16] and more recently reported in mammalian cells infected with viral or bacterial pathogens [17], pyroptosis typically engages the host inflammasome, a macromolecular complex that senses pathogen-associated molecular patterns (PAMPS) or danger-associated molecular patterns (DAMPS) to drive caspase-1 (and possibly caspase-11) dependent pro-inflammatory cascades [18]. Recently, these ideas have been extended to interactions between fungal pathogens and their susceptible plant hosts [19]. For example, while there are no *bona fide* caspases in plants, plant vacuolar processing enzyme gamma (VPEg) exhibits caspase 1-like activity in plant systems [20], [21]. Plant VPE protease activity is suppressed by caspase-1 specific inhibitors (xVAD-fmk) and is necessary for cell death mediation from a wide range of plant pathogens. The *Fusarium verticilliodes* toxin Fumonisin B1 requires the vacuolar processing enzyme (VPE) for PCD, as VPE mutants prevent mycotoxin-triggered death [22]. Similarly, the model oomycete pathogen,*Hyaloperonospora arabidopsidis,* induces VPE activity during infection of *Arabidopsis* [23]. This finding is somewhat surprising, as *Hpa* is a biotroph. It was proposed that increased VPE activity may benefit the pathogen by mediating protein turnover and nutrient release. The plant and animal enzymes, while having similar model substrate specificity, clearly display differences. Unlike mammalian caspase 1 which is localized in the cytosol, VPE is localized in the plant vacuole. Therefore, activation of VPE signaling and proteolysis contribute to vacuolar breakdown and apoptosis.

Autophagy can also promote cell death during interactions between fungal pathogens and their susceptible plant hosts. Autophagy was originally understood to be a process whereby starved cells cannibalize their organelles and cytosolic components to promote their survival. However, autophagy as a means for cell survival and homeostasis is now also appreciated as a mechanism to control interactions between fungal pathogens and plants.

The role of plant autophagy in response to fungal pathogens has been investigated in several studies; however, the mechanistic details are incomplete. Autophagy-defective *Arabidopsis*plants were more resistant to the biotrophic fungus *Golovinomyces cichoracearum* in a salicylic acid (SA)-dependent manner [24]. Increased resistance to the virulent biotrophic bacterial pathogen *Pseudomonas syringae* pv. *tomato* was also observed in plants harboring mutations in autophagy proteins [25], [26]. These studies suggest a negative effect of plant autophagy in resistance responses toward biotrophic pathogens and are in contrast to N-mediated resistance to tobacco mosaic virus (TMV), which requires functional autophagic machinery for an effective (HR-PCD) defense response [27]. Thus, in various host pathogen interactions, the consequences of host autophagy-mediated PCD can be diverse.

Although the effect of autophagy on biotrophic interactions is complex and perhaps indirect in its relationship to disease, necrotrophic organisms feed off of dead cells and thus it might be expected that conditions favorable to cell death would promote nectrotrophic growth. Indeed,*Arabidopsis* ATG5, ATG10, and ATG18a mutants developed spreading necrosis and enhanced disease susceptibility upon infection with necrotrophic fungus *Alternaria brassicicola* [25], [26]. This includes elevated ROS, which appeared in noninfected regions as well. Similar runaway cell death symptoms were observed in the autophagy mutants after application of fumonisin B1. In contrast, no difference in growth characteristics was observed for the obligate biotroph *H. arabidopsidis* [25]. Thus, runaway (unrestricted) cell death in the mutants indicates that autophagy plays an anti-death role, perhaps by limiting ROS effects and restricting pathogen spread. Indeed, elevated spontaneous ROS accumulation is readily observed in autophagy mutants [28]. Further support for susceptibility to nectrotrophic lifestyles was shown by enhanced rates of infection by the necrotrophic pathogen *Botrytis cinerea* in AtATG18a mutants [29]. Not unexpectedly, the necrotrophic pathogen *B. cinerea* has been reported to benefit from PCD [30]. Thus, cell death induced by autophagy might be expected to result in enhanced fungal growth. It is reasonable that a necrotroph would benefit from a local environment of dead tissue. However, this is clearly not the case, as *B. cinerea* triggers autophagous cell death, resulting in a restricted growth phenotype (Dickman, unpublished observations). We interpret these observations to indicate that the process by which host cells die (i.e., the means by which death occurs) is key for determining the ultimate outcome of interactions between host and pathogen [31].

AND DEATH SHALL BE NO MORE; DEATH, THOU SHALT DIE [3]

The struggle between plant hosts and their fungal pathogens to control the mechanisms by which host cells die can have profound effects on the ultimate outcome of the interaction. With this idea in mind, it is notable that, to date, very few examples of fungal anti-death factors have been described. It has been shown that *B. cinerea* undergoes massive PCD following penetration of the host plant and establishment of a primary necrotic lesion [32]. The host defense molecule camalexin was shown to be capable of triggering PCD in *B. cinerea.* In response, the fungus suppresses this host-induced PCD to establish infection. Despite these intriguing observations, the anti-death factors that mediate the suppression of host PCD in this system have yet to be identified. The identification of factors that suppress host cell death in necrotrophic fungal pathosystems (e.g., see [33]) constitutes an important line of future investigation.

REFERENCES

1. Kerr JF, Wyllie AH, Currie AR (1972) Apoptosis: a basic biological phenomenon with wide-ranging implications in tissue kinetics. Br J Cancer 26 (4) 239–257.
2. Ellis RE, Yuan JY, Horvitz HR (1991) Mechanisms and functions of cell death. Annu Rev Cell Biol 7: 663–698. doi: 10.1146/annurev.cb.07.110191.003311
3. Donne J (1633) "Death Be Not Proud" (Holy Sonnet XI of XIX).
4. Christofferson DE, Yuan J (2010) Necroptosis as an alternative form of programmed cell death. Curr Opin Cell Biol 22 (2) 263–268. doi: 10.1016/j.ceb.2009.12.003
5. Kaczmarek A, Vandenabeele P, Krysko DV (2013) Necroptosis: the release of damage-associated molecular patterns and its physiological relevance. Immunity 38 (2) 209–223. doi: 10.1016/j.immuni.2013.02.003
6. Chaabane W, et al. (2013) Autophagy, apoptosis, mitoptosis and necrosis: interdependence between those pathways and effects on cancer. Arch Immunol Ther Exp (Warsz) 61 (1) 43–58. doi: 10.1007/s00005-012-0205-y

7. Zhou Z, Han V, Han J (2012) New components of the necroptotic pathway. Protein Cell 3 (11) 811–817. doi: 10.1007/s13238-012-2083-9

8. Dickman MB, Reed JC (2003) Paradigms for programmed cell death in animals and plants. In: Programmed Cell Death in Plants, ed. J Gray, pp.26–43. Blackwell, Oxford.

9. Dickman MB, Park YK, Oltersdorf T, Li W, Clemente T, French R (2001) Abrogation of disease development in plants expressing animal antiapoptotic genes. Proc Natl Acad Sci U S A 98: 6957–6962. doi: 10.1073/pnas.091108998

10. Carmona-Gutierrez D, Frohlich KU, Kroemer G, Madeo F (2010) Metacaspases are caspases. Doubt no more. Cell Death Differ 17: 377–78. doi: 10.1038/cdd.2009.198

11. Enoksson M, Salvesen GS (2010) Metacaspases are not caspases–always doubt. Cell Death Differ 17: 1221. doi: 10.1038/cdd.2010.45

12. Williams B, Dickman M (2008) Plant programmed cell death: can't live with it; can't live without it. Mol Plant Pathol 9: 531–544. doi: 10.1111/j.1364-3703.2008.00473.x

13. Dickman MB, Fluhr R (2013) Centrality of host cell death in plant-microbe interactions. Annu Rev Phytopathol 51: 25.1–25.28. doi: 10.1146/annurev-phyto-081211-173027

14. Wolpert TJ, Dunkle LD, Ciuffetti LM (2002) Host-selective toxins and avirulence determinants: what's in a name? Annu Rev Phytopathol 40: 251–285. doi: 10.1146/annurev.phyto.40.011402.114210

15. Lorang J, Lorang J, Kidarsa T, Bradford CS, Gilbert B, et al. (2012) Tricking the guard: exploiting plant defense for disease susceptibility. Science 338 (6107) 659–662. doi: 10.1126/science.1226743

16. Bergsbaken T, Fink SL, Cookson BT (2009) Pyroptosis: host cell death and inflammation. Nat Rev Microbiol 7 (2) 99–109. doi: 10.1038/nrmicro2070

17. Rosales-Reyes R, Pérez-López A, Sánchez-Gómez C, Hernández-Mote RR, Castro-Eguiluz D, et al. (2012) Salmonella infects B cells by macropinocytosis and formation of spacious phagosomes but does not induce pyroptosis in favor of its survival. Microb Pathog 52 (6) 367–374. doi: 10.1016/j.micpath.2012.03.007

18. Broz P, Ruby T, Belhocine K, Bouley DM, Kayagaki N, et al. (2012) Caspase-11 increases susceptibility to Salmonella infection in the absence of caspase-1. Nature 490 (7419) 288–291. doi: 10.1038/nature11419

19. Hara-Nishimura I, Hatsugai N (2011) The role of vacuole in plant cell death. Cell Death Differ 18 (8) 1298–1304. doi: 10.1038/cdd.2011.70

20. Hatsugai N, Kuroyanagi M, Yamada K, Meshi T, Tsuda S, et al. (2004) A plant vacuolar protease, VPE, mediates virus-induced hypersensitive cell death. Science 305 (5685) 855–858. doi: 10.1126/science.1099859

21. Rojo E, Martín R, Carter C, Zouhar J, Pan S, et al. (2004) VPEgamma exhibits a caspase-like activity that contributes to defense against pathogens. Curr Biol 14 (21) 1897–1906. doi: 10.1016/j.cub.2004.09.056

22. Zhang H, Dong S, Wang M, Wang W, Song W, et al. (2010) The role of vacuolar processing enzyme (VPE) from Nicotiana benthamiana in the elicitor-triggered hypersensitive response and stomatal closure. J Exp Bot 61 (13) 3799–3812. doi: 10.1093/jxb/erq189

23. Misas-Villamil JC, Toenges G, Kolodziejek I, Sadaghiani AM, Kaschani F, et al. (2013) Activity profiling of vacuolar processing enzymes reveals a role for VPE during oomycete infection. Plant J 73 (4) 689–700. doi: 10.1111/tpj.12062

24. Wang Y, Nishimura MT, Zhao T, Tang D (2011) ATG2, an autophagy-related protein, negatively affects powdery mildew resistance and mildew-induced cell death in Arabidopsis. Plant J 68 (1) 74–87. doi: 10.1111/j.1365-313x.2011.04669.x

25. Lenz HD, Haller E, Melzer E, Gust AA, Nurnberger T (2011) Autophagy controls plant basal immunity in a pathogenic lifestyle-dependent manner. Autophagy 7 (7) 773–774. doi: 10.4161/auto.7.7.15535

26. Lenz HD, Haller E, Melzer E, Kober K, Wurster K, et al. (2011) Autophagy differentially controls plant basal immunity to biotrophic and necrotrophic pathogens. Plant J 66 (5) 818–830. doi: 10.1111/j.1365-313x.2011.04546.x

27. Liu Y, Schiff M, Czymmek K, Talloczy Z, Levine B, Dinesh-Kumar SP (2005) Autophagy regulates programmed cell death during the plant innate immune response. Cell 20;121 (4) 567–77. doi: 10.1016/j.cell.2005.03.007

28. Yoshimoto K, Jikumaru Y, Kamiya Y, Kusano M, Consonni C, et al. (2009) Autophagy negatively regulates cell death by controlling NPR1-dependent salicylic acid signaling during senescence and the innate immune response in Arabidopsis. Plant Cell 21 (9) 2914–2927. doi: 10.1105/tpc.109.068635

29. Lai Z, Wang F, Zheng Z, Fan B, Chen Z (2011) A critical role of autophagy in plant resistance to necrotrophic fungal pathogens. Plant J 66 (6) 953–968. doi: 10.1111/j.1365-313x.2011.04553.x

30. Govrin EM, Levine A (2000) The hypersensitive response facilitates plant infection by the necrotrophic pathogen Botrytis cinerea. Curr Biol 10 (13) 751–757. doi: 10.1016/s0960-9822(00)00560-1

31. Kabbage M, Williams B, Dickman M (2013) Cell death control: The interplay of apoptosis and autophagy in the pathogenicity of Sclerotinia sclerotiorum. PLoS Pathog 9: e100328 doi:10.1371/journal.ppat.1003287.

32. Shlezinger N, Minz A, Gur Y, Hatam I, Dagdas YF, et al. (2011) Anti-apoptotic machinery protects the necrotrophic fungus Botrytis cinerea from host-induced apoptotic-like cell death during plant infection. PLoS Pathog 7 (8) e1002185 doi:10.1371/journal.ppat.1002185.

33. Zhu W, Wei W, Fu Y, Cheng J, Xie J, et al. (2013) A secretory protein of necrotrophic fungus Sclerotinia sclerotiorum that suppresses host resistance. PLoS ONE 8 (1) e53901 doi:10.1371/journal.pone.0053901.

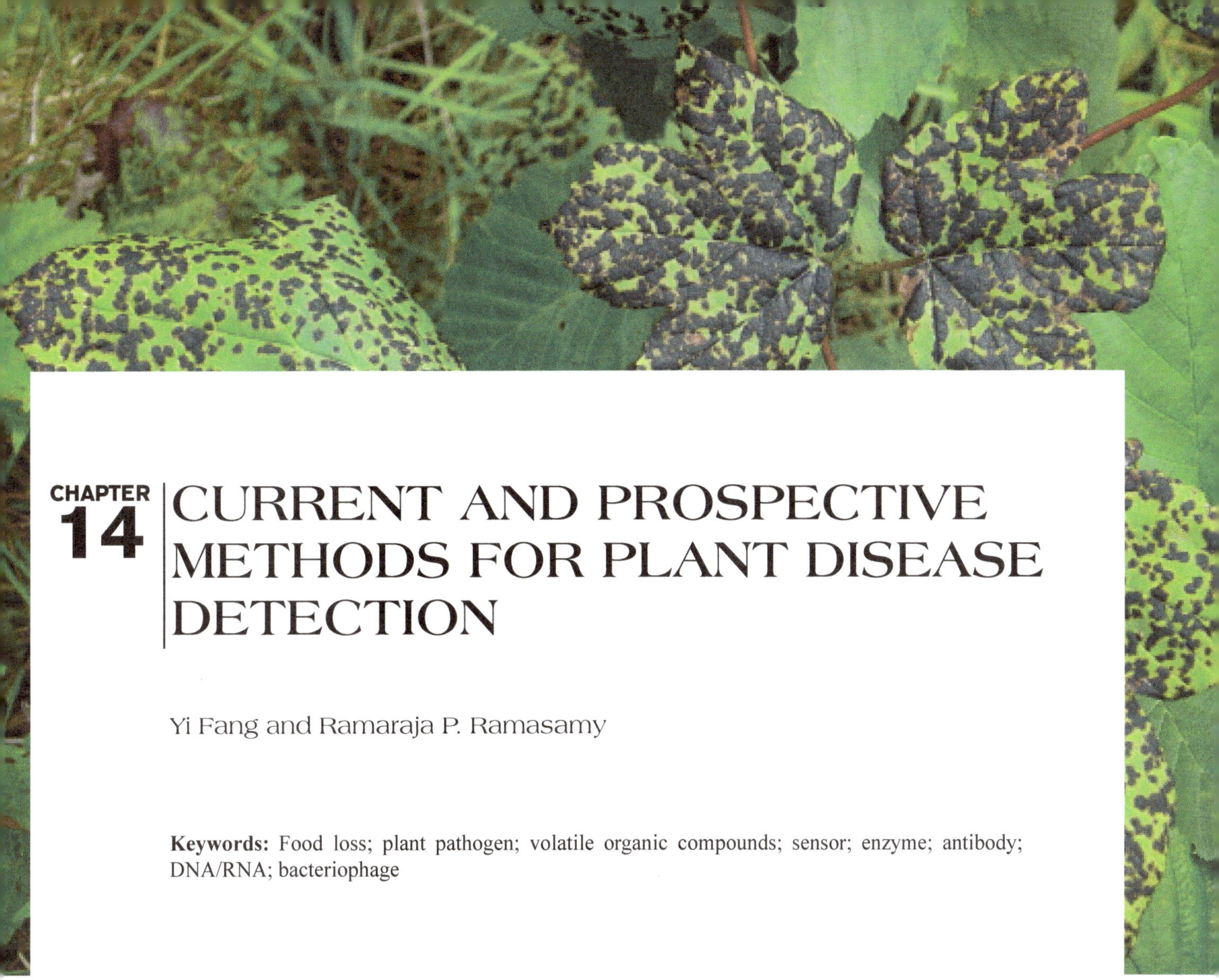

CHAPTER 14 CURRENT AND PROSPECTIVE METHODS FOR PLANT DISEASE DETECTION

Yi Fang and Ramaraja P. Ramasamy

Keywords: Food loss; plant pathogen; volatile organic compounds; sensor; enzyme; antibody; DNA/RNA; bacteriophage

INTRODUCTION

Global food security as determined by the balance of global food production and demand has become an important international issue in recent years [1,2,3]. In 2008, an increase in food prices brought about a global crisis that caused political and economic instability in some developing countries [4]. It was estimated that the demand for food will continue to increase for another 40 years due to the continuous increase in human population. The projections also indicate that an additional 70% of food production is required by 2050 to meet the needs [5]. Currently, over one billion people are suffering from different situations of malnutrition due to lack of food supply and approximately twice that population do not have access to sufficient nutrients or vitamins to meet their daily nutrition needs [6]. The situation can be attributed to the continuous decline in agricultural land area that causes a decrease in productivity. Although decrease in agricultural productivity can be attributed to a variety of reasons, damage caused by pests and pathogens plays a significant role in crop losses throughout the world. The losses in crop yield due to pathogen infections range between 20% and 40% [7]. On average, pathogen-induced losses of maize, barley, rice and soybean are estimated to be around 12%, groundnuts and potatoes are estimated to be around 24% and wheat and cotton are estimated to be around 50% and 80%, respectively [8]. Post-cultivation losses due to diseases and sub-standard quality are estimated to be 30%–40%. Overall, the economic losses due to infections are estimated at 40 billion dollars annually in the United States alone [9,10]. In order to minimize the disease induced damage in crops during growth, harvest and postharvest processing, as well as to maximize productivity and ensure agricultural sustainability, advanced disease detection and prevention in crops are highly important.

CURRENT METHODS FOR CROP DISEASE DETECTION

Detection and identification of diseases in crops could be realized via both direct and indirect methods. Direct detection of diseases includes molecular and serological methods that could be used for high-throughput analysis when large numbers of samples need to be analyzed. In these methods, the disease causing pathogens such as bacteria, fungi and viruses are directly detected to provide accurate identification of the disease/pathogen. On the other hand, indirect methods identify the plant diseases through various parameters such as morphological change, temperature change, transpiration rate change and volatile organic compounds released by infected plants.

Citation: Fang and Ramaraja P. Ramasamy, Current and Prospective Methods for Plant Disease Detection, Biosensors 2015, 4, 537-561; doi:10.3390/bios5030537

Direct Detection Methods

Table 1 compares the available direct detection methods for plant pathogens based on their limit of detection, advantages and limitations. Each of these methods is discussed in detail in the following paragraphs.

Polymerase Chain Reaction

In the years of 1984 and 1993, two Nobel prizes were awarded to J.F. Kohler and C. Milstein, and K. Mullis for development of monoclonal antibodies and amplification of nucleic acid sequences, respectively, using the technology of polymerase chain reaction (PCR). Based on the fidelity of DNA hybridization and replication, PCR was initially used for highly specific detection of diseases caused by bacteria and viruses [11]. Now it has been widely used for the detection of plant pathogens as well. In addition to the basic PCR technology, advanced PCR methods such as reverse-transcription PCR (RT-PCR) has also been used for plant pathogen identification due to its high sensitivity [12]. Multiplex PCR was proposed to enable simultaneous detection of different DNA or RNA by running a single reaction [13,14,15,16,17]. Real-time PCR platforms have also been used for on-site, rapid diagnosis of plant diseases based on the bacterial, fungal and viral nucleic acids [18,19]. Although PCR technique can provide high sensitivity and specificity due to the fidelity of DNA amplification, it is also limited by lack of operational robustness [20]. PCR depends on the efficacy of DNA extraction and the performance is affected by inhibitors present in the sample assay, polymerase activity, PCR buffer and concentration of deoxynucleoside triphosphate [20]. In addition, application of PCR for pathogen detection requires designing a primer to initiate DNA replication, which could limit the practical applicability of this technique for field sampling of diseases [19].

Table 1: Comparison of the current methods for detecting plant diseases resulting from bacterial pathogens

Techniques	Limit of Detection (CFU/mL) [12]	Advantages	Limitations
PCR	10^3–10^4	Mature and common technology, portable, easy to operate.	Effectiveness is subjected to DNA extraction, inhibitors, polymerase activity, concentration of PCR buffer and deoxynucleoside triphosphate.
FISH	10^3	High sensitivity.	Autofluorescence, photobleaching.
ELISA	10^5–10^6	Low cost, visual color change can be used for detection.	Low sensitivity for bacteria.
IF	10^3	High sensitivity, target distribution can be visualized.	Photobleaching.
FCM	10^4	Simultaneous measurement of several parameters, rapid detection.	High cost, overwhelming unnecessary information.

PCR: polymerase chain reaction; FISH: fluorescence in-situ hybridization; ELISA: enzyme-linked immunosorbent assay; IF: immunofluorescence; FCM: flow cytometry; CFU: colony forming unit.

Fluorescence in-situ Hybridization

Another type of molecular detection technique is fluorescence *in-situ* hybridization (FISH), which is applied for bacterial detection in combination with microscopy and hybridization of DNA probes and target gene from plant samples [21]. Due to the presence of pathogen-specific ribosomal RNA (rRNA) sequences in plants, recognizing this specific information by FISH can help detect the pathogen infections in plants. In addition to bacterial pathogens, FISH could also be used to detect fungi and viruses and other endosymbiotic bacteria that infect the plant [22,23]. The high affinity and specificity of DNA probes provide high single-cell sensitivity in FISH, because the probe will bind to each of the ribosomes in the sample. However, the practical limit of detection lies in the range of around 10^3 CFU/mL. In addition to the detection of culturable microorganisms that cause the plant diseases, FISH could also be used to detect yet-to-be cultured (so called unculturable) organisms in order to investigate complex microbial communities [24]. However, besides the advantages, FISH also has some pitfalls that compromise the technique's potency for plant disease detection [24,25,26]. For example, false positive results with autofluorescence materials are a common problem that often lowers the specificity. Accuracy and reliability of FISH is highly dependent on the specificity of the nucleotide probes. Insufficient penetration, higher order structure of target or probe (e.g., three-dimensional rRNA, loop and hairpin formation and rRNA-protein interactions), low rRNA content, photobleaching could also cause false negative results and hence compromise the limit of detection [24,25,26].

Enzyme-Linked Immunosorbent Assay

The enzyme-linked immunosorbent assay (ELISA) is another molecular method for identification of diseases based on antibodies and color change in the assay [27]. In this method, the target epitopes (antigens) from the viruses, bacteria and fungi are made to specifically bind with antibodies conjugated to an enzyme. The detection can be visualized based on color changes resulting from the interaction between the substrate and the immobilized enzyme. The performance of ELISA can be improved greatly with the application of specific monoclonal and recombinant antibodies which are commercially available [28,29]. Specific monoclonal antibodies have been used in ELISA to achieve lower limits of detection in the order of 10^5–10^6 CFU/mL [12]. For plant disease detection, tissue print-ELISA and lateral flow devices that enable detection have been fabricated for on-site detection. However, the sensitivity for bacteria is relatively low (10^5–10^6 CFU/mL, Table 1) making it useful only for the confirmation of plant diseases after visual symptoms appear but not for early detection before disease symptoms occur [12].

Immunofluorescence

Immunofluorescence (IF) is a fluorescence microscopy-based optical technique used for the analyses of microbiological samples. The technique can also be utilized to detect pathogen infections in plant tissues. For this technique, plant samples are fixed to microscope slides in thin tissue sections. Detection is achieved by conjugating a fluorescent dye to the specific antibody to visualize the distribution of target molecule throughout the sample [30]. IF has been used to detect onion crop infection by a fungus *Botrytis cinerea* [31]. IF has also been combined with other techniques such as FISH for *Solanum dulcamara* detection which causes crown rot in potatoes [32]. Similar to FISH and other fluorescence-based techniques, a significant problem with IF is photobleaching which results in false negative results. However, the decrease of sensitivity due to photobleaching can be controlled by reducing the intensity and duration of light exposure, increasing the concentration of fluorophores, and employing more robust fluorophores that are less sensitive to photobleaching.

Flow Cytometry

Flow cytometry (FCM) is a laser-based optical technique widely used for cell counting and sorting, biomarker detection and protein engineering. FCM is used for rapid identification of cells while cells pass through an electronic detection apparatus in a liquid stream. The advantage of this technology is the capability for simultaneous measurement of several parameters. The technique uses an incident laser beam and measures the scattering and fluorescence of the laser beam reflected from the sample. Although FCM has been primarily applied to study cell cycle kinetics and antibiotic susceptibility, to enumerate bacteria, to differentiate viable from non-viable bacteria, and to characterize bacterial DNA and fungal spores, it is still a relatively new technique for plant disease detection application [33]. FCM in combination with fluorescent probes has been applied for rapid detection of foodborne bacterial pathogens. Accurate detections within 30 min down to level of 10^4 colony forming units (CFU) per milliliter have been reported [33]. FCM has been proven to be efficient for detection of soil borne bacteria such as *Bacillus subtilis* in mushroom composts [34]. In addition to bacterial detection, FCM has also been reported for viability evaluation as well [35].

Indirect Detection Methods

In addition to the direct methods discussed above, indirect methods based on plant stress profiling and plant volatile profiling have also been used for the identification of biotic and abiotic stresses as well as pathogenic diseases in crops. In this regard, new types of optical sensors that detect biotic and abiotic stresses in plants have been developed and reported in the literature [36,37,38,39]. The optical sensors provide detailed information based on different electromagnetic spectra and thus, enable prediction of the plant health [40]. Thermography, fluorescence imaging and hyperspectral techniques are among the most favorable indirect methods for plant disease detection [41].

Thermography

Thermography allows imaging the differences in surface temperature of plant leaves and canopies. The emitted infrared radiation can be captured by thermographic cameras and color difference can be analyzed. Previous reports have demonstrated that the loss of water in plants regulated by stomata would be affected by phytopathogens [42,43,44,45,46]. The resulting disease can be monitored through thermographic imaging and the amount of water transpired can be determined, without the external temperature influences [44]. The temperature changes resulting from the plant pathogen infection have been reported by several research groups [42,43,44,45,46]. The thermographic detection of plant diseases can also be scaled up for direct detection of disease in plants [47]. Thermography is also a promising tool to monitor the heterogeneity in the infection of soilborne pathogens [48]. However, the practical applicability of thermography for disease monitoring is limited due to its high sensitivity to the change of environmental conditions during measurements. Additionally, thermographic detection lacks the specificity towards diseases, and therefore cannot be used to identify the type of infection or distinguish between diseases that produce similar thermographic patterns.

Fluorescence Imaging

In this technique, the chlorophyll fluorescence is measured on the leaves as a function of the incident light and the change in fluorescence parameters can be used to analyze pathogen infections, based on changes in the photosynthetic apparatus and photosynthetic electron transport reactions [49,50]. Using this technique, temporal and spatial variations of chlorophyll fluorescence were analyzed for precise detection of leaf rust and powdery mildew infections in wheat leaves at 470 nm [50]. Although fluorescence measurement provides sensitive detection of abnormalities in photosynthesis, the practical application of this technique in a field setting is limited [51,52,53].

Hyperspectral Techniques

Hyperspectral imaging can be used to obtain useful information about the plant health over a wide range of spectrum between 350 and 2500 nm. Hyperspectral imaging is increasingly being used for plant phenotyping and crop disease identification in large scale agriculture. The technique is highly robust and it provides a rapid analysis of the imaging data. Furthermore, hyperspectral imaging cameras facilitate the data collection in three dimension, with X- and Y- axes for spatial and Z- for spectral, which contributes to more detailed and accurate information about plant health across a large geographic area [40]. Hyperspectral techniques have been widely used for plant disease detection by measuring the changes in reflectance resulting from the biophysical and biochemical characteristic changes upon infection. *Magnaporthe grisea*infection of rice, *Phytophthora infestans* infection of tomato and *Venturia inaequalis* infection of apple trees have been identified and reported using hyperspectral imaging techniques [54,55,56].

Gas Chromatography

A completely different non-optical indirect method for plant disease detection involves the profiling of the volatile chemical signature of the infected plants. The pathogen infections of plants could result in the release of specific volatile organic compounds (VOCs) that are highly indicative of the type of stress experienced by plants [57]. An infection by*Phytophthora cactorum*, the fungus that causes crown rot diseases in strawberries, results in the release of *p*-ethylguaiacol and*p*-ethylphenol as characteristic VOCs from the infected portion of the strawberry plant/fruit [57,58]. VOCs are also produced when green leaf plants are damaged pathogenically and mechanically. For example, green leaf volatiles (GLVs) such as *cis*-3-hexenol, *cis*-hexenyl acetate and hexyl acetate are reported to have been produced during mechanical damage to plant leaves, *i.e.*, during herbivore invasion [59]. Profiling of such VOC could be used as a means to identify the type and nature of infection and accordingly be used for disease diagnostics and confirmation [57]. The volatile signature of plants could be analyzed using gas-chromatography (GC) technique to analyze the presence of the specific VOC that is indicative of a particular disease [60]. To enhance the performance of compound separation and analysis, the gas chromatography is often combined with mass spectrometry (GC-MS) to identify unknown compounds in the volatile sample [61,62,63]. In comparison to the optical imaging-based detection methods discussed above, GC/GC-MS can provide more accurate information about the plant disease due to its high specificity. It also allows the detection of diseases at different stages based on the quantitative information collected from the VOC sample. However, unlike the imaging system which can directly obtain the data on-field, GC/GC-MS requires sampling of pre-collected VOC for a longer time before data analysis, which severely limits its on-field application.

Although various methods for plant disease identification have already been developed and some have been implemented, their application is limited due to multiple reasons: they are either time consuming, destructive, demand a skilled technician, require laboratory set-up, do not provide real-time monitoring (e.g., FISH, ELISA, IF, FCM, GC-MS) and/or display low specificity (e.g., imaging techniques). Growers are interested in a solution that could help them identify pathogen infections in crops in a rapid, real-time and non-destructive fashion so that timely intervention and preventative treatments can be performed to contain the infection and minimize the crop losses. This would allow the growers to save millions of dollars in fungicide costs, by allowing them to localize sprayings and timely applications rather than preemptive spray massive regions of crop field.

DETECTION OF PLANT DISEASES USING PORTABLE SENSORS

A wide variety of sensors have been developed and commercialized for various applications including environmental monitoring and medical diagnostics. Depending on the operating principle of the sensor, the analytes could be detected using a sensor based on electrical, chemical, electrochemical, optical, magnetic or vibrational signals. The limit of detection could be enhanced by the use of nanomaterial matrices as transducers and the specificity could be enhanced by the use of bio-recognition elements such as DNA, antibody, enzymes *etc*.

Biosensor Platforms Based on Nanomaterials

Recent breakthroughs in nanotechnology enable the preparation of various nanoparticles and nanostructures with few technical hurdles. Nanoparticles display fascinating electronic and optical properties and can be synthesized using different types of materials for electronics and sensing applications [64]. For biosensing application, the limit of detection and the overall performance of a biosensor can be greatly improved by using nanomaterials for their construction. The popularity of nanomaterials for sensor development could be attributed to the friendly platform it provides for the assembly of bio-recognition element, the high surface area, high electronic conductivity and plasmonic properties of nanomaterials that enhance the limit of detection. Various types of nanostructures have been evaluated as platforms for the immobilization of a bio-recognition element to construct a biosensor. The immobilization of the biorecognition element, such as DNA, antibody and enzyme, can be achieved using various approaches including biomolecule adsorption, covalent attachment, encapsulation or a sophisticated combination of these methods. The nanomaterials used for biosensor construction include metal and metal oxide nanoparticles, quantum dots, carbon nanomaterials such as carbon nanotubes and graphene as well as polymeric nanomaterials. Nanoparticles have been utilized with other biological materials such as antibody for detecting*Xanthomonas axonopodis* that causes bacterial spot disease [65]. Gold nanoparticle-based optical immunosensors have been developed for detection of karnal bunt disease in wheat using surface plasmon resonance (SPR) [66]. In addition to single probe sensors, nano-chips made of microarrays which contain fluorescent oligo probes were also reported for detecting single nucleotide change in the bacteria and viruses with high sensitivity and specificity based on DNA hybridization [67]. Fluorescent silica nanoparticles (FSNPs) combined with antibody as a biomarker have been studied as the probe, which successfully detected plant pathogens such as *Xanthomonas axonopodis* pv. *Vesicatoria* that cause bacterial spot diseases in*Solanaceae* plant [65].

Quantum dots (QD) have also been used for biosensor construction for disease detection [68]. Due to their unique and advantageous optical properties, they have been used for disease detection using fluorescence resonance energy transfer (FRET) mechanism [69], which describes energy transfer between two light-reactive molecules. Many QD-FRET-based sensors have been developed for phytodisease detection such as the witches' broom disease of lime (WBDL) caused by*Candidatus* Phytoplasma aurantifolia (*Ca.* P. aurantifolia). The immunosensor constructed showed a high sensitivity and specificity of 100%, and a detection limit of 5 *ca.* P. aurantifolia/μL [70]. Additionally, QD-FRET was also reported to detect the disease vectors. For example, Rhizomania, which is the most destructive disease in sugar beet, is caused by *beet necrotic yellow vein virus* (BNYVV). *Polymyxa betae* (Keskin), the only known vector of BNYVV, for transmission of the virus to the plants was successfully reported to be detected by QD-FRET-based sensor [71].

In addition to QD-based sensors, the use of other novel materials for sensor fabrication have been explored in order to attain high sensitivity and low limits of detection [72,73,74]. Gold nanoparticles are widely used nanomaterials due to their high electroactivity and electronic conductivity for electron transfer [75,76]. Recently, nanomaterial-based electrochemical sensors have been reported for plant disease detection by our group [77]. The application of gold nanoparticle (AuNP) modified electrode has been reported by our group for the electrochemical detection of methyl salicylate, a key plant volatile organic compound released by plants during infections (Figure 1). [77]. Moreover, in addition to gold nanoparticles, semiconductive metal oxide nanoparticles have also been reported for VOC detection due to its advantages such as low cost, suitability for electron conduction for amperometric signal and the ease at which to obtain a desired size and shape. Our previous work has demonstrated the application of metal oxide nanoparticles such as SnO_2 and TiO_2 for VOC detection such as *p*-ethylguaiacol produced by infected strawberry (Figure 2) [57]. The limit of detection achieved was in the nanomolar concentration range [57]. In addition to the detection of plant VOC which is indicative of a particular disease, nanoparticles can also be used for detection of compounds released by the pathogen itself. Different types of phytopathogens—phytobacteria, viruses and fungi—have been detected through nanoparticle based amperometric biosensors [65,78,79].

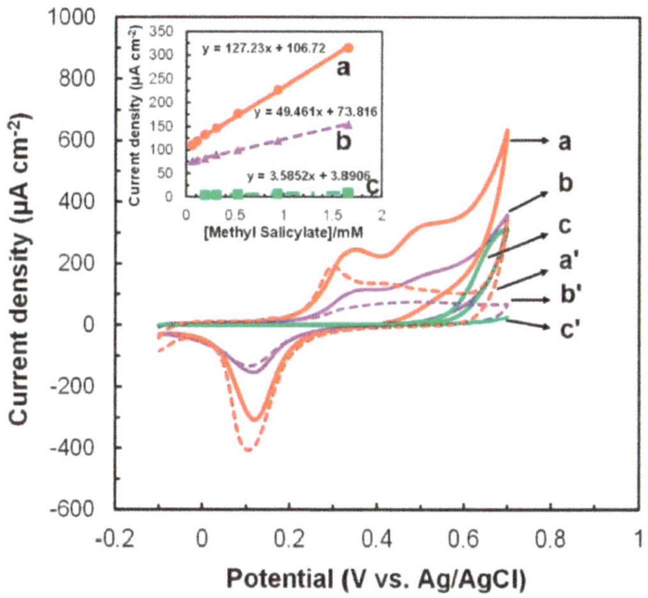

Figure 1: Cyclic voltammetry of (a) AuNP-Screen printed carbon electrode (SPCE), (b) planar gold and (c) SPCE in presence of 1.7 mM methyl salicylate. (a') AuNP-SPCE, (b') planar gold and (c') SPCE in the absence of methyl salicylate. The responses of current to methyl salicylate and sensitivity are shown in the inset. Figure is adopted from Ref. [77] with permission.

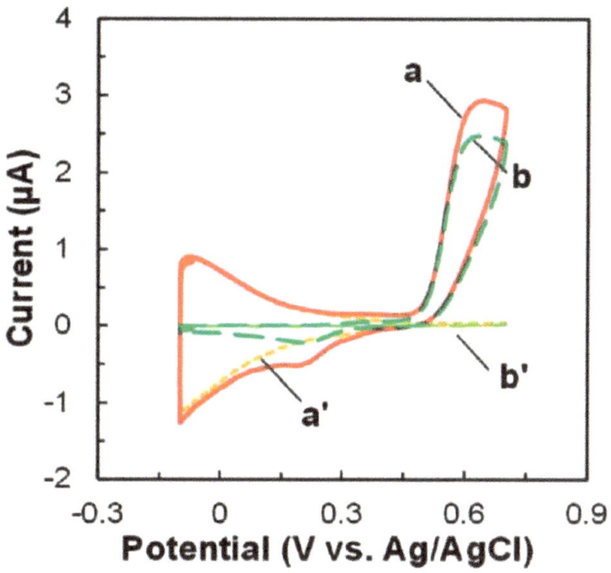

Figure 2: Cyclic voltammetry responses of (a and a') SnO_2-Screen printed (SP) and (b and b') TiO_2-SP (a and b) with and (a' and b') without the presence of 0.17 mM *p*-ethylguaiacol. Figure is adopted from Ref. [57] with permission.

Affinity Biosensors

Compared to the non-specific nanoparticle-based biosensors, inclusion of a bio-recognition element can greatly increase the specificity of the sensor. Consequently, other types of biosensors have been developed and among them affinity biosensors are popular. In affinity biosensors, the sensing is achieved based on the reaction of the bio-recognition element and the target analyte [80]. Affinity biosensors can be developed using antibody and DNA as recognition elements.

Antibody-Based Biosensors

Antibodies are versatile and are suitable for diverse immunosensing applications. Antibody-based biosensor allows rapid and sensitive detection of a range of pathogens especially for foodborne diseases and this technique has already been developed for food safety monitoring. The antibody-based biosensors provide several advantages such as fast detection, improved sensitivity, real-time analysis and potential for quantification. Antibody-based biosensors hold great value for agricultural plant pathogen detection [81]. The biosensors enable the pathogen detection in air, water and seeds with different platforms for greenhouses, on-field and postharvest storages of processors and distributors of crops and fruits [81].

The principle of establishing antibody-based immunosensors lies in the coupling of specific antibody with a transducer, which converts the binding event (the specific binding of antibody modified on the biosensor with the antigen, e.g., pathogen of interest) to a signal that can be analyzed (Figure 3a). Most antibody-based biosensors use one of the following types of electrochemical transducers: amperometric, potentiometric, impedimetric and conductometric. Amperometric biosensing platforms use electric current signal resulting from the specific binding event [82,83]. Potentiometric biosensors on the other hand convert the biorecognition of an analyte into a voltage signal

[82]. Impedimetric biosensors detect analyte by impedance change upon specific combination of the antibody and analyte. Based on the metabolic redox reactions of microorganisms, impedimetric biosensors are often used for biomass detection by microbial metabolism [84]. The conductometric biosensors are based on conductometric detection where the biological signal is converted to an electrical signal through a conductive polymer, such as polyacetylene, polypyrrole or polyaniline [80]. Other types of transducers (non-electrochemical) for affinity biosensing have been developed and reported including surface plasmon resonance (SPR), quartz crystal microbalance (QCM) and cantilever-based sensors. SPR-based sensors can measure the change in refractive index due to the attachment of the analyte to the metal surface (e.g., gold), which is usually modified with the recognition element of a conjugated ligand [85]. A QCM-based sensor detects a mass variation per unit area of the QCM crystal by measuring the change in frequency of a quartz crystal resonator. The QCM crystal is typically modified with a recognition element (e.g., antibodies) [86]. Similar to QCM-based sensors, cantilever-based sensors measure resonance frequency changes upon combination of the analytes and the sensor surface [81]. Based on the capability of detecting small analytes such as nucleic acid and proteins, cantilever-based sensors have been used for detection of pathogenic microorganisms [87].

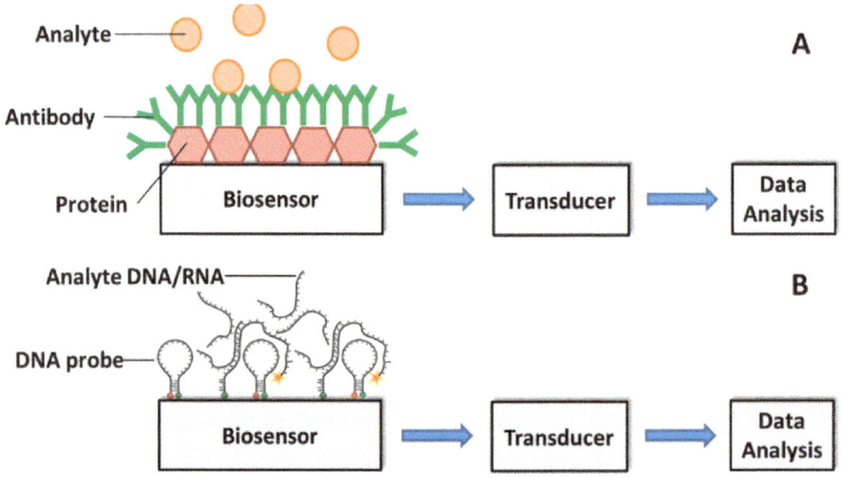

Figure 3: Schematic illustration of (**A**) antibody-based and (**B**) DNA/RNA-based biosensor for analyte detection. The specific combination of analyte and immobilized antibody (A) or DNA/RNA probe (B) produces a physicochemical change, such as mass, temperature, optical property or electrical potential. The change can be translated into a measurable signal for detection.

During the past decade, many articles have been published demonstrating the capability of antibody-based biosensors for detection of plant pathogens such as *Cowpea mosaic virus*, *Tobacco mosaic virus*, *Lettuce mosaic virus*, *Fusarium culmorum*,*Puccinia striiformis*, *Phytophthora infestans*, orchid viruses and *Aspergillus niger* [86,88,89,90,91,92,93,94]. 11-mercaptoundecanoic acid was self-assembled on gold surface and then crosslinked with anti-maize chlorotic mottle virus (anti-MCMV) for MCMV disease detection. SPR response to MCMV solutions was evaluated over time with different concentrations from 1–1000 ppb. The limit of detection was evaluated to be approximately 1 ppb [85]. In recent years, antibody-based biosensor technology has seen tremendous progress upon implementation of nanotechnology based approaches for the sensor fabrication. Gold nanorods (AuNRs) functionalized by antibodies have been used to detect*Cymbidium mosaic virus* (CymMV) or *Odontoglossum ringspot virus* (ORSV) for rapid diagnosis of viral infections. The limits of detection (LODs) for CymMV and ORSV were reported to be 48 and 42 pg/mL, respectively, in leaf saps [95]. The detection of CymMV and ORSV was not only reported by SPR technique, but also by QCM technique. The QCM technique was able to detect each of the orchid viruses as low as 1 ng [86]. Other nanomaterials made of polymers such as polypyrrole (PPy) nanoribbon modified chemiresistive sensors were fabricated by a lithographically patterned nanowire electrode position (LPNE) technique. The fabricated biosensor was investigated for the detection of *Cucumber mosaic virus* (CMV) and displayed excellent sensitivity with a detection limit of 10 ng/mL [96]. The limits of detection of the current antibody based biosensors are proved to be approximately two orders of magnitude higher than that of conventional ELISA methods [97]. Apart from the common abiotic material for antibody-based biosensor fabrication, biosensors based on living cells are characterized by low limit of detection, high specificity and rapid response time. A novel portable cell biosensor system for detection of *Potato virus Y* (PVY), *Cucumber mosaic virus* (CMV) and *Tobacco rattle virus* (TRV) was fabricated by immobilizing the vero cells carrying virus specific antibodies on their membranes. This study demonstrated an important step towards the development of a portable plant virus detection system suitable for on-field application [98].

Although the mechanisms, advantages and applications of antibody-based biosensors for plant disease detection have been highlighted, it is important to discuss the limitations. Since many biosensors based on antibodies focus on specific binding with a particular antigen, issues such as the exposure of a bacterial strain to environmental stress (pH and temperature), could cause errors in the measurement. In addition, the immobilization of large bacteria and fungi, whose diameters exceed the SPR range, might compromise the detection. More importantly, antibodies are vulnerable and are easy to get denatured, which requires specific environment (pH, temperature, *etc.*) for storage, otherwise, the behavior of antibody-based sensor will also be compromised due to the deterioration of antibody over time [84].

DNA/RNA-Based Affinity Biosensor

A recently developed new type of affinity biosensor uses nucleic acid fragments as elements for pathogen detection. The detection of specific DNA sequence is of significance in a variety of applications such as clinical human disease detection, environmental, horticulture and food analysis. Due to the possibility of detection at a molecular level, the DNA-based biosensor enables early detection of diseases before any visual symptoms appear. The application of specific DNA sequences has been widely used for detection of bacteria, fungi and genetically modified organisms. Based on the specific nucleic acid hybridization of the immobilized DNA probe on the sensor and the analyte DNA sequence, DNA-based biosensor allows rapid, simple and economical testing of genetic and infectious diseases. The most commonly adopted DNA probe is single stranded DNA (ssDNA) on electrodes with electroactive indicators to measure hybridization between probe DNA and the complementary DNA analyte [99]. There are four major types of DNA-based biosensors depending on their mode of transduction: optical, piezoelectric, strip type and electrochemical DNA biosensors. Optical DNA biosensors transduce the emission signal of a fluorescent label. The detection of DNA analyte is realized through a variation in physio-chemical properties such as mass, temperature, optical property and electrical property as a result of double-stranded DNA (dsDNA) hybridization occurs during the analyte recognition (Figure 3b). The optical DNA-based biosensors can be further classified to three subtypes—molecular beacons (MB), surface plasmon resonance (SPR) and quantum-dots. Unlike the optical type, piezoelectric DNA biosensors detect the analyte using a quartz crystal that oscillates at a specific frequency at an applied oscillating voltage. Detection of DNA hybridization can also be realized through nanoparticle-based colorimetric detection provided by strip type DNA biosensor. Electrochemical measurements are used for sequence-specific detection of analyte DNA by electrochemical DNA-based biosensors. The current change with constant applied potential can be monitored and used for interpretation of DNA hybridization in amperometric electrochemical DNA biosensor [99]. Bacterial pathogens are detectable by DNA-based biosensors due to their unique nucleic acid sequence, which can be specifically hybridized with the complementary DNA probe. The recognition of analyte DNA is dependent upon the formation of stable hydrogen bonds between the DNA probe and analyte DNA sequence. This is different from the antibody-based biosensors where hydrophobic, ionic and hydrogen bonds play a role in the stabilization of antigen-antibody complex.

In addition to DNA-DNA hybridization for bacterial detection, the specific hybridization of DNA and complementary RNA was also exploited for the detection of plant viruses by molecular beacons and QCM techniques. Two orchid viruses—*Cymbidium mosaic virus* (CymMV) and *Odontoglossum ringspot virus* (ORSV)—have been detected with specific oligonucleotide probes with a fluorescent moiety attached to one end of DNA while a quenching moiety attached to the opposite end. Four such molecules have been designed and this technique has been successfully applied to detect viral RNA of both orchid viruses with limits of detection as low as 0.5 ng of viral RNA in 100 mg orchid leaves (Figure 4) [99]. CymMV and ORSV were also detected by QCM DNA-based biosensor with a designed DNA probe modified with a mercaptohexyl group. The limits of detection of CymMV and ORSV were as low as approximately 1 ng in purified RNA and 10 ng in the crude sap (Figure 5) [100]. Although the application of DNA-based biosensors for plant disease detection is promising, PCR may have to be performed prior to the probing process due to the small quantity of nucleic acid present in the bacteria cells [101]. The limitations of DNA based biosensors include the requirement for the synthesis of specific DNA probe, amplification of DNA, high cost (DNA-based molecular beacons) and unsuitability for real-time detection (DNA-based piezoelectric biosensor).

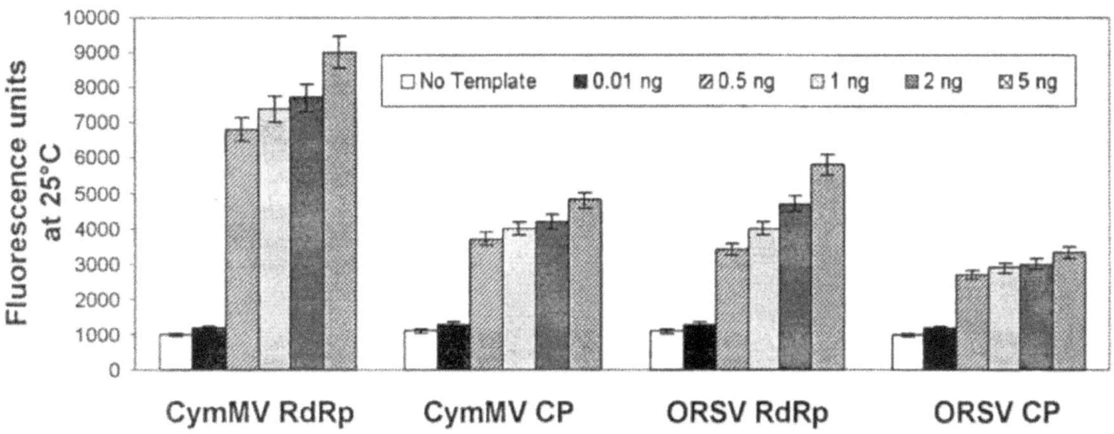

Figure 4: The fluorescence intensity of the four molecular beacons (CymMV RdRp, CymMV CP, ORSV RdRp and ORSV CP) with 0.01, 0.5, 1, 2, and 5 ng of purified viral RNA. The fluorescence intensity of all molecular beacons increased significantly and the limit of detection of purified viral RNA was estimated to be 0.5 ng. Figure has been adopted from Ref. [99].

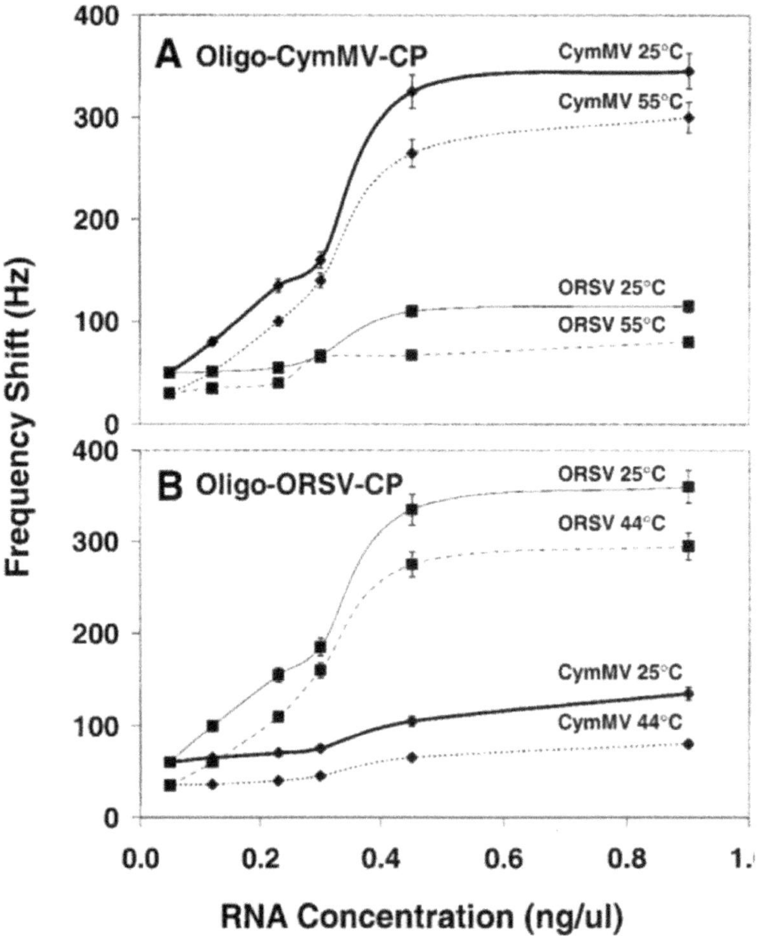

Figure 5: Sensitivity and specificity of (**A**) CymMV coat protein (CP) and (**B**) ORSV-CP QCM DNA biosensors upon incubation with increasing concentrations of viral RNA at different temperatures. Figure has been adopted from Ref. [100].

Enzymatic Electrochemical Biosensors

The use of enzyme as bio-recognition element can provide highly selective detection of the target analyte due to the high specificity of enzymes towards the analyte. An enzyme specific for the analyte of interest is immobilized on the nanomaterial modified-electrode. The amperometric detection is based on the bio-electrocatalytic reaction between the target analyte and electrode, which results in an electrical signal (current) that can be used for quantitative detection of the analyte. The amperometric signal can be obtained through either direct or mediated electron transfer based electrochemical reactions (Figure 6). Unlike other types of biosensors, which are not widely commercialized, the enzymatic electrochemical biosensors have been successfully commercialized, thanks to the invention of glucose biosensors, which are widely used in personal diabetes monitors [102]. A similar biosensing methodology can be adopted for plant pathogen detection, food quality detection and environmental monitoring [103]. For plant pathogen detection, enzymatic biosensors could be used if the target VOC could be collected in the form of a liquid sample. Previous studies have shown that several of phytohormones are catabolized by redox enzymes, offering prospects for using these enzymes for the development of highly selective enzyme-based biosensors for detecting plant chemicals [104]. Our previous work has already proved the detection of methyl salicylate with a bi-enzymatic system [105]. Many of the VOCs produced by infected crops are alcohols and aldehydes such as *cis*-3-hexen-1-ol and *trans*-2-hexanal, which can be catalyzed by alcohol dehydrogenase enzymes. Accordingly, these enzymes can be used for the development of biosensors for the detection of alcohol or aldehyde based VOCs which are specific to the infection. A summary of the different volatiles known to be released due to plant stresses are listed in Table 2 [60]. In addition to those specific volatile organic compounds, the common phytohormones such as auxin, cytokinins and gibberellins which are indicative of plant health could also be deactivated by oxidases. Gibberellin is deactivated by GA-2-oxidases which provides the potential for fabrication of gibberellin detection for plant disease prediction [104,106]. Although enzyme-based biosensors usually provide high sensitivity and specificity for the detection, stability of enzymes is of major concern. In addition, the enzyme catalysis varies with factors such as temperature and pH which compromises the accuracy of the biosensor.

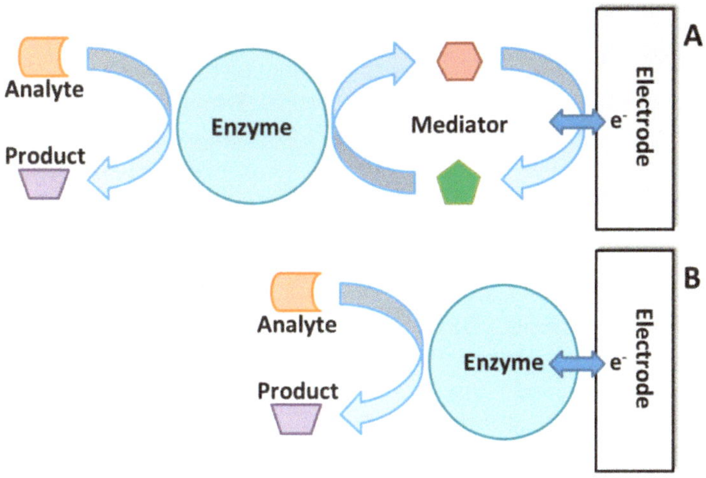

Figure 6: Schematic illustration of enzymatic biosensor based on (**A**) mediated electron transfer and (**B**) direct electron transfer (DET).

Table 2: VOCs emitted from whole, intact tomato plants or detached leaves, and biotic stress causing agents responsible for increases in VOC emissions [60].

Volatiles	Biotic Stress Causing Agents that Increase VOC Emissions
cis-3-hexen-1-ol	*Botrytis cinerea, Spodoptera littoralis, Lirimyza huidobrensis, Spodoptera exigua, Manduca sexta, Macrosiphum euphorbiae, Helicoverpa armigera*
trans-2-hexanal	*Botrytis cinerea, Spodoptera littoralis, Lirimyza huidobrensis, Spodoptera exigua, Manduca sexta, Helicoverpa armigera*
Methyl salicylate	*Botrytis cinerea, Spodoptera littoralis, Tetranychus urticae, Manduca sexta, Macrosiphum euphoria, Tobacco mosaic virus*

Bacteriophage-Based Biosensors

Bacteriophage is a virus, composed of protein capsid that encapsulates a DNA or RNA genome. It infects the bacteria and replicates within the bacteria and finally lyses the bacterial host to propagate. Being able to lyse the bacteria, bacteriophage has been widely studied and used in phage therapy to cure bacterial infections [107]. Phage therapy has been used for not only human diseases, but also plant disease control. In addition to phage therapy, bacteriophage is also emerging as a promising alternative for pathogen detection due to its high sensitivity, selectivity, low cost and higher thermostability [108,109,110]. Upon the interaction between the bacteriophage and the target analyte, the impedance of charge transfer reactions at the interface changes which is used as a signal for detection.

Bacteriophages have demonstrated to be successful in controlling plant pathogens recently such as *Dickeya solani*, the bacterial infecting of potatoes and tomatoes [111,112]. Tlili *et al.* reported the limit of detection for *E. coli* by using T4 phage and achieved 8×10^2 CFU/mL in less than 15 min and 10^2 CFU/mL within 40 min by impedimetric and loop-mediated isothermal amplification (LAMP) method respectively [113]. D. A. Schofield reported a phage-based diagnostic assay for detecting and identifying *Pseudomonas cannabina* pv. *Alisalensis* from cultures and diseased plant samples [114]. The bacterial*luxAB* reporter genes (encoding the luciferase) were integrated into *P. cannabina* pv *Alisalensis* phage PBSPCA1. In the presence of target pathogen cell, the *lux AB* can be expressed in the host cell due to specific infection through the phage. The expressed luciferase results in the light emission in presence of n-decanal, oxygen and flavin mononucleotide (Figure 7) [114]. In addition to *P. cannabina* pv. *alisalensi*, bacteriophages that specifically bind with other pathogens are also discovered and reported, such as *Pseudomonas syringae* pv. *Actinidiae*, which causes bacterial canker of kiwifruits, and *Ralstonia solanacearum*, a soilborne bacterium that is the causative agent of bacterial wilt in many important crops [115,116,117]. The progress in discovering more bacteriophages provides the possibility for fabrication of more bacteriophage-based sensors for plant disease detection.

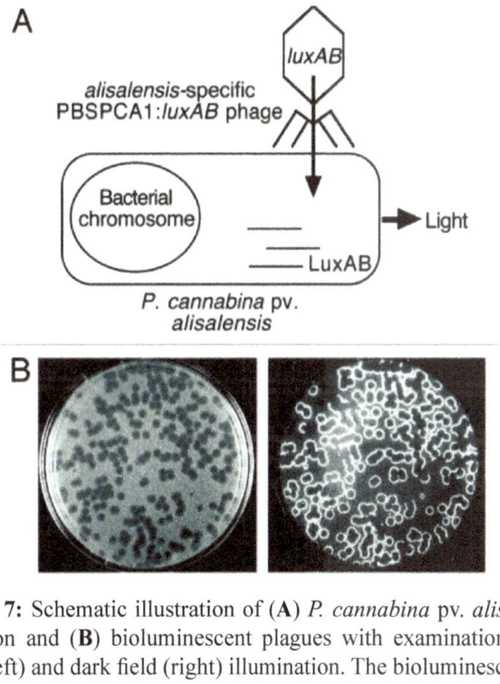

Figure 7: Schematic illustration of (**A**) *P. cannabina* pv. *alisalensis* detection and (**B**) bioluminescent plagues with examination under light (left) and dark field (right) illumination. The bioluminescence at the plague periphery (phage/cell interface) indicates phage-infected pathogens. Figure has been adopted from Ref. [114].

The advantages of using bacteriophage as the recognition element for biosensors are its high selectivity and low cost of the phage. Furthermore, compared to the antibody-based sensor, bacteriophage-based sensors are more thermostable which allows the detection in different temperature ranges and longer shelf life. Bacteriophage-based biosensors are also capable of differentiating the live and dead bacterial pathogens which decreases the false positive signals during measurement. However, measuring the pathogen in a real sample using phage-based biosensors could take a longer time because of the complex sample preparation requirement [113]. Apart from that, bacteriophage-based sensor can only be fabricated for detection of bacteria rather than fungi and viruses which severely limits its application for the majority of crops that are affected by fungal pathogens.

CHALLENGES AND FUTURE DIRECTIONS

Besides the unique advantages offered by the various disease detection methods for plant disease detection application, each method has its own limitations. When it comes to direct disease detection methods, PCR has displayed its ability in detecting plant pathogen with high sensitivity, however, it requires designing specific primers to amplify DNA for detecting different pathogens. The cost prohibitive procedure thus limits its application only to laboratory settings and high value target analytes [19]. In addition, parameters like polymerase activity, buffer concentration of deoxynucleoside triphosphate can bring uncertainty to the result [20]. On the other hand, as sophisticated equipment in the laboratory-based detection techniques, portable PCR and its specialized types such as RT-PCR, real-time PCR have been used for on-field detection. We anticipate the application of PCR will be continued in the future. Although the application of ELISA for plant disease detection is not much reported, fabrication of test strips based on ELISA will be an innovative strategy for plant disease detection in the future due to its visible color change signal. However, the application may only be confined to virus detection due to the excellent sensitivity of ELISA for viruses (Table 3). The application will be compromised for bacterial infections due to its poor sensitivity, on the other hand (Table 1). Although FISH and IF provide excellent sensitivity (Table 3), they are laboratory-based techniques which require skilled personnel to operate. Additionally, complex sample preparations and professional data analysis are required. The same holds true for GC-MS despite its ability to provide quantitative determination of VOCs produced by infected plants. While offering accurate data for disease detection, FCM provides overwhelming and sometimes unnecessary data which complicates the data analysis and requires professional and experienced technicians for interpreting the results of detection. Further, the expensive instrumentation makes it less likely for on-field application.

Among the indirect imaging techniques for plant disease detection, the hyperspectral technique is the most promising due to its robustness, high specificity and rapid data analysis. Hyperspectral imaging could be obtained using a camera fitted to an unmanned aerial vehicle (UAV) for collecting data across a wider area of agricultural field. Thermography and fluorescence imaging, being non-specific and susceptible to ambient environment, therefore are less suitable for on-field crop disease detection [51,52,53]. Affinity sensors, although extensively studied, still remain in the laboratory in most cases. This can be explained by the sensor's lack of the robustness, which is primarily due to the deterioration of antibody and DNA/RNA probe over time and the requirement for coupling with optical instruments. DNA-based molecular beacon biosensors could be potentially applied for on-field testing due to their properties such as real-time testing and low limit of detection (Table 3) given that the proper DNA probe is produced for this purpose. Also, the lower limit of detection provided by bacteriophage-based sensors is also promising, to be applied for live bacterial pathogen detection with the discovery of more bacteriophages [113]. However, the technique cannot be applied for fungal and viral pathogen detections. On the other hand, enzyme-based electrochemical biosensors constructed using nanomaterial platforms have higher specificity and applicability for real-time detection than other available technologies (Table 3). The electrochemical enzymatic biosensor also enables the fabrication of small-scale, portable and easy-to-use devices for detection. However, the current challenges reside in the identification, production and purification of enzymes specific to each target VOC produced by the infected plants. It is also promising that the enzymatic and affinity sensors can be fabricated for multiple pathogen detection, similar to electronic noses, rather than for single pathogen detection.

Table 3: Comparison of affinity biosensors, enzymatic biosensors, nanomaterial modified sensors and ELISA.

Techniques	Limit of Detection	Advantages	Disadvantages	Ref.
Nanomaterial sensor	35 nM (*p*-ethylguaiacol)	Real-time, easy to fabricate, stable	Low specificity compared to enzyme, DNA, antibody based biosensors	[57]
Enzymatic biosensor	0.98 μM (methyl salicylate)	High specificity, real-time	Interference from pH change	[105]
DNA-based piezoelectric biosensor	10 ng (virus)	Low cost	Not real-time, DNA may need to be amplified	[100]
DNA-based molecular beacon biosensor	0.5 ng (virus)	Real-time, low limit of detection	High cost, DNA may need to be amplified	[118]
Antibody-based quartz crystal biosensor	10 ng (virus)	Low cost	Not real-time	[86]
ELISA	2.5 ng (virus)	Low cost, low limit of detection for virus	Not real-time for bacteria, poor sensitivity for bacteria	[119]

CONCLUSIONS

In this article, we reviewed the currently existing methods for detection of plant diseases caused by pathogens such as bacteria, viruses and fungi. Although established methods such as PCR, FISH, ELISA, IF, FCM and GC-MS are already available and widely used for plant disease detection, they are relatively difficult to operate, require expert technicians and are time consuming for data analysis. In addition, most of these methods cannot provide real-time detection which makes them less suitable for on-field testing and early warning systems. On the other hand, imaging techniques such as thermography and fluorescence imaging, although they have been used on-field for disease detection, are proved to be susceptive to parameter change of the environment and lack of specificity of each type of disease. The untamed potential of various biosensors for plant disease detection has been comprehensively reviewed in this article. The advent of nanotechnology has resulted in the advancement of highly sensitive biosensors due to modern nanofabrication techniques. The specificity of the biosensors could be greatly enhanced by the use of enzymes, antibodies, DNA and bacteriophage as the specific recognition element.

ACKNOWLEDGMENTS

The authors acknowledge the National Science Foundation (CBET-1159540), National Institute of Food and Agriculture (Award # 2015-67021-23188) and ACS Herman Frasch Foundation (Grant # 045097-01) for financial support.

REFERENCES

1. Ingram, J. A food systems approach to researching food security and its interactions with global environmental change. *Food Secur.* **2011**, *3*, 417–431.
2. Keinan, A.; Clark, A.G. Recent explosive human population growth has resulted in an excess of rare genetic variants.*Science* **2012**, *336*, 740–743.
3. Senauer, B.; Vaclav, S. Feeding the World: A Challenge for the Twenty-First Century. Available online: http://onlinelibrary.wiley.com/doi/10.1111/j.1728-4457.2000.00827.x/pdf (accessed on 5 January 2015).
4. Rosset, P. Food sovereignty and the contemporary food crisis. *Development* **2008**, *51*, 460–463.
5. Godfray, H.C.J.; Beddington, J.R.; Crute, I.R.; Haddad, L.; Lawrence, D.; Muir, J.F.; Pretty, J.; Robinson, S.; Thomas, S.M.; Toulmin, C. Food security: The challenge of feeding 9 billion people. *Science* **2010**, *327*, 812–818.
6. Conway, G. *One Billion Hungry: Can We Feed the World?* Cornell University Press: Ithaca, NY, USA, 2012.
7. Savary, S.; Ficke, A.; Aubertot, J.; Hollier, C. Crop losses due to diseases and their implications for global food production losses and food security. *Food Secur.* **2012**, *4*, 519–537.
8. Oerke, E.-C. Crop losses to pests. *J. Agric. Sci.* **2006**, *144*, 31–43.
9. Pimentel, D.; Zuniga, R.; Morrison, D. Update on the environmental and economic costs associated with alien-invasive species in the United States. *Ecol. Econ.* **2005**, *52*, 273–288.
10. Roberts, M.J.; Schimmelpfennig, D.E.; Ashley, E.; Livingston, M.J.; Ash, M.S.; Vasavada, U. *The Value of Plant Disease Early-Warning Systems: A Case Study of USDA's Soybean Rust Coordinated Framework*; United States Department of Agriculture, Economic Research Service: Washington, DC, USA, 2006.
11. Cai, H.; Caswell, J.; Prescott, J. Nonculture molecular techniques for diagnosis of bacterial disease in animals a diagnostic laboratory perspective. *Vet. Pathol. Online* **2014**, *51*, 341–350.
12. López, M.M.; Bertolini, E.; Olmos, A.; Caruso, P.; Corris, M.T.; Llop, P.; Renyalver, R.; Cambra, M. Innovative tools for detection of plant pathogenic viruses and bacteria. *Int. Microbiol.* **2003**, *6*, 233–243.
13. James, D. A simple and reliable protocol for the detection of apple stem grooving virus by RT-PCR and in a multiplex PCR assay. *J. Virol. Methods* **1999**, *83*, 1–9.
14. Nassuth, A.; Pollari, E.; Helmeczy, K.; Stewart, S.; Kofalvi, S.A. Improved RNA extraction and one-tube RT-PCR assay for simultaneous detection of control plant RNA plus several viruses in plant extracts. *J. Virol. Methods* **2000**, *90*, 37–49.
15. Osiowy, C. Direct detection of respiratory syncytial virus, parainfluenza virus, and adenovirus in clinical respiratory specimens by a multiplex reverse transcription-PCR assay. *J. Clin. Microbiol.* **1998**, *36*, 3149–3154.
16. Pallisgaard, N.; Hokland, P.; Riishøj, D.C.; Pedersen, B.; Jørgensen, P. Multiplex reverse transcription-polymerase chain reaction for simultaneous screening of 29 translocations and chromosomal aberrations in acute leukemia. *Blood***1998**, *92*, 574–588.
17. Williams, K.; Blake, S.; Sweeney, A.; Singer, J.T.; Nicholson, B.L. Multiplex reverse transcriptase PCR assay for simultaneous detection of three fish viruses. *J. Clin. Microbiol.* **1999**, *37*, 4139–4141.
18. Lievens, B.; Brouwer, M.; Vanachter, A.C.R.C.; Cammue, B.P.A.; Thomma, B.P.H.J. Real-time PCR for detection and quantification of fungal and oomycete tomato pathogens in plant and soil samples. *Plant Sci.* **2006**, *171*, 155–165.
19. Schaad, N.W.; Frederick, R.D. Real-time PCR and its application for rapid plant disease diagnostics. *Can. J. Plant Pathol.* **2002**, *24*, 250–258.
20. Van der Wolf, J.; van Bechhoven, J.R.C.M.; Bonants, P.J.M.; Schoen, C.D. New technologies for sensitive and specific routine detection of plant pathogenic bacteria. In *Plant Pathogenic Bacteria*; Springer: Berlin, Germany, 2001; pp. 75–77.
21. Kempf, V.A.; Trebesius, K.; Autenrieth, I.B. Fluorescent in situ hybridization allows rapid identification of microorganisms in blood cultures. *J. Clin. Microbiol.* **2000**, *38*, 830–838.
22. Hijri, M. The use of Fluorescent in situ hybridisation in plant fungal identification and genotyping. In *Plant Pathology*; Springer: Berlin, Germany, 2009; pp. 131–145.
23. Kliot, A.; Kontsedalov, S.; Lebedev, G.; Brumin, M.; Cathrin, P.B.; Marubayashi, J.M.; Skaljac, M.; Belausov,

E.; Czosnek, H.; Ghanim, M. Fluorescence in situ hybridizations (FISH) for the localization of viruses and endosymbiotic bacteria in plant and insect tissues. *J. Vis. Exp.* **2014**, *84*, e51030.

24. Moter, A.; Göbel, U.B. Fluorescence in situ hybridization (FISH) for direct visualization of microorganisms. *J. Microbiol. Methods* **2000**, *41*, 85–112.
25. DeLong, E.F.; Wickham, G.S.; Pace, N.R. Phylogenetic stains: Ribosomal RNA-based probes for the identification of single cells. *Science* **1989**, *243*, 1360–1363.
26. Wallner, G.; Amann, R.; Beisker, W. Optimizing fluorescent in situ hybridization with rRNA-targeted oligonucleotide probes for flow cytometric identification of microorganisms. *Cytometry* **1993**, *14*, 136–143.
27. Clark, M.F.; Adams, A. Characteristics of the microplate method of enzyme-linked immunosorbent assay for the detection of plant viruses. *J. Gen. Virol.* **1977**, *34*, 475–483.
28. Gorris, M.T.; Alarcon, B.; Lopez, M.; Cambra, M. Characterization of monoclonal antibodies specific for *Erwinia carotovora* subsp. atroseptica and comparison of serological methods for its sensitive detection on potato tubers. *Appl. Environ. Microbiol.* **1994**, *60*, 2076–2085.
29. López, M.M.; Llop, P.; Cubero, J.; Penyalver, R.; Caruso, P.; Bertolini, E.; Penalver, J.; Gorris, M.T.; Cambra, M. Strategies for improving serological and molecular detection of plant pathogenic bacteria. In *Plant Pathogenic Bacteria*; Springer: Berlin, Germany, 2001; pp. 83–86.
30. Ward, E.; Foster, S.J.; Fraaije, B.A.; Mccartney, H.A. Plant pathogen diagnostics: Immunological and nucleic acid-based approaches. *Ann. Appl. Biol.* **2004**, *145*, 1–16.
31. Dewey, F.; Marshall, G. Production and use of monoclonal antibodies for the detection of fungi. In Proceeding of British Crop Protection Council Symposium, Farnham, UK, 18–21 November 1996.
32. Wullings, B.A.; VanBeuningen, A.R.; Janse, J.D.; Akkermans, A.D.L. Detection of *Ralstonia solanacearum*, which causes brown rot of potato, by fluorescent in situ hybridization with 23S rRNA-targeted probes. *Appl. Environ. Microbiol.***1998**, *64*, 4546–4554.
33. Chitarra, L.G.; van den Bulk, R.W. The application of flow cytometry and fluorescent probe technology for detection and assessment of viability of plant pathogenic bacteria. *Eur. J. Plant Pathol.* **2003**, *109*, 407–417.
34. Diaper, J.; Edwards, C. Flow cytometric detection of viable bacteria in compost. *FEMS Microbiol. Ecol.* **1994**, *14*, 213–220.
35. Porter, J.; Pickup, R.; Edwards, C. Evaluation of flow cytometric methods for the detection and viability assessment of bacteria from soil. *Soil Biol. Biochem.* **1997**, *29*, 91–100.
36. Bravo, C.; Moshou, D.; Oberti, R.; West, J.; McCartney, A.; Bodria, L.; Ramon, H. Foliar Disease Detection in the Field Using Optical Sensor Fusion. Available online: http://ecommons.cornell.edu/bitstream/handle/1813/10394/FP%2004%20008%20Bravo-Moshou%20Final%2022Dec2004.pdf?sequence=1&isAllowed=y (accessed on 2 March 2015).
37. Moshou, D.; Bravo, C.; Oberti, R.; West, J.; Bodria, A.; McCartney, A.; Ramon, H. Plant disease detection based on data fusion of hyper-spectral and multi-spectral fluorescence imaging using Kohonen maps. *Real Time Imaging* **2005**, *11*, 75–83.
38. Nilsson, H.-E. Remote sensing and image analysis in plant pathology. *Can. J. Plant Pathol.* **1995**, *17*, 154–166.
39. West, J.S.; Bravo, C.; Oberti, R.; Lemaire, D.; Moshou, D.; McCartney, H.A. The potential of optical canopy measurement for targeted control of field crop diseases. *Annu. Rev. Phytopathol.* **2003**, *41*, 593–614.
40. Mahlein, A.K.; Oerke, E.; Steiner, U.; Dehne, H. Recent advances in sensing plant diseases for precision crop protection. *Eur. J. Plant Pathol.* **2012**, *133*, 197–209.
41. Chaerle, L.; van der Straeten, D. Imaging techniques and the early detection of plant stress. *Trends Plant Sci.* **2000**, *5*, 495–501.
42. Chaerle, L.; Leinonen, I.; Jones, H.G.; Van Der Straeten, D. Monitoring and screening plant populations with combined thermal and chlorophyll fluorescence imaging. *J. Exp. Bot.* **2007**, *58*, 773–784.
43. Lindenthal, M.; Steiner, U.; Dehne, H.W.; Oerke, E.C. Effect of downy mildew development on transpiration of cucumber leaves visualized by digital infrared thermography. *Phytopathology* **2005**, *95*, 233–240.
44. Oerke, E.; Steiner, U.; Dehne, H.W.; Lindenthal, M. Thermal imaging of cucumber leaves affected by downy mildew and environmental conditions. *J. Exp. Bot.* **2006**, *57*, 2121–2132.
45. Oerke, E.-C.; Fröhling, P.; Steiner, U. Thermographic assessment of scab disease on apple leaves. *Precis. Agric.* **2011**,*12*, 699–715.
46. Stoll, M.; Schultz, H.R.; Baecker, G.; Berkelmann-Loehnertz, B. Early pathogen detection under different water status and the assessment of spray application in vineyards through the use of thermal imagery. *Precis. Agric.* **2008**, *9*, 407–417.
47. Lindenthal, M. *Visualisierung der Krankheitsentwicklung von Falschem Mehltau an Gurken durch Pseudoperonospora cubensis mittels Thermografie*; Universitäts-und Landesbibliothek Bonn: Bonn, Germany, 2005.
48. Hillnhütter, C.; Mahlein, A.K.; Sikora, R.A.; Oerke, E.-C. Remote sensing to detect plant stress induced by *Heterodera schachtii* and *Rhizoctonia solani* in sugar beet fields. *Field Crops Res.* **2011**, *122*, 70–77.
49. Bürling, K.; Hunsche, M.; Noga, G. Use of blue-green and chlorophyll fluorescence measurements for differentiation between nitrogen deficiency and pathogen infection in winter wheat. *J. Plant Physiol.* **2011**, *168*, 1641–1648.
50. Kuckenberg, J.; Tartachnyk, I.; Noga, G. Temporal and spatial changes of chlorophyll fluorescence as a basis for early and precise detection of leaf rust and powdery

mildew infections in wheat leaves. *Precis. Agric.* **2009**, *10*, 34–44.

51. Chaerle, L.; Lenk, S.; Leinonen, I.; Jones, H.G.; Van Der Straeten, D.; Dr., Buschmann, C. Multi-sensor plant imaging: Towards the development of a stress-catalogue. *Biotechnol. J.* **2009**, *4*, 1152–1167.
52. Cséfalvay, L.; Gaspero, G.D.; Matous, K.; Bellin, D.; Ruperti, B.; Olejnickova, J. Pre-symptomatic detection of*Plasmopara viticola* infection in grapevine leaves using chlorophyll fluorescence imaging. *Eur. J. Plant Pathol.* **2009**, *125*, 291–302.
53. Scholes, J.D.; Rolfe, S.A. Chlorophyll fluorescence imaging as tool for understanding the impact of fungal diseases on plant performance: A phenomics perspective. *Funct. Plant Biol.* **2009**, *36*, 880–892.
54. Delalieux, S.; van Aardt, J.; Keulemans, W.; Schrevens, E.; Coppin, P. Detection of biotic stress (*Venturia inaequalis*) in apple trees using hyperspectral data: Non-parametric statistical approaches and physiological implications. *Eur. J. Agron.* **2007**, *27*, 130–143.
55. Kobayashi, T.; Kanda, E.; Kitada, K.; Ishiguro, K.; Torigoe, Y. Detection of rice panicle blast with multispectral radiometer and the potential of using airborne multispectral scanners. *Phytopathology* **2001**, *91*, 316–323.
56. Zhang, M.; Qin, Z.; Liu, X.; Ustin, S.L. Detection of stress in tomatoes induced by late blight disease in California, USA, using hyperspectral remote sensing. *Int. J. Appl. Earth Observ. Geoinf.* **2003**, *4*, 295–310.
57. Fang, Y.; Umasankar, Y.; Ramasamy, R.P. Electrochemical detection of p-ethylguaiacol, a fungi infected fruit volatile using metal oxide nanoparticles. *Analyst* **2014**, *139*, 3804–3810.
58. Ellis, M.; Grove, G. Fruit rots cause losses in Ohio strawberries. *Ohio Rep. Res. Dev.* **1982**, *67*, 3–4.
59. Umasankar, Y.; Rains, G.C.; Ramasamy, R.P. Electroanalytical studies on green leaf volatiles for potential sensor development. *Analyst* **2012**, *137*, 3138–3145.
60. Jansen, R.M.; Hofstee, J.W.; Wildt, J.; Verstappen, F.W.; Bouwmeester, H.; van Henten, E.J. Induced plant volatiles allow sensitive monitoring of plant health status in greenhouses. *Plant Signal. Behav.* **2009**, *4*, 824–829.
61. Isidorov, V.; Zenkevich, I.; Ioffe, B. Volatile organic compounds in the atmosphere of forests. *Atmos. Environ.* **1985**, *19*, 1–8.
62. Kesselmeier, J.; Staudt, M. Biogenic volatile organic compounds (VOC): An overview on emission, physiology and ecology. *J. Atmos. Chem.* **1999**, *33*, 23–88.
63. Perera, R.M.; Marriott, P.J.; Galbally, I.E. Headspace solid-phase microextraction—Comprehensive two-dimensional gas chromatography of wound induced plant volatile organic compound emissions. *Analyst* **2002**, *127*, 1601–1607.
64. Shipway, A.N.; Katz, E.; Willner, I. Nanoparticle arrays on surfaces for electronic, optical, and sensor applications.*Chem. Phys. Chem.* **2000**, *1*, 18–52.
65. Yao, K.S.; Li, S.J.; Tzeng, K.C.; Cheng, T.C.; Chang, C.Y.; Chiu, C.Y.; Liao, C.Y.; Hsu, J.J.; Lin, Z.P. Fluorescence silica nanoprobe as a biomarker for rapid detection of plant pathogens. *Adv. Mater. Res.* **2009**, *79*, 513–516.
66. Singh, S.; Singh, M.; Agrawal, V.V.; Kumar, A. An attempt to develop surface plasmon resonance based immunosensor for Karnal bunt (*Tilletia indica*) diagnosis based on the experience of nano-gold based lateral flow immuno-dipstick test. *Thin Solid Films* **2010**, *519*, 1156–1159.
67. López, M.M.; Llop, P.; Olmos, A.; Marco-Noales, E.; Cambra, M.; Bertolini, E. Are molecular tools solving the challenges posed by detection of plant pathogenic bacteria and viruses? *Curr. Issues Mol. Biol.* **2009**, *11*, 13–46.
68. Frasco, M.F.; Chaniotakis, N. Semiconductor quantum dots in chemical sensors and biosensors. *Sensors* **2009**, *9*, 7266–7286.
69. Algar, W.R.; Krull, U.J. Quantum dots as donors in fluorescence resonance energy transfer for the bioanalysis of nucleic acids, proteins, and other biological molecules. *Anal. Bioanal. Chem.* **2008**, *391*, 1609–1618.
70. Rad, F.; Mohsenifar, A.; Tabatabaei, M.; Safarnejad, M.R.; Shahryari, F.; Safarpour, H.; Foroutan, A.; Mardi, M.; Davoudi, D.; Fotokian, M. Detection of Candidatus Phytoplasma aurantifolia with a quantum dots fret-based biosensor. *J. Plant Pathol.* **2012**, *94*, 525–534.
71. Safarpour, H.; Safarnejad, M.R.; Tabatabaei, M.; Mohsenifar, A.; Rad, F.; Basirat, M.; Shahryari, F.; Hasanzadeh, F. Development of a quantum dots FRET-based biosensor for efficient detection of Polymyxa betae. *Can. J. Plant Pathol.***2012**, *34*, 507–515.
72. Kuila, T.; Bose, S.; Khanra, P.; Mishra, A.K.; Kim, N.H.; Lee, J.H. Recent advances in graphene-based biosensors. *Biosens. Bioelectron.* **2011**, *26*, 4637–4648.
73. Pérez-López, B.; Merkoçi, A. Nanoparticles for the development of improved (bio) sensing systems. *Anal. Bioanal. Chem.* **2011**, *399*, 1577–1590.
74. Shiddiky, M.J.; Torriero, A.A. Application of ionic liquids in electrochemical sensing systems. *Biosens. Bioelectron.***2011**, *26*, 1775–1787.
75. Cao, X.; Ye, Y.; Liu, S. Gold nanoparticle-based signal amplification for biosensing. *Anal. Biochem.* **2011**, *417*, 1–16.
76. Mandler, D.; Kraus-Ophir, S. Self-assembled monolayers (SAMs) for electrochemical sensing. *J. Solid State Electrochem.***2011**, *15*, 1535–1558.
77. Umasankar, Y.; Ramasamy, R.P. Highly sensitive electrochemical detection of methyl salicylate using electroactive gold nanoparticles. *Analyst* **2013**, *138*, 6623–6631.
78. Boonham, N.; Glover, R.; Tomlinson, J.; Mumford, R. Exploiting generic platform technologies for the detection and identification of plant pathogens. In *Sustainable Disease Management in a European Context*; Springer: Berlin, Germany, 2008; pp. 355–363.

79. Chartuprayoon, N.; Rheem, Y.; Chen, W.; Myung, N. Detection of plant pathogen using LPNE grown single conducting polymer Nanoribbon. In *Meeting Abstracts*; The Electrochemical Society: Pennington, NJ, USA, 2010.

80. Sadanandom, A.; Napier, R.M. Biosensors in plants. *Curr. Opin. Plant Biol.* **2010**, *13*, 736–743.

81. Skottrup, P.D.; Nicolaisen, M.; Justesen, A.F. Towards on-site pathogen detection using antibody-based sensors. *Biosens. Bioelectron.* **2008**, *24*, 339–348.

82. Leonard, P.; Hearty, S.; Brennan, J.; Dunne, L.; Quinn, J.; Chakraborty, T.; O'Kennedy, R. Advances in biosensors for detection of pathogens in food and water. *Enzyme Microb. Technol.* **2003**, *32*, 3–13.

83. Palchetti, I.; Mascini, M. Electroanalytical biosensors and their potential for food pathogen and toxin detection. *Anal. Bioanal. Chem.* **2008**, *391*, 455–471.

84. Byrne, B.; Stack, E.; Gilmartin, N.; O'Kennedy, R. Antibody-based sensors: Principles, problems and potential for detection of pathogens and associated toxins. *Sensors* **2009**, *9*, 4407–4445.

85. Zeng, C.; Huang, X.; Xu, J.; Li, G.; Ma, J.; Ji, H.F.; Zhu, S.; Chen, H. Rapid and sensitive detection of maize chlorotic mottle virus using surface plasmon resonance-based biosensor. *Anal. Biochem.* **2013**, *440*, 18–22.

86. Eun, A.J.-C.; Huang, L.; Chew, F.-T.; Li, S.F.-Y.; Wong, S.-M. Detection of two orchid viruses using quartz crystal microbalance (QCM) immunosensors. *J. Virol. Methods* **2002**, *99*, 71–79.

87. Campbell, G.A.; Mutharasan, R. Detection of Bacillus anthracis spores and a model protein using PEMC sensors in a flow cell at 1 mL/min. *Biosens. Bioelectron.* **2006**, *22*, 78–85.

88. Candresse, T.; Lot, H.; German-Retana, S.; Krause-Sakate, R.; Thomas, J.; Souche, S.; Delaunay, T.; Lanneau, M.; Le Gall, O. Analysis of the serological variability of *Lettuce mosaic* virus using monoclonal antibodies and surface plasmon resonance technology. *J. Gen. Virol.* **2007**, *88*, 2605–2610.

89. Dickert, F.L.; Hayden, O.; Bindeus, R.; Mann, K.-J.; Blaas, D.; Waigmann, E. Bioimprinted QCM sensors for virus detection—Screening of plant sap. *Anal. Bioanal. Chem.* **2004**, *378*, 1929–1934.

90. Nugaeva, N.; Gfeller, K.Y.; Backmann, N.; Duggelin, M.; Lang, H.P.; Guntherodt, H.-J.; Hegner, M. An antibody-sensitized microfabricated cantilever for the growth detection of Aspergillus niger spores. *Microsc. Microanal.* **2007**, *13*, 13–17.

91. Skottrup, P.; Frokiaer, H.; Hearty, S.; O'Kennedy, R.; Hejgaard, J.; Nicolaisen, M.; Justesen, A.F. Monoclonal antibodies for the detection of *Puccinia striiformis* urediniospores. *Mycol. Res.* **2007**, *111*, 332–338.

92. Skottrup, P.; Hearty, S.; Frokiaer, H.; Leonard, P.; Hejgaard, J.; O'Kennedy, R.; Nicolaisen, M.; Justesen, A.F. Detection of fungal spores using a generic surface plasmon resonance immunoassay. *Biosens. Bioelectron.* **2007**, *22*, 2724–2729.

93. Torrance, L.; Ziegler, A.; Pittman, H.; Paterson, M.; Toth, R.; Eggleston, I. Oriented immobilisation of engineered single-chain antibodies to develop biosensors for virus detection. *J. Virol. Methods* **2006**, *134*, 164–170.

94. Zezza, F.; Pascale, M.; Mule, G.; Visconti, A. Detection of *Fusarium culmorum* in wheat by a surface plasmon resonance-based DNA sensor. *J. Microbiol. Methods* **2006**, *66*, 529–537.

95. Lin, H.-Y.; Huang, C.-H.; Lu, S.-H.; Kuo, I.-T.; Chau, L.-K. Direct detection of orchid viruses using nanorod-based fiber optic particle plasmon resonance immunosensor. *Biosens. Bioelectron.* **2014**, *51*, 371–378.

96. James, C. Polypyrrole nanoribbon based chemiresistive immunosensors for viral plant pathogen detection. *Anal. Methods* **2013**, *5*, 3497–3502.

97. Perdikaris, A.; Vassilakos, N.; Yiakoumettis, I.; Kektsidou, O.; Kintzios, S. Development of a portable, high throughput biosensor system for rapid plant virus detection. *J. Virol. Methods* **2011**, *177*, 94–99.

98. Kumar, A.; Dash, S.K.; Suman, D.P.S. DNA based biosensors for detection of pathogens. In *Plant Fungal Disease Management*, 1st ed.; Westville: New York, NY, USA, 2015; pp. 31–35.

99. Eun, A.J.-C.; Wong, S.-M. Molecular beacons: A new approach to plant virus detection. *Phytopathology* **2000**, *90*, 269–275.

100. Eun, A.J.-C.; Huang, L.; Chew, F.-T.; Li, S.F.-Y.; Wong, S.-M. Detection of two orchid viruses using quartz crystal microbalance-based DNA biosensors. *Phytopathology* **2002**, *92*, 654–658.

101. Ivnitski, D.; Abdel-Hamid, I.; Atanasov, P.; Wilkins, E.; Stricker, S. Application of electrochemical biosensors for detection of food pathogenic bacteria. *Electroanalysis* **2000**, *12*, 317–325.

102. Ronkainen, N.J.; Halsall, H.B.; Heineman, W.R. Electrochemical biosensors. *Chem. Soc. Rev.* **2010**, *39*, 1747–1763.

103. Haccoun, J.; Piro, B.; Tran, L.D.; Dang, L.A.; Pham, M.C. Reagentless amperometric detection of L-lactate on an enzyme-modified conducting copolymer poly (5-hydroxy-1,4-naphthoquinone-co-5-hydroxy-3-thioacetic acid-1,4-naphthoquinone). *Biosens. Bioelectron.* **2004**, *19*, 1325–1329.

104. Thomas, S.G.; Phillips, A.L.; Hedden, P. Molecular cloning and functional expression of gibberellin 2-oxidases, multifunctional enzymes involved in gibberellin deactivation. *Proc. Natl. Acad. Sci. USA* **1999**, *96*, 4698–4703.

105. Fang, Y.; Umasankar, Y.; Ramasamy, R.P. Plant Volatile Sensor: Enzymatic Transducer for Selective and Sensitive Determination of Methyl Salicyalte. In *Meeting Abstracts*; The Electrochemical Society: Pennington, NJ, USA, 2014.

106. Kulagina, N.V.; Shankar, L.; Michael, A.C. Monitoring glutamate and ascorbate in the extracellular space of brain tissue with electrochemical microsensors. *Anal. Chem.* **1999**, *71*, 5093–5100.

107. Mc Grath, S.; van Sinderen, D. *Bacteriophage: Genetics and Molecular Biology*; Horizon Scientific Press: Norfolk, UK, 2007.

108. Brigati, J.R.; Petrenko, V.A. Thermostability of landscape phage probes. *Anal. Bioanal. Chem.* **2005**, *382*, 1346–1350.

109. Kretzer, J.W.; Lehmann, R.; Schmelcher, M.; Banz, M.; Kim, K.-P.; Korn, C.; Loessner, M.J. Use of high-affinity cell wall-binding domains of bacteriophage endolysins for immobilization and separation of bacterial cells. *Appl. Environ. Microbiol.* **2007**, *73*, 1992–2000.

110. Neufeld, T.; Schwartz-Mittelmann, A.; Biran, D.; Ron, E.Z.; Rishpon, J. Combined phage typing and amperometric detection of released enzymatic activity for the specific identification and quantification of bacteria. *Anal. Chem.* **2003**,*75*, 580–585.

111. Adriaenssens, E.M.; van Vaerenbergh, J.; Vandenheuvel, D.; Dunon, V.; Ceyssens, P.-J.; de Proft, M.; Kropinski, A.M.; Noben, J.-P.; Maes, M.; Lavigne, R. T4-related bacteriophage LIMEstone isolates for the control of soft rot on potato caused by "Dickeya solani". *PLoS ONE* **2012**, *7*, e33227.

112. Fujiwara, A.; Fujisawa, M.; Hamasaki, R.; Kawasaki, T.; Fujie, M.; Yamada, T. Biocontrol of Ralstonia solanacearum by treatment with lytic bacteriophages. *Appl. Environ. Microbiol.* **2011**, *77*, 4155–4162.

113. Tlili, C.; Sokullu, E.; Safavieh, M.; Tolba, M.; Ahmed, M.U.; Zourob, M. Bacteria screening, viability, and confirmation assays using bacteriophage-impedimetric/loop-mediated isothermal amplification dual-response biosensors. *Anal. Chem.* **2013**, *85*, 4893–4901.

114. Schofield, D.A.; Bull, C.T.; Rubio, I.; Wechter, W.P.; Westwater, C.; Molineux, I.J. "Light-tagged" bacteriophage as a diagnostic tool for the detection of phytopathogens. *Bioengineered* **2013**, *4*, 50–54.

115. Askora, A.; Kawasaki, T.; Usami, S.; Fujie, M.; Yamada, T. Host recognition and integration of filamentous phage □RSM in the phytopathogen, Ralstonia solanacearum. *Virology* **2009**, *384*, 69–76.

116. Frampton, R.A.; Taylor, C.; Moreno, A.V.H.; Visnovsky, S.B.; Petty, N.K.; Pitman, A.R.; Fineran, P.C. Identification of bacteriophages for biocontrol of the kiwifruit canker phytopathogen *Pseudomonas syringae* pv. actinidiae. *Appl. Environ. Microbiol.* **2014**, *80*, 2216–2228.

117. Yamada, T.; Kawasaki, T.; Nagata, S.; Fujiwara, A.; Usami, S.; Fujie, M. New bacteriophages that infect the phytopathogen *Ralstonia solanacearum*. *Microbiology* **2007**, *153*, 2630–2639.

118. Eun, A.J.-C.; Wong, S.-M. Detection of cymbidium mosaic potexvirus and odontoglossum ringspot tobamovirus using immuno-capillary zone electrophoresis. *Phytopathology* **1999**, *89*, 522–528.

119. Hill, J.; Bryant, G.; Durand, D. Detection of plant virus by using purified IgG in ELISA. *J. Virol. Methods* **1981**, *3*, 27–35.

CHAPTER 15

PLANTS VERSUS FUNGI AND OOMYCETES: PATHOGENESIS, DEFENSE AND COUNTER-DEFENSE IN THE PROTEOMICS ERA

Abdelbasset El Hadrami, Ahmed F. El-Bebany, Zhen Yao, Lorne R. Adam, Ismailx El Hadrami and Fouad Daayf

Keywords: Biotrophs, counter-defenses, defenses, effectors, hemibiotrophs, necrotrophs, pathogenicity factors, plant-pathogen interactions, pathosystems, proteome, proteomics approach

INTRODUCTION

Plant-pathogen interactions are among the most complex and exciting associations to study. They involve living organisms and counterparts able to generate and evolve various infection and pathogenesis strategies. Through their co-evolution, both plants and their pathogens have shaped their strategies to survive and eventually win the fight against their respective invader or host. This occurs through a constant battle involving an array of pathways and signaling networks that shape pathogenesis, plant defenses and further pathogen counter-defenses [1,2]. Proteomic studies during plant-pathogen interactions are crucial because most of the pathways used either by plants or pathogens rely on protein synthesis and activity. These have been pursued for many decades at a slow pace and have led to the characterization of many families of the so-called Pathogenesis Related-proteins (PRs) involved in plant defenses [3], or others involved in pathogenesis [1]. In recent years, new bioinformatics mining tools have emerged, thereby accelerating the discovery and annotation of such proteins [2]. The 2-dimensional polyacrylamide gel electrophoresis (2D-PAGE) technology was combined with new analytical instruments such as nanoflow liquid chromatography coupled to tandem mass spectrometry (*i.e.*, LC-MS/MS) to establish proteome maps and explain specific aspects of host-pathogen interactions [4,5]. Other approaches that do not require the use of PAGE have been developed [6] and searchable databases have been increasingly updated with newly discovered proteins based on Molecular Weight (MW), Isoelectric Focusing point (IEF) and characteristic MS/MS spectra of*m/z* (mass over charge) [5]. Cellular and sub-cellular screenings and other *in silico* annotations or *in vitro* functional analyses were also conducted for peptides and proteins originating from the plant, the pathogen or their interactions [2]. Through proteomics, it is also possible to reveal post-translational modifications in proteins after their synthesis [7]. These include compartment signal peptide cleavages, glycosylation and deglycosylation, phosphorylation and dephosphorylation as well as other unfolding modifications. Many mechanisms of the interplay between signaling networks, on the other hand, remain to be elucidated to determine the nature of their antagonism or synergy [8,9]. Despite the significant progress made to date, currently available proteomic tools still lack the required sensitivity to detect peptides that are transient or in low abundance. Unfortunately, this applies to many signaling proteins, such as kinases and transcription factors, some of which have proven to act as key regulators during plant-pathogen interactions.

Citation: El Hadrami, Ahmed F. El-Bebany, Zhen Yao, Lorne R. Adam, Ismailx El Hadrami, and Fouad Daayf ,Plants versus Fungi and Oomycetes: Pathogenesis, Defense and Counter-Defense in the Proteomics Era, Int. J. Mol. Sci. 2012, 13(6), 7237-7259; doi:10.3390/ijms13067237

This review discusses the opportunities and challenges offered through the study of plant-pathogen interactions *in planta*using proteomic approaches to underline pathogen determinants of pathogenicity [4,10,11] and the resulting host defenses [12–14]. In particular, it highlights important findings in interactions involving fungi and oomycetes with biotrophic, necrotrophic, or hemibiotrophic lifestyles. Further to plant defenses, understanding pathogen counter-defenses could lead to better management of host resources towards durable resistance. Therefore, proteomic studies set to unravel the mechanisms that pathogens use to counter plant defenses are emphasized.

PROTEOMIC ANALYSIS OF PLANT DEFENSES AND PATHOGEN COUNTER-DEFENSES

Proteomic approaches have been used to examine both pathogenicity determinants and the multilayered facets of plant defenses in many economically important diseases [15]. Studying both sides of the same equation is a legitimate attempt to assemble a more complete picture of the role of the detected proteome during the interaction. Such studies faced many challenges, *i.e.*, how to distinguish between plants and pathogen proteins and between those that are constitutive from those resulting from the interaction. However, further approaches had some success using tag labeling, subtractive methods, and differential treatments [4,10,16,17].

Analysis of the proteome of plant-pathogen interactions may serve many purposes through basic questions: (i) what plant proteins were activated in response to the pathogen and *vice-versa*? (ii) are these activations a result of new transcriptions/translations or post-translational activations of pre-formed inactive products? (iii) does the pattern of activation vary among pathosystems? (iv) is there any trend that could co-evolutionary explain defenses and counter-defenses and provide future prospects for engineering durable resistance? (v) could bioagents be used to accurately manipulate such interactions?

Proteomics as a Tool to Characterize Plant Defenses

Two types of plant defenses were defined; the pathogen-associated molecular patterns PAMP-triggered or innate immunity (PTI), and the pathogen effector-triggered immunity (ETI) [18]. PTI defenses are associated with the recognition of the pathogen followed by a series of downstream events. These include the alteration of ion fluxes across the host's plasma membrane, the release of reactive oxygen species and nitric oxide as signaling molecules, the activation of MAP kinases, SA and JA/ET, and cross-talk between these pathways or with others such as ABA. ETI reactions, on the other hand, are elicited upon perception of effectors [19,20] released by the pathogen. Resistance gene-mediated defense follows, but often depends on the type of genes involved and the direct/indirect recognition of effectors [18]. Defense responses usually take place at sub-cellular levels, with cell structure rearrangements and transcriptional regulations that lead to the accumulation of defense-related molecules. The latter fortify the host cell wall [21–23], affect the integrity of the invading hyphae [24–26], and restrict progress of the pathogen [27,28]. Cell communication establishes among the tissues surrounding the initial infection sites and remote sites, hence leading to systemic acquired resistance (SAR) or induced systemic resistance (ISR) [29].

In host-pathogen interactions, proteomics have been truly complementary to functional genomics, and will continue to be so even more as entire genomes of plants and microorganisms are increasingly accessible. Vast Genbank databases with billions of sequences are daily being browsed, annotated or corrected/replaced by researchers from around the globe. It is interesting to note that proteomic investigation of plant defenses confirmed many of the results gathered using classical forward and reverse genetics, and biochemistry or molecular biology. It also provided a new generation of dataset, descriptive yet to be further analyzed and mastered for an overall view of the processes triggered in plants while fighting pathogens. It is possible through these approaches to now focus on an entire pathway involved in resistance and determine its interaction with other signaling and defense networks.

A number of pathosystems where proteomic approaches have been successfully used revealed either an increase or a decrease in protein abundance in response to infection by pathogens or to their pathogenicity factors. These proteins can be associated with photosynthesis and energy metabolism, anti-fungal activity, phytohormone homeostasis, ROS generation, detoxification and signaling, phenylpropanoid and lignin biosynthesis, and more (see below).

Proteomic Analysis of Plant Responses to Fungi and Oomycetes

Proteomic studies of plant-pathogen interactions *in planta* remain challenging. Data retrieval is important in determining proteins that accumulate differentially and are important for defense, pathogenicity, or counter-defense. Such studies often rely on (i) empirical comparison with databases and among pathosystems, (ii) complementary studies using the plant or pathogen individually as controls, or (iii) differential analysis using cultivars with various levels of resistance against highly aggressive isolates. Most studies, if not at a preliminary stage [4], clustered proteomes by nature, function, or sub-cellular localization of peptides [15,30], and reported on the technical challenges encountered in: (i) characterizing proteins in the absence of fully sequenced genomes of the host plants or their pathogens [31], and (ii) analyzing functions and regulation of the proteins encoded by the sequenced genomes.

To understand resistance of *Brassica carinata* to *Leptosphaeria maculans*, the causal agent of phoma leaf spot and stem canker, a proteomic investigation compared this species to a susceptible one, *Brassica napus* [16]. Forty-eight hours post-inoculation, 28 proteins accumulated differentially in the resistant phenotype, including superoxide dismutase, nitrate reductase, and carbonic anhydrase, as well as many photosynthesis-related enzymes. In cotton (*Gossypium hirsutum* L.), Coumans *et al.* [32] reported on proteins involved in plant defense against the black root rot fungus, *Thielaviopsis basicola*, especially PR-proteins and those implicated in ROS production and scavenging or in the isoprenoid biosynthesis pathway. In response to *Verticillium dahliae*, Wang *et al.* [33] also identified 51 upregulated and 17 downregulated proteins mainly involved in defense and stress responses, primary and secondary metabolism, lipid transport, and cytoskeleton organization. As such, analyzing proteomes during host-pathogen interactions will always generate differential expression of genes/proteins involved in disease or resistance. However, comparing proteomic research advances in pathosystems is complex, given the diversity of approaches used in proteome analysis and the genetic and biochemical backgrounds of the hosts or pathogens to be compared. Nonetheless, two categories can be distinguished, and involve either proteome data acquired only on the pathogen, or those gathered during host-pathogen interaction *in planta* [34]. One of the challenges relates to the distinction between proteins originating from the host and the pathogen. However, judicious experimental setups and recent technical and bioinformatics advances have been providing ways to better interpret such complex data. Data specifically related to the lifestyle of the pathogen, as biotroph, necrotroph or hemibiotroph are also valuable for proper result analysis and interpretation, but may bring their own complexities.

Pathosystems Involving Biotrophs

Biotrophs establish in their host after they reach the plant surface, adhere and grow penetrating hyphae or appressoria, then haustoria [1]. Using these structures, CWDEs and effectors are released, some into the cells through mechanisms including injection, secretion, or simply diffusion and exocytosis [1,19,20,35]. Proteomic approaches developed in recent years revealed the content of these secretomes in some pathosystems. A growing number of proteins harboring RxLR or other conserved motifs have been determined to be invasive effectors essential for the injection of virulence factors into the host cells, hence leading to disease expression [19]. Effectors with LysM motifs are, on the other hand, thought to be defensive effectors [20,36] for the protection of the fungal cell-wall chitin layer, and preservation of its integrity. These two types of effectors are believed to be widespread among fungi, especially biotrophs.

Probably one of the most studied pathosytems involving a biotroph is the interaction between rice (*Oryza sativa* L.) and the blast pathogen *Magnaporthe grisea*. Past studies examined the proteome patterns in leaves interacting with the pathogen and under various levels of nitrogen [37,38]. Among the noticeable changes, 12 proteins including a PR-5 were up-regulated upon infection and differentially accumulated with nitrogen supplementation [12,13].

In a biotrophic interaction involving wheat lines and leaf rust pathogen *Puccinia triticina*, Rampitsch *et al.* [14] found no noticeable differences between soluble proteomes (pH 4–8) of inoculated and non-inoculated leaves of cultivar Thatcher, a susceptible line and Thatcher + Lr1, a near isogenic resistant one carrying *Lr1*, a dominant resistance gene to rust. To explain this, they suggested a masking effect of healthy tissues over the localized hypersensitive response in the resistant cultivar. In a similar approach, Devos *et al.* [39] reported many proteins related to defense, and to metabolic and cell differentiation processes that increased in response to inoculation of *Arabidopsis thaliana* with *Plasmodiophora brassicae*, the causal agent of clubroot in crucifers. The authors hypothesized that the establishment of typical meristematic zones around the infection site could be due to some of the proteins involved in cell differentiation, hence leading to the creation of a sink to host auxins, carbohydrates, nitrogen and energy to initiate gall formation and sustain pathogen growth and development. While studying changes in *Brassica napus* L. roots in response to compatible infections by *P. brassicae*, Cao *et al.* [40] revealed a decrease in the relative abundance of adenosine kinase. The latter has a role in cytokinin homeostasis as well as during infection of crucifers by *P. brassicae*. Lignin biosynthesis-associated caffeoyl-CoA *O*-methyltransferase and ROS-associated enzymes also decrease after infection, suggesting a reduction in cell wall strengthening, thus enhancing susceptibility. So far, more studies are needed to confirm the functions of many important proteins in these pathosystems, with a caution not to systemically extrapolate function findings onto other pathosystems. It is also paramount to cautiously design experiments that compare the proteomes of healthy and infected tissues, in order to avoid dilution of information such as that involving small hypersensitive spots in large leaf surfaces, and the use of appropriate checks.

Proteomic approaches were also used in plant-oomycete interactions. In the case of sudden oak death caused by*Phytophthora ramorum*, Meijer *et al.* [41] characterized 17 proteins that were exclusively secreted, including putative mucins, glucoside hydrolases, transglutaminases, an annexin-like protein and a Kazal-type protease inhibitor. These results confirmed findings in other oomycetes such as *Phytophthora cinnamomi* and *Phytophthora infestans*, known to produce spores with storage proteins [42] and a cocktail of CWDEs and suppressors [1,43–45]. These proteins are deposited in vesicles that are released during spore germination, thereby triggering the suppression of cell death and the hypersensitive response [46], the inhibition of superoxide generation [47], or accumulation of phytoalexins [43,45,48,49].

Changes in protein patterns of *Medicago truncatula*'s roots interacting with *Aphanomyces euteiches* [50] included cell wall proteins and enzymes involved in the phenylpropanoid pathway, especially those associated with isoflavonoids and phytoalexin biosynthesis [51]. Similarly, Curto *et al.* [52] and Amey *et al.* [53] identified leaf proteins implicated in pea resistance to powdery and downy mildews, caused by *Erysiphe pisi* and *Peronospora viciae*, respectively. Therefore, many similarities exist between systems involving biotrophic fungi and oomycetes, which is not surprising due to their similar lifestyle.

Pathosystems Involving Necrotrophs

As with biotrophs, the data available so far for necrotrophs is scattered among different levels of tissues and experimental setups. Leaf proteome of rice after inoculation with *Rhizoctonia solani* showed an increase in the relative abundance of proteins involved in energy metabolism and photosynthesis, defense, antioxidant activity, as well as protein unfolding and degradation processes [54]. In maize embryos inoculated with *Fusarium verticillioides*, proteomic host defense responses included PRs, ROS-detoxifying enzymes, and others implicated in protein biosynthesis or folding and stabilization processes [55]. In wheat infected by *Fusarium graminearum*, which causes Fusarium head blight, Zhou *et al.* [17] ranged 33 plant proteins into defense-related or metabolism-related groups, and more have been identified more recently in wheat and barley interactions with this pathogen [56,57]. In response to *Alternaria brassicae*, *Brassica napus* L. produced enzymes involved in ROS production as well as auxin signal transduction and metabolic pathways [58], while application of oxalic acid, a major pathogenicity factor of many necrotrophic pathogens, including *Sclerotinia sclerotiorum*, resulted in down-regulation of SA-dependent pathways and up-regulation of JA/ET- or ABA-dependent ones [59,60].

On the pathogen side, the grey mold fungus *Botrytis cinerea* has been greatly investigated in recent years using an array of approaches, including proteomics [61], to identify its pathogenicity factors [62]. It was recently shown to manipulate the network of phytohormones and oxidative burst system of the host [1,63]. Comparison of proteomic profiles of two isolates of this fungus, with contrasting levels of virulence [61], showed their ability to differentially accumulate proteins, including isoforms of malate and glyceraldehyde-3-phosphate dehydrogenases.

Pathosystems Involving Hemibiotrophs

Fusarium oxysporum f.sp. *lycopersici* induced more defense-related proteins in the xylem sap of tomato plants in the compatible interaction than in the incompatible one and the non-infected checks [64]. Using the same model, Houterman *et al.* [31] identified 33 proteins, among which 21 were from tomato and seven from the pathogen, with 13 ascribed to response to infection, and others identified as CWDEs, peroxidases, and proteases.

Many of the proteins identified in such interactions may have homology in different plant species, but definitive confirmation always requires further functional analyses. Some level of difficulty may arise with regard to the fungal *versus*plant origin of identified proteins, and in functional analyses in systems where transformation is not straightforward, such as wheat.

Proteomics as a Tool to Examine Pathogenicity Determinants

Plant pathogens specialize in attacking their respective hosts and feeding on their tissues, thus causing disease and, potentially, epidemics. They possess a large number of features that allow them to infect and establish in/on plant tissues [1,65,66]. Upon infection, necrotrophs release an array of enzymes that allow them to dispose of the organic matter of dead tissues [67,68]. Biotrophs, on the other hand, require their host cells to remain alive and have evolved other pathogenicity determinants, such as haustoria, following host cell penetration [1]. Similarly, pathogens producing toxins may secrete these molecules at the forefront of the infection site to lay the ground for mycelial invasion [65,66,69–71]. Recently, advances in understanding the mechanisms of biotrophic, necrotrophic and hemibiotrophic lifestyles were reviewed [72], with special emphasis on the biotrophy-necrotrophy switch of hemibiotrophic pathogens. During the last few years, a new field of research focusing on pathogen effectors has emerged [19,20]. Pathogen effectors enable plant defense suppression and can reprogram the metabolic machinery of infected tissue to benefit the pathogen in its growth and development [73].

Effectors were deemed to be widespread among invading pathogens and can be either invasive or defensive [74,75]. Invasive effectors allow a given pathogen to attack, *i.e.*, polygalacturonases [4,68], whereas, the defensive ones protect it from degrading enzymes (*i.e.*, chitinases; [20]) and toxic metabolites synthesized by the host (*i.e.*, phytoalexins and defense-related metabolites; [2,21,22]). Suppressors of plant defenses are also among the pathogenicity determinants/effectors in fungi and oomycetes as they were documented in bacteria and viruses [1]. Biotrophs needing to establish a trophic relationship with their live host cells have evolved these type of mechanisms to temporarily turn off the plant guards and defenses and avoid detection [1]. For hemibiotrophs and nectrotrophs, more sophisticated strategies to suppress defense mechanisms have recently been reported. These include the manipulation of signaling pathways [2,63], the hijacking of essential defense molecules [76], the detoxification of major defense-related metabolites [21,22] and others yet to be determined. In a recent study involving *V. dahliae*, a fungus with multiple lifestyles, varying from necrotroph, to hemibiotroph, to endophyte, on a wide host range [77], El-Bebany *et al.* [4] established a proteomic map (pH 4–7) using *in vitro*-cultures of weakly- *versus* highly-aggressive isolates. The comparison revealed 25 differentially-regulated proteins, most of which were important determinants for host penetration and colonization, development of microsclerotia, and resistance to stress and antibiotics in the highly-aggressive isolate tested. The authors also reported on an increase in protein abundance of the isochorismatase hydrolase, thought to be implicated as a suppressor of plant defense responses and signaling [4]. Proteome maps are also being developed for other important fungi, such as *Cladosporium fulvum* and*Mycosphaerella fijiensis*, the causal agents of leaf mould in tomato and black Sigatoka in bananas and plantains, respectively. Four isolates of *Verticillium albo-atrum* from two geographical zones were used to compare the proteomes of two pathotypes on hop and establish the first mycelial proteome map of this fungus [10]. Among the 53 proteins identified, 17 were ascribed functions, and about 30 differentially accumulated in the mild *versus* the lethal pathotype. Among the proteins differentially abundant in the lethal pathotype were peroxiredoxin and ascorbate peroxidase, cytoskeleton components and regulators, and proteins involved in protein synthesis and energy metabolism.

In *L. maculans*, which causes economically important diseases of crucifers, including canola blackleg and stem canker [23,78], the mycelial proteome has recently been reported [79], showing several CWDEs previously identified in this and other necrotrophs. Proteomic analysis of mycelia provides important insights into the ability of pathogens to evolve new tactics to deal with plant defenses. However, there is still much to achieve regarding the spatio-temporal accumulation of peptides/proteins produced by pathogens throughout their disease cycle, and how such choices allow them to strategically take advantage of specific plant weaknesses, or surmount most sophisticated plant defenses. It is nonetheless obvious that integration of proteomic, transcriptomic, and metabolomic data over time is a major step towards understanding such intricate interactions. In their study on exoproteome of the notorious nectrotrophic fungus *Sclerotinia sclerotiorum*, Yajima and Kav [67] reported on 52 secreted proteins that included many CWDEs, known as pathogenicity factors in this pathogen, and highlighted the complementarity of proteomics to genomic approaches by identifying an α-l-arabinofuranosidase that was omitted in functional genomic studies involving ESTs.

Cao *et al.* [11] examined the proteome profile of virulent and avirulent isolates of the biotrophic *Pyrenophora tritici-repentis*, the causal agent of tan spot in wheat, known for its ability to produce several Ptr host-specific toxins. Virulent isolates usually produce the toxic chlorosis-inducing protein Ptr ToxB due to the expression of the *ToxB* gene. Avirulent isolates harbor a homolog of the gene but do not produce a functional protein [69,80]. Aside from *ToxB* gene, the virulent isolate actually exhibited a distinctive profile where several translated and accumulated proteins were up-regulated and ascribed to virulence, including an α-mannosidase and an exo-β-1,3-glucanase. In addition, saprophytic growth and development may account for differences in the proteome profile between virulent and avirulent isolates.

In other filamentous fungi, a proteome map has been created from mycelia collected during the dimorphic transition from budding to filamentous growth in *Ustilago maydis* [81]. In *Blumeria graminis* f.sp. *hordei*, proteic fractions collected from conidiospores were used to generate such a map. Among the identified proteins were enzymes involved in carbohydrate, lipid, and protein metabolism [82]. In *Cladosporium fulvum* interacting with tomato, three extracellularly secreted proteins, namely CfPhiA, Ecp6 and Ecp7, have been recently characterized [83]. Therefore, proteome characterization in this class remains sporadic and many studies are needed to fill the gaps for the different development stages of oomycetes and during their interactions with their hosts.

In oomycetes, especially *Phytophthora* spp., proteomics are informative in studying the proteic pool during various stages of growth and development, including hyphae, cysts, sporangia, zoospores, and appressoria [84–86].

Some work has been done on the use of exogenous supplementation of chemically-synthesized molecules. Shimizu *et al.*[87] used vanillin on *Phanerochaete chrysosporium* cultures and reported on the identification of up-regulated proteins involved in vanillin metabolism. A similar procedure was applied on the white rot fungus, *Phanerochaete chrysosporium*, using benzoic acid [88], allowing the characterization of a pool of mycelia proteins differentially responsive to the treatment. In general, not many fungal/oomycete pathogens have been extensively studied in this respect. A recent review reports on *B. cinerea*, *S. sclerotiorum*, and *F. graminearum* as the most studied plant pathogenic fungi in proteomics [89].

The studies available so far highlight the worth of proteomic approaches in determining pathogenicity factors and effectors of many pathogens, but also suggest that a great horizon is open to more studies to investigate the function and roles of individual microbial peptides and proteins in development stages and host-pathogen interactions.

THE FUTURE OF PROTEOMICS IN STUDYING PLANT-PATHOGEN INTERACTIONS

Major discoveries in plant-pathogen interactions using proteomics are yet to come, with the development of high throughput and sophisticated procedures to analyze, annotate, and compare proteomes [90], including translational plant proteomics [91]. Advances are to be expected with the development of new tools such as liquid-phase IEF, high resolution 2D, MRMs and iTRAQ quantification procedures [92,93]. Progress is also being made in mathematical algorithms and statistical procedures to cluster and analyze massive datasets. Accurate linking of functions to targets is being sought, and will still be, while studies are moving from pathosystems involving model species, such as the thale cress, to economically important crops and their diseases. Metaproteomics are also being developed to examine cohabitant microbial species proteomes, their interplays, and understand relationships among microorganism communities present either in the rhizosphere or the phyllosphere. This will lay the ground for the application of proteomic approaches in studying tripartite relationships, involving the host, its pathogen, and biocontrol agents or other biological entities. The accruing information about pathogens that evolve strategies to counteract plant defenses will certainly benefit from proteome studies and help understand intricacies of the co-evolution between pathogens and their hosts.

Going beyond the Model Systems

Proteomic approaches have not yet provided a full coverage of the entire proteome of plant species such as *A. thaliana* or the model fungus *C. fulvum*. More than 35,000 translated proteins were initially predicted in *A. thaliana* and currently only around 1/10 of that number has been characterized [94], likely due to challenges in separating and identifying all the proteins expressed under specific conditions. The latter relate to insolubility, low abundance, size, or pH. However, proteomics findings often feed back to genomics by complementing the annotation of ORFs and correcting gene structure predictions. In addition, proteomics in model plant species allows for the determination of recurrent peptides to be used as internal controls for large-scale proteomic studies. It also allows for the examination of post-translational processes that proteins go through, such as cleavage of signal peptides to allow compartmentalization, *N*-glycosylation, or phosphorylation. Such knowledge accelerates our understanding of other pathosystems involving economically important crops and destructive pathogens, hence increasing the likelihood of setup control management strategies that reduce their impact and restore yield and quality.

Carpentier *et al.* [95] discussed the challenges and opportunities facing proteomic approaches in non-model species, such as bananas and plantains, where the proteome analysis has flourished in absence of sequencing data on a large portion of the genome and with enormous challenges in conducting large scale gene expression. El Hadrami *et al.* [66] had examined the isoenzymatic profile and activities of a set of cytosolic and chloroplastic proteins involved in the scavenging and detoxification of ROS in a susceptible *versus* a partially resistant banana cultivar interacting with the toxin juglone, one of the pathogenicity factors of *M. fijiensis*. The main difference between the two tested phenotypes was in the quickness in triggering their antioxidant systems on perceiving the toxin, and how long the system remains active and coordinated.

It is clear that in the years to come, a continuous move will occur from studying plant models to studying non-model species. In spite of the lack of extensive genomic data on the latter and/or how they differ from model species, they will still benefit from the available genomics and proteomics datasets.

There is a need for appropriate extraction methods that minimize interferences with other analytes and maximize separation and quantification. As for identification and assigning functions, one would still rely on the background of model species unless other information can be gathered through complementary approaches.

Resolving Unambiguous Protein Identification

Significant progress has been made in proteomic studies in many pathosystems using differential models or comparing inoculated with mock-inoculated controls [4,10,14,33]. Separation of the proteome is conducted by 1D or 2D-PAGE *i.e.*, [4,10], which often lack sensitivity and reproducibility [10]. The advent of nanoflow-LC technology has made it possible to minimize some of these issues *i.e.*, [4]. To enhance the separation and gain an in-depth understanding, new approaches, such as liquid-phase IEF coupled to high-resolution 2D, have also been used in fungi [79]. Immense challenges are still encountered in mass spectrometry detection, identification, and quantification. Current methods use nanoflow-LC coupled to mass spectrometry technologies with triple or quadruple quads such as MALDI, MALDI-Q-TOF, ESI-MS/MS, ICP-MS/MS, as well as programs specialized in data acquisition, spectral analysis and protein identification through matches with onsite or online databases (*i.e.*, MASCOT, SEQUEST, X!TANDEM). Other quantification procedures using stable labeling of ^{13}C, ^{2}H, ^{15}N, or ^{18}O within techniques such as ICAT [96–99], SILAC [100], and iTRAQ™ [101] have also became popular and have greatly improved the quantitative determination of the detected proteins. iTRAQ MS/MS technology has been applied in profiling differentially translated and accumulated proteins from *F. graminearum* after an *in vitro*-induction of mycotoxins synthesis in comparison with a mock-treated control [102]. The presence of phosphorus and sulfur residues in proteins has also been used along with technologies such as ICP-MS to translate the molarity of these residues into absolute protein concentrations.

Resolving unambiguous identification of proteins among contrasting treatments remains an important challenge in proteomics. While one can conceivably manually compare various gels with a limited number of differentially up- or down-regulated protein spots, difficulties arise when proteome analysis is conducted using high-resolution 2D-PAGE, especially after the use of liquid-phase IEF or other pre-screening methods. Theoretically, on these gels, up to 15,000 protein spots could be detected but practically only a third of that number or less can accurately be resolved and analyzed. An extensive development of hardware and software to precisely analyze these images has been seen in recent years to maximize spot annotations, establish protein patterns, resolve coalescent spots, and provide an accurate comparison among tested treatments. To further minimize identification errors, spots can be excised from the PAGE and enzymatically digested and subjected to MS/MS and database matching. Matching against available databases is often hampered by an extensive presence of hypothetical proteins with no assigned function and/or with lack of full protein sequences. Combining transcriptomic and proteomic approaches often yield an accurate correlation between the variation in the gene expression and the changes in the levels of the detected proteins, allowing for an exact determination of identity and function [2]. These can also be complemented with an array of *in silico* analysis to predict compartment localization, structures, and function [2]. In addition, a better knowledge of the pathosystems considered for proteomic analysis through other molecular, biochemical, physiological, and epidemiological approaches allows and eases the unambiguous interpretation of the gathered data.

Assigning Functions to Detected Proteins

Searching PAMPS

Proteome analysis is currently becoming an important component in dissecting plant-pathogen interactions, and represents a substantial addition to the daily-basis research tool box. It helps identifying up- or down-regulated proteins and assigned functions that are linked to pathogenicity, defense, or counter-defense. The gap in access to pathogen genomes will eventually be filled with the arrival of faster new-generation sequencers. This will balance the former research trend that mainly focused on the elucidation of plant defenses. A number of effectors, Avr genes and proteins as well as PAMPs have already been characterized [18,20]. A new flow of information gathered from the characterization of these proteins is currently being used to assess their roles in pathogenicity, in inducing susceptibility or HR in the host, or even in manipulating host signaling cascades and overall defense-responses. Proteomic studies have recently characterized a new class of PAMPs, from the oomycete *P. parasitica* var. *nicotianae*, with a cell wall anchored position and an ability to bind cellulose [103]. Similarly, the approach revealed other PAMPs from oomycetes and other fungi [104,105].

Characterizing Membrane Proteins and Receptors

The development of proteomics had taken protein-protein interactions to a level above the usual yeast two-hybrid systems. It is now possible to isolate membrane proteins and general or specific receptors [90]. Membrane proteins and binding receptors achieve their physiological functions through interaction with other intracellular proteins and effectors. More exploration could lead to the characterization of multiple proteins and complexes interacting either directly or indirectly with intercellular effectors and proteins to trigger signaling cascades during plant-pathogen interactions. Santoni [106] described a two-step procedure to recover transmembrane proteins of the plasma membrane of *A. thaliana*. First, the author depleted the plasma membrane from its soluble cytosolic contaminants and other functionally associated proteins, using an alkaline treatment, then, solubilized the hydrophobic remaining proteins. Similarly, Sazuka *et al.* [107] used a proteomic approach on the same plant to identify membrane-bound proteins. Analysis of such proteins revealed 64 informative spots belonging to 42 protein families including transporters, channels, receptors, and membrane-anchored proteins such as GTPases. Elortza *et al.* [108] analyzed the membrane-embedded proteins involved in glycosylphosphatidylinositol metabolism and signaling pathways. These proteins are believed to play a role in cell adhesion and perception of effectors and in signal transduction during host-pathogen interactions. Biological membranes are predicted to contain 30% of proteins so far, but the difficulties arising from their poor solubility, separation and identification will greatly benefit from advances in proteomics [109], which may lead to a better understanding of their functions and roles in plant-pathogen interactions.

Unveiling Wound-Healing Processes

In comparison with the biomedical sector, there seem to be fewer opportunities from current knowledge to minimize plant infection and colonization through wound-healing based on proteome analysis. Studying membrane proteins using proteomic approaches is rather difficult, although new advances in analytical methods in the next few years may help overcome such challenges. A strategy to accelerate wound-healing would lead to limiting plant infection by some pathogens. Slicing a potato tuber triggers natural wound-healing, which makes this species a useful model that was explored using proteomics [110,111]. Proteomics revealed a substantial alteration of the proteome pattern [110], including suberization-associated anionic peroxidases, thought to play a major role reconstructing the damaged periderm. While analyzing the tuber epidermis proteome in wounded *versus* non-wounded tubers, Barel and Ginzberg [111] showed that wounding enhances the translation and accumulation of proteins with a role in tissue differentiation and proliferation and in the oxidative respiratory chain and energy metabolism. Among the proteins detected were caffeoyl-CoA *O*-methyltransferase and a series of peroxidases, known for their role in potato periderm suberization [112]. This knowledge could well be used for creating cultivars with accelerated wound-healing process that could minimize their exposure to attacks by necrotrophic or soil-borne pathogens.

Examining Further the Roles of Fungal and Oomycete Effectors

One of the fields that will tremendously benefit from the expansion of proteomics is the identification of effector proteins [20] and their receptors. Extracellular proteins (Ecps) have been purified and characterized from apoplastic fluids of tomato plants subjected to infection with *C. fulvum*. These proteins are widespread among tested pathogen isolates, and exhibit similar properties as race-specific Avr elicitors with regard to host perception [20]. Similarly, proteomic approaches have been developed to investigate the RxLR effectors and other similar motifs that are thought to be omnipresent across fungi and oomycetes. Using proteomics, it has been possible to show that a positive selection has targeted the *C*-terminal of these effectors, thus contributing to their evolution and diversity [113]. This is in agreement with the current knowledge about these effectors, where the *N*-terminal domain is known to be involved in their secretion, whereas the *C*-terminal domain is implicated in interference with plant defenses once they are inside the host cells. Another category of effectors that has been getting increasing attention in recent years is the so-called LysM, believed to be implicated in the protection of the chitin layer of the fungus being targeted by plant chitinases and in modulating plant defenses [20]. In the next few years, proteomics should provide a better understanding of their mode of action, their roles in pathogenicity, and their implications to interfere with plant defenses.

Deciphering Regulons and Operons

The notions of transcriptional operons and proteome regulons are currently gaining ground after the discovery, through genomic and proteomic approaches, of a vast interplay and connectivity between the genes, proteins and signaling pathways that operate in plant-pathogen interactions. Operons represent clusters of genes that are functionally interrelated and co-regulated by the same promoter [114]. Regulons are clusters of genes that are under the same regulatory mechanisms of a single or multiple proteins [115,116]. Gene duplication that occurred during evolution had led to the setting of these gene clusters into paralogous and orthologous dispositions either among the core- or the plastic-genome [117]. The current advances in proteomics, in parallel with genomics, and metabolite profiling, are in the process of providing a better understanding of the mechanisms of such complex networks, especially during plant-pathogen interactions [118].

Studying Tripartite Interactions

Pathosystems involving a host, a pathogen, and a biocontrol agent are more challenging to study in terms of discerning the role of each one of them in the interaction. Proteomics-based approaches have been suggested to study these complex systems [119] and unveil the three-way signaling cross-talks. Marra *et al.* [119] used an experimental set-up that allowed them to separate the individual proteomes from beans, *B. cinerea* or *Trichoderma solani* and the antagonistic biocontrol agent*Trichoderma atrviridae*. The approach was instrumental in showing an up-regulation of bean's defense-related proteins such as PRs and that *T. atroviride* qualitatively and quantitatively alters the protein profile of the plant regardless of infection by a pathogen. The study also revealed proteins in *T. atroviride* with homology to fungal hydrophobins and ABC transporters, and the up-regulation in the pathogen proteome of virulence factors such as cyclophilins, when interacting with the host alone or in the concomitant presence of the biocontrol antagonist as well. Many other studies need to be investigated in such an integrated mode in order to gather more valuable information that is closer to the events that happen in nature.

Characterizing Pathogen Counter-Defenses

Between pathogens and their hosts, the balance was for many decades leaning towards studying plant defenses. In recent years, with the advent of genomics-based technologies that made more sequenced pathogen genomes available, more studies burgeoned on characterization of pathogenicity factors and pathogen-based strategies that avoid host resistance. Such strategies included ways to avoid detection by the host, suppress or cope with host defenses, and strategically manipulate host pathways and signaling cascades, thereby leading to unnecessary and energy-costly defenses or ones more beneficial to the pathogen [1,2]. These seem to be evolutionary strategies for co-existence that allow the pathogens to make the best of their trophic interactions with the hosts. While many biotrophs seem to have adapted strategies to either remain undetectable by plant cognate guardees and PRRs or to cope with plant defenses by detoxifying phytoalexins and other defense-related metabolites [1,21,22], hemibiotrophs and necrotrophs tend to raise the bar in sophisticating their counter-defense strategies [1,2,63]. For instance, necrotrophic fungi such as *S. sclerotiorum* suppress the host's ROS signaling cascade*in planta* via catalases, superoxide dismutases, or other scavenging molecules, such as mannitol or oxalic acid [120,121], hence leading to an enhanced hypersensitive response locally and/or preventing signal transduction to remote healthy sites.*Botrytis cinerea* forces its host to produce ROS, which kill the tissues, thereby enabling their colonization [122,123]. *Fusarium oxysporum*, on the other hand, has been reported to use the COI1-mediated JA-signaling to promote disease development in*A. thaliana* [76]. This manipulation of signaling pathways has also been shown in potato interacting with *V. dahliae* and tomato with *B. cinerea* [2,63]. In potato, plant defenses include an increase in the synthesis and accumulation of rutin. This defense-related flavonol-glycoside represents a carbon source for the pathogen, but cleaving the sugar moiety, using enzymes such as glucosidases and rhamnosidases, exposes *V. dahliae* to quercetin, the rutin's aglycone. During its coevolution with its hosts, *V. dahliae* has evolved a mechanism that allows it to detoxify this flavonol using its quercetinases [2]. The byproduct of detoxification, the 2-PCPGCA, is composed of phloroglucinol, a potent antibiotic that theoretically expands the pathogen's competitiveness in the rhizosphere, and protocatechuates that get converted into benzoates and salicylates, potential triggers of the SA-pathway, known to interfere with the JA-pathway. In the years to come, these reports on counter-defense strategies evolved by hemibiotrophs and necrotrophs will benefit to unveil new mechanisms and interplays using proteomics-based technologies.

CONCLUDING REMARKS

In recent years, proteomic analysis has enhanced our knowledge about plants, pathogens and their interactions with each other and with other components of their respective environments. It also complemented information gathered through forward and reverse genetics, and functional genomics. The increasing availability of fully-sequenced and annotated genomes allowed the making of great strides towards identification and establishment of protein functions, which consequently feeds back on accurate annotation of ORFs and prediction of gene functions. Proteomics have also made it possible to study protein-protein interactions, characterize membrane-anchored proteins, and discover new PAMPs. In plant-pathogen interactions, proteomic approaches are instrumental in deciphering plant defense pathways and signaling cascades, and unveiling pathogenicity factors and pathogen effectors. The field is set to generate even more opportunities, but not without more complexity, in studies meant to better understand (i) tripartite interactions where the host plant interacts with both beneficial and pathogenic microbes, and (ii) the strategies evolved by pathogens to counteract their host defenses and other microbes' attacks. Hopefully, integration of this knowledge from complex interactions will set us a step ahead towards developing more novel and environmentally-safe disease management strategies.

ACKNOWLEDGMENTS

F.D. was supported by fundamental and applied research grants from NSERC-Canada and Manitoba ARDI. *I.E.* was supported by PROTARS-Morocco.

REFERENCES

1. El Hadrami, A.; El Hadrami, I.; Daayf, F. Suppression of Induced Plant Defense Responses by Fungal Pathogens. In*Molecular Plant-Microbe Interactions*; Bouarab, K., Brisson, N., Daayf, F., Eds.; CABI: Oxfordshire, UK, 2009; pp. 231–268.
2. El Hadrami, A.; Adam, L.R.; Daayf, F. Biocontrol treatments confer protection against *Verticillium dahliae* infection of potato by inducing anti-microbial metabolites. *Mol. Plant Microbe Interact* **2011**, *24*, 328–335.
3. Van Loon, L.C.; Rep, M.; Pieterse, C.M.J. Significance of inducible defense-related proteins in infected plants. *Annu. Rev. Phytopathol* **2006**, *44*, 135–162.
4. El-Bebany, A.F.; Rampitsch, C.; Daayf, F. Proteomic analysis of the phytopathogenic soilborne fungus *Verticillium dahliae* reveals differential protein expression in isolates that differ in aggressiveness. *Proteomics* **2010**, *10*, 289–303.
5. Bhadauria, V.; Bannizam, S.; Wang, L.X.; Wei, Y.D.; Peng, Y.L. Proteomic studies of phytopathogenic fungi, oomycetes and their interactions with hosts. *Eur. J. Plant Pathol* **2010**, *126*, 81–95.
6. Goodlett, D.R.; Yi, E.C. Proteomics without polyacrylamide: Qualitative and quantitative uses of tandem mass spectrometry in proteome analysis. *Funct. Integr. Genomics* **2002**, *2*, 138–153.
7. Xing, T.; Rampitsch, C.; Sun, S.; Romanowski, A.; Conroy, C.; Stebbing, J.A.; Wang, X. TAB2, a nucleoside diphosphate protein kinase, as a component of the tMEK2 disease resistance pathway in tomato. *Physiol. Mol. Plant Pathol* **2009**, *73*, 33–39.
8. Kunkel, B.N.; Brooks, D.M. Cross talk between signaling pathways in pathogen defense. *Curr. Opin. Plant Biol* **2002**, *5*, 325–331.
9. Spoel, S.H.; Koornneef, A.; Claessens, S.M.C.; Korzelius, J.P.; van Pelt, J.A.; Mueller, M.J.; Buchala, A.J.; Métraux, J.P.; Brown, R.; Kazan, K.; *et al.* NPR1 modulates cross-talk between salicylate- and jasmonate-dependent defense pathways through a novel function in the cytosol. *Plant Cell* **2003**, *15*, 760–770.
10. Mandelc, S.; Radisek, S.; Jamnik, P.; Javornik, B. Comparison of mycelial proteomes of two *Verticillium albo-atrum*pathotypes from hop. *Eur. J. Plant Pathol* **2009**, *125*, 159–171.
11. Cao, T.; Kim, Y.M.; Kav, N.N.V.; Strelkov, S.E. A proteomic evaluation of *Pyrenophora tritici-repentis*, causal agent of tan spot of wheat, reveals major differences between virulent and avirulent isolates. *Proteomics* **2009**, *9*, 1177–1196.
12. Kim, S.T.; Cho, K.S.; Yu, S.; Kim, S.G.; Hong, J.C.; Han, C.D.; Bae, D.W.; Nam, M.H.; Kang, K.Y. Proteomic analysis of differentially expressed proteins induced by rice blast fungus and elicitor in suspension-cultured rice cells. *Proteomics***2003**, *3*, 2368–2378.
13. Kim, S.T.; Kim, S.G.; Hwang, D.H.; Kang, S.Y.; Kim, H.J.; Lee, B.H.; Lee, J.J.; Kang, K.Y. Proteomic analysis

of pathogen responsive proteins from rice leaves induced by rice blast fungus, *Magnaporthe grisea*. *Proteomics* **2004**, *4*, 3569–3578.

14. Rampitsch, C.; Bykova, N.V.; McCallum, B.; Beimcik, E.; Ens, W. Analysis of the wheat and *Puccinia triticina* (leaf rust) proteomes during a susceptible hostpathogen interaction. *Proteomics* **2006**, *6*, 1897–1907.
15. Mehta, A.; Brasileiro, A.C.; Souza, D.S.; Romano, E.; Campos, M.A.; Grossi-de-Sa, M.F.; Silva, M.S.; Franco, O.L.; Fragoso, R.R.; Bevitori, R.; Rocha, T.L. Plant-pathogen interactions: What is proteomics telling us? *FEBS J* **2008**, *275*, 3731–3746.
16. Subramanian, B.; Bansal, V.K.; Kav, N.N. Proteome-level investigation of *Brassica carinata*-derived resistance to*Leptosphaeria maculans*. *J. Agric. Food Chem* **2005**, *53*, 313–324.
17. Zhou, W.; Eudes, F.; Laroche, A. Identification of differentially regulated proteins in response to a compatible interaction between the pathogen *Fusarium graminearum* and its host, *Triticum aestivum*. *Proteomics* **2006**, *6*, 4599–4609.
18. Jones, J.D.G.; Dangl, J.L. The plant immune system. *Nature* **2006**, *444*, 323–329.
19. Rehmany, A.P.; Gordon, A.; Rose, L.E.; Allen, R.L.; Armstrong, M.R.; Whisson, S.C.; Kamoun, S.; Tylere, B.M.; Birch, P.R.J.; Beynon, J.L. Differential recognition of highly divergent downy mildew avirulence gene alleles by RPP1 resistance genes from two *Arabidopsis* lines. *Plant Cell* **2005**, *17*, 1839–1850.
20. Stergiopoulos, I.; de Wit, P.J.G.M. Fungal effector proteins. *Ann. Rev. Phytopathol* **2009**, *47*, 233–263.
21. Daayf, F.; Schmitt, A.; Bélanger, R.R. Evidence of phytoalexins in cucumber leaves infected with powdery mildew following treatment with leaf extracts of *Reynoutria sacchalinensis*. *Plant Physiol* **1997**, *113*, 719–727.
22. Daayf, F.; Ongena, M.; Boulanger, R.; El Hadrami, I.; Bélanger, R.R. Induction of phenolic compounds in two cultivars of cucumber by treatment of healthy and powdery mildew-infected plants with extracts of *Reynoutria sachalinensis*. *J. Chem. Ecol* **2000**, *26*, 1579–1593.
23. El Hadrami, A.; Daayf, F. Hydroxycinnamates induced in canola in response to weakly aggressive isolates of*Leptosphaeria* spp. help prevent further infections by highly aggressive isolates. *Can. J. Plant Pathol* **2009**, *31*, 393–406.
24. Wang, X.; El Hadrami, A.; Adam, L.; Daayf, F. US-1 and US-8 genotypes of *Phytophthora infestans* differentially affect local, proximal and distal gene expression of phenylalanine ammonia-lyase and 3-hydroxy, 3-methylglutaryl CoA reductase in potato leaves. *Physiol. Mol. Plant Pathol* **2004**, *65*, 157–167.
25. Wang, X.; El Hadrami, A.; Adam, L.R.; Daayf, F. Genes encoding pathogenesis-related proteins PR-2, PR-3 and PR-9, are differentially regulated in potato leaves inoculated with isolates from US-1 and US-8 genotypes of *Phytophthora infestans* (Mont.) de Bary. *Physiol. Mol. Plant Pathol* **2005**, *67*, 49–56.
26. Wang, X.; El Hadrami, A.; Adam, L.R.; Daayf, F. Local and distal gene expression of *pr*-1 and *pr*-5 in potato leaves inoculated with isolates from the old (US-1) and the new (US-8) genotypes of *Phytophthora infestans* (Mont.) de Bary.*Environ. Exp. Bot* **2006**, *57*, 70–79.
27. El Hadrami, I.; Ramos, T.; El Bellaj, M.; El Idrissi-Tourane, A.; Macheïx, J.J. A sinapic derivative as an induced defense compound of date palm against *Fusarium oxysporum* f. sp. *albedinis*, the agent causing bayoud disease. *J. Phytopathol***1997**, *145*, 329–333.
28. Daayf, F.; El Bellaj, M.; El Hassni, M.; J'aiti, F.; El Hadrami, I. Elicitation of soluble phenolics in date palm (*Phoenix dactylifera* L.) callus by *Fusarium oxysporum* f. sp. *albedinis* culture medium. *Env. Exp. Bot* **2003**, *49*, 41–47.
29. Hammerschmidt, R.; Kúc, J. *Induced Resistance to Disease in Plants*; Kluwer Academic Publishers: Dordrecht, The Netherlands, 1995.
30. Quirino, B.F.; Candido, E.S.; Campos, P.F.; Franco, O.L.; Krüger, R.H. Proteomic approaches to study plant-pathogen interactions. *Phytochemistry* **2010**, *71*, 351–362.
31. Houterman, P.M.; Speijer, D.; Dekker, H.L.; de Koster, C.G.; Cornelissen, B.J.C.; Rep, M. The mixed xylem sap proteome of *Fusarium oxysporum*-infected tomato plants. *Mol. Plant Pathol* **2007**, *8*, 215–221.
32. Coumans, J.V.F.; Poljak, A.; Raftery, M.J.; Backhouse, D.; Pereg-Gerk, L. Analysis of cotton (*Gossypium hirsutum*) root proteomes during a compatible interaction with the black root rot fungus. *Proteomics* **2009**, *9*, 335–349.
33. Wang, F.X.; Ma, Y.P.; Yang, C.L.; Zhao, P.M.; Yao, Y.; Jian, G.L.; Luo, Y.M.; Xia, G.X. Proteomic analysis of the sea-island cotton roots infected by wilt pathogen *Verticillium dahliae*. *Proteomics* **2011**, *11*, 4296–4309.
34. Kav, N.N.V.; Srivastava, S.; Yajima, W.; Sharma, N. Application of proteomics to investigate plant-microbe interactions. *Curr. Proteomics* **2007**, *4*, 28–43.
35. Kamoun, S. A catalogue of the effector secretome of plant pathogenic oomycetes. *Annu. Rev. Phytopathol* **2005**, *44*, 41–60.
36. Knogge, W.; Scheel, D. LysM receptors recognize friend and foe. *Proc. Natl. Acad. Sci. USA* **2006**, *103*, 10829–10830.
37. Konishi, H.; Ishiguro, K.; Komatsu, S. A proteomics approach towards understanding blast fungus infection of rice grown under different levels of nitrogen fertilization. *Proteomics* **2001**, *1*, 1162–1171.
38. Rakwal, R.; Agrawal, G.K. Rice proteomics: Current status and future perspectives. *Electrophoresis* **2003**, *24*, 3378–3389.

39. Devos, S.; Laukens, K.; Deckers, P.; van der Straeten, D.; Beeckman, T.; Inze, D.; van Onckelen, H.; Witters, E.; Prinsen, E. A hormone and proteome approach to picturing the initial metabolic events during *Plasmodiophora brassicae*infection on *Arabidopsis. Mol. Plant Microbe. Interact* **2006**, *19*, 1431–1443.

40. Cao, T.; Srivastava, S.; Rahman, M.H.; Kav, N.N.V.; Hotte, N.; Deyholos, M.K.; Strelkov, S.E. Proteome-level changes in the roots of *Brassica napus* as a result of *Plasmodiophora brassicae* infection. *Plant Sci* **2008**, *174*, 97–115.

41. Meijer, H.J.G.; van de Vondervoort, P.J.I.; Yin, Q.Y.; de Koster, C.G.; Klis, F.M.; Govers, F.; de Groot, P.W.J. Identification of cell wall-associated proteins from *Phytophthora ramorum. Mol. Plant Microbe Interact* **2006**, *19*, 1348–1358.

42. Gubler, F.; Hardham, A.R. Secretion of adhesive material during encystment of *Phytophthora cinnamomi* zoospores, characterized by immunogold labeling with monoclonal antibodies to components of peripheral vesicles. *J. Cell Sci* **1988**, *90*, 225–235.

43. Doke, N.; Garas, N.; Kúc, J. Partial characterization and aspects of the mode of action of a hypersensitivity-inhibiting factor (HIF) from *Phytophthora infestans. Physiol. Plant Pathol* **1979**, *15*, 127–140.

44. Shiraishi, T.; Yamada, T.; Saitoh, K.; Kato, T.; Toyoda, K.; Yoshioka, H.; Kim, H.M.; Ichinose, Y.; Tahara, M.; Oku, H. Suppressor: Determinants of specificity produced by plant pathogens. *Plant Cell Physiol* **1994**, *35*, 1107–1119.

45. Andreu, A.; Tonón, C.; van Damme, M.; Huarte, M.; Daleo, G. Effect of glucans from different races of *Phytophthora infestans* on defense reactions in potato tuber. *Eur. J. Plant Pathol* **1998**, *104*, 777–783.

46. Storti, E.; Pelucchini, D.; Tegli, S.; Scala, A. A potential defense mechanism of tomato against the late blight disease is suppressed by germinating sporangia-derived substances from *Phytophthora infestans. J. Phytopathol* **1988**, *121*, 275–282.

47. Doke, N. Involvement of superoxide anion generation in the hypersensitive response of potato tuber tissue to infection with an incompatible race of *Phytophthora infestans* and to the hyphal wall components. *Physiol. Plant Pathol* **1983**, *23*, 345–357.

48. Ozeretskovskaya, O.L.; Vasyukova, N.I.; Perekhod, E.A.; Chalenko, G.I.; Il'inskaya, L.I.; Gerasimova, N.G. Plant resistance suppressors in the pathosystem formed by potato and the causal agent of late blight. *Appl. Biochem. Microbiol* **2001**, *37*, 506–511.

49. Wang, X.; El Hadrami, A.; Adam, L.R.; Daayf, F. Differential activation and suppression of potato defence responses by*Phytophthora infestans* isolates representing US-1 and US-8 genotypes. *Plant Pathol* **2008**, *57*, 1026–1037.

50. Colditz, F.; Nyamsuren, O.; Niehaus, K.; Eubel, H.; Braun, H.P.; Krajinski, F. Proteomic approach: Identification of*Medicago truncatula* proteins induced in roots after infection with the pathogenic oomycete *Aphanomyces euteiches.Plant Mol. Biol* **2004**, *55*, 109–120.

51. Arfaoui, A.; El Hadrami, A.; Mabrouk, Y.; Sifi, B.; Boudabous, A.; El Hadrami, I.; Daayf, F.; Chérif, M. Treatment of chickpea with Rhizobium isolates enhances the expression of phenylpropanoid defense-related genes in response to infection by *Fusarium oxysporum* f. sp. *ciceris. Plant Physiol. Biochem* **2007**, *45*, 470–479.

52. Curto, M.; Camafeita, E.; Lopez, J.A.; Maldonado, A.M.; Rubiales, D.; Jorrín, J.V. A proteomic approach to study pea (*Pisum sativum*) responses to powdery mildew (*Erysiphe pisi*). *Proteomics* **2006**, *6*, S163–S174.

53. Amey, R.C.; Schleicher, T.; Slinn, J.; Lewis, M.; Macdonald, H.; Neill, S.J.; Spencer-Phillips, P.T.N. Proteomic analysis of a compatible interaction between *Pisum sativum* (pea) and the downy mildew pathogen *Peronospora viciae. Eur. J. Plant Pathol* **2008**, *122*, 41–55.

54. Lee, S.J.; Kelley, B.S.; Damasceno, C.M.B.; John, B.S.; Kim, B.S.; Kim, B.D.; Rose, J.K.C. A functional screen to characterize the secretomes of eukaryotic pathogens and their hosts in planta. *Mol. Plant Microbe Interact* **2006**, *19*, 1368–1377.

55. Campo, S.; Carrascal, M.; Coca, M.; Abian, J.; San Segundo, B. The defence response of germinating maize embryos against fungal infection: A proteomics approach. *Proteomics* **2004**, *4*, 383–396.

56. Paper, J.M.; Scott-Craig, J.S.; Adhikari, N.D.; Cuomo, C.A.; Walton, J.D. Comparative proteomics of extracellular proteins *in vitro* and *in planta* from the pathogenic fungus *Fusarium graminearum. Proteomics* **2007**, *7*, 3171–3183.

57. Geddes, J.; Eudes, F.; Laroche, A.; Selinger, B. Differential expression of proteins in response to the interaction between the pathogen *Fusarium graminearum* and its host, *Hordeum vulgare. Proteomics* **2008**, *8*, 545–554.

58. Sharma, N.; Rahman, M.H.; Strelkov, S.E.; Thiagarajah, M.; Bansal, V.K.; Kav, N.N.V. Proteome-level changes in two*Brassica napus* lines exhibiting differential responses to the fungal pathogen *Alternaria brassicae. Plant Sci* **2007**, *172*, 95–110.

59. Liang, Y.; Srivastava, S.; Rahman, M.H.; Strelkov, S.E.; Kav, N.N.V. Proteome changes in leaves of *Brassica napus* L. as a result of *Sclerotinia sclerotiorum* challenge. *J. Agric. Food Chem* **2008**, *56*, 1963–1976.

60. Liang, Y.; Strelkov, S.E.; Kav, N.N.V. Oxalic acid mediated stress responses in *Brassica napus* L. *Proteomics* **2009**, *9*, 3156–3173.

61. Fernández-Acero, F.J.; Jorge, I.; Calvo, E.; Vallejo, I.; Carbú, M.; Camafeita, E.; Garrido, C.; López, J.A.; Jorrin, J.; Cantoral, J.M. Proteomic analysis of phytopathogenic fungus *Botrytis cinerea* as a potential tool for identifying pathogenicity factors, therapeutic targets and for basic research. *Arch. Microbiol* **2007**, *187*, 207–215.

62. El Oirdi, M.; Bouarab, K. Plant signalling components EDS1 and SGT1 enhance disease caused by the necrotrophic pathogen *Botrytis cinerea. New Phytol* **2007**, *175*, 131–139.

63. El Oirdi, M.; Abd El Rahman, T.; Rigano, L.; El Hadrami, A.; Rodriguez, M.; Daayf, F.; Vojnov, A.; Bouarab,

K. *Botrytis cinerea* manipulates the antagonistic effects between plant immune pathways to restore its disease. *Plant Cell* **2011**, *23*, 2405–2421.

64. Rep, M.; Dekker, H.L.; Vossen, J.H.; de Boer, A.D.; Houterman, P.M.; Speijer, D.; Back, J.W.; de Koster, C.G.; Cornelissen, B.J. Mass spectrometric identification of isoforms of PR proteins in xylem sap of fungus-infected tomato.*Plant Physiol* **2002**, *130*, 904–917.
65. El Hadrami, A.; El Idrissi-Tourane, A.; El Hassni, M.; Daayf, F.; El Hadrami, I. Toxin-based *in vitro* selection and its potential application to date palm for resistance to the bayoud Fusarium wilt. *C. R. Biol* **2005**, *328*, 732–744.
66. El Hadrami, A.; Kone, D.; Lepoivre, P. Effect of juglone on active oxygen species and antioxidants in susceptible and partial resistant banana cultivars to black leaf streak disease. *Eur. J. Plant Pathol* **2005**, *113*, 241–254.
67. Yajima,W.;Kav,N.N.Theproteomeofthephytopathogenic fungus *Sclerotinia sclerotiorum*. *Proteomics* **2006**, *6*, 5995–6007.
68. Tudzynski, P.; Kokkelink, L. *Botrytis cinerea*: Molecular Aspects of a Necrotrophic Life Style. In *The Mycota: Plant Relationships*, 2nd ed; Springer: Berlin, Germany, 2009; Volume 5, pp. 29–50.
69. Strelkov, S.E.; Lamari, L.; Ballance, G.M. Characterization of a host-specific protein toxin (Ptr ToxB) from *Pyrenophora tritici-repentis*. *Mol. Plant Microbe Interact* **1999**, *12*, 728–732.
70. Wolpert, T.J.; Dunkle, L.D.; Ciuffetti, L.M. Host-selective toxins and avirulence determinants: What's in a name? *Annu Rev. Phytopathol* **2002**, *40*, 251–285.
71. Friesen, T.L.; Chu, C.-G.; Liu, Z.H.; Xu, S.S.; Halley, S.; Faris, J.D. Host-selective toxins produced by *Stagonospora nodorum* confer disease susceptibility in adult wheat plants under field conditions. *Theor. Appl. Gen* **2008**, *118*, 1489–1497.
72. Horbach, R.; Navarro-Quesada, A.R.; Knogge, W.; Deising, H.B. When and how to kill a plant cell: Infection strategies of plant pathogenic fungi. *J. Plant Physiol* **2011**, *168*, 51–62.
73. Koeck, M.; Hardham, A.R.; Dodds, P.N. The role of effectors of biotrophic and hemibiotrophic fungi in infection. *Cell Microbiol* **2011**, *13*, 1849–1857.
74. De Wit, P.J.G.M.; Mehrabi, R.; van den Burg, H.A.; Stergiopoulos, I. Fungal effector proteins: Past, present and future.*Mol. Plant Pathol* **2009**, *10*, 735–747.
75. Walton, J.D.; Avis, T.J.; Alfano, J.R.; Gijzen, M.; Spanu, P.; Hammond-Kosack, K.; Sánchez, F. Effectors, effectors *et encore des* effectors: The XIV international congress on molecular-plant microbe interactions, Quebec. *Mol. Plant Microb Interact* **2009**, *22*, 1479–1483.
76. Thatcher, L.F.; Manners, J.M.; Kazan, K. *Fusarium oxysporum* hijicks COI1-mediated jasmonate signaling in order to promote disease development in *Arabidopsis*. *Plant J* **2009**, *58*, 927–939.
77. Alkher, H.; El Hadrami, A.; Rashid, K.Y.; Adam, L.R.; Daayf, F. Cross-pathogenicity of *Vertcillium dahliae* between potato and sunflower. *Eur. J. Plant Pathol* **2009**, *124*, 505–519.
78. El Hadrami, A.; Fernando, W.G.D.; Daayf, F. Variations in relative humidity modulate *Leptospaheria* spp. pathogenicity and interfere with canola mechanisms of defence. *Eur. J. Plant Pathol* **2010**, *126*, 187–202.
79. Vincent, D.; Balesdent, M.H.; Gibon, J.; Claverol, S.; Lapaillerie, D.; Lomenech, A.M.; Blaise, F.; Rouxel, T.; Martin, F.; Bonneu, M.; *et al.* Hunting down fungal secretomes using liquid-phase IEF prior to high resolution 2-DE.*Electrophoresis* **2009**, *30*, 4118–4136.
80. Lamari, L.; Bernier, C.C. Genetics of tan necrosis and extensive chlorosis in tan spot of wheat caused by *Pyrenophora tritici-repentis*. *Phytopathology* **1991**, *81*, 1092–1095.
81. Böhmer, M.; Colby, T.; Bohmer, C.; Brautigam, A.; Schmidt, J.; Boker, M. Proteomic analysis of dimorphic transition in the phytopathogenic fungus *Ustilago maydis*. *Proteomics* **2007**, *7*, 675–685.
82. Noir, S.; Colby, T.; Harzen, A.; Schmidt, J.; Panstruga, R. A proteomic analysis of powdery mildew (*Blumeria graminis*f.sp. *hordei*) conidiospores. *Mol. Plant Pathol* **2008**, *10*, 223–236.
83. Bolton, M.D.; van Esse, H.P.; Vossen, J.H.; de Jonge, R.; Stergiopoulos, I.; Stulemeijer, I.; van den Berg, G.C.; Borrás-Hidalgo, O.; Dekker, H.L.; de Koster, C.G.; *et al.* The novel *Cladosporium fulvum* lysin motif effector Ecp6 is a virulence factor with orthologues in other fungal species. *Mol. Microbiol* **2008**, *69*, 119–136.
84. Krämer, R.; Freytag, S.; Schmelzer, E. *In vitro* formation of infection structures of *Phytophthora infestans* is associated with synthesis of stage specific polypeptides. *Eur. J. Plant Pathol* **1997**, *103*, 43–53.
85. Shepherd, S.J.; van West, P.; Gow, N.A.R. Proteomic analysis of asexual development of *Phytophthora palmivora*. *Mycol. Res* **2003**, *107*, 395–400.
86. Ebstrup, T.; Saalbach, G.; Egsgaard, H. A proteomics study of *in vitro* cyst germination and appressoria formation in*Phytophthora infestans*. *Proteomics* **2005**, *5*, 2839–2848.
87. Shimizu, M.; Yuda, N.; Nakamura, T.; Tanaka, H.; Wariishi, H. Metabolic regulation at the tricarboxylic acid and glyoxylate cycles of the lignin-degrading basidiomycetes *Phanerochaete chrysosporium* against exogenous addition of vanillin. *Proteomics* **2005**, *5*, 3919–3931.
88. Matsuzaki, F.; Shimizu, M.; Wariishi, H. Proteomic and metabolomic analyses of the white-rot fungus *Phanerochaete chrysosporium* exposed to exogenous benzoic acid. *J. Proteome Res* **2008**, *7*, 2342–2350.
89. Gonzalez-Fernandez, R.; Jorrin-Novo, J.V. Contribution of proteomics to the study of plant pathogenic fungi. *J. Proteome Res* **2012**, *11*, 3–16.
90. Cordwell, S.J.; Thingholm, T.E. Technologies for plasma membrane proteomics. *Proteomics* **2010**, *10*, 611–627.
91. Agrawal, G.K.; Pedreschi, R.; Barkla, B.J.; Bindschedler, L.V.; Cramer, R.; Sarkar, A.; Renaut, J.; Job, D.; Rakwal,

R. Translational plant proteomics: A perspective review article. *J. Proteomics* **2012**. in press.

92. Anderson, L.; Hunter, C.L. Quantitative mass spectrometric multiple reaction monitoring assays for major plasma proteins. *Mol. Cell. Proteomics* **2006**, *5*, 573–588.
93. Casado-Vela, J.; Martínez-Esteso, M.J.; Rodriguez, E.; Borrás, E.; Elortza, F.; Bru-Martínez, R. iTRAQ-based quantitative analysis of protein mixtures with large fold change and dynamic range. *Proteomics* **2010**, *10*, 343–347.
94. Baginsky, S.; Gruissem, W. Current Status of *Arabidopsis thaliana* Proteomics. In *Plant Proteomics*; Šamaj, J., Thelen, J., Eds.; Springer: Berlin, Germany, 2007; pp. 105–120.
95. Carpentier, S.C.; Panis, B.; Vertommen, A.; Swennen, R.; Sergeant, K.; Renaut, J.; Laukens, K.; Witters, E.; Samyn, B.; Devreese, B. Proteome analysis of nonmodel plants: A challenging but powerful approach. *Mass. Spectrom. Rev* **2008**,*27*, 354–377.
96. Gygi, S.P.; Rist, B.; Gerber, S.A.; Turecek, F.; Gelb, M.H.; Aebersold, R. Quantitative analysis of complex protein mixtures using isotope-coded affinity tags. *Nat. Biotechnol* **1999**, *17*, 994–999.
97. Turecek, F. Mass spectrometry in coupling with affinity capture-release and isotope-coded affinity tags for quantitative protein analysis. *J. Mass Spectrom* **2002**, *37*, 1–14.
98. Hansen, K.C.; Schmitt-Ulms, G.; Chalkley, R.J.; Hirsch, J.; Baldwin, M.A.; Burlingame, A.L. Mass spectrometric analysis of protein mixtures at low levels using cleavable ^{13}C-isotope-coded affinity tag and multidimensional chromatography. *Mol. Cell Proteom* **2003**, *2*, 299–314.
99. Barnidge, D.R.; Hall, G.D.; Stocker, J.L.; Muddiman, D.C. Evaluation of a cleavable stable isotope labeled synthetic peptide for absolute protein quantification using LC-MS/MS. *J. Proteome Res* **2004**, *3*, 658–661.
100. Ong, S.E.; Blagoev, B.; Kratchmarova, I.; Kristensen, D.B.; Steen, H.; Pandey, A.; Mann, M. Isotope labeling by amino acids in cell culture, SILAC, as a simple and accurate approach to expression proteomics. *Mol. Cell. Proteomics* **2002**, *1*, 376–386.
101. Ross, P.L.; Huang, Y.N.; Marchese, J.N.; Williamson, B.; Parker, K.; Hattan, S.; Khainovski, N.; Pillai, S.; Dey, S.; Daniels, S.; *et al.* Multiplexed protein quantitation in *Saccharomyces cerevisiae* using amine-reactive isobaric tagging reagents. *Mol. Cell. Proteomics* **2004**, *3*, 1154–1169.
102. Taylor, R.D.; Saparno, A.; Blackwell, B.; Anoop, V.; Gleddie, S.; Tinker, N.A.; Harris, L.J. Proteomic analyses of*Fusarium graminearum* grown under mycotoxin-inducing conditions. *Proteomics* **2008**, *8*, 2256–2265.
103. Gaulin, E.; Dramé, N.; Lafitte, C.; Torto-Alalibo, T.; Martinez, Y.; Ameline-Torregrosa, C.; Khatib, M.; Mazarguil, H.; Villalba-Mateos, F.; Kamoun, S.; *et al.* Cellulose binding domains of a *Phytophthora* cell wall protein are novel pathogen-associated molecular patterns. *Plant Cell* **2006**, *18*, 1766–1777.
104. Brunner, F.; Rosahl, S.; Lee, J.; Rudd, J.J.; Geiler, C.; Kauppinen, S.; Rasmussen, G.; Scheel, D.; Nürnberger, T. Pep-13, a plant defense-inducing pathogen-associated pattern from *Phytophthora* transglutaminases. *EMBO J* **2002**, *21*, 6681–6688.
105. Halim, V.A.; Hunger, A.; Macioszek, V.; Landgraf, P.; Nürnberger, T.; Scheel, D.; Rosahl, S. The oligopeptide elicitor Pep-13 induces salicylic acid-dependent and – independent defense reactions in potato. *Physiol. Mol. Plant Pathol* **2004**,*64*, 311–318.
106. Santoni, V. Plant plasma membrane protein extraction and solubilization for proteomic analysis. *Meth. Mol. Biol* **2006**,*355*, 93–109.
107. Sazuka, T.; Keta, S.; Shiratake, K.; Yamaki, S.; Shibata, D. A proteomic approach to identification of transmembrane proteins and membrane-anchored proteins of *Arabidopsis thaliana* by peptide sequencing. *DNA Res* **2004**, *11*, 101–113.
108. Elortza, F.; Nühse, T.S.; Foster, L.J.; Stensballe, A.; Peck, S.C.; Jensen, O.N. Proteomic analysis of glycosylphosphatidylinositol-anchored membrane proteins. *Mol. Cell. Proteomics* **2003**, *2*, 1261–1270.
109. Tan, S.; Tan, H.T.; Chung, M.C.M. Membrane proteins and membrane proteomics. *Proteomics* **2008**, *8*, 3924–3932.
110. Chaves, I.; Pinheiro, C.; Paiva, J.A.P.; Planchon, S.; Sergeant, K.; Renaut, J.; Graça, J.A.; Costa, G.; Coelho, A.V.; Pinto Ricardo, C.P. Proteomic evaluation of wound-healing processes in potato (*Solanum tuberosum* L.) tuber tissue.*Proteomics* **2009**, *9*, 4154–4175.
111. Barel, G.; Ginzberg, I. Potato skin proteome is enriched with plant defence components. *J. Exp. Bot* **2008**, *59*, 3347–3357.
112. Bernards, M.A.; Fleming, W.D.; Llewellyn, D.B.; Priefer, R.; Yang, X.; Sabatino, A.; Plourde, G.L. Biochemical characterization of the suberization-associated anionic peroxidase of potato. *Plant Physiol* **1999**, *121*, 135–146.
113. Win, J.; Morgan, W.; Bos, J.; Krasileva, K.V.; Cano, L.M.; Chaparro-Garcia, A.; Ammar, R.; Staskawicz, B.J.; Kamoun, S. Adaptive evolution has targeted the *C*-terminal domain of the RXLR effectors of plant pathogenic Oomycetes. *Plant Cell* **2007**, *19*, 2349–2369.
114. Osbourn, A.E.; Field, B. Operons. *Cell. Mol. Life Sci* **2009**, *66*, 3755–3775.
115. Williams, R.M.; Primig, M.; Washburn, B.K.; Winzeler, E.A.; Bellis, M.; Sarrauste de Menthiere, C.; Davis, R.W.; Esposito, R.E. The Ume6 regulon coordinates metabolic and meiotic gene expression in yeast. *Proc. Natl. Acad. Sci. USA* **2002**, *99*, 13431–13436.
116. Pelechano, V.; Jimeno-González, S.; Rodríguez-Gil, A.; García-Martínez, J.; Pérez-Ortín, J.E.; Chávez, S. Regulon-specific control of transcription elongation across the yeast genome. *PLoS Gen* **2009**, *5*. [CrossRef]
117. Haas, B.J.; Kamoun, S.; Zody, M.C.; Jiang, R.H.; Handsaker, R.E.; Cano, L.M.; Grabherr, M.; Kodira, C.D.; Raffaele, S.; Torto-Alalibo, T.; *et al.* Genome sequence and analysis of the Irish potato famine pathogen *Phytophthora infestans*.*Nature* **2009**, *461*, 393–398.

118. Thurston, G.; Regan, S.; Rampitsch, C.; Xing, T. Proteomic and phosphoproteomic approaches to understand plant-pathogen interactions. *Physiol. Mol. Plant Pathol* **2005**, *66*, 3–11.

119. Marra, R.; Ambrosino, P.; Carbone, V.; Vinale, F.; Woo, S.L.; Ruocco, M.; Ciliento, R.; Lanzuise, S.; Ferraioli, S.; Soriente, I.; *et al*. Study of the three-way interaction between *Trichoderma atroviride*, plant and fungal pathogens by using a proteomic approach. *Curr. Genet* **2006**, *50*, 307–321.

120. Jennings, D.B.; Ehrenshaft, M.; Mason Pharr, D.; Williamson, J.D. Roles for mannitol and mannitol dehydrogenase in active oxygen-mediated plant defense. *Proc. Natl. Acad. Sci. USA* **1998**, *95*, 15129–15133.

121. Cessna, S.G.; Sears, V.E.; Dickma, M.B.; Low, P.S. Oxalic acid, a pathogenicity factor for *Sclerotinia sclerotiorum*, suppresses the oxidative burst of the host plant. *Plant Cell* **2000**, *12*, 2191–2199.

122. Govrin, E.M.; Levine, A. The hypersensitive response facilitates plant infection by the necrotrophic pathogen *Botrytis cinerea*. *Curr. Biol* **2000**, *10*, 751–757.

123. Mayer, A.M.; Staples, R.C.; Gilad, N.L. Mechanisms of survival of necrotrophic fungal plant pathogens in hosts expressing the hypersensitive response. *Phytochemistry* **2001**, *58*, 33–41.

CHAPTER 16

BOTRYTIS CINEREA VIRULENCE FACTORS: NEW INSIGHTS INTO A NECROTROPHIC AND POLYPHAGEOUS PATHOGEN

Mathias Choquer, Elisabeth Fournier, Caroline Kunz, Caroline Levis, Jean-Marc Pradier, Adeline Simon and Muriel Viaud

Keywords: Botrytisgray moldtoxinssignal transductionhost adaptation

PHYLOGENETIC INTRODUCTION TO THE GRAY MOLD AGENT

Botrytis cinerea is an ascomycete responsible for gray mould on hundreds of dicot plants (Elad *et al.*, 2004). In grapevine, conidia can contaminate leaves or inflorescences, but the fungus develops mainly in the autumn on ripe grape berries. The wide variety of symptoms on different organs and plants may suggest that *B. cinerea* has a large 'arsenal of weapons' to attack its host plants. Among the necrotrophic and polyphageous fungi, the gray mold agent is one of the most-studied models (van Kan, 2006). One challenge is to elucidate whether *B. cinerea* has specific virulence factors compared with other fungal pathogens with a narrower host range. Another question to address is whether all strains of*B. cinerea* share a common arsenal of weapons or whether there are strain-specific adaptations. This review focuses on some of the latest inputs in the knowledge of *B. cinerea* biology and highlights the interest of combining population genetics data, gene knock-out approaches and high-throughput genomic tools. Recent data confirm that the gray mold fungus shares conserved virulence factors with other fungal plant pathogens, but they also reveal unique features and a high variability between strains that may be linked to its necrotrophic and polyphagous behaviour.

Recent phylogenetic studies yielded a better knowledge of the *B. cinerea* (teleomorph *Botryotinia fuckeliana*) species. The genus *Botrytis* Persoon (Ascomyceta, *Leotiomycetes, Sordariaceae*) comprises two clades (Staats *et al.*, 2005): one including four species that attack only Eudicot plants, the other including the 18 remaining species that attack mainly Monocot plants. Apart from the gray mold agent, most *Botrytis* species have a very narrow host range. Recently, multiloci gene genealogies showed that the *B. cinerea* morphological species were subdivided in two genetically isolated groups, named Group I and Group II, that were therefore proposed to be phylogenetic species. Group I (also referred as '*Botrytis pseudocinerea*') and Group II (*B. cinerea sensu stricto*) strains exhibit differences in their ecology and their resistance pattern to fungicides (Fournier *et al.*, 2003, 2005). Moreover, they are unable to cross with each other (D. Fortini & R. Fritz, pers. commun.). Functional studies presented in this review were performed using *B. cinerea* Group II model strains.

Main characteristics of the infection process

The infection process of *B. cinerea* is usually described by the following stages: penetration of the host surface, killing of

Citation: Mathias Choquer, Elisabeth Fournier, Caroline Kunz, Caroline Levis, Jean-Marc Pradier, Adeline Simon and Muriel Viaud, Botrytis cinerea virulence factors: new insights into a necrotrophic and polyphageous pathogen, DOI:10.1111/j.1574-6968.2007.00930.x

host tissue/primary lesion formation, lesion expansion/tissue maceration and sporulation (van Kan, 2006). Gene inactivation approaches, based on fungal protoplast transformation and homologous recombination at target locus (Tudzynski & Siewers, 2004), allowed the identification of genes participating in all the stages of the disease cycle (Table 1).

Table 1: *Botrytis cinerea* virulence genes characterized by gene inactivation approaches

Biological process	Function	Gene	Strain	*In vitro* mutant defects	*In planta* mutant phenotype (Tested host)	References
Appressorium penetration	Tetraspanin	*Bcpls1*	T4	ND	No penetration ^{C} (Rose petals, bean and tomato leaves)	Gourgues *et al.* (2004)
Fungal cell wall integrity	Chitin synthase	*Bcchsl*	Bd90	Chitin content reduction; cell wall weakening	Reduced colonization ^{M} (Grapevine leaves)	Soulie *et al.* (2003)
		BcchsIIIa	Bd90	Chitin content reduction; radial growth reduction; increase in hyphal ramification	Reduced colonization ^{M} (Grapevine and thale cress leaves)	Soulie *et al.* (2006)
Toxin biosynthesis	P450 monooxygenase	*Bcbot1**	T4 SAS56	No botrydial secretion	Reduced colonization ^{C} (Tomato fruits, bean and tomato leaves) in T4 mutant only	Siewers *et al.* (2005)
H_2O_2 generation	Superoxide dis-mutase	*Bcsod1*	B05.10	Increased sensitivity to oxidative stress	Reduced colonization ^{C} (Bean leaves)	Rolke *et al.* (2004)
Detoxification	ABC transporter	*BcAtrB*	B05.10	Increased sensitivity to resveratrol and fenpiclonil	Reduced colonization ^{C} (Grapevine leaves)	Schoonbeek *et al.* (2001)
Plant cell wall degradation	Pectin methyl-esterase	*Bcpme1**	Bd90 B05.10	Growth reduction on pectin medium observed in Bd90 mutant	Reduced colonization ^{M} (Apple fruits, grapevine and thale cress leaves) in Bd90 mutant only	Valette-Collet *et al.* (2003) Kars *et al.* (2005b)
	Endo-polygalacturonase	*Bcpg1*	B05.10	Growth reduction on pectin medium	Reduced colonization ^{C} (Apple fruits, tomato leaves and fruits)	ten Have *et al.* (1998)
		Bcpg2	B05.10	ND	Delay in secondary lesion formation; reduced colonization ^{C} (Tomato and bean leaves)	Kars *et al.* (2005a)
	Endo-β-1,4-xylanase	*Xyn11A*	B05.10	Reduced endo-β-1,4-xylanase activity	Delay in primary lesion formation; Reduced colonization ^{C} (Grape berries and tomato leaves)	Brito *et al.* (2006)
Signal transduction	Gα subunits of G-proteins	*Bcg1*	B05.10	Altered colony morphology; defect in botrydial and proteases secretion	Infection stopped after primary lesions formation ^{C} (Bean leaves, tomato fruits and leaves)	Schulze Gronover *et al.* (2001)
		Bcg2	B05.10	ND	Reduced colonization ^{C} (Bean leaves, tomato fruits and leaves)	Schulze Gronover *et al.* (2001)
		Bcg3	B05.10	Defect in germination and sporulation; excessive clerotia formation	Delay in primary lesion formation ^{C} (Tomato leaves)	Doehlemann *et al.* (2006a)
	Adenylate cyclase	*Bac*	B05.10	Reduced germination, growth and sporulation	Reduced colonization; no conidiation ^{C} (Bean leaves)	Klimpel *et al.* (2002); Doehlemann *et al.* (2006a)
	MAP kinase	*Bmp1*	B05.10	Defect in germination on hydrophobic surfaces; reduced vegetative growth	No appressoria; no penetration; no colonization ^{C} (Tomato leaves, carnation flowers)	Zheng *et al.* (2000); Doehlemann *et al.* (2006a)
		Bmp3	B05.10	Reduced vegetative growth and conidiation; no sclerotia; Sensitivity to low osmolarity, to oxidative stress and to phenylpyrrole	Reduced penetration efficiency; reduced colonization ^{C} (Gerbera flower petals, tomato leaves)	Rui & Hahn (2007)
		Bcsak1	B05.10	Defect in conidiation; increased sclerotial formation; Sensitivity to osmotic and oxidative stresses	No penetration; reduction of secondary lesions expansion ^{M} (Bean leaves)	Segmuller *et al.* (2007)
	Histidine kinase	*Bos1*	USW111	Osmosensitivity; resistance to dicarboximides and phenylpyrroles; defect in conidiation	Reduced colonization ^{M} (Apple fruits, bean and tomato leaves)	Viaud *et al.* (2006)
Protein folding	Cyclophilin A	*Bcp1*	T4	Resistance to cyclosporin A	Reduced colonization ^{C} (Tomato and bean leaves)	Viaud *et al.* (2003)
(Peptidyl-prolyl isomerisation)	FK506 binding protein (FKBP)	*Bcpic5**	T4 B05.10	Resistance to FK506 drug	Reduced colonization ^{C} (Bean and tomato leaves) in T4 mutant only	Gioti *et al.* (2006) Mey *et al.*, (pers. comm.)

*Strain dependant virulence factor.
ND, no detected defect; C, conidia inoculum; M, mycelium inoculum.

Botrytis cinerea develops different infection structures

Conidia of *B. cinerea* are considered to be the main produced and dispersed inoculum (Holz *et al.*, 2004). Their germination and adhesion on plant surfaces represent crucial steps preceding host penetration and colonization, whereby sensing and recognition of the host surface characteristics, including hydrophobicity and sugar sources, seem essential (Doehlemann *et al.*, 2005, 2006b). Germlings secrete an extracellular matrix that serves in their attachment through hydrophobic interactions but also release several extracellular enzymes that might facilitate their penetration by breaching the physical barriers of the plant (Doss, 1999). *Botrytis cinerea* is thought to enter the host mainly by producing degrading enzymes and causing an oxidative burst rather than using physical pressure. Appressorium-like structures depicted as swollen tips have been found frequently in *B. cinerea*(Fourie & Holz, 1995; Tenberge, 2004). These structures (Fig. 1a) are not highly melanized and not always separated from the germ tube by a septum, which would be necessary for the generation of an osmotic pressure as high as the one observed in appressoria of the rice blast fungus *Magnaporthe grisea* (Howard & Valent, 1996). However, both fungi share a conserved membrane protein, tetraspanin, required for the appressorium-mediated penetration into host plants (Gourgues *et al.*, 2004). When plants are inoculated with mycelium, a different form of penetration was observed, whereby hyphae growing on the plant surface heavily ramify into short bulbous cell aggregates (Fig. 1band d) described earlier as 'claw-like' structures (Kunz *et al.*, 2006). These structures are similar to the 'infection cushions' or 'complex appressoria' of the closely related fungus *Sclerotinia sclerotiorum*(Hegedus & Rimmer, 2005; Jurick & Rollins, 2007). The Importance of infection cushions in fungal penetration has not yet been molecularly investigated in *B. cinerea*, but they were observed on different organs and plants (Sharman & Heale, 1977; Garcia-Arenal & Sagasta, 1980; Backhouse & Willetts, 1987; Fullerton *et al.*, 1999).

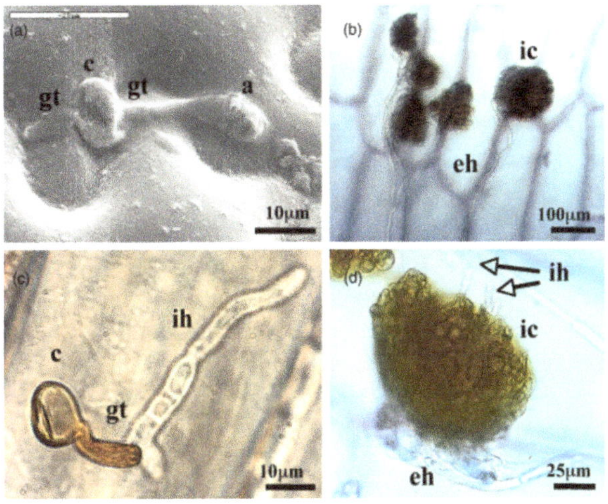

Figure 1: *Botrytis cinerea* develops different penetration structures that are all accumulating H_2O_2. Conidial suspensions from K1 strain were inoculated on bean leaves (a). Pictures were taken 12 h postinoculation (hpi) using scanning electron microscopy (Kindly provided by Brigitte Gelly, INRA). Mycelium plugs (b and d) or conidia (c) from Bd90 strain were inoculated on the upper side of onion epidermis. At 10 hpi, the epidermal strips were transferred onto a solution containing 1 mg L^{-1} of DAB and were left overnight. Pictures were taken 24 hpi. Inoculated conidia (indicated by c) form germ tubes (gt) and appressorium-like structures (a), allowing penetration of an infectious hypha (ih), whereas inoculated mycelium (indicated by eh for external hyphae) forms infectious cushions (ic) as a penetration structure.

Host cell death occurs through fungal toxins and an oxidative burst generated by both the pathogen and the host

As a necrotrophic and polyphageous pathogen, *B. cinerea* secretes nonspecific phytotoxins to kill cells from a large spectrum of plants. Among the numerous metabolites isolated from fermentation broths (Collado *et al.*, 2000, 2007), the most well known is the sesquiterpen botrydial (Fig. 2). Botrydial is produced during plant infection (Deighton *et al.*, 2001), and induces chlorosis and cell collapse, which seems to facilitate both penetration and colonization (Colmenares *et al.*, 2002). Botrydial biosynthetic pathway genes are organized into a physical cluster, coregulated and overexpressed *in planta* as shown by macro-array studies (Viaud *et al.*, 2003; Gioti *et al.*, 2006). Inactivation of one of these genes (*CND5/Bcbot1*) in three different strains demonstrated that botrydial is a strain-dependant virulence factor (Siewers *et al.*, 2005). In addition to botrydial and its derivatives, aggressive isolates of *B. cinerea*also produce a second class of toxins derived from botcinic acid (Reino *et al.*, 2004; Tani *et al.*, 2006) (Fig. 2), which may explain the strain-dependent effect of *Bcbot1* inactivation. In addition to these toxins,*B. cinerea* produces reactive oxygen species (ROS) during infection (Schouten *et al.*, 2002; Tenberge, 2004). The hydrogen peroxide (H_2O_2) accumulates in the early steps of infection, both in germinating spores (Fig. 1c) and in the infection cushions (Fig. 1d). Moreover, gene inactivation of the superoxide dismutase (SOD) indicated that this H_2O_2-generating enzyme is a virulence factor (Rolke *et al.*, 2004) contributing to the accumulation of phytotoxic H_2O_2 levels.

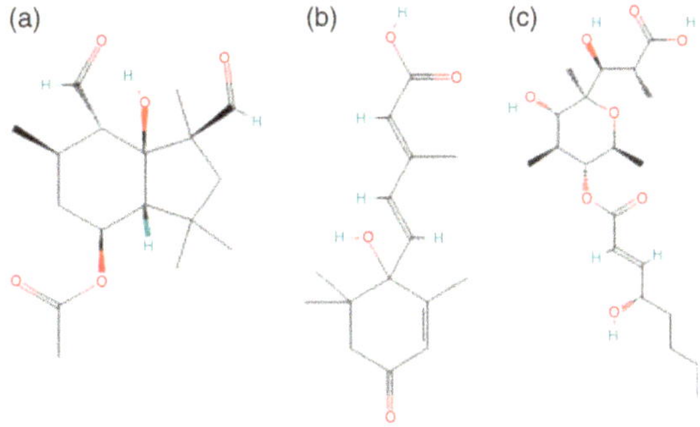

Figure 2: *Botrytis cinerea* produces a wide range of secondary metabolites. Biochemical structures have been determined for the sesquiterpenes botrydial (a. PubChem compound: CID: 185781) and abscissic acid (b. CID: 5375200) and for the polyketide called botcinic acid (c. CID: 11509607). Gene inactivation studies demonstrated that the phytoxin botrydial (Collado *et al.*, 2000, 2007) is a strain-dependent virulence factor (Siewers *et al.*, 2005) but did allow one to determine whether the vegetal hormone abscissic acid has a role in virulence (Siewers *et al.*, 2004). The genes coding for the biosynthesis of the phytotoxic botcinic acid and its botcinin derivatives (Tani *et al.*, 2006) have not been identified yet, but a genomic approach provided a list of 20 candidate genes that are coding for polyketide synthases (Kroken *et al.*, 2003).

Plant cell death also occurs through an oxidative burst produced by the host itself in reaction to *B. cinerea* attack (Govrin & Levine, 2000). This rapid production of ROS by the plant was originally known to be required for defense gene expression and a hypersensitive reaction (HR), a type of programmed cell death (PCD) thought to limit access of the pathogen to water and nutrients. Nevertheless, in some host–pathogen interactions, PCD has a clear role in promoting pathogen growth (Greenberg & Yao, 2004). Studies in *Arabidopsis thaliana* and tobacco suggested that *B. cinerea* may even need the HR to achieve full pathogenicity (Govrin & Levine, 2000; Dickman *et al.*, 2001). Indeed, *A. thaliana* mutants with a delayed or reduced cell death response are generally more resistant to *B. cinerea* infection, whereas mutants in which cell death was accelerated were more susceptible (van Baarlen *et al.*, 2007). There is also increasing evidence that fungi generate PCD-promoting molecules as virulence factors (Greenberg & Yao, 2004). For example, *B. cinerea* endo-polygalacturonase 1 (BcPG1), first known as a virulence factor (ten Have *et al.*, 1998), elicits defense responses in grape including the production of ROS (Poinssot *et al.*, 2003; Vandelle *et al.*, 2006). Recently, a nonpathogenic mutant overproducing BcPG1 was shown to provoke an HR-like reaction on grape and bean leaves that correlates with a cessation of tissue colonization (Kunz *et al.*, 2006). Independent studies also correlated low virulence strains with strong HR in bean leaf disks (Unger *et al.*, 2005). Therefore, PCD has an important role in*B. cinerea* virulence but, depending on its timing and strength and depending on the host, PCD may also correlate with resistance.

Enzymatic degradation arsenal relies on multi-genic families

The capacity for degrading the host cuticle and cell wall has been extensively studied in *B. cinerea* (ten Have *et al.*, 2002; Kars & van Kan, 2004). As a necrotroph, this fungus is notably equipped with multiple cell wall-degrading enzymes (CWDEs) that allow plant tissue colonization and the release of carbohydrates for consumption. Pectin, the major host cell wall component, can be degraded by a set of fungal pectinases. Among them, the polygalacturonase genes (*BcPG1–6*) were shown to be differentially expressed, depending on the stage of infection and the host, which suggested specialized functions for these pectinolytic activities (ten Have *et al.*, 2001). Gene inactivation indicated that BcPG2 activity is involved in the penetration step by breaching the pectin network localized at the anticlinal cell wall (Kars *et al.*, 2005a), whereas BcPG1 activity is required during colonization to breach the pectin network through the middle lamella (ten Have *et al.*, 1998). Pectin methylesterases are assumed to facilitate the action of PGs by demethylating pectin to pectate, and inactivation of the *bcpme1* gene resulted in a strong reduction of virulence in several plant hosts (Valette-Collet *et al.*, 2003). Nonpectinolytic CWDEs, like cellulases or hemicellulases, also facilitate *B. cinerea* infection as illustrated by inactivation of an endo-β-1,4-xylanase gene that led to a reduced virulence (Brito *et al.*, 2006).

Most of the degrading enzymes are encoded by multigenic families and some may have partly redundant functions. This would explain why inactivation of several genes encoding CWDEs (Espino *et al.*, 2005; Kars *et al.*, 2005b; van Kan, 2006) or cutinolytic enzymes, i.e. cutinase A and lipase 1 (van Kan *et al.*, 1997; Reis *et al.*, 2005), did not affect fungal virulence. Taken together, the multiplicity of the reported degrading activities and the reduction of virulence observed for several mutants impaired in degrading enzymes strongly support a major role in pathogenicity for the enzymatic degradation arsenal of *B. cinerea*.

Signal transduction regulating the infection process shows specific features in *B. cinerea*

Signal transduction cascades regulating fungal development and virulence are remarkably conserved between distantly related fungi, especially those involving the cAMP-dependant protein kinase and mitogen-activated protein kinases (MAPK) (Lengeler *et al.*, 2000; Xu, 2000). Different regulations among pathogens could then be generated by differences in upstream inputs (environmental signals and receptors), connection between pathways and their downstream outputs (target genes). Recent studies highlight specific roles in the pathogenicity of some signaling components of *B. cinerea* when compared with other fungal pathogens.

Heterotrimeric GTP-binding proteins form a family of regulators that receive signals directly from receptors and are required for the infection process in many fungal models (Lengeler *et al.*, 2000). In*B. cinerea*, inactivation of *Bcg1*, one of the three Gα subunit encoding genes, led to a severe pleitrophic phenotype (Schulze Gronover *et al.*, 2001; Siewers *et al.*, 2005). The mutant has an abnormal colony morphology, a defect in protease and botrydial secretion and is blocked after primary lesion development. Some of the defects (colony morphology) can be restored by the addition of cAMP, which demonstrated that BCG1 is involved in the cAMP-dependant pathway. The importance of this common pathway in vegetative growth and pathogenicity was confirmed by the inactivation of the adenylate cyclase gene, *bac* (Klimpel *et al.*, 2002). Interestingly, the *bcg1* mutant also has defects (proteases and botrydial secretion) that cannot be restored by cAMP, suggesting that BCG1 controls another signaling pathway besides the cAMP one. A suppression subtractive hybridization (SSH) approach confirmed that BCG1 controls the expression of secondary metabolism genes including botrydial (*Bcbot1* gene), protease and CWDE genes (Schulze Gronover *et al.*, 2004). Another expression study, using macroarrays, has shown that the botrydial gene cluster (including *Bcbot1*) is controlled by the calcineurin-dependent pathway (Viaud *et al.*, 2003). Calcineurin phosphatase has an essential role in fungal morphogenesis and virulence (Fox & Heitman, 2002) and was shown to be involved in *M. grisea*and *B. cinerea* appressorium formation (Viaud *et al.*, 2002, 2003). The observation that the botrydial gene *Bcbot1* is regulated by both BCG1 and calcineurin suggests a common signaling pathway that regulates infection-related functions including the production of secondary metabolites (J. Schumacher, M. Viaud & B. Tudzynski, pers. commun.).

Signaling networks of filamentous fungi usually contain three MAPK cascades (Xu, 2000). The corresponding MAPK, BMP1, BcSAK1 and BMP3 have been genetically characterized in *B. cinerea*. BMP1 is the ortholog of the yeast FUS3/KSS1 and of the *M. grisea* virulence factor PMK1. *bmp1* null mutants can germinate on plant surfaces, but they do not produce any appressoria. Even on wounded plants, they are unable to colonize the host tissues (Zheng *et al.*, 2000; Doehlemann *et al.*, 2006a). This phenotype shows that, as in other pathogenic fungi, the widely conserved BMP1 pathway is essential for host surface sensing, infection structure morphogenesis and colonization.

BcSAK1 is the ortholog of the yeast HOG1 involved in osmoregulation. In filamentous fungi, the corresponding pathway allows responses to osmotic and oxidative stresses. Null mutants, obtained in the phytopathogens *M. grisea* and *Colletotrichum lagenarium*, were still fully pathogenic (Dixon *et al.*, 1999; Kojima *et al.*, 2002). In contrast, *bcSak1* mutants are severely impaired in pathogenic development (Segmuller *et al.*, 2007). The putative upstream sensor of the BcSAK1 pathway could be BOS1 (S. Liu & S. Fillinger, pers. commun.), a member of the class III histidine-kinases (HKs) that are involved in osmoregulation and resistance to dicarboximide and phenylpyrrole fungicides (Catlett *et al.*, 2003). As for the *bcSak1* mutants, *bos1* null mutants are impaired in osmotic stress, conidiation and plant tissue colonization (Viaud *et al.*, 2006). This was the first HK reported to be a virulence factor in a phytopathogenic fungus. In *M. grisea* and in the seed-borne necrotrophic pathogen *Alternaria brassicicola*, inactivation of a homologous

HK gene did not impair pathogenicity (Motoyama *et al.*, 2005). The differential functions of the class III HKs and HOG1 orthologs among different pathogens may be linked to different fungal strategies for the management of osmotic and oxidative stresses during plant infection, or to differences in the level of stress encountered. Indeed, *B. cinerea* needs to resist the oxidative burst that occurs in the early stages of infection, which could explain the high importance of the BOS1 and BcSAK1 pathway(s). Macro-array approaches are in progress to investigate the downstream-regulated genes (S. Fillinger, S. Liu & M. Viaud, pers. commun.).

The third MAPK characterized in *B. cinerea* is BMP3 that is the homologue of the yeast SLT2 involved in cell wall integrity. However, *bmp3* null mutants do not seem to be affected in cell wall integrity as expected from the data on other fungi. The phenotype suggests that BMP3 is involved in saprophytic growth, response to low osmolarity, conidiation, surface sensing, host penetration and lesion formation. Therefore, as for BcSAK1, BMP3 presents unique features in *B. cinerea*.

The recent inputs into the complex signaling network-controlling pathogenicity in *B. cinerea* has revealed both conserved patterns with other fungi and specific features for this necrotrophic pathogen. Other signaling components not mentioned in this review and cross-talks between them are under investigation (Tudzynski & Schulze Gronover, 2004).

A POSSIBLE HOST ADAPTATION OF *B. CINEREA* STRAINS?

Botrytis cinerea populations display a significant phenotypic variability in their level of aggressiveness, the oxidative burst occurring during infection (Unger *et al.*, 2005) and toxin production (Reino *et al.*, 2004). The genetic bases for this variability are not yet elucidated, but different nonexclusive hypotheses are under investigation. A first source of variability can be the occurrence of mycoviruses among wild populations. Indeed, a double-stranded RNA virus isolated from *B. cinerea* was shown to confer hypovirulence-associated traits i.e. a reduced conidiation rate and a defect in bean leaf colonization (Castro *et al.*, 2003). A second source of variability could be differences in gene content and gene variability among *B. cinerea* strains. This hypothesis is supported by some of the functional studies mentioned above, showing that inactivation of the same gene in different *B. cinerea* Group II strains can lead to different virulence phenotypes. These strain-dependent virulence factors are, for example, the botrydial biosynthesis gene *Bcbot1* and the CWDE gene *Bcpme1* (Table 1). Depending on the genetic background, the null mutants were affected or not in their virulence, suggesting that differences in gene content, gene regulation and enzyme or toxin production might occur between strains.

These observed variability and strain-dependant virulence factors may suggest that *B. cinerea*populations have developed strain-dependent infection strategies in which each toxin or CWDE has a more or less important role. This raises the question: does the host plant play a role in the evolution of virulence factors within *B. cinerea* populations? To the authors' knowledge, this research question has not yet been directly addressed, but some indirect evidence for the role of host plant in the adaptation of*B. cinerea* populations already exists. First, conidia from *B. cinerea* strains isolated from tomato adhered and germinated faster on tomato leaves than strains isolated from grape (Cotoras & Silva, 2005). Moreover, the genotyping with microsatellite markers (Fournier *et al.*, 2002) of 10 French populations of *B. cinerea* Group II sampled on grapevine and bramble showed that gene flow was slightly but significantly reduced between populations from the two hosts, even in strict sympatry (E. Fournier & T. Giraud, pers. commun.). Hence, *B. cinerea* populations may be slightly but efficiently adapted to their different host plants, rather in terms of host preference than true host specialization. Indeed, hosts and their parasites are involved in an evolutionary arms race characterized by adaptation and counter-adaptation of mechanisms of host defense and parasite attack. The selective pressures exerted by plants may thus impact the evolution of virulence factors at the population level, and the strain-dependent effects observed in many functional studies of *B. cinerea* virulence factors may be a consequence of this process.

The effect of a host plant on pathogen evolution may also be investigated at an interspecific level: the reciprocal selective pressures leave signatures of positive or balancing selection at the molecular level that can be detected between closely related species specialized on different hosts. In the *Botrytis*genus, the evolutionary forces acting on two paralogous necrosis and ethylene-inducing like proteins (NLPs) have been studied (Staats *et al.*, 2007). NLPs constitute a family of secreted phytotoxic proteins that elicit plant cell death and directly interact with plant molecules. Among the *Botrytis* species, the evolution of a small number of residues of both NEP paralogs were shown to be driven by positive selection whereas the remaining parts of the proteins seemed to be evolutionarily constrained. Although no correlation was found with host range or type of host attacked, these results show that genes involved in the infectious process may be the target of various selective pressures that may have driven the evolution of related species on different hosts.

CONCLUSION – *BOTRYTIS* ENTERS THE GENOMIC AREA

The number of sequenced fungal genomes from both saprophytic and pathogenic species is increasing significantly (Galagan *et al.*, 2005; Xu *et al.*, 2006), which provides new comparative and functional genomics approaches to investigation of virulence. The neighbor species *B. cinerea* and *S. sclerotinium*are the first Leotiomycetes, but also some of the first polyphageous and necrotrophic fungi, to be sequenced, by the Broad Institute (http://www.broad.mit.edu/annotation/fgi/) and the Genoscope (http://www.genoscope.cns.fr/) (Fillinger *et al.*, 2007). Comparing the gene content and organization among them and with other sequenced fungi will highlight the differences that could be associated with their necrotrophic lifestyles and their broad host ranges. Moreover, two strains of *B. cinerea* were sequenced (B05.10 and T4), which may contribute to an understanding of the phenotypic and genotypic variability of this species and the occurrence of strain-dependent virulence factors.

These genomic resources will stimulate the development of high-throughput functional analyses such as transcriptomics and proteomics. In parallel, gene inactivation analysis will benefit from a genetically modified strain of *B. cinerea* (*ku* null mutant), which has a high rate of homologous recombination that should facilitate the inactivation of further candidate genes (M. Choquer & M. Viaud, pers. commun.).

DNA chips using the whole set of *B. cinerea* genes are under development. Such transcriptomic approaches have already been performed, to a lesser scale, by the use of expressed sequences tags (Viaud *et al.*, 2005) spotted on high-density filters. This way, the expression of 3032 genes was studied in different genetic backgrounds or physiological conditions. Twenty-seven genes were shown to be overexpressed during the infection of *Arabidopsis thaliana* (Gioti *et al.*, 2006). Among them, some had already been described as virulence genes (*Bcpg1, Bcbot1*) but most had unknown functions. For five of them, the corresponding null mutants were affected during colonization (A. Gioti & C. Levis, pers. commun.), which highlights the utility of genomic analysis for the identification of new virulence factors that may be specific to *B. cinerea*. The functional study of the interaction between the gray mould agent and its host plants should also greatly benefit from the sequencing of the grape genome (Jaillon *et al.*, 2007). A kinetic view of both complete transcriptomes during infection may give a first integrative view of the interaction and insights into the molecular dialogue occurring between the partners.

Statement

A pathogen profile of *Botrytis cinerea* was published while this Minireview was in press: Williamson B., Tudzynski B., Tudzynski P., Van Kan I.A.L. (2007) *Botrytis cinerea*: the cause of grey mould disease. *Mol Plant Pathol* **8:** 561–580.

ACKNOWLEDGEMENTS

The author's work is funded by an INRA 'Jeune Equipe' Grant.

REFERENCES

1. Backhouse D., Willetts H.J. (1987) Development and structure of infection cushions of Botrytis cinerea. Trans Br Mycol Soc 89: 89–95.
2. Brito N., Espino J.J., Gonzalez C. (2006) The endo-beta-1,4-xylanase xyn11A is required for virulence in Botrytis cinerea. Mol Plant Microbe Interact 19: 25–32.
3. Castro M., Kramer K., Valdivia L., Ortiz S., Castillo A. (2003) A double-stranded RNA mycovirus confers hypovirulence-associated traits to Botrytis cinerea. FEMS Microbiol Lett 228: 87–91.
4. Catlett N.L., Yoder O.C., Turgeon B.G. (2003) Whole-genome analysis of two-component signal transduction genes in fungal pathogens. Eukaryot Cell 2: 1151–1161.
5. Collado I.G., Aleu J., Hernandez-Galan R., Duran-Patron R. (2000) Botrytis species: an intriguing source of metabolites with wide ranges of biological activities. Structure, chemistry and bioactivity of metabolites isolated from Botrytis species. Curr Organic Chem 4: 1261–1286.
6. Collado I.G., Sánchez A.J., Hanson J.R. (2007) Fungal terpene metabolites: biosynthetic relationships and the control of the phytopathogenic fungus Botrytis cinerea. Nat Prod Rep 24: 674–686.
7. Colmenares A.J., Aleu J., Duran-Patron R., Collado I.G., Hernandez-Galan R. (2002) The putative role of botrydial and related metabolites in the infection mechanism of Botrytis cinerea. J Chem Ecol 28: 997–1005.
8. Cotoras M., Silva E. (2005) Differences in the initial events of infection of Botrytis cinerea strains isolated from tomato and grape. Mycologia 97: 485–492.
9. Deighton N., Muckenschnabel I., Colmenares A.J., Collado I.G., Williamson B. (2001) Botrydial is produced in plant tissues infected by Botrytis cinerea. Phytochemistry 57: 689–692.
10. Dickman M.B., Park Y.K., Oltersdorf T., Li W., Clemente T., French R. (2001) Abrogation of disease development in plants expressing animal antiapoptotic genes. Proc Natl Acad Sci USA 98: 6957–6962.
11. Dixon K.P., Xu J.R., Smirnoff N., Talbot N.J. (1999) Independent signaling pathways regulate cellular turgor during hyperosmotic stress and appressorium-mediated plant infection by Magnaporthe grisea. Plant Cell 11: 2045–2058.
12. Doehlemann G., Molitor F., Hahn M. (2005) Molecular and functional characterization of a fructose specific transporter from the gray mold fungus Botrytis cinerea. Fungal Genet Biol 42: 601–610.
13. Doehlemann G., Berndt P., Hahn M. (2006a) Different signalling pathways involving a Galpha protein, cAMP and a MAP kinase control germination of Botrytis cinerea conidia. Mol Microbiol 59: 821–835.
14. Doehlemann G., Berndt P., Hahn M. (2006b) Trehalose metabolism is important for heat stress tolerance and spore germination of Botrytis cinerea. Microbiology 152: 2625–2634.
15. Doss R.P. (1999) Composition and enzymatic activity of the extracellular matrix secreted by germlings of Botrytis cinerea. Appl Environ Microbiol 65: 404–408.
16. Elad Y., Williamson B., Tudzynski P., Delen N. (2004) Botrytis spp. and diseases they cause in agricultural systems – an introduction. Botrytis: Biology, Pathology and Control (Elad Y., Williamson B., Tudzynski P., Delen N., eds), pp. 1–6. Kluwer Academic Publishers, Dordrecht, the Netherlands.
17. Espino J.J., Brito N., Gonzalez C. (2005) Botrytis cinerea endo-1,4-glucanase Cel5A is expressed during infection but is not required for pathogenesis. Physiol Mol Plant Pathol 66: 213–221.
18. Fillinger S., Amselem J., Artiguenave F., et al. (2007) The genome projects of the plant pathogenic fungi Botrytis cinerea and Sclerotinia sclerotiorum. Macromolecules of Grape and Wines (Jeandet P., Clément C., Conreux A., eds), pp. 125–133. Editions TEC & DOC, Paris, France.
19. Fourie J.F., Holz G. (1995) Initial infection processes by Botrytis cinerea on nectarine and plum fruit and the development of decay. Phytopathology 85: 82–87.
20. Fournier E., Giraud T., Loiseau A., Vautrin D., Estoup A., Solignac M., Cornuet J.M., Brygoo Y. (2002) Characterization of nine polymorphic microsatellite loci in the fungus Botrytis cinerea (Ascomycota). Mol Ecol Notes 2: 253–255.

21. Fournier E., Lévis C., Fortini D., Giraud T., Leroux P., Brygoo Y. (2003) Characterization of Bc-hch, the Botrytis cinerea homolog of the Neurospora crassa het-c vegetative incompatibility locus, and its use as a population marker. Mycologia 95: 951–961.

22. Fournier E., Giraud T., Brygoo Y. (2005) Partition of the Botrytis cinerea complex in France using multiple gene genealogies. Mycologia 97: 1251–1267.

23. Fox D.S., Heitman J. (2002) Good fungi gone bad: the corruption of calcineurin. Bioessays 24: 894–903.

24. Fullerton R.A., Harris F.M., Hallet I.C. (1999) Rind distortion of lemon caused by Botrytis cinerea Pers. NZ J Crop Horticult Sci 27: 205–214.

25. Galagan J.E., Henn M.R., Ma L.J., Cuomo C.A., Birren B. (2005) Genomics of the fungal kingdom: insights into eukaryotic biology. Genome Res 15: 1620–1631.

26. Garcia-Arenal F., Sagasta E.M. (1980) Scanning electron microscopy of Botrytis cinerea penetration of bean (Phaseolus vulgaris) hypocotyls. Phytopathol Z 99: 37–42.

27. Gioti A., Simon A., Le Pecheur P., Giraud C., Pradier J.M., Viaud M., Levis C. (2006) Expression profiling of Botrytis cinerea genes identifies three patterns of up-regulation in planta and an FKBP12 protein affecting pathogenicity. J Mol Biol 358: 372–386.

28. Gourgues M., Brunet-Simon A., Lebrun M.H., Levis C. (2004) The tetraspanin BcPls1 is required for appressorium-mediated penetration of Botrytis cinerea into host plant leaves. Mol Microbiol 51: 619–629.

29. Govrin E.M., Levine A. (2000) The hypersensitive response facilitates plant infection by the necrotrophic pathogen Botrytis cinerea. Curr Biol 10: 751–757.

30. Greenberg J.T., Yao N. (2004) The role and regulation of programmed cell death in plant-pathogen interactions. Cell Microbiol 6: 201–211.

31. Hegedus D.D., Rimmer S.R. (2005) Sclerotinia sclerotioru: when "to be or not to be" a pathogen? FEMS Microbiol Lett 251: 177–184.

32. Holz G., Coertze S., Williamson B. (2004) The ecology of Botrytis on plant surfaces. Botrytis: Biology, Pathology and Control (Elad Y., Williamson B., Tudzynski P., Delen N., eds), pp. 9–27. Kluwer Academic Publishers, Dordrecht, the Netherlands.

33. Howard R.J., Valent B. (1996) Breaking and entering: host penetration by the fungal rice blast pathogen Magnaporthe grisea. Annu Rev Microbiol 50: 491–512.

34. Jaillon O., Aury J.M., Noel B., et al. (2007) The grapevine genome sequence suggests ancestral hexaploidization in major angiosperm phyla. Nature, doi: .

35. Jurick W.M. II., Rollins J.A. (2007) Deletion of the adenylate cyclase (sac1) gene affects multiple developmental pathways and pathogenicity in Sclerotinia sclerotiorum. Fungal Genet Biol 44: 521–530.

36. Kars I., Van Kan J.A. (2004) Extracellular enzymes and metabolites involved in pathogenesis of Botrytis. Botrytis: Biology, Pathology and Control (Elad Y., Williamson B., Tudzynski P., Delen N., eds), pp. 99–118. Kluwer Academic Publishers, Dordrecht, the Netherlands.

37. Kars I., Krooshof G.H., Wagemakers L., Joosten R., Benen J.A., Van Kan J.A. (2005a) Necrotizing activity of five Botrytis cinerea endopolygalacturonases produced in Pichia pastoris. Plant J 43: 213–225.

38. Kars I., McCalman M., Wagemakers L., Van Kan J.A. (2005b) Functional analysis of Botrytis cinerea pectin methylesterase genes by PCR-based targeted mutagenesis: Bcpme1 and Bcpme2 are dispensable for virulence of strain B05-10. Mol Plant Pathol 6: 641–652.

39. Klimpel A., Schulze Gronover C., Williamson B., Stewart J.A., Tudzynski B. (2002) The adenylate cyclase (BAC) in Botrytis cinerea is requiered for full pathogenicity. Mol Plant Pathol 3: 439–450.

40. Kojima K., Kikuchi T., Takano Y., Oshiro E., Okuno T. (2002) The mitogen-activated protein kinase gene MAF1 is essential for the early differentiation phase of appressorium formation in Colletotrichum lagenarium. Mol Plant Microbe Interact 15: 1268–1276.

41. Kroken S., Glass N.L., Taylor J.W., Yoder O.C., Turgeon B.G. (2003) Phylogenomic analysis of type I polyketide synthase genes in pathogenic and saprobic ascomycetes. Proc Natl Acad Sci USA 100: 15670–15675.

42. Kunz C., Vandelle E., Rolland S., et al. (2006) Characterization of a new, nonpathogenic mutant of Botrytis cinerea with impaired plant colonization capacity. New Phytol 170: 537–550.

43. Lengeler K.B., Davidson R.C., D'Souza C., Harashima T., Shen W.C., Wang P., Pan X., Waugh M., Heitman J. (2000) Signal transduction cascades regulating fungal development and virulence. Microbiol Mol Biol Rev 64: 746–785.

44. Motoyama T., Kadokura K., Ohira T., Ichiishi A., Fujimura M., Yamaguchi I., Kudo T. (2005) A two-component histidine kinase of the rice blast fungus is involved in osmotic stress response and fungicide action. Fungal Genet Biol 42: 200–212.

45. Poinssot B., Vandelle E., Bentejac M., Adrian M., Levis C., Brygoo Y., Garin J., Sicilia F., Coutos-Thevenot P., Pugin A. (2003) The endopolygalacturonase 1 from Botrytis cinerea activates grapevine defense reactions unrelated to its enzymatic activity. Mol Plant Microbe Interact 16: 553–564.

46. Reino J.L., Hernandez-Galan R., Duran-Patron R., Collado I.G. (2004) Virulence-toxin production relationship in isolates of the plant pathogenic fungus Botrytis cinerea. J Phytopathol 152: 563–566.

47. Reis H., Pfiffi S., Hahn M. (2005) Molecular and functional characterization of a secreted lipase from Botrytis cinerea. Mol Plant Pathol 6: 257–267.

48. Rolke Y., Liu S., Quidde T., Williamson B., Schouten A., Weltring K.M., Siewers V., Tenberge K.B., Tudzynski B., Tudzynski P. (2004) Functional analysis of H2O2-generating systems in Botrytis cinerea: the major Cu–Zn-superoxide dismutase (BCSOD1) contributes to virulence on French bean, whereas a glucose oxidase (BCGOD1) is dispensable. Mol Plant Pathol 5: 17–27.

49. Rui O., Hahn M. (2007) The Sit 2-type MAP kinase Bmp3 of Botrytis cinerea is required for normal saprotrophic growth, conidiation, plant surface sensing and host tissue colonization. Mol Plant Pathol 8: 173–184.

50. Schoonbeek H.J., Del Sorbo G., De Waard M.A. (2001) The ABC transporter BcatrB affects the sensitivity of Botrytis cinerea to the phytoalexin resveratrol and the fungicide fenpiclonil. Mol Plant Microbe Interact 14: 562–571.

51. Schouten A., Wagemakers L., Stefanato F.L., Van Der Kaaij R.M., Van Kan J.A. (2002) Resveratrol acts as a natural profungicide and induces self-intoxication by a specific laccase. Mol Microbiol 43: 883–894.

52. Schulze Gronover C., Schorn C., Tudzynski B. (2004) Identification of Botrytis cinerea genes up-regulated during infection and controlled by the Galpha subunit BCG1 using suppression subtractive hybridization (SSH). Mol Plant Microbe Interact 17: 537–546.

53. Schulze Gronover C.S., Kasulke D., Tudzynski P., Tudzynski B. (2001) The role of G protein alpha subunits in the infection process of the gray mold fungus Botrytis cinerea. Mol Plant Microbe Interact 14: 1293–1302.

54. Segmuller N., Ellendorf U., Tudzynski B., Tudzynski P. (2007) BcSAK1, a stress-activated mitogen-activated protein kinase, is involved in vegetative differentiation and pathogenicity in Botrytis cinerea. Eukaryot Cell 6: 211–221.

55. Sharman S., Heale J.B. (1977) Penetration of carrot roots by the grey mould fungus Botrytis cinerea Pers. Physiol Plant Pathol 10: 63–71.

56. Siewers V., Smedsgaard J., Tudzynski P. (2004) The P450 monooxygenase BcABA1 is essential for abscisic acid biosynthesis in Botrytis cinerea. Appl Environ Microbiol 70: 3868–3876.

57. Siewers V., Viaud M., Jimenez-Teja D., Collado I.G., Gronover C.S., Pradier J.M., Tudzynski B., Tudzynski P. (2005) Functional analysis of the cytochrome P450 monooxygenase gene bcbot1 of Botrytis cinerea indicates that botrydial is a strain-specific virulence factor. Mol Plant Microbe Interact 18: 602–612.

58. Soulie M.C., Piffeteau A., Choquer M., Boccara M., Vidal-Cros A. (2003) Disruption of Botrytis cinerea class I chitin synthase gene Bcchs1 results in cell wall weakening and reduced virulence. Fungal Genet Biol 40: 38–46.

59. Soulie M.C., Perino C., Piffeteau A., Choquer M., Malfatti P., Cimerman A., Kunz C., Boccara M., Vidal-Cros A. (2006) Botrytis cinerea virulence is drastically reduced after disruption of chitin synthase class III gene (Bcchs3a). Cell Microbiol 8: 1310–1321.

60. Staats M., Van Baarlen P., Van Kan J.A. (2005) Molecular phylogeny of the plant pathogenic genus Botrytis and the evolution of host specificity. Mol Biol Evol 22: 333–346.

61. Staats M., Van Baarlen P., Schouten A., Van Kan J.A., Bakker F.T. (2007) Positive selection in phytotoxic protein-encoding genes of Botrytis species. Fungal Genet Biol 44: 52–63.

62. Tani H., Koshino H., Sakuno E., Cutler H.G., Nakajima H. (2006) Botcinins E and F and botcinolide from Botrytis cinerea and structural revision of botcinolides. J Nat Prod 69: 722–725.

63. Ten Have A., Mulder W., Visser J., Van Kan J.A. (1998) The endopolygalacturonase gene Bcpg1 is required for full virulence of Botrytis cinerea. Mol Plant Microbe Interact 11: 1009–1016.

64. Ten Have A., Breuil W.O., Wubben J.P., Visser J., Van Kan J.A. (2001) Botrytis cinerea endopolygalacturonase genes are differentially expressed in various plant tissues. Fungal Genet Biol 33: 97–105.

65. Ten Have A., Tenberge K.B., Benen J.A., Tudzynski P., Visser J., Van Kan J.A. (2002) The contribution of cell wall degrading enzymes to pathogenesis of fungal plant pathogens. The Mycota XI, Agricultural Applications (Kempken F., ed), pp. 341–358. Springer-Verlag, Berlin, Germany.

66. Tenberge K.B. (2004) Morphology and cellular organisation in Botrytis interactions with plants. Botrytis: Biology, Pathology and Control (Elad Y., Williamson B., Tudzynski P., Delen N., eds), pp. 67–84. Kluwer Academic Publishers, Dordrecht, the Netherlands.

67. Tudzynski B., Schulze Gronover C. (2004) Signalling in Botrytis cinerea. Botrytis: Biology, Pathology and Control (Elad Y., Williamson B., Tudzynski P., Delen N., eds), pp. 85–97. Kluwer Academic Publishers, Dordrecht, the Netherlands.

68. Tudzynski P., Siewers V. (2004) Approaches to molecular genetics and genomics of Botrytis. Botrytis: Biology, Pathology and Control (Elad Y., Williamson B., Tudzynski P., Delen N., eds), pp. 53–60. Kluwer Academic Publishers, Dordrecht, the Netherlands.

69. Unger C., Kleta S., Jandl V.G., Tiedemann A. (2005) Suppression of the defence-related oxidative burst in bean leaf tissue and bean suspension cells by the necrotrophic pathogen Botrytis cinerea. J Phytopathol 153: 15–26.

70. Valette-Collet O., Cimerman A., Reignault P., Levis C., Boccara M. (2003) Disruption of Botrytis cinerea pectin methylesterase gene Bcpme1 reduces virulence on several host plants. Mol Plant Microbe Interact 16: 360–367.

71. Van Baarlen P., Woltering E.J., Staats M., Van Kan J.A. (2007) Histochemical and genetic analysis of host and non-host interactions of Arabidopsis with three Botrytis species: an important role for cell death control. Mol Plant Pathol 8: 41–54.

72. Van Kan J.A. (2006) Licensed to kill: the lifestyle of a necrotrophic plant pathogen. Trends Plant Sci 11: 247–253.

73. Van Kan J.A., Van't Klooster J.W., Wagemakers C.A., Dees D.C., Van Der Vlugt-Bergmans C.J. (1997) Cutinase A of Botrytis cinerea is expressed, but not essential, during penetration of gerbera and tomato. Mol Plant Microbe Interact 10: 30–38.

74. Vandelle E., Poinssot B., Wendehenne D., Bentejac M., Alain P. (2006) Integrated signaling network involving

calcium, nitric oxide, and active oxygen species but not mitogen-activated protein kinases in BcPG1-elicited grapevine defenses. Mol Plant Microbe Interact 19: 429–440.

75. Viaud M., Brunet-Simon A., Brygoo Y., Pradier J.M., Levis C. (2003) Cyclophilin A and calcineurin functions investigated by gene inactivation, cyclosporin A inhibition and cDNA arrays approaches in the phytopathogenic fungus Botrytis cinerea. Mol Microbiol 50: 1451–1465.

76. Viaud M., Legeai F., Pradier J.M., et al. (2005) Expressed sequence tags from the phytopathogenic fungus Botrytis cinerea. Eur J Plant Pathol 111: 139–146.

77. Viaud M., Fillinger S., Liu W., Polepalli J.S., Le Pecheur P., Kunduru A.R., Leroux P., Legendre L. (2006) A class III histidine kinase acts as a novel virulence factor in Botrytis cinerea. Mol Plant Microbe Interact 19: 1042–1050.

78. Viaud M.C., Balhadere P.V., Talbot N.J. (2002) A Magnaporthe grisea cyclophilin acts as a virulence determinant during plant infection. Plant Cell 14: 917–930.

79. Xu J.R. (2000) Map kinases in fungal pathogens. Fungal Genet Biol 31: 137–152.

80. Xu J.R., Peng Y.L., Dickman M.B., Sharon A. (2006) The dawn of fungal pathogen genomics. Annu Rev Phytopathol 44: 337–366.

81. Zheng L., Campbell M., Murphy J., Lam S., Xu J.R. (2000) The BMP1 gene is essential for pathogenicity in the gray mold fungus Botrytis cinerea. Mol Plant Microbe Interact 13: 724–732.

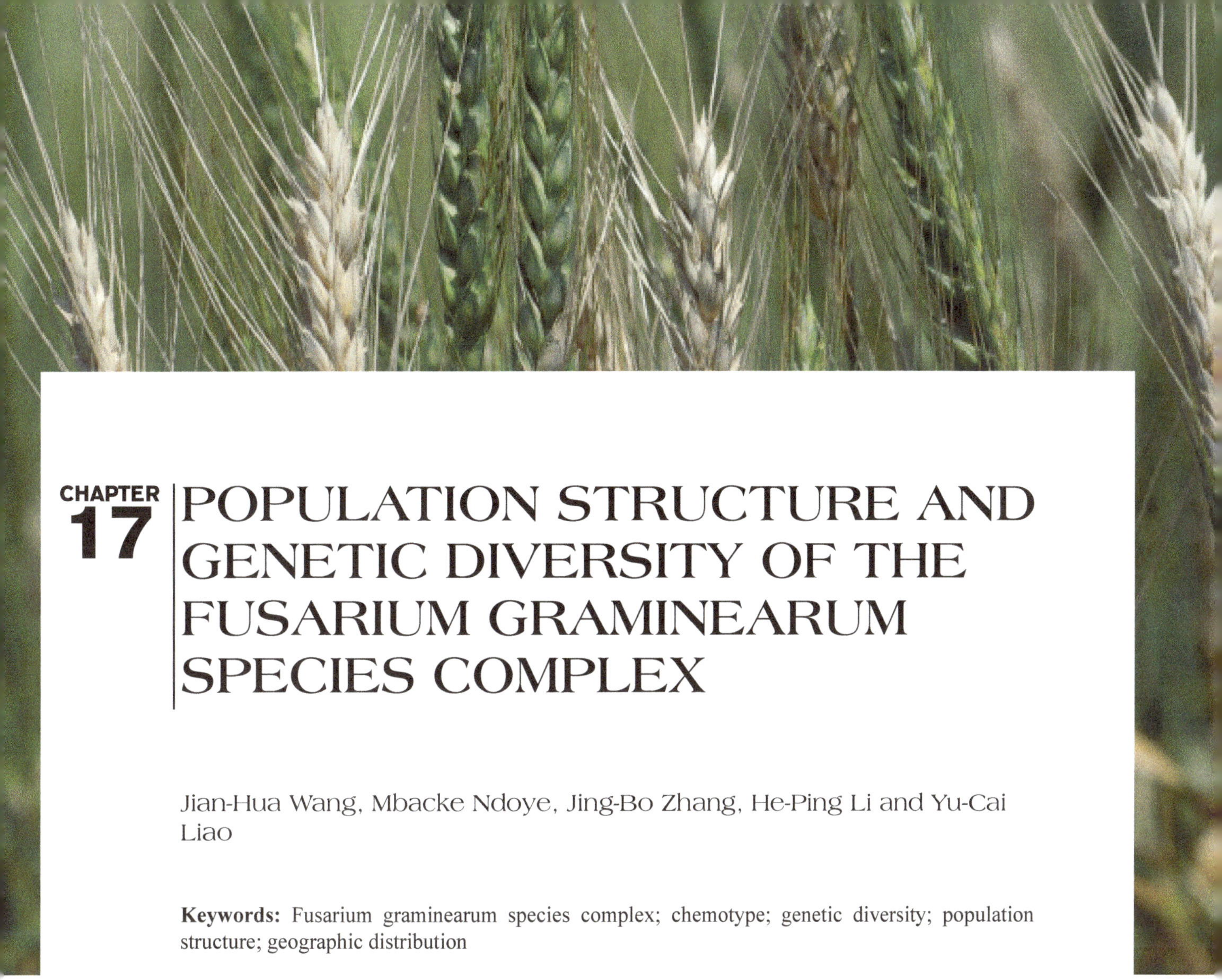

CHAPTER 17

POPULATION STRUCTURE AND GENETIC DIVERSITY OF THE FUSARIUM GRAMINEARUM SPECIES COMPLEX

Jian-Hua Wang, Mbacke Ndoye, Jing-Bo Zhang, He-Ping Li and Yu-Cai Liao

Keywords: Fusarium graminearum species complex; chemotype; genetic diversity; population structure; geographic distribution

INTRODUCTION

Filamentous fungi within the *Fusarium graminearum* species complex (*Fg* complex) are the most important etiological agents of *Fusarium* head blight (FHB) on wheat and other cereal grains worldwide [1,2,3]. Over the past decade, the *Fg*complex has emerged as one of the most destructive pathogens of cereals resulting in losses valued at billions of dollars to the grain industry [4,5]. Besides reducing yield, these fungi can cause further losses through contamination of crops with mycotoxins such as deoxynivalenol, nivalenol and zearalenone [6,7], which pose a serious threat to animal health and food safety.

Members of the *Fg* complex appear to have evolved allopatrically [8,9,10,11]. Biogeographic data suggests that the majority of species within the *Fg* complex have evolutionary origins in the Southern Hemisphere [2,3] and Asia [9]. During recent years, changed agronomy, trade and shipment of cereal commodities, and global climate change have aggravated the spread and severity of FHB northward in China [12,13,14] and also have resulted in the reemergence of FHB to epidemic levels in the United States, Canada and Europe, causing losses of about $3 billion for the United States alone between 1998 and 2000 [15,16]. It seems likely that some *Fg* complex species have been introduced into non-indigenous areas relatively recently, especially in areas that rely extensively on agricultural imports [2,3,17,18,19,20,21,22,23,24,25].

In view of the importance of diseases caused by members of the *Fg* complex, procedures for effective management of these pathogens are urgently needed. In this context, information on the population genetic diversity and dispersal of the *Fg*complex is, consequently, of importance. Various genotypic and phenotypic approaches have been in use over the past few years in research pertaining to the population genetic structure of the *Fg* complex [1,26,27,28,29,30,31,32,33].

This review seeks to outline some of the findings of genetic diversity studies on the *Fg* complex that have been conducted over the past two decades. It is not meant to be a comprehensive review of the literature but rather a reflection on some of the common approaches, assumptions, and findings in the field. The genetic diversity, population structure and geographic distribution of the *Fg* complex that have been the focus of these studies will be reviewed. In addition, potential topics for future research will be discussed.

Citation: Jian-Hua Wang, Mbacke Ndoye, Jing-Bo Zhang, He-Ping Li, and Yu-Cai Liao, Population Structure and Genetic Diversity of the Fusarium graminearum Species Complex, Toxins 2011, 3(8), 1020-1037; doi:10.3390/toxins3081020

POPULATION STRUCTURE AND GEOGRAPHIC DISTRIBUTION OF THE F_G COMPLEX

Natural occurrence of the *Fg* complex causing FHB on cereal crops have been reported in different continents, e.g., Africa [11,34,35,36,37], America [3,4,7,17,20,25,33,38,39,40,41,42,43,44,45,46,47,48,49], Asia [9,13,14,17,21,22,23,24,28,30,50,51,52,53,54,55,56,57], Europe [50,58,59,60], and Oceania [18,61,62,63,64]. Members of the *Fg*complex were considered a single cosmopolitan species, due to the failure of morphological species recognition to accurately assess species limits for this group. O'Donnell *et al.* [1] first identified seven phylogenetic lineages within the *Fg*complex, employing genealogical concordance phylogenetic species recognition (GCPSR) [65]. Most recently, employing a high-throughput multilocus genotyping (MLGT) assay of portions of 13 housing keeping genes, combined with GCPSR and molecular marker technologies, 13 phylogenetically distinct, cryptic species have been identified within the *Fg* complex [1,2,3,9,11]. The species designation *Fusarium graminearum* will therefore be *sensu stricto*, and other *Fg* complex species designations will be used. These 13 species include: *F. austroamericanum*, *F. meridionale*, *F. boothii*, *F. mesoamericanum*, *F. acaciae-mearnsii*, *F. asiaticum*, *F. graminearum sensu stricto*, *F. cortaderiae*, *F. brasilicum*, *F. gerlachii*, *F. vorosii*, *F. aethiopicum*, and*F. ussurianum* [1,2,3,9,11]. In addition, a putative new species within the *Fg* complex was identified in Nepal by Chandler *et al.* [10] and later characterized by Desjardins and Proctor [66] as the "Nepal lineage".

A biogeographic hypothesis suggests that the basal most species within the *Fg* complex may be endemic to the southern hemisphere while the derived species evolved within the northern hemisphere [3]. According to the surveys to date, *F. austroamericanum*, *F. meridionale*, *F. cortaderiae*, and *F. brasilicum* appear to be endemic to South America [3,46]; *F. acaciae-mearnsii* to Australia or less likely Africa [2]; *F. asiaticum*, *F. vorosii* and *F. ussurianum* to Asia [2]; *F. aethiopicum* to Africa [11];*F. boothii* and *F. mesoamericanum* to Central America [2,11]; *F. gerlachii* to the US [3].

In the surveys conducted worldwide to date, *F. graminearum sensu stricto* (*F. graminearum s.s.*) is cosmopolitan in distribution and has been found in Asia, Africa, America, Europe, and Oceania, while another species, *F. asiaticum*, is widespread in Asia [1,2,59,41,66,54]. FHB was initially reported in 1936 in China, and since then FHB epidemics have become more severe and frequent in the middle and lower regions of the Yangtze River, and in Heilongjiang province in northeastern China [12]. More recently FHB frequently occurs in areas along the Yangtze River, a region with conditions favorable for infection by the pathogen [13,14,21,22,23,24]. *F. asiaticum* is the predominant species in this region [13,14,30,21,22,23,24]. Based on the climate data of 30 y during 1970–1999, our group assayed the geographical distribution patterns of *F. graminearum s.s.* and *F. asiaticum* in China. Our results indicated that *F. graminearum s.s.* was mainly obtained from cooler regions where the annual average temperature was 15 °C or lower. In contrast, 95% of *F. asiaticum* isolates were collected from warmer regions [14,21]. These results suggest ecological factors may have had a significant effect on the distribution of these species [13,14,46,51,54,66]. In Japan, *F. graminearum s.s.* and *F. asiaticum* are the predominant species, with *F. asiaticum* predominating in the southern regions as in China [52]. A preliminary report of the *F. graminearum* clade isolated from maize in Korea indicated that *F. graminearum s.s.* was dominant, accounting for 74% of strains, while three additional species, *F. meridionale*, *F. boothii*, and *F. asiaticum*, were also present at low frequencies [67]. However, *F. asiaticum*,*F. meridionale* and *F. boothii* were the major causal agents of *Gibberella* ear rot of maize in Nepal, whereas *F. graminearum s.s.*, which dominates in maize elsewhere in Asia and worldwide, was not detected [66]. Similarly, *F. meridionale* and *F. boothii*play a substantial role in the infection and trichothecene contamination of maize in Argentina. *F. graminearum s.s.* was not represented among the isolates examined [46]. In contrast, a previous survey of the *Fg* complex collected from wheat in the center of Argentina identified all 113 isolates as *F. graminearum s.s.* [41]. In Korea, 80% of the isolates collected from rice belonged to *F. asiaticum*, which is dramatically different from the results on maize [54]. Although the underlying factors for species distribution are unknown, the occurrence of *F. asiaticum* in Louisiana closely overlaps with rice-growing areas in Louisiana [25].

CHEMOTYPE DISTRIBUTION OF THE F_G COMPLEX

Every species within the *Fg* complex is capable of producing type B trichothecenes in planta [11]. Based on the chemical structure, including the acetylation position, three type B trichothecene chemotypes have been identified within the *Fg*complex: deoxynivalenol and 3-acetyldeoxy-nivalenol (3-AcDON), deoxynivalenol and 15-acetyldeoxynivalenol (15-AcDON), and nivalenol and 4-acetylnivalenol (NIV) [32,68,69].

In the limited surveys to date, strains of several newly discovered species were found to represent only a single chemotype: 3-AcDON for the *F. ussurianum* strains [9], 15-AcDON for *F. aethiopicum*, *F. boothii* and *F. vorosii* strains [3,9,11,20,70], and NIV for *F. meridionale* and *F. gerlachii* strains [3,11,20,46,70]. 3-AcDON, 15-AcDON and NIV producers were identified in both *F. graminearum s.s.* and *F. asiaticum*. The remaining 5 species, *F. acaciae-mearnsii*, *F. mesoamericanum*, *F. austroamericanum*, *F. cortaderiae*, and *F. brasilicum* can produce both 3-AcDON and NIV [20].

Dramatic differences in the geographic distribution of *Fg* complex chemotypes were reported by Gale *et al.* [30], Zhang *et al.* [21] and Qu *et al.* [13] in China, and by Karugia *et al.* [56] in Japan. Results of these surveys show that Chinese populations of *F. graminearum s.s.* on wheat and barley produce 15-AcDON predominately or exclusively, in contrast to populations of *F. asiaticum*, which predominately produce 3-AcDON or NIV. While in Japan, out of a total of 179 *F. asiaticum*isolates, 80 isolates were of the NIV chemotype, and the remaining 99 isolates were of 3-AcDON producers [56]. NIV chemotype was predominant among the *Fg* complex strains isolated from Korean rice [54,57]. In contrast, the predominant chemotype among the corn isolates varied according to geographic region [57]. In Nepal, NIV producers of the *Fg* complex were isolated more than twice as frequently as DON producers from maize seed samples [66,71].

Direct competitive enzyme-linked immunosorbent assay (ELISA) indicated that 75% of grain samples from Kenya were contaminated with DON [35]. Eleven *Fusarium*-related mycotoxins were quantified by chemical analysis, the results demonstrated that DON was detected at the highest frequency (69%) [36]. The trichothecene chemotype of 560 South African isolates of the *Fg* complex collected from diseased wheat, barley and maize were determined using MLGT assay. The results showed that 535 isolates possessed the 15-AcDON chemotype, 22 isolates represented the NIV chemotype, and only 3 isolates were of the 3-AcDON chemotype [37]. *F. aethiopicum*, a novel species reported only from Ethiopia, belonged to 15-AcDON chemotype [11,37].

In the surveys conducted to date in Europe, the 15-AcDON chemotype of the *Fg* complex was the predominant, followed by the 3-AcDON chemotype and a few isolates of the NIV chemotype. For example, in UK [72], Germany [73], and Luxembourg [74], the 15-AcDON chemotype represented more than 90% of the population, and DON producers also were in high portion in Netherlands [58]. In New Zealand, however, Lauren *et al.* [62] and Monds *et al.* [18] found that among the*F. graminearum s.s.* strains, NIV chemotypes were as common as DON chemotypes, with the DON chemotypes producing primarily 15-AcDON. All the isolates of *F. cortaderiae*, the other *Fg* clade species found in New Zealand, were NIV producers [18].

Previous studies indicated that *F. graminearum s.s.* isolates with a 15-AcDON chemotype represented the predominant cause of FHB in North America [68,75,76]. However, a significant increasing frequency of 3-AcDON chemotype from west to east occurred in Canada in the last decade [20,77]. Schmale *et al.* [78] evaluated the trichothecene genotypes of 998 *Gibberella zeae* (*G. zeae*) isolates gathered from 39 winter wheat fields in the eastern US, and found that 92% percent of the isolates were 15-AcDON, 7% were 3-AcDON, and 1% was NIV, with an increasing trend of 3-AcDON producers from south to north. Starkey *et al.* [3] identified 2 NIV-chemotype isolates (NRRL 38371 and NRRL 38383) within 6 *F. graminearum s.s.* strains in Louisiana. A recent survey by Gale *et al.* [25] suggests that NIV-type *F. graminearum s.s.* populations represented a high proportion (79%) of isolates in Louisiana. It was suggested that the NIV chemotype of *F. graminearum s.s.* was rare in the United States, and that the only way for the NIV chemotype with adequate fitness to become established in North America would be by migration and introgression into the resident DON populations [71]. In addition, chemotype variation has been maintained by natural selection and appears to have adaptive potential for FHB pathogens [32].

In Argentina, within the *Fg* complex strains, 15-AcDON chemotype was the most common [44,47], and trichothecene chemotype diversity among the isolates examined was directly tied to species differences [46]. Two studies of the wheat kernel samples in southern Brazil showed that 15-AcDON chemotype was predominant (>92%), followed by the NIV chemotype, and no 3-AcDON chemotype was observed [43,49]. A recent survey of barley grain samples revealed that, 4% of the strains were 3-AcDON chemotype, 66% were 15-AcDON chemotype and 29% were NIV chemotype [48].

Broad geographic surveys for trichothecene chemotype of the *Fg* complex from cereal crops are important to establish baselines for these genotypes against which any future shifts in populations can be assessed [3,20,25,42,77,78].

MOLECULAR MARKERS FOR F_G COMPLEX GENETIC ANALYSES

High genotypic diversity has been revealed within the Fg complex, even within a species [30,33,38,56,79,80], using molecular markers. Molecular markers such as sequence characterized amplified regions (SCAR), single strand conformational polymorphism (SSCP), randomly amplified polymorphic DNA (RAPD), amplified fragment length polymorphism (AFLP), restriction fragment length polymorphism (RFLP), sequence related amplified polymorphism (SRAP), single nucleotide polymorphism (SNP), and variable number of tandem repeat (VNTR) have been used extensively to study genetic variability of the Fg complex. Recently, a high resolution, SNP-based multilocus genotyping assay was used in numerous investigations of diversity within the Fg complex, and several novel species have been identified.

SCAR (Sequence Characterized Amplified Regions) or SSCP (Single Strand Conformational Polymorphism)

SCAR analysis using Fg16F/R primer was previously used to distinguish *F. graminearum s.s.* and *F. asiaticum* isolates from China [14]. However, differentiation of *F. asiaticum* from *F. meridionale* on the basis of SCAR fragment size was not reliable because the Fg16F/R amplicon size is identical (497 bp) [29]. However, because the sequence of the *F. meridionale* and *F. asiaticum* amplicons differ, single-strand conformational polymorphism (SSCP) enabled unequivocal resolution of *F. meridionale* from *F. asiaticum* [13].

SSCP analysis of the Fg16F/R PCR amplicon differentiated *F. graminearum s.s.*, *F. asiaticum* and *F. meridionale* from China and revealed three haplotypes among sequence-characterized amplified region (SCAR) type 1 *F. graminearum s.s.* isolates: 1A, 1B and 1C, while the remaining SCAR types displayed the same SSCP patterns with no polymorphism [13]. All the *F. graminearum s.s.* isolates from China and Europe, and 5 of the seven isolates from the US were haplotype 1A. The remaining two American isolates (A2 and A5) were haplotype 1B [13]. DNA sequencing of the Fg16F/R product confirmed that haplotype 1B differed from that of 1A at nucleotide 348 (C-T) and 412 (T-A) [81]. SSCP assays of a further 440 isolates from China indicated that 90 were *F. graminearum* (haplotype 1A) and 350 were *F. asiaticum* (haplotype 5) with no polymorphism within the species [81]. This high level of uniformity contrasts markedly with isolates from elsewhere. For example, three haplotypes (haplotypes 3, 4 and 5) were observed among *F. asiaticum* isolates from Nepal [29], while two haplotypes were observed among the limited number of isolates from Europe (haplotypes 1A and 6) and the US (haplotypes 1A and 1B).

Investigators have used a variety of methods to improve the resolving power of SSCP, however, differentiation among polymorphic molecules on a polyacrylamide matrix is not entirely predictable, and the method can result in false negatives, ambiguous results and experimental artifacts. The low haplotype diversity detected by the SSCP markers indicates that it is too conserved for population genetic studies of the *Fg* complex [14].

RAPD (Randomly Amplified Polymorphic DNA)

RAPD technique is a PCR procedure involving very short (10 or fewer bases) arbitrary primers at a low annealing temperature (30–38 °C). RAPD analysis detects two types of genetic variation: (i) in the length of DNA between the two primer binding sites; and (ii) in sequence variation at the priming regions. When the amplified products from such a reaction are analyzed on an electrophoresis gel, unique banding patterns are seen. These patterns may reflect the differences characteristic of certain species or strains. The RAPD method is simple, sensitive, rapid and has been used for distinguishing several pathogens [82]. A high level of genotypic diversity in *Fg* complex isolates in Canada, via RAPD markers, was demonstrated by Dusabenyagasani *et al.* [83], and 90.56% of the genetic variability was explained by within-region variation. RAPD analysis of 42 isolates, which originated from northwest Europe, the US and Nepal, identified three groups, two of which contained the isolates from Nepal, and a third contained the isolates from Europe and the US [29]. A considerable genetic resemblance was found by RAPD between 34 isolates from northeastern and northwestern China. The isolate grouping was not related to pathogenicity or to host cultivar [84]. Similar findings were obtained in the US, where *F. graminearum s.s.* is the predominant FHB species [3,38,42]. However, relatively low amount of genetic diversity were revealed using RAPD analysis of isolates in Canada [85] and Brazil [86].

RAPD has the major disadvantage that it is very sensitive to the reaction conditions, DNA quality and PCR temperature profiles [87], which limits its application. This problem can be minimized if strains under study are treated identically. RAPD has been criticized for lack of reproducibility [88].

AFLP (Amplified Fragment Length Polymorphism)

AFLP is a DNA fingerprinting technique that detects genomic restriction fragments and thus resembles the RFLP technique. The method allows for the specific co-amplification of high numbers of restriction fragments. Typically 50–100 restriction fragments are amplified and detected on denaturing polyacrylamide gels.

The advantage of AFLP over other techniques is that multiple bands are derived from all over the genome. This prevents overinterpretation or misinterpretation due to point mutations or single-locus recombination, which may affect other genotypic characteristics. Different enzymes and/or selective extension nucleotides can be used to create new sets of markers. Therefore, AFLP can provide an almost limitless set of genetic markers, and as many as 50–75 AFLP markers can be resolved per reaction [89]. The AFLP markers showed a higher degree of resolution for discriminating closely related fungi. In *Fusarium* taxonomy, AFLP was first applied to discriminate between different strains of the *Fg* complex [90]. It was used for assessing the population structure of the *Fg* complex and identified considerable diversity within the clade [14,91,92,93]. Bowden *et al.* [94] using AFLP analysis to determine genetic diversity in the *Fg* complex from Kansas and North Dakota, found a high degree of homogeneity between subpopulations. The analysis of 59 isolates from Australia by AFLP showed that 56 isolates had distinct haplotypes, and the spatial diversity within the *F. graminearum s.s.* strains was high within a single field [63]. The AFLP analysis by Monds *et al.* [18] showed that the *Fg* complex was represented by *F. graminearum s.s.* and *F. cortaderiae* in New Zealand, and that these two species originated from at least two regions, and probably on at least two hosts. More diversity of the *Fg* complex was detected by AFLP analysis of populations from Europe, the US or Nepal [14]. Based on AFLP analyses, 270 bands were detected among 333 *F. asiaticum* isolates from the five populations in Korea, and 36% of the AFLP bands were polymorphic [54]. Among the 270 multilocus haplotypes, 225 were represented by a single strain, and 45 haplotypes were detected more than once. Genotypic diversity varied by population, and the number of polymorphic loci in an individual population ranged from 32 to 56 [54]. A sample of 103 *F. graminearum s.s.* isolates from Brazil were assessed employing AFLP. Astolfi *et al.* [49] found that isolates with the same haplotype were rare and genotypic diversity was uniformly high (≥98% of the count) in all three subpopulations analyzed.

In our experience, AFLP is not highly reproducible. When controlled experiments are conducted over 10% of the bands are not reproduced. Also scoring band size is exceedingly difficult. Besides, alleles are not easily recognized [95], even when a capillary DNA sequencer is employed [96].

SRAP (Sequence Related Amplified Polymorphism)

Sequence-related amplified polymorphism (SRAP) is a novel molecular marker technique based on two-primer amplification that preferentially amplifies open reading frames (ORFs) [97]. The forward primers preferentially amplify exonic regions, and the reverse primers preferentially amplify intronic regions and regions with promoters. The observed polymorphism originates in the variation in the length of these extrons, introns, promoters, and spacers, both among individuals and among species [97]. With this unique primer design, SRAP markers are more reproducible, more stable, and less complex than RAPD and AFLPs. This technique has been used to investigate genetic diversity of the *Fg* complex in Canada [40]. SRAP results are in agreement with findings of high diversity of the *Fg* complex isolates from the United States and eastern Canada based on RAPD and AFLP markers [98]. It could help to assess whether different management practices affect the genetic diversity of the *Fg* complex populations [98].

The disadvantage of this technique is that both dominant and co-dominant markers can be produced which makes the interpretation of the data a little bit more complicated.

SNP (Single Nucleotide Polymorphism)

A SNP is a single base pair mutation at a specific locus, usually consisting of two alleles that can be amplified by two pairs of primers in one PCR reaction. Cuomo *et al.* [99] identified more than 10,000 SNPs from a comparison of two *F. graminearum s.s.* strains (strains PH-1 and GZ3639), which were frequently located near telomeres and within other discrete chromosomal segments. *F. graminearum* genes specifically expressed during plant infection (including predicted secreted proteins, major facilitator transporters, amino acid transporters, and cytochrome P450s) are all overrepresented in high SNP density regions, which may allow the fungus to adapt rapidly to changing environments or hosts. Perhaps the SNP have the greatest potential in elucidating the dynamics of host pathogen interactions.

Yang *et al.* [22] characterized *Fusarium* isolates at the species level by a robust set of diagnostic primers based on SNPs among members of the *Fg* complex. In addition, numerous SNPs-based trichothecene chemotype assays were developed from the trichothecene biosynthetic pathway genes. Two sets of multiplex primers specific to individual chemotypes were designed from the *Tri3* and *Tri12* genes [20,32]. Both multiplex assays were widely used to predict the chemotype of the *Fg*complex strains [3,25,43,48,49,78,100]. Primers derived from the *Tri13* and *Tri7* genes [10,58], which are responsible for the oxygenation and acetylating of the C-4 atom of the trichothecene molecule [101], were used to discriminate between DON and NIV producers [10,43,51,53,58,72,100]. The primer pairs Tri303F/R and Tri315F/R derived from the *Tri3* gene is used to predict if a DON chemotype is a 3-AcDON or 15-AcDON producer [21,51,72]. Very recently, Zhang *et al.* [23] also developed an effective set of diagnostic primers based on SNPs among three different chemotype isolates and determined that there was a recent shift to the 3-AcDON chemotype. The high genetic diversity of this group of genes suggests that the fungus has great capacity for adaptability and genetic change during its interaction with a host species. The richness in SNPs in the fungal genome combined with real-time PCR assays would provide a powerful tool for development of markers for various identifications within the *Fg* complex.

VNTR (Variable Number of Tandem Repeat)

VNTR markers have been developed for *Fg* complex species and have been effectively used for population analyses [20,55,70,102]. VNTR may be more desirable than RFLP markers for population genetic analysis due to their ability to generate accurate polymorphic data and due to their codominance. Generally, the development of VNTR markers is costly and burdensome and therefore few VNTR markers have been reported to date for the *Fg* complex [102,103]. A relative small genome size, 36.1 Mb, of *F. graminearum s.s.* (strain PH-1, NRRL 31084) sequenced recently by the Center for Genome Research, Cambridge, MA, USA [99], may carry a low level of variation within tandem repeat sequences in the genome. However, the availability of the whole genome sequence may allow a bioinformatic approach for the further development of VNTR markers in these species. Zhang *et al.* [23], using seven highly informative VNTR markers on 448 *F. asiaticum* isolates from Chinese barley, showed a significant degree of population subdivision ($P<0.001$) among populations from the upper, middle, and lower valleys of the Yangtze River, with little gene flow ($Nm = 1.210$). A comprehensive VNTR analysis of 1106*F. asiaticum* isolates collected from western China, centre and eastern provinces, indicated that the genetic diversity of the species showed a clear substructure even within provinces [24]. Ten VNTR markers were used to determine the genetic diversity of an older population ($n = 115$ isolates) of *G. zeae* collected from 1997 to 2000 with a newer population ($n = 147$ isolates) collected in 2008. High gene diversity and genotypic diversity were revealed, but the linkage disequilibrium were low in both populations, and the differentiation between the two populations was low [100].

RFLP (Restriction Fragment Length Polymorphism)

The RFLP method is used to assess the extent of natural variation or relatedness in the genomes of different species, biotypes, strains or races of fungal pathogens. Deletions or insertions in the DNA sequences may lead to variation (polymorphisms) in fragment sizes. The sensitivity and reliability of the assay may be enhanced by employing highly repetitive DNA sequences as probes, as the signal is present in multiple copies.

Members of the *Fg* complex and *F. pseudograminearum* are morphologically very similar, and were previously designated Group 1 and Group 2 of *F. graminearum* [104]. RFLP [105], DNA sequence [106], and isozyme [107] data indicated that these two groups were phylogenetically distinct and should be regarded as separate species. RFLP analysis of the *Fg* complex collected from eastern China showed a population mean gene diversity of 30.6% to 36.4% and identified 65 distinct haplotypes among 225 isolates [30].

An alternative PCR-RFLP marker system based on single-copy nuclear polymorphic (SCNP) sequences in *Fg* complex was developed [25]. These markers can be useful when VNTR analysis is technologically not possible, mutation rates differ, or selective neutrality is a concern [25]. This method not only substantially reduced screening time and cost but also allowed for the strategic positioning of markers across the whole genome. Based on the moderate number of alleles per locus and the low number of shared genotypes, this PCR-RFLP system allows for excellent genotypic resolution without excessive polymorphisms [25]. Therefore it is suitable for population level analysis of the *Fg* complex in the future.

GCPSR (Genealogical Concordance Phylogenetic Species Recognition) and MLGT (Multilocus Genotyping Assay)

GCPSR was first used to investigate species diversity within the Fg complex by O'Donnell et al. [1]. A global collection of FHB strains were investigated by Ward et al. [32], O'Donnell et al. [2], and Starkey et al. [3], phylogenetic analyses of DNA sequences from portions of 13 independent loci, including translation elongation factor (EF-1α), α-tubulin (α-tub), β-tubulin (β-tub), histone H3 (HIS), phosphate permease (PHO), reductase (RED), trichothecene 3-O-acetyltransferase (Tri101), ammonium ligase (URA), ITS/28S rDNA locus, and four adjacent mating-type (MAT) genes (MAT1-1-3, MAT1-1-2, MAT1-1-1, and MAT1-2-1) plus the corresponding intergenic regions (A, B, and C), totaling 16.33 kb, provided strong evidence that this morphologically defined species comprises at least 13 phylogenetically distinct species.

MLGT was developed by Ward et al. [20]. Forty-one oligonucleotide probes targeting species or trichothecene chemotype-specific variation within 6 genes were developed base on SNPs [20], which facilitated the species and chemotype assays of the Fg complex. Based on the MLGT assays, two novel species within the Fg complex, F. aethiopicum [11] and F. ussurianum [9], were recently identified in Ethiopia and the Russian Far East, respectively. Four novel probes, 2 for F. aethiopicum and 2 for F. ussurianum, were developed from variation identified in the Tri101 and RED genes for identification of these two species [9,11]. MLGT is based on direct interrogation of SNPs. It provides a rapid, efficient and accurate framework for studies to understand the population dynamics, trichothecene chemotype distribution, and ecology of the Fg complex [9,11,20,25,37,45].

CONCLUSIONS

Members of the *Fg* complex were thought to represent the single species *F. graminearum* (teleomorph *G. zeae*) [11]. Thus multiple *Fg* complex were reported under the name *F. graminearum* in previous studies. Morphological variation within the*Fg* complex is insufficient for discerning species identity. However, by employing GCPSR, a MLGT assay, and different molecular marker technologies, 13 phylogenetically distinct species have been identified within the *Fg* complex [1,2,3,9,11,65]. These findings indicate that diversity within the *Fg* species complex is much greater than previously believed [46].

As FHB is a serious threat to cereal production worldwide, information on the global distribution, aggressiveness, and toxin accumulation of *Fg* complex strains will be critical to identifying and implementing pathogen control strategies. Some updated molecular marker technologies, such as VNTR and PCR-RFLP, will play an important role in population genetic analyses of the *Fg* complex. The increasing database of global surveillance information is providing a detailed understanding of the geographic distribution, host range, aggressiveness, epidemiology, and evolutionary dynamics of the FHB pathogens and their toxin potential, all of which are critical for monitoring changes in this economically devastating disease worldwide [11].

ACKNOWLEDGMENTS

This work was supported by the National Basic Research Program of China (2009CB118806), the National Natural Science Foundation of China (30571160, 30771337), the Ministry of Agriculture of China (2008ZX08002-001,2009ZX08002-001B), and the Ministry of Education of China (20090146120013). The authors are thankful for a Doctorial fellowship from the Chinese Exchange program (to M.N.).

REFERENCES

1. O'Donnell, K.; Kistler, H.C.; Tacke, B.K.; Casper, H.H. Gene genealogies reveal global phylogeographic structure and reproductive isolation among lineages of *Fusarium graminearum*, the fungus causing wheat scab. *Proc. Natl. Acad. Sci. USA* **2000**, *97*, 7905–7910.
2. O'Donnell, K.; Ward, T.J.; Geiser, D.M.; Kistler, H.C.; Aoki, T. Genealogical concordance between mating type locus and seven other nuclear genes supports formal recognition of nine phylogenetically distinct species within the*Fusarium graminearum* clade. *Fungal Genet. Biol.* **2004**, *41*, 600–623.
3. Starkey, D.E.; Ward, T.J.; Aoki, T.; Gale, L.R.; Kistler, H.C.; Geiser, D.M.; Suga, H.; Tóth, B.; Varga, J.; O'Donnell, K. Global molecular surveillance reveals novel *Fusarium* head blight species and trichothecene toxin diversity. *Fungal Genet. Biol.* **2007**, *44*, 1191–1204.
4. Clear, R.M.; Patrick, S.K. *Fusarium* head blight pathogens isolated from *Fusarium*-damaged kernels of wheat in western Canada, 1993 to 1998. *Can. J. Plant Pathol.* **2000**, *22*, 51–60.
5. Windels, C.E. Economic and social impacts of *Fusarium* head blight: Changing farms and rural communities in thenorthern great plains. *Phytopathology* **2000**, *90*, 17–21.
6. Snijders, C.H.A. *Fusarium* head blight and mycotoxin contamination of wheat, a review. *Eur. J.Plant Pathol.* **1990**, *96*, 187–198.
7. Tuite, J.; Shaner, G.; Everson, R.J. Wheat scab in soft red winter wheat in Indiana in 1986 and its relation to some quality measurements. *Plant Dis.* **1990**, *74*, 959–962.
8. Giraud, T.; Refregier, G.; Gac, I.L.; De Vienne, D.M.; Hood, M.E. Speciation in fungi. *Fungal Genet. Biol.* **2008**, *45*, 791–802.
9. Yli-Mattila, T.; Gagkaeva, T.; Ward, T.J.; Aoki, T.; Kistler, H.C.; O'Donnell, K. A novel Asian clade within the *Fusarium graminearum* species complex includes a newly discovered cereal head blight pathogen from the Russian Far East.*Mycologia* **2009**, *101*, 841–852.
10. Chandler, E.A.; Simpson, D.R.; Thomsett, M.A.; Nicholson, P. Development of PCR assays to *Tri7* and *Tri13*trichothecene biosynthetic genes, and characterisaton of chemotypes of *Fusarium graminearum*, *Fusarium culmorum* and*Fusarium cerealis*. *Physiol. Mol. Plant Pathol.* **2003**, *62*, 355–367.
11. O'Donnell, K.; Ward, T.J.; Aberra, D.; Kistler, H.C.; Aoki, T.; Orwig, N.; Kimura, M.; Bjornstad, A.; Klemsdal, S.S. Multilocus genotyping and molecular phylogenetics resolve a novel head blight pathogen within the *Fusarium graminearum* species complex from Ethiopia. *Fungal Genet. Biol.* **2008**, *45*, 1514–1522.
12. Chen, L.F.; Bai, G.H.; Desjardins, A.E. Recent Advances in Wheat Head Scab Research in China. Proceedings of the International Symposium on Wheat Improvement for Scab Resistance, Suzhou and Nanjing, Jiangsu, China, 2000; pp. 258–273. Available online: http://www.nal.usda.gov/pgdic/WHS/whsindex.html (accessed on 15 August 2011).
13. Qu, B.; Li, H.P.; Zhang, J.B.; Xu, Y.B.; Huang, T.; Wu, A.B.; Zhao, C.S.; Carter, J.; Nicholson, P.; Liao, Y.C. Geographic distribution and genetic diversity of *Fusarium graminearum* and *F. asiaticum* on wheat spikes throughout China. *Plant Pathol.* **2008**, *57*, 15–24.
14. Qu, B.; Li, H.P.; Zhang, J.B.; Huang, T.; Carter, J.; Liao, Y.C.; Nicholson, P. Comparison of genetic diversity and pathogenicity of *Fusarium* head blight pathogens from

China and Europe by SSCP and seedling assays on wheat.*Plant Pathol.* **2008**, *57*, 642–651.

15. Nganje, W.E.; Johnson, D.D.; Wilson, W.W.; Leistritz, F.L.; Bangsund, D.A.; Tiapo, N.M. Economic Impacts of*Fusarium*Head Blight in Wheat and Barley: 1998–2000. Proceedings of Agribusiness and Applied Economics Report, North Dakota State University, Fargo, ND, USA, 2001; pp. 1–41. Available online: http://ageconsearch.umn.edu/bitstream/23515/1/aer464.pdf (accessed on 15 August 2011).
16. Nganje, W.E.; Bangsund, D.A.; Leistritz, F.L.; Wilson, W.W.; Tiapo, N.M. Estimating the Economic Impact of a Crop Disease: The Case of *Fusarium* Head Blight in U.S. Wheat and Barley. National Fusarium Head Blight Forum Proceedings, Okemos, MI, USA, 2002; pp. 275–281. Available online: http://scab.pw.usda.gov/pdfs/forum_02_proc.pdf#page=288 (accessed on 15 August 2011).
17. Lee, Y.W.; Jeon, J.J.; Kim, H.; Jang, I.Y.; Kim, H.S.; Yun, S.H.; Kim, J.G. Lineage Composition and Trichothecene Production of *Gibberella Zeae* Population in Korea. In *New Horizons of Mycotoxicology for Assuring Food Safety*; Yoshizawa, T., Ed.; Kagawa, Japan, 2004; pp. 117–122.
18. Monds, R.D.; Cromey, M.G.; Lauren, D.R.; Marshall, J. *Fusarium graminearum, F. cortaderiae* and *F. pseudograminearum*in New Zealand: Molecular phylogenetic analysis, mycotoxin chemotypes and co-existence of species. *Mycol. Res.***2005**, *109*, 410–420.
19. Roux, J.; Steenkamp, E.T.; Marasas, W.F.O.; Wingfield, M.J.; Wingfield, B.D. Characterization of *Fusarium graminearum*from Acacia and Eucalyptus using β-tubulin and histone gene sequences. *Mycologia* **2001**, *93*, 704–711.
20. Ward, T.J.; Clear, R.M.; Rooney, A.P.; O'Donnell, K.; Gaba, D.; Patrick, S.; Starkey, D.E.; Gilbert, J.; Geiser, D.M.; Nowicki, T.W. An adaptive evolutionary shift in *Fusarium* head blight pathogen populations is driving the rapid spread of more toxigenic *Fusarium graminearum* in North America. *Fungal Genet. Biol.* **2008**, *45*, 473–484.
21. Zhang, J.B.; Li, H.P.; Dang, F.J.; Qu, B.; Xu, Y.B.; Zhao, C.S.; Liao, Y.C. Determination of the trichothecene mycotoxin chemotypes and associated geographical distribution and phylogenetic species of the *Fusarium graminearum* clade from China. *Mycol.Res.* **2007**, *111*, 967–975.
22. Yang, L.; van der Lee, T.; Yang, X.; Yu, D.; Waalwijk, C. *Fusarium* populations on Chinese barley show a dramatic gradient in mycotoxin profiles. *Phytopathology* **2008**, *98*, 719–727.
23. Zhang, H.; Zhang, Z.; Van der Lee, T.; Chen, W.Q.; Xu, J.; Xu, J.S.; Yang, L.; Yu, D.; Waalwijk, C.; Feng, J. Population genetic analyses of *Fusarium asiaticum* populations from barley suggest a recent shift favoring 3ADON producers in southern China. *Phytopathology* **2010**, *100*, 328–336.
24. Zhang, Z.; Zhang, H.; Van der Lee, T.; Chen, W.Q.; Arens, P.; Xu, J.; Xu, J.S.; Yang, L.J.; Yu, D.Z.; Waalwijk, C.; Feng, J. Geographic substructure of *Fusarium asiaticum* isolates collected from barley in China. *Eur. J. Plant Pathol.* **2010**, *127*, 239–248.
25. Gale, L.R.; Harrison, S.A.; Ward, T.J.; O'Donnell, K.; Milus, E.A.; Gale, S.W.; Kistler, H.C. Nivalenol-type populations of *Fusarium graminearum* and *F. asiaticum* are prevalent on wheat in southern Louisiana. *Phytopathology* **2011**, *101*, 124–134.
26. Bowden, R.L.; Leslie, J.F. Diversity of *Gibberella zeae* at small spatial scales. *Phytopathology* **1994**, *84*, 1140.
27. Bowden, R.L.; Leslie, J.F. Sexual recombination in *Gibberella zeae*. *Phytopathology* **1999**, *89*, 182–188.
28. Carter, J.P.; Rezanoor, H.N.; Desjardins, A.E.; Nicholson, P. Variation in *Fusarium graminearum* isolates from Nepal associated with their host of origin. *Plant Pathol.* **2000**, *49*, 1–10.
29. Carter, J.P.; Rezanoor, H.N.; Holden, D.; Desjardins, A.E.; Plattner, R.D.; Nicholson, P. Variation in pathogenicity associated with the genetic diversity of *Fusarium graminearum*. *Eur. J. Plant Pathol.* **2002**, *108*, 573–583.
30. Gale, L.R.; Chen, L.F.; Hernick, C.A.; Takamura, K.; Kistler, H.C. Population analysis of *Fusarium graminearum* from wheat fields in eastern China. *Phytopathology* **2002**, *92*, 1315–1322.
31. Miedaner, T.; Schilling, A.G.; Geiger, H.H. Molecular genetic diversity and variation for aggressiveness in populations of *Fusarium graminearum* and *Fusarium culmorum* sampled from wheat fields in different countries. *J. Phytopathol.* **2001**,*149*, 641–648.
32. Ward, T.J.; Bielawski, J.P.; Kistler, H.C.; Sullivan, E.; O'Donnell, K. Ancestral polymorphism and adaptive evolution in the trichothecene mycotoxin gene cluster of phytopathogenic *Fusarium*. *Proc. Natl. Acad. Sci. USA* **2002**, *99*, 9278–9283.
33. Zeller, K.A.; Bowden, R.L.; Leslie, J.F. Diversity of epidemic populations of *Gibberella zeae* from small quadrats in Kansas and North Dakota. *Phytopathology* **2003**, *93*, 874–880.
34. Onyike, N.B.N.; Nelson, P.E. *Fusarium* species associated with millet grain from Nigeria, Lesotho, and Zimbabwe.*Mycologia* **1991**, *83*, 708–712.
35. Muthomi, J.W.; Ndung'u, J.K.; Gathumbi, J.K.; Mutitu, E.W.; Wagacha, J.M. The occurrence of*Fusarium* species and mycotoxins in Kenyan wheat. *Crop Prot.* **2008**, *27*, 1215–1219.
36. Wagacha, J.M.; Steiner, U.; Dehne, H.-W.; Zuehlke, S.; Spiteller, M.; Muthomi, J.; Oerke, E.-C. Diversity in mycotoxins and fungal species infecting wheat in Nakuru district, Kenya. *J. Phytopathol.* **2010**, *158*, 527–535.
37. Boutigny, A.-L.; Ward, T.J.; Van Coller, G.J.; Flett, B.; Lamprecht, S.C.; O'Donnell, K.; Viljoen, A. Analysis of the*Fusarium graminearum* species complex from wheat, barley, and maize in South Africa provides evidence of species-specific differences in host preference. *Fungal Genet. Biol.* **2011**, *48*, 914–920.
38. Zeller, K.A.; Bowden, R.L.; Leslie, J.F. Population differentiation and recombination in wheat scab

populations of*Gibberella zeae* from the United States. *Mol. Ecol.* **2004**, *13*, 563–571.

39. Geraldo, M.R.F.; Tessmann, D.J.; Kemmelmeier, C. Production of mycotoxins by *Fusarium graminearum* isolated from small cereals (wheat, triticale and barley) affected with scab disease in Southern Brazil. *Braz. J. Microbiol.* **2006**, *37*, 58–63.

40. Fernando, W.G.D.; Zhang, J.X.; Dusabenyagasani, M.; Guo, X.W.; Ahmed, H.; McCallum, B. Genetic diversity of*Gibberella zeae* isolates from Manitoba. *Plant Dis.* **2006**, *90*, 1327–1342.

41. Ramirez, M.L.; Reynoso, M.M.; Farnochi, M.C.; Torres, A.M.; Leslie, J.F.; Chulze, S.N. Population genetic structure of*Gibberella zeae* isolated from wheat in Argentina. *Food Addit. Contam.* **2007**, *24*, 1115–1120.

42. Gale, L.R.; Ward, T.J.; Balmas, V.; Kistler, H.C. Population subdivision of *Fusarium graminearum sensu stricto* in the Upper Midwestern United States. *Phytopathology* **2007**, *97*, 1434–1439.

43. Scoz, L.B.; Astolfi, P.; Reartes, D.S.; Schmale, D.G., III; Moraes, M.G.; Del Ponte, E.M. Trichothecene mycotoxin genotypes of *Fusarium graminearum sensu stricto* and *Fusarium meridionale* in wheat from Southern Brazil. *Plant Pathol.***2009**, *58*, 344–351.

44. Alvarez, C.L.; Azcarate, M.P.; Pinto, V.F. Toxigenic potential of *Fusarium graminearum sensu stricto* isolates from wheat in Argentina. *Int. J. Food Microbiol.* **2009**, *135*, 131–135.

45. Sampietro, D.A.; Marín, P.; Iglesias, J.; Presello, D.A.; Vatuone, M.A.; Catalan, C.A.N.; Gonzalez Jaen, M.T. A molecular based strategy for rapid diagnosis of toxigenic *Fusarium* species associated to cereal grains from Argentina.*Fungal Biol.* **2010**, *114*, 74–81.

46. Sampietro, D.A.; Díaz, C.G.; Gonzalez, V.; Vattuone, M.A.; Ploper, L.D.; Catalan, C.A.; Ward, T.J. Species diversity and toxigenic potential of *Fusarium graminearum* complex isolates from maize fields in northwest Argentina. *Int. J. Food Microbiol.* **2011**, *145*, 359–364.

47. Reynoso, M.M.; Ramirez, M.L.; Torres, A.M.; Chulze, S.N. Trichothecene genotypes and chemotypes in *Fusarium graminearum* strains isolated from wheat in Argentina. *Int. J. Food Microbiol.* **2011**, *145*, 444–448.

48. Astolfi, P.; Dos Santos, J.; Schneider, L.; Gomes, L.B.; Silva, C.N.; Tessmann, D.J.; Del Ponte, E.M. Molecular survey of trichothecene genotypes of *Fusarium graminearum* species complex from barley in southern Brazil. *Int. J. Food Microbiol.* **2011**, *148*, 197–201.

49. Astolfi, P.; Reynoso, M.M.; Ramirez, M.L.; Chulze, S.N.; Alves, T.C.A.; Tessmann, D.J.; Del Ponte, E.M. Genetic population structure and trichothecene genotypes of *Fusarium graminearum* isolated from wheat in southern Brazil.*Plant Pathol.* **2011**.

50. Gagkaeva, T.Y.; Yli-Mattila, T. Genetic diversity of *Fusarium graminearum* in Europe and Asia. *Eur. J. Plant Pathol.* **2004**,*110*, 550–562.

51. Ji, L.; Cao, K.; Wang, S. Determination of deoxynivalenol and nivalenol chemotypes of *Fusarium graminearum* from China by PCR assay. *J. Phytopathol.* **2007**, *155*, 505–512.

52. Suga, H.; Karugia, G.W.; Ward, T.; Gale, L.R.; Tomimura, K.; Nakajima, T.; Miyasaka, A.; Koizumi, S.; Kageyama, K.; Hyakumachi, M. Molecular characterization of the *Fusarium graminearum* species complex in Japan. *Phytopathology***2008**, *98*, 159–166.

53. Haratian, M.; Sharifnabi, B.; Alizadeh, A.; Safaie, N. PCR analysis of the *Tri13* gene to determine the genetic potential of *Fusarium graminearum* isolates from Iran to produce nivalenol and deoxynivalenol. *Mycopathologia* **2008**, *166*, 109–116.

54. Lee, J.; Chang, I.Y.; Kim, H.; Yun, S.H.; Leslie, J.; Lee, Y.W. Genetic diversity and fitness of *Fusarium graiminearum*populations from rice in Korea. *Appl. Environ. Microbiol.* **2009**, *75*, 3289–3295.

55. Karugia, G.W.; Suga, H.; Gale, L.R.; Nakajima, T.; Ueda, A.; Hyakumachi, M. Population structure of *Fusarium asiaticum* from two Japanese regions and eastern China. *J. Gen. PlantPathol.* **2009**, *75*, 110–118.

56. Karugia, G.W.; Suga, H.; Gale, L.R.; Nakajima, T.; Tomimura, K.; Hyakumachi, M. Population structure of the*Fusarium graminearum* species complex from a single Japanese wheat field sampled in two consecutive years. *Plant Dis.* **2009**, *93*, 170–174.

57. Lee, S.-H.; Lee, J.; Nam, Y.J.; Lee, S.; Ryu, J.-G.; Lee, T. Population Structure of *Fusarium graminearum* from maize and rice in 2009 in Korea. *Plant Pathol.* **2010**, *26*, 321–327.

58. Waalwijk, C.; Kastelein, P.; de Vries, I.; Kerenyi, Z.; van der Lee, T.; Hesselink, T.; Köhl, J.; Kema, G. Major changes in*Fusarium spp.* in wheat in the Netherlands. *Eur. J. Plant Pathol.* **2003**, *109*, 743–754.

59. Láday, M.; Juhász, Á.; Mulè, G.; Moretti, A.; Szécsi, Á.; Logrieco, A. Mitochondria DNA diversity and lineage determination of European isolates of *Fusarium graminearum* (*Gibberella zeae*). *Eur. J. Plant Pathol.* **2004**, *110*, 545–550.

60. Tóth, B.; Mesterházy, Á.; Horváth, Z.; Bartók, T.; Varga, M.; Varga, J. Genetic variability of central European isolates of the *Fusarium graminearum* species complex. *Eur. J. Plant Pathol.* **2005**, *113*, 35–45.

61. Sayer, S.T.; Lauren, D.R. *Fusarium* infection in New Zealand grain. *N. Z. J. Crop Hortic. Sci.* **1991**, *19*, 143–148.

62. Lauren, D.R.; Sayer, S.T.; Di Menna, M.E. Trichothecene production by *Fusarium* species isolated from grain and pasture throughout New Zealand. *Mycopathologia* **1992**, *120*, 167–176.

63. Akinsanmi, O.A.; Backhouse, D.; Simpfendorfer, S.; Chakraborty, S. Genetic diversity of Australian *Fusarium graminearum* and *F. pseudograminearum*. *Plant Pathol.* **2006**, *55*, 494–504.

64. Wright, D.G.; Thomas, G.J.; Loughman, R.; Fuso-Nyarko, J.; Bullock, S. Detection of *Fusarium graminearum* in wheat grains in Western Australia. *Australas. Plant Dis. Notes* **2010**, *5*, 82–84.

65. Taylor, J.W.; Jacobson, D.J.; Kroken, S.; Kasuga, T.; Geiser, D.M.; Hibbett, D.S.; Fisher, M.C. Phylogenetic species recognition and species concepts in Fungi. *Fungal Genet. Biol.* **2000**, *31*, 21–32.

66. Desjardins, A.E.; Proctor, R.H. Genetic diversity and trichothecene chemotypes of the *Fusarium graminearum* clade isolated from maize in Nepal and identification of a putative new lineage. *Fungal Biol.* **2011**, *115*, 38–48.

67. Jeon, J.J.; Kim, H.; Kim, H.S.; Zeller, K.A.; Lee, T.; Yun, S.H.; Bowden, R.L.; Leslie, J.F.; Lee, Y.W. Genetic diversity of*Fusarium graminearum* from maize in Korea. *Fungal Genet. Newsl.* **2003**, *50*, S142.

68. Miller, J.D.; Greenhalgh, R.; Wang, Y.Z.; Lu, M. Trichothecene chemotypes of three *Fusarium* species. *Mycologia* **1991**,*83*, 121–130.

69. Moss, M.O.; Thrane, U. *Fusarium* taxonomy with relation to trichothecene formation. *Toxicol. Lett.* **2004**, *153*, 23–28.

70. Goswami, R.S.; Kistler, H.C. Pathogenicity and in planta mycotoxin accumulation among members of the *Fusarium graminearum* species complex on wheat and rice. *Phytopathology* **2005**, *95*, 1397–1404.

71. Desjardins, A.E.; Busman, M.; Manandhar, G.; Jarosz, A.M.; Manandhar, H.K.; Proctor, R.H. Distribution of the mycotoxins nivalenol and deoxynivalenol in naturally and experimentally infected maize. *J. Agric. Food Chem.* **2008**,*56*, 5428–5436.

72. Jennings, P.; Coates, M.E.; Walsh, K.; Turner, J.A.; Nicholson, P. Determination of deoxynivalenol- and nivalenol-producing chemotypes of *Fusarium graminearum* isolated from wheat crops in England and Wales. *Plant Pathol.* **2004**,*53*, 643–652.

73. Talas, F.; Parzies, H.K.; Miedaner, T. Diversity in genetic structure and chemotype composition of *Fusarium graminearum sensu stricto* populations causing wheat head blight in individual fields in Germany. *Eur. J. Plant Pathol.***2011**, *1*, 39–48.

74. Pasquali, M.; Giraud, F.; Brochot, C.; Cocco, E.; Hoffmann, L.; Bohn, T. Genetic *Fusarium* chemotyping as a useful tool for predicting nivalenol contamination in winter wheat. *Int. J. Food Microbiol.* **2010**, *137*, 246–253.

75. Abramson, D.; Clear, R.M.; Smith, D.M. Trichothecene production by *Fusarium* spp. isolated from Manitoba grain.*Can. J. Plant Pathol.* **1993**, *15*, 147–152.

76. Abramson, D.; Clear, R.M.; Gaba, D.; Smith, D.M.; Patrick, S.K.; Saydak, D. Trichothecene and moniliformin production by *Fusarium* species from western Canadian wheat. *J. Food Prot.* **2001**, *64*, 1220–1225.

77. Guo, X.W.; Fernando, W.G.D.; Seow-Brock, H.Y. Population structure, chemotype diversity, and potential chemotype shifting of *Fusarium graminearum* in wheat fields of Manitoba. *Plant Dis.* **2008**, *92*, 756–762.

78. Schmale, D.G., III; Wood-Jones, A.K.; Cowger, C.; Bergstrom, G.C.; Arellano, C. Trichothecene genotypes of *Gibberella zeae* from winter wheat fields in the eastern USA. *Plant Pathol.* **2011**.

79. Burlakoti, R.R.; Ali, S.; Secor, G.A.; Neate, S.M.; McMullen, M.P.; Adhikari, T.B. Genetic relationships among populations of *Gibberella zeae* from barley, wheat, potato, and sugar beet in the Upper Midwest of the United States.*Phytopathology* **2008**, *98*, 969–976.

80. Schmale, D.G., III; Leslie, J.F.; Zeller, K.A.; Saleh, A.A.; Shields, E.J.; Bergstrom, G.C. Genetic structure of atmospheric populations of *Gibberella zeae*. *Phytopathology* **2006**, *96*, 1021–1026.

81. Qu, B. Genetic Diversity of *Fusarium Graminearum* in China and its Comparison with the Isolates from of Nepal, Europe, and USA. PhD thesis, Hauzhong Agricultural University, Wuhan, China, September 2002.

82. Narayanasamy, P. *Molecular Biology in Plant Pathogenesis and Disease Management: Microbial Plant Pathogens*, 1st ed; Springer: New York, NY, USA, 2008; Volume 1, pp. 175–178.

83. Dusabenyagasani, M.; Dostaler, D.; Hamelin, R.C. Genetic diversity among *Fusarium graminearum* strains from Ontario and Quebec. *Can. J. Plant Pathol.* **1999**, *21*, 308–314.

84. Liu, W.-C.; Xi, J.-H.; Li, H.-Y.; Pan, H.-Y.; Hu, H.-Q.; Guo, Y.-L.; Bai, R.-L. RAPD analysis of isolates from *Fusarium*spp. causing wheat head blight in northeast China. *Mycosystema* **2002**, *21*, 63–70.

85. Quellet, T.; Seifert, K.A. Genetic characterization of *Fusarium graminearum* strains using RAPD and PCR amplification.*Phytopathology* **1993**, *83*, 1003–1007.

86. Busso, C.; Kaneshima, E.N.; De Assis Franco, F.; Querol, C.B.; De Castro-Prado, M.A. Vegetative compatibility and molecular characterization of *Fusarium graminearum* isolates from the State of Paraná, Brazil. *Cienc. Rural.* **2007**, *37*, 1813–1816.

87. Riedy, M.F.; Hamilton, W.J.; Aquadro, C.F. Excess non-parental bands in offspring from known pedigrees assayed using RAPD PCR. *Nucleic Acid Res.* **1992**, *20*, 918.

88. Williams, J.G.; Kubelik, A.R.; Livak, K.J.; Rafalski, J.A.; Tingey, S.V. DNA polymorphisms amplified by arbitrary primers are useful as genetic markers. *Nucleic Acids Res.* **1990**, *18*, 6531–6535.

89. Blears, M.J.; de Grandis, S.A.; Lee, H.; Trevors, J.T. Amplified fragment length polymorphism (AFLP): A review of the procedure and its applications. *J. Ind. Microbiol. Biotechnol.* **1998**, *21*, 99–114.

90. Leissner, C.E.W.; Niessen, M.L.; Vogel, R.F. Use of the AFLP technique for the identification and discrimination of*Fusarium graminearum*. *Cer. Res. Com.* **1997**, *25*, 555–556.

91. Abd-Elsalam, K.A.; Schnieder, F.; Verreet, J.A. Population analysis of *Fusarium* species. *Phytomedizin* **2002**, *3*, 18–19.

92. Kiprop, E.K.; Baudoin, J.P.; Mwangombe, A.W.; Kimani, P.M.; Mergeai, G.; Maquet, A. Characterization of Kenyan isolates of *Fusarium udum* from Pigeonpea [*Cajanus cajan* (L.) Millsp.] by cultural characteristics, aggressiveness and AFLP analysis. *J. Phytopathol.* **2002**, *150*, 517–527.

93. Sivaramakrishnan, S.; Kannan, S.; Singh, S.D. Genetic variability of *Fusarium* wilt pathogen isolates of chickpea (*Cicer arietinum* L.) assessed by molecular markers. *Mycopathologia* **2002**, *155*, 171–178.

94. Bowden, R.L.; Zeller, K.A.; Leslie, J.F. Population structure of *Gibberella zeae* in the Great Plains of North America. In Proceedings of the International Symposium on Wheat Improvement for Scab Resistance, Suzhou and Nanjing, Jiangsu, China, 2000; pp. 211–213.

95. Majer, D.; Lewis, B.G.; Mithen, R. Genetic variation among field isolates of *Pyrenopeziza brassicae*. *Plant Pathol.* **1998**, *47*, 22–28.

96. O'Donnell, K.; Sutton, D.A.; Rinaldi, M.G.; Magnon, K.C.; Cox, P.A.; Revankar, S.G.; Sanche, S.; Geiser, D.M.; Juba, J.H.; van Burik, J.-A.H.; Padhye, A.; Anaissie, E.J.; Francesconi, A.; Walsh, T.J.; Robinson, J.S. Genetic diversity of human pathogenic members of the *Fusarium oxysporum* complex inferred from multilocus DNA sequence data and amplified fragment length polymorphism analyses: Evidence forthe recent dispersion of a geographically widespread clonal lineage and nosocomial origin. *J. Clin. Microbiol.* **2004**, *42*, 5109–5120.

97. Li, G.; Quiros, C.F. Sequence related amplified polymorphism (SRAP), a new system based on a simple PCR reaction: Its application to mapping and gene tagging in *Brassica*. *Theor. Appl. Genet.* **2001**, *103*, 455–461.

98. Mishra, P.K.; Tewari, J.P.; Clear, R.M.; Turkington, T.K. Molecular genetic variation and geographical structuring in*Fusarium graminearum*. *Ann. Appl. Biol.* **2004**, *145*, 299–307.

99. Cuomo, C.A.; Güldener, U.; Xu, J.R.; Trail, F.; Turgeon, B.G.; Pietro, A.D.; Walton, J.D.; Ma, L.J.; Baker, S.E.; Rep, M.; *et al.* The *Fusarium graminearum* genome reveals a link between localized polymorphism and pathogen specialization.*Science* **2007**, *317*, 1400–1402.

100. Burlakoti, R.R.; Neate, S.M.; Adhikari, T.B.; Gyawali, S.; Salas, B.; Steffenson, B.J.; Schwarz, P.B. Trichothecene profiling and population genetic analysis of *Gibberella zeae* from barley in North Dakota and Minnesota. *Phytopathology* **2011**, *101*, 687–695.

101. Kimura, M.; Tokai, T.; Takahashi-Ando, N.; Ohsato, S.; Fujimura, M. Molecular and genetic studies of*Fusarium*trichothechne biosynthesie: Pathways, genes, and evolution. *Biosci. Biotechnol. Biochem.* **2007**, *71*, 2105–2123.

102. Suga, H.; Gale, L.R.; Kistler, H.C. Development of VNTR markers for two *Fusarium graminearum* clade species. *Mol. Ecol. Notes* **2004**, *4*, 468–470.

103. Giraud, T.; Fournier, E.; Vautrin, D.; Solignac, M.; Vercken, E.; Bakan, B.; Brygoo, Y. Isolation of eight polymorphic microsatellite loci, using an enrichment protocol, in the phytopathogenic fungus *Fusarium culmorum*. *Mol. Ecol. Notes***2002**, *2*, 121–123.

104. Francis, R.G.; Burgess, L.W. Characteristics of two populations of *Fusarium roseum* " *Graminearum*" in eastern Australia. *Trans. Br. Mycol. Soc.* **1977**, *68*, 421–427.

105. Benyon, F.H.L.; Burgess, L.W.; Sharp, P.J. Molecular genetic investigations and reclassification of *Fusarium* species in sections *Fusarium* and *Roseum*. *Mycol. Res.* **2000**, *104*, 1164–1174.

106. Aoki, T.; O'Donnell, K. Morphological and molecular characterization of *Fusarium pseudograminearum* sp. nov., formerly recognised as the group 1 population of *F. graminearum*. *Mycologia* **1999**, *91*, 597–609.

107. Láday, M.; Bagi, F.; Mesterházy, A.; Szécsi, Á. Isozyme evidence for two groups of *Fusarium graminearum*. *Mycol. Res.***2000**, *104*, 788–793.

CHAPTER 18 MOLECULAR DEFENSE RESPONSES IN ROOTS AND THE RHIZOSPHERE AGAINST FUSARIUM OXYSPORUM

Yi Chung Chen, Brendan N Kidd, Lilia C Carvalhais and Peer M Schenk

Keywords: Arabidopsis, Fusarium oxysporum, plant defense, transcription factor

Biotic stresses in the roots and rhizophere include plant disease-causing pathogens, such as fungi, oomycetes, bacteria and viruses, as well as insects, nematodes, other herbivores and parasitic plant interactions. Fungal pathogens can be simply classified into categories, biotrophic or necrotrophic, depending on how they acquire nutrients from the host plant. Many biotrophs live in the intercellular region between mesophyl cells, and some produce haustoria as a feeding structure which allows them to absorb nutrients from the host plant while evading recognition or defense responses. Salicylic acid (SA)-, reactive oxygen species (ROS)- and localized programmed cell death, known as the hypersensitive response, provide an effective mechanism to isolate infection with obligate biotrophic pathogens.3 In contrast, necrotrophic pathogens produce enzymes to degrade cells and often toxins to kill host tissues to absorb nutrients. Plants are able to effectively defend against these pathogens via the jasmonic acid (JA) and ethylene (ET) pathways that produce pathogenesisrelated (PR)-proteins and secondary metabolites while suppressing ROS production (Fig. 1). However, it is important to note that some pathogens may behave as both a necrotroph and a biotroph depending on the different stage of the infection cycle. These hemi-biotrophic pathogens typically act as biotrophs at an early stage, feeding on living cells and establishing infection before shifting to a necrotrophic phase to complete their life cycle.

Fusarium oxysporum is a hemi-biotrophic root pathogen that infects a number of plants, including cotton, tomato, banana and Arabidopsis. F. oxysporum is believed to be able to manipulate defense pathways, and previous research has identified that some Arabidopsis-infecting isolates hijack the JA pathway, a pathway typically used for induced systemic resistance (ISR) by beneficial rhizophere microbes or against necrotrophic pathogens or insects. Impaired JA-signaling mutants, myc2, coi1 and pft1/med5, although more susceptible to other pathogens, showed strong resistance against F. oxysporum coi1-mediated F. oxysporum resistance was independent of both JA- and SA- dependent defense gene expression, but correlated with compromised nondefensive aspects of JA-dependent responses in this mutant, suggesting that F. oxysporum hijacks non-defensive aspects of the JAsignaling pathway to enhance disease. Pantelides et al. (2013) also suggested that F. oxysprourn hijacks ETR1-mediated ET signaling to promote disease development in Arabidopsis ETR1 is an ET receptor and acts upstream of the ET pathway regulator CTR1.12 ET and JA work tightly together to confer resistance against necrotrophic pathogens.

Citation: Chung Chen, Brendan N Kidd, Lilia C Carvalhais & Peer M Schenk (2014) Molecular defense responses in roots and the rhizosphere against Fusarium oxysporum, Plant Signaling & Behavior, 9:12, e977710, DOI: 10.4161/15592324.2014.977710

However, many F. oxysporum strains live in the soil and often colonize roots as endophytes without causing damage or even provide benefits against pathogens. A wealth of literature has reported that Fusarium wilt can be controlled by non-pathogenic strains of F. oxysporum in numerous crops, such as banana, asparagus, basil, chickpea, watermelon, cyclamen, tomato, spinach, cucumber, and melon (for a comprehensive review see15). Although there are a few reports that these protective F. oxysporum strains were effective against other pathogens including Meloidogyne incognita, Phytophtora capsici, and Pythium ultimum, they generally exhibit biological control attributes against pathogenic F. oxysporum. However, other well-known biocontrol agents against F. oxysporum, including Trichoderma and Pseudomonas, inhibit a wider range of pathogens. Recent studies have shown that plants are able to recruit certain microbes in their rhizosphere during defense signaling. These "soldier" microbes are known to produce antimicrobial and insecticidal compounds and differ from those under normal growth condition where most beneficial microbes assist with nutrient supply and plant growth promotion. It would interesting to study whether protective F. oxysporum strains can be actively recruited during plant defense signaling.

Apart from using biological control agents, genetic manipulation to increase resistance to F. oxysporum has made progress on several fronts. It was already mentioned above that desensitizing plants for JA signaling provided resistance in Arabidopsis. McGrath et al. (2005) identified a transcription factor-encoding gene, ERF4, which is a negative defense gene regulator controlled by JA and fungal pathogens. Inactivation of ERF4 showed an increase resistance to F. oxysporum. Recently, Chen et al. (2014) reported that erf72 Arabidopsis mutants showed increased resistance to F. oxysporum. ERF72 has a role in ROS response, but erf72 plants did not show a noticeable altered oxidative burst in the root region. Chen et al. (2014) also examined the pub22/23/24 U-box type E3 ubiqutin ligase triple mutant, which is known to possess enhanced ROS production in response to pathogen challenge. However pub 22/23/24 mutants showed increased resistance to F. oxysporum, suggesting that ROS homeostasis and heightened innate immune responses provide protection against soil-borne F. oxysporum.

CONCLUSIONS

Although soil-borne diseases result in significant yield losses world-wide, relatively little is known how plants are able to defend themselves against these pathogens. Root-specific defense gene expression, structural responses, secondary metabolite and root exudate production in the rhizosphere may all assist in fighting off soil-borne pathogens and pests, and perhaps also promote colonization of plant defense–assisting beneficial microbes. Soil-born *F. oxysporum* appears to be an "accidental" pathogen that disguises as a beneficial microbe to activate the JA pathway to gain entry into the cell as a hemibiotrophic pathogen. Further studies of the early steps leading to *F. oxysporum* colonization of root cells are needed to elucidate the underlying mechanisms of recognition and early signaling during this interesting plant-microbe interaction.

REFERENCES

1. Vogele TT, Mendgen K. Rust haustoria: nutrient uptake and beyond. New Phytol 2003; 159:93-100; http://dx.doi.org/10.1046/j.1469-8137.2003.00761.x
2. Ellis J, Catanzariti AM, Dodds P. The problem of how fungal and oomycete avirulence proteins enter plant cells. Trends Plant Sci 2006; 11:61-3; PMID:16406302; http://dx.doi.org/10.1016/j.tplants.2005.12.008
3. Pieterse CM, Leon-Reyes A, Van der Ent S, Van Wees SC. Networking by small-molecule hormones in plant immunity. Nat Chem Biol 2009; 5:308-16; PMID:19377457; http://dx.doi.org/10.1038/nchembio.164
4. Agrios GN. Plant Pathology. Fifth edition, Amsterdam: Elsevier Academic Press, 2005.
5. Brown JF, Ogle HJ. Plant Pathogens and Plant Diseases. Amidale: Rockvale Publications, 1997.
6. Berrocal-Lobo M, Molina A. Arabidopsis defense response against Fusarium oxysproum. Trends Plant Sci 2008; 13:145-50; PMID:18289920; http://dx.doi.org/10.1016/j.tplants.2007.12.004
7. Thatcher L, Manners JM, Kazan K. Fusarium oxysporum hijacks COI1-mediated jasmonate signaling to promote disease development in Arabidopsis. Plant J. 2009; 58:927-39; PMID:19220788; http://dx.doi.org/10.1111/j.1365-313X.2009.03831.x
8. Cole SJ, Yoon AJ, Faull KF, Diener AC. Host perception of jasmonates promotes infection by Fusarium oxysporum formae speciales that produce isoleucine- and leucine-conjugated jasmonates. Mol Plant Pathol 2014; 15:589-600; PMID:24387225; http://dx.doi.org/10.1111/mpp.12117
9. Anderson JP, Badruzsaufari E, Schenk PM, Manners JM, Desmond OJ, Ehlert C,Maclean DJ, Ebert PR, Kazan K. Antagonistic interaction between abscisic acid and jasmonate-ethylene signaling pathways modulates defense gene expression and disease resistance in Arabidopsis. Plant Cell 2004; 16:3460-79; PMID:15548743; http://dx.doi.org/10.1105/tpc.104.025833
10. Kidd B, Edgar C, Kumar K, Aitken EA, Schenk PM, Manners JM, Kazan K. The mediator complex subunit PFT1 is a key regulator of jasmonate-dependent defense in Arabidopsis. Plant Cell 2009; 21:2237-52; PMID:19671879; http://dx.doi.org/10.1105/tpc.109.066910
11. Pantelides IS, Tjamos SE, Pappa S, Kargakis M, Paplomatas EJ. The ethylene receptor ETR1 is required for Fusarium oxysporum pathogenicity. Plant Pathol 2013; 62:1302-09; PMID:21813131; http://dx.doi.org/10.1111/ppa.12042
12. Guo HW, Ecker JR. The ethylene signaling pathway: new insights. Curr Opin Plant Biol2004; 7:40-9; PMID:14732440; http://dx.doi.org/10.1016/j.pbi.2003.11.011
13. Lorenzo O, Piqueras R, Sanchez-Serrano JJ, Solano R. ETHYLENE RESPONSE FACTOR1 integrates signals from ethylene and jasmonate pathways in plant defense.

Plant Cell2003; 15:165-78; PMID:12509529; http://dx.doi.org/10.1105/tpc.007468

14. Swarupa V, Ravishankar K, Rekha A. Plant defense response against Fusarium oxysporum and strategies to develop tolerant genotypes in banana. Planta 2014; 239:735-51; In press; PMID:24420701; http://dx.doi.org/10.1007/s00425-013-2024-8

15. Alabouvette C, Olivain C, Migheli Q, Steinberg C. Microbiological control of soil-borne phytopathogenic fungi with special emphasis on wilt-inducing Fusarium oxysporum. New Phytol 2009; 184:529-44; PMID:19761494; http://dx.doi.org/10.1111/j.1469-8137.2009.03014.x

16. Benhamou N, Garand C, Goulet A. Ability of non-pathogenic Fusarium oxysporum Fo47 to induce resistance against Pythium ultimum infection in cucumber. Appl Env Microbiol 2002; 68:4044-60; PMID:12147506; http://dx.doi.org/10.1128/AEM.68.8.4044-4060.2002

17. Silvar C, Merino F, Diaz J. Resistance in pepper plants induced by Fusarium oxysporum f. sp. lycopersici involves different defence-related genes. Plant Biol 2009; 11:68-74; PMID:19121115; http://dx.doi.org/10.1111/j.1438-8677.2008.00100.x

18. Dababat AEFA, Sikora RA. Induced resistance by the mutualistic endophyte, Fusarium oxysporum strain 162, toward Meloidogyne incognita on tomato. Biocontr Sci Technol 2007; 17:969-75; http://dx.doi.org/10.1080/09583150701582057[Taylor & Francis Online], [Web of Science ®]

19. Carvalhais LC, Dennis PG, Badri DV, Tyson GW, Vivanco JM, Schenk PM. Activation of the jasmonic acid plant defence pathway alters the composition of rhizosphere bacterial communities. PLoS One 2013; 8:e56457; PMID:23424661; http://dx.doi.org/10.1371/journal.pone.0056457

20. McGrath KC, Dombrecht B, Manners JM, Schenk PM, Edgar CI, Maclean DJ, ScheibleWR, Udvardi MK, Kazan K. Repressor-and activator-type ethylene response factors functioning in jasmonate signalling and resistance identified via a genome-wide screen of Arabidopsis transcription factor gene expression. Plant Physiol 2005; 139:949-59; PMID:16183832; http://dx.doi.org/10.1104/pp.105.068544

21. Chen YC, Wong CL, Muzzi F, Vlaardingerbroek I, Kidd BN, Schenk PM. Root defense analysis against Fusarium oxysporum reveals new regulators confer resistance. Sci Rep 2014; 4:5584; PMID:24998294[PubMed], [Web of Science ®]

22. Pan L, Kawai M, Yu LH, Kim KM, Hirata A, Umeda M, Uchimiya H. The Arabidopsis thaliana ethylene-responsive element binding protein (AtEBP) can function as a dominant suppressor of Bax-induced cell death of yeast. FEBS Letts 2001; 508:375-8; PMID:11728455; http://dx.doi.org/10.1016/S0014-5793(01)03098-8

23. Ogawa T, Pan LH, Kawai-Yamada M, Yu L, Yamamura S, Koyama T, Kitajima S, Ohme-Takagi M, Sato F, Uchimiya H. Functional analysis of Arabidposis ethylene-responsive elemen binding protein conferring resistance to Bax and abiotic stress-induced plant cell death. Plant Physiol 2005; 138:1436-45; PMID:15980186; http://dx.doi.org/10.1104/pp.105.063586

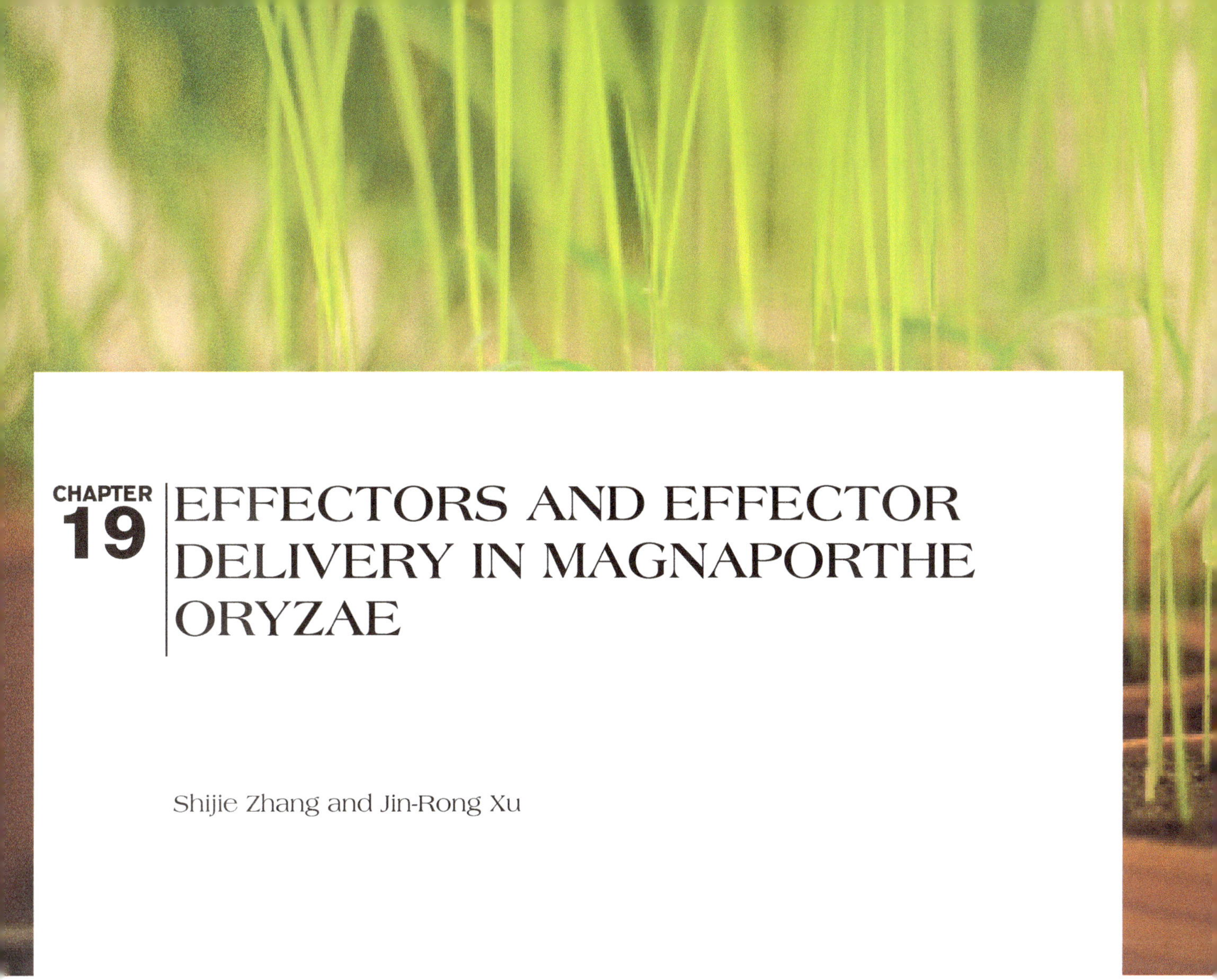

CHAPTER 19

EFFECTORS AND EFFECTOR DELIVERY IN MAGNAPORTHE ORYZAE

Shijie Zhang and Jin-Rong Xu

Rice blast caused by *Magnaporthe oryzae* is one of the most destructive fungal diseases of rice and a model for studying fungal-plant interactions. The fungus penetrates plant cells with appressoria and develops the narrow primary invasive hyphae (IH) and, subsequently, the bulbous secondary IH. As a hemibiotrophic pathogen, biotrophic IH are enclosed in the extra-invasive-hyphal membrane (EIHM) produced by the plant cells [1]. Like many other fungal pathogens [2], *M. oryzae* secretes effector proteins to manipulate plant immunity and physiology to promote infection.

AVR EFFECTORS

Avr proteins, a special class of effectors encoded by the avirulence (*AVR*) genes, could be recognized by corresponding R proteins and lead to the race-specific recognition [3]. In the rice–*M. oryzae* interaction, over 40 *AVR* genes have been identified. Among the *AVR* genes that have been cloned, all but *ACE1* encode secreted proteins expressed in IH. *ACE1* is specifically expressed in appressoria, and it encodes an intracellular hybrid PKS-NRPS protein[4]. Avr-Piz-t and Avr-Pita are the other two Avr proteins with known biochemical functions. Avr-Piz-t functions to suppress pathogen-associated molecular pattern (PAMP)-triggered immunity by inhibiting the ubiquitin ligase activity of the rice RING E3 ubiquitin ligase APIP6 [5]. *AVR-Pita*encodes a putative neutral zinc metalloprotease [6] and it belongs to a gene family with at least two additional members [7].

PWL1, *PWL2*, *PWL3*, and *PWL4* are members of another *AVR* gene family [8]. Pwl effectors are small, glycine-rich proteins that are present in rice pathogens and function as avirulence proteins in infection of weeping lovegrass and finger millet. In contrast, *AVR1*-CO39 was cloned from a weeping lovegrass isolate. Its specific expression in IH triggers hypersensitive response (HR) and resistance in cultivars carrying the Pi-CO39 R gene [9]. Avr-Pia, Avr-Pik/km/kp, and Avr-Pii were identified in the same re-sequencing study of strain Ina168 [10]. Avr-Pia directly interacts with Rga5-A, which also interacts with Avr-CO39 [11].

NON-AVR EFFECTORS

The best characterized non-Avr effector in *M. oryzae* is Slp1, a secreted LysM protein that is accumulated at the interface between IH and EIHM [12]. Slp1 is dispensable for appressorium penetration but required for invasive growth *in planta*. It competes with the chitin elicitor binding protein CEBiP for binding to chitin oligosaccharides. Thus, Slp1

Citation: Zhang S, Xu J-R (2014) Effectors and Effector Delivery in Magnaporthe oryzae. PLoS Pathog 10(1): e1003826. doi:10.1371/journal.ppat.1003826

functions to suppress chitin-induced plant immune responses, including generation of reactive oxygen species and expression of defense-related genes [12].

Using an expression profiling approach, a number of low molecular weight secreted proteins specifically expressed or highly induced in biotrophic invasive hyphae were identified, including 58 candidate effectors that were up-regulated over 10-fold during plant infection. Four of them,*BAS1–BAS4*, were confirmed to be fungal biotrophy-associated secreted (BAS) proteins [13]. MC69 was identified by systematic disruption of *in planta* expressed secreted protein genes[14]. The 54-aa Mc69 protein was essential for IH development, although it was not translocated into rice cytoplasm. In a separate study, five secreted protein genes named*MoCDIP1–5* were found to induce plant cell death in a transient expression assay with rice protoplasts. Four of them also induced cell death in *Nicotiana benthamiana* [15].

LOCALIZATION OF FUNGAL EFFECTORS IN PLANT CELLS

M. oryzae effectors can be divided into two distinct types based on their localization in plant cells (Figure 1) [16]. Cytoplasmic effectors, including Avr-Pita, Pwl1, Pwl2, Bas2, and Avr-Piz-T, are preferentially accumulated in the biotrophic interfacial complex (BIC) before being delivered into plant cells. The BIC is a distinct plant-derived, membrane-rich structure developed at the tip of primary IH by *M. oryzae*. In each newly invaded rice cell, effectors are first secreted into BICs before delivery. The BICs are persistent and left behind when the primary IH differentiates into the secondary IH. In addition, the fungus continues to secrete effectors into BICs even after IH have grown extensively as pseudohyphae and invaded neighboring plant cells [17].

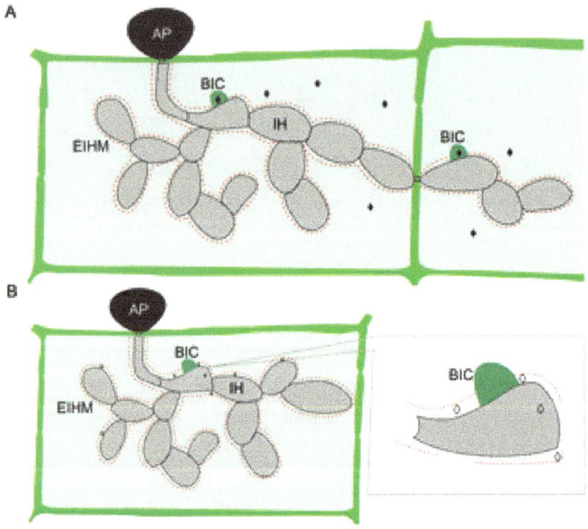

Figure 1: Localization of *M. oryzae* **effectors during plant infection. A.** Cytoplasmic effectors (♦) are secreted into the biotrophic interfacial complex (BIC) before being translocated into plant cells. **B.** Apoplastic effectors (◊) are secreted into the space between the fungal cell wall and extra-invasive-hyphal membrane (EIHM). Like IH, effector proteins may move from cell to cell via plasmodesmata. AP, appressorium.

On the other hand, apoplastic effectors such as Bas4, Avr1-CO39, and Slp1 are not associated with the BIC. After secretion, they are dispersed in the extracellular space between the fungal cell wall and EIHM. In the rice cells colonized by transformants expressing the Bas4-GFP fusion protein, GFP signals appeared to outline the IH [13], consistent with its extracellular localization. To date, it is not clear how the apoplastic effectors are recognized by surface receptors or translocated into plant cells to interact with their intracellular targets. Furthermore, no specific protein motifs or sequences have been identified in the cytoplasmic or apoplastic effectors that are responsible for their localization in plant cells after being secreted. Therefore, it is impossible to predict whether an effector is apoplastic or cytoplasmic solely based on its amino acid sequence.

EFFECTOR SECRETION SYSTEMS

Consistent with two types of effectors, two distinct effector secretion systems have been identified in *M. oryzae* (Figure 2) [16]. Cytoplasmic effectors are delivered into plant cells via the BIC, which is independent of the Golgi-dependent secretory system. Instead, secretion into the BIC is associated with a novel form of secretion involving the exocyst complex and t-SNAREs. Targeted deletions of the exocyst components *SEC5* and *EXO70* resulted in impaired secretion of cytoplasmic effectors and pathogenicity defects but had no effect on the secretion of apoplastic effectors [16]. Mutants deleted of the t-SNARE component *SSO1* were defective in BIC development and pathogenesis [16].

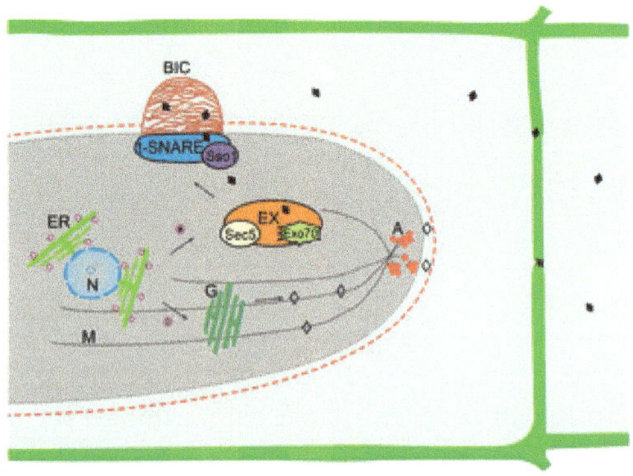

Figure 2: Two effector secretion mechanisms identified in *M. oryzae*. The apoplastic effectors (◊) utilize the conventional ER-Golgi secretion pathway for secretion. Disruption of the cytoskeleton by benzimidazoleand LatA treatments interferes with the secretion of apoplastic effectors. In contract, the secretion of cytoplasmic effectors (♦) into the BIC is independent of the Golgi bodies and cytoskeleton but involves the exocyst complex and t-SNAREs. A, cortical actin; ER, endoplasmic reticulum; EX, exocyst complex; G, Golgi bodies; M, microtubules; N, Nucleus.

The apoplastic effectors are secreted from IH by the conserved ER to Golgi secretory pathway that is independent of the BIC. Treatment with Brefeldin A that interferes with Golgi-dependent secretion inhibited the secretion of apoplastic effectors such as Bas4 and Slp1 but had no effect on the localization of cytoplasmic effectors Pwl2, Bas1, and Bas107 to the BIC [16]. Furthermore, actin and microtubules essential for vesicle trafficking are required only for the secretion of apoplastic effectors but not for cytoplasmic effectors. Many genes involved in the ER to Golgi secretory pathway and post-translational modifications also are likely important for the secretion of apoplastic effectors in *M. oryzae*. For example, the *LHS1* ER chaperone gene is required for the translocation and secretion of apoplastic effectors such as Slp1 [18].

TRANSLOCATION OF EFFECTORS INTO PLANT CELLS

Several cytoplasmic effectors, including Pwl2, Bas1, and Avr-Piz-t, are translocated into the cytoplasm of rice cells [5], [13]. Whereas some apoplastic effectors may be recognized by host plant surface receptors, the others that have intracellular targets, such as Avr1-CO39, also enter plant cells. In rice leaf sheath cells penetrated by fungal transformants expressing the Pwl2 and Avr-Piz-t proteins tagged with NLS and mRFP sequences, red fluorescence was observed in the nucleus of plant cells [17]. However, the underlying molecular mechanism responsible for effector translocation into plant cells is not clear. In experiments with purified recombinant proteins, the native Avr1-CO39 protein was translocated into rice protoplasts, indicating that it can enter plant cells independent of fungal factors [19]. In Oomycete pathogens such as *Phytophthora* species, the RXLR sequence is important for the translocation of effector proteins into plant cells [20]. Nevertheless, there is no evidence for the presence of functional RXLR sequences in *M. oryzae* effector proteins.

MOVEMENT OF EFFECTORS IN PLANT CELLS

In the rice leaf sheath cells penetrated by *M. oryzae*, the cytoplasmic effectors Pwl2 and Bas1, but not apoplastic effector Bas4, were moved to neighboring cells ahead of the invasive hyphae[17]. Red fluorescence was observed in the nucleus of rice leaf sheath cells penetrated by a transformant expressing the *PWL2*-mCherry-NLS fusion construct and a number of surrounding cells without IH. Fungal effectors entering un-colonized plant cells may be able to suppress host defense responses or elicit susceptibility. In fact, it takes a shorter time (2 h) for IH to move through the subsequently invaded cells than in the first colonized cells (12 h) [17].

Interestingly, movement of cytoplasmic effectors from vein-associated cells into neighboring cells was rare and of low efficiency in comparison with effector movement among regular epidermal cells [17], suggesting that effector trafficking depends on rice cell types. In addition, unlike the 39.3 kD Pwl2-mCherry fusion, the 68.3 kD Pwl2-tdTomato protein was defective in the movement from the penetrated cells to surrounding cells [17]. Therefore, cell-to-cell effector translocation also depends on the protein size. These results indicate that *Magnaporthe*effectors may be moved symplastically through plasmodesmata [17]. The movement of effector proteins and invasion of neighboring cells by IH may be co-regulated because IH of *M. oryzae*preferentially contacted and crossed the plant cell wall at the pit fields [1].

Although more and more fungal effectors are being discovered, our understanding of effector delivery and cell-to-cell movement *in planta* is relatively limited in comparison with that of bacterial effectors. Studies in *M. oryzae* showed that some effectors may be secreted via an unconventional protein secretion system to the BIC [16]. The BIC-like structures may also exist in other plant pathogenic fungi, particularly other hemibiotrophic pathogens. To date, no common plant entry sequence has been identified in fungal effectors, indicating that fungi may utilize a variety of mechanisms for effector translocation. *M. oryzae* and other fungal pathogens may have conserved mechanisms to recognize plasmodesmata for the movement of effector proteins and spreading of IH.

REFERENCES

1. Kankanala P, Czymmek K, Valent B (2007) Roles for rice membrane dynamics and plasmodesmata during biotrophic invasion by the blast fungus. Plant Cell 19: 706–724. doi: 10.1105/tpc.106.046300
2. Djamei A, Kahmann R (2012) Ustilago maydis: Dissecting the molecular interface between pathogen and plant. PLoS Pathog 8: e1002955 doi:10.1371/journal.ppat.1002955.
3. de Wit P, Mehrabi R, van den Burg HA, Stergiopoulos I (2009) Fungal effector proteins: past, present and future. Mol Plant Pathol 10: 735–747. doi: 10.1111/j.1364-3703.2009.00591.x
4. Bohnert HU, Fudal I, Dioh W, Tharreau D, Notteghem JL, et al. (2004) A putative polyketide synthase peptide synthetase from Magnaporthe grisea signals pathogen attack to resistant rice. Plant Cell 16: 2499–2513. doi: 10.1105/tpc.104.022715
5. Park C, Chen S, Shirsekar G, Zhou B, Khang C, et al. (2012) The Magnaporthe oryzae effector Avrpiz-t targets the ring E3 ubiquitin ligase apip6 to suppress pathogen-associated molecular pattern-triggered immunity in rice. Plant Cell 24: 4748–4762. doi: 10.1105/tpc.112.105429
6. Jia Y, McAdams SA, Bryan GT, Hershey HP, Valent B (2000) Direct interaction of resistance gene and avirulence gene products confers rice blast resistance. EMBO J 19: 4004–4014. doi: 10.1093/emboj/19.15.4004
7. Khang CH, Park S-Y, Lee Y-H, Valent B, Kang S (2008) Genome organization and evolution of the AVR-Pita avirulence gene family in the Magnaporthe grisea species complex. Mol Plant Microbe Interact 21: 658–670. doi: 10.1094/mpmi-21-5-0658
8. Kang SC, Sweigard JA, Valent B (1995) The PWL host specificity gene family in the blast fungus Magnaporthe grisea. Mol Plant Microbe Interact 8: 939–948. doi: 10.1094/mpmi-8-0939
9. Peyyala R, Farman ML (2006) Magnaporthe oryzae isolates causing gray leaf spot of perennial ryegrass possess a functional copy of the AVR1-CO39 avirulence gene. Mol Plant Pathol 7: 157–165. doi: 10.1111/j.1364-3703.2006.00325.x
10. Yoshida K, Saitoh H, Fujisawa S, Kanzaki H, Matsumura H, et al. (2009) Association genetics reveals three novel avirulence genes from the rice blast fungal pathogen Magnaporthe oryzae. Plant Cell 21: 1573–1591. doi: 10.1105/tpc.109.066324
11. Cesari S, Thilliez G, Ribot C, Chalvon V, Michel C, et al. (2013) The rice resistance protein pair RGA4/RGA5 recognizes the Magnaporthe oryzae effectors Avr-Pia and Avr1-CO39 by direct binding. Plant cell 25: 1463–1481. doi: 10.1105/tpc.112.107201
12. Mentlak TA, Kombrink A, Shinya T, Ryder LS, Otomo I, et al. (2012) Effector-mediated suppression of chitin-triggered immunity by Magnaporthe oryzae is necessary for rice blast disease. Plant Cell 24: 322–335. doi: 10.1105/tpc.111.092957
13. Mosquera G, Giraldo MC, Khang CH, Coughlan S, Valent B (2009) Interaction transcriptome analysis

identifies Magnaporthe oryzae BAS1-4 as biotrophy-associated secreted proteins in rice blast disease. Plant Cell 21: 1273–1290. doi: 10.1105/tpc.107.055228

14. Saitoh H, Fujisawa S, Mitsuoka C, Ito A, Hirabuchi A, et al. (2012) Large-scale gene disruption in Magnaporthe oryzae identifies mc69, a secreted protein required for infection by monocot and dicot fungal pathogens. PLoS Pathog 8: e1002711 doi:10.1371/journal.ppat.1002711.
15. Chen S, Songkumarn P, Venu RC, Gowda M, Bellizzi M, et al. (2013) Identification and characterization of in planta-expressed secreted effector proteins from Magnaporthe oryzae that induce cell death in rice. Mol Plant Microbe Interact 26: 191–202. doi: 10.1094/mpmi-05-12-0117-r
16. Giraldo MC, Dagdas YF, Gupta YK, Mentlak TA, Yi M, et al. (2013) Two distinct secretion systems facilitate tissue invasion by the rice blast fungus Magnaporthe oryzae. Nat Commun 4: 1996. doi: 10.1038/ncomms2996
17. Khang CH, Berruyer R, Giraldo MC, Kankanala P, Park S-Y, et al. (2010) Translocation of Magnaporthe oryzae effectors into rice cells and their subsequent cell-to-cell movement. The Plant Cell Online 22: 1388–1403. doi: 10.1105/tpc.109.069666
18. Mentlak TA, Talbot NJ, Kroj T (2011) Effector translocation and delivery by the rice blast fungus Magnaporthe oryzae. Effectors in Plant-Microbe Interactions 219–241. doi: 10.1002/9781119949138.ch9
19. Ribot C, Cesari S, Abidi I, Chalvon V, Bournaud C, et al. (2013) The Magnaporthe oryzae effector AVR1CO39 is translocated into rice cells independently of a fungal-derived machinery. Plant J 74: 1–12. doi: 10.1111/tpj.12099
20. Jiang RH, Tripathy S, Govers F, Tyler BM (2008) RXLR effector reservoir in two Phytophthora species is dominated by a single rapidly evolving superfamily with more than 700 members. Proc Natl Acad Sci USA 105: 4874–4879. doi: 10.1073/pnas.0709303105

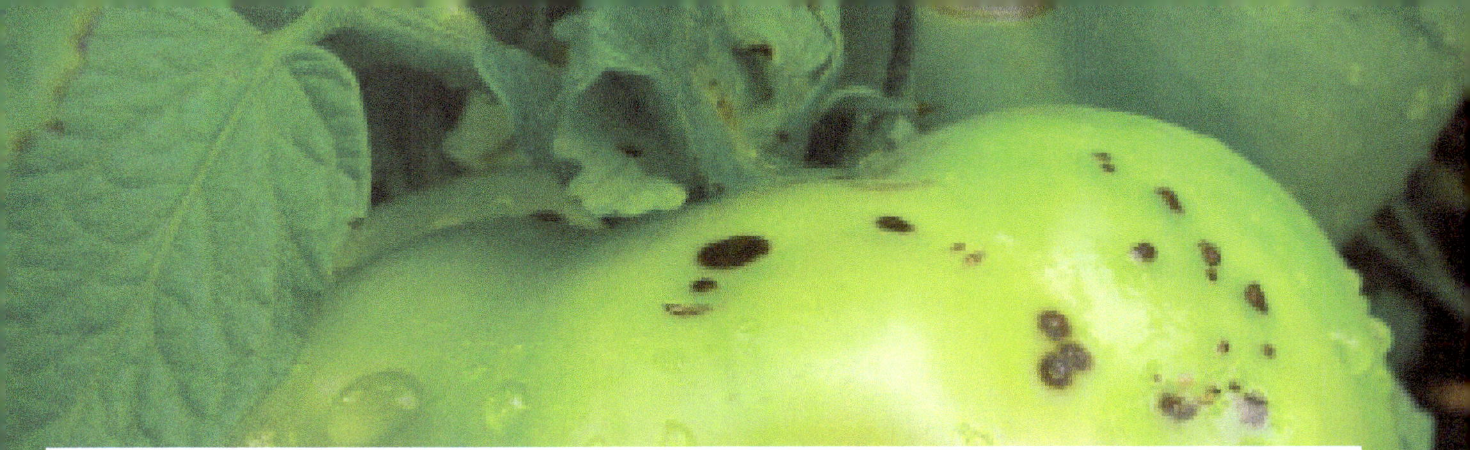

CHAPTER 20 COMPUTATIONAL MODELS IN PLANT-PATHOGEN INTERACTIONS

Andrés Pinzón, Emiliano Barreto, Adriana Bernal, Luke Achenie, Andres F González Barrios, Raúl Isea and Silvia Restrepo

BACKGROUND

Control and management of plant diseases and the identification of factors that contribute to the spread a given plant pathogen attack are at the basis of phytopathology. Mathematical models and computational simulations have been used, along with molecular and physiological approaches, to solve these and other issues.

In the early 1990s the use of stochastic models in plant pathology was reviewed [1, 2], mostly focused on epidemics. In this work we update topics not fully covered in previous reviews as well as associated experimental approaches that characterize the systems biology era [3]. Most of the review will focus on thePhytophthora infestans - Solanum tuberosum pathosystem, but its discussion will be general enough as to be applicable to any other plant pathogen system. A brief discussion of boolean networks and how this approach could drive the modeling of the compatible interaction between P. infestans and S. tuberosum is also introduced.

Citation: Andrés Pinzón, Emiliano Barreto, Adriana Bernal, Luke Achenie, Andres F González Barrios, Raúl Isea, and Silvia Restrepo, Computational models in plant-pathogen interactions, DOI: 10.1186/1742-4682-6-24

Experimental approaches to the study of molecular plant-pathogen interactions in*Phytophthora* species

Plants use various strategies to resist infection by a particular pathogen [4]. These strategies are part of the plant›s innate immune system and can be grouped into two broad categories [5]. The first recognizes common pathogen-associated molecular patterns (PAMPs), and acts as an early plant warning to potential infection [6]. This recognition leads to the induction of a basal plant defense, which in some cases includes a hypersensitive response (HR). The HR is characterized by the rapid death of cells surrounding the infected region and commonly leads to a broad spectrum plant response, the Systemic Acquired Resistance [7].

A second defense system in plants involves pairs of gene products, an effector molecule from the pathogen and an associated resistance protein (R) from the host, which recognizes it. This defense mechanism is highly specific and is triggered once a given effector is recognized by its associated R defense protein [5].

Plants with the capacity for protection from a pathogen attack are considered as resistants and a pathogen that lacks the ability to infect it is referred to as avirulent on that plant [4]. In this case, the host-pathogen interaction is considered incompatible. On the other hand, when a compatible interaction occurs, the pathogen becomes virulent and a plant that is incapable of resisting the attack is considered non-resistant.

Plant pathogens have developed several strategies to evade such plant defense responses and to become virulent. For some of these pathogens the evasion mechanisms are at least partially known, as in the case of bacteria such as Pseudomonas syringae. However, for most plant pathogen species, these evasion mechanisms are almost completely unknown. This is the case for P. infestans, the causal agent of late blight of potato, a disease that affects S. tuberosum and some other species in the Solanaceae family [8]. Oomycetes from the genus Phytophthora are plant pathogens devastating for agriculture and natural ecosystems [9]. For instance, in the United States alone, P. infestans causes estimated losses that exceed $US 5 billion annually [10].

Despite its economic importance, the fundamental molecular mechanisms underlying the pathogenicity of P. infestans are poorly understood. It was not until recent years that information crucial to the understanding of its genomics and infectious mechanisms was accessible to the research community [11]. For example, in 2006, the first effort to classify the secretome of plant pathogenic Oomycetes was carried out by Kamoun et al. Furthermore, although the general molecular events associated with the interaction between P. infestans andS. tuberosum were already known in 1991 [12], it was not until last year (2008) that all the known molecular and cytological processes underlying plant-pathogen interactions in various Phytophthora species were revised [9].

From the biological strategies used so far to study the processes underlying plant-pathogen interactions, three are most suitable as basis for a computational systems biology approach: (a) gene expression, (b) structural and comparative genomics and (c) protein-protein interactions.

Gene expression

Gene expression approaches constitute a starting point from which to determine the best strategy for building a computational model of a plant disease. Host-expressed molecules give insights into the underlying defense mechanisms, whereas identification of the pathogen counterparts allows us to ascertain possible mechanisms of attack and/or avoidance mechanisms used to establish a disease.

Differential expression of particular genes

A common strategy in gene expression analysis is to identify a particular gene of interest, and then to study or characterize its expression profile in different hosts and/or treated tissues. For instance, based on the findings that during the early phases of the interaction between P. infestans and potato, the genes ipiB and ipiO are expressed at high levels, Pieterse et al. hypothesized that these genes played an important role in the early stages of the infection process [13]. Both genes were isolated and their expression studied in various host tissues and different host plants. The results showed that the expression of these genes was activated in compatible, incompatible and non-host interactions. In the case of ipiO, it was revealed that a motif on the promoter region functioned as a glucose repression element in yeast. This observation helped to generate hypotheses about its behavior in cultivars with different resistance levels. The authors concluded that perhaps a variable nutrient environment could trigger the expression of ipiO and ipiB depending on the host and/or the expressing tissue.

Most of the crucial P. infestans protein elicitors known to-date [14] have also been revealed by this approach. This is the case for the Avr3a avirulence gene, the first to be cloned from P. infestans. Subsequently, this gene was the subject of the first report of cell death suppression from a filamentous plant pathogen [15, 16].

Differential expression of particular genes has also been used to study Systemic Acquired Resistance (SAR) and HR in challenged plants [17, 18] to test, for instance, the correlation between the expression of basal SAR marker genes with resistance to P. infestans [19].

High throughput differential gene expression

This approach focuses in the identification of all the genes expressed in a cell under a particular condition. Since this approach allows us to differentiate clearly between the expression profiles of cells under different conditions, its application is of special interest in plant-pathogen interactions, allowing us to solve research questions such as: which genes are expressed in a compatible interaction that are not expressed in a compatible one? Or, are there any sub-regulations, positive or negative feed-backs, present in one case but not the other?

Different techniques such as DNA microarrays [20, 21, 22, 23], serial analysis of gene expression [24, 25] and differential display [26, 27] have been used to study high throughput differential gene expression.

In the case of P. infestans, genes expressed in host cells challenged by this pathogen have been screened on compatible [28, 29, 30] and incompatible interactions [31, 32, 33], elucidating important issues about the mechanisms of interaction with its hosts.

For instance, gene regulation was revealed in a DNA microarray analysis of 7680 potato cDNA clones, representing approximately 5000 unique sequences expressed during a compatible interaction [30]. This work focused on the role of gene suppression in the compatible interaction, and its profile was obtained from microarray data evaluated at five time points. From this study, suppression of genes involved in the jasmonic acid (JA) defense pathway was revealed [34], as well as a severe down-regulation of the carbonic anhydrase (CA) gene, responsible for the reversible hydration of carbon dioxide to bicarbonate. Further analysis showed that CA was first down-regulated and then up-regulated during the incompatible interaction, clearly differentiating susceptibility from resistance, opening questions about the mechanisms that lead to its rapid suppression and the possibility of a connection between CA suppression and the overall down-regulation of the JA defense pathway.

Differential expression has also been studied on the pathogen side in P. infestans [35, 21, 23] and otherPhytophthora species [21, 36], revealing differential expression of e.g. the hsp70 and hsp90 genes, under distinct pathogen developmental stages and pathogenicity structures [37, 36].

Although still fragmented, this approach provides a systemic view of the pathogenicity process, considering gene expression as a network and helping us to develop strategies to control or prevent the disease by manipulation of either the pathogen or the host.

Structural and comparative genomics

Along with differential gene expression analysis, this is the most common modern approach to studying plant pathogen interactions, mostly due to the proteomic techniques as well as data mining and functional genomics tools available nowadays.

To date, one nuclear and six chloroplast genomes have been sequenced and two more nuclear genome sequencing projects are in progress in Solanaceous species (Additional file 1). On the pathogen side, five Oomycete genomes have been sequenced [11] and several studies at the genome scale have been carried out thanks to the availability of genomic information on these Oomycetes [38, 39, 40] and their hosts.

Therefore, the possibility of performing comparisons between different organisms at the sequence level [40] has allowed agronomically important resistance genes in potato to be isolated [41], pathogen avirulence genes [42] and gene families [10] to be identified, and novel proteins implicated in a given interaction to be identified [43]. For example, in the case of S. tuberosum, comparative analysis has revealed a physical co-localization between resistance loci in tomato, tobacco and pepper [44].

This approach has also revealed how two widely divergent microorganisms, P. infestans and the human malaria parasite Plasmodium falciparum, use equivalent host-targeting signals to deliver virulence and avirulence gene products into their hosts [45]. These products have been characterized by a particular protein motif, leading to the hypothesis of pathogenicity mechanisms conserved between both organisms [46]. This motif is the host-targeting (HT) signal of P. falciparum, centered on an RxLx core, revealed after the discovery of the RxLR host translocation motif of Oomycete effectors [47, 48, 49]. Owing to the availability of such data, it has been shown that although Plasmodium and Phytophthora are divergent eukaryotes, they share leader sequences, which suggests a conserved machinery for transport of effector proteins, a finding otherwise hard to achieve.

Protein-protein interactions

One approach to study protein-protein interactions is by using yeast two hybrid screening, co-immunoprecipitation [50] or surface plasmon resonance. This is arguably the most important approach towards a broad understanding of any plant pathogen interaction. It enables some mechanisms for the suppression of host defense in several organisms, such as the fungal pathogen Septoria lycopersici [51] or the Oomycete Phytophthora sojae [52], to be revealed.

In the case of P. infestans, relevant host defense suppression molecules have been also identified by this approach, such as the extracellular protease inhibitors EPI1 [53], EPI10 - the first protease inhibitor reported in any plant-associated pathogen, which suppresses tomato defense by targeting - the P69B subtilisin-like serine protease [54], and the EPIC family of secreted proteins that target the extracellular cysteine protease PIP1 (Phytophthora Inhibited Protease 1) [55].

Protein-protein interactions play an important role in recognition between plant pathogens and their hosts. This recognition has been studied at two levels: recognition of the host by the pathogen and recognition of the pathogen by the host [56, 57]. During an interaction, host resistance (R) and pathogen avirulence (Avr) proteins interact in a gene-for-gene manner. Proteins encoded by R alleles recognize the products of corresponding Avr alleles, thus triggering disease resistance. Using an association genetics approach [58], theP. infestans Avr3a effector was shown to be recognized in tomato cytoplasm by R3a (a member of the R3complex locus on chromosome 11). R3a was isolated by positional cloning the same year [41].

Together, these and other studies [59, 23], along with computational chemistry and/or computational modeling and prediction of protein-protein interactions [60], provide valuable information about the recognition mechanisms in S. tuberosum - P. infestans R-Avr interactions and could lead to the identification of metabolic and/or signaling pathways underlying incompatible interactions.

Quantitative models in plant pathology

In cases where experimental data for a biological system start to accumulate, it is feasible and convenient to integrate all the information gathered into a quantitative model. This approach allows us to obtain a mathematical and networked framework for a descriptive model of the biological phenomenon [61]. This type of model strengthens the predictive capacity of future responses, for instance under different conditions, and it also helps to broaden our view of the potential interactions that could take place in any molecular reaction [62].

In order to capture time-dependent dynamic phenomena, a systems biology approach should allow us to integrate various ranges of spatial and temporal biological scales, as well as processing of different signals, genotypic variation and responses to external perturbations. As seen in the previous section, typical experiments describing the interaction between P. infestans and its hosts are clearly related to each of these characteristics.

Functional genomics and proteomic approaches produce the most suitable data for the development of a theoretical model [61]. For instance, microarray-based differential expression analysis evaluates expression patterns at different times [30], under different conditions [21, 33] with host and pathogen genotypic variation. On the other hand, gene expression and host targeting of protease inhibitors work at different levels of signaling and at different spatial and temporal scales [54, 53].

Data gathered from such plant-pathogen interaction approaches, along with the development of interaction, pathways and metabolism databases [63, 64], as well as standardized systems biology languages [65, 66] andin silico research platforms [67, 68], have opened the door to modern computational model approaches at the molecular level in several organisms, including Oomycetes.

Predominantly, phytopathologists have used computational and quantitative modeling approaches to describe the temporal dynamics of plant diseases. Consistently, the bulk of the literature written in this field has been focused on the epidemiology of the disease, so research on the modeling of plant-pathogen molecular interactions is under-represented.

Quantitative modeling of plant-pathogen epidemiology

Deterministic approaches

In 1969, Waggoner and Horsfall published Epidem, the first computer simulation of a plant disease [69]. Epidem was mainly a simulator of potato and tomato blights. Since then, models used in the plant-pathogen field have often belonged to the family of logistic equations.

The fundamental logistic model was proposed in 1963 by VanderPlank [70, 71] and it describes the rate at which a disease spreads over time (Table 1).

Table 1: Solanaceous genome projects.

Species	Genome	Status	reference
Nicotianatabacum	mitochondrion	Finished	[106]
Nicotianatomentosiformis	chloroplast	Finished	[107]
Solanum tuberosum	chloroplast	Finished	[108]
Solanum bulbocastanum	chloroplast	Finished	[109]
Solanum lycopersicum	chloroplast	Finished	[110]
Nicotianasylvestris	chloroplast	Finished	[111]
Atropa belladonna	chloroplast	Finished	[112]
Solanum tuberosum	Nuclear	In progress	12984*
Solanum lycopersicum	Nuclear	In progress	9509*

*NCBI's genome project identification number.

In this model [71], y_t is the proportion of diseased tissue (severity) at time t and λ is the rate of change of diseased tissue unit in a given unit time. The term $(1\text{-}y_t)$ indicates that new infections occur only in non-infected tissue. The slope of the disease curve depends on the infection rate (λ) and the inoculum y_t. At a higher infection rate, the curve rise more steeply.

However, this model assumes that a lesion always remains infectious and also neglects the lag between the time at which an infection occurs and the time it becomes infectious (latent period). As such, the so-called generalized model considers both a latent period ($p > 0$) and an infectious period (i) [71, 72] (Table 1). These values can range from $p < 7$, $i<65$ days in Puccinia recondita [73] to $p < 2$, $i<8$ days in P. infestan s [74].

Other relevant characteristics of a plant disease are its spatial pattern and the arrangement of disease entities (i.e., spores). The spatial patterns are influenced by dispersal of disease entities [75]. In cases where spore dispersal is not carried out directly from infected tissue but by environmental factors such as wind, a model assuming a constant source of inoculum, such as a monomolecular model, is more appropriate [71].

In this case the infected tissue is not part of the source, so the shape of the disease curve depends solely on the rate of infection. In the case of Phytophthora, five potential mechanisms of dispersal have been described for some major species [75]; for P. infestans, P. cinnamoni and P. syringae, a real mechanism of dispersal could be represented correctly by this model.

Some other models have been derived from the general logistic model. For instance, the Gompertz model is similar to the general logistic one and can be seen as a logarithmic form of it. When different data sets are compared, it is appropriate to use a model that allows us to make such comparisons; for those cases a Weibull model should be considered [76].

The spread of disease has also been modeled [77, 78]. Since the early 1980s, the epidemic wave velocity of P. infestans has been measured by several means [79, 80, 81, 82].

Although widely used, deterministic models do not represent the underlying biological process in a proper way. Spore germination is a good example of a stochastic process; for instance, examination of a single spore will reveal stochastic behavior, which can only be inferred by the examination of a significant number of units. Thus, in these cases, the process under study is better described by a probability function [2].

Stochastic approaches

Stochastic modeling of epidemics has been studied since the early 1960s. Most of the stochastic approaches carried out at that time were also concerned with the progress of the infection over time, represented by the so-called general stochastic epidemic model [83]:

$$\Pr\{I(\tau+\delta)=\chi+1, S(\tau+\delta_\tau)=\gamma-1 \mid I(\tau)=\chi, S(\tau)=\gamma\}=\beta_{\chi\gamma}\delta_\tau+\varphi(\delta_\tau)$$

Where $(\tau, \tau+\delta_\tau)$ is the time interval. Here I(τ) represents the number of infectives, S(τ) the number of susceptibles and R(τ) the number of removals at time $\tau \geq 0$. The removal of infected tissue is also considered probabilistic and it will occur with the following probability [83]:

$$\Pr\{I(\tau+\delta)=\chi-1, S(\tau+\delta_\tau)=\gamma \mid I(\tau)=\chi, S(\tau)=\gamma\}=\gamma_{\chi\gamma}\delta_\tau+\varphi(\delta_\tau)$$

in the same time interval, where $\gamma > 0$; $\chi = 0, 1, ..., N$,. Since a given removal does not depend on previous ones, a removal is considered independent [83]:

$$\Pr\{I(\tau+\delta)=\chi, S(\tau+\delta_\tau)=\gamma \mid I(\tau)=\chi, S(\tau)=\gamma\}=1-\gamma\chi\delta_\tau-\beta_{\chi\gamma}+\delta_\tau+\varphi(\delta_\tau)$$

Transition probabilities are given as [83]:

$$P_{\chi\gamma}(\tau)=\Pr\{I(\tau)=\chi, S(\tau)=\gamma \mid I(0)=i, S(0)=s=N-i\}$$

In non-stochastic models, stochasticity can be approached by adding randomness to state variables. For instance, Vanderplank's model was used in the description of the zucchini yellow mosaic virus disease [84]. In this case, stochasticity was achieved by adding a "brownian motion term to the growth rate parameter". As the authors stated, a significant difference between a stochastic and a deterministic version of the same model can be seen only if large data sets are employed. This observation could explain why in recent years, when biological data acquisition has grown faster than ever, the use of stochastic models has become more popular.

Stochastic modeling in plant pathology has also been applied to processes at different levels of biological organization, such as at the organ level, crops [85], spatial patterns, evolution [2] and aerial spread [86]. For instance, the spatial spread of disease in race-specific and race-nonspecific cultivar mixtures was studied using a spatially explicit stochastic model [87]. This model was based on the assumption that disease can be significantly higher in monocultures than in cultivar mixtures and it only considered stochastic variation of spore dispersal at constant sporulation rate, although there exist many other sources of stochastic variation (such as genotypic variation) [2].

No matter whether they are stochastic or deterministic, the models described above have been focused on higher scales of plant pathogen interactions, such as the population, organ or ecological level. Nevertheless, in any plant-pathogen disease, the molecular level of the interaction (i.e., protein-protein, protein-DNA, regulatory and metabolic network regulation) is intrinsically involved and surely accounts for much of the variation observed at other levels. Therefore, genetic processes in an organism can be seen as networks thatbridge the gap between genotype and phenotype [88]. A good example of this situation can be found in the collective behavior of bacteria in Quorum Sensing (QS) mechanisms.

QS is a common strategy used by several plant-pathogenic bacteria to assess local population density and/or physical confinement. In a recent publication, a model describing the Ti plasmid quorum-sensing gene network was constructed [89]. It was shown that it could operate as an «on-off» gene expression switch that is sensitive to the environment, allowing the question about how bacteria really behave or respond to be answered in QS.

Although this topic is absent in some plant-pathogen organisms, such as P. infestans, the characteristics of quantitative modeling of molecular mechanisms could elucidate several questions in phytopathology.

In silico modeling of plant-pathogen molecular interactions

Plants resist pathogen attacks by shifting their defense mechanisms, as reflected in quantitative and kinetics enhancements [62]. The mechanism that controls host defense activation consists of a highly interconnected network, in which host defense genes interact with each other as well as with effector proteins present in the cell [90, 91, 92]. The availability of high-throughput gene expression and proteomics data has generated an unprecedented opportunity for comprehensive study of these types of biological networks [89, 93].

Since an important phase in host-pathogen interactions involves protein-protein recognition [94, 91], efforts to elucidate networks of such interactions are of special interest in phytopathology. For example, a whole-genome computational strategy to infer protein interactions was applied to ten pathogens, including species of Mycobacterium, Apicomplexa and Kinetoplastida [91]. This work started with the identification of pairs of matching proteins known to interact between the host and the pathogen, and by assessing the likelihood of this interaction by means of structural modeling, expression properties and subcellular location. As a result, an enriched candidate set of proteins is obtained, suitable for experimental study.

With the current genome sequence information for several Phytophthora genomes (Additional file 1) and those under sequencing [11, 95], this approach could be applicable to a Phytophthora-Solanaceae model and thus enhance our limited knowledge about the molecular interactions in these genera.

Another good theoretical framework to start working with is a space of interconnected operators such as a boolean network (Figure 1). Boolean networks present some advantages when compared to similar strategies such as hidden Markov models [96, 97]. For instance, it is possible to perform a simulation while «avoiding the statistical basis around them, provides the option to perform simpler computational simulations, insert additional regulators or quantitative and biochemical data parameters into the model when available» [98].

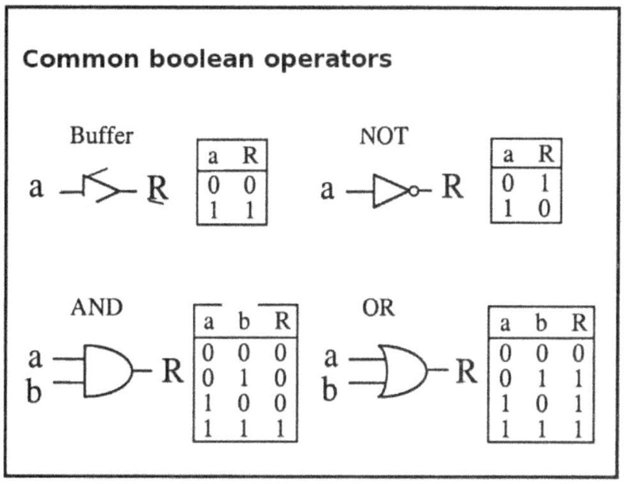

Figure 1: Boolean formalism. Adapted from [98] The most frequent types of boolean operators are the buffer, NOT, AND and OR gates. Tables adjacent to each of these gates are known as "true" tables, where "a" and "b" represent the input (or stimuli) and R the output (or response).

Towards a boolean description of the *P. infestans - Solanum tuberosum* interaction

Although clustering analysis can be used to infer gene function from expression data, the detailed interaction between genes within or between clusters cannot be deduced by this approach [99]. In order to deduce such interactions, data from differential expression analysis can be represented in a boolean formalism. This representation can be achieved in a typical boolean binary form, where repression and/or induction of a given gene can be expressed by an on or off switch and thus translated into a network structure and simulated by computational analysis. This approach has been successfully implemented in the simulation of plant defense signaling networks in Arabidopsis thaliana in response to different treatments with salicylic acid, jasmonic acidand ethylene [98] (figure 2). For example, genes up-regulated to the same level in both treatments can be expressed by an OR operator, as in the case of phyA and PhyB in figure 2, thus leading to different possible initial states (input domains) as represented by panels A and B in the same figure. Data from differential expression analysis can also be represented by three possible boolean states (Table 2). This approach has been successfully used in the inference of gene regulatory networks [100] where «-1» was also introduced to address the negative interaction between components in the network. Experimental information on the compatible interaction between P. infestans and S. tuberosum is being approached by our laboratory using a similar strategy, in order to hypothesize the network space of carbonic anhydrase in this interaction.

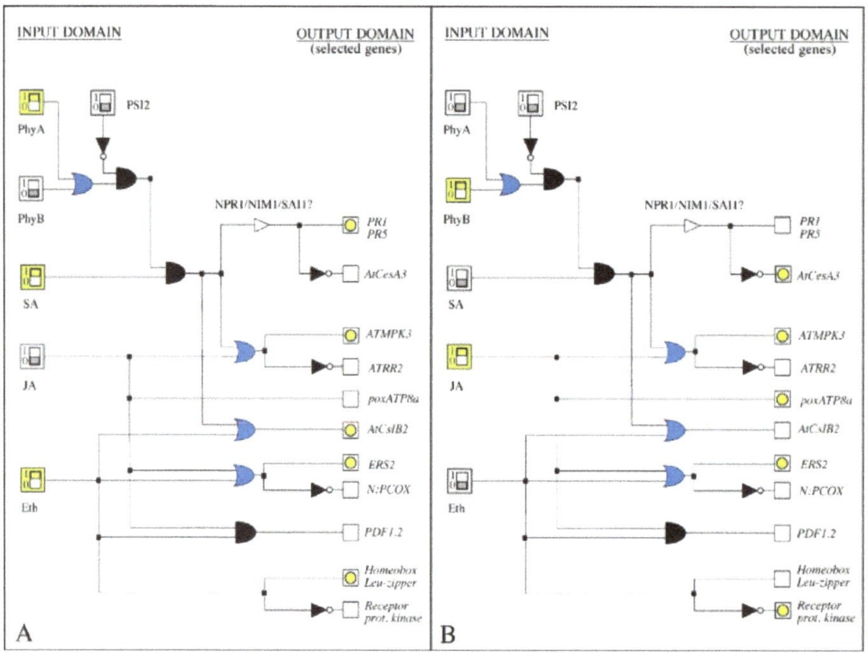

Figure 2: Boolean representation of a signaling network. Adapted from [98] Boolean representation of the signal transduction network controlling the plant's defense response against pathogens in Arabidopsis thaliana, represented by a series of output genes selected from microarray data. The activated switches are represented in yellow. Diode symbols in yellow indicate the induced genes. Empty squares correspond to no significant expression. A and B represent two of the various possible outputs given the input.

Table 2: Boolean representation of defense-related genes expressed during a compatible interaction between P. infestans and S. tuberosum.

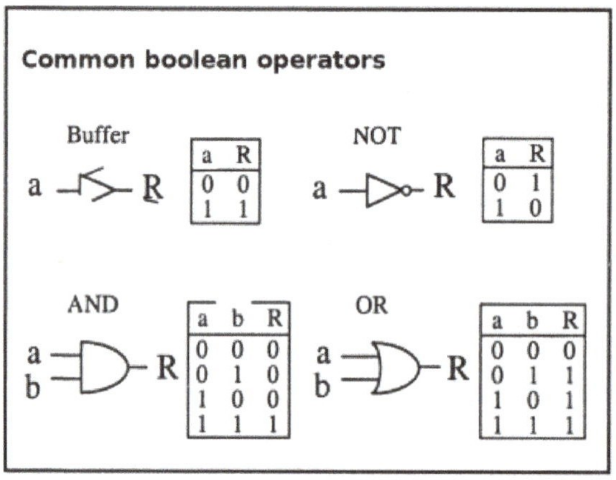

Adapted from [30]. Inductions, quantified as ratios greater than 1.5-fold, are represented in bold font. Repressions, quantified as ratios lower than 1.5-fold, are represented in bold and italic font. Grey columns contain the boolean representation for each gene expression level. Inductions here are represented by 1, repressions by -1 and non-significant expression by zero. These boolean values can further be formalized and simulated into a computational model.

This approach can also be used in systems lacking biological information, by gathering data common to other organisms or from related species. This possibility opens the door to implementation in other species of Oomycetes where lack of information is typical.

CONCLUSION

The idea of the stochastic modeling of biological systems is not new, although traditionally, the mathematical frameworks used to represent and study these processes have been deterministic. This situation can be explained by taking into consideration the fact that quantitative and computational modeling usually require the availability of important computational resources. These resources increase proportionally with the number of variables involved in the model; then apart from restrictions on the availability of biological information included in the model, there also exist restrictions on the availability of computational resources to perform a given simulation. In recent years, computational resources have become less restrictive. Moreover, stochastic processes are probabilistic in nature and thus require the use of more data as confidence in calculations depends on them. More data means more calculations and interconnections between variables. Thus, the availability of computational resources and biological data has restricted the use of stochastic approaches for decades, not only in plant pathology, but in biological processes in general.

Any top level biological observation implies an underlying molecular process, which is not isolated from its environment.

This situation has always been evident to plant pathologists, as reflected in the conversion of environmental factors such water or humidity into model variables. Epidemiological research will find molecular plant-pathogen interaction models an important tool for describing disease spread and dynamics from its roots. In turn, molecular plant pathogen interactions cannot be modeled in isolation from environmental variables, largely analyzed by deterministic approaches since the early 1960s. Therefore, in order to reflect the real biological phenomena, it is crucial to take this information into account when a molecular interaction is considered.

Biological information that implies a networked structure can be represented by a boolean formalism. This approach avoids the immediate necessity of chemical kinetics and the use of sets of equations (for example differential equations) to run a simulation. Thus, boolean networks are a viable and ideal strategy for the computational modeling of protein-protein interactions, metabolic networks and differential expression data available today for organisms for which molecular kinetic information is not available.

To date, differential gene expression data, protein-protein interaction and functional comparative analysis represent the only information available, not only for the majority of Oomycetes and their hosts but also for several other organisms. Here we argue that, due to the type of information available - although hidden Markov models, neural networks and flux balance analysis have recently been used - a boolean representation of plant-pathogen interactions between P. infestans and S. tuberosum is one of the most suitable approaches for computational modeling; an ongoing effort in our laboratory, based on microarray data for a compatible interaction between these organisms. Once available, boolean networks will allow kinetic information to be put back into the model and thus complement it with new information as it becomes available.

Quantitative representation and computational simulation of biological data is an important tool for understanding complex biological networks and interactions. To date, the availability of efficient algorithms, biological information and computational resources have opened the door to new insights into the analysis of such information. Bioinformatics, systems biology and its most representative tool, computational modeling, allow us to study complex plant pathogen interactions in a way unreachable to scientists two decades ago. Understanding of plant-pathogen interactions at the deterministic and stochastic, molecular and population levels requires a holistic approach, where any piece of available information is important. Today we are facing an integrative era of biological information, which approaches biological phenomena not from their individual parts but from their interactions. Without a doubt, this approach reflects biological reality in a more convenient and realistic way, but it also brings new challenges as well as the necessity for new tools, which cover not only the biological sciences field but also engineering and mathematics.

ACKNOWLEDGEMENTS

The authors thank the Vicerrectoria de Investigaciones at Los Andes University, Colombia for its support.

REFERENCES

1. Rouse D: Stochastic modeling of plant disease epidemic processes. CSA. 1991, 1: 647-665.
2. Shaw M: Modeling stochastic processes in plant pathology. Annual Reviews in Phytopathology. 1994, 32: 523-544.
3. Kitano H: Systems biology: a brief overview. Science. 2002, 295 (5560): 1662-1664.
4. Cook R: The molecular mechanisms responsible for resistance in plant-pathogen interactions of the gene-for-gene type function more broadly than previously imagined. Proc Natl Acad Sci USA. 1998, 95 (17): 9711-9712.
5. Jones J, Dangl J: The plant immune system. Nature. 2006, 444 (7117): 323-329.
6. Ingle R, Carstens M, Denby K: Pamp recognition and the plant-pathogen arms race. BioEssays. 2006, 28: 880-889.
7. Wang D, Weaver N, Kesarwani M, Dong X: Induction of protein secretory pathway is required for systemic acquired resistance. Science. 2005, 308 (5724): 1036-1040.
8. Ribeiro D, Ribeiro O, Erwin D: Phytophthora diseases worldwide. Amer Phytopathol Society, St. Paul, MN. 1996, 561-
9. Attard A, Gourgues M, Galiana E, Panabières F, Ponchet M, Keller H: Strategies of attack and defense in plant-oomycete interactions, accentuated for Phytophthora parasitica dastur (syn. p. nicotianae breda de haan). J Plant Physiol. 2008, 165 (1): 83-94.
10. Avrova A, Whisson S, Pritchard L, Venter E, De Luca S, Hein I, Birch P: A novel non-protein-coding infection-specific gene family is clustered throughout the genome of Phytophthora infestans. Microbiology. 2007, 153 (Pt 3): 747-759.
11. Tyler BM, Tripathy S, Zhang X, Dehal P, Jiang RH, Aerts A, Arredondo FD, Baxter L, Bensasson D, Beynon JL, Chapman J, Damasceno CM, Dorrance AE, Dou D, Dickerman AW, Dubchak IL, Garbelotto M, Gijzen M, Gordon SG, Govers F, Grunwald NJ, Huang W, Ivors KL, Jones RW, Kamoun S, Krampis K, Lamour KH, Lee MK, McDonald WH, Medina M, Meijer HJ, Nordberg EK, Maclean DJ, Ospina-Giraldo MD, Morris PF, Phuntumart V, Putnam NH, Rash S, Rose JK, Sakihama Y, Salamov AA, Savidor A, Scheuring CF, Smith BM, Sobral BW, Terry A, Torto-Alalibo TA, Win J, Xu Z, Zhang H, Grigoriev IV, Rokhsar DS, Boore JL: Phytophthora genome sequences uncover evolutionary origins and mechanisms of pathogenesis. Science. 2006, 313 (5791): 1261-1266.
12. de Wit P, Pieterse M, Govers F: Molecular aspects of the potato - Phytophthora infestans interaction. European Journal of Plant Pathology. 1992, 98 (2):
13. Pieterse C, Derksen A, Folders J, Govers F: Expression of the Phytophthora infestans ipib and ipio genes in planta and in vitro. Mol Gen Genet. 1994, 244 (3): 269-277.

14. Kamoun S, van West P, Jong A, Groot K, Vleeshouwers V, Govers F: A gene encoding a protein elicitor of Phytophthora infestans is down-regulated during infection of potato. Mol Plant Microbe Interact. 1997, 10 (1): 13-20.
15. Bos J, Kanneganti T, Young C, Cakir C, Huitema E, Win J, Armstrong M, Birch P, Kamoun S: The c-terminal half ofPhytophthora infestans rxlr effector avr3a is sufficient to trigger r3a-mediated hypersensitivity and suppress infl-induced cell death in Nicotianabenthamiana. Plant J. 2006, 48 (2): 165-176.
16. Kamoun S: Groovy times: filamentous pathogen effectors revealed. Curr Opin Plant Biol. 2007, 10 (4): 358-365.
17. Halim V, Eschen-Lippold L, Altmann S, Birschwilks M, Scheel D, Rosahl S: Salicylic acid is important for basal defense of Solanum tuberosum against Phytophthora infestans. Mol Plant Microbe Interact. 2007, 20 (11): 1346-1352.
18. Anand A, Uppalapati S, Ryu C, Allen S, Kang L, Tang Y, Mysore K: Salicylic acid and systemic acquired resistance play a role in attenuating crown gall disease caused by Agrobacterium tumefaciens. Plant Physiol. 2008, 146 (2): 703-715.
19. Vleeshouwers V, Dooijeweert W, Govers F, Kamoun S, Colon LT: The hypersensitive response is associated with host and nonhost resistance to Phytophthora infestans. Planta. 2000, 210: 853-864.
20. Richmond T, Somerville S: Chasing the dream: plant est microarrays. Curr Opin Plant Biol. 2000, 3 (2): 108-116.
21. Moy P, Qutob D, Chapman B, Atkinson I, Gijzen M: Patterns of gene expression upon infection of soybean plants byPhytophthora sojae. Mol Plant Microbe Interact. 2004, 17 (10): 1051-1062.
22. Silvar C, Merino F, Díaz J: Differential activation of defense-related genes in susceptible and resistant pepper cultivars infected with Phytophthora capsici. J Plant Physiol. 2009, 165 (10): 1120-1124.
23. Tian Z, Liu J, Wang B, Xie C: Screening and expression analysis of Phytophthora infestans induced genes in potato leaves with horizontal resistance. Plant Cell Rep. 2006, 25 (10): 1094-1103.
24. Wang G, Vega-Sanchez M, Gowda M: Tag-based approaches for deep transcriptome analysis in plants. Plant Science. 2007, 173: 371-380.
25. Cerami E, Bader G, Gross B, Sander C: cpath: open source software for collecting, storing, and querying biological pathways. BMC Bioinformatics. 2006, 7: 497-
26. Truesdell G, Dickman M: Isolation of pathogen/stress-inducible cdnas from alfalfa by mrna differential display. Plant Mol Biol. 1993, 33 (4): 737-743.
27. Zhi Wu, Sheng Li, Yu Fen, Zi Sheng, Wu Xue, Xue Bao, Li F, Fu F, Sha L, Li W: Identification of drought-responsive genes from maize inbred lines. Journal of plant physiology and molecular biology. 2006, 33 (6): 607-611.
28. Boller T, Collinge M, Beyer K, Binder A: Identification of potato genes induced during colonization by Phytophthora infestans. Molecular Plant Pathology. 2001, 2: 125-134.
29. Evers D, Ghislain M, Hausman J, Dommes J: Differential gene expression in two potato lines differing in their resistance to Phytophthora infestans. J Plant Physiol. 2003, 160 (6): 709-712.
30. Restrepo S, Myers K, del Pozo O, Martin G, Hart A, Buell C, Fry W, Smart C: Gene profiling of a compatible interaction between Phytophthora infestans and Solanum tuberosum suggests a role for carbonic anhydrase. Mol Plant Microbe Interact. 2005, 18 (9): 913-922.
31. Duncan J, Lyon G, Birch P, Avrova A, Toth R: Isolation of potato genes that are induced during an early stage of the hypersensitive response to Phytophthora infestans. MPMI. 1999, 12: 356-361.
32. Ronning CM, Stegalkina SS, Ascenzi RA, Bougri O, Hart AL, Utterbach TR, Vanaken SE, Riedmuller SB, White JA, Cho J, Pertea GM, Lee Y, Karamycheva S, Sultana R, Tsai J, Quackenbush J, Griffiths HM, Restrepo S, Smart CD, Fry WE, Hoeven Van Der R, Tanksley S, Zhang P, Jin H, Yamamoto ML, Baker BJ, Buell CR: Comparative analyses of potato expressed sequence tag libraries. Plant Physiol. 2003, 131 (2): 419-29.
33. Tian Z, Song B, Wang B, Liu J, Xie C: Monitoring the expression patterns of potato genes associated with quantitative resistance to late blight during Phytophthora infestans infection using cdna microarrays. Plant Science. 2005, 169 (6): 1155-1167.
34. Glauser G, Grata E, Dubugnon L, Rudaz S, Farmer E, Wolfender J: Spatial and temporal dynamics of jasmonate synthesis and accumulation in Arabidopsis in response to wounding. Biol Chem. 2008, 283 (24): 16400-7.
35. Randall T, Lam S, Binder A, Boller T, Beyer K, Jimenez S, Collinge M: Characterization of Phytophthora infestans genes regulated during the interaction with potato. Molecular Plant Pathology. 2002, 3 (6): 473-485.
36. Le Berre J, Engler G, Panabières F: Exploration of the late stages of the tomato-Phytophthora parasitica interactions through histological analysis and generation of expressed sequence tags. New Phytol. 2008, 177 (2): 480-492.
37. Avrova A, Venter E, Birch P, Whisson S: Profiling and quantifying differential gene transcription in Phytophthora infestans prior to and during the early stages of potato infection. Fungal Genet Biol. 2003, 40 (1): 4-14.
38. Judelson H: Genomics of the plant pathogenic oomycete phytophthora: insights into biology and evolution. Adv Genet. 2007, 57: 97-141.
39. Lamour K, Win J, Kamoun S: Oomycete genomics: new insights and future directions. FEMS Microbiol Lett. 2007, 274 (1): 1-8.
40. Martin F, Bensasson D, Tyler B, Boore L: Mitochondrial genome sequences and comparative genomics of Phytophthora ramorum and p. sojae. Curr Genet. 2007, 51 (5): 285-296.
41. Huang S, Vossen van der E, Kuang H, Vleeshouwers V, Zhang N, Borm T, van Eck H, Baker B, Jacobsen E, Visser R: Comparative genomics enabled the isolation

of the r3a late blight resistance gene in potato. Plant J. 2005, 42 (2): 251-261.

42. Whisson S, Torto T, Ochwo M, Birch P, Kamoun S, Jorunn B, Miles B, Armstrong M: Intraspecific comparative genomics to identify avirulence genes from phytophthora. New Phytol. 2003, 159: 63-72.

43. Kamoun S: A catalogue of the effector secretome of plant pathogenic oomycetes. Annu Rev Phytopathol. 2006, 44: 41-60.

44. Gebhardt C, Valkonen J: Organization of genes controlling disease resistance in the potato genome. Annu Rev Phytopathol. 2001, 39: 79-102.

45. Haldar K, Kamoun S, Hiller N, Bhattacharje S, van Ooij C: Common infection strategies of pathogenic eukaryotes. Nat Rev Microbiol. 2006, 4 (12): 922-931.

46. Bhattacharjee S, Hiller N, Liolios K, Win J, Kanneganti T, Young C, Kamoun S, Haldar K: The malarial host-targeting signal is conserved in the irish potato famine pathogen. PLoS Pathogens. 2006, 2 (5): e50-

47. Rehmany A, Gordon A, Rose L, Allen R, Armstrong M, Whisson S, Kamoun S, Tyler B, Birch P, Beynon J: Differential recognition of highly divergent downy mildew avirulence gene alleles by rpp1 resistance genes from two arabidopsislines. Plant Cell. 2005, 17 (6): 1839-1850.

48. Birch P, Rehmany A, Pritchard L, Kamoun S, Beynon J: Trafficking arms: oomycete effectors enter host plant cells. Trends Microbiol. 2006, 14 (1): 8-11.

49. Morgan W, Kamoun S: Rxlr effectors of plant pathogenic oomycetes. Curr Opin Microbiol. 2007, 10 (4): 332-338.

50. Hermann J, Westermann B, Neupert W: Analysis of protein-protein interactions in mitochondria by coimmunoprecipitation and chemical cross-linking. Methods Cell Biol. 2001, 65: 217-230.

51. Bouarab K, Melton R, Peart J, Baulcombe D, Osbourn A: A saponin-detoxifying enzyme mediates suppression of plant defenses. Nature. 2002, 418 (6900): 889-892.

52. Darvill A, Rose J, Ham K, Albersheim P: Molecular cloning and characterization of glucanase inhibitor proteins. Plant Cell. 2003, 14: 1329-1345.

53. Tian M, Huitema E, Cunha L, Torto-Alalibo T, Kamoun S: A kazal-like extracellular serine protease inhibitor fromPhytophthora infestans targets the tomato pathogenesis-related protease p69b. J Biol Chem. 2004, 279 (25): 26370-26377.

54. Benedetti B, Tian M, Kamoun S: A second kazal-like protease inhibitor from Phytophthora infestans inhibits and interacts with the apoplastic pathogenesis-related rotease p69b of tomato. Plant Physiology. 2005, 138: 1785-1793.

55. Tian M, Win J, Song J, Hoorn van der R, Knaap van der E, Kamoun S: A Phytophthora infestans cystatin-like protein targets a novel tomato papain-like apoplastic protease. Plant Physiol. 2007, 143 (1): P364-377.

56. Tyler B: Molecular basis of recognition between Phytophthora pathogens and their hosts. Annu Rev Phytopathol. 2002, 40: 137-167.

57. Lee Van der T, Robold A, Testa A, Klooster J, Govers F: Mapping of avirulence genes in Phytophthora infestans with amplified fragment length polymorphism markers selected by bulked segregant analysis. Genetics. 2001, 157 (3): 949-956.

58. Armstrong M, Whisson S, Pritchard L, Bos J, Venter E, Avrova A, Rehmany A, Ulrike Bohme, Brooks K, Cherevach I, Hamlin N, White B, Fraser A, Lord A, Quail M, Churcher C, Hall N, Berriman M, Huang S, Kamoun S, Beynon J, Birch P: An ancestral oomycete locus contains late blight avirulence gene avr3a, encoding a protein that is recognized in the host cytoplasm. Proc Natl Acad Sci USA. 2005, 102 (21): 7766-7771.

59. Bent A, Mackey D: Elicitors, effectors, and r genes: the new paradigm and a lifetime supply of questions. Annu Rev Phytopathol. 2007, 45: 399-436.

60. Dyer M, Murali M, Sobral B: Computational prediction of host-pathogen protein-protein interactions. Bioinformatics. 2007, 23 (13): i159-i166.

61. Aggarwal K, Lee K: Functional genomics and proteomics as a foundation for systems biology. Briefings in functional genomics and proteomics. 2003, 2: 175-184.

62. Katagiri F: A global view of defense gene expression regulation-a highly interconnected signaling network. Curr Opin Plant Biol. 2004, 7 (5): 506-511.

63. Cubero J, Lastra B, Salcedo C, Piquer J, López M: Systemic movement of Agrobacterium tumefaciens in several plant species. J Appl Microbiol. 2006, 101 (2): 412-421.

64. Masoudi-Nejad A, Goto S, Endo T, Kanehisa M: Kegg bioinformatics resource for plant genomics research. Methods Mol Biol. 2007, 406: 437-458.

65. Evans T, Gillespie C, Wilkinson D: The sbml discrete stochastic models test suite. Bioinformatics. 2008, 24 (2): 285-286.

66. Finney A, Hucka M: Systems biology markup language: Level 2 and beyond. Biochem Soc Trans. 2003, 31 (Pt 6): 1472-1473.

67. Albrecht M, Huthmacher C, Tosatto S, Lengauer T: Decomposing protein networks into domain-domain interactions. Bioinformatics. 2005, 21 (Suppl 2): ii220-ii221.

68. Bell G, Lewitter F: Visualizing networks. Methods Enzymol. 2006, 411: 408-421.

69. Horsfall J, Waggoner P: Epidem a simulator of plant disease written for a computer. Conn agric Exp Stn Bull. 1969, 698: 80-

70. Zadocks J, Rainey R, Mollison D, Tyrrell D, Fitt B: Twenty-five years of botanical epidemiology (and discussion). Philosophical Transactions of the Royal Society of London. 1988, 321: 377-387.

71. Billard L, Dayananda P, Chakraborty S: Estimation of rate parameter and its relationship with latent and infectious periods in plant disease epidemics. Biometrics. 1995, 51: 284-292.

72. Billard L: Some statistical opportunities in the agricultural sciences. The canadian Journal of statistics. 1998, 3 (26): 392-404.

73. Metah Y, Zadocks J: uredospore production and sporulation period of Puccinia recondita fsp. triticina on primary leaves of wheat. Netherlands journal of plant pathology. 1970, 76: 267-276.
74. Vleeshouwers V, Van Dooijeweert W, Govers F, Kamoun S, Colon L: Does basal pr gene expression in Solanum species contribute to non-specific resistance to Phytophthora infestans?. Physiological and Molecular Plant Pathology. 2000, 57: 35-42.
75. Ristaino J, Gumpertz M: New frontiers in the study of dispersal and spatial analysis of epidemics caused by species in the genus phytophthora. Annu Rev Phytopathol. 2000, 38: 541-576.
76. Madden H, Hughes G, Bosch F: The study of plant disease epidemics. The American Phytopathological Society. 2007, APS Press St. Paul, Minnesota
77. Fry W, Minoque K: Models for the spread of plant disease: model description. Phytopathology. 1983, 73: 1168-73.
78. Fry W, Minoque K: Models for the spread of plant disease: some experimental results. Phytopathology. 1983, 73: 1173-76.
79. Ferrandino F: Dispersive epidemic waves: I. focus expansion in a linear planting. Phytopathology. 1993, 798 (83): 802-
80. Bosch van den F, Zadoks J: On the spread of plant disease: a theory on foci. Annu Rev Phytopathol. 1994, 32: 503-21.
81. Sherm H: On the velocity of epidemic waves in model plant disease epidemics. Ecol Model. 1996, 87: 217-22.
82. Bolker B: Analytic models for the patchy spread of plant disease. Bull Math Biol. 1999, 61: 849-874.
83. Gani J, Jerwood D: The cost of a general stochastic epidemic. Journal of Applied Probability. 1972, 9 (2): 257-269.
84. Marcus R: Deterministic and stochastic logistic models for describing increase of plant disease. Crop Protection. 1991, 10: 155-159.
85. Grand L, Benson D, Gottwald T, Vernia C: Temporal and spatial epidemiology of phytophthora root rot in fraser fir plantations. Phytopathology. 2006, 90 (9): 1171-1180.
86. Mayton H, Aylor D, Fry W, Andrade-Piedra J: Quantifying the rate of release and escape of Phytophthora infestanssporangia from a potato canopy. Phytopathology. 2001, 91 (12): 1189-1196.
87. Xu M, Ridout X: Stochastic simulation of the spread of race-specific and race-nonspecific aerial fungal pathogens in cultivar mixtures. Plant Pathology. 2000, 49 (2): 207-218.
88. Alm E, Arkin A: Biological networks. Curr Opin Struct Biol. 2003, 13 (2): 193-202.
89. Aguda BD, Goryachev AB: From pathways databases to network models of switching behavior. PLoS Comput Biol. 2007, 3 (9): 1674-1678.
90. Vidal M: Interactome modeling. FEBS Lett. 2005, 579 (8): 1834-1838.
91. Davis F, Barkan D, Eswar N, McKerrow J, Sali A: Host pathogen protein interactions predicted by comparative modeling. Protein Sci. 2007, 16 (12): 2585-2596.
92. Li P, Zhang C, Perkins E, Gong P, Deng Y: Comparison of probabilistic boolean network and dynamic bayesian network approaches for inferring gene regulatory networks. BMC Bioinformatics. 2007, 8 (Suppl 7): S13-
93. Colizzaa V, Maritanb A, Flamminia A, Vespignani A: Characterization and modeling of protein-protein interaction networks. Physica A: Statistical Mechanics and its Applications. Physics Applied to Biological Systems, Physics Applied to Biological Systems. 2005, 35: 1-27.
94. Freitas R, Santos , Chen Z, Batista M, Duarte J, Borges A, Texixeira A, Ferreira R, Monteiro S: The role of plant defense proteins in fungal pathogenesis. Molecular Plant Pathology. 2007, 8 (5): 677-700.
95. Govers F, Gijzen M: Phytophthora genomics: the plant destroyers' genome decoded. Mol Plant Microbe Interact. 2006, 19 (12): 1295-1301.
96. Gupta M, Qu P, Ibrahim J: A temporal hidden markov regression model for the analysis of gene regulatory networks. Biostatistics. 2007, 8 (4): 805-820.
97. Won K, Sandelin a, Marstrand T, Krogh A: Modeling promoter grammars with evolving hidden markov models. Bioinformatics. 2008, 24 (15): 1669-1675.
98. Genoud T, Trevino M, Métraux J: Numeric simulation of plant signaling networks. Plant Physiol. 2001, 126 (4): 1430-1437.
99. Honda H, Hakamada K, Hanai T, Kobayashi T: A preprocessing method for inferring genetic interaction from gene expression data using boolean algorythm. Journal of Bioscience and Bioengineering. 2004, 98: 457-463.
100. Gupta R, Achenie L: A network model for gene regulation. Computers and Chemical Engineering. 2006, 31 (8): 950-961.
101. Campbell C: Interpretation and uses of diseases progress curves for root diseases. Plant Disease Epidemiology: Population Dynamics of Management. Edited by: Jones AL, Leonard KJ, Fry WE. 1986, New York: MacMillan, 1: 38-54.
102. Kishalappa A, Ludwig A: Calculation of apparent infection rate in plant diseases: Development of a method to correct for host growth. Phytopathology. 1982, 72: 1373-1377.
103. Kester D, Fenton C, Kuniyuski C: Models for noninfectious bud failure in almonds. Phytopathology. 1988, 78: 139-143.
104. Vander Plank J: Plant diseases: Epidemic and control. 1963, Academic Press. New York
105. Vander Plank JE: Host-pathogen interactions in plant disease. 1982, Academic Press. New York
106. Sugiyama Y, Watase Y, Nagase M, Makita N, Yagura S, Hirai A, Sugiura M: The complete nucleotide sequence and multipartite organization of the tobacco mitochondrial genome: comparative analysis of

mitochondrial genomes in higher plants. Mol Genet Genomics. 2005, 272 (6): 603-615.

107. Yukawa M, Tsudzuki T, Sugiura M: The chloroplast genome of Micotiana sylvestris and Nicotiana tomentosiformis: complete sequencing confirms that the Nicotiana sylvestris progenitor is the maternal genome donor of Nicotiana tabacum. Mol Genet Genomics. 2006, 275: 367-373.

108. Vezzi A, Scotti N, Gray JC, Valle G, Grillo S, Gargano D, Cardi T: The complete nucleotide sequence of potato (Solanum tuberosum cv.desiree) chloroplast dna. Abstracts 2nd Solanaceae genome workshop. 2005

109. Grevich J, Saski C, Quesada-Vargas T, Guda C, Tomkins J, Jansen RK, Daniell H, Lee SB: Complete chloroplast genome sequences of Solanum bulbocastanum, Solanum lycopersicum and comparative analyses with other solanaceae genomes. Theor Appl Genet. 2006, 8: 112-

110. Kahlau S, Aspinall S, Gray J, Bock R: Sequence of the tomato chloroplast dna and evolutionary comparison of solanaceous plastid genomes. J Mol Evol. 2006, 63 (2): 194-207.

111. Yukawa M, Tsudzuki T, Sugiura M: The chloroplast genome of Nicotiana sylvestris and Nicotiana tomentosiformis: complete sequencing confirms that the Nicotiana sylvestris progenitor is the maternal genome donor of Nicotiana tabacum. Mol Genet Genomics. 2006, 275 (4): 367-373.

112. Schmitz C, Regel R, Du T, Hupfer H, Herrmann R, Maier R: The plastid chromosome of Atropa belladonna and its comparison with that of Nicotiana tabacum: the role of rna editing in generating divergence in the process of plant speciation. Mol Biol Evol. 2002, 19 (9): 1602-1612.

CHAPTER 21

PUSHING THE BOUNDARIES OF RESISTANCE: INSIGHTS FROM BRACHYPODIUM-RUST INTERACTIONS

Melania Figueroa, Claudia V. Castell-Miller, Feng Li, Scot H. Hulbert and James M. Bradeen

Keywords: Brachypodium, rust fungi, Puccinia, plant immunity, non-host resistance

INTRODUCTION

Cereals, which are classified within the grass family Poaceae, also known as Gramineae, are essential worldwide commodities (Chapman, 1996). The importance of these plant species ranges from economic to ecological, and stems from their multiple applications in industry, food production, livestock feed, generation of renewable and sustainable energy among others. The critical role of cereals in human nutrition is highlighted by the fact that wheat, corn, and rice provide nearly two thirds of the global caloric intake (Tilman et al., 2011). Other crops, like oat and millet, exemplify the extraordinary nutritional value that cereals can provide (Bewley et al., 2012;Shukla and Srivastava, 2014). In a different context, the cultivation of grasses such as switchgrass, sweet sorghum, *Miscanthus*, and especially sugarcane and corn, are the main basis of biofuel production (Somerville, 2007) increasing the value of these plant species.

According to predictions of worldwide population growth, agricultural production must increase approximately twofold to meet the dietary needs of nine billion people in the year 2050 (US Census Bureau, 2015; Tilman et al., 2011). In addition to this, crop production systems need to consider a dietary bias toward higher consumption of meat and dairy, and the demands of increasing biomass to support biofuel production (Ray et al., 2013). Increasing crop yields and boosting agricultural production are among the greatest challenges humanity faces, as research suggests current crop yield trends are not on track to reach our target food production (Ray et al., 2013). Unpredictable and recurrent epidemics caused by plant pathogens are factors that have hampered productivity of crops since the birth of agriculture about 10,000 years ago (Oerke, 2006). The mid 1800s Irish potato famine caused by the oomycete *Phytophthora infestans* (Woodham-Smith, 1991) and other similar epidemics demonstrate the influence of plant diseases in history and society (Schumann, 1991). Currently, pre-harvest plant disease causes an approximately 15% loss of global crop production (Popp and Hantos, 2011).

The global production of small grains is severely affected by rust fungi (Maier et al., 2003). Some of these economically important pathogens include members of the genus *Puccinia* such as *P. graminis*, which affects production of wheat (*Triticum aestivum* and *T. durum*), barley (*Hordeum vulgare*) and oat (*Avena sativa*), *P. triticina*, the causal agent of wheat leaf rust, *P. striiformis*, which causes stripe rust on wheat, and *P. coronata*, the causal agent of oat crown rust (Cummins, 1971). In fact, the recurrent outbreaks

Citation: Figueroa M, Castell-Miller CV, Li F, Hulbert SH and Bradeen JM (2015) Pushing the boundaries of resistance: insights from Brachypodium-rust interactions. Front. Plant Sci. 6:558. doi: 10.3389/fpls.2015.00558

of *P. coronata* f. sp. *avenae* and the rapid emergence of hypervirulent races of the wheat stem rust fungus, *P. graminis* f. sp. *tritici* (i.e., TTKSK) are two examples that illustrate the potential devastating economic effects of rust fungi on crops around the globe (Pretorius et al., 2000; Leonard and Martinelli, 2005; Singh et al., 2006; Park, 2008;Pennisi, 2010). This review discusses the biological attributes of rust fungi, and how those characteristics can contribute to expand our knowledge in plant immunity. We focus on the development of *Brachypodium*-rust pathosystems, which are important at two fronts, first helping us to understand the principles of non-host resistance (NHR) and second holding promise to enhance disease resistance engineering technologies.

THE TWO FACES OF RUST FUNGI—DANGEROUS PATHOGENS AND GOOD MODELS

Rust fungi exhibit complex lifecycles with both sexual and asexual (clonal) cycles of reproduction. In the case of the previously listed *Puccinia* species, sexual reproduction is completed on an alternate host as these pathogens undergo a heteroecious life cycle (Al-Kherb et al., 1987; Jin et al., 2010). In contrast, other rust fungi known to be autoecious, like the causal agent of flax rust*Melampsora lini*, depend solely on one host to complete both asexual and sexual reproduction (Lawrence et al., 2007). An interesting biological characteristic of rust fungi is that they generally show high levels of host specificity, although there is variability in host range exhibited by these pathogens. For instance, species like *P. triticina* infect a limited number of plant species (Roelfs et al., 1992), while others like *P. graminis* and *P. coronata* have a host range in the order of hundreds of related species (Anikster, 1984; Wahl et al., 1984; Leonard and Szabo, 2005). However, in these cases, the rust species can generally be classified into different physiological variants known as formae speciales (ff. spp.), each of them having a much more restricted host range (Anikster, 1984). For example, *P. graminis* f. sp. *avenae* affects only oat production, while *P. graminis* f. sp. *tritici*targets barley and wheat (Leonard and Szabo, 2005). The investigation of the evolutionary and physiological factors determining host range and specificity of rust species and formae speciales is of interest as those factors could provide insights into the mechanisms that drive pathogen jumps from one host species to another and the emergence of new diseases (Bettgenhaeuser et al., 2014). Unraveling the riddles of sexual reproduction in rust fungi, the biological contribution of alternate hosts in introducing genetic variation, particularly virulence alterations, can also help understand the rise of new pathogens or pathotypes.

Rust fungi have co-evolved with their respective host(s) for millions of years, a process that has resulted in the adoption of complex and effective infection strategies involving specialized cell types and morphological structures (Harder, 1984; Harder and Chong, 1984; Wahl et al., 1984). Such complexity is manifested by the production of up to five type of spores: basidiospores, pycniospores, aeciospores, urediniospores, and teliospores. However, most research in developmental biology of rust fungi has been directed to the infection mediated by urediniospores as they are responsible for multiple infection cycles in the gramineous host with agronomic value (Leonard and Szabo, 2005). *P. graminis* f. sp. *tritici* is one of the best characterized pathogenic rust fungi and has been widely used to describe the morphological differentiation processes associated with rust diseases of cereals. The elaborate developmental program followed by rust fungi is initiated by the recognition of the host environment, where chemical, thigmotropic and topographical characteristics of the leaf surface prompt hyphal growth and orientation (Hoch et al., 1987; Heath, 1997; Brand and Gow, 2012). Moisture on the leaf surface induces the germination of urediniospores, a process that is followed by the elongation of a germ tube and location of a stoma, which serves as a cue to form an appressorium and gain access to internal cavities of the plant (Staples and Macko, 2004; Leonard and Szabo, 2005). The course of infection continues with the differentiation of a substomatal vesicle and the emergence of a primary hypha that quickly extends in the plant mesophyll space. During more advanced stages of infection, feeding structures, referred to as haustoria, penetrate the walls of living plant cells, and invaginate the host plasma membrane forming an intimate association with the plant (Harder and Chong, 1984; Staples and Macko, 2004). Finally, the colonization leads to sporulation and the generation of uredial pustules erupting through the epidermis of the leaf (Leonard and Szabo, 2005). Management of diseases caused by rust fungi is often a difficult task as the pathogen grows asymptomatically during colonization and disease does not manifest until the fungus is undergoing sporulation (Leonard and Szabo, 2005). In this context, furthering our understanding of biochemical and molecular events particularly during early stages of infection should be regarded as a research priority.

The effectiveness of introducing disease resistance into crops to counteract the negative effect of rust fungi was established in the very early days of plant breeding with the identification of a resistance gene effective against *P. striiformis* (Biffen, 1907). Early studies in the flax-*M. lini*pathosystem played an important role in establishing the concepts of plant immunity, especially Flor's research leading to the "gene-for-gene" model, and demonstrate the usefulness of rust fungi to uncover the most fundamental principles in plant pathology (Flor, 1971; Lawrence et al., 2007;Ellis et al., 2014). It is now evident that the genetically determined plant immune system constitutes the underlying basis of breeding crop plants for pathogen resistance and dissecting the factors that govern the onset of rust disease can bring innovation to crop protection programs (Dodds and Rathjen, 2010). The distinctive morphological features that rust fungi adopt during infection, as well as the extended period of time that these pathogens spend in contact with their host, enable ideal experimental systems to elucidate how the fungus copes with the plant defenses and resolve molecular changes in relationship to infection stages at a high-resolution.

THE PLANT IMMUNE SYSTEM AND THE PURSUIT OF CROP HEALTH

In nature plants are constantly threatened by different microorganisms; however, the development of disease as an outcome of those interactions is rather uncommon (Dangl and Jones, 2001; Lipka et al., 2010). Whether a plant is suitable as a host to a particular would-be pathogen is partly conferred by a battery of physiological and biochemical immune responses that result in a wide array of phenotypes ranging from complete immunity (incompatibility) to full susceptibility (compatibility). Plant disease resistance has been classified historically as either host resistance, describing resistance that is specific to a plant cultivar, variety or accession to a pathogen that is able to infect other host genotypes of that species (Flor, 1971; Jones and Dangl, 2006; Dodds and Rathjen, 2010), or NHR, referring to a type of resistance that is present across all genotypes of a plant species and prevents infection by any genetic variants (i.e., formae speciales, races, isolates) of a given non-adapted pathogen (Heath, 1981, 2000; Mysore and Ryu, 2004; Bettgenhaeuser et al., 2014). Considering that most of the naturally occurring disease resistance is attributed to NHR, this type of plant immunity holds great promise to agricultural practices as means to provide durable and broad-spectrum resistance (Mysore and Ryu, 2004; Ellis, 2006; Schulze-Lefert and Panstruga, 2011). Our present understanding of the mechanisms underlying NHR is much less extensive than the accumulated body of knowledge that conveys the molecular basis of host resistance. Recent evidence suggests that components mediating host resistance are likely important determinant factors in NHR, and thus both types of resistance may share some of the same inducible responses (Huitema et al., 2003; Mysore and Ryu, 2004; Douchkov et al., 2014; Lee et al., 2014). Consequently, the distinction between host and NHR may not truly reflect mechanistic differences among the two types of resistance, as both phenomena may exploit the same means of pathogen perception and defense.

Microbial recognition follows a two-tier system, enabling the plant to mount a defense response that may or may not be effective depending on the physiological characteristics of the would-be pathogen. First, the plant is equipped to recognize conserved pathogen-associated molecular patterns (PAMPs) such as the fungal cell wall component chitin (Jones and Dangl, 2006; Dodds and Rathjen, 2010). This phenomenon, known as PAMP-triggered immunity (PTI), is thought to be particularly relevant in NHR as it effectively prevents infection by non-adapted pathogens (Schulze-Lefert and Panstruga, 2011). PTI operates via plant cell surface-associated receptors (pattern recognition receptors, PRRs) that activate various basal defense responses such as callose deposition, production of reactive oxygen species, activation of pathogenesis-related proteins, etc. (Monaghan and Zipfel, 2012; Fu and Dong, 2013). The second tier of plant defense, known as effector-triggered immunity (ETI), exploits the backbone of many pathogens' virulence strategies. Pathogens secrete proteins and small molecules, known as effectors, which among other functions can suppress or inhibit PTI-associated defense responses to allow infection to proceed (Hogenhout et al., 2009; Koeck et al., 2011). During ETI, the plants recognize effectors via intracellular receptor proteins of the nucleotide binding leucine rich repeat (NB-LRR) class, turning on complex cellular responses in order to restrict pathogen growth. These receptors are encoded by resistance (*R*) genes, and, in general, embody the traditional "gene-for-gene" concept initially described in the flax rust pathosystem (Flor, 1971; Jones and Dangl, 2006; Dodds and Rathjen, 2010). The recognition of effectors is typically associated with a rapid and localized cell death reaction at the site of infection, known as the hypersensitive response (HR); however, HR-like cell death can also occur during PTI (Khatib et al., 2004). The ETI concept is generally considered to encompass recognition of effectors delivered/translocated into the host cell; however, some effectors accumulate in apoplastic space and exert their function extracellularly (Hogenhout et al., 2009). Some of these apoplastic effectors are recognized by plasma membrane-associated receptors in the plant via a mechanism that differs from ETI and PTI (Stotz et al., 2014). This type of resistance, named effector-triggered defense (ETD), is commonly associated with apoplastic fungal leaf pathogens and can be manifested as slow developing cell death (Stotz et al., 2014). Regardless of the terminology and mechanistic differences between ETD and ETI, ETD should be regarded as part of the second tier of microbial recognition as its fundamental principle is still recognition of effector molecules. Successful infection by rust fungi requires haustorium-mediated delivery and translocation of effectors into the invaded plant cells (Catanzariti et al., 2006, 2007; Panstruga and Dodds, 2009;Garnica et al., 2014), and while the secretion of effectors via other structures has not been demonstrated it is possible that those events take place and are relevant to the rust pathogen–plant interaction outcomes. The presence of apoplastic effectors produced by rust fungi has not been demonstrated either, but it seems reasonable to think that this class of effectors could be important during early stages of rust infection, prior the differentiation of the first haustorium. Barley-*P. graminis* interactions mediated by the *Rpg1* gene clearly begin before haustorial formation since the RPG1 protein is phosphorylated within 5 min of spores from avirulent pathotypes contacting the leaf surface (Nirmala et al., 2010). According to these findings, two effector proteins required for the phosphorylation and subsequent degradation of RPG1 are present on the spore surface and probably other fungal structures (Nirmala et al., 2011).

In addition to immune system-induced resistance, genetic protection from disease can occur through the action of resistance genes that have other effects. These genes often provide broad-spectrum resistance against multiple pathogens, and, although frequently referred to as resistance genes, do not necessarily encode receptor-like proteins like NB-LRR proteins. An example of this type of resistance in wheat is the response conditioned by the gene *Lr34*, which encodes a putative adenosine triphosphate-binding cassette (ABC) transporter protein, and confers partial resistance against leaf, stem and stripe rust fungi, as well as barley yellow dwarf virus and powdery mildew (Dyck et al., 1966; Singh, 1993; Krattinger et al., 2009; Risk et al., 2013). In summary, this broad array of defense mechanisms represents targets that may be exploited in various strategies to prevent and minimize crop losses.

BRACHYPODIUM DISTACHYON: A MODEL SYSTEM THAT DEFIES RUST DISEASES

Throughout research history, the use of experimental model organisms such as *Escherichia coli,Saccharomyces cerevisiae*, *Neurospora crassa*, and *Caenorhabditis elegans* have accelerated scientific discovery and led to significant breakthroughs (Miklos and Rubin, 1996; Guarente and Kenyon, 2000; Davis and Perkins, 2002; Replansky et al., 2008). Research in wheat, oat, and other monocots often faces long-standing challenges associated with the ploidy of the plants and the large size of their genomes (Chawade et al., 2010; Mayer et al., 2014). The use of representative model species can help circumvent those problems and accelerate the pace of research. In addition, leveraging basic knowledge from model organisms for applied purposes, such as disease resistance, can be extremely advantageous to crop improvement. Dicotyledonous species such as *Arabidopsis thaliana* (thale cress or mouse-ear cress) and *Solanum lycopersicum* (tomato) have served as excellent models to elucidate the plant immune system, understand the basis of plant–microbe interactions and have made significant contributions to crop improvement (Piquerez et al., 2014). However, dicotyledonous and monocotyledonous plants diverged at least 140–150 million years ago (Chaw et al., 2004; Davies et al., 2004; Anderson et al., 2005) and the substantial phylogenetic distance between cereals and dicots can often hinder downstream applications of knowledge gained from dicot models to cereal research and breeding. Thus, there is need for developing appropriate systems to study monocotyledonous plants. Acknowledging this necessity and the role of cereals in food security, rice (*Oryza sativa* L.) was launched as a genetic and genomic model several years ago (Goff et al., 2002; Project, 2005). These efforts facilitated advances in comparative genomics, and demonstrated that characterization of the genic content and landscape of one grass species could be informative and applicable to other related species (Gale and Devos, 1998; Phillips and Freeling, 1998). Rice also turned out to be a useful model to study plant immunity and the existing genomic resources and tools in this species are highly valuable to plant scientists and breeders (Phillips and Freeling, 1998; Chen and Ronald, 2011; Chen et al., 2013). Nevertheless, basic biological and physiological differences between *Oryza* spp. and temperate grasses can hinder rice's utility to inform on certain processes, like rust resistance, in cool-season crops (Phillips and Freeling, 1998;Draper et al., 2001; Vogel and Bragg, 2009; Chen et al., 2013). Disease resistance in rice to important cereal pathogens is not always genetically tractable. While rice acts as a non-host to all rust species, the variation in NHR phenotypes across cultivars can be subtle and often influenced by environmental factors; limiting the value of rice in rust research (Ayliffe et al., 2011, 2013).

To tackle some of these challenges, the small wild grass *Brachypodium distachyon* (Pooideae subfamily), also known as purple false brome, was developed as an experimental model to investigate temperate grass biology (Draper et al., 2001; Vogel and Bragg, 2009; Brkljacic et al., 2011; Mur et al., 2011). *B. distachyon* diverged from the wheat phylogenetic lineage approximately 35–40 million years ago indicating it is more closely related to Triticeae than rice (Bossolini et al., 2007). As a species, *B. distachyon* has no agronomic value; however, it possesses biological attributes advantageous for its use as a model system (Draper et al., 2001; Vogel and Bragg, 2009;Mur et al., 2011). Similar to *A. thaliana*, *B. distachyon* exhibits a small plant size, compact genome (~270–300 Mb), short life cycle (2–3 months), genetic tractability, and minimal growth requirements. There are several species within the genus *Brachypodium* and their taxonomic classification has been recently resolved (Catalán et al., 2012; López-Alvarez et al., 2012). *B. distachyon* and *B. stacei* include diploid plants with 5 and 10 pairs of chromosomes, respectively; whereas *B. hybridum* is an allotetraploid species that resulted from a hybridization event between*B. distachyon* and *B. stacei* (López-Alvarez et al., 2012). The different ploidies within the genus offers opportunity to conduct evolutionary, gene inheritance and expression studies in a system that is easier to handle than related crops like wheat and oat. The embracement of *Brachypodium*as model has resulted in a wealth of genetic, genomic, and experimental assets that set the stage to answer many interesting questions in plant biology and pathology. Some of the established resources include large natural germplasm collections and several families of recombinant inbred lines available to the scientific community (Jenkins et al., 2003; Vogel et al., 2006a, 2009; Vogel and Hill, 2008; Scoles et al., 2009). In addition, there is access to bacterial artificial chromosome (BAC) and expressed sequence tag (EST) libraries (Vogel et al., 2006b; Huo et al., 2008), genetic and physical maps (Febrer et al., 2010), simple sequence repeat (SSR) microsatellite markers (Azhaguvel et al., 2009; Vogel et al., 2009; Garvin et al., 2010), a high quality Sanger-based reference genome assembly of *B. distachyon* inbred line Bd21 and deep sequencing data corresponding to additional genotypes[1] (International Brachypodium Initiative, 2010; Gordon et al., 2013). The implementation of efficient *Agrobacterium tumefaciens*-mediated transformation methods allowed the creation of two large banks of T-DNA insertion mutants in *B. distachyon*[2,3], which are also a great resource available to the community (Alves et al., 2009; Bragg et al., 2012;Thole et al., 2012).

From the perspective of research in plant immunity, *B. distachyon* is emerging as a suitable system to investigate plant–microbe interactions (Fitzgerald et al., 2015). In contrast to rice,*Brachypodium* species can act as a host to *Puccinia brachypodii*, an adapted pathogen of *B. sylvaticum*, thus better positioning this system to investigate various aspects of rust compatibility (Barbieri et al., 2011). Initial pathogenicity tests on various ecotypes of *B. distachyon* showed the potential of the species to investigate the molecular basis of resistance against pathogens like*Blumeria graminis*, *Magnaporthe oryzae*, and various rust fungi such as *P. hordei*, *P. triticina*, *P. striiformis* f. sp. *tritici* and *P. striiformis* f. sp. *hordei* (Draper et al., 2001; Routledge et al., 2004;Peraldi et al., 2011). Since these early reports, the number of studies examining the interaction of *B. distachyon* with pathogenic fungi, some of which affect food and biofuel crops, is growing. *B. distachyon* has been established as a pathosystem to study important grain diseases such as to*Fusarium* head blight, caused by *Fusarium graminearum* (Peraldi et al., 2011), *Rhizoctonia* root rot in wheat, caused by *Rhizoctonia solani* (Schneebeli et al., 2014), spot blotch and common root rot caused by *Cochliobolus sativus* (Zhong et al., 2014), Septoria tritici blotch disease, caused by*Zymoseptoria tritici* (O'Driscoll et al., 2015), take-all caused by *Gaeumannomyces graminis*(Sandoya and de Oliveira Buanafina, 2014), Stagonospora nodorum blotch and net blotch of barley caused by *Pyrenophora teres* (Falter and Voigt, 2014), as well as other grass diseases caused by*Sclerotinia homoeocarpa* and *Ophiosphaerella agrostis* (Sandoya and de Oliveira Buanafina, 2014).

The role of *Brachypodium* spp. as a non-host to formae speciales of rust fungi, notably *P. graminis*,*P. triticina*, and *P. striiformis*, is now very well-established (Garvin, 2011; Ayliffe et al., 2013;Figueroa et al., 2013). Most recently, *B. distachyon* was also recognized as a non-host to *P. emaculata*, the causal agent of switchgrass rust (Gill et al., 2015). Challenging *Brachypodium* spp. with non-adapted rust fungi results in symptoms that are remarkably different than typical symptoms in compatible interactions (Figure 1A; Barbieri et al., 2011; Ayliffe et al., 2013; Figueroa et al., 2013; Gill et al., 2015). The density and appearance of these symptoms varies greatly. Certain*Brachypodium*-rust interactions result in immunity, whereas others can display symptoms at a macroscopic scale, ranging from small necrotic flecks to spreading necrotic and/or chlorotic lesions that may surround small sporulating pustules, responses that are absent in fully-compatible interactions (Figures 1A–E; Ayliffe et al., 2013; Figueroa et al., 2013). In cases where macroscopic symptoms are completely absent due to either failure of the fungus to accomplish plant colonization and/or reach sporulation stages, only microscopic examination of inoculated tissue can allow symptom visualization. *B. distachyon* possesses sufficient chemical, topographical, and thigmotropic signals to induce differentiation of appressoria and in some cases, substomatal vesicles, by most *Puccinia* species examined; although orientation of the germ tube can be affected as in the case of *P. emaculata* (Gill et al., 2015). If fungal growth occurs in the mesophyll space, the infection can advance to the formation of haustoria and sporulating pustules (Figure 2; Ayliffe et al., 2013; Figueroa et al., 2013). Interestingly, the formation of haustoria and pustules was not detected in the *P. emaculata–B. distachyon* interaction (Gill et al., 2015). However, such observations could be specific to the *P. emaculata* isolate used in their study. The resistance of *B. distachyon* to *P. graminis* f. sp. *tritici* has been observed as prehaustorial, as most of the infection sites failed to display growth in the plant mesophyll (Figueroa et al., 2013), and post-haustorial, as different races of stem rust fungus can accomplish the formation of colonies that vary in size (Figures 2C–E; Ayliffe et al., 2013). Most importantly for genetic studies, resistance in*Brachypodium* to rust fungi varies substantially across accessions or inbred lines. This phenotypic variability makes it possible to employ genetic-based approaches to map and clone genes governing stem rust resistance in *Brachypodium* species. In fact, differential phenotypes of *B. distachyon*inbred lines Bd3-1 and Bd1-1 in response to *P. brachypodii* allowed the identification of three distinctive rust resistance quantitative trait loci QTLs (Barbieri et al., 2011, 2012). Assessment of the inheritance of the resistance to *P. striiformis* f. sp. *tritici* in two different *B. distachyon*mapping families from parental lines with contrasting phenotypes, BdTR13k × Bd21 and Bd10TR10h × TEK4, suggested that major differences in NHR are simply inherited (Ayliffe et al., 2013). Similar differential responses to *P. graminis* f. sp. *tritici* have been detected in multiple *B. distachyon* inbred lines and accessions (Ayliffe et al., 2013; Figueroa et al., 2013), encouraging comparable resistance inheritance studies against this pathogen.

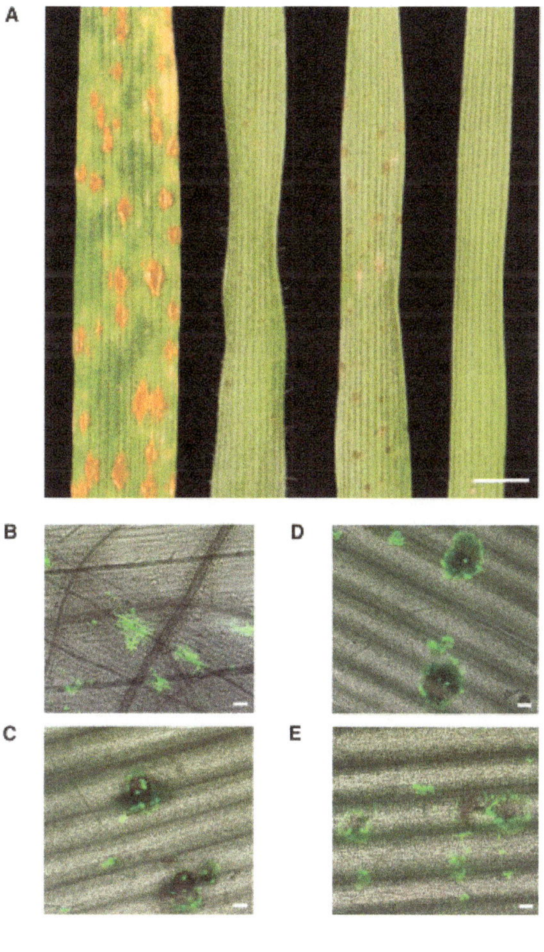

Figure 1: Symptoms induced by *P. graminis* **f. sp.** *tritici* **in wheat and** *B. distachyon*. (A) Left to right, macroscopic symptoms of *P. graminis* f. sp. *tritici* (CRL 75-36-700-3) in susceptible *Triticum aestivum* cv. McNair 701, *B. distachyon* inbred lines Bd1-1, Bd2-3, Bd21 at 12 days after inoculation, scale = 2 mm. **(B–E)** Microscopic symptoms induced by *P. graminis* f. sp. *tritici* and various inbred lines of *B. distachyon* at 96 h post-inoculation. Inoculations were performed as previously reported (Figueroa et al., 2013). Fungal tissue was stained with wheat germ agglutinin conjugated to fluorescein isothiocyanate (WGA-FITC; Sigma-Aldrich, St. Louis, MO, USA) as previously described (Ayliffe et al., 2013). **(B)** Susceptible *Triticum aestivum* cv. McNair 701. **(C)** Inbred line Bd1-1. **(D)** Inbred line Bd2-3. **(E)** Inbred line Bd21. Images were generated by merging bright and fluorescence fields captured using an Olympus IX70 Inverted Fluorescence Microscope, scale = 100 μm.

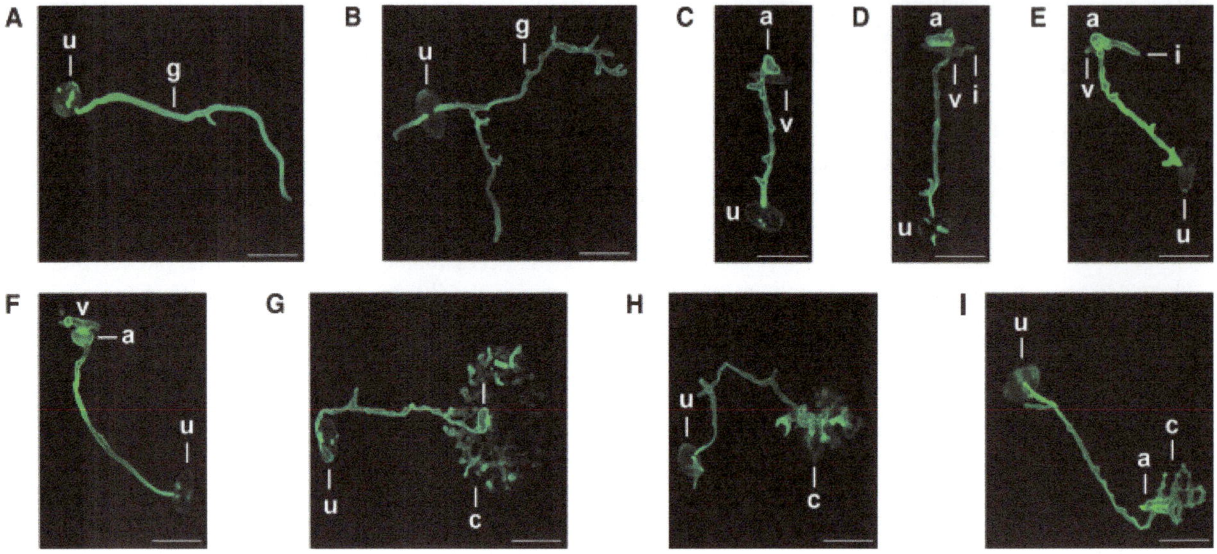

Figure 2: Differentiation of morphological structures associated with the infection of *P. graminis* **f. sp.** *tritici* **on** *B. distachyon*. *P. graminis* f. sp. *tritici*(CRL 75-36-700-3) was used to inoculate various *B. distachyon* inbred lines following procedures previously reported (Figueroa et al., 2013). Fungal tissue was stained WGA-FITC as previously described (Ayliffe et al., 2013). **(A,B)** Germination of urediniospores (u) on the surface of leaves of *B. distachyon* inbred line Bd1-1, micrographs illustrate the different morphological features of germ tubes as they elongate to locate a stoma. **(C–E)** Differentiation of an appressorium (a) and substomatal vesicle (v) in*B. distachyon* inbred lines Bd1-1, Bd2-3, Bd21, respectively, at 24 hpi (10 h of light exposure). Notice differences in the growth of primary infection hyphae (i) as it emerges from the substomatal vesicle. **(F)** Example of an infection site in *B. distachyon* inbred line Bd1-1 that failed to continue growth after forming the appressorium and substomatal vesicle. Image was capture at 96 hpi. **(G–I)** Presence of rust fungal colonies in *B. distachyon* inbred lines Bd1-1, Bd2-3, Bd21, respectively, at 96 hpi. Images were captured using epifluorescence in a Nikon A1 Spectral Confocal Microscope, scale = 50 μm.

Our understanding of the genomic architecture controlling NHR in *B. distachyon* is in its infancy. Microscopic analyses and transcriptional profiling have started to shed light into the mechanisms that could mediate resistance in *B. distachyon* against non-adapted rust pathogens. Accumulation of callose and papillae formation has been observed at infection sites during penetration (substomatal vesicles) in interactions with *P. graminis* ff. spp. *tritici*, *lolii*, and *phleipratensis*, *P. striiformis* f. sp. *tritici*, and *P. triticina* (Ayliffe et al., 2013; Figueroa et al., 2013). Interestingly, comparisons of callose deposition at infection sites where the growth of rust fungi was arrested in both host and non-host scenarios, suggests that both types of mechanisms of resistance share common responses and supports the role of PTI in NHR (Ayliffe et al., 2013). Quantification of transcript abundance when three different *B. distachyon* inbred lines were challenged with *P. emaculata* suggest a mixture of responses linked to jasmonic acid, ethylene, and salicylic acid signaling pathways, as well as induction of HRs (Gill et al., 2015). In contrast, evaluation of salicylic acid levels indicated no change in response to *P. graminis* f. sp. *tritici* (Ayliffe et al., 2013). While cell death and dark pigmentation in the rust-infected tissue of various *Brachypodium* accessions and lines is commonly observed (Figures 1C–E; Ayliffe et al., 2013; Figueroa et al., 2013), microscopic analyses indicated that such necrotic responses are not typically associated with autofluorescence, suggesting that hypersensitive cell death is not playing a role in NHR (Ayliffe et al., 2013).

The relative contributions of ETI and PTI to NHR is a long-standing question. It is thought that evolutionary distance between host and non-host plays a major role in determining the relative importance of ETI and PTI processes in these plant–microbe interactions, with ETI favored in non-hosts that are closely related to the natural host (Schulze-Lefert and Panstruga, 2011;Bettgenhaeuser et al., 2014). The phenotypic outcome of*Brachypodium*-rust interactions appear to be highly influenced by the genotype of both plant and fungus, which suggests a relatively strong ETI component to these NHR interactions (Ayliffe et al., 2013; Figueroa et al., 2013;Bettgenhaeuser et al., 2014; Gill et al., 2015). Members of *P. graminis* which are pathogens of plants in the tribe Aveneae and Poeae (ff. spp. *avenae*, *lolii*, *phalaridi*, *phleipratensis*) are more likely to form sporulating pustules on *Brachypodium* accessions than *P. graminis* f. sp. *tritici*(Ayliffe et al., 2013; Figueroa et al., 2013), which supports the view that such interactions resemble more host resistance responses (Ayliffe et al., 2013; Bettgenhaeuser et al., 2014). Future studies are required to carefully evaluate whether ETI is playing a role in the resistance of *B. distachyon*against non-adapted rust pathogens. The mode-of-action of R genes and other broad-spectrum resistance genes is often poorly understood, and therefore, the efficacy of their use in breeding programs is likely undermined. Elucidation of the mechanisms of action of these genes is an area of investigation that could be strengthened by basic research in model plants like *B. distachyon*.

The *B. distachyon* T-DNA insertion mutant resources provide an excellent platform to enable reverse genetics and functional genomic approaches to identify major components of NHR against rust fungi. In addition, the various levels of disease resistance in *Brachypodium* against rust isolates can be exploited to conduct genetic-based approaches to identify, map and clone genes that govern NHR. Recent genomic advances and catalogs of high confidence predicted effectors in North American and Australian stem rust isolates (Duplessis et al., 2011; Upadhyaya et al., 2015) have positioned *P. graminis* f. sp. *tritici* as an ideal organism to investigate the role of ETI during NHR. In combination with the resources available for *B. distachyon*, *P. graminis* f. sp. *tritici* makes a powerful system to investigate the strategies by which obligate biotrophs can be able to reproduce in the face of less suited "host" environments.

Additionally, the *Brachypodium* NHR system could be regarded as a potential source of genetic disease resistance for cereals. The effectiveness of *B. distachyon* derived-genes in transgenic cereals as a strategy for crop protection has not been evaluated yet; however, the high genome collinearity and synteny among the species (International Brachypodium Initiative, 2010) suggests that such genes have high probability of being effective in cereal crops. On the other hand, advances in unraveling the genomic structure and organization of *Hordeum vulgare* (International Barley Genome Sequencing Consortium et al., 2012) and *Triticum aestivum* (Mayer et al., 2014), and the seed transcriptome assembly of *Avena sativa* (Gutierrez-Gonzalez et al., 2013) will enable comparisons with *B. distachyon*, which could be useful to support engineering of disease resistance programs via genome editing techniques.

In addition to its utility in studying NHR, it may be possible to utilize *B. distachyon* for experimentation essentially as a host if sufficiently compatible accession-pathotype combinations can be found. Natural infections of *P. striiformis* resulting in large uredinia have been observed on several accessions grown in nurseries in the Pacific Northwest, where pathotype variation is very high (E. Elmore, Pers. Comm.). Further development of this pathosystem by identifying the most compatible pathotype-accession combination, or possibly generating new lines by crossing the most susceptible accessions, would enable researchers to take advantage of *B. distachyon*'s amenability to transformation. High density marker systems for cereals are enabling candidate R genes to be identified relatively easily, but verification of the candidates are still lacking because transformation systems for wheat and barley are very inefficient. A compatible *B. distachyon* stripe rust system could be used to test candidate stripe rust genes or other components of resistance for function in stably transformed plants. Resistance signaling pathways could be examined by introducing heterologous R genes with, or without, other required donor species components or by using *B. distachyon* mutants. Novel approaches to engineering resistance could be tested in *B. distachyon* transgenics. An example would be RNAi constructs that appear to silence essential rust fungus genes and confer resistance but have only been tested by transient assays to date (Panwar et al., 2013; Yin et al., 2014). Fungal genes, like those encoding putative effectors, could be stably expressed in *B. distachyon* to examine their effects on host physiology to provide insight into their function.

CONCLUSIONS AND REMARKS

Current trends in food and bioenergy production systems compel the development of transformative approaches to combat plant disease in a sustainable and environmentally safe manner. Rust fungi are among the most important pathogens of grasses and cereals. Understanding the biology of these organisms, especially their mechanisms of virulence and host specificity, is essential to understand emergence of new pathogens or races. Exploring non-host and host pathogen interactions within a single rust species offers unique opportunities to also investigate how new pathogens emerge. Such fundamental knowledge can aid in planning and designing effective and long lasting crop protection strategies. The prospect of exploiting NHR as a natural durable approach to decrease crop yield losses due to biotic stress is well-supported. Defining the molecular and genetic mechanisms conferring NHR is not particularly straightforward given its presumed polygenic nature. However, the amenability of *Brachypodium*-rust fungi as a model pathosystem may accelerate the discovery of molecular and genetic mechanisms underpinning NHR. One avenue to uncover the genetic factors that control disease outcome is the screening of mutagenized *Brachypodium* populations to identify mutants that show susceptibility/compatibility to rust fungi. Combining genetic and genomic tools in both crops and model systems can result in foundational knowledge to support translational plant research. Model-to-crop studies can strengthen plant genetic engineering programs and their potential benefits to agriculture outweigh the time and scale of investments that such studies require. Genome editing techniques and transgenic technologies have quickly emerged as powerful research tools that can enhance traditional plant breeding techniques (Belhaj et al., 2013; Dangl et al., 2013). In spite of regulatory debates and controversial views, these approaches offer great potential to reduce crop yield losses and enable engineering of durable and sustainable plant disease resistance as exemplified by the commercial use of the genetically engineered multiviral resistant squash (Tricoll et al., 1995) and ringspot virus resistant papaya (Lius et al., 1997; Ferreira et al., 2002). Thus, it is worthwhile to invest efforts and research funding to build experimental systems that are meaningful to address immediate societal needs.

CONFLICT OF INTEREST STATEMENT

The authors declare that the research was conducted in the absence of any commercial or financial relationships that could be construed as a potential conflict of interest.

ACKNOWLEDGMENTS

We acknowledge the support from start-up funds provided by the University of Minnesota. We thank Grant Barthel at the University Imaging Center (http://uic.umn.edu/), and Matthew Martin at the Department of Plant Pathology for their assistance during collection of photographs, David Garvin for sharing seed stocks of *B. distachyon* inbred lines and the Cereal Disease Laboratory, USDA-ARS, St. Paul, MN for providing the isolate of *P. graminis* f. sp. *tritici* (CRL 75-36-700-3). We also thank the Stakman-Borlaug Center (SBC) for Sustainable Plant Health at the University of Minnesota for their supportive role to continue this research. The SBC brings together diverse multidisciplinary teams to tackle problems with both national and international relevance and is especially interested in innovative approaches to minimize the impact of rust pathogens in wheat, oat, and other crops which comprise economically and socially significant agricultural threats.

REFERENCES

1. Al-Kherb, S. M., Roelfs, A., and Groth, J. (1987). Diversity for virulence in a sexually reproducing population of *Puccinia coronata*. *Can. J. Bot.* 65, 994–998. doi: 10.1139/b87-137

2. Alves, S. C., Worland, B., Thole, V., Snape, J. W., Bevan, M. W., and Vain, P. (2009). A protocol for *Agrobacterium*-mediated transformation of *Brachypodium distachyon* community standard line Bd21. *Nat. Protoc.* 4, 638–649. doi: 10.1038/nprot.2009.30

3. Anderson, C. L., Bremer, K., and Friis, E. M. (2005). Dating phylogenetically basal eudicots using rbcL sequences and multiple fossil reference points. *Am. J. Bot.* 92, 1737–1748. doi: 10.3732/ajb.92.10.1737

4. Anikster, Y. (1984). "The formae speciales," in *The Cereals Rusts*, Vol 1, eds W. R. Bushnell and A. P. Roelfs (Orlando, FL: Academic Press, Inc.), 124–137.

5. Ayliffe, M., Singh, D., Park, R., Moscou, M. J., and Pryor, T. (2013). The infection of *Brachypodium distachyon* with selected grass rust pathogens. *Mol. Plant Microbe Interact.* 26, 946–957. doi: 10.1094/MPMI-01-13-0017-R

6. Ayliffe, M. A., Jin, Y., Kang, Z., Persson, M., Steffenson, B., Wang, S., et al. (2011). Determining the basis of nonhost resistance in rice to cereal rusts. *Euphytica* 179, 33–40. doi: 10.1007/s10681-010-0280-2

7. Azhaguvel, P., Li, W., Rudd, J. C., Gill, B. S., Michels, G., and Weng, Y. (2009). Aphid feeding response and microsatellite-based genetic diversity among diploid *Brachypodium distachyon* (L.) *Beauv accessions. Plant Gen. Res.* 7, 72–79. doi: 10.1017/S1479262108994235

8. Barbieri, M., Marcel, T. C., and Niks, R. E. (2011). Host status of false brome grass to the leaf rust fungus *Puccinia brachypodii*and the stripe rust fungus *P. striiformis. Plant Dis.* 95, 1339–1345. doi: 10.1094/PDIS-11-10-0825

9. Barbieri, M., Marcel, T. C., Niks, R. E., Francia, E., Pasquariello, M., Mazzamurro, V., et al. (2012). QTLs for resistance to the false brome rust *Puccinia brachypodii* in the model grass *Brachypodium distachyon* L. *Genome* 55, 152–163. doi: 10.1139/g2012-001

10. Belhaj, K., Chaparro-Garcia, A., Kamoun, S., and Nekrasov, V. (2013). Plant genome editing made easy: targeted mutagenesis in model and crop plants using the CRISPR/Cas system. *Plant Methods* 9, 39. doi: 10.1186/1746-4811-9-39

11. Bettgenhaeuser, J., Gilbert, B., Ayliffe, M., and Moscou, M. J. (2014). Nonhost resistance to rust pathogens–a continuation of continua. *Front. Plant Sci.* 5:664. doi: 10.3389/fpls.2014.00664

12. Bewley, J., Bradford, K., Hilhorst, H., and Nonogaki, H. (2012). *Seeds: Physiology of Development, Germination and Dormancy*. New York: Springer-Verlag.

13. Biffen, R. H. (1907). Studies in the inheritance of disease-resistance. *J. Agric. Sci.* 2, 109–128. doi: 10.1017/S0021859600001234

14. Bossolini, E., Wicker, T., Knobel, P. A., and Keller, B. (2007). Comparison of orthologous loci from small grass genomes*Brachypodium* and rice: implications for wheat genomics and grass genome annotation. *Plant J.* 49, 704–717. doi: 10.1111/j.1365-313X.2006.02991.x

15. Bragg, J. N., Wu, J., Gordon, S. P., Guttman, M. E., Thilmony, R., Lazo, G. R., et al. (2012). Generation and characterization of the western regional research center *Brachypodium* T-DNA insertional mutant collection. *PLoS ONE* 7:e41916. doi: 10.1371/journal.pone.0041916

16. Brand, A., and Gow, N. A. (2012). "Tropic orientation responses of pathogenic fungi," in *Morphogenesis and Pathogenicity in Fungi*, eds J. Pérez-Martín and A. Di Pietro (Heidelberg: Springer), 21–41. doi: 10.1007/978-3-642-22916-9_2

17. Brkljacic, J., Grotewold, E., Scholl, R., Mockler, T., Garvin, D. F., Vain, P., et al. (2011). *Brachypodium* as a model for the grasses: today and the future. *Plant Physiol.* 157, 3–13. doi: 10.1104/pp.111.179531

18. Catalán, P., Müller, J., Hasterok, R., Jenkins, G., Mur, L. A., Langdon, T., et al. (2012). Evolution and taxonomic split of the model grass *Brachypodium distachyon. Ann. Bot.* 109, 385–405. doi: 10.1093/aob/mcr294

19. Catanzariti, A.-M., Dodds, P. N., and Ellis, J. G. (2007). Avirulence proteins from haustoria-forming pathogens. *FEMS Microbiol. Lett.* 269, 181–188. doi: 10.1111/j.1574-6968.2007.00684.x

20. Catanzariti, A. M., Dodds, P. N., Lawrence, G. J., Ayliffe, M. A., and Ellis, J. G. (2006). Haustorially expressed secreted proteins from flax rust are highly enriched for avirulence elicitors. *Plant Cell* 18, 243–256. doi: 10.1105/tpc.105.035980

21. Chapman, G. P. (1996). *The Biology of Grasses*. Salisbury: CAB International.

22. Chaw, S.-M., Chang, C.-C., Chen, H.-L., and Li, W.-H. (2004). Dating the monocot–dicot divergence and the origin of core eudicots using whole chloroplast genomes. *J. Mol. Evol.* 58, 424–441. doi: 10.1007/s00239-003-2564-9

23. Chawade, A., Sikora, P., Bräutigam, M., Larsson, M., Vivekanand, V., Nakash, M. A., et al. (2010). Development and characterization of an oat TILLING-population and identification of mutations in lignin and β-glucan biosynthesis genes. *BMC Plant Biol.* 10:86. doi: 10.1186/1471-2229-10-86

24. Chen, H., He, H., Zhou, F., Yu, H., and Deng, X. W. (2013). Development of genomics-based genotyping platforms and their applications in rice breeding. *Curr. Opin. Plant Biol.* 16, 247–254. doi: 10.1016/j.pbi.2013.04.002

25. Chen, X., and Ronald, P. C. (2011). Innate immunity in rice. *Trends Plant Sci.* 16, 451–459. doi: 10.1016/j.tplants.2011.04.003

26. Cummins, G. B. (1971). *The Rust Fungi of Cereals, Grasses, and Bamboos*. New York, NY: Spinger-Verlag.

27. Dangl, J. L., Horvath, D. M., and Staskawicz, B. J. (2013). Pivoting the plant immune system from dissection to deployment.*Science* 341, 746–751. doi: 10.1126/science.1236011

28. Dangl, J. L., and Jones, J. D. (2001). Plant pathogens and integrated defence responses to infection. *Nature* 411, 826–833. doi: 10.1038/35081161

29. Davies, T. J., Barraclough, T. G., Chase, M. W., Soltis, P. S., Soltis, D. E., and Savolainen, V. (2004). Darwin's abominable mystery: insights from a supertree of the angiosperms. *Proc. Natl. Acad. Sci. U.S.A.* 101, 1904–1909. doi: 10.1073/pnas.0308127100

30. Davis, R. H., and Perkins, D. D. (2002). *Neurospora*: a model of model microbes. *Nat. Rev. Genet.* 3, 397–403. doi: 10.1038/nrg797

31. Dodds, P. N., and Rathjen, J. P. (2010). Plant immunity: towards an integrated view of plant–pathogen interactions. *Nat. Rev. Genet.* 11, 539–548. doi: 10.1038/nrg2812

32. Douchkov, D., Lueck, S., Johrde, A., Nowara, D., Himmelbach, A., Rajaraman, J., et al. (2014). Discovery of genes affecting resistance of barley to adapted and non-adapted powdery mildew fungi. *Genome Biol.* 15, 518. doi: 10.1186/s13059-014-0518-8

33. Draper, J., Mur, L. A. J., Jenkins, G., Ghosh-Biswas, G. C., Bablak, P., Hasterok, R., et al. (2001). *Brachypodium distachyon*. A new model system for functional genomics in grasses. *Plant Physiol.* 127, 1539–1555. doi: 10.1104/pp.010196

34. Duplessis, S., Cuomo, C. A., Lin, Y. C., Aerts, A., Tisserant, E., Veneault-Fourrey, C., et al. (2011). Obligate biotrophy features unraveled by the genomic analysis of rust fungi. *Proc. Natl. Acad. Sci. U.S.A.* 108, 9166–9171. doi: 10.1073/pnas.1019315108

35. Dyck, P., Samborski, D., and Anderson, R. (1966). Inheritance of adult-plant leaf rust resistance derived from the common wheat varieties Exchange and Frontana. *Can. J. Genet. Cytol.* 8, 665–671. doi: 10.1139/g66-082

36. Ellis, J. (2006). Insights into nonhost disease resistance: can they assist disease control in agriculture? *Plant Cell* 18, 523–528. doi: 10.1105/tpc.105.040584

37. Ellis, J. G., Lagudah, E. S., Spielmeyer, W., and Dodds, P. N. (2014). The past, present and future of breeding rust resistant wheat. *Front. Plant Sci.* 5:641. doi: 10.3389/fpls.2014.00641

38. Falter, C., and Voigt, C. A. (2014). Comparative cellular analysis of pathogenic fungi with a disease incidence in *Brachypodium distachyon* and *Miscanthus × giganteus*. *Bioenergy Res.* 7, 958–973. doi: 10.1007/s12155-014-9439-3

39. Febrer, M., Goicoechea, J. L., Wright, J., Mckenzie, N., Song, X., Lin, J., et al. (2010). An integrated physical, genetic and cytogenetic map of *Brachypodium distachyon*, a model system for grass research. *PLoS ONE* 5:e13461. doi: 10.1371/journal.pone.0013461

40. Ferreira, S., Pitz, K., Manshardt, R., Zee, F., Fitch, M., and Gonsalves, D. (2002). Virus coat protein transgenic papaya provides practical control of papaya ringspot virus in Hawaii. *Plant Dis.* 86, 101–105. doi: 10.1094/PDIS.2002.86.2.101

41. Figueroa, M., Alderman, S., Garvin, D. F., and Pfender, W. F. (2013). Infection of *Brachypodium distachyon* by formae speciales of *Puccinia graminis*: early infection events and host-pathogen incompatibility. *PLoS ONE* 8:e56857. doi: 10.1371/journal.pone.0056857

42. Fitzgerald, T. L., Powell, J. J., Schneebeli, K., Hsia, M. M., Gardiner, D. M., Bragg, J. N., et al. (2015). *Brachypodium* as an emerging model for cereal–pathogen interactions. *Ann. Bot.* 115, 717–731. doi: 10.1093/aob/mcv010

43. Flor, H. (1971). Current status of the gene-for-gene concept. *Annu. Rev. Phytopathol.* 9, 275–296. doi: 10.1146/annurev.py.09.090171.001423

44. Fu, Z. Q., and Dong, X. (2013). Systemic acquired resistance: turning local infection into global defense. *Annu. Rev. Plant Biol.*64, 839–863. doi: 10.1146/annurev-arplant-042811-105606

45. Gale, M., and Devos, K. (1998). Plant comparative genetics after 10 years. *Science* 282, 656–659. doi: 10.1126/science.282.5389.656

46. Garnica, D. P., Nemri, A., Upadhyaya, N. M., Rathjen, J. P., and Dodds, P. N. (2014). The ins and outs of rust haustoria. *PLoS Pathog.* 10:e1004329. doi: 10.1371/journal.ppat.1004329

47. Garvin, D. F. (2011). "Investigating rust resistance with the model grass *Brachypodium*," in *Proceedings of the 2011 Borlaug Global Rust Initiative Technical Workshop, June 13–19*, ed. R. McIntosh (St. Paul, MN), 89–91.

48. Garvin, D. F., Mckenzie, N., Vogel, J. P., Mockler, T. C., Blankenheim, Z. J., Wright, J., et al. (2010). An SSR-based genetic linkage map of the model grass *Brachypodium distachyon*. *Genome* 53, 1–13. doi: 10.1139/G09-079

49. Gill, U. S., Uppalapati, S. R., Nakashima, J., and Mysore, K. S. (2015). Characterization of *Brachypodium distachyon* as a nonhost model against switchgrass rust pathogen *Puccinia emaculata*. *BMC Plant Biol.* 15:113. doi: 10.1186/s12870-015-0502-9

50. Goff, S. A., Ricke, D., Lan, T.-H., Presting, G., Wang, R., Dunn, M., et al. (2002). A draft sequence of the rice genome (*Oryza sativa* L. ssp. *japonica*). *Science* 296, 92–100. doi: 10.1126/science.1068275

51. Gordon, S., Priest, H., Des Marais, D., Schackwitz, W., Figueroa, M., Martin, J., et al. (2013). Genome diversity in*Brachypodium distachyon*: deep sequencing of highly diverse inbred lines *Plant J.* 79, 361–374. doi: 10.1111/tpj.12569

52. Guarente, L., and Kenyon, C. (2000). Genetic pathways that regulate ageing in model organisms. *Nature* 408, 255–262. doi: 10.1038/35041700

53. Gutierrez-Gonzalez, J. J., Tu, Z. J., and Garvin, D. F. (2013). Analysis and annotation of the hexaploid oat seed transcriptome.*BMC Genomics* 14:471. doi: 10.1186/1471-2164-14-471

54. Harder, D. (1984). "Developmental ultrastructure of hyphae and spores," in *The Cereal Rusts*, Vol 1. eds W. R. Bushnell and A. P. Roelfs (Orlando, FL: Academic Press, Inc.) 333–373.

55. Harder, D. E., and Chong, J. (1984). "Structure and Physiology of Haustoria," in *The Cereal Rusts*, Vol 1, W. R. Bushnell and A. P. Roelfs (Orlando, FL: Academic Press, Inc.), 416–460. doi: 10.1016/b978-0-12-148401-9.50020-0

56. Heath, M. C. (1981). A generalized concept of host-parasite specificity. *Phytopathology* 71, 1121–1123. doi: 10.1094/Phyto-71-1121

57. Heath, M. C. (1997). Signalling between pathogenic rust fungi and resistant or susceptible host plants. *Ann. Bot.* 80, 713–720. doi: 10.1006/anbo.1997.0507

58. Heath, M. C. (2000). Nonhost resistance and nonspecific plant defenses. *Curr. Opin. Plant Biol.* 3, 315–319. doi: 10.1016/S1369-5266(00)00087-X

59. Hoch, H. C., Staples, R. C., Whitehead, B., Comeau, J., and Wolf, E. D. (1987). Signaling for growth orientation and cell differentiation by surface topography in *Uromyces*. *Science* 235, 1659–1662. doi: 10.1126/science.235.4796.1659

60. Hogenhout, S. A., Van Der Hoorn, R. A., Terauchi, R., and Kamoun, S. (2009). Emerging concepts in effector biology of plant-associated organisms. *Mol. Plant Microbe Interact.* 22, 115–122. doi: 10.1094/MPMI-22-2-0115

61. Huitema, E., Vleeshouwers, V. G., Francis, D. M., and Kamoun, S. (2003). Active defence responses associated with non-host resistance of *Arabidopsis thaliana* to the oomycete pathogen *Phytophthora infestans*. *Mol. Plant Pathol* 4, 487–500. doi: 10.1046/j.1364-3703.2003.00195.x

62. Huo, N., Lazo, G. R., Vogel, J. P., You, F. M., Ma, Y., Hayden, D. M., et al. (2008). The nuclear genome of *Brachypodium distachyon*: analysis of BAC end sequences. *Funct. Integr. Genomics* 8, 135–147. doi: 10.1007/s10142-007-0062-7

63. International Barley Genome Sequencing Consortium, Mayer, K. F., Waugh, R., Brown, J. W., Schulman, A., Langridge, P., et al. (2012). A physical, genetic and functional sequence assembly of the barley genome. *Nature* 491, 711–716. doi: 10.1038/nature11543

64. International Brachypodium Initiative. (2010). Genome sequencing and analysis of the model grass *Brachypodium distachyon.Nature* 463, 763–768. doi: 10.1038/nature08747

65. Jenkins, G., Hasterok, R., and Draper, J. (2003). "Building the molecular cytogenetic infrastructure of a new model grass," in*Application of Novel Cytogenetic and Molecular Techniques in Genetics and Breeding of the Grasses*, eds Z. Zwierzykowski, M. Surma, and P. Kachlicki (Poznan: Polish Academy of Sciences), 77–84.

66. Jin, Y., Szabo, L. J., and Carson, M. (2010). Century-old mystery of *Puccinia striiformis* life history solved with the identification of *Berberis* as an alternate host. *Phytopathology* 100, 432–435. doi: 10.1094/PHYTO-100-5-0432

67. Jones, J. D., and Dangl, J. L. (2006). The plant immune system. *Nature* 444, 323–329. doi: 10.1038/nature05286

68. Khatib, M., Lafitte, C., Esquerré-Tugayé, M. T., Bottin, A., and Rickauer, M. (2004). The CBEL elicitor of *Phytophthora parasitica* var. *nicotianae* activates defence in *Arabidopsis thaliana* via three different signalling pathways. *New Phytol.* 162, 501–510. doi: 10.1111/j.1469-8137.2004.01043.x

69. Koeck, M., Hardham, A. R., and Dodds, P. N. (2011). The role of effectors of biotrophic and hemibiotrophic fungi in infection.*Cell. Microbiol.* 13, 1849–1857. doi: 10.1111/j.1462-5822.2011.01665.x

70. Krattinger, S. G., Lagudah, E. S., Spielmeyer, W., Singh, R. P., Huerta-Espino, J., Mcfadden, H., et al. (2009). A putative ABC transporter confers durable resistance to multiple fungal pathogens in wheat. *Science* 323, 1360–1363. doi: 10.1126/science.1166453

71. Lawrence, G. J., Dodds, P. N., and Ellis, J. G. (2007). Rust of flax and linseed caused by *Melampsora lini*. *Mol. Plant Pathol.* 8, 349–364. doi: 10.1111/j.1364-3703.2007.00405.x

72. Lee, H. A., Kim, S. Y., Oh, S. K., Yeom, S. I., Kim, S. B., Kim, M. S., et al. (2014). Multiple recognition of RXLR effectors is associated with nonhost resistance of pepper against *Phytophthora infestans*. *New Phytol.* 203, 926–938. doi: 10.1111/nph.12861

73. Leonard, K., and Martinelli, J. (2005). Virulence of oat crown rust in Brazil and Uruguay. *Plant Dis.* 89, 802–808. doi: 10.1094/PD-89-0802

74. Leonard, K. J., and Szabo, L. J. (2005). Pathogen profile. Stem rust of small grains and grasses caused by *Puccinia graminis.Mol. Plant Pathol.* 6, 489–489. doi: 10.1111/j.1364-3703.2005.00299.x

75. Lipka, U., Fuchs, R., Kuhns, C., Petutschnig, E., and Lipka, V. (2010). Live and let die–*Arabidopsis* nonhost resistance to powdery mildews. *Eur. J. Cell Biol.* 89, 194–199. doi: 10.1016/j.ejcb.2009.11.011

76. Lius, S., Manshardt, R. M., Fitch, M. M., Slightom, J. L., Sanford, J. C., and Gonsalves, D. (1997). Pathogen-derived resistance provides papaya with effective protection against papaya ringspot virus. *Mol. Breed.* 3, 161–168. doi: 10.1023/A:1009614508659

77. López-Alvarez, D., López-Herranz, M. L., Betekhtin, A., and Catalán, P. (2012). A DNA barcoding method to discriminate between the model plant *Brachypodium distachyon* and its close relatives *B. stacei* and *B. hybridum* (Poaceae). *PloS ONE*7:e51058. doi: 10.1371/journal.pone.0051058

78. Maier, W., Begerow, D., Weiß, M., and Oberwinkler, F. (2003). Phylogeny of the rust fungi: an approach using nuclear large subunit ribosomal DNA sequences. *Can. J. Bot.* 81, 12–23. doi: 10.1139/b02-113

79. Mayer, K. F., Rogers, J., Doležel, J., Pozniak, C., Eversole, K., Feuillet, C., et al. (2014). A chromosome-based draft sequence of the hexaploid bread wheat (*Triticum aestivum*) genome. *Science* 345, 1251788. doi: 10.1126/science.1251788

80. Miklos, G. L. G., and Rubin, G. M. (1996). The role of the genome project in determining gene function: insights from model organisms. *Cell* 86, 521–529. doi: 10.1016/S0092-8674(00)80126-9

81. Monaghan, J., and Zipfel, C. (2012). Plant pattern recognition receptor complexes at the plasma membrane. *Curr. Opin. Plant Biol.* 15, 349–357. doi: 10.1016/j.pbi.2012.05.006

82. Mur, L. A., Allainguillaume, J., Catalan, P., Hasterok, R., Jenkins, G., Lesniewska, K., et al. (2011). Exploiting the *Brachypodium*tool box in cereal and grass research. *New Phytol.* 191, 334–347. doi: 10.1111/j.1469-8137.2011.03748.x

83. Mysore, K. S., and Ryu, C. M. (2004). Nonhost resistance: how much do we know? *Trends Plant Sci.* 9, 97–104. doi: 10.1016/j.tplants.2003.12.005

84. Nirmala, J., Drader, T., Chen, X., Steffenson, B., and Kleinhofs, A. (2010). Stem rust spores elicit rapid RPG1 phosphorylation.*Mol. Plant Microbe Interact.* 23, 1635–1642. doi: 10.1094/MPMI-06-10-0136

85. Nirmala, J., Drader, T., Lawrence, P. K., Yin, C., Hulbert, S., Steber, C. M., et al. (2011). Concerted action of two avirulent spore effectors activates Reaction to *Puccinia graminis* 1 (Rpg1)-mediated cereal stem rust resistance. *Proc. Natl. Acad. Sci. U.S.A.*108, 14676–14681. doi: 10.1073/pnas.1111771108

86. O'Driscoll, A., Doohan, F., and Mullins, E. (2015). Exploring the utility of *Brachypodium distachyon* as a model pathosystem for the wheat pathogen *Zymoseptoria tritici*. *BMC Res. Notes* 8:132. doi: 10.1186/s13104-015-1097-9

87. Oerke, E.-C. (2006). Crop losses to pests. *J. Agric. Sci.* 144, 31–43. doi: 10.1017/S0021859605005708

88. Panstruga, R., and Dodds, P. N. (2009). Terrific protein traffic: the mystery of effector protein delivery by filamentous plant pathogens. *Science* 324, 748–750. doi: 10.1126/science.1171652

89. Panwar, V., Mccallum, B., and Bakkeren, G. (2013). Endogenous silencing of*Puccinia triticina* pathogenicity genes through in planta-expressed sequences leads to the suppression of rust diseases on wheat. *Plant J.* 73, 521–532. doi: 10.1111/tpj.12047

90. Park, R. (2008). Breeding cereals for rust resistance in Australia. *Plant Pathol.* 57, 591–602. doi: 10.1111/j.1365-3059.2008.01836.x

91. Pennisi, E. (2010). Armed and dangerous. *Science* 327, 804–805. doi: 10.1126/science.327.5967.804

92. Peraldi, A., Beccari, G., Steed, A., and Nicholson, P. (2011). *Brachypodium distachyon*: a new pathosystem to study *Fusarium*head blight and other *Fusarium* diseases of wheat. *BMC Plant Biol.* 11, 100–114. doi: 10.1186/1471-2229-11-100

93. Phillips, R. L., and Freeling, M. (1998). Plant genomics and our food supply: an introduction. *Proc. Natl. Acad. Sci. U.S.A.* 95, 1969–1970. doi: 10.1073/pnas.95.5.1969

94. Piquerez, S. J. M., Harvey, S. E., Beynon, J. L., and Ntoukakis, V. (2014). Improving crop disease resistance: lessons from research on *Arabidopsis* and tomato. *Front. Plant Sci.* 5:671. doi: 10.3389/fpls.2014.00671

95. Popp, J., and Hantos, K. (2011). The impact of crop protection on agricultural production. *Stud. Agric. Econ.* 113, 47–66. doi: 10.7896/j.1003

96. Pretorius, Z. A., Singh, R. P., Wagoire, W. W., and Payne, T. S. (2000). Detection of virulence to wheat stem rust resistance gene Sr31 in *Puccinia graminis*. f. sp. *tritici* in Uganda. *Plant Dis.* 84, 203. doi: 10.1094/PDIS.2000.84.2.203B

97. Project, I. R. G. S. (2005). The map-based sequence of the rice genome. *Nature* 436, 793–800. doi: 10.1038/nature03895

98. Ray, D. K., Mueller, N. D., West, P. C., and Foley, J. A. (2013). Yield trends are insufficient to double global crop production by 2050. *PLoS ONE* 8:e66428. doi: 10.1371/journal.pone.0066428

99. Replansky, T., Koufopanou, V., Greig, D., and Bell, G. (2008). *Saccharomyces sensu* stricto as a model system for evolution and ecology. *Trends Ecol. Evol.* 23, 494–501. doi: 10.1016/j.tree.2008.05.005

100. Risk, J. M., Selter, L. L., Chauhan, H., Krattinger, S. G., Kumlehn, J., Hensel, G., et al. (2013). The wheat Lr34 gene provides resistance against multiple fungal pathogens in barley. *Plant Biotech. J.* 11, 847–854. doi: 10.1111/pbi.12077

101. Roelfs, A. P., Singh, R. P., and Saari, E. E. (eds). (1992). *Rust Diseases of Wheat: Concepts and Methods of Disease Management*. Mexico, DF: CIMMYT.

102. Routledge, A. P., Shelley, G., Smith, J. V., Talbot, N. J., Draper, J., and Mur, L. A. (2004). *Magnaporthe grisea* interactions with the model grass *Brachypodium distachyon* closely resemble those with rice (*Oryza sativa*). *Mol. Plant Pathol.* 5, 253–265. doi: 10.1111/j.1364-3703.2004.00224.x

103. Sandoya, G. V., and de Oliveira Buanafina, M. M. (2014). Differential responses of *Brachypodium distachyon* genotypes to insect and fungal pathogens. *Physiol. Mol. Plant Pathol.* 85, 53–64. doi: 10.1016/j.pmpp.2014.01.001

104. Schneebeli, K., Mathesius, U., and Watt, M. (2014). *Brachypodium distachyon* is a pathosystem model for the study of the wheat disease rhizoctonia root rot. *Plant Pathol.* 64, 91–100. doi: 10.1111/ppa.12227

105. Schulze-Lefert, P., and Panstruga, R. (2011). A molecular evolutionary concept connecting nonhost resistance, pathogen host range, and pathogen speciation. *Trends Plant Sci.* 16, 117–125. doi: 10.1016/j.tplants.2011.01.001

106. Schumann, G. L. (1991). *Plant Diseases: Their Biology and Social Impact*. St. Paul, MN: APS Press.

107. Scoles, G., Filiz, E., Ozdemir, B., Budak, F., Vogel, J., Tuna, M., et al. (2009). Molecular, morphological, and cytological analysis of diverse *Brachypodium distachyon* inbred lines. *Genome* 52, 876–890. doi: 10.1139/G09-062

108. Shukla, K., and Srivastava, S. (2014). Evaluation of finger millet incorporated noodles for nutritive value and glycemic index. *J. Food Sci. Technol.* 51, 527–534. doi: 10.1007/s13197-011-0530-x

109. Singh, R. (1993). Genetic association of gene Bdv1 for tolerance to barley yellow dwarf virus with genes Lr 34 and Yr18 for adult plant resistance to rusts in bread wheat. *Plant Dis.* 77, 1103–1106. doi: 10.1094/PD-77-1103

110. Singh, R. P., Hodson, D. P., Jin, Y., Huerta-Espino, J., Kinyua, M., Wanyera, R., et al. (2006). Current status, likely migration and strategies to mitigate the threat to wheat production from race Ug99 (TTKS) of stem rust pathogen. *CAB Rev.* 1, 54. doi: 10.1079/PAVSNNR20061054

111. Somerville, C. (2007). Biofuels. *Curr. Biol.* 17, R115–R119. doi: 10.1016/j.cub.2007.01.010

112. Staples, R. C., and Macko, V. (2004). "Germination of urediniospores and differentiation of infection structures," in *The Cereal Rusts*, Vol. 1, eds W. R. Bushell and A. P. Roelfs (Academic Press, Inc.), 250–282.

113. Stotz, H. U., Mitrousia, G. K., De Wit, P. J., and Fitt, B. D. (2014). Effector-triggered defence against apoplastic fungal pathogens. *Trends Plant Sci.* 19, 491–500. doi: 10.1016/j.tplants.2014.04.009

114. Thole, V., Peraldi, A., Worland, B., Nicholson, P., Doonan, J. H., and Vain, P. (2012). T-DNA mutagenesis in *Brachypodium distachyon*. *J. Exp. Bot.* 63, 567–576. doi: 10.1093/jxb/err333

115. Tilman, D., Balzer, C., Hill, J., and Befort, B. L. (2011). Global food demand and the sustainable intensification of agriculture.*Proc. Natl. Acad. Sci. U.S.A.* 108, 20260–20264. doi: 10.1073/pnas.1116437108

116. Tricoll, D. M., Carney, K. J., Russell, P. F., Mcmaster, J. R., Groff, D. W., Hadden, K. C., et al. (1995). Field evaluation of transgenic squash containing single or multiple virus coat protein gene constructs for resistance to cucumber mosaic virus, watermelon mosaic virus 2, and zucchini yellow mosaic virus. *Nat. Biotechnol.* 13, 1458–1465. doi: 10.1038/nbt1295-1458

117. Upadhyaya, N. M., Garnica, D. P., Karaoglu, H., Sperschneider, J., Nemri, A., Xu, B., et al. (2015). Comparative genomics of Australian isolates of the wheat stem rust pathogen *Puccinia graminis* f. sp. *tritici* reveals extensive polymorphism in candidate effector genes. *Front. Plant Sci.* 5:759. doi: 10.3389/fpls.2014.00759

118. US Census Bureau. (2015). *International Data Base.* Available at:http://www.census.gov/population/international/data/worldpop/table_population.php

119. Vogel, J., and Bragg, J. (2009). "*Brachypodium distachyon*, a new model for the Triticeae," in *Genetics and Genomics of the Triticeae*, eds C. Feuillet and G. J. Muehlbauer (New York: Springer), 427–449. doi: 10.1007/978-0-387-77489-3_16

120. Vogel, J., and Hill, T. (2008). High-efficiency *Agrobacterium*-mediated transformation of *Brachypodium distachyon* inbred line Bd21-3. *Plant Cell Rep.* 27, 471–478. doi: 10.1007/s00299-007-0472-y

121. Vogel, J. P., Garvin, D. F., Leong, O. M., and Hayden, D. M. (2006a). *Agrobacterium*-mediated transformation and inbred line development in the model grass *Brachypodium distachyon*. *Plant Cell Tissue Organ. Cult.* 84, 199–211. doi: 10.1007/s11240-005-9023-9

122. Vogel, J. P., Gu, Y. Q., Twigg, P., Lazo, G. R., Laudencia-Chingcuanco, D., Hayden, D. M., et al. (2006b). EST sequencing and phylogenetic analysis of the model grass *Brachypodium distachyon*. *Theor. Appl. Genet.* 113, 186–195. doi: 10.1007/s00122-006-0285-3

123. Vogel, J. P., Tuna, M., Budak, H., Huo, N., Gu, Y. Q., and Steinwand, M. A. (2009). Development of SSR markers and analysis of diversity in Turkish populations of *Brachypodium distachyon*. *BMC Plant Biol.* 9:88. doi: 10.1186/1471-2229-9-88

124. Wahl, I., Anikster, Y., Manisterski, J., and Segal, A. (1984). "Evolution at the center of origin," in *The Cereal Rusts*, Vol. 1,*Origins, Specificity, Structure, and Physiology*, eds W. R. Bushnell and A. P. Roelfs (Orlando: Academic Press, Inc), 39–77.

125. Woodham-Smith, C. (1991). *The Great Hunger: Ireland 1845–1849*. London: Penguin books.

126. Yin, C., Park, J.-J., Gang, D. R., and Hulbert, S. H. (2014). Characterization of a tryptophan 2-monooxygenase gene from*Puccinia graminis* f. sp. *tritici* involved in auxin biosynthesis and rust pathogenicity. *Mol. Plant Microbe Interact.* 27, 227–235. doi: 10.1094/MPMI-09-13-0289-FI

127. Zhong, S., Ali, S., Leng, Y., Wang, R., and Garvin, D. F. (2014). *Brachypodium distachyon–Cochliobolus sativus* pathosystem is a new model for studying plant–fungal interactions in cereal crops. *Phytopathology* 105, 482–489. doi: 10.1094/PHYTO-08-14-0214-R

CHAPTER 22 FUNGAL DEVELOPMENT OF THE PLANT PATHOGENUSTILAGO MAYDIS

Evelyn Vollmeister, Kerstin Schipper, Sebastian Baumann, Carl Haag, Thomas Pohlmann, Janpeter Stock and Michael Feldbrugge

Keywords: Effector; endosome; homeodomain transcription factor; microtubule; mRNA transport; post-transcriptional regulation

INTRODUCTION

Ustilago maydis is a basidiomycete that infects corn (Fig. 1a) and serves as an excellent model for plant pathogenicity (Banuett, 1995; Nadal et al., 2008; Brefort et al., 2009). However, it is also known to be a delicacy in Central America for hundreds of years and has even found its place in the food industry (Fig. 1b–d). Extensive research over the last couple of decades established this microorganism as a model for a number of important cellular processes such as signalling, transcriptional and post-transcriptional regulation, molecular transport, cell cycle regulation, as well as DNA recombination and repair (Bolker, 2001; Kahmann & ¨ Kamper, 2004; Perez-Martin ¨ et al., 2006; Feldbrugge ¨ et al., 2008; Steinberg & Perez-Martin, 2008; Brefort et al., 2009).

The research on DNA recombination and repair is a superb example documenting how the mechanistic insights gained in U. maydis promote research into human health (Llorente & Modesti, 2009). The foundation for work on DNA recombination was laid by R. Holliday more than 40 years ago. He introduced the concept of the now famous Holliday junction of DNA recombination while studying U. maydis (Holliday, 1964, 2004).

More recent research on DNA repair identified the key factor Brh2, a BRCA2 family protein (Kojic et al., 2002, 2011). Its founding member is a human tumour suppressor encoded by a predisposition gene of hereditary breast cancer. Brh2 from U. maydis catalyses the assembly of active recombinase complexes (Yang et al., 2005) and acts in double-strand repair to reunite broken ends (Mazloum & Holloman, 2009b). In addition, it might even be involved in unconventional strand invasion during the repair of defective replication forks (Mazloum & Holloman, 2009a). Based on these studies, mammalian BRCA2 research has gained tremendous momentum in resolving its molecular function. The protein is now considered a universal recombinase regulator that is conserved from lower to higher eukaryotes (Thorslund & West, 2007; Thorslund et al., 2010).

One of the major breakthroughs for the U. maydis research community was the public release of the genome sequence combined with a profound manual annotation that resulted in high-quality data currently curated in the database MUMDB at MIPS (Kamper ¨ et al., 2006; MIPS U. maydis database; and the Munich information centre for protein sequences, respectively; http://mips.helmholtzmuenchen.de/genre/proj/ustilago/). This spurred the application of new approaches such as transcriptome-wide DNA microarrays (Scherer et al., 2006;

Citation: Evelyn Vollmeister, Kerstin Schipper, Sebastian Baumann, Carl Haag, Thomas Pohlmann, Janpeter Stock an Michael Feldbrugge, Fungal development of the plant pathogenUstilago maydis, DOI:10.1111/j.1574-6976.2011.00296.x

Zarnack et al., 2008; Heimel et al., 2010b), proteome-wide approaches (Bohmer ¨ et al., 2007), gene replacement strategies for efficient knockouts or more sophisticated promoter and gene fusions (Brachmann et al., 2004; Kamper, 2004; Garcia-Pedrajas ¨ et al., 2008). Establishing fluorescence proteins such as eGfp, mRfp, mCherry, Yfp, Cfp, photoactivatable Gfp and splitYfp facilitated comprehensive in vivo localization, colocalization and interaction studies (Steinberg & Perez-Martin, 2008; Heimel et al., 2010a; Schuster et al., 2011b).

Figure 1: The good and the evil: Ustilago maydis as a delicacy and a pathogen. (a) Corn field close to Marburg, Germany. Infected plants can be seen in the foreground and enlarged at the bottom (picture taken by R. Kellner). (b) Quesadilla cuitlacoche (open) from the street restaurant 'Bazaar Sabado' in ´ Mexico City. (c) Menu of the restaurant (2009). Arrowhead indicates Quesadilla cuitlacoche. (d) Tin filled with U. maydis-infected corn sold as 'cream of corn truffle' (arrowhead). (e, f) Haploid cells proliferate by budding (arrowheads indicate buds, two different stages are shown; size bar = 5 mm). (g) Primary (1) and secondary (2) septum were stained with Calcofluor white (inverted image).

In this review, we will use the sexual life cycle of U. maydis as a blue print to describe recent findings focusing on the postgenomic era. Complementary information can be found in earlier reviews (Kahmann & Kamper, 2004; ¨ Feldbrugge ¨ et al., 2006) and in the special issue of Fungal Genetics & Biology dedicated to U. maydis (Kronstad, 2008).

For the sake of clarity, the life cycle is divided into four distinct phases: proliferation of haploid cells, mating, filamentation and plant infection. However, the results described here indicate that key players of the underlying cellular events are often important at several additional stages during the life cycle.

SYNOPSIS OF THE PLANT-DEPENDENT LIFE CYCLE

The saprophytic phase can be considered as the default state and starts with meiosis during the germination of diploid teliospores. The resulting haploid cells proliferate by budding (Fig. 1e–g). A prerequisite for infection is the mating of two compatible haploid cells, preferentially on the plant surface (Fig. 2). They recognize each other using a pheromone receptor system that consists of seven-transmembrane domain receptors and small lipopeptide pheromones (pra1/2 and mfa1/2, respectively) encoded at the biallelic a mating-type locus (Fig. 3; Bolker ¨ et al., 1992; Spellig et al., 1994; Szabo´ et al., 2002). Active pheromone signalling results in the formation of conjugation tubes, which orient their growth along the pheromone gradient of the mating partner. Thereby, compatible partners approach each other and fuse at their tips, initiating plasmogamy (Fig. 2; Snetselaar et al., 1996). Characteristic for basidiomycetes, plasmogamy and karyogamy are separated in time (Kruzel & Hull, 2010). A stable dikaryon is formed that grows filamentously with a defined axis of polarity. Hyphae expand at the apical growth cone (Fig. 2) and insert retraction septa at regular intervals at the basal pole. These septa separate the viable compartment from hyphal segments devoid of a cytoplasm, resulting in the formation of regularly spaced empty sections (Lehmler et al., 1997; Steinberg et al., 1998). The nuclei travel to the centre of the cell and are positioned in a defined distance of about 10 mm from each other as well as about 50 mm from the tip (Steinberg et al., 1998; Fuchs et al., 2005).

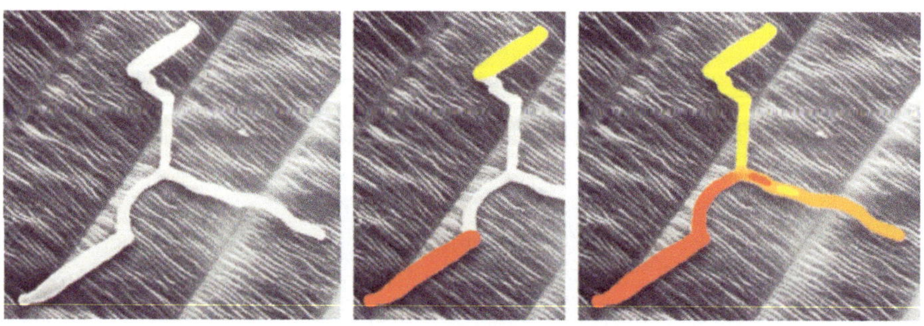

Figure 2: Mating on the plant surface. Two haploid cells (coloured yellow and red in the middle panel) landed in close proximity on the plant surface. Conjugation tubes (coloured yellow and red in the right panel) grew towards each other and fused at their tips. The resulting filament

(coloured orange in the right panel) with two nuclei (coloured yellow and red in the right panel; note that their position is an approximation) grows on the plant surface [the uncoloured figure on the left is reprinted from Kamper ¨ et al. (2006) with permission from the Nature Publishing Group].

The developmental switch resulting in hyphal growth is genetically controlled by the action of homeodomain transcription factors encoded as two separate subunits, bWest (bW) and bEast (bE), at the multiallelic b mating-type locus (Kronstad & Leong, 1990; Gillissen et al., 1992). The activity of the transcription factor is elegantly coupled to plasmogamy because it can only function as a heterodimer with subunits derived from different mating partners (e.g. bW1/ bE19, bW5/ bE8, etc.; Kamper ¨ et al., 1995). At present, 19 different b alleles that promote outbreeding are known (J. Kamper, pers. commun.; Barnes ¨ et al., 2004).

The dikaryotic filament is the infectious form of the fungus. It grows in close contact with the plant and is able to sense surface signals that trigger the formation of appressoria (Mendoza-Mendoza et al., 2009a; Lanver et al., 2010). These are specialized infection structures that enable the pathogen to enter the plant. Initially, hyphae grow intracellularly by invagination of the plant plasma membrane, establishing a tight interaction zone with colonized host cells (Snetselaar & Mims, 1993; Doehlemann et al., 2008b). At later stages, proliferation also occurs intercellularly and the dikaryotic mycelium grows towards bundle sheets (Doehlemann et al., 2008b). Subsequently, massive proliferation and hyphal fragmentation occurs. This is accompanied by an irregular division of host cells, resulting in the formation of tumours at all aerial parts of the plants. After karyogamy, black diploid teliospores develop within the tumours. Ripe tumours rupture and mature spores are spread by the wind. Under favourable conditions, teliospores germinate and release haploid cells. The life cycle begins anew.

PROLIFERATION OF HAPLOID CELLS AND CYTOKINESIS

Saprophytic cells exhibit a defined cylindrical cell shape comparable to Schizosaccharomyces pombe. However, they do not divide by insertion of a central septum, but proliferate by budding like Saccharomyces cerevisiae (Fig. 1e–g). During proliferation, the bud expands by polar tip growth, presumably by actin-mediated secretion of remodelling enzymes and building blocks of the cell wall (Banuett & Herskowitz, 2002; Weber et al., 2006). This notion is supported by the observation that an actin-dependent class V myosin localizes to regions of polar growth and is important for cell morphology (Weber et al., 2003). After the daughter cell reaches a certain size, the nucleus migrates from the centre of the mother into the bud where division occurs (Holliday, 1974; O'Donnell & McLaughlin, 1984). At the mother/bud neck region, the nuclear envelope is striped off, nuclear pore complexes are disassembled and chromosomes are released, enabling an open mitosis that takes place within the daughter cell (Steinberg et al., 2001; Straube et al., 2005; Theisen et al., 2008). This process is most probably regulated by a conserved signalling pathway containing small GTPase Ras3 and dual-function germinal centre kinase Don3 (Straube et al., 2005; Sandrock et al., 2006).

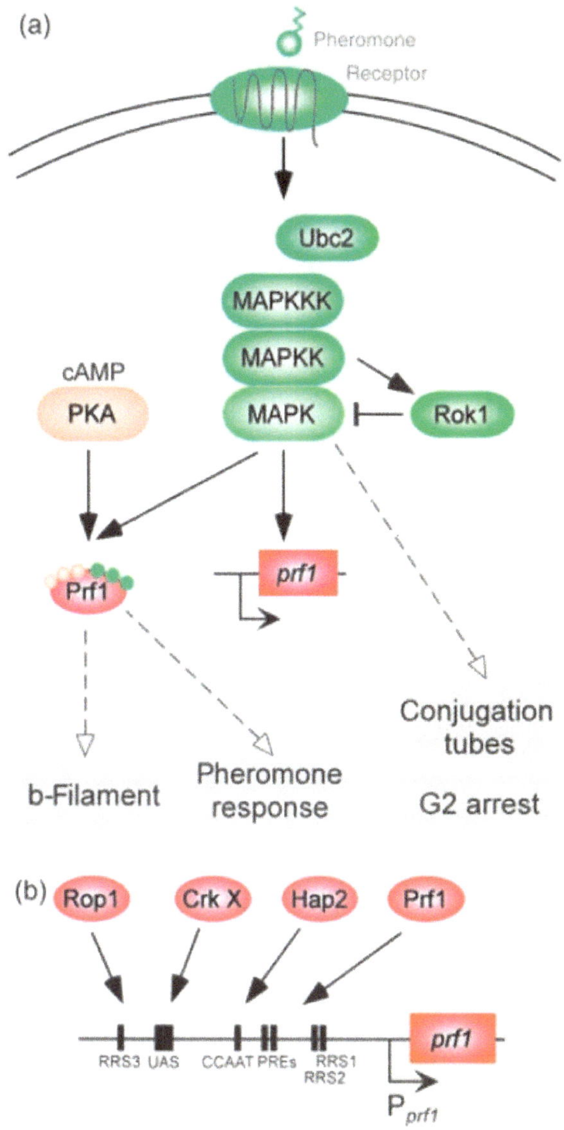

Figure 3: Networks of signalling and transcription factors regulate the pheromone response. (a) Schematic representation of the pheromone signalling network. Lipopeptide pheromone (a1 or a2, encoded by mfa1 or mfa2, respectively) activates its cognate seven-transmembrane domain receptor (Pra2 or Pra1, respectively). The activated receptor triggers a signalling pathway containing the scaffold protein Ubc2 and a pheromoneresponsive MAPK module (MAPKKK Kpp4, MAPKK Fuz7 and MAPK Kpp2). The MAPK regulates Prf1 via phosphorylation and at the transcriptional level. In addition, the MAPK module regulates two Prf1-independent events: conjugation tube formation and cell cycle arrest in G2. MAPK activity is counteracted by phosphatase Rok1, which is activated at the transcriptional level by the MAPK module, resulting in negative feedback regulation. Crosstalk occurs at the level of cAMP signalling. cAMPdependent PKA phosphorylates Prf1. This is apparently a prerequisite for signalling and a connection to environmental signals (see text). (b) At least four different transcription factors, Rop1, Crk X (yet unknown Crk1- dependent transcription factor) Hap2 and Prf1 (shown as ovals), recognize defined cis-active elements in the prf1 promoter (RRS1–3, Rop1-responsive sequence; CCAAT, CAT box; PREs, pheromone response elements; UAS, upstream activating sequence; bent arrow, transcriptional start site). Thereby, various environmental and nutritional signals could be funnelled into the pheromone response.

Chromosome segregation as well as nuclear movement depend on astral microtubules and involve the molecular motor dynein (Fink et al., 2006), a minus-end-directed motor that is split untypically into two subunits Dyn1 and Dyn2 in U. maydis (Straube et al., 2001).

Nuclear and cytoplasmic microtubule organizing centres (MTOCs, minus-ends of microtubules) nucleate dynamic microtubules that are organized in antiparallel bundles traversing the length of unbudded cells (Straube et al., 2003). During cell division, MTOCs are transported by dynein towards the neck region, where they polarize the microtubule cytoskeleton (Fink & Steinberg, 2006). Growing plus-ends reach into the distal pole of the mother as well as towards the growing pole of the daughter cell (Straube et al., 2003).

One transport function of microtubules is the movement of endosomes containing the t-SNARE-like protein Yup1 (Wedlich-Soldner ¨ et al., 2000). These endosomes shuttle bidirectionally along microtubules. Their active transport is mediated by the concerted action of dynein and the plusend-directed molecular motor Kin3, a Kinesin-3 type motor related to UNC104 and KIF1A from higher eukaryotes (Wedlich-Soldner ¨ et al., 2002). Strains carrying a temperature-sensitive mutation in yup1 are altered in cell shape, indicating that yup1-positive endosomes are important for morphology (Wedlich-Soldner ¨ et al., 2000). However, inhibition of microtubule function does not cause such morphological alterations (Fuchs et al., 2005), indicating that endosomes might exhibit a microtubule-independent function in regulating morphology during cell growth. Two lines of evidence suggest that endosomal transport is necessary for septation. Firstly, endosomes accumulate at the site of septum formation (Wedlich-Soldner ¨ et al., 2000; Schink & Bolker, 2009) and secondly ¨ kin3D strains are defective in cytokinesis due to aberrant septum formation (WedlichSoldner ¨ et al., 2002).

To complete the cell division, mother and daughter cells need to form two septa consecutively (Fig. 1e–g). Between these septa, a vacuolar fragmentation zone is formed, defining the location of the breakdown of the connecting cell wall and the physical separation (O'Donnell & McLaughlin, 1984; Banuett & Herskowitz, 2002; Weinzierl et al., 2002).

Mutants in this process were identified by their aberrant donut-shaped colonies on plates, hence their name don1 and don3 (Weinzierl et al., 2002; Feldbrugge ¨ et al., 2006). In a liquid medium, don mutants exhibit a distinct cell separation defect. They form tree-like cell clusters due to their failure to form a secondary septum (Weinzierl et al., 2002). Don1 is a guanine nucleotide exchange factor (GEF) that specifically activates the small G-protein Cdc42 (Weinzierl et al., 2002; Mahlert et al., 2006; Hlubek et al., 2008). Don3 is a dualfunction germinal centre kinase that acts parallel to the Cdc42 signalling pathway (Weinzierl et al., 2002; Sandrock et al., 2006). Chemical genetics revealed that Don3 kinase triggers the dynamic rearrangement of septins from hourglass-shaped collars into ring-like structures, depending on a contractile actomyosin ring (Bohmer ¨ et al., 2008, 2009). Thus, a signalling network containing small G-protein Cdc42 and kinase Don3 is crucial in determining septin and actin-ring organization during secondary septum formation.

Substantial progress has recently been made by connecting components of this signalling network to microtubule-dependent transport (Schink & Bolker, 2009). The ¨ authors demonstrate that Don1 is targeted to Yup1-positive endosomes via the specific interaction of the lipid-binding FYVE domain of Don1 with phosphatidylinositol 3-phosphate, a lipid present on these endosomes. Strains expressing a don1 variant with mutations in the FYVE domain are defective in secondary septum formation like kin3D strains. The data suggest that the exchange factor Don1 is targeted towards sites of secondary septum formation by microtubule-dependent transport of endosomes. Thereby, local activation of its cognate G-protein Cdc42 results in a signalling cascade orchestrating secondary septum formation and separation (Schink & Bolker, 2009). Apart from the ¨ recycling of pheromone receptors (Steinberg, 2007b), this is a new function for these endosomes in U. maydis and fits the general view of endosomes operating as multipurpose platforms (Gould & Lippincott-Schwartz, 2009).

MATING AND PHEROMONE SIGNALLING

Pheromone signalling determining fungal mating belongs to the best-characterized biological signal transduction phenomena studied in eukaryotes (Elion, 2000; Saito, 2010). The main framework was elucidated in S. cerevisiae and, in comparison with various other fungi such as S. pombe and Cryptococcus neoformans, pheromone signalling proved to be an evolutionarily conserved mechanism (Davey, 1998; Bahn et al., 2007). Key components are G-protein-coupled pheromone receptors that recognize cognate peptide pheromones, resulting in the activation of an evolutionarily conserved MAPK module consisting of three kinases (mitogen-activated protein kinase module; MAPKKK, MAPKK and MAPK). The terminal MAPK phosphorylates, among others, transcription factors that determine pheromoneregulated gene expression. A common feature is that the central MAPK pathways crosstalk with other signalling routes to discern appropriate mating conditions such as the nutritional status (Davey, 1998; Saito, 2010).

In U. maydis mating of two haploid cells is a prerequisite for infection (Fig. 2), a process that is regulated by the a and b mating-type loci. The lipopeptide pheromone of one mating partner is perceived by the cognate receptor of the compatible partner (Bolker ¨ et al., 1992; Spellig et al., 1994; Szabo´ et al., 2002). This elicits an intracellular signal transduction pathway leading to the activation of a defined MAPK module (Fig. 3a; MAPKKK Kpp4, MAPKK Fuz7 and MAPK Kpp2; summarized in Feldbrugge ¨ et al., 2006). Direct phosphorylation of the central HMG box transcription factor Prf1 (high mobility group; Hartmann et al., 1996; Muller ¨ et al., 1999; Kaffarnik et al., 2003) triggers the expression of a defined set of genes containing pheromone response elements in their regulatory regions (Urban et al., 1996b; Zarnack et al., 2008). Among these are the a matingtype genes, mfa1/2 and pra1/2, enhancing the pheromone response. This feed-forward regulation is further supported by autoactivation of the prf1 gene at the transcriptional level (Fig. 3b; Hartmann et al., 1999).

Genome-wide transcriptional profiling revealed that MAPK regulation of Prf1 is necessary for the expression of a defined subset of pheromone-responsive genes. Analysis of strains expressing Prf1 variants with mutated MAPK sites showed that in the majority of cases, pheromone responsiveness of target genes was completely lost while in the remaining cases regulation was alleviated. Among the latter are genes encoding the key transcription factors for filamentous growth bW/bE as well as Rbf1 (Scherer et al., 2006; Heimel et al.,

2010b). Thus, target genes respond differentially to MAPK phosphorylation of Prfl (Zarnack et al., 2008) and MAPK signalling via Prfl prepares the main transcriptional cascade that regulates filamentous growth and further pathogenic development (Heimel et al., 2010b; Wahl et al., 2010b).

In addition, the pheromone-responsive MAPK module is involved in the regulation of the pheromone-induced cell cycle arrest in the G2 phase, which is a Prfl-independent process (Fig. 3a; Garcia-Muse et al., 2003). This ensures synchronization of the cell cycle of both nuclei after cell fusion. Pheromone stimulation as well as transcriptional activation of a constitutively active MAPKK triggers reduced expression of several genes encoding cell cycle regulators such as B-type cyclin Clb1 or orthologues of polo kinases (Perez-Martin et al., 2006; Zarnack et al., 2008). It is noteworthy that MAPK signalling also promotes a negative feedback loop by inducing the expression of Rok1, a dualspecificity phosphatase that inactivates MAPK Kpp2 (Fig. 3a; Di Stasio et al., 2009).

Moreover, the MAPK module is responsible for eliciting the formation of conjugation tubes, another Prfl-independent event (Fig. 3a; Muller ¨ et al., 2003). During the polar growth of these filaments, the pheromone receptor localizes to their tips. This specific localization as well as receptor turnover by endocytosis might be important for the orientation within the pheromone gradient. Thereby, conjugation tubes of compatible mating partners grow towards each other and fuse at their tips (Snetselaar et al., 1996; Fuchs et al., 2006). After plasmogamy, pheromone signalling is switched off by an active bW/bE heterodimer (Laity et al., 1995; Urban et al., 1996b; Heimel et al., 2010a).

Mating on the surface of the plant is crucial for infection. Thus, cell recognition and fusion are influenced by environmental factors such as nutrient availability, hydroxy fatty acids or hydrophobic surfaces (Hartmann et al., 1999; Mendoza-Mendoza et al., 2009a). Two molecular mechanisms for the underlying signalling integration event have been identified. Firstly, an evolutionarily conserved cAMP signalling pathway that consists of a receptor-coupled heterotrimeric G-protein, adenylate cyclase and protein kinase A (PKA) influences mating (Feldbrugge ¨ et al., 2006; Brefort et al., 2009). An important molecular mechanism is the direct phosphorylation of Prfl by cAMP-activated PKA, a regulatory event that is apparently a prerequisite for pheromone response (Fig. 3a; Kaffarnik et al., 2003). PKA-mediated activation of Prfl is also sufficient to increase the expression of mating-type genes, indicating that upregulation of the intracellular cAMP level favours mating (Kaffarnik et al., 2003; Zarnack et al., 2008). Crosstalk of the PKA and MAPK signalling is also important at later stages of the pathogenic development (Feldbrugge ¨ et al., 2006; Brefort et al., 2009).

Secondly, prfl expression at the transcriptional level is tightly controlled by various regulatory elements in the promoter region (Fig. 3b). For example, the upstream activating sequence (UAS) element integrates different carbon sources (Hartmann et al., 1999) and the unusual MAPK Crk1 signals via this element (Garrido et al., 2004). In addition, the HMG box transcription factor Rop1 as well as the CCAAT-box-binding protein Hap2 are crucial for the activation of prfl expression via binding sites in the prfl promoter (Brefort et al., 2005; Mendoza-Mendoza et al., 2009b). Thus, at least four different transcriptional activators (Prfl, yet unknown UAS-binding protein X, Rop1, Hap2) that integrate potentially different signals converge at the prfl promoter to accurately control the expression of this key transcription factor (Fig. 3b). This is an excellent example of how a central signalling node can integrate various external and internal cues to determine optimal mating conditions.

FILAMENTOUS GROWTH AND TRANSCRIPTIONAL NETWORKING

Homeodomain transcription factors constitute crucial players in developmental programmes not only in fungi, but also in plants and animals (Pearson et al., 2005; Hay & Tsiantis, 2010). They often function as master regulators eliciting transcriptional cascades that include additional transcription factors. In U. maydis, the homeodomain transcription factor bW/bE is necessary and sufficient to regulate filamentous growth. This Prfl-activated key regulator connects sexual and pathogenic development. Right after its discovery at the b mating-type locus, it was proposed that, like homeodomain transcription factors from higher eukaryotes, it triggers a defined transcriptional programme essential for development (Kronstad & Leong, 1990; Schulz et al., 1990; Gillissen et al., 1992). However, it turned out to be more difficult than expected to identify the responsible target genes. Only very recently a major breakthrough was achieved with the advent of genome-wide transcriptional profiling (Kahmann & Kamper, 2004; Heimel ¨ et al., 2010b; Wahl et al., 2010b).

Initially, it was discovered that the activity of the transcription factor was regulated by heterodimerization of subunits derived from different mating partners (Kamper ¨ et al., 1995). This information was used to demonstrate that an active bW/ bE heterodimer is sufficient to elicit pathogenic development. Haploid strains were generated that express active versions under the control of their own promoters. These strains are called solopathogenic because they infect plants independent of mating (Bolker ¨ et al., 1995).

For the identification of functionally important components of filamentous growth, various comparative methods were applied investigating haploid cells and b-induced filaments. Several genes were identified that exhibited significantly increased expression during filamentous growth, but unfortunately, these were all dispensable for infection (Schauwecker et al., 1995; Wosten ¨ et al., 1996; Romeis et al., 2000).

A significant improvement was the generation of haploid strains expressing active bW2/bE1 variants under the control of carbon- or nitrogen-source-regulated promoters. For the first time, this allowed a defined time-resolved monitoring of the underlying regulatory cascade (Brachmann et al., 2001). Using RNA fingerprinting, 10 additional targets could be identified (Brachmann et al., 2001). For example, the involvement of MAPK Kpp6 in the early steps of infection such as plant penetration could be demonstrated (Brachmann et al., 2003).

The same strains were the foundation for applying highdensity microarrays covering around 90% of all predicted genes (Heimel et al., 2010b). Performing extensive timecourse experiments with carbon- and nitrogen-sourcemediated activation of the bW/bE heterodimer resulted in the detection of 345 bW/bE-responsive genes (206 induced and 139 repressed;

Heimel et al., 2010b). Global analysis of the respective gene functions revealed that 20 genes involved in cell wall remodelling, such as chitin synthases, exochitinases, chitin deacetylases as well as glucanases, are induced, indicating active remodelling of the fungal cell wall during filamentous growth (Heimel et al., 2010b). A number of clustered genes encoding potential effector proteins for plant interaction were also activated (Kamper ¨ et al., 2006; Heimel et al., 2010b). Among the downregulated genes are several encoding cell cycle regulators such as cyclins Cln1, Clb1 and Clb2. For Clb1, it has been shown that repression leads to G2 cell cycle arrest (Garcia-Muse et al., 2004). Apparently, the bW/bE heterodimer is responsible for maintaining the G2 cell cycle arrest until the filament resumes the cell cycle after plant penetration (Perez-Martin et al., 2006).

One of the most rewarding results of this analysis was the identification of the downstream master regulator Rbf1. This zinc finger transcription factor, which is an immediate early target of the bW/bE heterodimer at the transcriptional level, is responsible for the vast majority of the gene expression programme (Fig. 4a; Heimel et al., 2010b). Accordingly, overexpression of Rbfl is sufficient to trigger filamentous growth. Because it is difficult to envision that hundreds of different bW/bE heterodimers trigger the same transcriptional programme, the presence of such a master regulator was no surprise. Interestingly, rbfl is also a direct Prfl target and its expression is pheromone induced via MAPK signalling (Fig. 4b; Zarnack et al., 2008). Thus, during the initial phase of mating, Rbfl activity must be inhibited by a currently unknown mechanism.

Rbfl amplifies the transcriptional cascade by the activation of various transcriptional regulators such as Bizl and Foxl (Fig. 4a; Heimel et al., 2010b). Bizl is a zinc fingercontaining factor involved in cell cycle regulation and is important for appressoria formation and plant penetration (Flor-Parra et al., 2006). Foxl regulates the expression of numerous effector proteins that appear to be involved in the suppression of plant defence (Zahiri et al., 2010).

Using a temperature-sensitive mutant, it was demonstrated that the active bW/bE transcription factor is not only needed to establish the transcriptional cascade of pathogenic development, but is also crucial at later stages. Interfering with the activity of bW/bE in planta results in an unusual activation of pheromone signalling, defects in cell cycle synchronization and reduced expression of effector proteins. This aberrant regulation correlates with impaired infection (Wahl et al., 2010b).

Another important direct target of the bW/bE transcription factor is clp1 (Fig. 4b). It encodes a regulator that is required for clamp formation, a special structure needed for nuclear sorting during mitotic division of the dikaryon in planta (Scherer et al., 2006). Recently, the level of complexity was added to by demonstrating that Clp1 interacts with bW and counteracts bW/bE-dependent functions such as G2 cell cycle arrest and filamentous growth (Fig. 4b). Clp1 also interacts with Rbfl during the repression of pheromone signalling (Heimel et al., 2010a). Furthermore, Clp1 binds Cib1, a bZIP transcription factor required for pathogenic development, whose activity is not regulated at the transcriptional level, but at the level of alternative splicing (Fig. 5b; Heimel et al., 2010a). Thus, besides the simple linear transcription cascade Prfl 4 bW/bE4 Rbfl (Fig. 4a), there is an underlying complicated network of transcription factors and protein interactions that mediate intensive feedback and feedforward regulation to orchestrate pathogenic development (Fig. 4b).

FILAMENTOUS GROWTH AND CYTOSKELETAL FUNCTIONS

In general, fungal hyphae exhibit a characteristic growth mode. At their apical pole, a growth cone is responsible for the rapid expansion of membrane and cell wall. In subapical regions, filaments are partitioned by septa forming molecular barriers (Harris, 2006; Steinberg, 2007a; Fischer et al., 2008; Zarnack & Feldbrugge, 2010). Polar growth is ¨ mediated by intensive membrane dynamics, i.e. local exoand endocytosis that is supported by vesicle trafficking. This might be promoted by distinct membrane microdomains at the apical pole that are rich in specific lipids such as sphingolipids and ergosterol (Harris et al., 2005; Steinberg, 2007b). Two juxtaposed macromolecular units are involved in growth cone function: the Spitzenkorper and the polari- ¨ some. The latter is a multiprotein complex consisting of landmark proteins, signalling molecules such as small GTPases and formins that nucleate actin cables for shortdistance transport (Harris et al., 2005; Sudbery & Court, 2007). The Spitzenkorper is thought to function as a supply ¨ centre for exocytotic vesicles (Gierz & Bartnicki-Garcia, 2001; Bartnicki-Garcia, 2002), for example, specialized vesicles, so-called chitosomes, transport and export chitin synthases as well as building blocks for cell wall synthesis (Bartnicki-Garcia, 2006; Riquelme et al., 2007). Active transport along the actin and microtubule cytoskeleton is mediated by molecular motors for short- and long-distance delivery of vesicles, respectively. This process is streamlined for efficiency and any disturbances lead to reduced growth rates or a complete blockage (Harris, 2006; Steinberg, 2007a).

In U. maydis, a number of recent observations confirm that dikaryotic filaments grow like typical hyphae (Steinberg, 2007a). For example, the landmark protein Spa2 localizes to the tip (Carbo & Perez-Martin, 2008) and sphingolipid biosynthesis is important for polar growth. These lipids are most likely part of membrane microdomains at the hyphal tip (Canovas & Perez-Martin, 2009) and could function to enrich signalling components with membrane anchors such as the small G protein Rac1 (Mahlert et al., 2006). Septins might serve as diffusion barriers organizing these membrane subdomains. Strains carrying corresponding loss-of-function mutations are affected in establishing the axis of polarity, whereas septins are dispensable during the maintenance of filamentous growth (AlvarezTabares & Perez-Martin, 2010).

Analysis of eight different chitin synthases revealed that Chs7 was particularly important for polar growth and localizes to the hyphal tip like its related family members Chs5 and Mcs1 (Weber et al., 2006). The chitin synthase Mcs1 contains a unique domain with a similarity to myosin motors (Weber et al., 2006). This domain is not required for motility, but apparently for different functions (Treitschke et al., 2010). Mcs1 might be important in early infection stages. It is conceivable that a defective Mcs1 results in altered cell wall composition. This could trigger the plant's defence response (Treitschke et al., 2010).

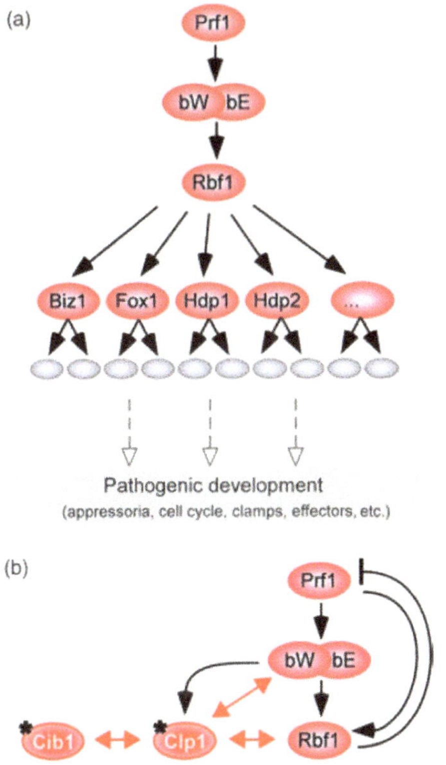

Figure 4: Intensive transcriptional networking determines pathogenicity. (a) Schematic representation showing a cascade of transcription factors (red ovals) during the regulation of filamentous growth and further pathogenic development. Various target genes encoding regulatory, structural and enzymatic factors are indicated by grey ovals. (b) The central transcription factor cascade Prf14 bW/bE4 Rbf1 is subject to intensive regulation. At the transcriptional level, Prf1 activates Rbf1 during pheromone response, whereas Rbf1 mediates the repression of Prf1 at later stages. The clp1 gene is one of the few direct targets of bW/bE. Moreover, Clp1 regulates the activity of bW/bE and Rbf1 at the level of protein/protein interaction (red arrowheads). Clp1 activity is influenced by its interaction with Cib1. Another level of complexity (symbolized by asterisks) is the regulation of Cib1 and Clp1 expression at the post-transcriptional level.

Short-distance transport of vesicles that, for example, contain chitin synthases is mediated by the actin cytoskeleton, which is essential for filamentous growth (Fuchs et al., 2005; Steinberg, 2007a). The actin-dependent motor Myo5 is potentially involved in polar growth. Accordingly, the motor accumulates at plus-ends of actin cables at the poles of filaments, and loss of Myo5 results in severe defects during polar growth (Weber et al., 2003; Schuchardt et al., 2005).

Long-distance transport of molecular cargo is mediated by the microtubule cytoskeleton (Steinberg, 2007c). Although microtubules are dispensable for filamentous growth, they are needed to determine the axis of polarity. In their absence, growth is retarded and defects in the insertion of basal retraction septa occur (Fuchs et al., 2005; Steinberg, 2007a). One of the main functions of microtubules during filamentous growth appears to be the bidirectional transport of Yup1-positive vesicles that also colocalize with the endosomal marker Rab5, a small GTPase (Fuchs et al., 2006). Although the knowledge about the function of endosomal shuttling is currently scarce, significant progress has been made in analysing and modelling the mechanism of transport by molecular motors (Steinberg, 2007c; Ashwin et al., 2010; Schuster et al., 2011a). Study of plus-endbinding proteins revealed that microtubules are highly polarized, with plus-ends localizing to both poles (Lenz et al., 2006). Similar to haploid cells, Kin3 and split Dyn1/2 mediate anterograde and retrograde transport of endosomes, respectively (Lenz et al., 2006; Schuster et al., 2011b). Dynein as well as its regulators dynactin and Lis1 accumulate at the plus-ends of microtubules, where they might form a reservoir for retrograde transport (Lenz et al., 2006; Schuster et al., 2011a). Apparently, dynein can be loaded onto endosomes either at plus-ends of microtubules or during its retrograde movement (Schuster et al., 2011a, b). Conventional kinesin Kin1 (Lehmler et al., 1997) is responsible for recycling dynein to the microtubule plusends for additional rounds of minus-end-directed transport (Becht et al., 2006; Lenz et al., 2006). Notably, kin1 was initially named kin2 in a study describing kin1 and kin2 encoding a Kinesin-7 and Kinesin-1 type motor, respectively (Lehmler et al., 1997). In order to apply a standardized kinesin nomenclature, kin1 and kin2 were subsequently renamed kin7a and kin1, respectively (Lawrence et al., 2004; Schuchardt et al., 2005).

As mentioned above, knowledge of the biological function of endosomal transport during polar growth is still limited. It has been speculated that these vesicles constitute early endosomes that might be important for endocytotic membrane recycling or act during long-distance signalling to the nucleus (Fuchs et al., 2006; Steinberg, 2007a). However, additional work is required to substantiate these assumptions.

FILAMENTOUS GROWTH AND POSTTRANSCRIPTIONAL REGULATION

Post-transcriptional regulation is essential in determining the quantitative and spatiotemporal expression of proteins. Key factors are RNA-binding proteins that function, for example, during the splicing, localization, stability and translation of mRNAs (Dreyfuss et al., 2002). These processes can also be coupled. Localization and translation of mRNAs result, for example, in the defined subcellular localization of the encoded protein, a cellular mechanism that is conserved from bacteria to humans (St Johnston, 2005; Du et al., 2007; Nevo-Dinur et al., 2011). mRNA localization is mainly achieved by active transport along the actin or the microtubule cytoskeleton. In S. cerevisiae, ASH1 mRNA is actively transported from the mother cell towards the distal pole of the daughter cell. Local translation at the pole ensures that the encoded transcription factor specifi- cally localizes to the nucleus of the daughter cell in order to ascertain a distinct transcriptional programme (Jansen, 2001; Zarnack & Feldbrugge, 2007). ¨ ASH1 mRNA transport is mediated by two RNA-binding proteins, She2p and She3p, that cooperatively bind RNA elements, so-called zipcodes, in the target mRNA. This is essential for actindependent transport mediated by myosin (Bohl ¨ et al., 2000; Muller ¨ et al., 2011). Recently, it was reported that She3pmediated ASH1 mRNA transport is also operational in filaments of Candida albicans, most likely to regulate hyphal morphology and invasive growth (Elson et al., 2009).

Microtubule-dependent mRNA transport is mainly studied in higher eukaryotes. Important processes regulated at this level are oocyte and embryo development in Drosophila melanogaster as well as neuronal processes in mammals (St Johnston, 2005; Dahm et al., 2007). Although the RNAbinding proteins and the transported target mRNAs are evolutionarily not conserved, the general concept holds true for all investigated transport processes to date: specific RNAbinding proteins recognize zipcodes forming large mRNPs (mRNA-containing ribonucleoprotein complexes) that are transported along microtubules by kinesin(s) and dynein (Holt & Bullock, 2009).

The importance of such control at the post-transcriptional level is an emerging theme for U. maydis. This additional layer of regulation functions in concert with the extensive regulation at the level of transcription described above (Feldbrugge ¨ et al., 2008; Vollmeister & Feldbrugge, ¨ 2010). For example, although clp1 is a direct and immediate early target gene of bW/bE, the protein accumulates only at later stages during plant penetration, suggesting translational control (Scherer et al., 2006). Moreover, precise expression of the Clp1 interaction partner Cib1 appears to be regulated by alternative splicing (Heimel et al., 2010a).

Investigating various RNA-binding proteins in U. maydis revealed that two of them, Khd4 and Rrm4, are important for filamentous growth (Becht et al., 2005). The loss of Khd4 causes pleiotropic effects such as altered cell and filament morphology as well as defects in pheromone response and plant infection (Becht et al., 2005). The protein contains at least five potential RNA-binding domains of the KH type (hnRNP K homology domain, KH1-5). It constitutes the founding member of a multi-KH domain family of fungal proteins with representatives in other pathogens such as C. neoformans and C. albicans. Khd4 recognizes the sequence AUACCC via the two central tandem-KH domains (KH3 and KH4; Vollmeister et al., 2009). Interestingly, this motif is enriched in the 30 untranslated region of mRNAs and a number of transcripts, which exhibit altered expression in khd4D strains, contain this binding site. Because most mRNAs show an increased abundance in deletion strains, Khd4 might promote mRNA instability (Vollmeister et al., 2009; Vollmeister & Feldbrugge, 2010). ¨ Direct target mRNAs, whose expression is regulated by Khd4 and whose function can be related to the mutant phenotypes, are yet to be identified. To this end, we are currently conducting in vivo UV crosslinking experiments (Konig ¨ et al., 2010).

The deletion of rrm4 causes specific defects during filamentous growth, namely bipolar growth and failure to insert retraction septa (Fig. 5a–c; Becht et al., 2005, 2006). Interestingly, this mutant phenotype is reminiscent of growth defects of strains affected in microtubule functions (Lehmler et al., 1997; Steinberg et al., 1998; Fuchs et al., 2005; Schuchardt et al., 2005; Schuster et al., 2011a). Rrm4 contains three N-terminal RNA recognition motifs and a Cterminal MLLE domain that mediate RNA binding (Becht et al., 2006; Konig ¨ et al., 2007) and protein/protein interactions, respectively (Kozlov et al., 2010). Subcellular localization revealed that Rrm4 shuttles in mRNPs along microtubules (Fig. 5d–e; Supporting Information, Movies S1 and S2). RNA binding as well as the formation of motile mRNPs are essential for protein function (Becht et al., 2006; Konig ¨ et al., 2007). The deletion of kin1 results in the accumulation of Rrm4 at the poles. This is consistent with the hypothesis that split dynein mediates the retrograde transport of mRNPs, because conventional kinesin mediates anterograde transport of split dynein to the plus-ends of microtubules (Becht et al., 2006; S. Baumann, T. Pohlmann & M. Feldbrugge, unpublished data). ¨ In vivo UV crosslinking revealed that Rrm4 binds a distinct set of mRNAs that contain a potential CA-rich binding motif (Konig ¨ et al., 2009). These mRNAs encode proteins of cytotopically related groups such as proteins involved in translation, mitochondrial proteins or polarity factors. Important examples are ubi1 and rho3 encoding a natural fusion of ubiquitin with ribosomal protein Rpl40 and a small GTPase, respectively (Konig ¨ et al., 2009). RNA live imaging revealed that these mRNAs are molecular cargos of the Rrm4 transport unit. Remarkably, the CA-rich 30 UTR of ubi1 functions as a cis-active region in increasing the amount and processivity of trafficking (Konig ¨ et al., 2009). This is a characteristic feature of so-called mRNA zipcodes. Rho3 accumulates at retraction septa, suggesting a regulatory role during this process. These data led to the current model that Rrm4 functions in long-distance transport of mRNAs, a process that is conserved throughout evolution (Fig. 5f–g; Zarnack & Feldbrugge, 2007, 2010). The local translation of Rrm4 ¨ target mRNAs appears to be important for the subcellular localization of encoded proteins (Fig. 5f–g). Intriguingly, we could recently show that Rrm4-containing mRNPs are cotransported with Yup1-positive endosomes, suggesting yet another function of microtubule-dependent transport of endosomes (discussed above; S. Baumann, T. Pohlmann & M. Feldbrugge, unpublished data).

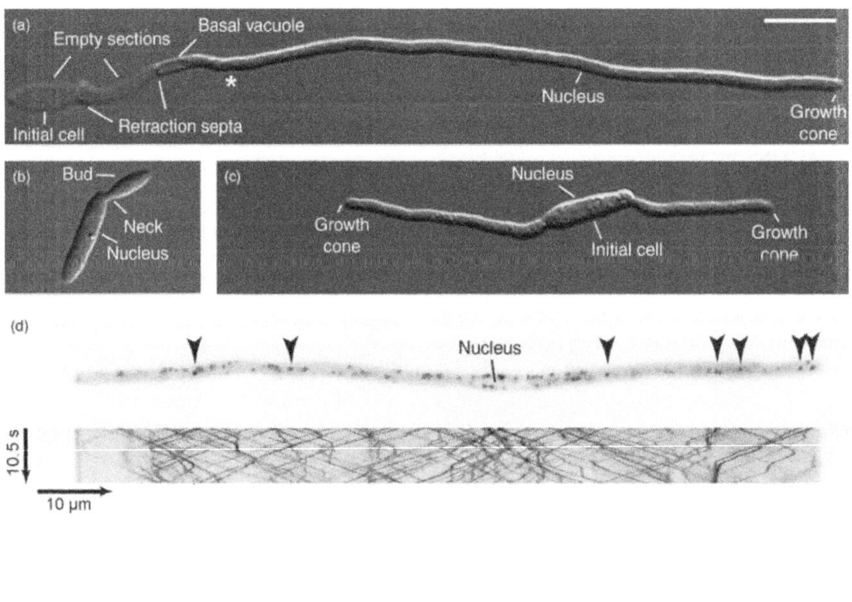

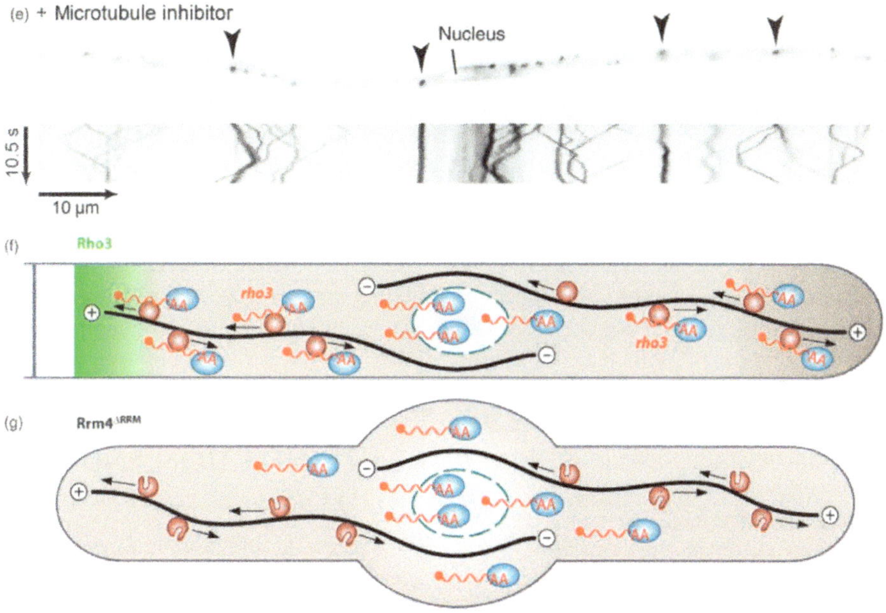

Figure 5: Microtubule-dependent shuttling of Rrm4-containing mRNPs is important for filamentous growth. (a) Filament of a monokaryotic strain expressing active bW2/bE1 variants under the control of a nitrogen-source-regulated promoter. Filaments are grown for 6 h under inducing conditions. Asterisk marks the predicted position of the next retraction septum (size bar = 10 mm). (b) rrm4D cells show no aberrant growth phenotype. (c) rrm4D filament exhibits defects in the formation of a single axis of polarity. The initial cell forms two growth cones and fails to insert a retraction septum (Becht et al., 2006). (d) Filament expressing Rrm4 fused to Gfp. The fusion protein accumulates in defined cytoplasmic particles (arrowheads in the inverted image detecting Gfp fluorescence) that shuttle bidirectionally along microtubules (kymograph in the lower part). In the kymograph, time is plotted vs. distance. Thus, motion of Rrm4 particles is visible as defined tracks (note the reversal of shuttling at the poles). Picture and kymograph correspond to Movie S1. (e) Filament expressing Rrm4 fused to Gfp was treated with the microtubule inhibitor benomyl (Fuchs et al., 2005). Presented as described in (d). Picture and kymograph correspond to Movie S2. (f) Model depicting microtubule-dependent transport of mRNAs. Rrm4-containing particles (dark red circles) transport mRNAs (red wavy lines) carrying the poly(A)-binding protein (blue ovals) bidirectionally along microtubules (black lines). The transport of rho3 mRNA might promote the accumulation of Rho3 at the retraction septum. (g) Deletion of RRM1 to RRM3 in Rrm4 causes a loss-offunction phenotype. Filaments grow mostly bipolar and target mRNAs are no longer transported along microtubules. The mutant Rrm4 is still part of the shuttling units, indicating that Rrm4 is an integral component of the transport machinery and does not just hitchhike like the poly(A)-binding protein [(f), (g) are reprinted from Zarnack & Feldbr ¨ugge (2010) with permission from the American Society of Microbiology].

FILAMENTOUS GROWTH AND MITOCHONDRIAL INHERITANCE

Sexual reproduction in eukaryotes goes along with the combination of nuclear and organelle genomes, such as mitochondrial genomes, in the zygote. This allows recombination events to accelerate evolution. However, a common process in eukaryotic sex is uniparental mitochondrial inheritance, a process that results in the asexual inheritance of the organelle genome. A possible explanation is the avoidance of evolutionary conflicts caused by a heteroplasmic situation. For example, mitochondria with increased replication rates, but decreased functional performance might dominate the population, resulting in disadvantages for the zygote (Partridge & Hurst, 1998; Xu, 2005). Possible mechanisms for uniparental mitochondrial inheritance include unequal size and mitochondrial numbers of gametes (Xu, 2005). An extreme example is human oocytes carrying about 10 000 more mitochondrial genomes than sperm cells, providing for a confident head start of maternal mitochondria in the population. Other mechanisms operating postfusion of the gametes include the interplay with nuclearencoded genes (Basse, 2010). For example, the mating-type specific homeodomain genes control uniparental inheritance in C. neoformans, suggesting that respective target genes encode proteins that determine uniparental inheritance (Xu, 2005; Yan et al., 2007).

In U. maydis, the inheritance of mitochondria is determined at late stages of filamentous growth when the fungus penetrates the plant. This coincides with the start of cell division of the dikaryon, and mitochondria are passed on to the next cell. Importantly, uniparental mitochondrial inheritance exists in U. maydis and the a2 mitotype is predominantly inherited (Fedler et al., 2009). How is this achieved at the molecular level? Initially, it was observed that the a2 mating-type locus, but not its a1 counterpart contains two genes, lga2 and rga2 (left and right genes of the a2 locus), which are only found in U. maydis and close relatives such as the head smut fungus Sporisorium reilianum (Urban et al., 1996a; Schirawski et al., 2005b). Because Lga2 contains a mitochondrial import signal, it was proposed that it might be involved in uniparental mitochondrial inheritance (Urban et al., 1996a). Work accomplished in the last few years has supported this hypothesis and has shed some light on the underlying mechanism (Basse, 2010).

During mating conjugation tubes fuse, plasmogamy occurs and cellular contents including the mitochondria are mixed (Fig. 2). How is it possible to separate the two mitochondrial populations to ensure uniparental inheritance? The key player is Lga2, a protein that is attached to the outside of the mitochondria (Bortfeld et al., 2004; Mahlert et al., 2009). This protein serves two main functions. Firstly, it inhibits the fusion of the mitochondria (Fig. 6) to prevent the mobilization of homing introns and secondly, Lga2 triggers mitochondrial fragmentation and mtDNA degradation (Bortfeld et al., 2004; Mahlert et al., 2009). Rga2 also localizes to the mitochondria even though a clear targeting signal is missing (Bortfeld et al., 2004). It appears to counteract Lga2 by protecting the mitochondria of the a2 mitotype (Fedler et al., 2009; Basse, 2010).

Based on the expression of these genes, the following model can be proposed (Basse, 2010). Like in all genes of the a locus, lga2 and rga2 expression is activated upon pheromone signalling (Urban et al., 1996b). Thereby, a2 mitotype mitochondria are preloaded with Lga2 and Rga2 before cell fusion. In contrast to mfa1/2 and pra1/2, which are repressed by an active bW/bE heterodimer, lga2 is one of the few directly activated targets of the bW/bE regulator, and rga2 expression is maintained constant (Romeis et al., 2000; Brachmann et al., 2001). Consistently, early during plasmogamy, Lga2 prevents the fusion of mitochondria (Fig. 6) and over time Lga2 eliminates those mitochondria that are not sufficiently protected by Rga2, namely mitochondria of the a1 mitotype (Basse, 2010).

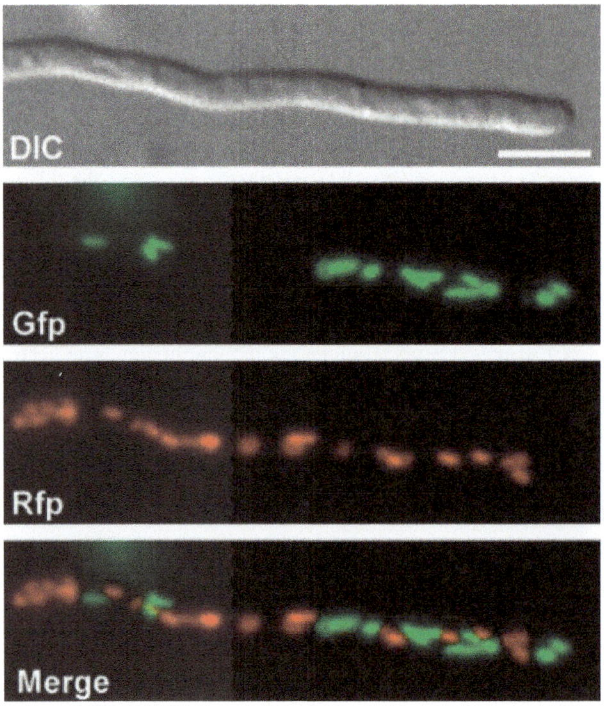

Figure 6: Separate parental populations of mitochondria in dikaryotic filaments. Tip of dikaryon resulting from the mating of strain FB1 (a1b1) with GF5 (a2b13). These strains express Gfp and Rfp targeted to the mitochondria, respectively (mtGfp and mtRfp). Epifluorescence images show green and red fluorescence as well as a merged picture at the bottom, indicating that mitochondria of the two parental cells do not fuse during filamentous growth [figure reprinted from Fedler et al. (2009) with permission from the Genetics Society of America].

Possible downstream effectors for the two functions of Lga2 are dynamin-related GTPase Dnm1 and the mitochondrial protein Mrb1. Dnm1 is crucial in preventing mitochondrial fusion in an Lga2-dependent manner (Mahlert et al., 2009). Mrb1, a regulator of the p32 family, interacts with Rga2 and appears to counteract the function of Lga2 (Bortfeld et al., 2004; Basse, 2010). This is based on the following observation. While a2mrb1D strains fail to infect plants, virulence can be restored by the deletion of lga2. Importantly, dikaryons of a1/a2mrb1D strains are impaired in virulence because they are arrested at the early phase of infection when uniparental inheritance should occur (Bortfeld et al., 2004). However, its precise mode of action is currently unclear.

This is now the first glimpse of this process, but as so often, the story is more complex. For example, the dominance of a2 mitotypes is not 100%, indicating additional mechanisms independent of the Lga2/Rga2 system (Bortfeld et al., 2004; Fedler et al., 2009). Moreover, in S. reilianum, there are three mating types: a1, a2 and a3. Lga2 and rga2 are present on a2, but successful mating and potential uniparental mitochondrial inheritance is possible between a1 and a3 (Schirawski et al., 2005b).

PLANT INFECTION AND FUNGAL EFFECTOR PROTEINS

For successful infection, mating should take place on the plant surface, such as on the leaves, stems or part of the flower. Therefore, it is not surprising that plant-derived signals stimulate the formation of filaments as well as appressoria (Fig. 7). For example, corn lipids trigger filamentous growth in liquid culture (Klose et al., 2004). Moreover, hydrophobicity acts as an inducer of filamentation and appressorium differentiation on artificial surfaces and the latter is further enhanced by the addition of hydroxy fatty acids. Because both signals are present on the plant surface, they are most likely also operational under natural conditions (Fig. 7; Mendoza-Mendoza et al., 2009a).

Two signalling components, the tetraspan membrane protein Sho1 and its interaction partner the single transmembrane mucin Msb2, appear to be involved in sensing the hydrophobic signal. Accordingly, the deletion of the corresponding genes strongly reduces appressoria formation. Sho1 interacts specifically with MAPK Kpp6, suggesting that Sho1/Msb2 operate upstream of a Kpp6-containing MAPK module during early infection (Lanver et al., 2010).

The perception of host signals is not the only crucial event that is a prerequisite for the development of infection structures on the plant surface. Additional processes involve actin-mediated functions (Berndt et al., 2010) as well as components of the N- and O-glycosylation pathways (Schirawski et al., 2005a; Fernandez-Alvarez et al., 2010). Unfortunately, insights into the underlying mechanisms that influence early infection are currently missing.

Once biotrophic fungi enter the plant, they depend on living material for growth and nutrition, so that many biotrophs form specialized feeding structures termed haustoria (Mendgen & Hahn, 2002). During plant invasion of U. maydis fungal hyphae invaginate the plasma membrane of infected host cells, leading to the formation of a biotrophic interface without extensive feeding structures. Instead, hyphae orient their growth towards vascular bundles, where they might feed on transported sugars (Doehlemann et al., 2008a; Brefort et al., 2009). In line with this, a novel sucrose transporter, Srt1, was recently described as a major component ensuring fungal nutrition (Fig. 7; Wahl et al., 2010a). Interestingly, this transporter has a higher substrate affinity than a comparable plant transporter, suggesting a redirection of sugar flow towards the parasite (Wahl et al., 2010a).

The invasion and proliferation of pathogens within a plant are often accompanied by host recognition, usually triggering defence. Thus, for survival, a successful biotroph should avoid or suppress plant immune responses. Indeed, transcriptional profiling of infected maize indicated that U. maydis is recognized by the plant within the first 24 h after infection as, for example, pathogenesis-related genes are activated. At later stages, however, plant defence appears to be specifically shut down (Fig. 7; Doehlemann et al., 2008a). A possible mechanism would be the interference with plant signalling. Conceivably, jasmonic acid signalling, which appears to be induced in infected tissue, antagonizes salicylic acid signalling, a key component of the plant defence against biotrophs (Doehlemann et al., 2008a).

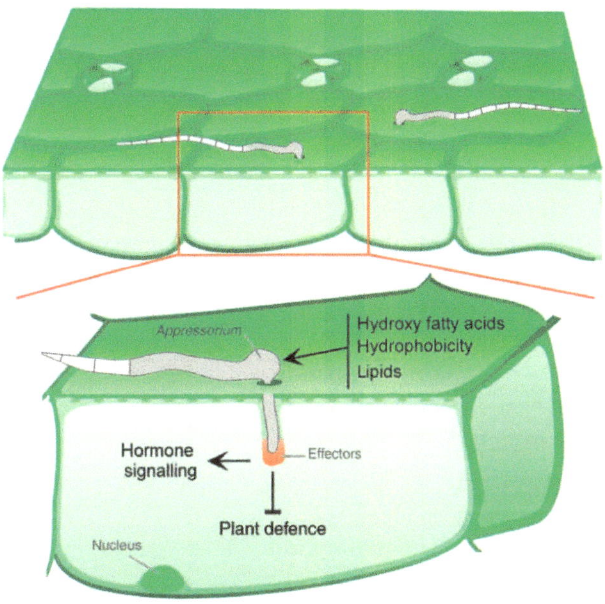

Figure 7: Infectious hyphae enter the plant. A model of the early infection process on the corn epidermis with a single plant cell shown in detail (below). Dikaryotic filaments sense plant surface signals, form appressoria and penetrate the surface. Fungal effectors (red), which are thought to counteract plant defence and alter hormone signalling, are secreted into the biotrophic interface.

In accordance with the current views, the fungal pathogen uses secreted effectors to trigger alterations of plant responses during biotrophic growth. Effectors are delivered to the biotrophic interface that constitutes a hotspot for plant–pathogen communication. One important finding that went along with the completion of the genome sequence was the discovery of gene clusters coding for novel secreted proteins that are crucial for infection (Fig. 7; Kamper ¨ et al., 2006). In five of 12 cases, the deletion of distinct gene clusters led to disease-associated phenotypes, indicating that these clusters contain virulence factors (Kamper ¨ et al., 2006). A common feature is that clustered effector genes exhibit strong transcriptional activation during biotrophic growth (Kamper ¨ et al., 2006), and many of these genes with virulence-associated phenotypes are regulated in a bW/bE-dependent manner (Wahl et al., 2010b).

Apart from clustered effector genes, there are also single genes and gene families scattered throughout the genome that encode potential effectors (Kamper ¨ et al., 2006; Doehlemann et al., 2009; Khrunyk et al., 2010). Currently, the best-studied example is Pep1, an effector that is essential for biotrophic development. pep1 deletion mutants arrest during penetration without causing disease symptoms (Doehlemann et al., 2009). They fail to induce jasmonic acid signalling and do not suppress the plant defence mechanism during the early stage of infection. This is accompanied by an accumulation of toxic substances and papillae formation at the penetration sites (Doehlemann et al., 2009).

Remarkably, it appears that specific effector subsets are important during different stages of pathogenic development, as the block of corresponding mutants differed with respect to the infection stage (Kamper ¨ et al., 2006). This finding was recently expanded by providing evidence for organ-specific effector functions. Because plant tissues and organs differ strongly in their physiology, the fungus might adapt to each environment using a special, spatiotemporalregulated assembly of effectors (Skibbe et al., 2010). Recently, the genome of S. reilianum, a related corn pathogen, was sequenced and annotated (http://mips.helmholtz-muenchen. de/genre/proj/sporisorium; Schirawski et al., 2010). Genome mining revealed that many of the genes for novel secreted proteins found in U. maydis are also present. This is likely due to the fact that they parasitize on the same host. However, most shared effectors are highly divergent, which may be attributed to the differing infection strategies of U. maydis and S. reilianum (Schirawski et al., 2010). In essence, during development in planta U. maydis relies on numerous, thoroughly regulated effectors to induce the host to foster its development.

CONCLUSIONS

Our tour following the fungal development of U. maydis revealed quite a number of remarkable recent findings. Highlights are for example (1) the endosomal transport of signalling components during cytokinesis (Schink & Bolker, ¨ 2009); (2) the discovery of Rbf1 and the underlying extensive regulation at the transcriptional and post-transcriptional level (Heimel et al., 2010a, b); (3) uniparental mitochondrial inheritance (Fedler et al., 2009); (4) microtubule-dependent mRNA transport (Konig ¨ et al., 2009); and (5) fungal effectors during infection (Kamper ¨ et al., 2006; Brefort et al., 2009).

In the future, resolving the underlying molecular mechanisms will be essential in answering the following key questions: how do endosomes function during membrane recycling, signalling and microtubule-dependent transport? How is the expression of transcription factors regulated at the post-transcriptional level? What are the precise molecular functions of effectors?

The results obtained will further advance U. maydis as a simple eukaryotic model for various aspects of basic cellular functions. Within the last few years, a few scientists have spearheaded distinct research areas covering several aspects of cell biology and plant pathogenicity. Now appears to be the right time for expansion.

ACKNOWLEDGEMENTS

We acknowledge lab members and Dr V. Gohre for critically ¨ reading the manuscript. Our deepest gratitude is due to R. Kellner, Dr H. Leal-Lara, Dr K.M. Sneetselar and Dr C. Basse for images shown in Figs 1a, d, 2 and 6, respectively. Special thanks are due to Drs G. Guerra-Sanchez and J. Pablo ´ Pardo-Vazquez for introducing M.F. to Quesadilla cuitlacoche. We are grateful to Dr R. Kahmann and the Max Planck Institute for Terrestrial Microbiology in Marburg for their generous support. S.B. and T.P. are members of the International Max-Planck Research School for Environmental, Cellular and Molecular Microbiology, C.H. is a member of the HHUD graduate school Molecules of Infection and J.S. participates in the graduate school of the CLIB Graduate Cluster Industrial Biotechnology. Our research was in part financed by grants from the German Science Foundation (DFG Fe448/3 and DFG FOR1334).

REFERENCES

1. Alvarez-Tabares I & Perez-Martin J (2010) Septins from the phytopathogenic fungus Ustilago maydis are required for proper morphogenesis but dispensable for virulence. PLoS One 5: e12933.
2. Ashwin P, Lin C & Steinberg G (2010) Queueing induced by bidirectional motor motion near the end of a microtubule. Phys Rev E Stat Nonlin Soft Matter Phys 82: 051907.
3. Bahn YS, Xue C, Idnurm A, Rutherford JC, Heitman J & Cardenas ME (2007) Sensing the environment: lessons from fungi. Nat Rev Microbiol 5: 57–69.
4. Banuett F (1995) Genetics of Ustilago maydis, a fungal pathogen that induces tumors in maize. Annu Rev Genet 29: 179–208.
5. Banuett F & Herskowitz I (2002) Bud morphogenesis and the actin and microtubule cytoskeletons during budding in the corn smut fungus, Ustilago maydis. Fungal Genet Biol 37: 149–170.
6. Barnes CW, Szabo LJ, May G & Groth JV (2004) Inbreeding levels of two Ustilago maydis populations. Mycologia 96: 1236–1244.
7. Bartnicki-Garcia S (2002) Hyphal tip growth: outstanding questions. Molecular Biology of Fungal Development (Osiewacz HD, ed), pp. 29–58. Dekker (Marcel), New York.
8. Bartnicki-Garcia S (2006) Chitosomes: past, present and future. FEMS Yeast Res 6: 957–965.
9. Basse CW (2010) Mitochondrial inheritance in fungi. Curr Opin Microbiol 13: 712–719.
10. Becht P, Vollmeister E & Feldbrugge M (2005) Role for RNA- ¨ binding proteins implicated in pathogenic development of Ustilago maydis. Eukaryot Cell 4: 121–133.
11. Becht P, Konig J & Feldbr ¨ ugge M (2006) The RNA-binding ¨ protein Rrm4 is essential for polarity in Ustilago maydis and shuttles along microtubules. J Cell Sci 119: 4964–4973.
12. Berndt P, Lanver D & Kahmann R (2010) The AGC Ser/Thr kinase Aga1 is essential for appressorium formation and maintenance of the actin cytoskeleton in the smut fungus Ustilago maydis. Mol Microbiol 78: 1484–1499.
13. Bohl F, Kruse C, Frank A, Ferring D & Jansen RP (2000) She2p, a ¨ novel RNA-binding protein tethers ASH1 mRNA to the Myo4p myosin motor via She3p. EMBO J 19: 5514–5524.
14. Bohmer C, B ¨ ohmer M, B ¨ olker M & Sandrock B (2008) Cdc42 and ¨ the Ste20-like kinase Don3 act independently in triggering cytokinesis in Ustilago maydis. J Cell Sci 121: 143–148.
15. Bohmer C, Ripp C & B ¨ olker M (2009) The germinal centre kinase ¨ Don3 triggers the dynamic rearrangement of higher-order septin structures during cytokinesis in Ustilago maydis. Mol Microbiol 74: 1484–1496.
16. Bohmer M, Colby T, B ¨ ohmer C, Brautigam A, Schmidt J & B ¨ olker ¨ M (2007) Proteomic analysis of dimorphic transition in the phytopathogenic fungus Ustilago maydis. Proteomics 7: 675–685.
17. Bolker M (2001) ¨ Ustilago maydis – a valuable model system for the study of fungal dimorphism and virulence. Microbiology 147: 1395–1401.
18. Bolker M, Urban M & Kahmann R (1992) The ¨ a mating type locus of U. maydis specifies cell signaling components. Cell 68: 441–450.
19. Bolker M, Genin S, Lehmler C & Kahmann R (1995) Genetic ¨ regulation of mating, and dimorphism in Ustilago maydis. Can J Bot 73: 320–325.
20. Bortfeld M, Auffarth K, Kahmann R & Basse CW (2004) The Ustilago maydis a2 mating-type locus genes lga2 and rga2 compromise pathogenicity in the absence of the mitochondrial p32 family protein Mrb1. Plant Cell 16: 2233–2248.
21. Brachmann A, Weinzierl G, Kamper J & Kahmann R (2001) ¨ Identification of genes in the bW/bE regulatory cascade in Ustilago maydis. Mol Microbiol 42: 1047–1063.
22. Brachmann A, Schirawski J, Muller P & Kahmann R (2003) An ¨ unusual MAP kinase is required for efficient penetration of the plant surface by Ustilago maydis. EMBO J 22: 2199–2210.
23. Brachmann A, Konig J, Julius C & Feldbr ¨ ugge M (2004) A reverse ¨ genetic approach for generating gene replacement mutants in Ustilago maydis. Mol Genet Genomics 272: 216–226.
24. Brefort T, Muller P & Kahmann R (2005) The high-mobility- ¨ group domain transcription factor Rop1 is a direct regulator of prf1 in Ustilago maydis. Eukaryot Cell 4: 379–391.
25. Brefort T, Doehlemann G, Mendoza-Mendoza A, Reissmann S, Djamei A & Kahmann R (2009) Ustilago maydis as a pathogen. Annu Rev Phytopathol 47: 423–445.
26. Canovas D & Perez-Martin J (2009) Sphingolipid biosynthesis is required for polar growth in the dimorphic phytopathogen Ustilago maydis. Fungal Genet Biol 46: 190–200.
27. Carbo N & Perez-Martin J (2008) Spa2 is required for morphogenesis but it is dispensable for pathogenicity in the phytopathogenic fungus Ustilago maydis. Fungal Genet Biol 45: 1315–1327.
28. Dahm R, Kiebler M & Macchi P (2007) RNA localisation in the nervous system. Semin Cell Dev Biol 18: 216–223.

29. Davey J (1998) Fusion of a fission yeast. Yeast 14: 1529–1566.

30. Di Stasio M, Brefort T, Mendoza-Mendoza A, Munch K & ¨ Kahmann R (2009) The dual specificity phosphatase Rok1 negatively regulates mating and pathogenicity in Ustilago maydis. Mol Microbiol 73: 73–88.

31. Doehlemann G, Wahl R, Horst RJ et al. (2008a) Reprogramming a maize plant: transcriptional and metabolic changes induced by the fungal biotroph Ustilago maydis. Plant J 56: 181–195.

32. Doehlemann G, Wahl R, Vranes M, de Vries RP, Kamper J & ¨ Kahmann R (2008b) Establishment of compatibility in the Ustilago maydis/maize pathosystem. J Plant Physiol 165: 29–40.

33. Doehlemann G, van der Linde K, Assmann D, Schwammbach D, Hof A, Mohanty A, Jackson D & Kahmann R (2009) Pep1, a secreted effector protein of Ustilago maydis, is required for successful invasion of plant cells. PLoS Pathog 5: e1000290.

34. Dreyfuss G, Kim VN & Kataoka N (2002) Messenger-RNAbinding proteins and the messages they carry. Nat Rev Mol Cell Bio 3: 195–205.

35. Du TG, Schmid M & Jansen RP (2007) Why cells move messages: the biological functions of mRNA localization. Semin Cell Dev Biol 18: 171–177.

36. Elion EA (2000) Pheromone response, mating and cell biology. Curr Opin Microbiol 3: 573–581.

37. Elson SL, Noble SM, Solis NV, Filler SG & Johnson AD (2009) An RNA transport system in Candida albicans regulates hyphal morphology and invasive growth. PLoS Genet 5: e1000664.

38. Fedler M, Luh KS, Stelter K, Nieto-Jacobo F & Basse CW (2009) The a2 mating-type locus genes lga2 and rga2 direct uniparental mitochondrial DNA (mtDNA) inheritance and constrain mtDNA recombination during sexual development of Ustilago maydis. Genetics 181: 847–860.

39. Feldbrugge M, B ¨ olker M, Steinberg G, K ¨ amper J & Kahmann R ¨ (2006) Regulatory and structural netwoks orchestrating mating, dimorphism, cell shape, and pathogenesis in Ustilago maydis. The Mycota I (Kues U & Fischer R, eds), pp. 375–391. ¨ Springer-Verlag, Berlin.

40. Feldbrugge M, Zarnack K, Vollmeister E, Baumann S, Koepke J, ¨ Konig J, M ¨ unsterk ¨ otter M & Mannhaupt G (2008) The ¨ posttranscriptional machinery of Ustilago maydis. Fungal Genet Biol 45: S40–S46.

41. Fernandez-Alvarez A, Elias-Villalobos A & Ibeas JI (2010) The requirement for protein O-mannosylation for Ustilago maydis virulence seems to be linked to intrinsic aspects of the infection process rather than an altered plant response. Plant Signal Behav 5: 412–414.

42. Fink G & Steinberg G (2006) Dynein-dependent motility of microtubules and nucleation sites supports polarization of the tubulin array in the fungus Ustilago maydis. Mol Biol Cell 17: 3242–3253.

43. Fink G, Schuchardt I, Colombelli J, Stelzer E & Steinberg G (2006) Dynein-mediated pulling forces drive rapid mitotic spindle elongation in Ustilago maydis. EMBO J 25: 4897–4908.

44. Fischer R, Zekert N & Takeshita N (2008) Polarized growth in fungi – interplay between the cytoskeleton, positional markers and membrane domains. Mol Microbiol 68: 813–826.

45. Flor-Parra I, Vranes M, Kamper J & Perez-Martin J (2006) Biz1, a ¨ zinc finger protein required for plant invasion by Ustilago maydis, regulates the levels of a mitotic cyclin. Plant Cell 18: 2369–2387.

46. Fuchs U, Manns I & Steinberg G (2005) Microtubules are dispensable for the initial pathogenic development but required for long-distance hyphal growth in the corn smut fungus Ustilago maydis. Mol Biol Cell 16: 2746–2758.

47. Fuchs U, Hause G, Schuchardt I & Steinberg G (2006) Endocytosis is essential for pathogenic development in the corn smut fungus Ustilago maydis. Plant Cell 18: 2066–2081.

48. Garcia-Muse T, Steinberg G & Perez-Mart ´ ´ın J (2003) Pheromone-induced G2 arrest in the phytopathogenic fungus Ustilago maydis. Eukaryot Cell 2: 494–500.

49. Garcia-Muse T, Steinberg G & Perez-Mart ´ ´ın J (2004) Characterization of B-type cyclins in the smut fungus Ustilago maydis: roles in morphogenesis and pathogenicity. J Cell Sci 117: 487–506.

50. Garcia-Pedrajas MD, Nadal M, Kapa LB, Perlin MH, Andrews DL & Gold SE (2008) DelsGate, a robust and rapid gene deletion construction method. Fungal Genet Biol 45: 379–388.

51. Garrido E, Voß U, Muller P, Kahmann R & P ¨ erez-Mart ´ ´ın J (2004) The induction of sexual development and virulence in the smut fungus Ustilago maydis on Crk1, a novel MAPK protein. Gene Dev 18: 3117–3130.

52. Gierz G & Bartnicki-Garcia S (2001) A three-dimensional model of fungal morphogenesis based on the vesicle supply center concept. J Theor Biol 208: 151–164.

53. Gillissen B, Bergemann J, Sandmann C, Schroeer B, Bolker M & ¨ Kahmann R (1992) A two-component regulatory system for self/non-self recognition in Ustilago maydis. Cell 68: 647–657.

54. Gould GW & Lippincott-Schwartz J (2009) New roles for endosomes: from vesicular carriers to multi-purpose platforms. Nat Rev Mol Cell Bio 10: 287–292.

55. Harris SD (2006) Cell polarity in filamentous fungi: shaping the mold. Int Rev Cytol 251: 41–77.

56. Harris SD, Read ND, Roberson RW, Shaw B, Seiler S, Plamann M & Momany M (2005) Polarisome meets Spitzenkorper: ¨ microscopy, genetics, and genomics converge. Eukaryot Cell 4: 225–229.

57. Hartmann HA, Kahmann R & Bolker M (1996) The pheromone ¨ response factor coordinates filamentous growth and pathogenicity in Ustilago maydis. EMBO J 15: 1632–1641.

58. Hartmann HA, Kruger J, Lottspeich F & Kahmann R (1999) ¨ Environmental signals controlling sexual development of the corn smut fungus Ustilago maydis

through the transcriptional regulator Prf1. Plant Cell 11: 1293–1306.

59. Hay A & Tsiantis M (2010) KNOX genes: versatile regulators of plant development and diversity. Development 137: 3153–3165.

60. Heimel K, Scherer M, Schuler D & Kamper J (2010a) The ¨ Ustilago maydis Clp1 protein orchestrates pheromone and b-dependent signaling pathways to coordinate the cell cycle and pathogenic development. Plant Cell 22: 2908–2922.

61. Heimel K, Scherer M, Vranes M, Wahl R, Pothiratana C, Schuler D, Vincon V, Finkernagel F, Flor-Parra I & Kamper J (2010b) ¨ The transcription factor Rbf1 is the master regulator for bmating type controlled pathogenic development in Ustilago maydis. PLoS Pathog 6: e1001035.

62. Hlubek A, Schink KO, Mahlert M, Sandrock B & Bolker M (2008) ¨ Selective activation by the guanine nucleotide exchange factor Don1 is a main determinant of Cdc42 signalling specificity in Ustilago maydis. Mol Microbiol 68: 615–623.

63. Holliday R (1964) A mechanism for gene conversion in fungi. Genet Res 5: 282–304.

64. Holliday R (1974) Ustilago maydis. Handbook of Genetics (King RC, ed), pp. 575–595. Plenum Press, New York.

65. Holliday R (2004) Early studies on recombination and DNA repair in Ustilago maydis. DNA Repair 3: 671–682.

66. Holt CE & Bullock SL (2009) Subcellular mRNA localization in animal cells and why it matters. Science 326: 1212–1216.

67. Jansen RP (2001) mRNA localization: message on the move. Nat Rev Mol Cell Bio 2: 247–256.

68. Kaffarnik F, Muller P, Leibundgut M, Kahmann R & Feldbr ¨ ugge ¨ M (2003) PKA and MAPK phosphorylation of Prf1 allows promoter discrimination in Ustilago maydis. EMBO J 22: 5817–5826.

69. Kahmann R & Kamper J (2004) ¨ Ustilago maydis: how its biology relates to pathogenic development. New Phytol 164: 31–42.

70. Kamper J (2004) A PCR-based system for highly efficient ¨ generation of gene replacement mutants in Ustilago maydis. Mol Genet Genomics 271: 103–110.

71. Kamper J, Reichmann M, Romeis T, B ¨ olker M & Kahmann R ¨ (1995) Multiallelic recognition: nonself-dependent dimerization of the bE and bW homeodomain proteins in Ustilago maydis. Cell 81: 73–83.

72. Kamper J, Kahmann R, B ¨ olker M ¨ et al. (2006) Insights from the genome of the biotrophic fungal plant pathogen Ustilago maydis. Nature 444: 97–101.

73. Khrunyk Y, Munch K, Schipper K, Lupas AN & Kahmann R ¨ (2010) The use of FLP-mediated recombination for the functional analysis of an effector gene family in the biotrophic smut fungus Ustilago maydis. New Phytol 187: 957–968.

74. Klose J, de Sa MM & Kronstad JW (2004) Lipid-induced filamentous growth in Ustilago maydis. Mol Microbiol 52: 823–835.

75. Kojic M, Kostrub CF, Buchman AR & Holloman WK (2002) BRCA2 homolog required for proficiency in DNA repair, recombination, and genome stability in Ustilago maydis. Mol Cell 10: 683–691.

76. Kojic M, Zhou Q, Fan J & Holloman WK (2011) Mutational analysis of Brh2 reveals requirements for compensating mediator functions. Mol Microbiol 79: 180–191.

77. Konig J, Julius C, Baumann S, Homann M, G ¨ oringer HU & ¨ Feldbrugge M (2007) Combining SELEX and yeast three- ¨ hybrid system for in vivo selection and classification of RNA aptamers. RNA 13: 614–622.

78. Konig J, Baumann S, Koepke J, Pohlmann T, Zarnack K & ¨ Feldbrugge M (2009) The fungal RNA-binding protein Rrm4 ¨ mediates long-distance transport of ubi1 and rho3 mRNAs. EMBO J 28: 1855–1866.

79. Konig J, Zarnack K, Rot G, Curk T, Kayikci M, Zupan B, Turner ¨ DJ, Luscombe NM & Ule J (2010) iCLIP reveals the function of hnRNP particles in splicing at individual nucleotide resolution. Nat Struct Mol Biol 17: 909–915.

80. Kozlov G, Menade M, Rosenauer A, Nguyen L & Gehring K (2010) Molecular determinants of PAM2 recognition by the MLLE domain of poly(A)-binding protein. J Mol Biol 397: 397–407.

81. Kronstad JW (2008) Mining the genome of the biotrophic fungal pathogen Ustilago maydis. Fungal Genet Biol 45 (suppl 1): S1–S2.

82. Kronstad JW & Leong SA (1990) The b mating-type locus of Ustilago maydis contains variable and constant regions. Gene Dev 4: 1384–1395.

83. Kruzel EK & Hull CM (2010) Establishing an unusual cell type: how to make a dikaryon. Curr Opin Microbiol 13: 706–711.

84. Laity C, Giasson L, Campbell R & Kronstad J (1995) Heterozygosity at the b mating-type locus attenuates fusion in Ustilago maydis. Curr Genet 27: 451–459.

85. Lanver D, Mendoza-Mendoza A, Brachmann A & Kahmann R (2010) Sho1 and Msb2-related proteins regulate appressorium development in the smut fungus Ustilago maydis. Plant Cell 22: 2085–2101.

86. Lawrence CJ, Dawe RK, Christie KR et al. (2004) A standardized kinesin nomenclature. J Cell Biol 167: 19–22.

87. Lehmler C, Steinberg G, Snetselaar KM, Schliwa M, Kahmann R & Bolker M (1997) Identification of a motor protein required ¨ for filamentous growth in Ustilago maydis. EMBO J 16: 3464–3473.

88. Lenz JH, Schuchardt I, Straube A & Steinberg G (2006) A dynein loading zone for retrograde endosome motility at microtubule plus-ends. EMBO J 25: 2275–2286

89. Llorente B & Modesti M (2009) Fungal BRCA2 ortholog Brh2 brings 50 end strand invasion back on stage. Mol Cell 36: 539–540

90. Mahlert M, Leveleki L, Hlubek A, Sandrock B & Bolker M (2006) ¨ Rac1 and Cdc42 regulate hyphal growth and cytokinesis in the dimorphic fungus Ustilago maydis. Mol Microbiol 59: 567–578.

91. Mahlert M, Vogler C, Stelter K, Hause G & Basse CW (2009) The a2 mating-type-locus gene lga2 of Ustilago maydis interferes with mitochondrial dynamics and fusion, partially in dependence on a Dnm1-like fission component. J Cell Sci 122: 2402–2412.

92. Mazloum N & Holloman WK (2009a) Brh2 promotes a template-switching reaction enabling recombinational bypass of lesions during DNA synthesis. Mol Cell 36: 620–630.

93. Mazloum N & Holloman WK (2009b) Second-end capture in DNA double-strand break repair promoted by Brh2 protein of Ustilago maydis. Mol Cell 33: 160–170.

94. Mendgen K & Hahn M (2002) Plant infection and the establishment of fungal biotrophy. Trends Plant Sci 7: 352–356.

95. Mendoza-Mendoza A, Berndt P, Djamei A, Weise C, Linne U, Marahiel M, Vranes M, Kamper¨ J & Kahmann R (2009a) Physical–chemical plant-derived signals induce differentiation in Ustilago maydis. Mol Microbiol 71: 895–911.

96. Mendoza-Mendoza A, Eskova A, Weise C, Czajkowski R & Kahmann R (2009b) Hap2 regulates the pheromone response transcription factor Prfl in Ustilago maydis. Mol Microbiol 72: 683–698.

97. Muller¨ M, Heym RG, Mayer A, Kramer K, Schmid M, Cramer P, Urlaub H, Jansen RP & Niessing D (2011) A cytoplasmic complex mediates specific mRNA recognition and localization in yeast. PLoS Biol 9: e1000611.

98. Muller¨ P, Aichinger C, Feldbrugge¨ M & Kahmann R (1999) The MAP kinase Kpp2 regulates mating and pathogenic development in Ustilago maydis. Mol Microbiol 34: 1007–1017.

99. Muller¨ P, Weinzierl G, Brachmann A, Feldbrugge¨ M & Kahmann R (2003) Mating and pathogenic development of the smut fungus Ustilago maydis are regulated by one mitogen-activated protein kinase cascade. Eukaryot Cell 2: 1187–1199.

100. Nadal M, Garcia-Pedrajas MD & Gold SE (2008) Dimorphism in fungal plant pathogens. FEMS Microbiol Lett 284: 127–134.

101. Nevo-Dinur K, Nussbaum-Shochat A, Ben-Yehuda S & Amster-Choder O (2011) Translation-independent localization of mRNA in E. coli. Science 331: 1081–1084.

102. O'Donnell KL & McLaughlin DJ (1984) Postmeiotic mitosis, basidiospore development, and septation in Ustilago maydis. Mycologia 76: 486–502.

103. Partridge L & Hurst LD (1998) Sex and conflict. Science 281: 2003–2008.

104. Pearson JC, Lemons D & McGinnis W (2005) Modulating Hox gene functions during animal body patterning. Nat Rev Genet 6: 893–904.

105. Perez-Martin J, Castillo-Lluva S, Sgarlata C, Flor-Parra I, Mielnichuk N, Torreblanca J & Carbo N (2006) Pathocycles: Ustilago maydis as a model to study the relationships between cell cycle and virulence in pathogenic fungi. Mol Genet Genomics 276: 211–229.

106. Riquelme M, Bartnicki-Garc´ıa S, Gonzalez´-Prieto JM, Sanchez´-Leon´ E, Verd´ın-Ramos JA, Beltran´-Aguilar A & Freitag M (2007) Spitzenkorper¨ localization and intracellular traffic of green fluorescent protein-labeled CHS-3 and CHS-6 chitin synthases in living hyphae of Neurospora crassa. Eukaryot Cell

107. 6: 1853–1864.

108. Romeis T, Brachmann A, Kahmann R & Kamper¨ J (2000) Identification of a target gene for the bE–bW homeodomain protein complex in Ustilago maydis. Mol Microbiol 37: 54–66.

109. Saito H (2010) Regulation of cross-talk in yeast MAPK signaling pathways. Curr Opin Microbiol 13: 677–683.

110. Sandrock B, Bohmer C & Bolker¨ M (2006) Dual function of the germinal centre kinase Don3 during mitosis and cytokinesis in

111. Ustilago maydis. Mol Microbiol 62: 655–666. Schauwecker F, Wanner G & Kahmann R (1995) Filament-specific expression of a cellulase gene in the dimorphic fungus Ustilago maydis. Biol Chem H-S 376: 617–625.

112. Scherer M, Heimel K, Starke V & Kamper¨ J (2006) The Clp1 protein is required for clamp formation and pathogenic development of Ustilago maydis. Plant Cell 18: 2388–2401.

113. Schink KO & Bolker¨ M (2009) Coordination of cytokinesis and cell separation by endosomal targeting of a Cdc42-specific guanine nucleotide exchange factor in Ustilago maydis. Mol Biol Cell 20: 1081–1088.

114. Schirawski J, Bohnert¨ HU, Steinberg G, Snetselaar K, Adamikowa L & Kahmann R (2005a) Endoplasmic reticulum glucosidase II is required for pathogenicity of Ustilago maydis. Plant Cell 17: 3532–3543.

115. Schirawski J, Heinze B, Wagenknecht M & Kahmann R (2005b) Mating type loci of Sporisorium reilianum: novel pattern with three a and multiple b specificities. Eukaryot Cell 4: 1317–1327.

116. Schirawski J, Mannhaupt G, Munch¨ K et al. (2010) Pathogenicity determinants in smut fungi revealed by genome comparison. Science 330: 1546–1548.

117. Schuchardt I, Assmann D, Thines E, Schuberth C & Steinberg G (2005) Myosin-V, Kinesin-1, and Kinesin-3 cooperate in hyphal growth of the fungus Ustilago maydis. Mol Biol Cell 16: 5191–5201.

118. Schulz B, Banuett F, Dahl M, Schlesinger R, Schafer¨ W, Martin T, Herskowitz I & Kahmann R (1990) The b alleles of U. maydis, whose combinations program pathogenic development, code for polypeptides containing a homeodomain-related motif. Cell 60: 295–306.

119. Schuster M, Kilaru S, Ashwin P, Lin C, Severs NJ & Steinberg G (2011a) Controlled and stochastic retention concentrates dynein at microtubule ends to keep endosomes on track. EMBO J 30: 652–664.

120. Schuster M, Lipowsky R, Assmann MA, Lenz P & Steinberg G (2011b) Transient binding of dynein controls bidirectional long-range motility of early endosomes. P Natl Acad Sci USA 109: 3618–3623.

121. Skibbe DS, Doehlemann G, Fernandes J & Walbot V (2010) Maize tumors caused by Ustilago maydis require organ-specific genes in host and pathogen. Science 328: 89–92.

122. Snetselaar KM & Mims CW (1993) Infection of maize stigmas by Ustilago maydis: light and electron microscopy. Phytopathology 83: 843–850.

123. Snetselaar KM, Bolker¨ M & Kahmann R (1996) Ustilago maydis mating hyphae orient their growth toward pheromone sources. Fungal Genet Biol 20: 299–312.

124. Spellig T, Bolker¨ M, Lottspeich F, Frank RW & Kahmann R (1994) Pheromones trigger filamentous growth in Ustilago maydis. EMBO J 13: 1620–1627.

125. Steinberg G (2007a) Hyphal growth: a tale of motors, lipids, and the Spitzenkorper¨. Eukaryot Cell 6: 351–360.

126. Steinberg G (2007b) On the move: endosomes in fungal growth and pathogenicity. Nat Rev Microbiol 5: 309–316.

127. Steinberg G (2007c) Tracks for traffic: microtubules in the plant pathogen Ustilago maydis. New Phytol 174: 721–733.

128. Steinberg G & Perez-Martin J (2008) Ustilago maydis, a new fungal model system for cell biology. Trends Cell Biol 18: 61–67.

129. Steinberg G, Schliwa M, Lehmler C, Bolker¨ M, Kahmann R & McIntosh JR (1998) Kinesin from the plant pathogenic fungus Ustilago maydis is involved in vacuole formation and cytoplasmic migration. J Cell Sci 111: 2235–2246.

130. Steinberg G, Wedlich-Soldner¨ R, Brill M & Schulz I (2001) Microtubules in the fungal pathogen Ustilago maydis are highly dynamic and determine cell polarity. J Cell Sci 114: 609–622.

131. St Johnston D (2005) Moving messages: the intracellular localization of mRNAs. Nat Rev Mol Cell Bio 6: 363–375.

132. Straube A, Enard W, Berner A, Wedlich-Soldner¨ R, Kahmann R & Steinberg G (2001) A split motor domain in a cytoplasmic dynein. EMBO J 20: 5091–5100.

133. Straube A, Brill M, Oakley BR, Horio T & Steinberg G (2003) Microtubule organization requires cell cycle-dependent nucleation at dispersed cytoplasmic sites: polar and perinuclear microtubule organizing centers in the plant pathogen Ustilago maydis. Mol Biol Cell 14: 642–657.

134. Straube A, Weber I & Steinberg G (2005) A novel mechanism of nuclear envelope break-down in a fungus: nuclear migration strips off the envelope. EMBO J 24: 1674–1685.

135. Sudbery P & Court H (2007) Polarised growth in fungi. The Mycota VIII (Howard RJ & Gow NAR, eds), pp. 137–166. Springer-Verlag, Berlin.

136. Szabo´ Z, Tonnis¨ M, Kessler H & Feldbrugge¨ M (2002) Structure– function analysis of lipopeptide pheromones from the plant pathogen Ustilago maydis. Mol Genet Genomics 268: 362–370.

137. Theisen U, Straube A & Steinberg G (2008) Dynamic rearrangement of nucleoporins during fungal 'open' mitosis. Mol Biol Cell 19: 1230–1240.

138. Thorslund T & West SC (2007) BRCA2: a universal recombinase regulator. Oncogene 26: 7720–7730.

139. Thorslund T, McIlwraith MJ, Compton SA, Lekomtsev S, Petronczki M, Griffith JD & West SC (2010) The breast cancer tumor suppressor BRCA2 promotes the specific targeting of RAD51 to single-stranded DNA. Nat Struct Mol Biol 17: 1263–1265.

140. Treitschke S, Doehlemann G, Schuster M & Steinberg G (2010) The myosin motor domain of fungal chitin synthase V is dispensable for vesicle motility but required for virulence of the maize pathogen Ustilago maydis. Plant Cell 22: 2476–2494.

141. Urban M, Kahmann R & Bolker¨ M (1996a) The biallelic a mating type locus of Ustilago maydis: remnants of an additional pheromone gene indicate evolution from a multiallelic ancestor. Mol Gen Genet 250: 414–420.

142. Urban M, Kahmann R & Bolker¨ M (1996b) Identification of the pheromone response element in Ustilago maydis. Mol Gen Genet 251: 31–37.

143. Vollmeister E & Feldbrugge¨ M (2010) Posttranscriptional control of growth and development in Ustilago maydis. Curr Opin Microbiol 13: 693–699.

144. Vollmeister E, Haag C, Zarnack K, Baumann S, Konig¨ J, Mannhaupt G & Feldbrugge¨ M (2009) Tandem KH domains of Khd4 recognize AUACCC and are essential for regulation of morphology as well as pathogenicity in Ustilago maydis. RNA 15: 2206–2218.

145. Wahl R, Wippel K, Goos S, Kamper¨ J & Sauer N (2010a) A novel high-affinity sucrose transporter is required for virulence of the plant pathogen Ustilago maydis. PLoS Biol 8: e1000303.

146. Wahl R, Zahiri A & Kamper J (2010b) The Ustilago maydis b mating type locus controls hyphal proliferation and expression of secreted virulence factors in planta. Mol Microbiol 75: 208–220.

147. Weber I, Gruber C & Steinberg G (2003) A class-V myosin required for mating, hyphal growth, and pathogenicity in the dimorphic plant pathogen Ustilago maydis. Plant Cell 15: 2826–2842.

148. Weber I, Assmann D, Thines E & Steinberg G (2006) Polar localizing class V myosin chitin synthases are essential during early plant infection in the plant pathogenic fungus Ustilago maydis. Plant Cell 18: 225–242.

149. Wedlich-Soldner¨ R, Bolker¨ M, Kahmann R & Steinberg G (2000) A putative endosomal t-SNARE links exo- and endocytosis in the phytopathogenic fungus Ustilago maydis. EMBO J 19:1974–1986.

150. Wedlich-Soldner¨ R, Straube A, Friedrich MW & Steinberg G (2002) A balance of KIF1A-like kinesin and dynein organizes early endosomes in the fungus Ustilago maydis. EMBO J 21: 2946–2957.

151. Weinzierl G, Leveleki L, Hassel A, Kost G, Wanner G & Bolker¨ M (2002) Regulation of cell separation in the dimorphic fungus Ustilago maydis. Mol Microbiol

45: 219–231. Wosten¨ HA, Bohlmann R, Eckerskorn C, Lottspeich F, Bolker¨ M & Kahmann R (1996) A novel class of small amphipathic peptides affect aerial hyphal growth and surface hydrophobicity in Ustilago maydis. EMBO J 15: 4274–4281.

152. Xu J (2005) The inheritance of organelle genes and genomes: patterns and mechanisms. Genome 48: 951–958.

153. Yan Z, Sun S, Shahid M & Xu J (2007) Environment factors can influence mitochondrial inheritance in the fungus Cryptococcus neoformans. Fungal Genet Biol 44: 315–322. Yang H, Li Q, Fan J, Holloman WK & Pavletich NP (2005) The BRCA2 homologue Brh2 nucleates RAD51 filament formation at a dsDNA–ssDNA junction. Nature 433: 653–657.

154. Zahiri A, Heimel K, Wahl R, Rath M & Kamper¨ J (2010) The Ustilago maydis forkhead transcription factor Fox1 is involved in the regulation of genes required for the attenuation of plant defenses during pathogenic development. Mol Plant Microbe In 23: 1118–1129.

155. Zarnack K & Feldbrugge¨ M (2007) mRNA trafficking in fungi. Mol Genet Genomics 278: 347–359

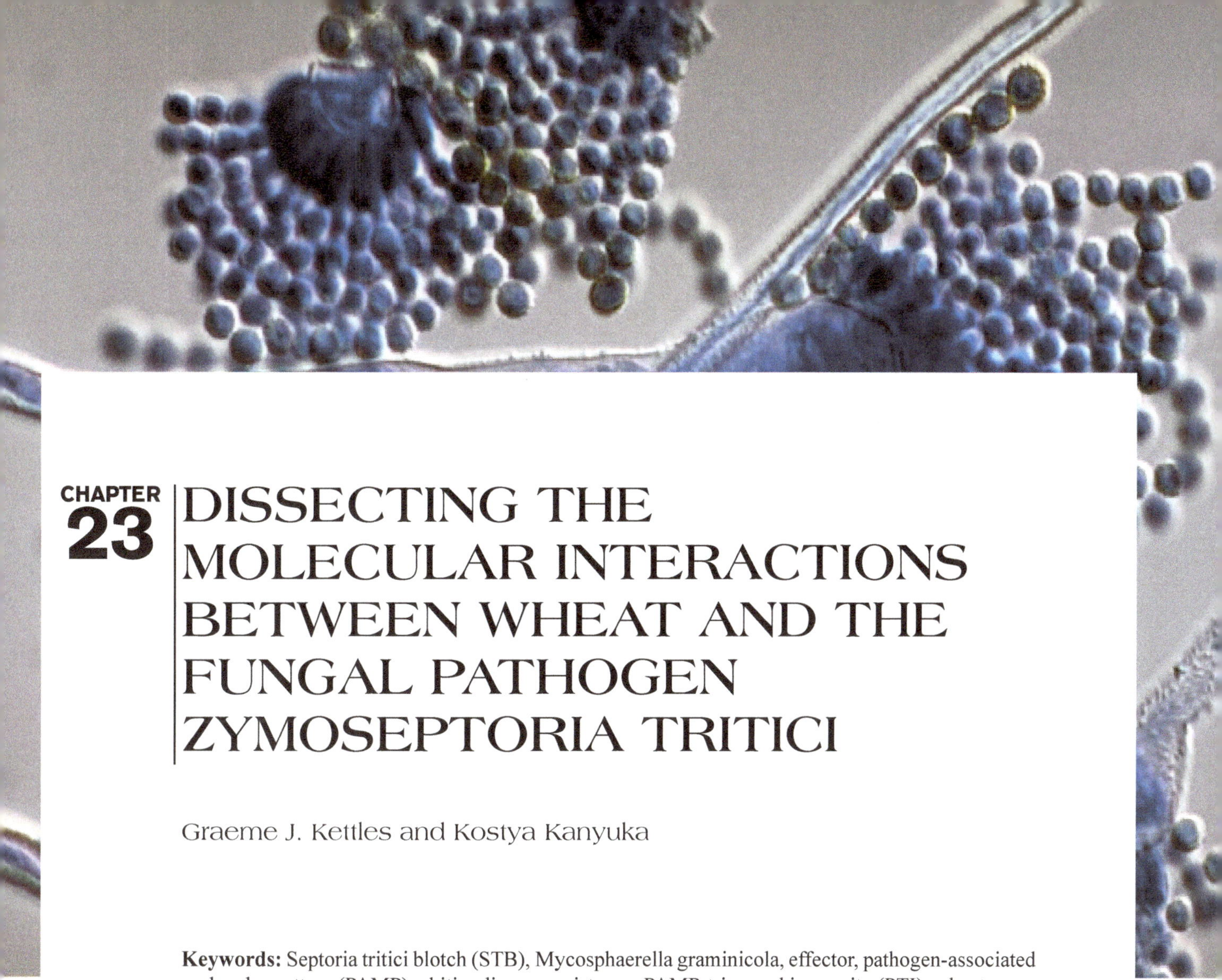

CHAPTER 23

DISSECTING THE MOLECULAR INTERACTIONS BETWEEN WHEAT AND THE FUNGAL PATHOGEN ZYMOSEPTORIA TRITICI

Graeme J. Kettles and Kostya Kanyuka

Keywords: Septoria tritici blotch (STB), Mycosphaerella graminicola, effector, pathogen-associated molecular pattern (PAMP), chitin, disease resistance, PAMP-triggered immunity (PTI), wheat

The Dothideomycete fungus *Zymoseptoria tritici* (previously known as *Mycosphaerella graminicola* and *Septoria tritici*) is the causative agent of *Septoria tritici* leaf blotch (STB) disease of wheat (*Triticum aestivum* L.). In Europe, STB is the most economically damaging disease of wheat, with an estimated ~€1 billion per year in fungicide expenditure directed toward its control. Here, an overview of our current understanding of the molecular events that occur during *Z. tritici* infection of wheat leaves is presented. On the host side, this includes the contribution of (1) the pathogen-associated molecular pattern-triggered immunity (PTI) layer of plant defense, and (2) major *Stb* loci for resistance against *Z. tritici*. On the pathogen side of the interaction, we consolidate evidence from recent bioinformatic, transcriptomic and proteomic studies that begin to explain the contribution of *Z. tritici* effector proteins to the biphasic lifestyle of the fungus. This includes the discovery of chitin-binding proteins in the *Z. tritici* secretome, which contribute to evasion of immune surveillance by this pathogen, and the possible existence of 'necrotrophic' effectors from *Z. tritici*, which may actively stimulate host recognition in a manner similar to related necrotrophic fungal pathogens. We finish by speculating on how some of these recent fundamental discoveries might be harnessed to help improve resistance to STB in the world's second largest food crop.

THE *ZYMOSEPTORIA TRITICI* LIFE CYCLE AND PATHOGENESIS PROGRAM

Zymoseptoria tritici is an ascomycete fungus belonging to the family Mycosphaerellaceae in the class Dothideomycetes. The *Z. tritici* lifestyle is described as hemibiotrophic, with two distinct phases of infection. Following inoculation onto the leaf surface by rain splash, spores germinate and the fungus invades exclusively through the stomata before undergoing a prolonged asymptomatic phase of very slow hyphal growth in the apoplastic space between mesophyll cells. There is no evidence of host cell penetration or formation of specialized feeding structures such as haustoria (Kema et al., 1996). This phase has also often been refered to as 'biotrophic.' Recent transcriptomic and metabolic profiling of *Z. tritici* infection of susceptible wheat, however, indicated that the fungus' own lipids and fatty acids are most likely used as the main energy sources during this phase. That said, a number of fungal genes encoding secreted cutinases and lipases were found to be significantly up-regulated during the asymptomatic phase, and

Citation: Kettles GJ and Kanyuka K (2016) Dissecting the Molecular Interactions between Wheat and the Fungal Pathogen Zymoseptoria tritici. Front. Plant Sci. 7:508. doi: 10.3389/fpls.2016.00508

it has been proposed that host lipids (e.g., leaf surface waxes) may also be utilized (Rudd et al., 2015). This first infection phase is also characterized by the absence, or very weak defense response (Keon et al., 2007; Yang et al., 2013b; Rudd et al., 2015) suggesting that *Z. tritici* is able to successfully suppress or avoid immune elicitation. However, the ratio of fungal to wheat biomass at this stage is extremely low and therefore any local host defense response (if present) may be diluted out and be undetectable by current methods. The asymptomatic phase typically lasts 7–10 days following inoculation depending on the particular cultivar-isolate combination, after which there is a rapid transition to the symptomatic phase, which is frequently referred to as 'necrotrophic.' This second infection phase is typified by the large-scale reprogramming of both host and pathogen transcriptomes, a strong activation of host defense responses culminating in apoptotic-like rather than necrotic cell death and release of nutrients into the leaf apoplast and, as a consequence, a substantial build-up of fungal biomass (Keon et al., 2007; Yang et al., 2013b; Rudd et al., 2015). Macroscopically visible dead leaf areas then expand to form irregularly-shaped blotches (lesions) in which fungal asexual sporulation structures called pycnidia develop. These erupt through stomatal openings and release pycnidiospores that may initiate further rounds of infection when spread to healthy tissues via rain splash.

Z. TRITICI EFFECTORS AND HOST CELL SURFACE IMMUNE RECEPTORS PLAY KEY ROLES DURING THE ASYMPTOMATIC INFECTION PHASE

The principle function of the immune system is to recognize non-self molecules that betray the presence of an invader. Non-self molecules that trigger innate immune responses are considered pathogen-associated molecular patterns (PAMPs) (Jones and Dangl, 2006). Several fungal PAMPs have been described, including chitin (polymer of long chains of *N*-acetylglucosamine), β-glucans (polymers of glucose), mannans (polymers derived from mannose, galactose, and glucose) derived from cell walls, and ergosterol found in cell membranes (Klarzynski et al., 2000; Lochman and Mikes, 2006; McGreal et al., 2006; Shetty et al., 2009; Sánchez-Vallet et al., 2014). Successful immune activation following PAMP elicitation leads to PAMP-triggered immunity (PTI) (Jones and Dangl, 2006).

For a successful pathogen to usurp host immunity, the invader must deploy secreted effector proteins, metabolites or other mechanisms to overcome host defenses triggered during PTI. Of particular importance to fungal-plant interactions (including the *Z. tritici* – wheat interaction) is the PAMP chitin, which represents one of the major components of fungal cell walls and is naturally absent in plants (Sánchez-Vallet et al., 2014). Chitin functions as an elicitor of defense in both dicots and monocots, including a number of model and crop species, such as rice and wheat. Preventing or suppressing chitin-triggered defense is therefore likely to be important for any fungal pathogen. The Ecp6 effector from the fungus *Cladosporium fulvum* that causes the tomato leaf mold disease (Bolton et al., 2008) was demonstrated to have ultrahigh chitin-binding affinity via cooperation of two out of its three lysin motifs (LysM) (de Jonge et al., 2010). *Z. tritici* encodes three LysM-containing proteins (Mg1LysM, Mg3LysM, and MgxLysM) with Mg3LysM being the functional ortholog of *C. fulvum* Ecp6. It has been recently demonstrated that both Mg1LysM and Mg3LysM are transcriptionally highly upregulated during symptomless colonization of wheat and both bind chitin (Marshall et al., 2011). However, only Mg3LysM is capable of blocking chitin-induced elicitation of wheat defenses (Figure 1). Curiously, both Mg3LysM and Mg1LysM also have a unique protective effect for *Z. tritici*, shielding the fungal cell wall from digestion by host chitinases (Marshall et al., 2011). This activity has not been observed for LysM effectors from other fungi including Ecp6. These results demonstrate the importance of chitin-binding effectors during the early stage of plant colonization.

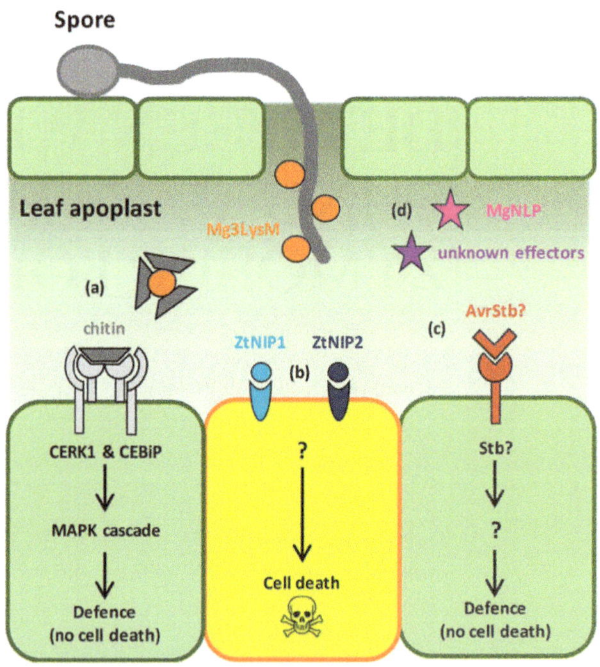

Figure 1: Molecular events in the *Zymoseptoria tritici*–wheat interaction. (a)Fungal PAMP chitin is recognized by the host receptors Chitin Elicitor Binding Protein (CEBiP) and Chitin Elicitor Receptor Kinase 1 (CERK1), triggering MAP kinase cascades and immune activation. Multi-functional LysM-domain containing effector Mg3LysM scavenges chitin to suppress immunity and protects fungal hyphae from wheat chitinases. **(b)**'Necrotrophic' effectors, Necrosis-Inducing Protein 1/2 (ZtNIP1/2), induce host cell death via an unknown mechanism. **(c)** *Stb* gene-specified resistance presumably triggered following recognition of cognate fungal effectors (AvrStb) secreted into the apoplast. This results in arrest of pathogen growth via an unknown mechanism that does not involve HR. **(d)** The Necrosis and Ethylene-Inducing Peptide 1 (NEP1)-like effector protein MgNLP has an unknown function(s) during wheat infection, but triggers cell death in dicotyledonous plants. Unknown effectors – *Z. tritici* effectors predicted from bioinformatic analyses.

Molecular recognition of chitin fragments in cereals requires the Chitin Elicitor Binding Protein (CEBiP) and the Chitin Elicitor Receptor Kinase 1 (CERK1) (Kaku et al., 2006; Shimizu et al., 2010). CEBiP was originally identified as a chitin receptor in rice (OsCEBiP) (Kaku et al., 2006) and is a membrane-anchored extracellular protein that is predicted to be heavily glycosylated (Kaku et al., 2006). The nature and importance of these glycosylations to OsCEBiP function are currently unknown. OsCEBiP recognizes and binds chitin via its extracellular LysM-containing domain but contains no intracellular signaling domain, and is therefore unable to initiate immune signaling alone. OsCERK1 is a transmembrane protein with an extracellular LysM-containing domain and an intracellular kinase domain (Shimizu et al., 2010). OsCERK1 is essential for the generation of a Mitogen-Activated Protein (MAP) kinase signaling cascade following chitin elicitation (Shimizu et al., 2010). The precise mechanism of chitin binding by OsCEBiP has recently been elucidated and shown to require receptor dimerization (Hayafune et al., 2014).

Wheat contains one *CEBiP* and two sequence-related *CERK1* genes that show high degrees of similarity to their rice orthologs (81 and 86%, respectively) (Lee et al., 2014). Whilst wheat *CEBiP*and *CERK1* are understudied relative to their rice counterparts, their importance during the *Z. tritici* asymptomatic infection phase have been revealed in a recent study (Lee et al., 2014). It was demonstrated that the *Z. tritici* mutant lacking a functional Mg3LysM gene, which normally shows dramatically reduced virulence, was able to regain nearly wild-type virulence on wheat plants in which either *CEBiP* or *CERK1* was silenced through Virus-Induced Gene Silencing (VIGS) (Lee et al., 2014, 2015). This is consistent with the role of the Mg3LysM effector in chitin sequestration (Marshall et al., 2011).

Z. TRITICI EFFECTORS POTENTIALLY CONTRIBUTING TO THE SECOND, SYMPTOMATIC INFECTION PHASE

Bioinformatics-based genome-wide approaches have identified several hundred *Z. tritici* genes encoding candidate secreted proteins that bear hallmarks of effectors identified from other pathosystems, i.e., small, cysteine rich, and expressed *in planta* (do Amaral et al., 2012; Gohari et al., 2015). Approximately 100 of these candidate effector genes have been shown to peak in expression either during the switch from asymptomatic to symptomatic infection phase, or during the symptomatic phase (Yang et al., 2013b; Rudd et al., 2015). Some of these effectors also become quite abundant during this second phase and can be detected in apoplastic fluids from infected leaf tissues using proteomics analysis (Yang et al., 2013a).

One of the *Z. tritici* effectors transcriptionally upregulated toward the late stage of symptomless colonization by *Z. tritici* is MgNLP, a member of the Necrosis and Ethylene-inducing Peptide 1 (NEP1)-like (NLP) protein family identified and characterized from a number of bacteria, fungi and oomycetes, including many plant pathogens (Gijzen and Nürnberger, 2006). Whilst NLPs often occur in large families, thus making functional redundancy problematic, *Z. tritici* harbors only a single NEP1-like protein (Motteram et al., 2009). Similar to other NLPs, MgNLP was found to induce defense-related responses and cell death in leaves of dicotyledonous plants (Motteram et al., 2009). Curiously, MgNLP does not induce defense genes or trigger cell death in wheat, and a targeted *MgNLP* deletion mutant of *Z. tritici* is not compromised in its ability to infect wheat (Motteram et al., 2009). Therefore, MgNLP may be functionally redundant and its exact function during wheat infection remains elusive (Figure 1).

It has been recently shown that liquid *Z. tritici* culture filtrates contain proteinaceous effectors able to induce cell death when injected into wheat leaves (M'Barek et al., 2015). This study identified two putative 'Necrosis-Inducing Proteins' designated as ZtNIP1 and ZtNIP2 through fast protein liquid chromatography and liquid-chromatography mass spectrometry fractionation of culture filtrates. Syringe-infiltration of heterologously expressed ZtNIP1 and ZtNIP2 into wheat leaves induced cell death in a cultivar-dependent fashion (M'Barek et al., 2015). Whilst the specific cell death-inducing activity of these proteins is as yet unknown, these findings point to the existence of interactions between the *Z. tritici* 'necrotrophic' effectors and the corresponding host susceptibility factors (Figure 1), similar to those described for necrotrophic pathogens which infect wheat,*Phaeosphaeria nodorum* (syn. *Stagonospora nodorum*) and *Pyrenophora tritici-repentis* (Tan et al., 2015).

Functional characterisation of other candidate phase-specific *Z. tritici* effectors identified through bioinformatic, proteomic and genome-wide transcriptomic analyses (do Amaral et al., 2012; Gohari et al., 2015; Rudd et al., 2015) is currently underway in several laboratories around the globe including ours, and appears a rich source of future discoveries.

MANAGING *SEPTORIA TRITICI* BLOTCH DISEASE IN WHEAT

The control of STB disease in wheat crops currently relies on the intensive use of fungicides with mixtures of azole and succinate dehydrogenase inhibitor (SDHI) fungicides being most effective (Torriani et al., 2015). Resistance to STB is also an important target in wheat breeding, recently catalyzed by new EU regulations restricting the use of some of the most potent fungicide products as well as increasingly frequent reports of fungicide resistance in field populations of *Z. tritici*[1]. In comprehensive worldwide wheat germplasm screening programs, a number of exotic as well as synthetic wheat genotypes have been identified as good sources of STB resistance. At least 20 distinct genetic loci have been identified (*Stb* loci) that confer qualitative, often isolate-specific resistance to *Z. tritici*. In addition, a large number of quantitative trait loci (QTL), which make smaller contributions to the *Z. tritici* resistance phenotype, have been mapped genetically. For a comprehensive up-to-date review of both *Stb* loci and resistance QTLs see Brown et al. (2015).

Although major STB resistance loci (designated as *Stb1* through to *Stb18*, *StbSm3*, and *StbWW*) have been identified in hexaploid wheat, all of these with the exception of *Stb16q* provide protection against individual or small groups of *Z. tritici* isolates. *Stb6* is one of the better-characterized resistance genes. *Stb6*-based resistance conforms to the gene-for-gene hypothesis, whereby avirulent *Z. tritici* isolates (such as the reference isolate IPO323 with fully sequenced genome;Goodwin et al., 2011) carry the matching *AvrStb6* gene (Brading et al., 2002). However, the identity of *AvrStb6* is not yet known and so the frequency of this avirulence gene in current *Z. tritici* field populations remains to be determined. Interestingly, *Stb6* is found in many commercial wheat cultivars and breeding lines originating from Europe, China, Israel, and the United States (US) that are known sources of *Z. tritici* resistance (Chartrain et al., 2005). This gene appears to contribute some resistance to STB in the field at least in the United Kingdom (UK) (Arraiano et al., 2009) indirectly suggesting that *AvrStb6* may be relatively well represented in regional field populations of *Z. tritici*. Frequencies of avirulence toward other known *Stb* genes in field populations of the fungus are unknown. However, it is plausible that these may be sufficiently high for some individual *Stb* genes to provide a certain degree of resistance durability. With this in mind, it is worth noting that US wheat breeders have been actively deploying two major resistance genes, *Stb1*and *Stb4*, since the early 1970s (Goodwin, 2007). *Stb1* provided long-lasting resistance to wheat in the central USA (Adhikari et al., 2004), and *Stb4* remained effective in California for about 15 years until the effectiveness of this gene decreased (Jackson et al., 2000). Efficacy and durability of the other known *Stb* genes against natural fungal populations is not known. Importantly, some wheat genotypes used as major sources of resistance to STB in current world breeding programs, for example, TE9111, Kavkaz-K4500 L.6.A.4 and Veranopolis, all contain ≥3 *Stb* genes (Chartrain et al., 2004; Kollers et al., 2013). This strongly suggests that pyramiding *Z. tritici* isolate-specific resistance genes may be an effective strategy for developing wheat cultivars with high levels of field resistance.

Stb16q is particularly interesting from the disease resistance breeding point of view because this gene appears to confer broad-spectrum seedling stage resistance, as so far no resistance-breaking*Z. tritici* isolates have been found. *Stb16q* has been identified in synthetic hexaploid wheat and was shown to originate from one of its parents, the diploid wild wheat species *Aegilops tauschii* (Tabib Ghaffary et al., 2012). Interestingly, broad-spectrum resistance to *Z. tritici* has also been identified in the diploid wheat *Triticum monococcum* and was shown to be controlled by a single genetic locus *TmStb1* (Jing et al., 2008). Efforts are currently underway to introgress *TmStb1* into hexaploid wheat[2]. It would be interesting to evaluate the usefulness of *Stb16q* and *TmStb1* for potentially achieving durable broad-spectrum resistance to *Z. tritici* in bread wheat as well as to explore further the full potential of synthetic hexaploid wheat and wild relatives of wheat as novel sources of resistance to *Z. tritici.*

The functional characterisation of several of the known *Stb* genes would greatly increase fundamental understanding of the wheat immune system and may facilitate the identification of novel candidate resistance genes using bioinformatic approaches.

FUTURE STRATEGIES FOR STB DISEASE RESISTANCE

Exploiting PTI to Enhance Resistance to *Z. tritici*

The battle that exists in the apoplastic space for chitin binding goes some way to determining the overall outcome of the *Z. tritici*-wheat interaction. The comparative chitin-binding affinity of the wheat receptors CEBiP and CERK1 relative to chitin-scavenging ability of the fungal effector Mg3LysM is unknown. It is not unreasonable to assume that pathogen effectors such as Mg3LysM, under high evolutionary pressure, might be optimized for chitin binding affinity. One novel strategy, therefore, would be to engineer wheat immune receptors to match or exceed the chitin affinity of the competing fungal effector proteins. Such a tactic might allow 'supercharging' the PTI layer of the wheat immune system. Indeed, such a resource might already exist in nature. However, no study has yet examined the variation in chitin sensitivity across commercial wheat cultivars or their wild relatives.

Furthermore, chitin is likely not the only PAMP present in the apoplastic space during *Z. tritici*infection. β-glucans are another major components of the fungal cell wall and preparations of these compounds from *Z. tritici* have been shown to elicit defense responses in wheat and to confer some resistance to a normally susceptible cultivar (Shetty et al., 2009). Understanding of how plants perceive β-glucans is immature in comparison to perception of chitin. Indeed, it is not known if a dedicated membrane-bound immune receptor exists. Nonetheless, enhancement of multiple immune receptors concurrently might offer the same benefits as has been widely discussed with the stacking of dominant isolate- or race-specific resistance (*R*) genes.

Exploiting Major Resistance Genes for Enhanced Resistance to *Z. tritici*

Whilst considerable scientific and breeding efforts have been directed toward the identification of useful sources of *Z. tritici* resistance and development of elite resistant wheat, fully resistant varieties are not yet available to farmers. Taking the UK as an example, the Agriculture and Horticulture Development Board (AHDB) 2016/17 recommended list for winter wheat provides information on the level of disease resistance for current commercially available wheat cultivars[3]. For *Z. tritici* resistance, the vast majority of cultivars (34 of 36) score 6 or less, on a scale of 0–9 (whereby 0 indicates high level of susceptibility and 9 indicates high level of resistance). Only two cultivars (Graham and KWS Siskin) scored as high as 7. This highlights the need for development of novel strategies in breeding for *Z. tritici* resistance.

Single major *Stb* genes can confer complete resistance to *Z. tritici*, albeit against a relatively narrow range of isolates harboring the corresponding avirulence (*Avr*) genes (Figure 1). Deployment of single resistance genes rarely provides durable disease resistance as avirulent isolates of a target pathogen need only to escape this single recognition event in order to regain virulence. Stacking of multiple *Stb* genes against *Z. tritici* either through marker assited breeding or via cysgenics will likely provide more effective resistance as evidenced by the prevalence of multiple *Stb* genes in highly resistant germplasm. Biochemical and molecular characterisation of *Stb* gene products will aid the design of optimal stacking strategies and may also facilitate identification of novel *Stb*alleles.

Effector-Assisted Screening for Novel Resistances in Wheat

For biotrophic pathogens, recognition of an effector by a corresponding major host resistance protein often leads to induction of the hypersensitive response (HR) and localized cell death that correlates with limited pathogen spread beyond the primary inoculated loci. By contrast, recognition of some 'necrotrophic' pathogen effectors triggers induction of host cell death that leads to host susceptibility (Faris et al., 2013). Notable examples are the toxic effectors from the pathogens *P. nodorum* and *P. tritici-repentis* (*Ptr*) that cause glume blotch and tan spot disease of wheat, respectively (Faris et al., 2013; Vleeshouwers and Oliver, 2014). For *Ptr* ToxA, effector sensitivity correlates with disease susceptibility across a collection of wheat cultivars (reviewed inFaris et al., 2013). ToxA sensitivity is dependent on the wheat gene *Tsn1*, although there is no evidence of a direct ToxA-Tsn1 interaction (Faris et al., 2010). Nonetheless, screening of Australian wheat germplasm using recombinant ToxA protein has helped direct breeding programs toward the elimination of *Ptr* susceptibility. Effector-directed breeding reduced the area of ToxA-sensitive wheat grown in Western Australia by nearly 50% over the first three growing seasons since the program was initiated in 2009 (Vleeshouwers and Oliver, 2014). Similar programs may be initiated to breed out susceptibility to *Z. tritici*. However, knowledge of *Z. tritici* susceptibility genes in wheat and identification of 'necrotrophic' effectors potentially contributing to disease is less advanced than for pathogens such as *Ptr*. Recent description of *Z. tritici* effectors ZtNIP1 and ZtNIP2 that induce cell death in some but not all wheat cultivars (M'Barek et al., 2015) suggest that sensitivity to 'necrotrophic' effectors may potentially contribute to *Z. tritici* susceptibility. If proven, this may be exploited for breeding resistant (or at least less susceptible) wheat in a similar fashion as discussed above for the wheat–*Ptr* interaction.

CONCLUSION AND FUTURE PERSPECTIVES

Molecular characterisation of the major events occurring during *Z. tritici* infection of wheat has enabled more thorough understanding of disease progression. Significant progress has been made, for example, in describing the candidate *Z. tritici* effector repertoire. However, how the >100 candidate effectors act in concert to induce disease remains unknown. Likewise, the existence of*Stb* loci conferring strong isolate-specific resistance has been known for some time, but only one has so far been cloned or functionally characterized (unpublished). Nonetheless, progress that has been made give reason to be optimistic that application of more advanced (and rapid) molecular tools will allow fuller understanding of STB disease development, taking it on par with model pathogens.

Where perhaps a greater problem lies, however, is the translation of fundamental findings into real-world impact. But even here, trails have been blazed using other pathosystems that may point the way for *Z. tritici* research. The implementation of an effector-directed breeding program against *Ptr* (discussed above) is a prime example of how fundamental research can lead to low-cost, time-saving tools for the benefit of agriculture.

It is speculated that introduction of *Z. tritici* susceptibility into breeding programs came as a result of breeding for higher yields combined with tight association of gene loci conditioning susceptibility and those for yield potential (Torriani et al., 2015). In the event that culprit susceptibility gene loci are identified and characterized biochemically in wheat, genome editing technologies such as CRISPR-Cas9 (Wang et al., 2014) might be used in their modification or inactivation without the necessary yield loss penalties demanded by agriculture.

FUNDING

Research was funded by the Institute Strategic Programme Grant "20:20 Wheat®" (grant no. BB/J/00426X/1) from the Biotechnology and Biological Sciences Research Council of the UK (BBSRC) and by the Rothamsted Institute Fellowship Programme.

CONFLICT OF INTEREST STATEMENT

The authors declare that the research was conducted in the absence of any commercial or financial relationships that could be construed as a potential conflict of interest.

ACKNOWLEDGMENTS

We would like to thank Kim E. Hammond-Kosack for critical reading of this manuscript and helpful comments.

REFERENCES

1. Adhikari, T. B., Yang, X., Cavaletto, J. R., Hu, X., Buechley, G., Ohm, H. W., et al. (2004). Molecular mapping of Stb1, a potentially durable gene for resistance to *Septoria tritici* blotch in wheat. *Theor. Appl. Genet.* 109, 944–953. doi: 10.1007/s00122-004-1709-6
2. Arraiano, L. S., Balaam, N., Fenwick, P. M., Chapman, C., Feuerhelm, D., Howell, P., et al. (2009). Contributions of disease resistance and escape to the control of *Septoria tritici* blotch of wheat. *Plant Pathol.* 58, 910–922. doi: 10.1111/j.1365-3059.2009.02118.x
3. Bolton, M. D., van Esse, H. P., Vossen, J. H., de Jonge, R., Stergiopoulos, I., Stulemeijer, I. J., et al. (2008). The novel*Cladosporium fulvum* lysin motif effector Ecp6 is a virulence factor with orthologues in other fungal species. *Mol. Microbiol.*69, 119–136. doi: 10.1111/j.1365-2958.2008.06270.x
4. Brading, P. A., Verstappen, E. C. P., Kema, G. H. J., and Brown, J. K. M. (2002). A gene-for-gene relationship between wheat and *Mycosphaerella graminicola*, the *Septoria tritici* blotch pathogen. *Phytopathology* 92, 439–445. doi: 10.1094/PHYTO.2002.92.4.439
5. Brown, J. K., Chartrain, L., Lasserre-Zuber, P., and Saintenac, C. (2015). Genetics of resistance to *Zymoseptoria tritici* and applications to wheat breeding. *Fungal Genet. Biol.* 79, 33–41. doi: 10.1016/j.fgb.2015.04.017
6. Chartrain, L., Brading, P. A., and Brown, J. K. M. (2005). Presence of the Stb6 gene for resistance to *Septoria tritici* blotch (*Mycosphaerella graminicola*) in cultivars used in wheat-breeding programmes worldwide. *Plant Pathol.* 54, 134–143. doi: 10.1111/j.1365-3059.2005.01164.x
7. Chartrain, L., Brading, P. A., Makepeace, J. C., and Brown, J. K. M. (2004). Sources of resistance to *Septoria tritici* blotch and implications for wheat breeding. *Plant Pathol.* 53, 454–460. doi: 10.1111/j.1365-3059.2004.01052.x
8. de Jonge, R., van Esse, H. P., Kombrink, A., Shinya, T., Desaki, Y., Bours, R., et al. (2010). Conserved fungal LysM effector Ecp6 prevents chitin-triggered immunity in plants. *Science* 329, 953–955. doi: 10.1126/science.1190859
9. do Amaral, A. M., Antoniw, J., Rudd, J. J., and Hammond-Kosack, K. E. (2012). Defining the predicted protein secretome of the fungal wheat leaf pathogen *Mycosphaerella graminicola*. *PLoS ONE* 7:e49904. doi: 10.1371/journal.pone.0049904

10. Faris, J. D., Liu, Z., and Xu, S. S. (2013). Genetics of tan spot resistance in wheat. *Theor. Appl. Genet.* 126, 2197–2217. doi: 10.1007/s00122-013-2157-y

11. Faris, J. D., Zhang, Z., Lu, H., Lu, S., Reddy, L., Cloutier, S., et al. (2010). A unique wheat disease resistance-like gene governs effector-triggered susceptibility to necrotrophic pathogens. *Proc. Natl. Acad. Sci. U.S.A.* 107, 13544–13549. doi: 10.1073/pnas.1004090107

12. Gijzen, M., and Nürnberger, T. (2006). Nep1-like proteins from plant pathogens: recruitment and diversification of the NPP1 domain across taxa. *Phytochemistry* 67, 1800–1807. doi: 10.1016/j.phytochem.2005.12.008

13. Gohari, A. M., Ware, S. B., Wittenberg, A. H., Mehrabi, R., M'Barek, S. B., Verstappen, E. C. P., et al. (2015). Effector discovery in the fungal wheat pathogen *Zymoseptoria tritici*. *Mol. Plant Pathol.* 16, 931–945. doi: 10.1111/mpp.12251

14. Goodwin, S. B. (2007). Back to basics and beyond: increasing the level of resistance to *Septoria tritici* blotch in wheat. *Australas. Plant Pathol.* 36, 532–538. doi: 10.1071/AP07068

15. Goodwin, S. B., M'barek, S. B., Dhillon, B., Wittenberg, A. H. J., Crane, C. F., Hane, J. K., et al. (2011). Finished genome of the fungal wheat pathogen *Mycosphaerella graminicola* reveals dispensome structure, chromosome plasticity, and stealth pathogenesis. *PLoS Genet.* 7:e1002070. doi: 10.1371/journal.pgen.1002070

16. Hayafune, M., Berisio, R., Marchetti, R., Silipo, A., Kayama, M., Desaki, Y., et al. (2014). Chitin-induced activation of immune signaling by the rice receptor CEBiP relies on a unique sandwich-type dimerization. *Proc. Natl. Acad. Sci. U.S.A.* 111, E404–E413. doi: 10.1073/pnas.1312099111

17. Jackson, L. F., Dubcovsky, J., Gallagher, L. W., Wennig, R. L., Heaton, J., Vogt, R. L., et al. (2000). 2000 regional barley and common and durum wheat performance tests in California. *UC Davis Agron. Prog. Rep.* 272, 1–56.

18. Jing, H. C., Lovell, D., Gutteridge, R., Jenk, D., Kornyukhin, D., Mitrofanova, O. P., et al. (2008). Phenotypic and genetic analysis of the *Triticum monococcum–Mycosphaerella graminicola* interaction. *New Phytol.* 179, 1121–1132. doi: 10.1111/j.1469-8137.2008.02526.x

19. Jones, J. D. G., and Dangl, J. L. (2006). The plant immune system. *Nature* 444, 323–329. doi: 10.1038/nature05286

20. Kaku, H., Nishizawa, Y., Ishii-Minami, N., Akimoto-Tomiyama, C., Dohmae, N., Takio, K., et al. (2006). Plant cells recognize chitin fragments for defense signaling through a plasma membrane receptor. *Proc. Natl. Acad. Sci. U.S.A.* 103, 11086–11091. doi: 10.1073/pnas.0508882103

21. Kema, G. H. J., Yu, D. Z., Rijkenberg, H. J., Shaw, M. W., and Baayen, R. P. (1996). Histology of the pathogenesis of*Mycosphaerella graminicola* in wheat. *Phytopathology* 86, 777–786. doi: 10.1094/Phyto-86-777

22. Keon, J., Antoniw, J., Carzaniga, R., Deller, S., Ward, J. L., Baker, J. M., et al. (2007). Transcriptional adaptation of*Mycosphaerella graminicola* to programmed cell death (PCD) of its susceptible wheat host. *Mol. Plant Microbe Interact.* 20, 178–193. doi: 10.1094/MPMI-20-2-0178

23. Klarzynski, O., Plesse, B., Joubert, J. M., Yvin, J. C., Kopp, M., Kloareg, B., et al. (2000). Linear beta-1,3 glucans are elicitors of defense responses in tobacco. *Plant Physiol.* 124, 1027–1038. doi: 10.1104/pp.124.3.1027

24. Kollers, S., Rodemann, B., Ling, J., Korzun, V., Ebmeyer, E., Argillier, O., et al. (2013). Genetic architecture of resistance to*Septoria tritici* blotch (*Mycosphaerella graminicola*) in European winter wheat. *Mol. Breed.* 32, 411–423. doi: 10.1007/s11032-013-9880-6

25. Lee, W. S., Rudd, J. J., Hammond-Kosack, K. E., and Kanyuka, K. (2014). *Mycosphaerella graminicola* LysM effector-mediated stealth pathogenesis subverts recognition through both CERK1 and CEBiP homologues in wheat. *Mol. Plant Microbe Interact.*27, 236–243. doi: 10.1094/MPMI-07-13-0201-R

26. Lee, W. S., Rudd, J. J., and Kanyuka, K. (2015). Virus induced gene silencing (VIGS) for functional analysis of wheat genes involved in *Zymoseptoria tritici* susceptibility and resistance. *Fungal Genet. Biol.* 79, 84–88. doi: 10.1016/j.fgb.2015.04.006

27. Lochman, J., and Mikes, V. (2006). Ergosterol treatment leads to the expression of a specific set of defence-related genes in tobacco. *Plant Mol. Biol.* 62, 43–51. doi: 10.1007/s11103-006-9002-5

28. Marshall, R., Kombrink, A., Motteram, J., Loza-Reyes, E., Lucas, J., Hammond-Kosack, K. E., et al. (2011). Analysis of two in planta expressed LysM effector homologs from the fungus *Mycosphaerella graminicola* reveals novel functional properties and varying contributions to virulence on wheat. *Plant Physiol.* 156, 756–769. doi: 10.1104/pp.111.176347

29. M'Barek, B. S., Cordewener, J., Ghaffary, S. M. T., Lee, T. V. D., Liu, Z., Gohari, A. M., et al. (2015). FPLC and liquid-chromatography mass spectrometry identify candidate necrosis-inducing proteins from culture filtrates of the fungal wheat pathogen *Zymoseptoria tritici*. *Fungal Genet. Biol.* 79, 54–62. doi: 10.1016/j.fgb.2015.03.015

30. McGreal, E. P., Rosas, M., Brown, G. D., Zamze, S., Wong, S. Y., Gordon, S., et al. (2006). The carbohydrate-recognition domain of Dectin-2 is a C-type lectin with specificity for high mannose. *Glycobiology* 16, 422–430. doi: 10.1093/glycob/cwj077

31. Motteram, J., Küfner, I., Deller, S., Brunner, F., Hammond-Kosack, K. E., Nürnberger, T., et al. (2009). Molecular characterization and functional analysis of MgNLP, the sole NPP1 domain-containing protein, from the fungal wheat leaf pathogen *Mycosphaerella graminicola*. *Mol. Plant Microbe Interact.* 22, 790–799. doi: 10.1094/MPMI-22-7-0790

32. Rudd, J. J., Kanyuka, K., Hassani-Pak, K., Derbyshire, M., Andongabo, A., Devonshire, J., et al. (2015). Transcriptome and metabolite profiling the infection cycle of *Zymoseptoria tritici* on wheat (*Triticum aestivum*) reveals a biphasic interaction with plant immunity involving differential pathogen chromosomal

contributions, and a variation on the hemi-biotrophic lifestyle definition. *Plant Physiol.* 167, 1158–1185.

33. Sánchez-Vallet, A., Mesters, J. R., and Thomma, B. P. (2014). The battle for chitin recognition in plant-microbe interactions.*FEMS Microbiol. Rev.* 39, 171–183. doi: 10.1093/femsre/fuu003

34. Shetty, N. P., Jensen, J. D., Knudsen, A., Finnie, C., Geshi, N., Blennow, A., et al. (2009). Effects of beta-1,3-glucan from*Septoria tritici* on structural defence responses in wheat. *J. Exp. Bot.* 60, 4287–4300. doi: 10.1093/jxb/erp269

35. Shimizu, T., Nakano, T., Takamizawa, D., Desaki, Y., Ishii-Minami, N., Nishizawa, Y., et al. (2010). Two LysM receptor molecules, CEBiP and OsCERK1, cooperatively regulate chitin elicitor signaling in rice. *Plant J.* 64, 204–214. doi: 10.1111/j.1365-313X.2010.04324.x

36. Tabib Ghaffary, S. M., Faris, J. D., Friesen, T. L., Visser, R. G., van der Lee, T. A., Robert, O., et al. (2012). New broad-spectrum resistance to *Septoria tritici* blotch derived from synthetic hexaploid wheat. *Theor. Appl. Genet.* 124, 125–142. doi: 10.1007/s00122-011-1692-7

37. Tan, K. C., Phan, H. T., Rybak, K., John, E., Chooi, Y. H., Solomon, P. S., et al. (2015). Functional redundancy of necrotrophic effectors - consequences for exploitation for breeding. *Front. Plant Sci.* 6:501. doi: 10.3389/fpls.2015.00501

38. Torriani, S. F. F., Melichar, J. P. E., Mills, C., Pain, N., Sierotzki, H., and Courbot, M. (2015). *Zymoseptoria tritici*: a major threat to wheat production, integrated approaches to control. *Fungal Genet. Biol.* 79, 8–12. doi: 10.1016/j.fgb.2015.04.010

39. Vleeshouwers, V. G., and Oliver, R. P. (2014). Effectors as tools in disease resistance breeding against biotrophic, hemibiotrophic, and necrotrophic plant pathogens. *Mol. Plant Microbe Interact.* 27, 196–206. doi: 10.1094/MPMI-10-13-0313-IA

40. Wang, Y., Cheng, X., Shan, Q., Zhang, Y., Liu, J., Gao, C., et al. (2014). Simultaneous editing of three homoeoalleles in hexaploid bread wheat confers heritable resistance to powdery mildew. *Nat. Biotechnol.* 32, 947–951. doi: 10.1038/nbt.2969

41. Yang, F., Li, W., Derbyshire, M., Larsen, M. R., Rudd, J. J., and Palmisano, G. (2013a). Unraveling incompatibility between wheat and the fungal pathogen *Zymoseptoria tritici* through apoplastic proteomics. *BMC Genomics* 16:362. doi: 10.1186/s12864-015-1549-6

42. Yang, F., Li, W., and Jørgensen, H. J. L. (2013b). Transcriptional reprogramming of wheat and the hemibiotrophic pathogen*Septoria tritici* during two phases of the compatible interaction. *PLoS ONE* 8:e81606. doi: 10.1371/journal.pone.0081606

INDEX